Antimicrobial Stewardship: Principles and Practice

FSC
www.fsc.org
MIX
Paper from
responsible sources
FSC® C013604

Antimicrobial Stewardship: Principles and Practice

Edited by

Kerry L. LaPlante, PharmD, FCCP

Cheston B. Cunha, MD

Haley J. Morrill, PharmD

Louis B. Rice, MD

and

Eleftherios Mylonakis, MD, PhD, FIDSA

CABI

CABI is a trading name of CAB International

CABI
Nosworthy Way
Wallingford
Oxfordshire OX10 8DE
UK

Tel: +44 (0)1491 832111
Fax: +44 (0)1491 833508
E-mail: info@cabi.org
Website: www.cabi.org

CABI
745 Atlantic Avenue
8th Floor
Boston, MA 02111
USA

Tel: +1 (617)682-9015
E-mail: cabi-nao@cabi.org

A catalogue record for this book is available from the British Library, London, UK.

Library of Congress Cataloging-in-Publication Data

Names: LaPlante, Kerry, editor. | C.A.B. International, publisher.
Title: Antimicrobial stewardship : principles and practice / senior editors, Kerry LaPlante, Eleftherios Mylonakis, and Louis Rice ; section editors, Kimberle Chapin, Panagiotis Ziakis, Cheston Cunha, Christy Varughese, and Haley Morrill.
Description: Wallingford, Oxfordshire ; Boston, MA : CABI, [2017] | Includes bibliographical references.
Identifiers: LCCN 2016022765 (print) | LCCN 2016023845 (ebook) | ISBN 9781780644394 (alk. paper) | ISBN 9781780644400 (ePDF) | ISBN 9781786390530 (ePub)
Subjects: | MESH: Anti-Bacterial Agents--administration & dosage | Drug Resistance, Microbial | Inappropriate Prescribing--prevention & control
Classification: LCC RM267 (print) | LCC RM267 (ebook) | NLM QV 350 | DDC 615.7/922--dc23
LC record available at https://lccn.loc.gov/2016022765

ISBN-13: 978 1 78064 439 4

Commissioning editor: Caroline Makepeace
Editorial assistant: Emma McCann
Production editor: James Bishop

Typeset by SPi, Pondicherry, India
Printed and bound by CPI Group (UK) Ltd, Croydon, CR0 4YY

Contents

Editors and Contributors

Cesar A. Arias, MD, PhD, Molecular Genetics and Antimicrobial Resistance Unit, Universidad El Bosque, Bogota, and Division of Infectious Diseases, University of Texas Medical School at Houston, 6431 Fannin, MSB 1.150, Houston, TX 77030, US. E-mail: cesar.arias@uth.tmc.edu

Rafael Araos, MD, MMSc, Assistant Professor of Medicine, Infectious Diseases Unit, Department of Medicine, Facultad de Medicina Clinica Alemana—Universidad del Desarrollo, Avenida Vitacura 5951, Vitacura, Santiago, 7650568, Chile. E-mail: rafaaraos@gmail.com

Armen Arshakyan, MD, Saban Community Clinic, 8405 Beverly Blvd, Los Angeles, CA 90048, US. E-mail: armenarshakyan@yahoo.com

Allison H. Bartlett, MD, MS, Assistant Professor of Pediatrics, Section of Infectious Diseases, Quality Chief, Department of Pediatrics, Associate Medical Director, Infection Control and Antimicrobial Stewardship Programs, Fellowship Director, Pediatric Infectious Diseases, The University of Chicago Medicine, 5841 S. Maryland Ave, Rm C-638A, MC6054 Chicago, IL 60637, US. E-mail: abartlett@peds.bsd.uchicago.edu

Andrew Berti, PharmD, PhD, Research Fellow, School of Pharmacy, University of Wisconsin-Madison, 777 Highland Avenue, Room 1204, Madison, WI 53705-2222, US. E-mail: andrew.berti@wisc.edu

April M. Bobenchik, PhD, D(ABMM), Assistant Professor of Pathology and Laboratory Medicine, The Warren Alpert Medical School of Brown University, 293 Eddy St., APC 11, Providence, Rhode Island 02903, US. E-mail: april_bobenchik@brown.edu

Hugh G. Boothe, PharmD, Clinical Supervisor of Inpatient Pharmacy, Jacobi Medical Center, 1400 Pelham Parkway South, Bronx, NY 10461. E-mail: hughgmboothe@gmail.com

Carey-Ann D. Burnham, PhD, D(ABMM), Associate Professor of Pathology and Immunology, Molecular Microbiology, and Pediatrics, Washington University in St. Louis School of Medicine, 660 S Euclid Ave, Campus Box 8118, St. Louis, MO 63110, US. E-mail: cburnham@path.wustl.edu

Michael S. Calderwood, MD, MPH, FIDSA, Regional Hospital Epidemiologist, Assistant Professor of Medicine, Infectious Disease & International Health, Dartmouth-Hitchcock Medical Center, One Medical Center Drive, Lebanon, NH 03756, US. E-mail: michael.s.calderwood@hitchcock.org

Michael S. Calderwood, MD, MPH, FIDSA, Regional Hospital Epidemiologist, Assistant Professor of Medicine, Infectious Disease & International Health, Dartmouth-Hitchcock Medical Center, One Medical Center Drive, Lebanon, NH 03756, US.

Gerald A. Capraro, MS, PhD, D(ABMM), Medical Director, Clinical Microbiology Laboratory, Carolinas Pathology Group, Carolinas Healthcare System, 5040 Airport Center Parkway, Building H, Suite A, Charlotte, NC 28208, US. E-mail: gerald.capraro@carolinashealthcare.org

Joseph J. Carreno, PharmD, Assistant Professor, Pharmacy Practice, Albany College of Pharmacy and Health Sciences, 106 New Scotland Ave, Albany, NY 12208-3492, US. E-mail: Joseph.Carreno@acphs.edu

Kimberle C. Chapin, MD, ABMM, FCAP, Director of Microbiology and ID Molecular Diagnostics, Department of Pathology, Lifespan Academic Medical Centers, Professor of Medicine and Pathology, The Warren Alpert Brown University Medical School of Brown University, 222 Richmond St, Providence, RI 02903, US. E-mail: kchapin@lifespan.org

Christopher W. Crank, PharmD, MS, BCPS AQ ID, Associate Director, Clinical Pharmacy Services, Clinical Specialist, Infectious Diseases, Rush University Medical Center, 1653 W. Congress Pkwy, Chicago, IL 60612, US. E-mail:christopher_w_crank@rush.edu

Cheston B. Cunha, MD, Assistant Professor of Medicine, Medical Director, Antimicrobial Stewardship Program, Rhode Island Hospital and Miriam Hospital, Division of Infectious Disease, Rhode Island Hospital, 593 Eddy Street, Physician's Office Building Suite # 328, Providence, RI 02903, US. E-mail: ccunha@lifespan.org

George L. Daikos, First Department of Medicine, Medical School, National and Kapodistrian University of Athens, Agiou Thoma 17, Goudi 11527, Athens, Greece, E-mail: gldaikos@gmail.com

Robert S. Daum, MD, CM, Professor of Pediatrics, Microbiology, and Molecular Medicine, Director, The University of Chicago Medicine, MRSA Research Center, Department of Pediatrics, Section of Infectious Diseases, 5841 South Maryland Avenue, MC 6054, Chicago, IL 60637-1470, US. E-mail: rdaum@peds.bsd.uchicago.edu

Susan L. Davis, PharmD, Associate Professor (Clinical), Pharmacy Practice, Wayne State University and Infectious Diseases Pharmacy Specialist, Henry Ford Hospital, Detroit. Address: Eugene Applebaum College of Pharmacy and Health Sciences, Wayne State University, 259 Mack Ave, Detroit, MI 48201, US. E-mail: sldavis@wayne.edu

Thomas J. Dilworth, PharmD, BCPS-AQ ID, Clinical Specialist, Infectious Diseases, Department of Pharmacy Services, Aurora St. Luke's Medical Center, 2900 W. Oklahoma Ave, Milwaukee, WI 53215, US. E-mail: thomas.dilworth@aurora.org

Christopher D. Doern, PhD, Assistant Professor of Pathology, Director of Clinical Microbiology, Department of Pathology, School of Medicine, Virginia Commonwealth University, PO Box 980662, 1101 E. Marshall Street, Richmond, VA 23298-0662, US. E-mail: cdoern@mcvh-vcu.edu

Gary V. Doern, PhD, Emeritus Professor of Pathology, Department of Pathology, University of Iowa Carver College of Medicine, 200 Hawkins Drive, Iowa City, IA 52242, US. E-mail: gdoern@nc.rr.com

Curtis J. Donskey, MD, Professor of Medicine, Division of Infectious Diseases and HIV Medicine, Case Western University, Cleveland, Ohio; and Staff Physician; Infectious Diseases Section, Cleveland Ohio Geriatric Research, Education, and Clinical Center (GRECC), Louis Stokes Cleveland Veterans Affairs Medical Center, 10701 East Blvd, Cleveland, OH 44106, US. E-mail: curtisd123@yahoo.com

Monica Dorobisz, PharmD, Infectious Diseases Pharmacist and Chair of the Kent Antimicrobial Stewardship Program, Department of Pharmacy, Kent Hospital, 455 Toll Gate Rd, Warwick, RI 02886, US. E-mail: mdorobisz@kentri.org

Ramy H. Elshaboury, PharmD, BCPS AQ-ID, Clinical Pharmacy Coordinator—Infectious Diseases, Massachusetts General Hospital, 55 Fruit Street GRB005, Boston, MA 02114, US. E-mail: relshaboury@partners.org

Sandy J. Estrada, PharmD, BCPS AQ-ID, Pharmacist Specialist, Infectious Diseases, Department of Pharmacy, Lee Memorial Hospital, PO Box 2218, Fort Myers, FL 33902, US. E-mail: Sandy.Estrada@LeeMemorial.org

Islam M. Ghazi, PharmD, BCPS, BCACP, Pharmacotherapy Research Fellow, Center for Anti-Infective Research and Development, Hartford Hospital, 80 Seymour Street, Hartford, CT 06102, US. E-mail: islam.ghazi@hhchealth.org

Jennifer L. Goldman, MD, MS, Assistant Professor of Pediatrics, University of Missouri-Kansas City, and Medical Director, Antimicrobial Stewardship Program, Children's Mercy Hospital, Kansas City, 2401 Gillham Road, Kansas City, MO 64108, US. E-mail: jlgoldman@cmh.edu

Blanca E. Gonzalez, MD, Center for Pediatric Infectious Diseases, Cleveland Clinic Children's Hospital, Assistant Professor of Pediatrics, Cleveland Clinic Lerner College of Medicine of Case Western Reserve University, 9500 Euclid Avenue/S25, Cleveland, OH 44195, US. E-mail: gonzalb3@ccf.org

Christopher J. Graber, MD, MPH, FIDSA, Associate Clinical Professor, David Geffen School of Medicine at UCLA (University of California, Los Angeles), and Infectious Diseases Section, VA Greater Los Angeles Healthcare System, 11301 Wilshire Blvd, 111-F, Los Angeles, CA 90073, US. E-mail: christopher.graber@va.gov

Inge C. Gyssens, MD, PhD, Professor of Infectious Diseases, Research Group of Immunology and Biochemistry, Faculty of Medicine, Hasselt University, Hasselt, Belgium and Department of Medicine, Radboud University Medical Centre, Geert Grooteplein 10, 6525 GA Nijmegen, The Netherlands. E-mail: inge.gyssens@radboudumc.nl

Inge C. Gyssens, MD, PhD, Professor of Infectious Diseases, Research Group of Immunology and Biochemistry, Faculty of Medicine, Hasselt University, Hasselt, Belgium; Department of Medicine, Radboud University Nijmegen Medical Centre and Canisius-Wilhelmina Ziekenhuis (Hospital), Weg door Jonkerbos 100 6532 SZ Nijmegen, The Netherlands. E-mail: i.gyssens@aig.umcn.nl

Sajeev Handa, MD, SFHM, Director, Division of Hospital Medicine, Rhode Island Hospital. Clinical Assistant Professor of Medicine and Clinical Assistant Professor of Neurology, The Warren Alpert Medical School of Brown University, 222 Richmond St, Providence, RI 02903, US. E-mail: shanda@lifespan.org

Amy Hanson, PharmD, BCPS, AQ-ID, Clinical Pharmacy Specialist, Infectious Diseases, Rush University Medical Center, 1653 W. Congress Pkwy, Chicago, IL 60612, US. E-mail: Amy_Hanson@rush.edu

Emily L. Heil, PharmD, BCPS AQ ID, Infectious Diseases Specialist and Assistant Professor, University of Maryland School of Pharmacy, 20 N Pine St, Baltimore, MD 21201, US. E-mail: eheil@rx.umaryland.edu

Elizabeth D. Hermsen, PharmD, MBA, BCPS ID, Head, Global Antimicrobial Stewardship, Global Population Health, Merck & Co., Inc., 2000 Galloping Hill Road, Kenilworth, NJ 07033, US. E-mail: elizabeth.hermsen@merck.com

Styliani Karanika, MD, Postdoctoral Research Associate, Infectious Diseases Division, The Warren Alpert Medical School of Brown University, Rhode Island Hospital, 593 Eddy Street, POB, 3rd Floor, Suite 328/330, Providence, RI 02903, US. E-mail: Styliani_karanika@brown.edu

Keith S. Kaye, MD, MPH, Professor of Medicine, Corporate Vice President of Quality and Patient Safety, Corporate Medical Director, Infection Prevention, Epidemiology and Antimicrobial Stewardship, Detroit Medical Center, Wayne State University, University Health Center, 4201 Saint Antoine, Suite 2B, Box 331, Detroit, MI 48201, US. E-mail: KKaye@dmc.org

Powel Kazanjian, MD, PhD, Professor of Medicine, Chief, Infectious Disease Division, Taubman Center, University of Michigan Health System, Floor 3 Reception D, 1500 E Medical Center Dr SPC 5352, Ann Arbor, MI 48109, US. E-mail: pkazanji@med.umich.edu

Joseph L. Kuti, PharmD, Associate Director, Clinical and Economic Studies, Center for Anti-Infective Research and Development, Hartford Hospital, 80 Seymour Street, Hartford, CT 06102, US. E-mail: joseph.kuti@hhchealth.org

Kerry L. LaPlante, PharmD, Associate Professor of Pharmacy, University of Rhode Island, Kingston, Adjunct Clinical Associate Professor of Medicine, Brown University, Providence, Rhode Island, Director of the Rhode Island Infectious Diseases Research Program (RIID), Veterans Affairs Medical Center, Providence, Rhode Island and Infectious Diseases Pharmacotherapy Specialist, University of Rhode Island College of Pharmacy, 7 Greenhouse Road Suite 295A, Kingston RI 02881, US. E-mail: kerrylaplante@uri.edu

Thomas P. Lodise, PharmD, PhD, Associate Professor, Pharmacy Practice, Albany College of Pharmacy and Health Sciences, 106 New Scotland Ave, Albany, NY 12208-3492, US. E-mail: thomas.lodise@acphs.edu

Nicole M. Mahoney, PhD, Director, Global Regulatory Policy, Global Regulatory Affairs, Merck & Co., Inc., 1700 Rockville Pike, Suite 525, Rockville MD 20852, US. E-mail: nicole.mahoney@merck.com

Antonis Markogiannakis, BPharm, MPH, PhD, Department of Pharmacy, Laikon General Hospital, Athens, Greece. E-mail: mark@laiko.gr

Aileen Martinez, PharmD, BCPS, Pharmacist Specialist, Infectious Diseases, Martin Health System, 200 SE Hospital Ave., Stuart, FL 34994, US. E-mail: Aileen.Martinez@martinhealth.org

Dmitriy Martirosov, PharmD, Clinical Pharmacy Specialist, Henry Ford Hospital, 2799 W. Grand Blvd, Detroit, MI 48202, US. E-mail: dmartir1@hfhs.org

Dayna McManus, PharmD, BCPS, Clinical Pharmacy Specialist-Infectious Diseases, Inova Fairfax Hospital, 3300 Gallows Road, Falls Church, VA 22042, US. E-mail: dayna.mcmanus@inova.org or dayna.mcmanus@gmail.com

John McNelis, MD FACS FCCM, Professor of Clinical Surgery, Department of Surgery, Albert Einstein College of Medicine, and Chairman, Department of Surgery, North Bronx Health Network, Jacobi Medical Center, North Central Bronx Hospital, Rm 513, 1400 Pelham Parkway South, Bronx, NY 10461, US. E-mail: john.mcnelis@nbhn.net

Renée-Claude Mercier, PharmD, BCPS AQ ID, Professor of Pharmacy and Medicine, College of Pharmacy, University of New Mexico, 2502 Marble Ave NE, Albuquerque, NM 87131, US. E-mail: rmercier@salud.unm.edu

Haley J. Morrill, PharmD, Department of Pharmacy Practice, College of Pharmacy, University of Rhode Island, Kingston, RI 02881, US, and Infectious Diseases Research Program, Veterans Affairs Medical Center, 830 Chalkstone Ave, Providence, RI 028908, US. E-mail: haley.morrill@va.gov

Jacob Morton, PharmD, MBA, BCPS, Postdoctoral Outcomes in Antimicrobial Stewardship Fellow, Providence VA Medical Center, 830 Chalkstone Ave, Providence, RI 02908, US. E-mail: mjacob.unc@gmail.com

Jose M. Munita, MD, Center for Antimicrobial Resistance and Microbial Genomics, Department of Internal Medicine, Division of Infectious Diseases, University of Texas Medical School at Houston, 6431 Fannin, MSB 1.150, Houston, TX 77030, US. E-mail: jose.m.munita@uth.tmc.edu

Eleftherios Mylonakis, MD, PhD, FIDSA, Dean's Professor of Medical Science (Medicine, and Molecular Microbiology and Immunology), Chief, Infectious Diseases Division, The Warren Alpert Medical School of Brown University, Rhode Island Hospital, 593 Eddy Street, POB, 3rd Floor, Suite 328/330, Providence, RI 02903, US. E-mail: emylonakis@lifespan.org

Jason G. Newland, MD, MEd, Associate Professor of Pediatrics, Washington University School of Medicine in St. Louis, Director, Antimicrobial Stewardship Program, St. Louis Children's Hospital, Campus Box 8116, 660 S. Euclid Ave., St. Louis, MO 63110, US. E-mail: jgnewland@wustl.edu

David P. Nicolau, PharmD, FCCP, FIDSA, Center for Anti-Infective Research and Development, Hartford Hospital, 80 Seymour Street, Hartford, CT 06102, US. E-mail: david.nicolau@hhchealth.org

Jessica K. Ortwine, PharmD, Infectious Diseases Clinical Pharmacy Specialist, Parkland Hospital and Health System, Clinical Assistant Professor in Internal Medicine, University of Texas Southwestern Medical School, 5200 Harry Hines Boulevard, Dallas, TX 75235, US. E-mail: Jessica.ortwine@phhs.org

Suresh Paudel, MD, Postdoctoral Research Associate, Infectious Diseases Division, The Warren Alpert Medical School of Brown University, Rhode Island Hospital, 593 Eddy Street, POB, 3rd Floor, Suite 328/330, Providence, RI 02903, US. E-mail: suresh_paudel@brown.edu

Evangelia-Theophano Piperaki, Department of Microbiology, Medical School, National and Kapodistrian University of Athens, Mikras Asias 75, Athens 11527, Greece. E-mail: vagge71@yahoo.com or epiper@med.uoa.gr

Jason M. Pogue, PharmD, Clinical Pharmacist, Infectious Diseases, Sinai-Grace Hospital, Detroit Medical Center, Clinical Assistant Professor of Medicine, Wayne State University School of Medicine, 6071 W. Outer Drive, Detroit, MI 48235, US. E-mail: jpogue@dmc.org

Louis B. Rice, MD, Joukowsky Family Professor and Chair, Department of Medicine, and Physician-in Chief, Division of Infectious Diseases, Rhode Island and The Miriam Hospitals, The Warren Alpert Medical School of Brown University, 593 Eddy Street, Providence, RI 02903, US. E-mail: lrice@lifespan.org

Warren Rose, PharmD, Associate Professor, School of Pharmacy, University of Wisconsin-Madison, 777 Highland Avenue, Room 4123, Madison, WI 53705-2222, US. E-mail: warren.rose@wisc.edu

John C. Rotschafer, PharmD, FCCP, Department of Experimental and Clinical Pharmacology, University of Minnesota College of Pharmacy, Weaver-Densford Hall 7-189, 308 Harvard Street SE, Minneapolis, MN 55455, US. E-mail: rotsc001@umn.edu

Michael A. Ruggero, PharmD, BCPS, Clinical Pharmacist, Hospital of the University of Pennsylvania, 3400 Spruce St, Philadelphia, PA 19104, US. E-mail: michael.ruggero@uphs.upenn.edu or Ruggero.michael@gmail.com

Michael Samarkos, MD, Assistant Professor, First Department of Internal Medicine, Medical School, National and Kapodistrian University of Athens and University Hospital of Athens "Laikon", Agiou Thoma 17, Athens 11527, Greece, E-mail: msamarkos@otenet.gr or msamarkos@gmail.com

Nikolaos V. Sipsas, MD, FIDSA, Associate Professor, Infectious Disease Unit, Pathophysiology Department, Medical School, National and Kapodistrian University of Athens and University Hospital of Athens "Laikon", Mikras Asias 75, Athens 11527, Greece. E-mail: nsipsas@med.uoa.gr

Harold C. Standiford, MD (retired), formerly Professor of Medicine, University of Maryland School of Medicine, and Medical Director for Antimicrobial Stewardship/Infection Control and Antimicrobial Effectivess, University of Maryland Medical Center, 22 S. Greene Street, Baltimore, MD 21201, US. E-mail: hstandiford02@gmail.com

Melvin E. Stone Jr., MD, Associate Professor of Clinical Surgery, Albert Einstein College of Medicine, and Director Surgical Intensive Care Unit, Jacobi Medical Center,1400 Pelham Parkway South, Bronx, NY 10461, US. E-mail: melvin.stone@nbhn.net

Mahlet Tadele, MD, Assistant Professor of Medicine, Department of Medicine/Division of Infectious Disease, Albert Einstein College of Medicine, and Jacobi Medical Center, 1400 Pelham Parkway South, Bronx, NY 10461, US. E-mail: mahlet.tadele@nbhn.net

Philip Toltzis, MD, Division of Pediatric Critical Care, Rainbow Babies and Children's Hospital, Professor of Pediatrics, Case Western Reserve University School of Medicine, Cleveland, OH 44106, US. E-mail: Philip.Toltzis@UHHospitals.org

Jeffrey E. Topal, MD, Section of Infectious Disease, Yale School of Medicine, Yale New Haven Hospital, New Haven, CT 06504, US. E-mail Jeffrey.Topal@ynhh.org

Damary C. Torres, PharmD, BCOP, Associate Clinical Professor, Clinical Health Professions, College of Pharmacy and Health Sciences, St. John's University, Queens, New York, and Clinical Pharmacy Specialist, Department of Pharmacy, Winthrop-University Hospital, State University of New York at Stony Brook, 259 First Street, Mineola, NY 11501, US. E-mail: castanhd@stjohns.edu

Leonidas Tzouvelekis, Department of Microbiology, Medical School, National and Kapodistrian University of Athens, Mikras Asias 75, Athens 11527, Greece. E-mail: lstzouvelekis@gmail.com

Christy A. Varughese, PharmD, BCPS, Antimicrobial Stewardship Pharmacist, Department of Pharmacy, Rush University Medical Center, 1653 W. Congress Parkway, Chicago, IL 60612, US. E-mail: Christy_A_Varughese@Rush.edu

Jamie L. Wagner, PharmD, Clinical Assistant Professor, Department of Pharmacy Practice, University of Mississippi School of Pharmacy, 2500 N. State Street, Jackson, MS 39216, US. E-mail: jwagner@umc.edu

Carla Walraven, PharmD, BCPS AQ ID, Infectious Diseases/Antimicrobial Stewardship Pharmacist, University of New Mexico Hospitals, Department of Pharmacy Services, Albuquerque, NM 87106, US. E-mail: cwalraven@salud.unm.edu

Theoklis E. Zaoutis, MD, MSCE, Professor of Pediatrics and Epidemiology, Department of Biostatistics and Epidemiology at the University of Pennsylvania Perelman School of Medicine, and Chief, Division of Infectious Diseases, Children's Hospital of Philadelphia (CHOP), 34th & Civic Center Blvd., CHOP North Room 1514, Philadelphia, PA 19104-6021. E-mail: zaoutis@email.chop.edu

Panayiotis D. Ziakas, MD, PhD, Research Associate in Medicine; Division of Infectious Diseases, The Warren Alpert School of Medicine of Brown University, Providence, Rhode Island, and Rhode Island Hospital, 593 Eddy St., POB #328, Providence, RI 02903, US. E-mail: Panagiotis_Ziakas@brown.edu

Preface/Introduction

The recent attention to the millions of people infected yearly with antimicrobial-resistant organisms, and the likely tens of thousands of associated deaths (>23,000 in the US alone), have resulted in a call to action for acute care hospitals and (more recently) long-term care facilities to implement antimicrobial stewardship programs (ASPs). Clinicians, pharmacists, administrators, information technology, microbiology laboratories and a number of other groups were asked to develop an integrated strategy in order to improve antimicrobial usage, enhance patient outcomes, reduce antimicrobial cost, and minimize side effects, including drug resistance and nosocomial infections. However, this strategy has not been clearly defined, and so it has been left to each institution to develop a program that works. With this book, we try to assist these different groups, and to understand the background and the complex dynamic that are required for a successful ASP. Certainly, in this endeavour, we all learned from mistakes and from colleagues. We hope that we were able to pass some of our hard-learned lessons on to the readers of this book.

The book is a joint effort led by five editors, Kerry L. LaPlante, Cheston B. Cunha, Haley J. Morrill, Louis B. Rice, and Eleftherios Mylonakis, with a team of section editors and the authors, who have brought together 38 chapters that describe the principles and practice of antimicrobial stewardship, mainly with reference to the US.

There are eight sections. The first provides an overview of antibiotic stewardship, and it looks at the principles and history of, clinical perspectives on, and importance of education in antimicrobial stewardship. The second, on the principles of antibiotic resistance in antibiotic stewardship, covers the mechanisms of antibiotic resistance, the roles of selection, induction, and colonization in its emergence, and its association with antibiotic usage strategies. Section III, on the role of the microbiology laboratory, considers active surveillance, the antibiogram, selective reporting, and new diagnostics. The fourth section, on infection control, reviews the epidemiology of a number of important antimicrobial-resistant organisms and discusses the role of the hospital epidemiologist in antibiotic stewardship. Section V, on the pharmacokinetic and pharmacodynamic aspects of antibiotic dosing, deals with the principles of pharmacokinetics/pharmacodynamics, the optimal use of several classes of antibiotics, and switching from intravenous to oral dosing. The sixth section, on the role of the pharmacy department, examines the role of pharmacists, formulary management, benchmarking, and stewardship interventions and technologic support for these. Section VII deals with the measurement of outcomes of stewardship programs, and includes discussion of the guidelines and milestones, economic aspects, adverse events, and stewardship in areas of increased resistance. The final section explores antimicrobial stewardship at different practice sites, looking at how it converges with the unique populations involved.

Acknowledgements

The editors would like to express their gratitude to the authors, who have tolerated both our deadlines and our comments. We challenged their patience, but hope that they will be pleased with the result. We also express our gratitude to the people behind the scenes who worked tirelessly to put this book together. In particular, we thank Jennifer DeAngelis, and Kathryn Daffinee for their assistance with the coordination of the book; their natural organizational skills, and their patience with the contributors, are greatly appreciated.

We would also like to add our own personal acknowledgments to our colleagues and families.

KLL: I would like to thank my students and fellows who continue to inspire me to work hard, stay inquisitive, and become better a better clinician and researcher. I would also like to thank my amazing husband and children for their support love and encouragement, as such an undertaking always has an impact on family life. Their support has encouraged and sustained me throughout this project.

CBC: I would like to thank my exceptional wife, my partner in all things, for all her support, love, and understanding. I also thank my father for being an outstanding dad, role model, and mentor, providing me a shining example of what it means to be an excellent person and a masterful teacher–clinician.

HJM: A huge thank you to my colleagues for their continual encouragement and for inspiring me daily. To my friends and family, many thanks for the love and support!

LBR: I would like to thank all of the colleagues and trainees who inspire me every day, and my family for understanding my strange fascination with this arcane subject.

EM: A big thank you to the teachers, colleagues, students, and administrators for tolerating my presence and to my family for tolerating my absence.

We hope that the book will provide guidance for hospitals on starting antibiotic stewardship programs and provide researchers with a valuable resource on the subject.

Kerry L. LaPlante
Cheston B. Cunha
Haley J. Morrill
Louis B. Rice
Eleftherios Mylonakis
June 2016

1 Principles of Antimicrobial Stewardship

CHESTON B. CUNHA*

Rhode Island Hospital, Providence, Rhode Island, US

Introduction

Antimicrobial Stewardship (or Antibiotic Stewardship) Programs (ASPs) have become the mechanism to optimize antimicrobial therapy within hospitals. There are many components of an ASP and these require the support and enthusiastic participation of the Infectious Disease Division, one or more Infectious Disease-trained Doctors of Pharmacy (PharmD), the Pharmacy Department, Microbiology Laboratory, and Infection Control (IC). These components should be organized under an ASP Program Director, an Infectious Disease clinician, with the requisite interpersonal, diplomatic, and organizational skills to assure ASP implementation and coordination to achieve its goals (Doron and Davidson, 2011; Hand, 2013) There is no pro forma structure for ASP programs as each hospital has its own ASP challenges. Each ASP, under the leadership and guidance of the Infectious Disease ASP Director, should tailor the ASP to the needs of the institution, i.e., some hospitals have problems with multidrug-resistant (MDR) Gram-negative bacilli, others have methicillin-resistant *Staphylococcus aureus* (MRSA) or vancomycin-resistant enterococci (VRE) concerns, and still others have problems with *Clostridium difficile* infections (Cunha *et al.*, 2013). An essential element in a successful ASP is medical staff education. Medical staff education efforts need to be ongoing and periodically to focus on different topics so as to continually reinforce basic ASP principles. Aside from the major ASP problem areas among all institutions, each hospital, in this era of limited economic resources, certainly needs and can justify an ASP on the basis of cost savings alone. The economic advantages of a well configured and executed ASP cannot be overestimated. Therefore, whether the hospital has major ASP challenges with the nosocomial problems mentioned or not, the economic benefits of ASPs should not be underestimated (Fraser *et al.*, 1997).

ASP Principles of Optimal Antibiotic Therapy

There are several key tenets of ASPs, which begin with optimal antibiotic utilization. The inappropriate or unnecessary utilization of antibiotics to treat either nonbacterial infections, e.g., viral infections, or those with fever and leukocytosis and mimicking a bacterial infection. Many antibiotic days are wasted treating "fever and leukocytosis" of nonbacterial origin and this is a needless waste of institutional resources. Aside from wasting valuable hospital resources, unnecessary antibiotic treatment also comes with the potential perils of unwanted side effects, e.g., hematologic adverse events (AEs) or antibiotic-related complications, e.g., *C. difficile*. Another area where antimicrobial therapy is unwarranted and potentially harmful is in the unnecessary treatment of "colonizers" in respiratory secretions, nonpurulent wounds, or urine in those with indwelling urinary catheters. Treating colonization is unnecessary and, in general, it is more difficult to eradicate than infection due to the same organism. The problem with needlessly "covering" colonizers in body fluids is that these organisms are often of the MDR variety and the prolonged treatment associated with trying to eradicate such pathogens is frequently complicated by the subsequent development of further antimicrobial resistance (Lutters *et al.*, 2004).

The next consideration in selecting appropriate antibiotics is to take into account the spectrum of

*E-mail: ccunha@lifespan.org

activity of the antibiotic against the known or presumed pathogen, which is related to the flora of the anatomical site of infection. All too often, clinicians use "broad spectrum" antimicrobial therapy in a "shotgun" approach regardless of anatomical location. The pathogens responsible for various intra-abdominal infections (IAIs) depend on the resident flora, which becomes the pathogenic flora in different locations in the gastrointestinal (GI) tract. Gastric and small bowel pathogens are different from biliary pathogens, which are yet again different from liver/colon pathogens. The antibiotic chosen should have the appropriate spectrum of activity and a high degree of activity against the presumed pathogens from the anatomical site of infection. The use of an antibiotic with an incorrect spectrum of activity results in suboptimal therapy or in the selection of organisms not covered by the antibiotic, e.g., MRSA, VRE (Weiss *et al.*, 2011).

Pharmacokinetic (PK) principles are important in dosing, but are critical in assessing antibiotic penetration at the site of infection. Tissue serum PK principles, i.e., serum/tissue concentration gradient, lipid solubility, pK_a, local pH, and estimated local tissue concentrations should be considered by the Infectious Disease consultant. Infectious disease consultation should be obtained in such difficult cases because of the interaction of multifactorial factors based upon PK principles. However, prescribing without the knowledge of the importance of PK factors may predispose to either therapeutic failure or antibiotic resistance at the site of infection (Cunha *et al.*, 2013).

It should be obvious that the shortest duration of antimicrobial therapy that eliminates the infection should be used, although in practice quite the opposite frequently occurs. All too often antibiotics are continued for additional days after infection has resolved. Shorter durations of therapy are clearly associated with decreased costs to the institution. In patient terms, a shorter course of therapy means a shorter length of stay (LOS) with less antibiotic exposure and less potential for AEs as well as increased antibiotic resistance potential (Pinzone *et al.*, 2014; Pogue *et al.*, 2014).

ASP Strategies for Optimal Antibiotic Therapy

Restricted formulary

The single most important component in optimizing antimicrobial therapy is formulary restriction. Formulary restriction limits antibiotic selection to preferred antibiotics based on their "low resistance potential," safety profile, and *C. difficile* potential. It is well known that the use of certain antibiotics predisposes to certain pathogens, e.g., intravenous vancomycin exerts a selective pressure on the enterococcal fecal flora, resulting in the emergence of VRE. Because there are fewer options to treat VRE than vancomycin-susceptible enterococci (VSE), it makes sense in an ASP program to educate the staff in minimizing the use of antibiotics that promote the emergence of VRE in the fecal flora at the expense of a decreased VSE population. Similarly, with MRSA, the use of some antibiotics is associated with an increased MRSA colonization, e.g., ceftazidime, and the control of MRSA begins by avoiding antibiotics that predispose to increased MRSA prevalence. The other part of MRSA containment in hospital is based on effective IC containment measures. Patients with MRSA admitted from the community introduce MRSA to hospital, and containing the intrahospital spread of MRSA from colonized/infectious patients depends entirely upon effective IC containment measures (Hayman and Sbravati, 1985; Pulcini and Gyssens, 2013; Reed *et al.*, 2013).

Minimizing emergence of multidrug-resistant Gram-negative bacilli

Another important ASP goal is to minimize the emergence of MDR Gram-negative bacilli (GNBs). Different hospitals have different problems with different organisms and the approach should be tailored to local epidemiologic concerns. In general though, the ASP operating principle is that antibiotics with a "low resistance potential," i.e., the development of resistance is fairly independent of volume and duration of use, are relatively unlikely to result in MDR GNBs. In contrast, other antibiotics with a "high resistance potential," even with limited use, have been associated with the emergence of MDR GNBs, e.g., imipenem, ceftazidime, gentamicin/tobramycin, ciprofloxacin. Within each antibiotic class, there are one or more "low resistance potential" alternatives for use by medical staff, e.g., instead of imipenem, meropenem, doripenem, or ertapenem may be used, in place of ceftazidime, cefepime or amikacin may be used, and in place of ciprofloxacin, levofloxacin or moxifloxacin may be utilized (Cunha, 1998, 2000, 2003; Pulcini *et al.*, 2014).

Major Problems of ASPs

Antibiotic resistance

Major control of antibiotic resistance depends not on antibiotic class, antibiotic volume (tonnage), or duration of antibiotic use, but rather primarily on the widespread use of "high resistance potential" antibiotics, e.g., ciprofloxacin (re: *Streptococcus pneumoniae, Pseudomonas aeruginosa*) vs. levofloxacin or moxifloxacin; imipenem (re: *P. aeruginosa*) vs. meropenem, doripenem, gentamicin, or tobramycin (re: *P. aeruginosa*) vs. amikacin; ceftazidime (re: *P. aeruginosa*) vs. cefepime or other 3rd generation cephalosporins; macrolides (re: *S. pneumoniae*) vs. doxycycline.

The antibiotics with "low resistance potential" cited above should be used preferentially over their "high resistance potential" counterparts. Another aspect of controlling resistance has to do with untoward collateral effects not causing resistance per se but related to causing changes in the flora from susceptible to more resistant organisms; e.g., the use of vancomycin does not cause an increase of enterococci resulting from colonization by VRE, but the intravenous use of vancomycin inhibits/eliminates VSE, which are normally the predominant species in bowel flora, so resulting in an increase in the number or emergence of more VRE. Similarly, the widespread use of some antibiotics predisposes to other unrelated resistant organisms; e.g., the use of ceftazidime inhibits/decreases methicillin-susceptible *S. aureus* (MSSA) in respiratory secretions in ventilated patients, leading to colonization of respiratory secretions by MRSA (Cunha, 2000, 2003).

It is a common misconception that antibiotic resistance is related to a high volume or duration of use of common antibiotics. This misconception of resistance is that over time, e.g., years of use, resistance is inevitable, leading to a lack of effective antibiotics over time. Even modest use or modest use of antibiotics over decades will lead to resistance or worsening resistance, but this occurs only with "high resistance potential" antibiotics. High volume use over years of "low resistance potential" antibiotics, e.g., ceftriaxone, doxycycline, amikacin, or nitrofurantoin, has not led to meaningful resistance problems. Unless these concepts are understood and implemented, the success of any ASP in trying to control antibiotic resistance will be limited accordingly (Cunha, 1998, 2001, 2003). Formulary restriction helps selective antibiotic prescribing to minimize the emergence of resistant pathogens, i.e., VRE, MRSA, and MDR GNBs. If, in the rare cases when there are situations when only a "high resistance potential" antibiotic is requested for a specific one time use, then this is within the purview of the ASP Stewardship Director.

If the approach of the ASP to try to control antibiotic resistance is by restricting the use of certain drug classes, e.g., 3rd generation cephalosporins, carbapenems, or quinolones, such efforts are doomed to failure unless it is understood that individual agents, within each antibiotic class, are "high resistance potential" antibiotics, and that control of resistance depends on limiting the use of these antibiotics. The restriction of "high resistance potential" antibiotics is best achieved by formulary restriction, and the control of resistant organisms within the hospital depends on effective IC containment measures (Cunha *et al.*, 2013; Landelle *et al.*, 2014).

Clostridium difficile

Perhaps the most difficult ASP challenge is the control of *C. difficile*, i.e., the causative organism of diarrhea and/or colitis. As with antibiotic resistance, unless the determinants of *C. difficile* infections are understood, control measures will be correspondingly ineffective. *C. difficile* is often community acquired, but when such patients are hospitalized, the bacteria become an IC contaminant problem. Drugs that are likely to induce the production of *C. difficile* toxin include a few, but not most, antibiotics. Little can be done to alter chemotherapy regimens and change antidepressants, but intervention aimed at avoiding the use of proton pump inhibitors (PPIs) in patients receiving quinolones is easily accomplished (Kelly and LaMont, 2008; Feazel *et al.*, 2014). (Note also that PPIs themselves have been implicated in *C. difficile* infection, e.g., Janarthanan *et al.*, 2012.) Medicine staff members often think that "all antibiotics" cause *C. difficile* infection. In fact, nearly all cases of infection by *C. difficile*, when due to antibiotic use, are limited to relatively few antibiotics, e.g., β-lactams (excluding ceftriaxone) quinolones (administered with PPIs) and clindamycin. Therefore, in patients with inflammatory bowel disease (IBD), or receiving chemotherapy, or with previous/recent/current *C. difficile* infection, every attempt should be made, if possible, to select an antibiotic that does not predispose to infection by *C. difficile*. If an antibiotic is needed in such cases, it is preferable to select one with an appropriate spectrum of activity that is protective against

C. difficile, e.g., doxycycline or tigecycline (Fowler *et al*., 2007; Valiquette *et al*., 2007).

Economic Benefits of ASPs

As mentioned previously, ASP programs should be implemented in all hospitals across the country for their economic benefits, irrespective of institutional problems with MRSA, VRE, MDR GNBs or *C. difficile*. In this era of cost containment and limited economic resources, hospital administrations should be enthusiastic in their support of ASPs, for medical as well as for their powerful economic benefits. As has already been mentioned, there are several ASP strategies to decrease antibiotic costs for the institution. The first intervention is that of decreased duration of therapy to decrease antibiotic costs. Decreased duration of therapy, aside from decreasing antibiotic costs, also minimizes the hospital's potential liability from unnecessary AEs associated with prolonged therapy. The concept of decreased duration of therapy is an important education initiative of ASPs so that the medical staff understand its rationale and basis and wholeheartedly support shorter, but effective, causes of therapy (Bantar *et al*., 2003; McQuillen *et al*., 2008).

The single most important cost-saving intervention by ASPs is the implementation of a program that switches from intravenous (IV) to oral (PO) antibiotic administration. First, at the same dose, PO antibiotics cost less than their IV counterparts. With the exception of linezolid, there are considerable cost savings in switching from an IV to a PO equivalent (same dose), e.g., levofloxacin or moxifloxacin. Second, IV to PO switch programs have other less conspicuous but even more important economic benefits than the antibiotic cost savings realized by using PO vs. IV doses of a given antibiotic. An important hospital cost savings component in IV to PO switch programs is the elimination of the cost of administering IV antibiotics. The hospital bears the related costs of the IV solution, IV tubing, antiseptics/dressings, and the pharmacy/nursing time associated with dispensing and administering IV antibiotics. When the same or an equivalent antibiotic is administered orally, obviously there are no associated IV antibiotic administration charges. These can be considerable with antibiotics that have a short half-life ($t_{1/2}$) and must be administered several times a day. The mean hospital IV administration charge per dose of IV antibiotics in the US is $10/IV dose. With antibiotics that are dosed q4h, q6h or q8h this represents, in addition to the acquisition cost of the antibiotic, an additional $60, $40, or $30/day from the IV administration charges alone. Also of critical economic importance in IV to PO switch programs is the virtual elimination of phlebitis and IV line bacteremias/sepsis, so clearly, without IV antimicrobial therapy, there are no IV-related antibiotic complications. It has been estimated that each episode of phlebitis costs the institution concerned $7500 and each episode of IV line infection costs it about $15,000. In addition, perhaps the most important, single economic benefit of IV to PO switch programs is the associated earlier hospital discharge and decreased hospital LOS. As hospital reimbursement is tied to LOS (depending upon the patient's illness), early hospital discharge with decreased LOS alone justify and pay for ASPs (Cunha, 2004, 2005, 2006).

Summary

ASPs are important because they are the mechanism to try to optimize antimicrobial therapy in all of its myriad aspects, as has been discussed. As with excellence in clinical care, excellence in antibiotic utilization, based on the aforementioned principles, should be done to improve patient care. Hospital administration should support ASP programs, not only to improve patient care, but for important economic reasons. The hospital stands to gain considerably from an effective ASP program in economic terms, particularly in those with a decreased duration of therapy and IV to PO switch components (Buyle *et al*., 2012; Eckmann *et al*., 2014). The key components of an ASP depend upon the diplomatic skills of the Infectious Disease Director, the enthusiastic support of the hospital administration, close collaboration of the Pharmacy Department, Microbiology Laboratory, and IC Department, and the education of the medical staff (Ohl and Luther 2014; Cunha, 2015). A restricted formulary is the key to success of any ASP and works best if implemented with an extensive educational program to explain the rationale of a restricted formulary. Each ASP should be tailored to the needs of the hospital and be reflective of the local epidemiology of resistance patterns and the incidence of *C. difficile* (Table 1.1) (Chung *et al*., 2013; Paño-Pardo *et al*., 2013; Lesprit *et al*., 2015).

ASP programs could be assessed by periodic audits that measure success or areas for improvement for future ASP initiative. All ASP programs must be

Table 1.1. Antibiotic stewardship programs (ASPs): overview of essential elements.

Key tenets of optimal antibiotic use	Consequences if ASP not implemented
ASP antibiotic therapeutic considerations	
• *Avoid "covering" or treating colonizers* in respiratory secretions, nonpurulent wounds and urine in those with indwelling urinary catheters	• Unnecessary wastes of antibiotic resources • Prolonged ineffective therapy of colonizers predisposes to resistance
• Select antibiotics *with the appropriate spectrum for the site of infection*	• If spectrum is inadequate or inappropriate, suboptimal therapy likely • Potential antibiotic failure if key pathogens missed
• *Apply pharmacokinetic (PK) principles relevant to the site of infection* to achieve effective local therapeutic concentrations	• Suboptimal tissue concentrations predisposes to resistance • Suboptimal tissue concentration may result in antibiotic failure
• Use the shortest duration of therapy that eliminates the infection	• Prolonged antibiotic therapy results in needless expense • Increased potential adverse events (AEs) and increased length of stay (LOS)
ASP optimal antibiotic therapy strategies	
• Minimize selection pressure on vancomycin-susceptible enterococci (VSE) *to avoid emergence of vancomycin-resistant enterococci (VRE)*	• Minimize use of antibiotics that promote emergence of VRE, e.g., vancomycin
• *Preferred use of "low resistance potential" antibiotics by a restrictive formulary*	• Avoid "high resistance potential" antibiotics, e.g., imipenem (vs. meropenem) and ceftazidime (vs. cefepime)
• *Avoid/minimize use of antibiotics that predispose* to *Clostridium difficile*	• Avoid β-lactams (excluding ceftriaxone), ciprofloxacin (especially with proton pump inhibitors, PPIs), and clindamycin
• *Minimize* intrahospital *C. difficile spread*	• Use effective infection control (IC) containment measures
• *Intrahospital* containment of methicillin-resistant *Staphylococcus aureus* MRSA, VRE, multi-drug resistant Gram-negative bacilli (MDR GNBs)	• Use effective IC containment measures
Economic benefits of ASPs	
• Decreased duration of therapy decreases antibiotic costs	• Depending on the infection site/type, use shortest duration of therapy that is effective
• Intravenous (IV) to oral (PO) switch programs decrease antibiotic costs	• PO antibiotics cost less than IV antibiotics cost (same dose)
• Eliminates IV administration cost/dose	• IV administration charge ($10/IV dose) is eliminated
• Permits earlier discharge with shorter LOS	• Considerable cost savings to hospital
• Eliminates phlebitis/IV line sepsis	• No phlebitis/IV line infections • Cost savings to hospital of phlebitis, $7500/episode and for IV line-related bacteremias $15,000/episode

individualized to conform to the needs of a particular institution. The easiest ASP intervention to implement is an IV to PO switch program; this alone pays for the ASP. However, in general, the two most difficult problems to contain are the control of MDR GNBs and *C. difficile*. Perhaps lacking is a more complete understanding of the factors that are important in the emergence of MDR GNBs and *C. difficile*. Most recently, it has been shown that formerly unsuspected drugs, e.g., antidepressants and PPIs, may predispose to infection by *C. difficile*. ASP is an ongoing process of evolution customized for each hospital and if needed, it can be modified. While the principles of ASP are clear, the implementation of principles remains problematic, i.e., the devil is still in the details (Cunha *et al.*, 2013).

References

Bantar, C., Sartori, B., Vesco, E., Heft, C., Saúl, M., Salamone, F., and Oliva, M.G. (2003) A hospitalwide intervention program to optimize the quality of antibiotic use: impact on prescribing practice, antibiotic consumption, cost savings, and bacterial resistance. *Clinical Infectious Diseases* 37, 180–186.

Buyle, F.M., Metz-Gercek, S., Mechtler, R., Kern, W.V., Robays, H., Vogelaers, D., and Struelens, M.J. on behalf of the Members of the Antibiotic Strategy International (ABS) Quality Indicators Team (2012) Prospective multicentre feasibility study of a quality of care indicator for intravenous to oral switch therapy with highly bioavailable antibiotics. *Journal of Antimicrobial Chemotherapy* 67, 2043–2046.

Chung, G.W., Wu, J.E., Yeo, C.L., Chan, D., and Hsu, L.Y. (2013) Antimicrobial stewardship: a review of prospective audit and feedback systems and an objective evaluation of outcomes. *Virulence* 4, 151–157.

Cunha, B.A. (1998) Antibiotic resistance. Control strategies. *Critical Care Clinics* 14, 309–327.

Cunha, B.A. (2000) Antibiotic resistance. *Medical Clinics of North America* 84, 1407–1429.

Cunha, B.A. (2001) Effective antibiotic-resistance control strategies. *The Lancet* 357, 1307–1308.

Cunha, B.A. (2003) Penicillin resistance in pneumococcal pneumonia. Antibiotics with low resistance potential are effective and pose less risk. *Postgraduate Medicine* 113, 42–44.

Cunha, B.A. (2004) Empiric oral monotherapy for hospitalized patients with community-acquired pneumonia: an idea whose time has come. *European Journal of Clinical Microbiology and Infectious Diseases* 23, 78–81.

Cunha, B.A. (2005) Oral antibiotic treatment of MRSA infections. *The Journal of Hospital Infection* 60, 88–90.

Cunha, B.A. (2006) Oral antibiotic therapy of serious systemic infections. *Medical Clinics of North America* 90, 1197–1222.

Cunha, B.A. (ed.) (2015) *Antibiotic Essentials*, 14th edn. JayPee Medical, New Delhi.

Cunha, C.B., Varughese, C.A., and Mylonakis, E. (2013) Antimicrobial stewardship programs (ASPs): the devil is in the details. *Virulence* 4, 147–149.

Doron, S. and Davidson, L. (2011) Antimicrobial Stewardship. *Mayo Clinic Proceedings* 86, 1113–1123.

Eckmann, C., Lawson, W., Nathwani, D., Solem, C.T., Stephens, J.M., Macahilig, C., Simoneau, D., Hajek, P., Charbonneau, C., Chambers, R. *et al.* (2014) Antibiotic treatment patterns across Europe in patients with complicated skin and soft-tissue infections due to methicillin [sic]-resistant *Staphylococcus aureus*: a plea for implementation of early switch and early discharge criteria. *International Journal of Antimicrobial Agents* 44, 56–64.

Feazel, L.M., Malhotra, A., Perencevich, E.N., Kaboli, P., Diekema, D.J., and Schweizer, M.L. (2014) Effect of antibiotic stewardship programmes on *Clostridium difficile* incidence: a systematic review and meta-analysis. *Journal of Antimicrobial Chemotherapy* 69, 1748–1754.

Fowler, S., Webber, A., Cooper, B.S., Phimister, A., Price, K., Carter, Y., Kibbler, C.C., Simpson, A.J.H., and Stone, S.P. (2007) Successful use of feedback to improve antibiotic prescribing and reduce *Clostridium difficile* infection: a controlled interrupted time series. *Journal of Antimicrobial Chemotherapy* 59, 990–995.

Fraser, G.L., Stogsdill, P., Dickens, J.D., Jr., and Prato, B.S. (1997) Antibiotic optimization: an evaluation of patient safety and economic outcomes. *Archives of Internal Medicine* 157, 1689–1994.

Hand, K. (2013) Antibiotic stewardship. *Clinical Medicine* 13, 499–503.

Hayman, J.N. and Sbravati, E.C. (1985) Controlling cephalosporin and aminoglycoside costs through pharmacy and therapeutics committee restrictions. *American Journal of Health-System Pharmacy* 42, 1343–1347.

Janarthanan, S., Ditah, I., Adler, D.G., and Ehrinpreis, M.N. (2012) *Clostridium difficile*-associated diarrhea and proton pump inhibitor therapy: a meta-analysis. *The American Journal of Gastroenterology* 107, 1001–1010.

Kelly, C.P. and LaMont, J.T. (2008) *Clostridium difficile*—more difficult than ever. *The New England Journal of Medicine* 359, 1932–1940.

Landelle, C., Marimuthu, K., and Harbarth, S. (2014) Infection control measures to decrease the burden of antimicrobial resistance in the critical care setting. *Current Opinion in Critical Care* 20, 499–506.

Lesprit, P., de Pontfarcy, A., Esposito-Farese, M., Ferrand, H., Mainardi, J.L., Lafaurie, M., Parize, P., Rioux, C., Tubach, F., and Lucet, J.C. (2015) Postprescription review improves in-hospital antibiotic use: a multicenter randomized controlled trial. *Clinical Microbiology, and Infection* 21: 180.e1–180.e7.

Lutters, M., Harbarth, S., Janssens, J.P., Freudiger, H., Herrmann, F., Michel, J.-P., and Vogt, N. (2004) Effect of a comprehensive multidisciplinary, educational program on the use of antibiotics in a geriatric university hospital. *Journal of the American Geriatrics Society* 52, 112–116.

McQuillen, D.P., Petrak, R.M., Wasserman, R.B., Nahass, R.G., Schll, J.A., and Martinelli, L.P. (2008) The value of infectious diseases specialists: non-patient care activities. *Clinical Infectious Diseases* 47, 1051–1063.

Ohl, C.A. and Luther, V.P. (2014) Health care provider education as a tool to enhance antibiotic stewardship practices. *Journal of Antimicrobial Chemotherapy* 28, 177–193.

Paño-Pardo, J.R., Campos, J., Natera Kindelán, C., and Ramos, A. (2013) Initiatives and resources to promote antimicrobial stewardship. *Enfermedades Infecciosas y Microbiología Clínica* 31(Suppl 4), 51–55.

Pinzone, M.R., Cacapardo, B., Abbo, L., and Nunnari, G. (2014) Duration of antimicrobial therapy in community acquired pneumonia: less is more. *The Scientific World Journal* 2014: 759138.

Pogue, J.M., Mynatt, R.P., Marchaim, D., Zhao, J.J., Barr, V.O., Moshos, J., Sunkara, B., Chopra, T., Chidurala, S., and Kaye, K.S. (2014) Automated alerts coupled with antimicrobial stewardship intervention lead to decreases in length of stay in patients with Gram-negative bacteremia. *Infection Control and Hospital Epidemiology* 35, 132–138.

Pulcini, C. and Gyssens, I.C. (2013) How to educate prescribers in antimicrobial stewardship practices. *Virulence* 4, 192–202.

Pulcini, C., Botelho-Nevers, E., Dyar, O.J., and Harbarth, S. (2014) The impact of infectious disease specialists on antibiotic prescribing in hospitals. *Clinical Microbiology and Infection* 20, 963–972.

Reed, E.E., Stevenson, K.B., West, J.E., Bauer, K.A., and Goff, D.A. (2013) Impact of formulary restriction with prior authorization by an antimicrobial stewardship program. *Virulence* 4, 158–162.

Valiquette, L., Cossette, B., Grant, M.P., Diab, H., and Pépin, J. (2007) Impact of a reduction in the use of high-risk antibiotics on the course of an epidemic of *Clostridium difficile*-associated disease caused by the hypervirulent NAPI/027 strain. *Clinical Infectious Diseases* 45(Supplément 2), S112–S121.

Weiss, K., Blais, R., Fortin, A., Lantin, S., and Gaudet, M. (2011) Impact of a multipronged education strategy on antibiotic prescribing in Quebec, Canada. *Clinical Infectious Diseases* 53, 433–439.

2 Clinical Perspectives on Antimicrobial Stewardship

Styliani Karanika, Suresh Paudel, and Eleftherios Mylonakis*

The Warren Alpert Medical School of Brown University, Rhode Island Hospital, Providence, Rhode Island, US

Introduction

In the US, about one third of inpatients and more than two thirds of critically ill patients are on antimicrobial therapy (Vincent *et al.*, 2009; Fridkin *et al.*, 1999), and antimicrobial agents have significantly increased average life expectancy and allowed medical procedures such as organ transplantation. However, when Alexander Fleming warned in 1945 of the lurking dangers from nonjudicious use of penicillin (Fleming, 1945), nobody could have imagined the recent worldwide challenges associated with multidrug-resistant infections. According to the 2013 report from the US Centers for Disease Control and Prevention (CDC), about 2 million patients are infected yearly with antibiotic-resistant organisms in the US and about 23,000 deaths are directly attributed to these infections (CDC, 2013).

The inappropriate use of broad-spectrum antimicrobial agents destroys the normal commensal flora, thereby compromising host immunity and leading to the development of various opportunistic infections (Corona *et al.*, 2010). The increasing incidence of these infections results in increasing mortality rates (Cosgrove *et al.*, 2002; CDC, 2016). Adding to this challenge is the reluctance of the pharmaceutical industry to invest in the development of new antimicrobial agents. Consequently, antimicrobial stewardship has risen to prominence as a potential strategy for preserving antimicrobial utility, and the CDC has announced that all acute care hospitals should urgently implement Antibiotic Stewardship Programs (ASPs) (Dellit *et al.*, 2007, Fridkin *et al.*, 2014). During the last 5 years, both the CDC (2015) and the World Health Organization (WHO, 2011) have launched campaigns in an effort to draw attention to the uneven battle against antibiotics and resistance; these have called to action all practitioners to assume their full responsibility of this urgent problem. Also, legislation such as that from the California Department of Public Health (CDPH), mandates all hospitals to implement ASPs (CDPH, 2014).

Although the approach adopted varies from institution to institution, the term "antimicrobial stewardship" describes an integrated strategy of optimizing the use of antimicrobial agents in order to improve patient outcomes, establish cost-effective approaches, and reduce side effects caused by antimicrobial use, including nosocomial infections and microbial resistance (MacDougall and Polk, 2005; Dellit *et al.*, 2007). As described in the 2007 Infectious Disease Society of America (IDSA)/Society for Healthcare Epidemiology of America (SHEA) guidelines (Dellit *et al.*, 2007), ASPs usually comprise an extensive team led by an infectious disease (ID) physician and a clinical pharmacist with ID training. This highly dynamic team also leans on the support of other groups, such as infection control (IC) professionals, hospital epidemiologists, clinical microbiologists and information system specialists.

The Development of an ASP

The initial step in the development of an ASP involves the identification of resistance patterns in the hospital

*Corresponding author. E-mail: emylonakis@lifespan.org

concerned and a review of the antimicrobial use through the combined efforts of IC professionals, and pharmacy and laboratory staff. The successive steps include an assessment of current resources, the identification of priority areas and intervention plans, the participation of hospital leaders in the program, the development of a business plan, and, finally, the implementation of the plan into action (Doron and Davidson, 2011). As noted above, an ASP should be customized and serve the special needs of an institution. However, there are some interventions that can be generalized. Thus, the best way to determine which strategy will work best for each institution is to assess weak spots, overall culture and mentality, and the estimated ease of implementation.

After the initial evaluation, further actions and the establishment of a multidisciplinary approach are needed. Prospective audit and feedback is one of the main core strategies that has been employed and found to be beneficial (Dellit *et al.*, 2007). This strategy includes case review by an expert, usually an ID clinician and/or an educated pharmacist, who provides feedback, especially if the antibiotics that have been selected have been considered to be unsuitable. This technique, although costly and labor intensive, appears to be more welcomed by physicians as they are free to choose to comply with the provided recommendation (Seto *et al.*, 1996), and it offers opportunities for education and can be easily tailored according to the existing resources and expertise, and the availability of personnel. Clinical studies have shown the direct efficacy of this type of implementation (Fraser *et al.*, 1997; Solomon *et al.*, 2001; Di Pentima and Chan, 2010; DiazGranados, 2012). DiazGranados *et al.* (2012) performed a prospective study in a hospital that evaluated the effect of prospective audit and feedback among 692 intensive care unit (ICU) patients, and found that this approach can influence the selection of antibiotics and the subsequent development of resistance. Di Pentima and Chan (2010) implemented real-time consultation on vancomycin regimens among 1709 pediatric patients, along with an automated report system on prescriptions, and found that this resulted in a reduction of vancomycin use ($P < 0.001$) and the avoidance of overuse errors ($P < 0.05$) (Di Pentima and Chan, 2010).

Formulary restriction or prior authorization

Another strategy in the development of an ASP includes formulary restrictions or prior authorizations. Many hospitals follow a selective formulary according to local susceptibility results in order to lead physicians to opt for particular antibiotics. Although it is common for hospitals to apply restrictions to certain antimicrobial agents, it should be highlighted that the acquired resistance is frequently not an antibiotic class phenomenon (Aldeyab *et al.*, 2012), and that the restrictions should focus primarily on agents with “high resistant potential” instead of on the whole class to avoid the phenomenon of “squeezing the balloon” (Cunha *et al.*, 2013). Additionally, prior authorization implementation requires a well-coordinated system in which the approvals should be obtained in a timely manner (Dancer *et al.*, 2013). Various studies have presented convincing data for the efficacy of these strategies (Quale *et al.*, 1996; White *et al.*, 1997). For example, Carling *et al.* (2003) prospectively studied this approach and showed that the restriction of high-risk antibiotics resulted in a significant reduction in the incidence of resistant enterobacteriaceae and a favorable impact on the rate of vancomycin-resistant enterococci.

Optimization of antimicrobial therapy

The optimization of antimicrobial therapy in everyday clinical practice is another field that physicians should focus upon as a part of ASP development. Data show that inappropriate antibiotic prescription or misuse can occur as much as 50% of the time (CDC, 2013) and the lack of knowledge of prescribing physicians is believed to be the primary reason for this (John and Fishman, 1997). Antimicrobial therapy is often started empirically and the reassessment of antimicrobial choices is a common omission. Providers should make sure that antimicrobial choices are reviewed frequently, especially 48 h after the start of treatment, and taking into account the newly available clinical and laboratory data, such as culture results (Stocker *et al.*, 2012). This review should include reassessment of the need for each antimicrobial agent, the use of the proper drug for the particular disease, and the evaluation of possibilities to de-escalate the regimen to more targeted antibiotics or monotherapy (Kaye, 2012, Stocker *et al.*, 2012; Pardo *et al.*, 2014). A reassessment should also be made of treatment duration, dose, route and potential for collateral damage.

Antibiotic cycling or rotation has been introduced as an ASP technique based on the hypothesis that alterations among particular antimicrobial agents or antimicrobial classes could deter the development of resistance. Raymond *et al.* (2001) made a prospective study of the rotation of empirical antibiotic treatments at a university medical center in which the rotation of the empirical antibiotic treatment took place every 3 months over a study period of 2 years. The study included 1456 ICU patients and found that this approach resulted in significant decrease in the incidence of some Gram-positive and Gram-negative infections, and decreased the infection-related mortality. Gruson *et al.* (2000) applied antibiotic cycling for 4 years and this prospective (before and after) study showed that among patients with ventilator-associated pneumonia (VAP, caused by potentially antibiotic-resistant microorganisms), the rates of occurrence of potentially antibiotic-resistant Gram-negative bacilli responsible for VAP decreased and the proportion of methicillin-sensitive *Staphylococcus aureus* (MSSA) increased from 40 to 60% of the total *S. aureus* (including methicillin-resistant *S. aureus*, MRSA). It should be noted though, that in a systematic review of studies of the efficacy of antibiotic rotation, methodological and standardization issues did not allow a clear assessment of the effectiveness of this method (Brown and Nathwani, 2005).

Pharmacodynamic dose optimization can maximize the effect of antibiotics with the minimum number of doses (Jenkins *et al.*, 2010; McCallum *et al.*, 2013), while reviewing possible changes from intravenous (IV) to oral (PO) antibiotic therapy according to bioavailability or the automatic pharmacy-ordered route change according to patient condition is another part of ASP development (Cyriac and James, 2014). More specifically, some antibiotics, such as fluoroquinolones and linezolid (Doron and Davidson, 2011), can be efficiently absorbed orally and have been suggested as decreasing the need for IV access, thus enhancing safety (Jenkins *et al.*, 2010) and resulting in significant cost savings (Ruttimann *et al.*, 2004). Also, the conversion from an IV to PO route in hemodynamically stable patients shortens the length of IV antibiotic therapy and is likely to shorten the duration of hospitalization (length of stay, LOS), reduce adverse effects (AEs), and decrease antibiotic-related expenditure (Paskovaty *et al.*, 2005).

Collaboration

The close collaboration between ID practitioners and pharmacists is one of the cornerstones of a successful ASP. Active physician participation, supported by a well-educated pharmacist, is essential for a successful de-escalation of antibiotics that promotes appropriate antibiotic use and decreases consumption, ultimately resulting in cost reduction (Vidaur *et al.*, 2005; Masterton, 2011). Also, members of the ASP can adjust and monitor the doses used by following the best tissue penetration patterns (including extended-infusion administration whenever necessary) (Avdic *et al.*, 2012; Canton and Bryan, 2012). For example, Bond and Raehl (2007) studied the management by pharmacists of 20 specific drugs through questionnaires for around 240,000 patients from 860 hospitals. They concluded that when vancomycin or aminoglycoside dosing was managed by a pharmacist using dosing nomograms, hospitals had significantly lower mortality ($P < 0.0001$), shorter LOS ($P < 0.0001$) and lower hospital costs ($P < 0.01$). Of note is that knowledge of pharmacokinetics and proper management of antimicrobial agents is particularly important among patients with organ dysfunction (Ulldemolins *et al.*, 2011).

Electronic surveillance

Electronic surveillance and other IT-support programs can assist in creating a variety of alerts or stops in many situations, with the ultimate benefit of reducing the use of antibiotics. A retrospective, cohort, multicenter study was conducted in Detroit by Pogue *et al.* (2014) on patients with monomicrobial Gram-negative bacteremia to determine the effect of active alerting of positive blood cultures on clinical parameters. The authors showed the significant efficacy of automated warnings on time to appropriate therapy (P = 0.014), length of stay (decrease of 2.2 days), and infection-related mortality. Such IT-based measures include time-limited stop orders for surgical (antibiotic) prophylaxis (Gómez *et al.*, 2006), alerts for drug interactions, and treatment decision algorithms for surgical prophylaxis that can impel physicians to follow guidelines according to the diagnosis (Hermsen *et al.*, 2008). Evans *et al.* (1998) made a prospective study of the effects of a computerized management (decision support) program for antibiotics and other anti-infective agents among 545 patients in a 12-bed ICU. When the results from implementation of the program were compared with the results from the 2 years prior to intervention, they found a significant reduction in drug prescriptions among patients with a history of drug allergy ($P < 0.01$), decreased discrepancies in the pattern of antibiotic susceptibilities

($P < 0.01$), reduced excess drug doses ($P < 0.002$), and reduced total days of treatment ($P < 0.001$).

Education

Education is an essential component of every successful ASP and should involve practicing clinicians, trainees, nurses, ancillary providers, and pharmacists in both inpatient and outpatient settings. Similar to the approach for ASPs in general, the educational and communication components of the program should be adapted and reflect the local flaws and priorities. An example of a large educational strategy is the campaign "Get Smart about Antibiotics" launched by the CDC. This campaign initially focused on outpatient healthcare providers and the general public, but later expanded to include inpatients and has been a great success (Ohl and Luther, 2014). Lectures with relevant topics, e-mail notifications with summary points, pocketbooks and in-training examinations are some of the direct ways that can used as parts of this educational effort (Chung *et al.*, 2013). Moreover, most of the above-mentioned components of an ASP, such as feedback/review, should also be seen as educational opportunities. Treatment algorithms can be part of such educational campaigns and their implementation can help physicians to standardize their approach to infections such as pneumonia, soft-tissue infections, infections by *Clostridium difficile*, and urinary tract infections. In all standardized methods, attention is needed to identify those patients who are not eligible for the implementation of such algorithms. For instance, patients with sepsis and those with multiple comorbidities require personalized clinical decisions beyond the provided streamlines.

Implementation

The implementation of an ASP should include monitoring the influence of interventions. Studies often use clinical outcomes, such as all-cause mortality, infection-related mortality, readmission rate, and duration of hospitalization in order to demonstrate that ASPs can improve patient outcomes (Pogue *et al.*, 2014). For example, encouraging outcomes on the length of hospital stay in the ASP intervention group compared with the control population have been demonstrated (Pasquale *et al.*, 2014; Pogue *et al.*, 2014). Additional, outcomes can be estimated using data obtained from patient records, the clinical microbiology laboratory, and infection control surveys. These outcomes are measured in terms of percentages of organisms resistant to certain antimicrobials, the percentage of multidrug-resistant organisms, or the number of infections due to specified organisms, and records of resistant organisms recovered, which is also a useful index (MacDougall and Polk, 2005; Schechner *et al.*, 2013). Similarly, the implementation of ASPs has resulted in increased and earlier ID consults that could rapidly improve patient outcomes, including mortality and readmission rates (Morrill *et al.*, 2014). *C. difficile* infections (CDIs) are directly associated with the use of some antibiotic classes. Thus, enhancing antimicrobial use could impact on CDI infection rate, thereby providing an indirect way to monitor and document the success of the ASP. Many hospitals already report this rate and in this way the monitoring of their antibiotic use can be ensured (Dancer *et al.*, 2013).

Measurement of antibiotic use

Additionally, it should be noted that there are two standardized statistical measures of antibiotic use—days of therapy (DOT), which is calculated per patient, and defined daily dose (DDD), which depicts total antibiotic consumption per institution. Both are popular in registering antibiotic data and the CDC has recently developed software by means of which every hospital with suitable expertise should be able to enter DOT data for further review (CDC, 2016). A significant reduction in the DDD of broad-spectrum antibiotics such as cephalosporins and fluoroquinolones has been reported by studies conducted in the US and European hospitals after the implementation of ASPs (Pate *et al.*, 2012; Borde *et al.*, 2014). Last, but not least, cost savings registration is another tool which realistically portrays the value of ASP performance (Nowak *et al.*, 2012). Clinical and microbiological outcomes, together with antimicrobial outcomes in terms of dosing, frequency, selection, timing, and route of administration, have been studied to determine the overall performance of ASPs (Wagner *et al.*, 2014).

Conclusion

In conclusion, ASPs have been expanded over the last decade and have had a positive impact on the use of antimicrobial agents and the subsequent development of antibiotic resistance. A growing number of ASP strategies, covering every aspect of a potentially beneficial intervention, have been developed and implemented (Johannsson *et al.*, 2011). Although institutions still face tough challenges in competing for this trend due to inadequate staffing and resources (Johannsson *et al.*, 2011), in future, ASPs will be proven to be worthy of physicians' expectations for combating antibiotic resistance.

References

Aldeyab, M.A., Harbarth, S., Vernaz, N., Kearney, M.P., Scott, M.G., Darwish Elhajji, F.W., Aldiab, M.A., and McElnay, J.C. (2012) The impact of antibiotic use on the incidence and resistance pattern of extended-spectrum beta-lactamase-producing bacteria in primary and secondary healthcare settings. *British Journal of Clinical Pharmacology* 74, 171–179.

Avdic, E., Cushinotto, L.A., Hughes, A.H., Hansen, A.R., Efird, L.E., Bartlett, J.G., and Cosgrove, S.E. (2012) Impact of an antimicrobial stewardship intervention on shortening the duration of therapy for community-acquired pneumonia. *Clinical Infectious Diseases* 54, 1581–1587.

Bond, C.A. and Raehl, C.L. (2007) Clinical and economic outcomes of pharmacist-managed antimicrobial prophylaxis in surgical patients. *American Journal of Health-System Pharmacy* 64, 1935–1942.

Borde, J.P., Kaier, K., Steib-Bauert, M., Vach, W., Geibel-Zehender, A., Busch, H., Bertz, H., Hug, M., de With, K., and Kern, W.V. (2014) Feasibility and impact of an intensified antibiotic stewardship programme targeting cephalosporin and fluoroquinolone use in a tertiary care university medical center. *BMC Infectious Diseases* 14:201.

Brown, E.M. and Nathwani, D. (2005) Antibiotic cycling or rotation: a systematic review of the evidence of efficacy. *Journal of Antimicrobial Chemotherapy* 55, 6–9.

Canton, R. and Bryan, J. (2012) Global antimicrobial resistance: from surveillance to stewardship. Part 2: stewardship initiatives. *Expert Review of Anti-infective Therapy* 10, 1375–1377.

Carling, P., Fung, T., Killion, A., Terrin, N., and Barza, M. (2003) Favorable impact of a multidisciplinary antibiotic management program conducted during 7 years. *Infection Control and Hospital Epidemiology* 24, 699–706.

CDC (2013) *Antibiotic Resistance Threats in the United States, 2013*. Centers for Disease Control and Prevention, Atlanta, Georgia. Available at: http://www.cdc.gov/drugresistance/pdf/ar-threats-2013-508.pdf (accessed 27 April 2016).

CDC (2015) Get Smart: Know When Antibiotics Work. Centers for Disease Control and Prevention, Atlanta, Georgia. Available at: http://www.cdc.gov/getsmart/community/index.html (accessed 27 April 2016).

CDC (2016) *Antimicrobial Use and Resistance (AUR) Module*. Centers for Disease Control and Prevention, Atlanta, Georgia. Available at: http://www.cdc.gov/nhsn/PDFs/pscManual/11pscAURcurrent.pdf (accessed 27 April 2016).

CDPH (2014) The California Antimicrobial Stewardship Program Initiative. California Department of Public Health, Sacramento, California. Available at: http://www.cdph.ca.gov/programs/hai/Pages/antimicrobialStewardship-ProgramInitiative.aspx (accessed 27 April 2016).

Chung, G.W., Wu, J.E., Yeo, C.L., Chan, D., and Hsu, L.Y. (2013) Antimicrobial stewardship: a review of prospective audit and feedback systems and an objective evaluation of outcomes. *Virulence* 4, 151–157.

Corona, A., Bertolini, G., Lipman, J., Wilson, A.P., and Singer, M. (2010) Antibiotic use and impact on outcome from bacteraemic critical illness: the BActeraemia Study in Intensive Care (BASIC). *Journal of Antimicrobial Chemotherapy* 65, 1276–1285.

Cosgrove, S.E., Kaye, K.S., Eliopoulous, G.M., and Carmeli, Y. (2002) Health and economic outcomes of the emergence of third-generation cephalosporin resistance in *Enterobacter* species. *Archives of Internal Medicine* 162, 185–190.

Cunha, C.B., Varughese, C.A., and Mylonakis, E. (2013) Antimicrobial stewardship programs (ASPs): the devil is in the details. *Virulence* 4, 147–149.

Cyriac, J.M. and James, E. (2014) Switch over from intravenous to oral therapy: a concise overview. *Journal of Pharmacology and Pharmacotherapeutics* 5, 83–87.

Dancer, S.J., Kirkpatrick, P., Corcoran, D.S., Christison, F., Farmer, D., and Robertson, C. (2013) Approaching zero: temporal effects of a restrictive antibiotic policy on hospital-acquired *Clostridium difficile*, extended-spectrum beta-lactamase-producing coliforms and methicillin [sic]-resistant *Staphylococcus aureus*. *International Journal of Antimicrobial Agents* 41, 137–142.

Dellit, T.H., Owens, R.C., McGowan, J.E., Jr., Gerding, D.N., Weinstein, R.A., Burke, J.P., Huskins, W.C., Paterson, D.L., Fishman, N.O., Carpenter, C.F. *et al.* (2007) Infectious Diseases Society of America and the Society for Healthcare Epidemiology of America guidelines for developing an institutional program to enhance antimicrobial stewardship. *Clinical Infectious Diseases* 44, 159–177.

Di Pentima, M.C. and Chan, S. (2010) Impact of antimicrobial stewardship program on vancomycin use in a pediatric teaching hospital. *Pediatric Infectious Disease Journal* 29, 707–711.

DiazGranados, C.A. (2012) Prospective audit for antimicrobial stewardship in intensive care: impact on resistance and clinical outcomes. *American Journal of Infection Control* 40, 526–529.

Doron, S. and Davidson, L.E. (2011) Antimicrobial stewardship. *Mayo Clinic Proceedings* 86, 1113–1123.

Evans, R.S., Pestotnik, S.L., Classen, D.C., Clemmer, T.P., Weaver, L.K., Orme, J.F., Jr., Lloyd, J.F., and Burke, J.P.

(1998) A computer-assisted management program for antibiotics and other antiinfective agents. *The New England Journal of Medicine* 338, 232–238.

Fleming, A. (1945) The penicillin finder assays its future. *New York Times* 26 June 1945, 21.

Fraser, G.L., Stogsdill, P., Dickens, J.D., Jr., Wennberg, D.E., Smith, R.P., Jr., and Prato, B.S. (1997) Antibiotic optimization. An evaluation of patient safety and economic outcomes. *Archives of Internal Medicine* 157, 1689–1694.

Fridkin, S.K., Steward, C.D., Edwards, J.R., Pryor, E.R., McGowan, J.E., Jr., Archibald, L.K., Gaynes, R.P., Tenover, F.C., and Project Intensive Care Antimicrobial Resistance Epidemiology (ICARE) Hospitals (1999) Surveillance of antimicrobial use and antimicrobial resistance in United States hospitals: Project ICARE Phase 2. *Clinical Infectious Diseases* 29, 245–252.

Fridkin, S.K., Baggs, J., Fagan, R., Magill, S., Pollack, L.A., Malpiedi, P., Slayton, R., Khader, K., Rubin, M.A., Jones, M. *et al.* (2014) Vital signs: improving antibiotic use among hospitalized patients. *Morbidity and Mortality Weekly Report (MMWR)* 63, 194–200.

Gómez, M.I., Acosta-Gnass, S.I., Mosqueda-Barboza, L., and Basualdo, J.A. (2006) Reduction in surgical antibiotic prophylaxis expenditure and the rate of surgical site infection by means of a protocol that controls the use of prophylaxis. *Infection Control and Hospital Epidemiology* 27, 1358–1365.

Gruson, D., Hilbert, G., Vargas, F., Valentino, R., Bebear, C., Allery, A., Bebear, C., Gbikpi-Benissan, G., and Cardina, J.-P. (2000) Rotation and restricted use of antibiotics in a medical intensive care unit. Impact on the incidence of ventilator-associated pneumonia caused by antibiotic-resistant Gram-negative bacteria. *American Journal of Respiratory and Critical Care Medicine* 162, 837–843.

Hermsen, E.D., Smith Shull, S., Puumala, S.E., and Rupp, M.E. (2008) Improvement in prescribing habits and economic outcomes associated with the introduction of a standardized approach for surgical antimicrobial prophylaxis. *Infection Control and Hospital Epidemiology* 29, 457–461.

Jenkins, T.C., Sabel, A.L., Sarcone, E.E., Price, C.S., Mehler, P.S., and Burman, W.J. (2010) Skin and soft-tissue infections requiring hospitalization at an academic medical center: opportunities for antimicrobial stewardship. *Clinical Infectious Diseases* 51, 895–903.

Johannsson, B., Beekmann, S.E., Srinivasan, A., Hersh, A.L., Laxminarayan, R., Polgreen, P.M., and The Infectious Diseases Society of America Emerging Infections Network (2011) Improving antimicrobial stewardship: the evolution of programmatic strategies and barriers. *Infection Control and Hospital Epidemiology* 32, 367–374.

John, J.F., Jr. and Fishman, N.O. (1997) Programmatic role of the infectious diseases physician in controlling antimicrobial costs in the hospital. *Clinical Infectious Diseases* 24, 471–485.

Kaye, K.S. (2012) Antimicrobial de-escalation strategies in hospitalized patients with pneumonia, intra-abdominal infections, and bacteremia. *Journal of Hospital Medicine* 7(Suppl 1), S13–21.

MacDougall, C. and Polk, R.E. (2005) Antimicrobial stewardship programs in health care systems. *Clinical Microbiology Reviews* 18, 638–656.

Masterton, R.G. (2011) Antibiotic de-escalation. *Critical Care Clinics* 27, 149–162.

McCallum, A.D., Sutherland, R.K., and Mackintosh, C.L. (2013) Improving antimicrobial prescribing: implementation of an antimicrobial i.v.-to-oral switch policy. *Journal of the Royal College of Physicians of Edinburgh* 43, 294–300.

Morrill, H.J., Gaitanis, M.M., and LaPlante, K.L. (2014) Antimicrobial stewardship program prompts increased and earlier infectious diseases consultation. *Antimicrobial Resistance and Infection Control* 3:12.

Nowak, M.A., Nelson, R.E., Breidenbach, J.L., Thompson, P.A., and Carson, P.J. (2012) Clinical and economic outcomes of a prospective antimicrobial stewardship program. *American Journal of Health-System Pharmacy* 69, 1500–1508.

Ohl, C.A. and Luther, V.P. (2014) Health care provider education as a tool to enhance antibiotic stewardship practices. *Infectious Disease Clinics of North America* 28, 177–193.

Pardo, J., Klinker, K.P., Borgert, S.J., Trikha, G., Rand, K.H., and Ramphal, R. (2014) Time to positivity of blood cultures supports antibiotic de-escalation at 48 hours. *Journal of Hospital Medicine* 48, 33–40.

Paskovaty, A., Pflomm, J.M., Myke, N., and Seo, S.K. (2005) A multidisciplinary approach to antimicrobial stewardship: evolution into the 21st century. *International Journal of Antimicrobial Agents* 25, 1–10.

Pasquale, T.R., Trienski, T.L., Olexia, D.E., Myers, J.P., Tan, M.J., Leung, A.K., Poblete, J.E., and File, T.M., Jr. (2014) Impact of an antimicrobial stewardship program on patients with acute bacterial skin and skin structure infections. *American Journal of Health-System Pharmacy* 71, 1136–1139.

Pate, P.G., Storey, D.F., and Baum, D.L. (2012) Implementation of an antimicrobial stewardship program at a 60-bed long-term acute care hospital. *Infection Control and Hospital Epidemiology* 33, 405–408.

Pogue, J.M., Mynatt, R.P., Marchaim, D., Zhao, J.J., Barr, V.O., Moshos, J., Sunkara, B., Chopra, T., Chidurala, S., and Kaye, K.S. (2014) Automated alerts coupled with antimicrobial stewardship intervention lead to decreases in length of stay in patients with Gram-negative bacteremia. *Infection Control and Hospital Epidemiology* 35, 132–138.

Quale, J., Landman, D., Saurina, G., Atwood, E., DiTore, V., and Patel, K. (1996) Manipulation of a hospital antimicrobial formulary to control an outbreak of vancomycin-resistant enterococci. *Clinical Infectious Diseases* 23, 1020–1025.

Raymond, D.P., Pelletier, S.J., Crabtree, T.D., Gleason, T.G., Hamm, L.L., Pruett, T.L., and Sawyer, R.G. (2001) Impact of a rotating empiric antibiotic schedule on infectious mortality in an intensive care unit. *Critical Care Medicine* 29, 1101–1108.

Ruttimann, S., Keck, B., Hartmeier, C., Maetzel, A., and Bucher, H.C. (2004) Long-term antibiotic cost savings from a comprehensive intervention program in a medical department of a university-affiliated teaching hospital. *Clinical Infectious Diseases* 38, 348–356.

Schechner, V., Temkin, E., Harbarth, S., Carmeli, Y., and Schwaber, M.J. (2013) Epidemiological interpretation of studies examining the effect of antibiotic usage on resistance. *Clinical Microbiology Reviews* 26, 289–307.

Seto, W.H., Ching, T.Y., Kou, M., Chiang, S.C., Lauder, I.J., and Kumana, C.R. (1996) Hospital antibiotic prescribing successfully modified by 'immediate concurrent feedback'. *British Journal of Clinical Pharmacology* 41, 229–234.

Solomon, D.H., Van Houten, L., Glynn, R.J., Baden, L., Curtis, K., Schrager, H., and Avorn, J. (2001) Academic detailing to improve use of broad-spectrum antibiotics at an academic medical center. *Archives of Internal Medicine* 161, 1897–1902.

Stocker, M., Ferrao, E., Banya, W., Cheong, J., Macrae, D., and Furck, A. (2012) Antibiotic surveillance on a paediatric intensive care unit: easy attainable strategy at low costs and resources. *BMC Pediatrics* 12:196.

Ulldemolins, M., Roberts, J.A., Lipman, J., and Rello, J. (2011) Antibiotic dosing in multiple organ dysfunction syndrome. *Chest* 139, 1210–1220.

Vidaur, L., Sirgo, G., Rodriguez, A.H., and Rello, J. (2005) Clinical approach to the patient with suspected ventilator-associated pneumonia. *Respiratory Care* 50, 965–974.

Vincent, J.L., Rello, J., Marshall, J., Silva, E., Anzueto, A., Martin, C.D., Moreno, R., Lipman, J., Gomersall, C., Sakr, Y. *et al.* (2009) International study of the prevalence and outcomes of infection in intensive care units. *Journal of the American Medical Association* 302, 2323–2329.

Wagner, B., Filice, G.A., Drekonja, D., Greer, N., MacDonald, R., Rutks, I., Butler, M., and Wilt, T.J. (2014) Antimicrobial stewardship programs in inpatient hospital settings: a systematic review. *Infection Control and Hospital Epidemiology* 35, 1209–1228.

White, A.C., Jr., Atmar, R.L., Wilson, J., Cate, T.R., Stager, C.E., and Greenberg, S.B. (1997) Effects of requiring prior authorization for selected antimicrobials: expenditures, susceptibilities, and clinical outcomes. *Clinical Infectious Diseases* 25, 230–239.

WHO (2011) World Health Day – 7 April 2011. Antimicrobial resistance: no action today, no cure tomorrow. World Health Organization, Geneva, Switzerland. Available at: http://www.who.int/world-health-day/2011/en/ (accessed 27 April 2016).

3 History of Antimicrobial Stewardship

Powel Kazanjian*

Taubman Center, University of Michigan Health System, Ann Arbor, Michigan, US

Introduction

Antimicrobial stewardship programs implemented in the 2000s are essentially an outgrowth of earlier efforts designed to control antibiotic overuse. Excessive usage of antibiotics was identified as a problem as far back as the 1950s when it was linked to the emergence of resistance. At that time, an effort was made to educate physicians about how resistance would accelerate unless they learned how to use antibiotics more judiciously. If overuse were left unchecked, physicians were told, it could hasten the demise of the antibiotic model itself. Decades later, it became clear that even though physicians were educated in the harms of overuse, they would not voluntarily restrain their indiscriminate antibiotic prescribing habits. By the 1970s, hospitals eventually resorted to implementing programs designed to limit antibiotic use through externally imposed restrictions. However, these programs remained ineffective because they lacked the authority to enforce antibiotic restrictions. The stewardship programs implemented in the 2000s sought to address these limitations by vesting members trained in infectious diseases with the authority to deny inappropriately prescribed antibiotics. These later programs offered the collateral promise of decelerating the burden of antibiotic-resistant organisms. Because stewardship programs cannot be disaggregated from antimicrobial resistance, it is necessary to tell their history together with that of resistance.

Resistance Accompanies Antimicrobial Use

Resistance was first recognized in the 1880s at the very inception of the development of antimicrobials. At that time, Robert Koch in Germany and Louis Pasteur in France were testing antimicrobials for the purpose of disinfecting germs that contaminated hospital environments (Bulloch, 1938). Pasteur and Koch carried out their studies during a time when animosities between their two nations festered in the aftermath of the 1871 Franco–Prussian War. The two researchers were in competition to prove that the new germ theory had practical value and to lay claim for their own nation to sensational new scientific discoveries (Gradman, 2009). They sought to apply bacteriologic techniques to identify the most potent "antiseptic' agent."

While conducting experiments in Paris in 1887, one of Pasteur's students, M.G. Kossiakoff, unexpectedly found that bacteria originally killed by boric acid and phenol and bichloride of mercury became "tolerant" to them (Kossiakoff, 1887). He coined the term "resistance" to describe the phenomenon of microbes becoming "acclimated" over time to increasing concentrations of disinfecting agents so that they were no longer destroyed by them. The French researcher concluded that resistance would become an issue of interest beyond the learned physicians of the day, and he asserted that it would someday become of interest to the laity as well.

Meanwhile, in Germany, Koch strove not to be eclipsed by his French counterparts. Paul Ehrlich, a student of Koch, began to test chemotherapy not only for disinfecting surfaces, but also for treating experimentally induced infections in laboratory animals (Bulloch, 1938). He tested the activity of dyes developed by German chemists for the purpose of decorating clothing sold by the thriving garment factories that fueled the German economy. Noting that certain dyes had differential staining affinity for particular microbes, Ehrlich postulated that they might also have specific killing activity. To test his hypothesis, Ehrlich studied the chemotherapeutic

*E-mail: pkazanji@med.umich.edu

activity of a variety of factory-obtained dyes in his trypanosomal mouse model (Ehrlich, 1909). Using a laborious trial and error method, Ehrlich identified both azo and trypan blue dyes that killed the trypanosomes (Ehrlich, 1909), but he also described that the microbes concomitantly developed resistance to one or the other dye in a specific fashion. Noting that mice with dye-resistant strains did not survive treatment, he postulated that specific resistance would be lethal unless the mice were treated with an alternate agent (Ehrlich, 1909). Later, another student of Koch, Julius Morgenroth, described resistance in a different setting. He showed in 1910 that resistance to the bacterium pneumococcus developed when infected mice were being treated with a derivative of quinine, ethylhydrocupreine (optochin) (Morgenroth and Levy, 1911). Thus, by 1910, German researchers had identified fundamental issues about resistance—its rapid onset, its ability to compromise antimicrobial efficacy, and its specificity, which could be averted by using an alternate drug.

Researchers also identified resistance as soon as antimicrobials were first used in humans. Optochin was licensed for use in humans in 1913, shortly after Morgenroth's animal experiments (Bulloch, 1938). Physicians immediately observed that not all humans receiving the drug responded to therapy. Seeking an explanation, Henry Moore and Alan Chesney, researchers at the Rockefeller Institute in New York, turned their attention to pneumococcal microbes recovered from respiratory specimens from humans who had failed therapy. They showed that these microbes required larger amounts of optochin than previously employed to inhibit growth in vitro, a phenomenon they called "fastness" (Moore and Chesney, 1917). They speculated that the pneumococcus organism could be "educated" to grow in the presence of a particular drug to which it had been exposed (Moore and Chesney, 1917). Optochin never gained widespread traction for use in humans, not because of resistance, but because it gave rise to troubling toxicities—including tinnitus, deafness, and blindness. In the absence of a safe alternate drug, physicians opted to use antisera to treat pneumococcal infections (Kazanjian, 2004). Although serum therapy was effective, it also had its own set of problems. Allergic reactions developed in response to horse sera and delays resulted from a need to identify a particular strain of pneumococcus and then match it with the strain of the pneumococcal isolate (Finland and Sutliff, 1933). Meanwhile, to circumvent these problems, investigators continued a vigilant search for an effective, nontoxic chemotherapeutic agent to treat human infections.

It was not until two decades later that newly discovered antimicrobial agents became available for use in humans. In 1936, Gerhard Domagk, a German pathologist, developed a chemically modified azo dye, prontosil—the first sulfa drug—which was effective when given to animals infected with streptococci (Domagk, 1935). Shortly thereafter, prontosil was shown to be effective in humans with pneumonia and urinary infections. Concomitant with its use in humans, resistance to prontosil was demonstrated in bacteria isolated from the sputum and urine of patients who failed therapy (MacLean *et al.*, 1939). As was the case for optochin, enthusiasm for using this sulfa drug was tempered not by the development of resistance, but by its array of toxic side effects, including rash and gastrointestinal intolerance. By 1941, however, a new antibiotic, penicillin, garnered immediate favor among physicians when it was first used to treat infections in soldiers wounded in battle during World War II (Abraham *et al.*, 1941). Unlike the earlier antimicrobials, enthusiasm for penicillin soared, in part because of its comparatively low toxicity profile. As had been the case with the earlier drugs, resistance among some bacteria, especially staphylococci, was described immediately following use of penicillin in humans (Rammelkamp and Maxon, 1942). Investigators in the 1940s noted a pattern—it seemed almost inevitable that a pathogen developed resistance soon after antibacterial drugs had been used in humans.

Is Resistance a Problem?

Did physicians view resistance as a clinical concern, or merely an insignificant by-product of antibiotic therapy? The dominant view among physicians up to the mid-20th century was that resistance was a trivial occurrence that could be overlooked (Abraham, 1953). This attitude, which may seem cavalier today, becomes comprehensible when it is considered in its historical context. Resistance emerged during the progressive era in the late 19th century, a time when belief in science as a means to deliver certain, dependable medical therapies was at its apogee (Diner, 1998). Anti infective therapies were the most noteworthy achievement of scientific medicine at the time, and acknowledging resistance as a problem could only tarnish the luster of the triumphant narrative of scientific medicine (Kazanjian, 2004).

The early faith in biomedicine was later reinforced when penicillin was introduced during World War II. At the time, a belief in advanced technology, including new weapons to defeat the enemy or antibiotics to keep the troops healthy after sustaining battle wounds, was viewed a key factor responsible for America's victory (Kazanjian, 2004). The metaphor of penicillin as a miracle drug that could vanquish microbial enemies was carried over to the postwar civilian period. During the 1950s, technologic advancements enabled mass production to overcome what was perceived to be penicillin's one limitation—its limited supply (Roueché, 1953). Now that the drug was available in surplus, people became captivated by the hope that potent antibiotics, together with effective vaccines, would someday render infectious diseases obsolete (Winslow, 1943). If physicians were to acknowledge resistance as a problem, it would only deter this confidence in biomedicine. Moreover, physicians could now witness firsthand the drama of restoring health to humans who had infections that would have previously been lethal. This memory would strengthen their already unwavering faith in penicillin.

Physicians became so confident in the powers of penicillin and the antibiotics that were to follow that they were willing to prescribe them without restraint. How often did physicians actually use antibiotics in circumstances where they had no demonstrated benefit? By the 1960s, a group of medical researchers studied physician prescribing habits and showed that physicians used antibiotics inappropriately in 60% of cases, including circumstances when the diagnosis was uncertain, or in situations where viral infections were more likely to be present (Scheckler and Bennett, 1970). As antimicrobials could potentially be beneficial, what was the downside of overprescribing drugs that physicians believed had magical powers and were now available in surplus?

In the 1950s, when optimism about the miracle drugs was at its pinnacle, a cadre of academic medical investigators in the US argued that physicians' casual use of antibiotics posed a grave problem. These physicians, led by Maxwell Finland at Harvard, Louis Weinstein at Tufts, Harry Dowling and Mark Lepper at Illinois, Ernest Jawetz at San Francisco, and Hobard Reimann at Jefferson, had begun their careers in the 1930s before antibiotic availability (Finland and Weinstein, 1953; Dowling *et al.*, 1955; Reimann, 1961). Acknowledging the success of antibiotics, they nonetheless focused their investigations on antibiotic misuse, which they witnessed firsthand, and the effects of resistance on populations. They linked antibiotic overuse with the hastening of resistance, which by then had accompanied aureomycin in the 1950s (Lepper, 1955). Finland (1960) showed that infections caused by the antibiotic-resistant organisms in the 1950s were not only inevitable, they were as lethal as those anteceding antibiotics. He was not the first to argue that antibiotic resistance was a medical concern; this had previously been noted following the use of optochin and sulfa drugs (Moore and Chesney, 1917; MacLean *et al.*, 1939). In addition, Alexander Fleming acknowledged that resistance to penicillin posed a problem in his Nobel Prize acceptance speech in 1945 (Fleming, 1999). Later, Rene Dubos, a plant biologist, predicted that drug-resistant organisms could threaten the antibiotic model itself (Dubos, 1959). However, Finland was the first investigator to use large-scale numbers to demonstrate that the increasing rates of resistance were associated with increased mortality (Finland, 1960). He showed that mortality from these infections, which had declined following the discovery and use of penicillin, had reverted in the 1950s to rates that preceded the introduction of antibiotics in the 1930s. Patients died of infections from drug-resistant microbes just as they had done before antibiotics were available.

The physician investigators argued that the unselected use of antibiotics in the new hospital environment of the 1950s facilitated the development of harmful resistance. New hospital structures, together with antibiotic overuse, compounded the acceleration of resistance (Finland, 1955). Changes in the hospital environment, the physicians showed, encouraged the rapid spread of resistant organisms among patients in intensive care units (ICUs) designed to provide life support for patients who had complications from heart attacks and to house ventilation systems for polio patients whose respiratory muscles were paralyzed (Dowling *et al.*, 1955). These units were equipped with devices that were used to prolong lives—mechanical ventilators to assist breathing through an endotracheal tube, cardiac monitors—including those with telemetry, external pacemakers, defibrillators, dialysis equipment for renal problems, equipment for constant monitoring of bodily functions, and an array of intravenous catheters, feeding tubes, suction pumps, and drains (Reimann, 1961). Physicians realized that the new hospital setting provided an environment that permitted the rapid transmission of bacteria from one person to another. Together, combination of the new nosocomial structures and imprudent physician prescribing habits fueled the increase in antibiotic resistance (Moser, 1956). The physician investigators argued in the 1950s that the problem of antibiotic-resistant organisms was developing so rapidly in hospitals that it offset any benefit that antimicrobials provided.

Finland and his colleagues were not content to simply observe this alarming trend. They decided to take action. How, they wondered, could physicians who had such unwavering faith in the miraculous drugs be persuaded to curb their indiscriminant and dangerous prescribing habits? The physician investigators resorted to the method they were most familiar with—education. They had faith that knowledge of the potential harm in encouraging resistance would be a sufficient deterrent for physicians to overuse antibiotics. The medical researcher–educators relied on the belief that knowledge would change physician behavior—but, as it would turn out, they underestimated the unwillingness of physicians to relinquish their habits of using antibiotics in circumstances where they were not absolutely necessary. The task of persuading physicians to voluntarily moderate their unrestrained antibiotic-prescribing behavior would prove daunting.

Physician Education

Finland and others told physicians in widely read journals in the 1950s that they had become too complacent and must correct their habit of giving antibiotics casually. The physician researcher–educators were telling physicians that they had been too optimistic about antibiotics (Reimann, 1961). They acknowledged that antibiotics definitely had favorable accomplishments in saving lives of people who would have otherwise died, but noted that they also created a problem of their own by enabling resistant infections (Finland, 1960). They also used the array of newly recognized side effects of the drugs, which became apparent by the 1950s following the licensing of penicillin, streptomycin, and aureomycin, to bolster their argument (Finland and Weinstein, 1953). They asserted that too much attention had been paid to the accomplishments of antibiotics as opposed to their dangers, which also included resistance and toxicities (Finland and Weinstein, 1953). In other words, they were using education as a method to persuade physicians to restrain their harmful habit of overprescribing antibiotics.

By the 1970s, the medical researchers elaborated on the previously recognized harms caused by antibiotic overuse. Clearly, antibiotics had not eliminated infections, as some had earlier predicted. Finland showed that overall incidence of hospital infections did not diminish in the antibiotic era (Finland, 1970). Rather, antibiotics simply shifted the bacteriologic patterns of serious infections. Finland noted that the widespread use of antibiotics at Boston City Hospital (Massachusetts) over the years was accompanied by a decline in bacteremias due to *Staphylococcus aureus* and an increase in much less frequent causes of bacteremias due to enterobacteriaceae and fungi. Finland concluded that the major factor responsible for the changing ecology of serious bacterial infections was the selective pressure of antibiotics that he claimed were "so widely used and overused" (Finland, 1970). Thus, infections continued to pose as much of a threat three decades into the antibiotic era as they had before antibiotics were available, but the problems that physicians had previously encountered had now become more varied and complex.

Finland grew anxious about this changing spectrum of infections that had emerged by the 1970s. He recognized that the problem with the new infections was that physicians did not have safe and effective drugs at hand to treat them. Consequently, patients were dying of infections just as they had done before the antibiotic era. He said that the increases in these bacteremias, then uninfluenced by licensed drugs, had "much more than compensated for the difference in number of cases and numbers of deaths" (Finland, 1970). He lamented that "the picture of serious bacterial infections, as reflected in blood stream invasion, is not a pleasant one to contemplate" (Finland, 1970). He admonished physicians for their recklessness. He said the "blame" for this "bad situation" is largely due to [physicians'] excessive and universal resort to antimicrobials" (Finland, 1970). That is, physicians, through their widespread use of antibiotics, had created a problem that was worse than the one they had started with.

By the 1970s, Finland was disconsolate that his attempts to change physician behavior through education had been ineffectual. He admitted that despite his fervent efforts, his educational tactic had had no impact whatsoever on reducing the emergence of resistant bacteria. Physicians, armed with knowledge of the dangers of antibiotic overuse, had, he acknowledged, not moderated their uncontrolled habits (Finland, 1979). Furthermore, antibiotic overuse had not abated at all over the past two decades despite his warnings, as physicians misused antibiotics at the same rate, 60%, as they had done earlier (Kunin *et al.*, 1973). Finland became resigned that promiscuous antibiotic use was an incorrigible habit which was not subject to voluntary restraint through education. He lamented that his attempts were becoming nothing more than "another jeremiad from an old and experienced physician, teacher, and clinical investigator reaching the 'end of his line'" (Finland, 1970). He then threw down the gauntlet to a new generation of physicians to become what he called "activists" to "reappraise our current efforts and to create novel methods of preventing and coping with the increasingly serious (hospital) infections" (Finland, 1970). The old guard of physician educators admitted that their tactics to resolve the problem of antibiotic overuse through education had failed.

Thus, a look into the period of the 1940s through 1970s from the perspective of antibiotic resistance shows that it was not the "golden era" antibiotic arcadia that current physicians view it as being. We have romanticized this era as being something that it never was. There was gold, Finland admitted—in curing infections in individual patients that had once been lethal—a dramatic and powerful experience for patients and doctors at their bedsides, but the gold had always been tarnished by the recognition that antibiotics had never diminished fatalities on a population level. Rather, antibiotics shifted the infections that were responsible for death in hospitals. Throughout this time we now view as golden, we hear the voices of physicians who were alarmed by the problems caused by antibiotic overuse and dismayed by their feckless attempts to reverse it. Resigned that physicians' behavior had been refractory to their outspokenness, they called for their trainees to take over the cause that they had begun.

Finland's trainees indeed dutifully accepted the invitation to become the activists that their mentor had requested, but like Finland, they too were researcher–educators who used the same tools of their mentors—education—to moderate excessive antibiotic prescribing. Operating under the same belief that education would favorably change physician behavior, the new breed of activists recast the argument to resonate with arguments made in other spheres at the time. They emphasized that antibiotics would be become a lost resource unless physicians learned to use them more cautiously. In the 1970s, they had aligned their argument of preserving a nonrenewable resource with discourses in other spheres that were garnering credibility and government action. At this time, environmentalists argued to reduce the consumption of fossil fuel use in order to preserve nonrenewable resources (A+E Networks, 2010). The environmentalist movement became mainstream in America in the 1970s when the government created the United States Environmental Protection Agency (US EPA), a body that was granted lawmaking powers, and America designated an annual Earth Day dedicated to conserving resources (US EPA, 2016). The new generation of antibiotic activists both reinforced and heightened their argument by aligning it with arguments made by environmentalists to conserve nonrenewable resources.

In addition to echoing themes of the environmentalists, the new generation of medical investigators sought an explanation for the ineffectualness of its forbear's arguments. Calvin Kunin, a trainee of Finland's, acknowledged "since the beginning of the antibiotic era," educational efforts had resulted in "no overall change in prescribing patterns" (Kunin *et al.*, 1973). Seeking to understand the reluctance to change habits from the perspective of the prescribing physician, Kunin speculated that physicians prescribe antibiotics during periods of uncertainty to avoid disastrous outcomes or out of fear of patient dissatisfaction and possibility of a lawsuit. He argued that clinicians had succumbed to perceived pressure to give priority in clinical settings to the immediate risk of individual patients over the long-term interest of microbial species to meet patients' expectations. Perhaps aware that he could not alter these factors, Kunin resorted to the same assertions of his predecessors, but delivered them more trenchantly. He stated acerbically, "overuse is to be condemned because it is wasteful and expensive" (Kunin *et al.*, 1973). Next, he warned that the problem increased the cost of medical care, and that one person's overuse affected all physicians (Dowling *et al.*, 1955). Finally, he asserted that antibiotic overuse would hasten the occurrence of the "antibiotic Armageddon" (Kunin, 1997). Kunin was stoking the latent fear of the finality of the antibiotic era as a deterrent to overuse, even though warnings of hastening the end of the antibiotic era had previously been insufficient to coax physicians to alter their habits.

Throughout the antibiotic era, the anxieties that physicians expressed about the ending of the antibiotic model have fluctuated in response to changing social and biological circumstances. The optimistic mood in the powers of biomedicine to terminate the world's epidemics in the 1950s overshadowed any consideration of resistance as a problem. This confidence had cooled by the 1980s, when AIDS shattered the false belief that infectious diseases would succumb to the laboratory's insights (Kazanjian, 2014). With a deluge of new antibiotics available for physicians to treat infections in the 1980s, the imperative to reduce overprescribing habits seemed less immediate as the number of agents available for physicians to use seemed plentiful (Powers, 2004). With an abundance of effective drugs to use, anxieties about the end of the antibiotic era had waned. By the 1990s, however, these apprehensions resurfaced as an array of microbes had rapidly become resistant to all the newly licensed antibiotics (Weber and Courvalin, 2005). Concerns about an antibiotic Armageddon were intensified in the 1990s with a curtailment of the development of new antibiotics in the pharmaceutical pipeline. At this time, increasing federal requirements to license antibiotics became a disincentive for pharmaceutical companies to produce them (Rex *et al.*, 2013). Specifically, regulations that required expensive development and low tolerance for risk on the part of regulatory agencies charged with the safety of the public were disincentives for pharmaceutical companies to develop new drugs (Laxminarayan *et al.*, 2013). As a consequence of expanding resistance and contraction of the availability of new drugs, the number of effective antibiotics that physicians had in their quivers dwindled (Weber and Courvalin, 2005). Because physicians would not voluntarily moderate their antibiotic-prescribing behavior during a period of concerns about the end of the antibiotic era, external agencies acted to impose restrictions on the physicians' liberty to prescribe.

Action: Regulation

In the 1990s, hospitals adopted specific programs to regulate antibiotic use in response to a heightened need to reform physician habits. Antibiotic regulation, however, was not an altogether a new strategy. Individual hospitals had put into place regulations and programs to restrict unnecessary antibiotics as far back as the 1950s. In an attempt to limit the increasing number of drug-resistant infections, for example, Hammersmith Hospital (London) began in 1957 to implement controls over the antibiotics that doctors prescribed. In this hospital, a policy that restricted the use of penicillin according to clearly defined indications, combined with an isolation policy, was effective (Barber *et al.*, 1960). The authors concluded that their "controlled antibiotic policy" was successful in checking the "ever increasing incidence of drug resistant staphylococcal infections" (Barber *et al.*, 1960). Later, in the 1970s, Boston City Hospital also reported the success of a regulated antibiotic policy (McGowan and Finland, 1974). Here, a system was created such that an antibiotic placed on a restricted list could be dispensed for the patient from the pharmacy only if authorized by a member of the infectious disease service (McGowan and Finland, 1974). Individual hospitals had effectively regulated antibiotic use on a small-scale and sporadic level as far back as the 1950s.

From the 1950s to 1970s, various national agencies sought to expand the scope of individual hospital control programs to a larger scale. The American Hospital Association's advisory Committee on Infections Within Hospitals, for instance, issued a requirement in 1958 that all hospitals conduct surveillance for nosocomial infections and establish an infection control program (American Hospital Association, 1958). Once established, these programs performed surveillance of nosocomial infections and formed a network, the Comprehensive Hospital Infections Project (CHIP), a research project funded by the US Centers for Disease Control (CDC), to report their findings. In response to CHIPs report on the widespread nature of nosocomial infections caused by antimicrobial-resistant organisms, the CDC recommended in 1970 that hospitals staff their programs with an infection control nurse and a hospital epidemiologist (Scheckler *et al.*, 1971). By 1976, the Joint Commission on Accreditation of Hospitals, JCAH) (now the Joint Commission on Accreditation of Healthcare Organizations, JCAHO) created infection control standards for these committees to evaluate and maintain records and they recommended surveillance personnel for each hospital (Hughes, 1987). However, JCAH fell short of providing specific aims or coordinated strategies on how to achieve these goals (Klapp and Ramphal, 1983). According to Harry Dowling, the committees were ineffectual because they lacked specific expertise and a specific strategy to limit unnecessary antibiotics, or they believed that no one should tell doctors how to treat their patients (Dowling, 1977). The committees recommended by the JCAH and CDC in the 1970s were not standardized in composition and they lacked specific goals.

To address these shortcomings, the Infectious Disease Society of America (IDSA) issued specific recommendations on committee expertise and aims in the 1980s. The IDSA gave guidelines on how to staff committees and how to adopt restriction policies for the use of selected antimicrobials (Marr *et al.*, 1988); it also addressed how to standardize programs that could be applied categorically across hospitals. It recommended that all hospitals assemble an infection control committee composed of individuals who had the necessary expertise to construct restriction lists—an infectious disease physician, infection control practitioner, a microbiologist and a pharmacist (Marr *et al.*, 1988). The IDSA gave guidance on how to create a restricted formulary, release restricted agents when predetermined protocols are followed, and provide direction in developing policies and guidelines to survey antibiotic use. Despite such explicit guidance on the structure and tasks of a committee, the IDSA stopped short of recommending that the committee be granted the authority to stop the use of antibiotics in situations where they were given without meeting the specified indication. For this reason, these programs did not succeed in stopping the spread of resistant organisms.

Following the urgent cry to save the threatened antibiotic model in the 1990s was a call for the creation of more stringent regulatory programs—the "antibiotic stewardship" programs of the 2000s. Unlike earlier restriction programs, these dedicated stewardship programs authorized physicians to deny antibiotics if they were prescribed without an accepted indication. In these programs, restricted antibiotics were first to be approved by a team of experts before they were released for use. By 1997, hospitals reported that prior authorization programs resulted in a shift from expensive broad-spectrum antibiotics to narrower agents, saved costs and increased susceptibilities to several drugs (White *et al.*, 1997). Furthermore, these programs did not compromise patient care or result in suboptimal antibiotic use, which had been a reservation of some infectious disease physicians (Dunagan and Medoff, 1993). In addition to reducing antibiotic expenditures, prior approval significantly lowered rates of infection caused by drug-resistant bacteria. (Frank *et al.*, 1997). Thus, these stewardship programs sought to improve antibiotic usage by involving administrative controls that granted experts the power to place limits on the latitude that physicians had in choosing antibiotics. Now possessing the power to discontinue an antibiotic chosen by a physician, stewardship programs had the teeth that previous programs lacked to restrict antibiotic overuse.

Notwithstanding their new regulatory authority, the stewardship programs were otherwise essentially a comprehensive program that encompassed multifaceted elements of previous restriction programs. These elements included educating healthcare workers about the importance of resistance and means to prevent or reduce its emergence and spread, creating a restricted formulary, linking improved diagnostics to identify the etiology of infections and help direct therapy, and helping to coordinate infection-control measures to prevent the transmission of resistant species (Fishman, 2006). Although the programs were intended to standardize these tasks, there was flexibility built into the actual mechanics of restricting antibiotics. These methods could vary from telephone approval, to antibiotic order forms (written justification), automatic stop orders, direct interactions, implementation of control categories, or simple chart entry. In short, they constitute a comprehensive program that systematizes and integrates the ad hoc activities of previous programs and also empowers the program to discontinue unnecessary antibiotics through a variety of mechanisms that can be customized to an individual hospital environment. In this sense, restriction programs are essentially an outgrowth of earlier restriction programs, with the exception that they have been designed to be more comprehensive and effective.

Conclusion

Today, antimicrobial stewardship programs are viewed as an urgent effort to correct a crisis of accelerating antibiotic resistance. A widely held view is that these programs have been created to address a pressing need to correct an acute problem so severe that it could jeopardize the longevity of the antibiotic model itself. This view implies that stewardship programs are the first coordinated attempt to resolve a newly recognized problem of antibiotic overuse and resistance. However, this chapter has shown that apprehensions about antimicrobial resistance posing a threat to the antibiotic model have, in fact, been longstanding. Medical researchers in the 1950s argued that the difficulties posed by resistant bacteria fueled by antibiotic overuse were even more worrisome than the problems that antibiotics were intended to solve. In the 1950s, these researchers sought to reform physician habits that if left unchecked could have hastened the end of the antibiotic era by encouraging resistance. It was only when their methods of education and reliance on the agency of the physician to voluntarily change their habits was shown to be ineffective that programs to externally impose restrictions on physician antibiotic choice were adopted. Antimicrobial stewardship programs represent the newest version of a series of attempts to rectify the chronic problem of excessive antibiotic use.

Will antibiotic stewardship programs be the last in the line of coordinated efforts to rectify antibiotic misuse? Arguably, the programs have already shown that they can reduce the expenditures of the latest most expensive antibiotics that hospitals have targeted as high priority. So hospitals have a financial incentive to continue their investments in stewardship programs. For this reason, stewardship programs will likely be long-lived. In actuality though, the overuse of costly antibiotics in the hospital setting represents only a fraction of the bulk of antibiotic overuse. Antibiotic stewardship programs cannot rectify the excessive use of less costly antimicrobials that takes place in a larger scale—the ambulatory and urgent-care settings. To date, programs targeting the restriction of all antibiotics in these loci have yet to be designed. Consequently, antibiotic use remains as imprudent in today's stewardship era as it has in the past when earlier generation restriction programs had been implemented. To rectify the problem of overuse on a larger scale, new strategies will need to address how to convince physicians to cease prescribing antibiotics pointlessly for conditions that are not infectious, or, in many circumstances, when the diagnosis is uncertain. Until programs that address these issues are designed and implemented on a wide-scale basis, the problem of accelerating antibiotic resistance will likely continue unabated in the antibiotic stewardship era.

References

Abraham, E.P. (1953) The development of drug resistance in microorganisms. In: Gale, E.F. and Davies, R. (eds) *Adaptation in Microorganisms. The Third Symposium of the Society for General Microbiology*. Cambridge University Press, Cambridge, UK, pp. 201–224.

Abraham, E.P., Chain, E., Fletcher, C.M., Florey, H.W., Gardner, A.D., Heatley, N.G., and Jennings, M.A. (1941) Further observations on penicillin. *The Lancet* 11, 177.

A+E Networks (2010) History: The 1970s. The Environmental Movement; History: Energy Crisis (1970s). A+E Networks, New York. Available at: http://www.history com/topics/1970s;http://www.history.com/topics/energy-crisis (accessed 28 April 2016).

American Hospital Association (1958) Prevention and Control of Staphylococcus Infections in Hospitals (Bulletin 1) May 21, 1958. Reprinted in: *Proceedings of the National Conference on Hospital-Acquired Staphylococcal Disease, Held at Atlanta, Georgia—September 15–17, 1958*.

Barber, M., Dutton, M.A., Beard, M.A., Elmes, P.C., and Williams, R. (1960) Reversal of antibiotic resistance in hospital staphylococcal infection. *British Medical Journal* 2, 11–17.

Bulloch, W. (1938) *The History of Bacteriology*. Dover, New York.

Diner, S. (1998) *America in the Progressive Era*. Hill and Wang, New York.

Domagk, G. (1935) Ein Beitrag zur Chemotherapie der bakteriellen Infektionen. *Deutsche Medizinische Wochenschrift* 61, 250–253.

Dowling, H. (1977) *Fighting Infection, Conquests of the Twentieth Century*. Harvard University Press, Cambridge, Massachusetts.

Dowling, H.F., Lepper, M.H., and Jackson, G.G. (1955) Clinical significance of antibiotic resistant bacteria. *Journal of the American Medical Association* 157, 327–337.

Dubos, R. (1959) *Mirage of Health*. Anchor Books, New York, pp. 1–235.

Dunagan, W.B. and Medoff, G. (1993) Formulary control of antimicrobial usage. What price freedom? *Diagnostic Microbiology and Infectious Disease* 16, 265–274.

Ehrlich, P. (1909) Ueber modern chemotherapie: Vortrag gehalten in der X; Tagung der Deutschen Dermatologischen Gesellschaft. *Beiträge zur Experimentellen Pathologie und Chemotherapie* 7, 53–69.

Finland, M. (1955) Emergence of antibiotic-resistant bacteria. *The New England Journal of Medicine* 253, 909–922, 969–978, 1019–1028.

Finland, M. (1960) Treatment of pneumonia and other serious infections. *The New England Journal of Medicine* 263, 207–221.

Finland, M. (1970) Changing ecology of bacterial infections as related to antibacterial therapy. *The Journal of Infectious Diseases* 122, 419–431.

Finland, M. (1979) Emergence of antibiotic resistance in hospitals, 1935–1975. *Reviews of Infectious Diseases* 1, 4–22.

Finland, M. and Sutliff, W. (1933) Immune reactions of human subjects to strains of pneumococci other than Types I, II, and III. *Journal of Experimental Medicine* 57, 95–103.

Finland, M. and Weinstein, L. (1953) Complications induced by antimicrobial agents. *The New England Journal of Medicine* 248, 220–226.

Fishman, N. (2006) Antimicrobial stewardship. *American Journal of Infection Control* 34, S55–S63.

Fleming, A. (1999) Penicillin, Nobel Lecture, December 11, 1945. In: *Nobel Lectures: Physiology or Medicine, 1942–1962*. World Scientific Publishing, Singapore, pp. 83–93.

Frank, M.O., Batteiger, B.E., Sorensen, S.J., Hartstein, A.I., Carr, J.A., McComb, J.S., Clark, C.D., Abel, S.R., Mikuta, J.M., and Jones, R.B. (1997) Decrease in expenditures and selected nosocomial infections following implementation of an antimicrobial-prescribing improvement program. *Clinical Performance and Quality Healthcare* 5, 180–188.

Gradman, C. (2009) *Laboratory Disease: Robert Koch's Medical Bacteriology*. Johns Hopkins University Press, Baltimore, Maryland.

Hughes, J. (1987) Nosocomial infection surveillance in the United States: historical perspective. *Infection Control* 8, 450–453.

Kazanjian, P.H. (2004) Changing interest among physicians towards pneumococcal vaccination throughout the twentieth century. *Journal of the History of Medicine and Allied Sciences* 59, 555–587.

Kazanjian, P. (2014) The AIDS pandemic in historic perspective. *Journal of the History of Medicine and Allied Sciences* 69, 351–382 (published online 22 October 2012).

Klapp, D.L. and Ramphal, R. (1983) Antibiotic restriction in hospitals associated with medical schools. *American Journal of Hospital Pharmacy* 40, 1957–1960.

Kossiakoff, M.G. (1887) De la propriété qui possedent les microbes de s'accomoder aux milieux antiseptiques. *Annales de l'Institut Pasteur. Microbiology* 1, 465–476.

Kunin, C. (1997) Antibiotic Armageddon. *Clinical Infectious Diseases* 25, 240–241.

Kunin, C.M., Tupasi, T., and Craig, W.A. (1973) Use of antibiotics: a brief exposition of the problem and some tentative solutions. *Annals of Internal Medicine* 79, 555–560.

Laxminarayan, R., Duse, A., Wattal, C., Zaidi, A.K.M., Wertheim, , H.F.L., Sumpradit, N., Vlieghe, E., Hara, G.L., Gould, I.M., Goossens, H. *et al.* (2013) Antibiotic resistance—the need for global solutions. *The Lancet Infectious Diseases* 13, 1057–1098.

Lepper, M. (1955) Microbial resistance to antibiotics. *Annals of Internal Medicine* 43, 299–315.

MacLean, I.H., Rogers, K.B., and Fleming, A. (1939) M. & B. 693 and pneumococci. *The Lancet* 1, 562.

Marr, J.J., Moffet, H.L., and Kunin, C.M. (1988) Guidelines for improving the use of antimicrobial agents in hospitals: a statement of the Infectious Diseases Society of America. *The Journal of Infectious Diseases* 157, 869–876.

McGowan, J.E., Jr. and Finland, M. (1974) Usage of antibiotics in a general hospital: effect of requiring justification. *The Journal of Infectious Diseases* 130, 165–168.

Moore, H. and Chesney, A. (1917) A study of ethylhydrocuprein (optochin) in the treatment of acute lobar pneumonia. *Archives of Internal Medicine* 19, 611–682.

Morgenroth, J. and Levy, R. (1911) Chemotherapie der Pneumokokkeninfection. *Berliner Klinische Wochenschrift* 48, 1560–1561.

Moser, R.H. (1956) Diseases of medical progress. *The New England Journal of Medicine* 255, 606–614.

Powers, J.H. (2004) Antimicrobial drug development – the past, the present, and the future. *Clinical Microbiology and Infection* 10(Suppl 4), 23–31.

Rammelkamp, C.H. and Maxon, T. (1942) Resistance of *Staphylococcus aureus* to action of penicillin. *Proceedings of the Society for Experimental Biology and Medicine* 51, 386–389.

Reimann, H.A. (1961) The misuse of antimicrobics. *Medical Clinics of North America* 45, 849–856.

Rex, J., Eisenstein, B., Alder, J., Goldberger, M., Meyer, R., Dane, A., Friedland, I., Knirsch, C., Sanhai, W.R., Tomayko, J. *et al.* (2013) A comprehensive regulatory framework to address the unmet need for new antibacterial treatments. *The Lancet Infectious Diseases* 13, 269–275.

Roueché, B. (1953) Something extraordinary. In: Roueché, B. *Eleven Blue Men*. Berkley Publishing Company, New York, pp. 139–158.

Scheckler, W.E. and Bennett, J.V. (1970) Antibiotic usage in seven community hospitals. *Journal of the American Medical Association* 213, 264–267.

Scheckler, W.E., Garner, J.S., Kaiser, A.B., and Bennett, J.V. (1971) Prevalence of infections and antibiotic usage in eight community hospitals. In: American Hospital Association. *Proceedings of the International Conference on Nosocomial Infections, August 3–6, 1970*. Centers for Disease Control, Atlanta, Georgia, pp. 299–305.

US EPA (2016) EPA History: Earth Day. US Environmental Protection Agency, Washington, DC. Available at: https://www.epa.gov/aboutepa/epa-history-earth-day (accessed 28 April 2016).

Weber, J.T. and Courvalin, P. (2005) An emptying quiver: antimicrobial drugs and resistance. *Emerging Infectious Diseases* 6, 791–793.

White, A.C., Atmar, R.L., Wilson, J., Cate, T.R., Stager, C.E., and Greenberg, S.B. (1997) Effects of requiring prior authorization for selected antimicrobials: expenditures, susceptibilities, and clinical outcomes. *Clinical Infectious Diseases* 25, 230–239.

Winslow, C.E.A. (1943) *The Conquest of Epidemic Disease*. University of Wisconsin Press, Madison, Wisconsin.

4 The Importance of Education in Antimicrobial Stewardship

INGE C. GYSSENS*

Radboud University Medical Center, Nijmegen, The Netherlands and Hasselt University, Hasselt, Belgium

Introduction

Antimicrobial resistance—the growing threat

The ability to treat infectious diseases with antimicrobials is regarded as an essential component of medical management. Antimicrobial drugs are among the most successful drugs, offering life-saving benefits in severe infections. In cases of sepsis, this is obtained by rapid administration of effective antibiotics, i.e., antibiotics to which the causative pathogen is still susceptible. Bacterial resistance to antibiotics is now considered a serious threat to the achievements of modern medicine by numerous agencies, the US Centers for Disease Control and Prevention (CDC, 2013), the European Center for Disease Prevention and Control (ECDC, 2014), World Health Organization (WHO, 2015), and governments (The White House, 2015). Few new antibiotics are in the research and development pipeline, particularly in the Gram-negative spectrum (Freire-Moran, 2011), and "old," potentially useful antibiotics are not marketed anymore in some countries (Pulcini *et al.*, 2012).

Effective antimicrobial drugs—a common good

Antimicrobials differ from other drugs in a particular way. They are the only drugs that do not directly target the patient but instead inhibit or kill invading pathogens and commensal microorganisms. Antimicrobial therapy is not only based on the characteristics of a patient and a drug, but also on the characteristics of the microorganisms causing the infection and the colonizing flora. A useful didactic tool that describes the complex interrelationships between humans, microorganisms, and antimicrobial drugs is the "pyramid of infectious diseases" (Fig. 4.1). The arrows in the pyramid illustrate the multiple interactions between the patient, the drug, the pathogen(s), and colonizing microflora. The activity of the antimicrobial is obtained at the cost of the development of resistance by the pathogen, and also of the colonizing flora. As is shown in the pyramid, the selection of the appropriate antimicrobial therapy is a complex decision, depending on knowledge of many different aspects of infectious diseases: immunological and genetic host factors, microbial virulence, and the pharmacokinetics and pharmacodynamics (PK/PD) of drugs.

Due to these complex interactions, prescribers of antimicrobial drugs experience an ethical dilemma when treating an individual patient. On the one hand, they want to choose optimal empirical therapy (i.e., broad-spectrum antimicrobials) to cover all potential pathogens, but this inevitably increases microbial resistance. On the other hand, prescribers have a responsibility toward that same patient, and other, yet unidentified patients, as well as future generations, to preserve the efficacy of antibiotics and minimize the development of resistance (Leibovici *et al.*, 2012). The first responsibility tends to promote overtreatment; the other responsibilities are usually overlooked. However, prudent antibiotic use is the only option to delay the emergence of resistance. As pointed out by Rice (2008), the simplest approach is to use fewer antibiotics and thereby apply less selective pressure to the commensal flora,

*E-mail: i.gyssens@aig.umcn.nl

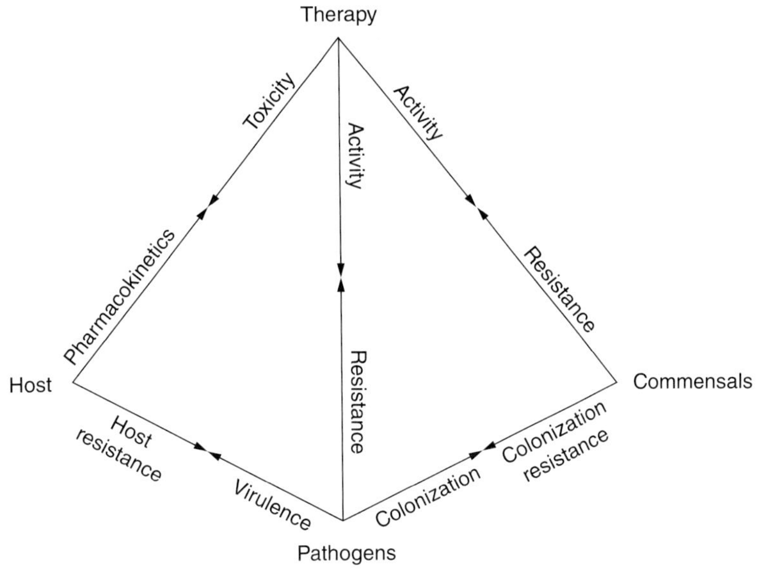

Fig. 4.1. The pyramid of infectious diseases.

and to use antibiotics to the minimum extent necessary to safely treat patients with serious infections.

Misuse of antimicrobial drugs

Antibiotics are the most extensively developed and prescribed among antimicrobial drugs. Misuse of antibiotics, i.e., unnecessary as well as inappropriate prescribing due to inadequate dosing and/or wrong duration of use is frequent; up to half of the antibiotic prescriptions, both in the community and in hospitals, are considered unjustified (Dellit *et al.*, 2007). International comparative studies show that the quantity and quality of antibiotic prescribing differs greatly between countries. In Europe, there is a trend to higher consumption from north to south and west to east (Adriaenssens *et al.*, 2011). The prevalence of resistant pathogens follows the same pattern (ECDC, 2014).

One of the major causes of overprescribing is insufficient knowledge of both infectious diseases and the responsible use of antimicrobial drugs. More than 20 years ago, in 1993, the British Society of Antimicrobial Chemotherapy (BSAC) Working Party on Antimicrobial Use considered training in Infectious Diseases (ID) and knowledge of prescribing of antimicrobial drugs insufficient in clinicians (Davey *et al.*, 1993). According to the Infectious Diseases Society of America (IDSA), clinician training and continuing education in appropriate antimicrobial use in the USA is "highly variable, non standardized, infrequent, and highly prone to bias, especially when conveyed or sponsored by pharmaceutical firms or their agents. Apart from initial training and, to a limited extent, in preparation for board recertification examinations, there is little if any compulsory training or education of physicians in antimicrobial stewardship" (Spellberg *et al.*, 2011). Conversely, focus on prescribing older, narrow-spectrum drugs in targeted therapy has been taught in medical schools and has been common practice in northern European countries such as Scandinavia and the Netherlands for several decades. (Halls, 1993). In low- and middle-income countries—settings with restricted access to medical literature—the role of up-to-date undergraduate and postgraduate education is even more important (Laxminarayan *et al.*, 2013).

Multifaceted strategies are recommended to promote prudent antibiotic use. One of the major strategies is to optimize the education of all healthcare professionals who prescribe antibiotics (WHO, 2001, 2012; Dellit *et al.*, 2007). In this chapter, all relevant aspect of the education of the public healthcare force are analyzed and discussed; there is a need for early and integrated education of all medical professionals. The principles of antimicrobial stewardship are reviewed. The time line and setting for this educational process and learning outcomes, the background of the teachers, the format of education, as well as its evaluation are also explored. Education of the public is considered to be of importance as well, as patient participation gets increasing attention in healthcare. However, in the Western world, the bulk of antibiotics are prescribed by healthcare practitioners, mostly physicians. It is, therefore logical to focus primarily on the education of the prescribers of antibiotics.

Principles of Antimicrobial Stewardship

According to the IDSA definition, antimicrobial stewardship includes optimizing the indication, selection, dosing, route of administration, and duration of antimicrobial therapy to maximize clinical cure or prevention of infection, whilst limiting the collateral damage of antimicrobial use, including toxicity, the selection of pathogenic organisms (such as *Clostridium difficile*), and the emergence of resistance (Dellit *et al.*, 2007). An important principle of prudent use is avoiding selection pressure of the antibiotic in the patient, both on the pathogen and the commensal, by choosing the least broad-spectrum antibiotic for directed therapy, adequate doses, a good timing, and the shortest possible duration of use (Fig. 4.1). The choice of the optimal broadness of spectrum of an empiric antibiotic in the era of increasing resistance is more difficult. The level of local in vitro resistance that would make an antibiotic obsolete in that setting is not known. For serious infections, a limit of 5–10% resistance is proposed, although the evidence for such a level is lacking. Antibiotic choices for empiric therapy should be guided by local antibiotic resistance and patient outcome data. In addition, in settings with a well-developed microbiology laboratory system, it is increasingly considered good practice to adapt the empiric therapy to a targeted therapy when culture results are known.

Streamlining or *de-escalating* antimicrobial therapy has become widely accepted, and intensive care physicians (Borgatta and Rello, 2014) and even hematologists (Averbuch *et al.*, 2013) have widely adopted this strategy. Targeted therapy decreases unnecessary broad-spectrum antimicrobial exposure and results in cost containment. De-escalation may also include discontinuation of empirical antimicrobial therapy based on clinical criteria and negative culture results (Dellit *et al.*, 2007).

Lack of application of these principles of infectious diseases and prudent antibiotic use may seriously hamper the quality of prescription. In this situation, the prescribing physician may prefer to err on the safe side, i.e., by prescribing maximal broad-spectrum treatment, instead of making a well-informed guess. A negative attitude, based on a lack of agreement with protocols or guidelines, will also affect prescribing. Likewise, a lack of self-efficacy, a lack of outcome expectancy, and inertia may lead to poor prescribing (Cabana *et al.*, 1999). Based on the available international recommendations for antimicrobial stewardship policies and on the literature (Davenport *et al.*, 2005; Dellit *et al.*, 2007), Table 4.1 presents the main principles for education in prudent antibiotic prescribing. The ultimate goal is to translate these principles into specific topics, concepts, disciplines, learning outcomes, and competencies for a core curriculum of medical doctors and other healthcare professionals. The program should cover the undergraduate level, the internship/foundation year, and specialty/professional training.

Education of the Professional in Antimicrobial Stewardship

In the community, antibiotics are prescribed universally by all medical doctors and dentists, unlike many other drugs, for which prescribing is kept within a specialty (for example, neuroleptic drugs). One could argue that in countries with a considerable over-the-counter (OTC) antibiotic use, the patients share this responsibility. However, even in the situation of OTC by the public, health professionals determine the purchasing behavior of the public, as patients and providers tend to copy the doctor's prescribing habits. In a survey in Indonesia, most prescribed and self-medicated antibiotics did not differ, but the self-medicated courses were shorter (Hadi *et al.*, 2008). In Europe, OTC antibiotic use by the public is low except for a few countries, e.g., Romania, Greece, Cyprus and Spain (Grigoryan *et al.*, 2010; European Commission, 2013; Guinovart *et al.*, 2015).

In hospitals, the majority of patients are treated for their infection by organ specialists. Only patients with complicated or severe infections will be referred to infectious diseases departments or medical microbiology/infectious diseases teams for consultation. On average, one patient out of three is treated with antibiotics during his/her hospital stay in an acute care facility. In the ECDC *Point Prevalence Survey on Healthcare-associated Infections and Antimicrobial Use in European Acute Care Hospitals* (ECDC, 2013),

Table 4.1. Elements of education on prudent antibiotic prescribing. Adapted from Pulcini and Gyssens (2013).

Topic	Concept, understanding	Field, discipline	Principles, learning outcomes, competencies
Bacterial resistance	Selection, mutation	(Micro)biology, genetics	• Extent and causes of bacterial resistance in pathogens (low antibiotic concentration, longtime exposure of microorganisms to antibiotics are driving resistance) • Extent, causes of bacterial resistance in commensals and the phenomenon of overgrowth (e.g. *Clostridium difficile* infection, yeast infection)
		Epidemiology	• Epidemiology of resistance, accounting for local variations and importance of surveillance (differences between wards, countries, etc.)
	Hygiene	Infection control – mostly microbiology	• Spread of resistant organisms
Antibiotics	Mechanisms of action of antibiotics/resistance Toxicity	Pharmacology	• Broad- versus narrow-spectrum antibiotics, preferred choice of narrow-spectrum drugs • Combination therapy (synergy, limiting emergence of resistance, broaden the spectrum)
	Costs	Ethics, public health, pharmacology	• Collateral damage of antibiotic use (toxicity, cost) • Consequences of bacterial resistance • Lack of development of new antibiotics (limited arsenal)
Diagnosis of infection	Infection/inflammation	Physiology /microbiology/ immunology/ infectious diseases	• Interpretation of clinical and laboratory biological markers • Fever and C-reactive protein (CRP) elevation are also a sign of inflammation, not per se of an infection
	Isolation, identification of bacteria, viruses and fungi	(Micro)biology	• Practical use of point-of-care tests (e.g. urine dipstick, streptococcal rapid antigen diagnostic test in tonsillitis, etc.) • Importance of taking microbiological samples for culture before starting antibiotic therapy
	Susceptibility to antibiotics	Microbiology/infectious diseases	• Interpretation of basic microbiological investigations (Gram stain, culture, PCR, serology, etc.)
Treatment of infection	Indication for antimicrobials	Clinical microbiology/ infectious diseases Organ specialty	• Definitions and indications of empiric/directed therapy vs. prophylaxis • Clinical situations when not to prescribe an antibiotic: ○ Colonization vs. infection (e.g., asymptomatic bacteriuria) ○ Viral infections (e.g., acute bronchitis) ○ Inflammation vs. infection (e.g., fever without a definite diagnosis in a patient with no severity criteria)
Prevention of infection		Pharmacotherapy, surgery, anesthesiology, clinical microbiology/infectious diseases	• Surgical antibiotic prophylaxis: indication, choice, duration (short), timing

Continued

Table 4.1. Continued.

Topic	Concept, understanding	Field, discipline	Principles, learning outcomes, competencies
Medical record keeping	Choice Duration Timing	Clinical medicine	• Documentation of antimicrobial indication in clinical notes • Recording (planned) duration or stop date
Prescribing antibiotics: initial steps	Empiric therapy (local guide, antibiotic booklet, etc.) Diagnostic uncertainty	Clinical microbiology/ infectious diseases/organ specialists Clinical pharmacology	• Best bacteriological guess for empiric therapy • Choice in case of prior use of antibiotics when selecting an antibiotic for empiric therapy • Choosing the dose and interval of administration (basic principles of pharmacokinetics (PK) and pharmacodynamics (PD) • Estimating the shortest possible adequate duration
Prescribing antibiotics: targeted therapy	Communication with the microbiology laboratory Value of specialist consultation in infectious diseases or microbiology	Clinical microbiology/ infectious diseases/ organ specialists Hospital pharmacy	• Reassessment of the antibiotic prescription around day 3 • Streamlining/de-escalation once microbiological results are known • IV-oral switch (bioavailability of antibiotics) • Therapeutic drug monitoring to ensure adequate drug levels (e.g. vancomycin)
Prescribing antibiotics: standard of care	Importance of guidelines in clinical practice Quality indicators of antibiotic use	Clinical medicine, organ specialists Quality Institute	• Prescribing antibiotic therapy according to national/local practice guidelines • Audit and feedback assessing prescribing practice using quality indicators
Communication skills	Discussion techniques	Psychology, clinical medicine	• Explaining to the patient the absence of an antibiotic prescription • Education of patients regarding prudent antibiotic use (comply with the doctors' prescription, no self-medication, etc.)

the overall European acute care hospital prevalence of antimicrobial use was 35%. The prevalence of use was highest in the intensive care units (ICUs) (56.5%) and in surgery (40.7%). Junior doctors of all specialties often prescribe the antibiotics under the supervision of their seniors. Therefore, both senior and junior doctors must be educated in order to change practice (De Souza *et al.*, 2006; Pulcini and Gyssens, 2013).

In some European countries (for example the UK and France), clinical pharmacists, midwives and nurses ("physician assistants" or "physician associates" in the UK, the Netherlands and Belgium) can also prescribe some antibiotics in selected clinical situations (Dryden *et al.*, 2009; NPC, 2012). In addition, pharmacists also play a key role by dispensing the drugs and advising the patients. Therefore, all healthcare professionals in contact with the patient must be educated with respect to knowledge on antimicrobial resistance, (lack of) evidence of benefit of antibiotics under different conditions and related beliefs, knowledge of management of symptoms, and the use of microbiology laboratory tests to guide antibiotic treatment. Education on the management of demanding patients is also required (Bond, 2005), and patients should receive consistent messages on correct and prudent antibiotic use when taking antibiotics (Finch *et al.*, 2004; Davey and Garner, 2007). Thus, antibiotic management requires effective teamwork among all health professions.

Finally, we must also consider the antibiotic prescriptions for animals made by the veterinarians and in agriculture, as strong positive associations have been found between the consumption of antibiotic classes and resistance. Conversely, voluntary discontinuation of cephalosporin use in pig production has significantly reduced extended-spectrum cephalosporinase-producing *Escherichia coli* in pigs (Agerso and Aarestrup, 2013). Behavioral research has shown that farmers rely mainly on veterinarians for their antibiotic use habits (Friedman *et al.*, 2007). Although there are similarities between antimicrobial stewardship for companion animals and the human sector (Garcia-Alvarez *et al.*, 2012), antibiotic use in livestock is further complicated by economic issues (Aarestrup, 2000). A modern veterinary education program should educate undergraduates on the ecology of antimicrobial resistance and the role of antimicrobial drug use (Fanning *et al.*, 2009).

Educational interventions targeted at professionals

Up to now, most initiatives on education in antimicrobial stewardship have been deployed in the postgraduate setting in hospitals. A considerable effort has been put into education that tries to change behavior. Many interventional programs to optimize antibiotic use in inpatients have been conducted worldwide, and to a lesser extent in primary care (Arnold and Straus, 2005; Davey *et al.*, 2005). Intervention strategies have been categorized as educational (Table 4.2), restrictive, and supportive (Davey *et al.*, 2005; Dellit *et al.*, 2007). Education is an essential element of any hospital program that targets antibiotic prescribing. Educational measures are usually more popular among clinicians than restrictive measures (Pulcini *et al.*, 2011). The limited success of inhospital education may be partly due to the rapid turnover of junior staff and the difficulty of maintaining a local continuous educational program.

Educational interventions can be categorized as passive or active. Passive education alone (lectures, educational events, leaflets, handouts), without active intervention, has been shown to be only marginally effective in changing antimicrobial prescribing practices and has no sustained impact (Dellit *et al.*, 2007). Antibiotic guidelines are intended to improve the quality of care, to support public health decisions, to diminish unwanted diversity of practice, and to increase transparency (for the healthcare worker and the public). Implementation of guidelines can be facilitated through provider education and feedback on antimicrobial use and patient outcomes (Dellit *et al.*, 2007). Clinical pathways have been successfully used to implement prudent antibiotic strategies, such as the de-escalation pathway described by Singh *et al.* (2000) to curb inappropriate antibiotic

Table 4.2. Educational interventions to influence and control physician behavior at hospital level.

Passive educational measures
- Developing/updating local antibiotic guidelines, clinical pathways
- Educational sessions, workshops/ local conferences

Active educational interventions
- Clinical rounds discussing cases
- Prospective audit with intervention and feedback
- Reassessment of antibiotic prescriptions, with streamlining and de-escalation of therapy
- Academic detailing, educational outreach visits

use for pulmonary infiltrates in the ICU. Face-to-face and one-to-one educational sessions provided by physicians are based on established principles of behavioral science, market research, and communications theory. This type of education has been used intensively and successfully by the pharmaceutical industry. In a hospital-based study by Solomon *et al.* (2001), daily academic detailing reminded prescribing physicians to stop the use of previously started but unnecessary antibiotics. Participative physician feedback and multidisciplinary interventions have also been found to be effective methods to increase the judicious use of antibiotics and reduce costs (Davey *et al.*, 2005).

Many national and international professional societies, organizations or governmental bodies have deployed educational antimicrobial stewardship activities at the postgraduate level. As an example, some programs are described here. Worldwide, the WHO's actions to fight healthcare-associated infections (HCAI) are building on the *WHO Global Strategy for Containment of Antimicrobial Resistance* (WHO, 2001) to further implement the World Health Assembly resolution WHA 51.17 on *Emerging and Other Communicable Diseases: Antimicrobial Resistance* (WHO, 1998). WHO chose antimicrobial resistance as the topic for World Health Day 2011, and put antimicrobial resistance on the health agenda of 129 WHO Member States. In the US, the IDSA published its report *Bad Bugs, No Drugs: As Antibiotic Discovery Stagnates ... A Public Health Crisis Brews* in 2004 and launched an advocacy campaign to spur government solutions. In 2007, the IDSA and Society for Healthcare Epidemiology of America (SHEA) issued guidelines for developing an institutional program to enhance antibiotic stewardship (Dellit *et al.*, 2007). Strategies were rated according to a grading system expressing the strength of the recommendation and evidence. The American Society for Microbiology (ASM) has been organizing workshops in the topic of antimicrobial policies/stewardship since 2004, and the program has been developed in collaboration with the European Society on Clinical Microbiology and Infection (ESCMID) Study Group on Antibiotic Policies (see below). The CDC started its campaign "Get Smart for Healthcare" in 2010 and focused on improving antibiotic use in inpatient healthcare facilities, starting with hospitals and then expanding to long-term care facilities. The goal of the campaign is to optimize the use of antimicrobial agents in inpatient healthcare settings by focusing on strategies to help hospitals and other inpatient facilities implement interventions to improve antibiotic use. Among other tools, the CDC provides slides, factsheets, and an annotated bibliography on the evidence base of outcomes on its website (http://www.cdc.gov/getsmart/healthcare/). The CDC is collaborating with SHEA to develop simple implementation tools and with the Institute for Healthcare Improvement (IHI, in Massachusetts) and SHEA to develop a driver diagram with practical antibiotic stewardship implementation strategies.

In Europe, the ECDC chose the hospital prescribers as target for their "European Antibiotic Awareness Day" (EAAD) campaign in 2010. A toolkit was developed to support efforts at a national level to increase prudent use of antibiotics in hospitals through the dissemination of evidence-based educational and information materials. The toolkit contains template materials and evidence-based key messages which may be adapted for use at national level, and suggests tactics for getting the messages on the prudent use of antibiotics through to the target audiences. The template toolkit materials and more information about the annual EAAD are available on the ECDC website (ECDC, 2016). In 2015, efforts on the prudent use of antibiotics in Europe were consolidated. The ESCMID Study Group on Antibiotic Policies (ESGAP) was created to provide a uniting European forum for those medical personnel and scientists actively involved in antibiotic stewardship at local, national and international levels. ESGAP provides an opportunity for training in the appropriate use of antibiotics (see www.escmid.org/esgap). Major activities are the postgraduate international education courses "Antimicrobial Stewardship: Measuring, Auditing and Improving" conducted bi-annually before the "European Conference on Microbiology and Infectious Diseases" (ECCMID). Up to now, seven courses have been organized, training over 400 medical doctors, scientists, and clinical pharmacists over the past decade. Another 1 day educational course is conducted as a workshop before the ASM's annual Interscience Conference on Antimicrobial Agents and Chemotherapy (ICAAC). For example, at the ICAAC in September 2015 at San Diego (California), the pre-ICAAC workshop was "Antimicrobial Stewardship in Hospitals." ESGAP has published an inventory of antimicrobial stewardship websites. (Pagani *et al.*, 2009).

On the postgraduate training track, most medical and surgical specialties are anatomically defined, but all have to deal with infections. In practice, each specialty thus has a certain degree of "claim" over antimicrobial prescribing in its field. As depicted in Table 4.1, the input of many disciplines requires training at the bedside, but the organ specialist, e.g., urologists, may not have the fully required background for implementing the general principles of prudent antibiotic prescribing in the microbiological diagnostic and therapeutic management of urological patients. In Canada, the Royal College of Physicians and Surgeons of Canada (CanMEDS) Physician Competency framework describes the knowledge, skills and abilities that specialist physicians need for better patient outcomes (CanMEDS, 2005). This model has been adapted around the world in the health profession and other professions. With resistance data increasing worldwide, multidisciplinary guideline development of national and local guidelines becomes paramount. The ADAPTE Process, a resource toolkit for guideline adaptation (The ADAPTE Collaboration, 2009), can be used to minimize barriers to the development and acceptance of guidelines (Cabana *et al.*, 1999). Guidelines must be evidence based, graded and developed by a multidisciplinary group, and involve all key stakeholders to foster acceptance and ownership. National or international guidelines should be adapted to the local context to ensure relevance for local practice and policies. Transparency is key to promoting confidence in the recommendations of the adapted guideline(s), and flexible, easily accessible formats must be used (an online tool, booklet, smartphone application, etc.). In hospitals, a multidisciplinary core group, including infectious diseases specialists, microbiologists and clinical pharmacists, and/or an antibiotic stewardship team must be involved in the development and implementation of a local educational program for residents and fellows on prudent antibiotic prescribing.

The format of internship/specialist training of medical doctors is very variable within Europe and in the world, both in the onset of exposure to patients, the duration and type of training, and the responsibilities allowed, which renders standardization of learning outcomes very difficult. This period is extremely crucial for shaping behavior, as juniors start to copy the behavior of their supervisors within their first weeks in the hospital (De Souza *et al.*, 2006). Competencies and learning outcomes must be clearly defined. The impact of learning sessions can be enhanced by measurement of current practice and the use of quality improvement strategies (Davey and Garner, 2007; McNulty *et al.*, 2012).

Recently, the training programs of graduated doctors into primary care or a specialty have been the subject of reforms in many countries. In Scotland, the Doctors Online Training System (DOTS), a mandatory web-based education resource for all foundation training doctors, was revised to highlight current issues in prudent antibiotic prescribing starting in 2008 (Nathwani *et al.*, 2011), and there is an interactive *Antimicrobial Stewardship Educational Workbook* available from the NHS (National Health Service) Education for Scotland (2015).

Education of the Public on Antimicrobials

The public is also a major stakeholder when it comes to prudent antibiotic use, and many efforts have been conducted to educate the adult population in the community. In the last decade, costly national and regional educational campaigns to educate the public have been conducted in various countries, e.g., France, Spain, and Belgium. Their effects are not always clear or sustainable (Huttner *et al.*, 2010; WHO, 2015). However, according to the last *Eurobarometer* (*Special Eurobarometer* 407) in 2013 (European Commission, 2013), knowledge of antibiotics is slowly improving in Europe, possibly as a result of the larger antibiotic awareness campaigns in Europe by the ECDC's annual EAAD, which has been held on 18 November since 2008 (EAAD, 2016 http://ecdc.europa.eu/en/eaad/pages/hame.aspx). Another example of sustained, targeted initiatives is the Canadian "Do Bugs Need Drugs (DBND)?" campaign, which was started in Alberta and now also operates in British Columbia (DBND, 2013), and in the US, the CDC's annual "Get smart about Antibiotics" week (http://www.cdc.gov/getsmart/). The intensity of the campaigns has varied widely, from simple Internet-based to expensive mass-media campaigns. All but one campaign targeted the public and physicians simultaneously (Huttner *et al.*, 2010).

Because of the major deficiencies found in elementary knowledge on antibiotics in surveys of adults throughout Europe (e.g., only 52% of respondents are aware that antibiotics are not effective against colds and flu) (European Commission, 2013) educating the public on this topic at an earlier age might be a good idea. Education at the primary and secondary level will prepare the individual who might later become a patient or, more importantly, the anxious parent of a sick toddler with, for example, acute otitis media. In addition, it provides the basic education of the future professional. In Fig. 4.2(a), a time line of opportunities for education on prudent antibiotic use of the public is depicted. Indeed, some groups have set up children's educational programs in the last decade and a lot of progress has been made.

Three major initiatives for children are outlined here. The first program, "Do Bugs Need Drugs?" started as a small 6 month pilot in Alberta, Canada, in 1998. It contained a kids' section dealing in a playful way with handwashing and the responsible use of antibiotics. The program evolved into a larger provincial program in Alberta and British Columbia and the website is regularly updated (http://www.dobugsneeddrugs.org/). Secondly, the "Microbes en Question" mobile children's health education campaign, supported by the French Caisse Nationale d'Assurance Maladie and the Pasteur Institute, was started as an innovative program in 2004. Finally, the European Union (EU)-funded antibiotic and hygiene teaching resource "e-Bug" was developed in the UK as an initiative of the Health Protection Agency. In preparation of the program, Lecky *et al.* (2014) examined the educational structure and educational resources or campaigns currently available by questionnaire in ten European countries. It appeared that the curricula in all countries covered the topic of human health and hygiene, but limited information was provided on antibiotics and their prudent use. The 9–11 and 12–15 year-old age groups were considered the most appropriate to aim the e-Bug resource at, each at an adapted level, in order to teach all children within compulsory education (Lecky and McNulty, 2011). The e-Bug program has recently developed a science road show to educate the parents too. The program has been helpful in increasing knowledge in adults and children (Lecky *et al.*, 2014).

Taking Education of the Prescribers to the Undergraduate Level

Similar to the difficult task of reeducating adults is changing the behavior of graduated professionals. A plethora of literature is available on the difficulties of changing the behavior of trained medical practitioners, the barriers being multiple, cultural among others (Cabana *et al.*, 1999; Hulscher *et al.*, 2010). Therefore, starting the teaching of prudent antibiotic use to undergraduate students might be more effective. In Fig. 4.2(b) a time line is depicted of opportunities for the education on prudent antibiotic use of prescribers of antimicrobials. Inclusion of antimicrobial chemotherapy in the undergraduate curriculum is logical, as doctors of all disciplines regularly treat infections.

Recently, the content, volume and quality of medical curricula teaching antimicrobial stewardship

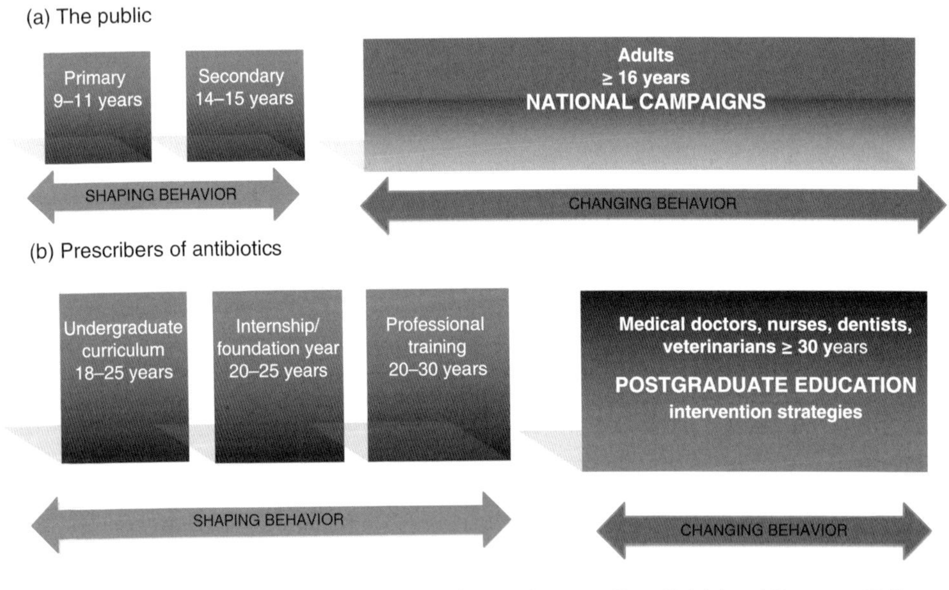

Fig. 4.2. Time line of education on prudent antibiotic use by age in years. From Pulcini and Gyssens, 2013.

principles and resistance has been studied in terms of imparting knowledge, attitude and behavior to medical students. There is much variation of curricula in the EU. Programs can either be quite detailed, such as the "Raamplan Artsopleiding 2009" in the Netherlands, in which general learning outcomes with keywords are specified; these include the prevention of infections and spread, aspects of guidelines and protocols, and the development of antibiotics. Other learning outcomes of university curricula on websites such as that of the University of Leuven (http://med.kuleuven.be/eng/education/meded) are very general and do not provide information on the topics covered. A recent cross-sectional survey on the teaching of prudent antibiotic prescribing in the undergraduate curriculum of medical schools was performed in 13 European countries. Thirty-five of 37 medical schools were included in the study. Prudent antibiotic use principles were taught in all but one, but only four of 13 countries had a national program. Interactive teaching formats were used less frequently than passive formats. The teaching was mandatory for 53% of the courses and started before clinical training in 71%. Wide variations in the exposure of students to important principles of prudent antibiotic use were found both among countries and within the same country. Some major principles were poorly covered (e.g., reassessment and duration of antibiotic therapy, communication skills). Whereas 77% of the (faculty) respondents fully agreed that the teaching of these principles should be prioritized, lack of time, mainly due to rigid curriculum policies, was the main barrier reported to implementation. The authors concluded that, given the study design, these are probably optimistic results (Pulcini *et al.*, 2015).

Indirect information on medical curricula is also available from a few surveys of medical students or junior doctors on the perceptions of, attitudes to, and knowledge of antimicrobial prescribing practices (Minen *et al.*, 2010; Pulcini *et al.*, 2011; Abbo *et al.*, 2013; Dyar *et al.*, 2014). In the US, Minen *et al.* (2010) surveyed the perceptions and attitudes of medical students about their training on antimicrobial use to identify gaps in medical education in a university hospital. Some 30% of medical students responded. The majority of third- and fourth-year medical students believed that antibiotics were overused in the hospital and in outpatient areas. The students recognized the importance of judicious antibiotic use and would have liked greater instruction on how to choose antibiotics appropriately. Over three quarters of the students would have liked more education on antibiotic selection, and 83% wanted this education to be during the third year of medical school. The resources that the students used the most for antibiotic selection included other physicians and handheld programs/apps such as Epocrates, but no clear resource emerged as the dominant preference (Minen *et al.*, 2010). Recently, Abbo *et al.* (2013) conducted a multicenter study (at the University of Miami, the Johns Hopkins University, and the University of Washington) that investigated medical students' knowledge, attitudes, and perceptions about antimicrobial stewardship, and their perceptions of their preparedness to prescribe antimicrobials appropriately. A 24-item electronic survey on antimicrobial prescribing and education was administered to fourth-year medical students. The response rate was high, 61%. Some 90% of the students said that they would like more education on the appropriate use of antimicrobials. Mean correct knowledge score (11 items) was 51%, with statistically significant differences between study sites and sources of information used to learn about antimicrobials. Differences were found between medical schools in educational resources used, perceived preparedness, and knowledge about antimicrobial use.

In Europe, Pulcini *et al.* (2011) conducted a survey on doctors who were still in their training years in a French and a Scottish university hospital. In both countries, doctors were first-year trainees at the same stage in their medical training, after 5 and 6 years of undergraduate training in Dundee and Nice, respectively. Overall, 30% of those surveyed stated that they had had no training in antibiotic prescribing in the past year, although 99% had prescribed an antibiotic within the last 6 months. A reassuring feature was that 91% (France) and 97% (Scotland) cited guidelines as a factor that influenced their prescribing.

In the above surveys, one of the striking differences between the US and European populations is that US students did not mention local guidelines as resource for prescribing (Minen *et al.*, 2010), whereas British students are explicitly stimulated to prescribe according to local guidelines (NPC, 2012). Also in Europe, final-year students at seven European medical schools were invited in 2012 to participate in an online survey to learn about medical students' knowledge of and perspectives on antibiotic prescribing and resistance, with the aim of helping to develop educational programs (Dyar *et al.*, 2014). The response rate was 35%. Most students (74%) wanted more education on choosing antibiotic treatments. Students at all schools felt least confident in choosing combination therapies, choosing the correct dose and interval of administration, and not prescribing in cases of diagnostic uncertainty. Most (66%) also felt that the antibiotics they would prescribe would contribute to resistance, and almost all (98%) felt that resistance would be a greater problem in the future (Dyar *et al.*, 2014).

The importance of undergraduate training in prudent prescribing of antibiotics has become increasingly recognized (Bond, 2005). In the UK and Scotland in particular, major efforts have been made to adapt and revise undergraduate education on antibiotics. In the UK, the Specialist Advisory Committee on Antimicrobial Resistance (SACAR) has proposed to undertake the development of learning outcomes, i.e., statements that indicate what a student should know, understand, and be able to do by the end of an educational program. Subsequently, the learning outcomes could be translated into competencies by the appropriate bodies (Davey and Garner, 2007). Prescribing is included as a component of the undergraduate program in the UK, and the importance of undergraduate training in prescribing is reflected in aspects of the General Medical Council's (GMC's) Tomorrows' Doctors (GMC, 2009), a document that has now been superseded (as of 1 January 2016) by two new documents: *Promoting Excellence: Standards for Medical Education and Training* (GMC, 2015a) and Outcomes for Graduates (GMC, 2015b). The GMC has also endorsed the NPC's report entitled A Single Competency Framework for All Prescribers, as in the UK nurses are allowed to prescribe antibiotics (NPC, 2012). In the UK, this competency is subjected to the standards for supervision for doctors in training set out in report The Trainee Doctor (GMC, 2011), a document that has also been superseded by one of the aforementioned new documents (GMC, 2015a). Statement competency examples given by the single competency framework report (NPC, 2012) include:

> Competency 1: knowledge. *Has up-to-date clinical, pharmacological and pharmaceutical knowledge relevant to own area of practice.* ... 11. Understands antimicrobial resistance and the roles of infection prevention, control and antimicrobial stewardship measures as outlined in the ARHAI [Advisory Committee on Antimicrobial Resistance and Healthcare Associated Infection] and PHE [Public Health England] Antimicrobial Prescribing and Stewardship Competences.
>
> Competency 7: the healthcare system. Understands and works within local and national policies, processes and systems that impact on prescribing practice. Sees how own prescribing impacts on the wider healthcare community. ... 59: understands and works within local frameworks for medicines use as appropriate (e.g. local formularies, care pathways, protocols and guidelines).

More specific learning outcomes or competencies can be developed that depart from this list.

Reexamining the principles and learning outcomes in Table 4.1, it is clear that much emphasis is needed on the transfer of basic scientific knowledge at an early stage. Similarly to the optimal periods and subjects identified to guarantee maximal exposure to children (Lecky and McNulty, 2011), the undergraduate curriculum and internship/foundation year seem optimal stages to build a solid knowledge base for later practice. For example: surgeons will have a much higher acceptance of prophylaxis guidelines if they have been exposed to the principles of guideline development and antibiotic prophylaxis when taught as core competencies in the third year of medical school. To reach this goal, a strong input is needed from academia to transfer the knowledge in the undergraduate curriculum. As depicted in Table 4.1, the curriculum is expected to deliver knowledge and shape the right attitude and behavior toward the basic principles of antimicrobial stewardship. A wide variety of disciplines must be involved, including epidemiology, ethics, and communication skills (working with guidelines, communicating with the patient). To link the undergraduate and postgraduate programs, in particular in the period of internship/foundation training, close collaboration between healthcare providers and academicians, and between hospitals and medical schools is needed.

Suitable formats of educational curricula

In the undergraduate curriculum, classical formal lectures are seldom considered to be a successful means of transferring knowledge. Over the past decade, problem-based learning has been introduced in many universities. This type of education allows for alternative formats of interactive learning in smaller student groups. It is important to identify the topics or concepts that benefit from a disease- (e.g., acute bronchitis) or problem- (e.g., antimicrobial resistance) oriented rather than a pathogen (e.g., methicillin-resistant *Staphylococcus aureus*, MRSA) oriented or a drug (e.g., antibiotic classes) oriented approach. Microbial resistance can be part of microbiology teaching, and information on antibiotics can be part of pharmacology and managing the demands of patients—in particular the parents of young children, and integrated into communication skills sessions. However, targeted "antibiotic" sessions in the format of problem-based learning are absolutely necessary to integrate all aspects of the topic (Bond, 2005). Apart from formal lectures, interactive learning with case vignettes, PowerPoint presentations, and role-play can be particularly appropriate for this topic. Elective rather than core modules are particularly suitable for discussions in small groups. Suitable topics are case studies, e.g., with questions and answer sessions, illustrating the evidence base of surgical prophylaxis.

Here are some examples. In the Netherlands, the University of Rotterdam has included a 1 week module on several concepts of antimicrobial resistance, hygiene, and prudent antibiotic prescribing in the core curriculum of the second year of medical school; the University of Nijmegen offers an additional elective, problem-based module on antibiotic policy for third year students, treating hygiene and infection control, antibiotic guidelines, principles of antibiotic prophylaxis, among others. In Scotland, an extensive range of e-learning resources have been developed to train both undergraduate and postgraduate healthcare professionals on prudent antibiotic prescribing (Nathwani *et al.*, 2011). Such training is mandatory for junior doctors. Also in the UK, the Prudent Antibiotic User (PAUSE), offers a website (http://www.pause-online.org.uk) of shared standardized teaching materials for prudent antimicrobial prescribing for use in the undergraduate medical curriculum. PAUSE provides standardized teaching aides for all educators of antibiotic prescribing based on patient-focused, reflective learning. Resources are designed to enable students to prepare for interactive sessions to compare standard patient vignettes with their own clinical experience. The structured preparation required of the students is a key to success. The interactive discussion sessions focus on learning prudent antibiotic prescribing through reflective practice. Prepared materials (in a PowerPoint format) in relation to structure and content for each interactive session are available for tutors to use. These resources include patient histories, clinical signs, investigations, and questions on diagnosis, assessing severity, appropriate prescribing, public health issues, and patient management. Best practice statements and core resource lists are also available. In addition, guideline answers to the questions with feedback are provided, including inappropriate responses and the corresponding reasons for them being inappropriate. The materials may be shared, reproduced, and modified as necessary.

Strong political support is necessary for a curriculum program to be successfully implemented. For instance, in the UK, the GMC requested in 2009 that all postgraduate deans and Royal Colleges ensure that infection prevention and control, and antimicrobial prescribing, become standard practice implemented in all clinical settings, and that these aspects are strongly emphasized in undergraduate and postgraduate medical training (McNulty *et al.*, 2012).

Conclusion

For over 30 years antibiotic policymakers have been trying to curb antibiotic resistance. The bulk of the efforts have mainly been conducted at the adult public and professional postgraduate level, aiming to change behavior. Changing the practice of professionals is extremely difficult and frustrating. Worldwide, only minimal investment has been put in antimicrobial stewardship education in undergraduate and fellowship training. It seems obvious that antimicrobial stewardship is likely to be more successful when the teaching is started much earlier, at the time when the knowledge, attitudes, and behavior of professionals are being shaped. There is a need for an undergraduate medical/professional curriculum that covers the principles of microbiology, infectious diseases, and clinical pharmacology, with emphasis on the principles of prudent prescribing in an adequate format. Appropriate curricula on antimicrobial stewardship are a joint responsibility of academia and the national ministries of Health and Education.

References

Aarestrup, F.M. (2000) Occurrence, selection and spread of resistance to antimicrobial agents used for growth promotion for food animals in Denmark. *APMIS (Acta Pathologica, Microbiologica et Immunologica Scandinavica)* 108 (Supplementum S101), 5–6.

Abbo, L.M., Cosgrove, S.E., Pottinger, P.S., Pereyra, M., Sinkowitz-Cochran, R., Srinivasan, A., Webb, D.J., and Hooton, T.M. (2013) Medical students' perceptions and knowledge about antimicrobial stewardship: how are we educating our future prescribers? *Clinical Infectious Diseases* 57, 631–638.

Adriaenssens, N., Coenen, S., Versporten, A., Muller, A., Minalu, G., Faes, C., Vankerckhoven, V., Aerts, M., Hens, N., Molenberghs, G., and Goossens, H. (2011) European Surveillance of Antimicrobial Consumption (ESAC): outpatient antibiotic use in Europe (1997–2009). *Journal of Antimicrobial Chemotherapy* 66(Suppl 6), vi3–12.

Agerso, Y. and Aarestrup, F.M. (2013) Voluntary ban on cephalosporin use in Danish pig production has effectively reduced extended-spectrum cephalosporinase-producing *Escherichia coli* in slaughter pigs. *Journal of Antimicrobial Chemotherapy* 68, 569–572.

Arnold, S.R. and Straus, S.E. (2005) Interventions to improve antibiotic prescribing practices in ambulatory care. *Cochrane Database of Systematic Reviews* 4: CD003539.

Averbuch, D., Orasch, C., Cordonnier, C., Livermore, D.M., Mikulska, M., Viscoli, C., Gyssens, I.C., Kern, W.V., Klyasova, G., Marchetti, O. *et al.* (2013) European guidelines for empirical antibacterial therapy for febrile neutropenic patients in the era of growing resistance: summary of the 2011 4th European Conference on Infections in Leukemia. *Haematologica* 98, 1826–1835.

Bond C. (2005) Education of patients and professionals. In: Gould, I.M., Van der Meer, J.W.M. (eds) *Antibiotic Policies Theory and Practice*. Kluwer/Plenum, New York.

Borgatta, B. and Rello, J. (2014) How to approach and treat VAP in ICU patients. *BMC Infectious Diseases* 14:211.

Cabana, M.D., Rand, C.S., Powe, N.R., Wu, A.W., Wilson, M.H., Abboud, P.A., and Rubin, H.R. (1999) Why don't physicians follow clinical practice guidelines? A framework for improvement. *JAMA (The Journal of the American Medical Association)* 282, 1458–1465.

CanMEDS (2005) *CanMEDS 2005 Framework*. Royal College of Physicians and Surgeons of Canada, Ottawa. Available at: http:// www.royalcollege.ca/portal/page/portal/rc/common/documents/canmeds/framework/the_7_canmeds_roles_e.pdf (accessed 30 April 2015).

CDC (2013) *Antibiotic Resistance Threats in the United States, 2013*. US Centers for Disease Control and Prevention, Atlanta, Georgia. Available at: http://www.cdc.gov/drugresistance/pdf/ar-threats-2013-508.pdf (accessed 5 May 2015).

Davenport, L.A., Davey, P.G., and Ker, J.S. (2005) An outcome-based approach for teaching prudent antimicrobial prescribing to undergraduate medical students: report of a Working Party of the British Society for Antimicrobial Chemotherapy. *Journal of Antimicrobial Chemotherapy* 56, 196–203.

Davey, P. and Garner, S. (2007) Professional education on antimicrobial prescribing: a report from the Specialist Advisory Committee on Antimicrobial Resistance (SACAR) Professional Education Subgroup. *Journal of Antimicrobial Chemotherapy* 60 (Suppl 1), i27–32.

Davey, P., Hudson, S., Ridgway, G., and Reeves, D. (1993) A survey of undergraduate and continuing medical education about antimicrobial chemotherapy in the United Kingdom. British Society of Antimicrobial Chemotherapy Working Party on Antimicrobial Use. *British Journal of Clinical Pharmacology* 36, 511–519.

Davey, P., Brown, E., Fenelon, L., Finch, R., Gould, I., Hartman, G., Holmes, A., Ramsay, C., Taylor, E., Wilcoxm M., and Wiffen, P. (2005) Interventions to improve antibiotic prescribing practices for hospital inpatients. *Cochrane Database of Systematic Review* 4: CD003543.

DBND (2013) Do Bugs Need Drugs? Alberta Health Services and the British Columbia Centre for Disease Control, Canada. Details available at http://www.dobugsneeddrugs.org/ (accessed 30 April 2016).

De Souza, V., MacFarlane, A., Murphy, A.W., Hanahoe, B., Barber, A., and Cormican, M. (2006) A qualitative study of factors influencing antimicrobial prescribing by non-consultant hospital doctors. *Journal of Antimicrobial Chemotherapy* 58, 840–843.

Dellit, T.H., Owens, R.C., McGowan, J.E., Jr., Gerding, D.N., Weinstein, R.A., Burke, J.P., Huskins, W.C., Paterson, D.L., Fishman, N.O., Carpenter, C.F., Brennan, P.J., Billeter, M. *et al.* (2007) Infectious Diseases Society of America and the Society for Healthcare Epidemiology of America guidelines for developing an institutional program to enhance antimicrobial stewardship. *Clinical Infectious Diseases* 44, 159–177.

Dryden, M.S., Cooke, J., and Davey, P. (2009) Antibiotic stewardship—more education and regulation not more availability? *Journal of Antimicrobial Chemotherapy* 64, 885–888.

Dyar, O.J., Pulcini, C., Howard, P., and Nathwani, D. (2014) European medical students: a first multicentre study of knowledge, attitudes and perceptions of antibiotic prescribing and antibiotic resistance. *Journal of Antimicrobial Chemotherapy* 69, 842–846.

ECDC (2013) *ECDC Surveillance Report. Point Prevalence Survey of Healthcare-associated Infections and Antimicrobial Use in European Acute Care Hospitals, 2011–2012*. European Center for Disease Prevention and Control, Stockholm. Available at: http://ecdc.europa.eu/en/publications/Publications/healthcare-associated-infections-antimicrobial-use-PPS.pdf (accessed 30 April 2016).

ECDC (2014) *Antimicrobial Resistance in Europe, 2013. Annual Report of the European Antimicrobial Resistance Surveillance Network (EARS-Net)*. European Center for Disease Prevention and Control, Stockholm. Available at: http://ecdc.europa.eu/en/publications/Publications/antimicrobial-resistance-surveillance-europe-2013.pdf (accessed 5 May 2015).

ECDC (2016) European Antibiotic Awareness Day. European Center for Disease Prevention and Control,

Stockholm. Available at: http://ecdc.europa.eu/en/EAAD/Pages/Home.aspx/ (accessed 30 April 2016).

European Commission (2013) *Special Eurobarometer 407. Antimicrobial Resistance: Report.* European Commission, Brussels. Available at: http://ec.europa.eu/health/antimicrobial_resistance/docs/ebs_407_en.pdf (accessed 5 May 2015).

Fanning, S., Whyte, P., and O'Mahony, M. (2009) Essential veterinary education on the development of antimicrobial and antiparasitic resistance: consequences for animal health and food safety and the need for vigilance. *Scientific and Technical Review* 28, 575–582.

Finch, R.G., Metlay, J.P., Davey, P.G., and Baker, L.J. (2004) Educational interventions to improve antibiotic use in the community: report from the International Forum on Antibiotic Resistance (IFAR) colloquium, 2002. *The Lancet Infectious Diseases* 4, 44–53.

Freire-Moran, L., Aronsson, B., Manz, C., Gyssens, I.C., So, A.D., Monnet, D.L., and ECDC-EMA Working Group (2011) Critical shortage of new antibiotics in development against multidrug-resistant bacteria—time to react is now. *Drug Resistance* 14, 118–124.

Friedman, D.B., Kanwat, C.P., Headrick, M.L., Patterson, N.J., Neely, J.C., and Smith, L.U. (2007) Importance of prudent antibiotic use on dairy farms in South Carolina: a pilot project on farmers' knowledge, attitudes and practices. *Zoonoses Public Health* 54, 366–375.

Garcia-Alvarez, L., Dawson, S., Cookson, B., and Hawkey, P. (2012) Working across the veterinary and human health sectors. *Journal of Antimicrobial Chemotherapy* 67(Suppl 1), i37–49.

GMC (2009) *Tomorrow's Doctors: Outcomes and Standards for Undergraduate Medical Education*, rev. 2013. General Medical Council, Manchester, UK. Available at: http://www.gmc-uk.org/Tomorrow_s_Doctors_1214.pdf_48905759.pdf (accessed 29 April 2016).

GMC (2011) *The Trainee Doctor*. General Medical Council, Manchester, UK. Available at: http://www.gmc-uk.org/The_Trainee_Doctor___to_be_withdrawn_on_01_01_2016.pdf_62052266.pdf (accessed 29 April 2016).

GMC (2015a) *Promoting Excellence: Standards for Medical Education and Training*. General Medical Council, Manchester, UK. Available at: http://www.gmc-uk.org/Promoting_excellence_standards_for_medical_education_and_training_0715.pdf_61939165.pdf (accessed 29 April 2016).

GMC (2015b) Outcomes for Graduates (Tomorrow's Doctors). General Medical Council, Manchester, UK. Available at: http://www.gmc-uk.org/static/documents/content/Outcomes_for_graduates_Jul_15.pdf (accessed 29 April 2016).

Grigoryan, L., Monnet, D.L., Haaijer-Ruskamp, F.M., Bonten, M.J., Lundborg, S., and Verheij, T.J. (2010) Self-medication with antibiotics in Europe: a case for action. *Current Drug Safety* 5, 329–332.

Guinovart, M.C., Figueras, A., Llop, J.C., and Llor, C. (2015) Obtaining antibiotics without prescription in Spain in 2014: Even easier now than 6 years ago. *Journal of Antimicrobial Chemotherapy* 70, 1270–1271.

Hadi, U., Duerink, D.O., Lestari, E.S., Nagelkerke, N.J., Werter, S., Keuter, M., Suwandojo, E., Rahardjo, E., van den Broek, P., and Gyssens, I.C. (2008) Survey of antibiotic use of individuals visiting public healthcare facilities in Indonesia. *International Journal of Infectious Diseases* 12, 622–629.

Halls, G.A. (1993) The management of infections and antibiotic therapy: a European survey. *Journal of Antimicrobial Chemotherapy* 31, 985–1000.

Hulscher, M.E., Grol, R.P., and van der Meer, J.W. (2010) Antibiotic prescribing in hospitals: a social and behavioural scientific approach. *The Lancet Infectious Diseases* 10, 167–175.

Huttner, B., Goossens, H., Verheij, T., and Harbarth, S. (2010) Characteristics and outcomes of public campaigns aimed at improving the use of antibiotics in outpatients in high-income countries. *The Lancet Infectious Diseases* 10, 17–31.

IDSA (2004) *Bad Bugs, No Drugs: As Antibiotic Discovery Stagnates ... A Public Health Crisis Brews*. Infectious Diseases Society of America, Alexandria, Virginia. Available at: http://www.idsociety.org/uploadedFiles/IDSA/Policy_and_Advocacy/Current_Topics_and_Issues/Advancing_Product_Research_and_Development/Bad_Bugs_No_Drugs/Statements/As%20Antibiotic%20Discovery%20Stagnates%20A%20Public%20Health%20Crisis%20Brews.pdf (accessed 29 April 2016).

Laxminarayan, R., Duse, A., Wattal, C., Zaidi, A.K., Wertheim, H.F., Sumpradit, N., Vlieghe, E., Hara, G.L., Gould, I.M., Goossens, H. *et al*. (2013) Antibiotic resistance – the need for global solutions. *The Lancet Infectious Diseases* 13, 1057–1098.

Lecky, D.M. and McNulty, C.A. (2011) e-Bug implementation in England. *Journal of Antimicrobial Chemotherapy* 66(Suppl 5), v63–v66.

Lecky, D.M., Hawking, M.K., Verlander, N.Q., and McNulty, C.A. (2014) Using interactive family science shows to improve public knowledge on antibiotic resistance: does it work? *PLoS ONE* 9(8):e104556.

Leibovici, L., Paul, M., and Ezra, O. (2012) Ethical dilemmas in antibiotic treatment. *Journal of Antimicrobial Chemotherapy* 67(1), 126.

McNulty, C.A., Cookson, B.D., and Lewis, M.A. (2012) Education of healthcare professionals and the public. *Journal of Antimicrobial Chemotherapy* 67(Suppl 1), i11–i8.

Minen, M.T., Duquaine, D., Marx, M.A., and Weiss, D. (2010) A survey of knowledge, attitudes, and beliefs of medical students concerning antimicrobial use and resistance. *Microbial Drug Resistance* 16, 285–289.

Nathwani, D., Sneddon, J., Malcolm, W., Wiuff, C., Patton, A., Hurding, S., Eastaway, A., Seaton, R.A., Watson, E., Gillies, E. *et al*. (2011) Scottish Antimicrobial Prescribing Group (SAPG): development and impact of the Scottish

National Antimicrobial Stewardship Programme. *International Journal of Antimicrobial Agents* 38, 16–26.

NHS Education for Scotland (2015) *Antimicrobial Stewardship Educational Workbook*. NHS (National Health Service) Education for Scotland, Edinburgh, UK. Available at: http://www.nes.scot.nhs.uk/education-and-training/by-theme-initiative/healthcare-associated-infections/training-resources/antimicrobial-stewardship-workbook.aspx (accessed 29 April 2016).

NPC (2012) *A Single Competency Framework for All Prescribers*, rev. 2014. National Prescribing Centre, National Institute for Health and Clinical Excellence (NICE), London. Available at: https://www.associationforprescribers.org.uk/images/Single_Competency_Framework.pdf (accessed 29 April 2016).

Pagani, L., Gyssens, I.C., Huttner, B., Nathwani, D., and Harbarth, S. (2009) Navigating the Web in search of resources on antimicrobial stewardship in health care institutions. *Clinical Infectious Diseases* 48, 626–632.

Pulcini, C. and Gyssens, I.C. (2013) How to educate prescribers in antimicrobial stewardship practices. *Virulence* 4, 192–202.

Pulcini, C., Williams, F., Molinari, N., Davey, P., and Nathwani, D. (2011) Junior doctors' knowledge and perceptions of antibiotic resistance and prescribing: a survey in France and Scotland. *Clinical Microbiology and Infection* 17, 80–87.

Pulcini, C., Bush, K., Craig, W.A., Frimodt-Møller, N., Grayson, M.L., Mouton, J.W., Turnidge, J., Harbarth, S., Gyssens, I.C., and ESCMID Study Group for Antibiotic Policies (2012) Forgotten antibiotics: an inventory in Europe, the United States, Canada, and Australia. *Clinical Infectious Diseases* 54, 268–274.

Pulcini, C., Wencker, F., Frimodt-Møller, N., Kern, W.V., Nathwani, D. and Rodríguez-Bano, J., Simonsen, G.S., Vlahović-Palčevski, V., and Gyssens, I.C. for the ESGAP Curriculum Working Group (2015) European survey on principles of prudent antibiotic prescribing teaching in undergraduate students. *Clinical Microbiology and Infection* 21, 354–361.

Rice, L.B. (2008) The Maxwell Finland Lecture: for the duration—rational antibiotic administration in an era of antimicrobial resistance and *Clostridium difficile*. *Clinical Infectious Diseases* 46, 491–496.

Singh, N., Rogers, P., Atwood, C.W., Wagener, M.M. and Yu, V.L. (2000) Short-course empiric antibiotic therapy for patients with pulmonary infiltrates in the intensive care unit. A proposed solution for indiscriminate antibiotic prescription. *American Journal of Respiratory and Critical Care Medicine* 162, 505–511.

Solomon, D.H., Van Houten, L., Glynn, R.J., Baden, L., Curtis, K., Schrager, H., and Avorn, J. (2001) Academic detailing to improve use of broad-spectrum antibiotics at an academic medical center. *Archives of Internal Medicine* 161, 1897–1902.

Spellberg, B., Blaser, M., Guidos, R.J., Boucher, H.W., Bradley, J.S., Eisenstein, B.I., Gerding, D., Lynfield, R., Reller, L.B., Rex, J. *et al.* (2011) Combating antimicrobial resistance: policy recommendations to save lives. *Clinical Infectious Diseases* 52(Suppl 5), S397–S428.

The ADAPTE Collaboration (2009) *The ADAPTE Process.* Guideline Adaptation: A Resource Toolkit, Version 2.0. Available from: http://www.g-i-n.net/document-store/working-groups-documents/adaptation/adapte-resource-toolkit-guideline-adaptation-2-0.pdf/view?searchterm=adapte (accessed 29 April 2016).

The White House (2015) *National Action Plan for Combating Antibiotic-resistant Bacteria*. The White House, Washington, DC. Available at: https://www.whitehouse.gov/sites/default/files/docs/national_action_plan_for_combating_antibotic-resistant_bacteria.pdf (accessed 5 May 2015).

WHO (1998) *Fifty-First World Health Assembly. Agenda Item 21.3. [Resolution] WHA 51.17, 16 May 1998. Emerging and Other Communicable Diseases: Antimicrobial Resistance*. World Health Organization, Geneva, Switzerland. Available at: http://apps.who.int/medicinedocs/index/assoc/s16334e/s16334e.pdf?ua=1 (accessed 29 April 2016).

WHO (2001) *WHO Global Strategy for Containment of Antimicrobial Resistance*. World Health Organization, Geneva, Switzerland. Available at: http://whqlibdoc.who.int/hq/2001/WHO_CDS_CSR_DRS_2001.2.pdf (accessed 5 May 2015).

WHO (2012) *The Evolving Threat of Antimicrobial Resistance: Options for Action*. World Health Organization, Geneva, Switzerland. Available at: http://whqlibdoc.who.int/publications/2012/9789241503181_eng.pdf (accessed 5 May 2015).

WHO (2015) *Worldwide Country Situation Analysis: Response to Antimicrobial Resistance*. World Health Organization, Geneva, Switzerland. Available at: http://apps.who.int/iris/bitstream/10665/163468/1/9789241564946_eng.pdf?ua=1 (accessed 5 May 2015).

5 Mechanisms of Resistance to Antibacterial Agents

Louis B. Rice*

The Warren Alpert Medical School of Brown University, Rhode Island Hospital, Providence, Rhode Island, US

Introduction

Clinically effective antibacterial agents work by mechanisms that inhibit essential processes in bacteria. This distinguishes them from antiseptics, for example, which often work by more general mechanisms and therefore have a greater potential for toxicity to human cells. Mechanisms by which bacteria develop resistance to antibacterial agents may be specific for the mechanism of action of that agent, or applicable to a range of antimicrobial classes. In this chapter, I will endeavor to detail the variety of resistance mechanisms commonly found in human pathogenic bacteria. For clarity, I will delineate mechanisms by antimicrobial class, although a breakdown by species would also be informative.

Resistance Mechanisms for Different Antimicrobial Classes

Common resistance mechanisms are summarized by antimicrobial (antibiotic) class in Table 5.1. These are further described—also by antimicrobial class—in the sections below.

Aminoglycosides

The aminoglycosides (amikacin, gentamicin, kanamycin, neomycin, netilmicin, paromomycin, streptomycin, and tobramycin) are hydrophilic antibiotics that are particularly active against aerobic, Gram-negative rods and their antimicrobial activity is concentration dependent. Their principal target is the 30S subunit of the ribosome, the binding of which prevents the elongation of the growing peptide chain through misreading or premature termination of peptide synthesis.

Aminoglycoside resistance is attributable to three mechanisms: (i) efflux pumps; (ii) target (ribosome) alterations; or (iii) enzymatic inactivation.

Resistance due to efflux

Several resistance–nodulation–cell division (RND)-type efflux pumps that remove aminoglycosides from the periplasmic space have been described in Gram-negative bacteria. *Pseudomonas aeruginosa* has an intrinsic inducible three-protein pump (MexXY–OprM) that effluxes aminoglycosides (Mao *et al.*, 2001; Jeannot *et al.*, 2005). In *Escherichia coli*, aminoglycosides are effluxed by the one-protein AcrD multidrug efflux transporter (Aires and Nikaido, 2005). Magnet *et al.* (2001) described an RND-type efflux pump in *Acinetobacter baumannii*.

Modification of the ribosome

Alterations in ribosomal proteins and in 16S rRNA, and enzymatic methylation of the rRNA, can affect aminoglycoside binding (Vakulenko and Mobashery, 2003; Maus *et al.*, 2005; Yamane *et al.*, 2005).

Aminoglycoside-modifying enzymes (AMEs)

Aminoglycoside inactivation by AMEs (phosphotransferases—APHs, nucleotidyltransferases or adenyltransferases—ANTs, and acetyltransferases—AACs) is the most frequent mechanism of resistance in clinical pathogens (Azucena and Mobashery, 2001). AMEs

*E-mail: lrice@lifespan.org

Table 5.1. Common mechanisms of resistance to different classes of antibiotics.

Antibiotic class	Type of resistance

Aminoglycosides	Decreased uptake
	Active efflux
	Enzymatic modification (aminoglycoside-modifying enzymes)
	  Phosphotransferase
	  Adenyltransferase
	  Acetyltransferase

β-Lactams	Altered penicillin-binding protein(s) Enzymatic degradation (β-lactamases)
Chloramphenicol	Enzymatic degradation Chloramphenicol acetyltranferases (CAT) Efflux

Glycopeptides	Altered target
	  D-Alanine (D-Ala)–D-Ala to D-Ala–D-lactate (D-Lac) or D-Ala–D-serine (D-Ser)
	Target overproduction
	  Excess peptidoglycan

Lincosamides	Altered target Ribosomal methylation

Macrolides	Altered target
	  Ribosomal methylation
	Efflux
	  Macrolide efflux pump (mef)

Oxazolidinones	Altered target
	  Ribosomal RNA mutation
	  Ribosomal methylation

Quinolones	Altered target
	  Quinolone-resistance-determining region (QRDR) mutations
	Efflux
	Protection from DNA binding
	  Quinolone resistance (Qnr) protein
	Enzymatic modification
	  Aminoglycoside acetyltransferase(6′)-Ib (AAC(6′)-Ib)

Streptogramins A	Enzymatic degradation Acetyltransferases
Streptogramins B	Altered target Ribosomal methylation
Rifampin	Altered target Mutations in RNA polymerase
Tetracyclines	Efflux Altered target Protection proteins
Sulfonamides	Altered target Altered dihydropteroate synthetase (DHPS) Acquisition of new low-affinity DHPS genes

Trimethoprim	Altered target
	  Substitutions in dihydrofolate reductase (DHFR)
	  Acquisition of new low-affinity DHFR genes
	Overproduction of target
	  Promoter mutation leading to overproduction of DHFR

covalently modify specific amino or hydroxyl groups, resulting in aminoglycosides that bind poorly (higher K_m, less affinity) to the ribosome. There are seven major phosphotransferases—APH (2″), APH (3′), APH(3″), APH(4), APH(6), APH(7″), and APH(9), four nucleotidyltransferases—ANT(2″), ANT(3″), ANT(4′), and ANT(6), and four acetyltransferases—AAC(1), AAC(2′), AAC(3), and AAC(6).

β-Lactam antibiotics

Penicillin-binding protein-mediated resistance

β-Lactam antibiotics (penicillins, cephalosporins, monobactams, and carbapenems) act by inhibiting cell wall synthesis enzymes known as penicillin-binding proteins (PBPs). The inhibition of PBPs interrupts cell wall synthesis and triggers the activity of autolysins, which rupture the cell and so lead to cell death (Tomasz, 1983). Susceptibility to inhibition by β-lactams varies between species, in some cases (such as *Enterococcus faecium*) due to the expression of PBPs with intrinsically low affinity. PBP-mediated resistance for normally susceptible bacteria takes several forms, including: (i) overproduction of a PBP—rare but documented in both enterococci and staphylococci (Fontana *et al.*, 1994; Henze and Berger-Bachi, 1995; Rice *et al.*, 2001); (ii) acquisition of a foreign PBP with low affinity—as in the gene for PBP 2a of *Staphylococcus aureus* (Chambers, 1997); (iii) recombination of a susceptible PBP with more resistant varieties—seen with the naturally transformable species *Streptococcus pneumoniae* and *Neisseria gonorrhoeae* (Barcus *et al.*, 1995; Hakenbeck and Coyette, 1998); and (iv) point mutations within PBPs that lower affinity for the β-lactam antibiotic—commonly seen in highly ampicillin-resistant *E. faecium* (Rybkine *et al.*, 1998).

β-Lactamase-mediated resistance

CHARACTERISTICS AND MECHANISM OF ACTION OF β-LACTAMASES β-Lactamases are members of a "superfamily" of active site serine proteases or D,D (D-alanyl-D-alanine)-peptidases that are related to PBPs (Matagne *et al.*, 1998). They disrupt the amide bond of a β-lactam, thus permanently inactivating the antibiotic. β-Lactamases are secreted into the periplasmic space in Gram-negative bacteria or into the surrounding medium by Gram-positive organisms. Genes encoding β-lactamases can be located on the bacterial chromosome or on plasmids, integrons, or transposons. Their expression may be constitutive or inducible. There are two accepted classification schemes for β-lactamases. The Bush–Jacoby scheme separates them based on functional characteristics, whereas the Ambler classification scheme is based on structural similarities (Bush, 1989a,b). Because of the simplicity of the Ambler system in placing the β-lactamases into four classes, A–D, this scheme is used for the rest of the chapter (Ambler, 1980). The reader is referred to the following website for updates (www.lahey.org/studies, in the process of transitioning, see http://www.ncbi.nlm.nih.gov/pathogens/submit_beta_lactamase/ for details).

CLASS A β-LACTAMASES The two commonly encountered class A β-lactamases found in the *Enterobacteriaciae* are designated TEM-1 and SHV-1. TEM-1 and SHV-1 β-lactamases are primarily penicillinases with relatively poor activity against cephalosporins. The origin of the TEM enzyme (which was named after the patient from whom it was isolated—Temoniera) is unclear, but the SHV enzyme (so named because sulfhydryl reagents had a variable effect on their substrate specificity) is the intrinsic chromosomal β-lactamase found in most *Klebsiella pneumoniae* isolates. Both can be located on plasmids and other mobile elements, and have been found in a wide variety of species. Both are also highly efficient enzymes whose expression confers very high levels of resistance to ampicillin. The growing prevalence of TEM- and SHV-mediated resistance was one of the inspirations for the development of the extended-spectrum cephalosporins, whose modifications diminished the activity of the TEM and SHV enzymes, but widespread use of these agents has led to the emergence and spread of TEM and SHV mutants (extended-spectrum β–lactamases, or ESBLs), which allow the hydrolysis of cephalosporins (Bradford, 2001a).

ESBLs are most commonly class A β-lactamases that have "expanded" or changed their substrate profile as a result of amino acid substitutions. Mutations at critical amino acids "extend the spectrum" of these enzymes and allow the hydrolysis of extended-spectrum cephalosporins (Phillipon *et al.*, 1989). In most cases, mutations in ESBLs render the enzymes more susceptible to inhibition by the more commonly employed inhibitors (clavulanic acid, sulbactam, and tazobactam). However, β-lactam–β-lactamase inhibitor combinations are rarely active against clinical ESBL-producing isolates, because many of the strains produce a variety of β-lactamases, thus effectively overwhelming the inhibitors.

Non-TEM-non-SHV ESBLs. CTX-M β-lactamases—so named for their greater activity against cefotaxime than other oxyimino-β-lactam substrates—are the most prevalent ESBLs worldwide. They can be divided into distinct clusters (see http://www.lahey.org/studies/) and are commonly found in *K. pneumoniae*, *E. coli*, typhoidal and nontyphoidal *Salmonella* spp., *Shigella* spp., *Citrobacter freundii*, *Enterobacter* spp., and *Serratia marcescens* (Paterson *et al.*, 2003; Kim *et al.*, 2005; Lartigue *et al.*, 2005). Many other clinically important non-TEM non-SHV ESBLs have been described (K1, GES-1, PER-1, PER-2, VEB-1, BES-1, IBI-1, IBI-2, and OXA-type) (Poirel *et al.*, 1999; Bradford, 2001a,b).

Serine carbapenemases of the Class A type. Class A carbapenemases are becoming increasingly common, especially in *Klebsiella pneumoniae* (Pitout *et al.*, 2015). Numerous studies are revealing that KPC (*Klebsiella pneumoniae*) β-lactamases are becoming endemic in many regions. These isolates are highly resistant to penicillins, cephalosporins, and commonly used β-lactam/β-lactamase inhibitor combinations, and show reduced susceptibility to carbapenems.

CLASS B β-LACTAMASES In contrast to the serine dependent β-lactamases (classes A, C, and D), class B β-lactamases are metalloenzymes in that they require a metal cofactor (usually zinc) for their activity. With few exceptions (see below), class B β-lactamases confer resistance to a wide range of β-lactam compounds, including cephamycins and carbapenems, and are resistant to inactivation by clavulanate, sulbactam, and tazobactam. The monobactam aztreonam is in many cases the only β–lactam available with activity against metalloenzyme-producing strains (Poirel *et al.*, 2000).

The metallo-β-lactamases (MBLs) of the VIM (Verona integron-encoded MBP) and IMP (imipenem) type are increasingly prevalent. The $bla_{\text{VIM-2}}$ gene has now spread to more than 20 countries (Rodriguez-Martinez *et al.*, 2010). IMP MBLs have been found as part of integrons in a variety of species (Walsh *et al.*, 2005). In recent years, metalloenzymes of the NDM (New Delhi MBL) variety have become increasingly prevalent (Poirel *et al.*, 2011). Although they originated in India, they can now be found in hospitals worldwide.

CLASS C β-LACTAMASES Ambler Class C chromosomal β-lactamases genes are present in almost all Gram-negative bacteria (*Salmonella*, *Klebsiella*, *Proteus mirabilis*, *Proteus vulgaris*, and *Stenotropomonas maltophilia* being the major exceptions) (Jacoby, 2009). AmpC (ampicillin C hydrolyzing) enzymes, which are chromosomally encoded through the *ampC* gene, are particularly important in clinical isolates of *C. freundii*, *Enterobacter aerogenes* and *E. cloacae*, *Morganella morganii*, *P. aeruginosa* and *S. marcescens*. Class C β-lactamases hydrolyze cephalosporins more effectively than they do penicillins, but their activity against penicillins is sufficient to confer clinically significant levels of resistance. Most class C enzymes are resistant to inhibition by clavulanate, sulbactam, and tazobactam, but are inhibited by the recently approved β-lactamase inhibitor avibactam.

Increases in *ampC* expression may result from the action of some β-lactam antibiotics (cefoxitin, clavulanic acid, and imipenem are specific examples) (Sanders and Sanders, 1988). Under these circumstances (induction), β-lactamase is produced for only as long as the antibiotic is present in the medium. Constitutive high-level production of AmpC β-lactamase most commonly results from a mutation in the *ampD* gene, reducing the quantity of (or eliminating) AmpD (cytosolic amidase) from the cytoplasm. Under these circumstances, a constant high level of anhydro-MurNAc-tripeptide (an inducer) is present in the cytoplasm, which complexes with the repressor AmpR (transcriptional regulator) to form an activator of *ampC* transcription (Jacobs *et al.*, 1997).

Plasmid-mediated class C β-lactamases have been described in many Gram-negative organisms. The loss of porin proteins in clinical isolates with plasmid-encoded AmpC enzymes may result in resistance to carbapenems (Bradford *et al.*, 1997; Stapleton *et al.*, 1999).

CLASS D β-LACTAMASES The OXA-type (oxacillin-hydrolyzing) β-lactamases have been most commonly described in the *Enterobacteriaceae*, *Acinetobacter* spp., and *P. aeruginosa* (Naas and Nordmann, 1999). Recent data suggest that these enzymes may have originated in *Acinetobacter* spp. (Périchon *et al.*, 2014). OXA enzymes confer resistance to a wide variety of penicillins and are only weakly inhibited by clavulanic acid. Their frequent location on mobile genetic elements (plasmids or integrons) facilitates spread (Vila *et al.*, 1997; Navia *et al.*, 2002; Poirel *et al.*, 2002).

Some OXA enzymes hydrolyze carbapenems. The first description of a serine carbapenemase in *A. baumannii* was ARI-1 (OXA-23) in 1985 (Paton *et al.*, 1993). Although OXA carbapenemases hydrolyze imipenem inefficiently, their presence in an organism with an active efflux pump or a porin mutation may confer clinically significant levels of resistance (Heritier *et al.*, 2005).

Chloramphenicol

ACETYLTRANSFERASES Chloramphenicol acts by binding to the ribosome and inhibiting bacterial protein synthesis. The most common mechanism of resistance to chloramphenicol is the production of chloramphenicol acetyltransferases (CATs) that confer extremely high levels of resistance. Relationships between the different CATs have been described in detail in a review by Schwarz *et al.* (2004). CATs are generally divided into two types: type A (classical) CATs and type B (xenobiotic) CATs (Schwarz *et al.*, 2004). In *S. aureus*, five structurally similar type A CATs (A, B, C, D, and one encoded by the prototypic plasmid pC194) have been described (Fitton and Shaw, 1979), with analogues of the type D enzyme being found also in *E. faecalis and S. pneumoniae*. CAT genes have also been identified in *Clostridium perfringens* and *C. difficile* (Schwarz *et al.*, 2004).

Three types of type A CATs (I, II, III) have been identified in Gram-negative bacteria. (Murray *et al.*, 1990; Murray and Shaw, 1997). Type B (xenobiotic) acetyltransferases (Murray and Shaw, 1997) are structurally unrelated to classic CATs, and those that have been demonstrated to acetylate chloramphenicol confer only low levels of chloramphenicol resistance, even when present in high copy number.

DECREASED ACCUMULATION OF CHLORAMPHENICOL Chloramphenicol serves as a substrate for many of the identified multidrug resistance (MDR) efflux pumps. As well as for some that are specific for chloramphenicol (Nikaido, 1998). Gram-negative bacteria also express efflux genes specific for both chloramphenicol and florfenicol (flo_{Pp}, flo_{St}) that are often reported from animal-derived *E. coli* and *Salmonella* isolates (Bolton *et al.*, 1999; White *et al.*, 2000).

Daptomycin

Daptomycin is a cyclic lipopeptide that acts in the presence of physiological concentrations of calcium to form pores in the cytoplasmic membranes of target Gram-positive bacterial cells. The end result is leakage of ions from the cell and cell death. Resistance to daptomycin in the clinical setting may arise associated with prolonged therapy. The precise mechanisms of resistance have not been defined, but overexpression of genes associated with increasing the positive charge of the cytoplasmic membrane and a disordering of phospholipids have been implicated (Mishra *et al.*, 2009; Arias *et al.*, 2011). Exposure to daptomycin induces the autoregulatory hVISA (heterogeneous vancomycin-intermediate *S. aureus*)-associated VraSR (vancomycin resistance associated sensor/regulator) two-component regulatory system in *S. aureus* (Muthaiyan *et al.*, 2008). Perhaps as a result of these common pathways, hVISA strains frequently express reduced susceptibility to daptomycin (Cui *et al.*, 2006), and strains expressing reduced susceptibility to daptomycin exhibit the hVISA phenotype (Camargo *et al.*, 2008).

Glycopeptides

Glycopeptide antibiotics (vancomycin and teicoplanin) inhibit cell wall synthesis by binding to the terminal D-alanyl–D-alanine of the pentapeptide peptidoglycan precursor molecule as it exits the cytoplasmic membrane, thus preventing the cross-linking of the cell wall. Glycopeptides are only active against Gram-positive bacteria because their size precludes entry through the outer membrane porins of Gram-negative bacteria.

Most glycopeptide resistance occurs through the acquisition of operons that encode the formation of altered peptidoglycan precursors. To date, several varieties of enterococcal glycopeptide resistance have been described (VanA through VanE, and VanG, VanM, and VanN). Of these, the most clinically important are VanA and VanB (Arthur *et al.*, 1996). VanA enterococci are phenotypically resistant to vancomycin and teicoplanin, whereas VanB strains are resistant to vancomycin but appear susceptible to teicoplanin. This susceptibility results from the fact that teicoplanin does not induce expression of resistance (Evers and Courvalin, 1996). Once the VanB operon is expressed, however, resistance to teicoplanin results.

VanA and VanB are encoded by similar operons in which three genes (*vanH*, *vanA*, *vanX* or $vanH_B$, *vanB*, $vanX_B$) whose transcription is regulated by two other genes (*vanS*, *vanR* or $vanS_B$, $vanR_B$) are required for expression of resistance (Arthur *et al.*, 1996). The ultimate outcome is a pentapeptide precursor terminating in D-alanine-D-lactate, in which the binding affinity of vancomycin to its target is reduced roughly 1000-fold (Hughes, 2003).

The majority of clinical vancomycin-resistant enterococci (VRE) strains are from the species *E. faecium*, a predilection that remains unexplained (Sahm *et al.*, 1999). The majority of vancomycin-resistant *E. faecium* strains that cause clinical infection are also resistant to ampicillin, owing to the expansion of a group of hospital-adapted clones (referred to as clonal complex 17) that have emerged around the world (Top *et al.*, 2008). Clonal complex 17 strains are now found worldwide and are characterized by their resistance to ampicillin and by the fact that they frequently harbor putative virulence determinants such as esp_{Efm} and hyl_{Efm} (Leavis *et al.*, 2004, 2007). Despite in vitro transfer of the VanA determinant to *S. aureus* (Noble *et al.*, 1992), and at least 11 instances in which VanA-expressing *S. aureus* have been described from clinical samples (Périchon and Courvalin, 2009), vancomycin-resistant *S. aureus* remain exceedingly rare.

The VanC operon is intrinsic to the cell wall synthesis machinery of the minor enterococcal species *E. casseliflavus* (including the biotype formerly classified as *E. flavescens)* and *E. gallinarum* (Vincent *et al.*, 1992; Dutka-Malen *et al.*, 1994). The peptidoglycan precursor in VanC strains terminates in D-alanine–D-serine, reducing vancomycin affinity about sevenfold and resulting in low levels of resistance.

Resistance to glycopeptides in *S. aureus* more commonly takes the form of reduced susceptibility. These strains, alternately called hVISA or hGISA (heterogeneous glycopeptide-intermediate *S. aureus*) express vancomycin minimum inhibitory concentration (MICs) in the 4–8 µg/ml range (Linares, 2001). However, within these cultures are smaller populations of cells that express higher levels of resistance. The resistance phenotype is characterized by a thickened cell wall, which may decrease glycopeptide susceptibility by providing an excess of false targets for glycopeptide binding. Animal studies suggest that the level of resistance expressed by hVISA strains will reduce the effectiveness of vancomycin therapy (Climo *et al.*, 1999). In *S. haemolyticus*, glycopeptide resistance has been reported associated with changes in the composition of the cross-links of the peptidoglycan (Billot-Klein *et al.*, 1996).

Vancomycin is, in general, a less bactericidal antibiotic than are the β-lactams. The bacteremia associated with *S. aureus* endocarditis, for example, takes roughly twice as much time to clear with vancomycin treatment than with treatment by the β-lactam oxacillin (Levine *et al.*, 1991), and vancomycin may be inferior to nafcillin in preventing failure and relapse when treating bacteremia due to methicillin-susceptible *S. aureus* (MSSA) (Chang *et al.*, 2003).

Linezolid

The oxazolidinone antibiotic linezolid inhibits bacterial protein synthesis by interacting with the *N*-formylmethionyl-tRNA–ribosome–mRNA ternary complex commonly referred to as the initiation complex. Resistance to linezolid has been described in both enterococci and staphylococci, but overall rates remain very low more than a decade after the first clinical use of this agent (Jones *et al.*, 2008). Resistance is most often associated with a G2576U (*E. coli* numbering scheme) point mutation in the 23S ribosomal RNA, although mutations at other positions may also contribute to resistance (Prystowsky *et al.*, 2001). As the 23S subunit genes exist as multiple copies, more than one copy must be mutated to confer resistance, with strains that have a higher percentage of mutated 23S genes expressing greater levels of resistance (Marshall *et al.*, 2002). Resistance to linezolid, macrolides and chloramphenicol has been attributed to a 6 bp deletion in the gene encoding riboprotein L4 in *S. pneumoniae* (Wolter *et al.*, 2005). Plasmid-mediated resistance to linezolid through the expression of the *cfr* (chloramphenicol–florfenicol resistance) rRNA methylase gene has been reported in staphylococci (Arias *et al.*, 2008) and enterococci (Diaz *et al.*, 2012), but the prevalence of this type of linezolid resistance remains quite low (Jones *et al.*, 2008).

Macrolides

Macrolides inhibit protein synthesis by binding reversibly to the peptidyl-tRNA binding region of the 50S ribosomal subunit. The most important mechanism of resistance to macrolides is through methylation of the ribosome (Weisblum, 1995), most commonly through expression of different *erm* (erythromycin ribosomal methylase) genes. Methylated ribosomes confer resistance to macrolides, clindamycin (a lincosamide), and the streptogramins B (MLS_B compounds), hence MLS_B resistance. Many *erm* genes have been described—*erm*(A) and the related *erm*(TR), and *erm*(B) and the related *erm*(AM), and resistance is frequently inducible by macrolides, but not by clindamycin ($iMLS_B$). Constitutively expressed *erm*-type resistance ($cMLS_B$) results in resistance to clindamycin as well.

The second major mechanism of resistance to macrolides (but not to clindamycin) is by the expression of efflux pumps encoded by *mef* genes (mef in Gram-positive bacteria and the membrane proteins AcrAB–TolC in *H. influenzae* and *E. coli*) (Zhong and Shortridge, 2000). Minor mechanisms of resistance to macrolides include esterases that hydrolyze the antibiotics, and point mutations within the genes encoding the 50S ribosomal subunit.

Quinupristin–dalfopristin

Quinupristin–dalfopristin is a mixture of semisynthetic streptogramins A and B. Resistance to this can result from resistance to streptogramin A alone and was first described in staphylococci conferred by genes encoding streptogramin A acetyltransferases—*vat*(A), *vat*(B) and *vat*(C), or ATP-binding efflux genes—*vga*(A) and *vga*(B). Two acetyltransferase-encoding resistance genes have been described that confer resistance to quinupristin–dalfopristin in *Enterococcus faecium*—*vat*(D) (previously *sat*(A)) and *vat*(E) (previously *sat*(G)). In most cases, these resistance genes are found along with an *erm* resistance gene that confers resistance to streptogramin B-type antibiotics (Soltani *et al.*, 2000).

Metronidazole

Metronidazole is prodrug whose activation depends upon reduction of its nitro group in the absence of oxygen. The activity of metronidazole appears to result in DNA damage and cell death (Edwards, 1993b). In some cases, decreased uptake and/or reduced rate of reduction are believed to be responsible for metronidazole resistance (Edwards, 1993a). Five *Bacteroides* genes, *nimA–E*, have been implicated in resistance to 5-nitroimidazole antibiotics (Carlier *et al.*, 1997; Leiros *et al.*, 2004). The enzyme thioredoxin reductase appears to be responsible for the reduction of metronidazole in *Trichomonas vaginalis* (Leitsch *et al.*, 2009).

Nitrofurantoin

Nitrofurantoin, 1-[(5-nitrofurfurylidene)amino]hydantoin, is a synthetic antibacterial agent used primarily in the treatment of urinary tract infections. The ability of nitrofurantoin to kill bacteria correlates with the presence of bacterial nitroreductases, which convert nitrofurantoin to highly reactive electrophilic intermediates (McOsker and Fitzpatrick, 1994). Strains of bacteria that are resistant to nitrofurantoin have been shown to possess diminished nitroreductase activity (Race *et al.*, 2005).

Polymyxin B and polymyxin E (colistin)

The polymyxins are bactericidal polycationic peptide antibiotics that bind to the Gram-negative bacterial cell membrane and increase its permeability, resulting in leakage of intracellular components. (Tam *et al.*, 2005; Tzeng *et al.*, 2005). Organisms that are resistant to polymyxins have cell walls that prevent access of the drug to the cell membrane. In general, polymyxins are bactericidal against *P. aeruginosa*, *Acinetobacter* spp., some *P. mirabilis* strains, and some strains of *S. marcescens*. *Proteus* spp., *Providentia* spp., *Neisseria* spp. and Gram-positive bacteria are resistant to polymyxins (Tam *et al.*, 2005; Tzeng *et al.*, 2005).

Polymyxin-resistant bacteria, including mutants, exhibit a modified lipopolysaccharide (LPS). LPS modifications that include alteration of the fatty acid content of lipid A, the addition of phosphoethanolamine (PEA) to the core and lipid A head groups, and the addition of 4-amino-4-deoxy-L-arabinose (Ara4N) to the core and lipid A regions have been well studied (Tzeng *et al.*, 2005). The *mtr* (multiple transferable resistance) gene complex encodes an energy-dependent efflux pump made up of the cell envelope proteins MtrC–MtrD–MtrE, and together with lipid A modification, and the type IV pilin secretion system, modulates the levels of polymyxin resistance in *N. meningitidis* (Tzeng *et al.*, 2005), while the polymyxin resistance (Pmr) two-component PmrAB system is involved in resistance to colistin in *A. baumanni* (Adams *et al.*, 2009). Recently, a novel transferable resistance to colistin has become widespread, predominantly in *E. coli*. Encoded by the *mcr-1* and -2 genes it has been found is strains isolated from both animals and humans. (Liu *et al.*, 2016)

Quinolones

Bacterial topoisomerases (DNA gyrase and topoisomerase IV) are essential enzymes that maintain negative supercoiling of DNA (DNA gyrase) or separate interlocked daughter DNA strands formed during replication, facilitating segregation into daughter cells (Topo IV). Quinolone antibiotics act by directly inhibiting DNA synthesis through their activity against DNA gyrase and Topo IV (Willmott *et al.*, 1994; Hiasa *et al.*, 1996; Shea and Hiasa, 1999). The end result is cellular death by mechanisms that are unclear. In general, the primary target of fluoroquinolones in Gram-negative bacteria is DNA gyrase, whereas in Gram-positive bacteria it is Topo IV (Blanche *et al.*, 1996; Pan and Fisher, 1999; Alovero *et al.*, 2000).

ALTERATIONS IN TARGET ENZYMES The most common mechanism of clinically significant levels of fluoroquinolone resistance is through alterations in regions of the topoisomerase enzymes known as the quinolone-resistance-determining region (QRDR) resulting from spontaneous mutations that occur within the respective genes. Particularly frequent sites for resistance-associated mutations are serine 83 and aspartate 87 of the GyrA subunit of DNA gyrase and serine 79 and aspartate 83 of the ParC subunit of Topo IV (Piddock, 1999).

The level of resistance conferred by a single-point mutation in the primary target enzyme will depend upon the reduction of enzyme affinity created by the mutation, as well as the affinity of the fluoroquinolone for the secondary target. Fluoroquinolone–species combinations for which single mutations result in significantly higher MICs (such as ciprofloxacin and *S. aureus* or *P. aeruginosa*) have readily selected out resistant mutants in the clinical setting (Coronado *et al.*, 1995). In the clinic, most highly resistant strains exhibit more than one mutation in both the GyrA and ParC subunits of the two enzymes. While single-point mutations conferring resistance to one fluoroquinolone may not result in "resistant" MICs for another, the MICs for the second fluoroquinolone will inevitably be increased. As such, resistance to fluoroquinolones is probably best considered to be a class resistance.

RESISTANCE DUE TO DECREASED INTRACELLULAR ACCUMULATION The most important mechanism by which the intracellular accumulation of fluoroquinolones is reduced is through the expression of MDR efflux pumps (Piddock, 1999). RND pumps in Gram-negative bacteria extend from the cytoplasmic membrane through the outer membrane, whereas Gram-positive pumps need only traverse the cytoplasmic membrane. Resistance results when expression of pumps is increased due to mutations within their regulatory genes (Ziha-Zarifi *et al.*, 1999). While resistance conferred by pumps is generally low level, the action of these pumps may augment the phenotypic resistance expressed by single topoisomerase point mutations, thereby facilitating selection in the clinic.

Three plasmid-mediated fluoroquinolone resistance gene classes have been described. The Qnr proteins (Jacoby, 2005; Nordmann and Poirel, 2005) protect DNA from quinolone binding (Tran and Jacoby, 2002; Tran *et al.*, 2005) and confer low levels of resistance. Six variants of Qnr have now been described (A, B, C, D, S, VC) (http://www.lahey.org/qnrStudies/). Plasmid-mediated ciprofloxacin and norfloxacin resistance can also be conferred by the AAC(6′)-Ib-cr protein, which is a mutant of the AAC(6′)-Ib aminoglycoside-modifying enzyme (Cattoir and Nordmann, 2009). One plasmid-mediated quinolone efflux pump (QepA) that effluxes the hydrophilic fluoroquinolones (ciprofloxacin, enrfloxacin, and norfloxacin) has been recognized among strains of *Enterobacteriaciae* (Cattoir and Nordmann, 2009).

Rifampin

Resistance to rifampin is generally due to point mutations in the chromosomal *rpoB* gene, which encodes the bacterial DNA-dependent RNA polymerase (Rpo) (Wehrli, 1983). These mutations reduce rifampin binding and generally occur at a high enough frequency to preclude the use of rifampin as a single agent for the treatment of bacterial infections.

Tetracyclines

Tetracyclines act by inhibiting attachment of aminoacyl-tRNA to the ribosome acceptor site and preventing the elongation of nascent protein peptide chains (Schnappinger and Hillen, 1996). Most tetracycline-resistance (Tet) determinants involve either efflux or ribosomal protection (see review by Chopra and Roberts, 2001).

Tetracycline efflux proteins are membrane-associated members of the major facilitator superfamily of proteins. Most efflux proteins (with the exception of Tet(B)) confer resistance to tetracyclines but do not affect minocycline activity (Chopra and Roberts, 2001). There are six groups of tetracycline-efflux proteins based on amino acid identity. Group 1 efflux proteins are found primarily in Gram-negative species (with the exception of Tet(Z)), while Group 2 efflux proteins (consisting only of Tet(K) and Tet(L)) are found primarily in Gram-positive species. Groups 3 through 6 efflux proteins are small groups consisting of one or two efflux proteins each.

The other major mechanism of tetracycline resistance results from the production of ribosome protection proteins. These act by binding to and altering the conformation of the ribosome in a way that inhibits tetracycline binding. Tet(M) and Tet(O) are the best characterized of these proteins. In some cases, their mobilization by transposons has led to widespread dissemination (Rice, 1998). The expression of the genes encoding efflux proteins and ribosomal protection proteins is increased in the presence of tetracycline (Su *et al.*, 1992; Hillen and Berens, 1994).

Many species are intrinsically resistant to tetracycline through the expression of RND efflux pumps. In *E. coli*, this resistance can be expressed through mutations in the *mar* (multiple-antibiotic-resistant) operon, which result in reduced outer membrane porin expression and expression of the AcrAB RND efflux pump (Alekshun and Levy, 1997). Several similar pumps, some of which are constitutively expressed, have been described in *P. aeruginosa* and other Gram-negative bacteria (Poole *et al.*, 1993).

Tigecycline

Tigecycline, a derivative of minocycline, owes its broad spectrum of activity to its resistance to the common efflux or ribosomal protection mechanisms that confer resistance to the older tetracyclines. *P. aeruginosa* and *Proteus* spp. are intrinsically resistant to tigecycline through the expression of RND-type efflux (Ruzin *et al.*, 2005a). Resistance to tigecycline in other Gram-negative species generally results from the activation of normally repressed AcrAB-type RND efflux pumps (Ruzin *et al.*, 2005b).

Trimethoprim–sulfamethoxazole

Sulfamethoxazole (a sulfonamide) and trimethoprim are inhibitors of two enzymes—dihydropteroic acid synthase (DHPS) and dihydrofolate reductase (DHFR), respectively—that act sequentially and synergistically in the manufacture of tetrahydrofolate.

INTRINSIC RESISTANCE Intrinsic resistance to trimethoprim–sulfamethoxazole is relatively rare and results from decreased access to the target enzymes (*P. aeruginosa*) (Then, 1982), low-affinity DHFR enzymes (*Neisseria* spp., *Clostridium* spp., *Brucella* spp., *Bacteroides* spp., *Moraxella catarrhalis* and *Nocardia* spp.) (Then and Angehrn, 1979) or the ability to absorb exogenous folate (*Enterococcus* spp., *Lactobacillus* spp.) (Zervos and Schaberg, 1985) or thymine (*Enterococcus* spp.) (Hamilton-Miller, 1988).

ACQUIRED RESISTANCE TO TRIMETHOPRIM Resistance to trimethoprim is most commonly attributable to the acquisition of low affinity *dhfr* genes, of which approximately 20 have been described (Huovinen, 2001). Mutational resistance to trimethoprim has also been described in several species, generally involving promoter mutations resulting in the overproduction of DHFR (in *E. coli*), *dhfr* point mutations (in *S. aureus* and *S. pneumoniae*), or both mechanisms (in *H. influenzae*) (Huovinen, 2001).

ACQUIRED RESISTANCE TO SULFONAMIDES Resistance to sulfonamides in different species has been attributed to point mutations or small insertions of DNA segments within chromosomal *dhps* genes (Huovinen, 2001; Enne *et al.*, 2002), as well as to transformation and recombination (Swedberg *et al.*, 1998; Skold, 2000). Plasmid-mediated, transferable resistance to sulfonamides has been reported in Gram-negative bacteria (Huovinen, 2001). In contrast to the diversity of *dhfr* genes, only three acquired low-affinity *dhps* genes (*sul*I, *sul*II, and *sul*III) have been described. Genes conferring resistance to sulfonamides are frequently incorporated into multiresistance integrons, which are themselves frequently integrated into transferable plasmids.

Conclusions

Bacteria employ a variety of mechanisms resistant to the inhibitory and bactericidal activity of different classes of antibiotics. Understanding the specific mechanisms of resistance can, at times, inform treatment choices to allow more effective therapy. Unfortunately, the most problematic of resistant bacteria causing problems in modern hospitals employ a variety of resistance mechanisms, some of which are fairly broad in their spectrum of activity. As such, they are likely to be active not just against older antibiotics, but also against newly developed antibiotics. Parsimonious use of these precious agents holds the most promise for preserving their activities for future generations.

References

Adams, M.D., Nickel, G.C., Bajaksouzian, S., Lavender, H., Murthy, A.R., Jacobs, M.R., and Bonomo, R.A. (2009) Resistance to colistin in *Acinetobacter baumannii* associated with mutations in the PmrAB two-component system. *Antimicrobial Agents and Chemotherapy* 53, 3628–3634.

Aires, J.R. and Nikaido, H. (2005) Aminoglycosides are captured from both periplasm and cytoplasm by the AcrD multidrug efflux transporter of *Escherichia coli*. *Journal of Bacteriology,* 187, 1923–1929.

Alekshun, M.N. and Levy, S.B. (1997) Regulation of chromosomally mediated multiple antibiotic resistance: the *mar* regulon. *Antimicrobial Agents and Chemotherapy,* 41, 2067–2075.

Alovero, F.L., Pan, X.S., Morris, J.E., Manzo, R.H., and Fisher, L.M. (2000) Engineering the specificity of antibacterial fluoroquinolones: benzenesulfonamide modifications at C-7 of ciprofloxacin change its primary target in *Streptococcus pneumoniae* from topoisomerase IV to gyrase. *Antimicrobial Agents and Chemotherapy* 44, 320–325.

Ambler, R.P. (1980) The structure of β-lactamases. *Philosophical Transactions of the Royal Society B: Biological Sciences* 289, 321–331.

Arias, C.A., Vallejo, M., Reyes, J., Panesso, D., Moreno, J., Castaneda, E., Villegas, M.V., Murray, B.E., and Quinn, J.P. (2008) Clinical and microbiological aspects of linezolid resistance mediated by the *cfr* gene encoding a 23S rRNA methyltransferase. *Journal of Clinical Microbiology* 46, 892–896.

Arias, C.A., Panesso, D., Mcgrath, D.M., Qin, X., Mojica, M.F., Miller, C., Diaz, L., Tran, T.T., Rincon, S., Barbu, E.M. *et al.* (2011) Genetic basis for in vivo daptomycin resistance in enterococci. *The New England Journal of Medicine* 365, 892–900.

Arthur, M., Reynolds, P., and Courvalin, P. (1996) Glycopeptide resistance in enterococci. *Trends in Microbiology* 4, 401–407.

Azucena, E. and Mobashery, S. (2001) Aminoglycoside-modifying enzymes: mechanisms of catalytic processes and inhibition. *Drug Resistance Updates,* 4, 106–117.

Barcus, V.A., Ghanekar, K., Yeo, M., Coffey, T.J., and Dowson, C.G. (1995) Genetics of high level penicillin resistance in clinical isolates of *Streptococcus pneumoniae*. *FEMS Microbiology Letters* 126, 299–303.

Billot-Klein, D., Gutmann, L., Bryant, D., Bell, D., van Heijenoort, J., Grewal, J., and Shlaes, D.M. (1996) Peptidoglycan synthesis and structure in *Staphylococcus haemolyticus* expressing increasing levels of resistance to glycopeptide antibiotics. *Journal of Bacteriology* 178, 4696–4703.

Blanche, F., Cameron, B., Bernard, F.X., Maton, L., Manse, B., Ferrero, L., Ratet, N., Lecoq, C., Goniot, A., Bisch, D., and Crouzet, J. (1996) Differential behaviors of *Staphylococcus aureus* and *Escherichia coli* type Ii DNA topoisomerases. *Antimicrobial Agents and Chemotherapy* 40, 2714–2720.

Bolton, L.F., Kelley, L.C., Lee, M.D., Fedorka-Cray, P.J., and Maurer, J.J. (1999) Detection of multidrug-resistant *Salmonella enterica* serotype *typhimurium* DT104 based on a gene which confers cross-resistance to florfenicol and chloramphenicol. *Journal of Clinical Microbiology* 37, 1348–1351.

Bradford, P.A. (2001a) Extended-spectrum beta-lactamases in the 21st century: characterization, epidemiology, and detection of this important resistance threat. *Clinical Microbiology Reviews* 14, 933–951.

Bradford, P.A. (2001b) What's new in beta-lactamases? *Current Infectious Diseases Reports* 3, 13–19.

Bradford, P.A., Urban, C., Mariano, N., Projan, S.J., Rahal, J.J., and Bush, K. (1997) Imipenem resistance in *Klebsiella pneumoniae* is associated with the combination of ACT-1, a plasmid-mediated AmpC beta-lactamase, and the loss of an outer membrane protein. *Antimicrobial Agents and Chemotherapy* 41, 563–569.

Bush, K. (1989a) Classification of β-lactamases: Groups 1, 2a, 2b and 2b′. *Antimicrobial Agents and Chemotherapy* 33, 264–270.

Bush, K. (1989b) Classification of β-lactamases: Groups 2c, 2d, 2e, 3 and 4. *Antimicrobial Agents and Chemotherapy* 33, 271–276.

Camargo, I.L., Neoh, H.M., Cui, L., and Hiramatsu, K. (2008) Serial daptomycin selection generates daptomycin-nonsusceptible *Staphylococcus aureus* strains with a heterogeneous vancomycin-intermediate phenotype. *Antimicrobial Agents and Chemotherapy* 52, 4289–4299.

Carlier, J.P., Sellier, N., Rager, M.N., and Reysset, G. (1997) Metabolism of a 5-nitroimidazole in susceptible and resistant isogenic strains of *Bacteroides fragilis*. *Antimicrobial Agents and Chemotherapy* 41, 1495–1499.

Cattoir, V. and Nordmann, P. (2009) Plasmid-mediated quinolone resistance in Gram-negative bacterial species: an update. *Current Medicinal Chemistry* 16, 1028–1046.

Chambers, H.F. (1997) Methicillin resistance in staphylococci: molecular and biochemical basis and clinical implications. *Clinical Microbiological Reviews* 10, 781–791.

Chang, F.Y., Peacock, J.E., Jr., Musher, D.M., Triplett, P., MacDonald, B.B., Mylotte, J.M., O'Donnell, A.,

Wagener, M.M., and Yu, V.L. (2003) *Staphylococcus aureus* bacteremia: recurrence and the impact of antibiotic treatment in a prospective multicenter study. *Medicine (Baltimore)* 82, 333–339.

Chopra, I. and Roberts, M. (2001) Tetracycline antibiotics: mode of action, applications, molecular biology, and epidemiology of bacterial resistance. *Microbiology and Molecular Biology Reviews* 65, 232–260.

Climo, M.W., Patron, R.L., and Archer, G.L. (1999) Combinations of vancomycin and beta-lactams are synergistic against staphylococci with reduced susceptibilities to vancomycin. *Antimicrobial Agents and Chemotherapy* 43, 1747–1753.

Coronado, V.G., Edwards, J.R., Culver, D.H., and Gaynes, R.P. (1995) Ciprofloxacin resistance among nosocomial *Pseudomonas aeruginosa* and *Staphylococcus aureus* in the United States. National Nosocomial Infections Surveillance (NNIS) System. *Infection Control and Hospital Epidemiology* 16, 71–75.

Cui, L., Iwamoto, A., Lian, J.Q., Neoh, H.M., Maruyama, T., Horikawa, Y., and Hiramatsu, K. (2006) Novel mechanism of antibiotic resistance originating in vancomycin-intermediate *Staphylococcus aureus*. *Antimicrobial Agents and Chemotherapy* 50, 428–438.

Diaz, L., Kiratisin, P., Mendes, R.E., Panesso, D., Singh, K.V., and Arias, C.A. (2012) Transferable plasmid-mediated resistance to linezolid due to *cfr* in a human clinical isolate of *Enterococcus faecalis*. *Antimicrobial Agents and Chemotherapy* 56, 3917–3922.

Dutka-Malen, S., Blaimont, B., Wauters, G., and Courvalin, P. (1994) Emergence of high-level resistance to glycopeptides in *Enterococcus gallinarum* and *Enterococcus casseliflavus*. *Antimicrobial Agents and Chemotherapy* 38, 1675–1677.

Edwards, D.I. (1993a) Nitroimidazole drugs – action and resistance mechanisms I. Mechanisms of action. *Journal of Antimicrobial Chemotherapy* 31, 9–20.

Edwards, D.I. (1993b) Nitroimidazole drugs—action and resistance mechanisms II. Mechanisms of resistance. *Journal of Antimicrobial Chemotherapy* 31, 201–210.

Enne, V.I., King, A., Livermore, D.M., and Hall, L.M. (2002) Sulfonamide resistance in *Haemophilus influenzae* mediated by acquisition of *sul2* or a short insertion in chromosomal *folP*. *Antimicrobial Agents and Chemotherapy* 46, 1934–1939.

Evers, S. and Courvalin, R. (1996) Regulation of VanB-type vancomycin resistance gene expression by the vanSB–VanRB two component regulatory system in *Enterococcus faecalis* V583. *Journal of Bacteriology* 178, 1302–1309.

Fitton, J.E. and Shaw, W.V. (1979) Comparison of chloramphenicol acetyltransferase variants in staphylococci. Purification, inhibitor studies and N-terminal sequences. *Biochemical Journal* 177, 575–582.

Fontana, R., Aldegheri, M., Ligozzi, M., Lopez, H., Sucari, A., and Satta, G. (1994) Overproduction of a low-affinity penicillin-binding protein and high-level ampicillin resistance in *Enterococcus faecium*. *Antimicrobial Agents and Chemotherapy* 38, 1980–1983.

Hakenbeck, R. and Coyette, J. (1998) Resistant penicillin-binding proteins. *Cellular and Molecular Life Science* 54, 332–340.

Hamilton-Miller, J.M. (1988) Reversal of activity of trimethoprim against Gram-positive cocci by thymidine, thymine and 'folates'. *Journal of Antimicrobial Chemotherapy,* 22, 35–39.

Henze, U.U. and Berger-Bachi, B. (1995) *Staphylococcus aureus* penicillin-binding protein 4 and intrinsic beta-lactam resistance. *Antimicrobial Agents and Chemotherapy* 39, 2415–2422.

Heritier, C., Poirel, L., Lambert, T., and Nordmann, P. (2005) Contribution of acquired carbapenem-hydrolyzing oxacillinases to carbapenem resistance in *Acinetobacter baumannii*. *Antimicrobial Agents and Chemotherapy* 49, 3198–3202.

Hiasa, H., Yousef, D.O., and Marians, K.J. (1996) DNA strand cleavage is required for replication fork arrest by a frozen topoisomerase-quinolone-DNA ternary complex. *The Journal of Biological Chemistry* 271, 26424–26429.

Hillen, W. and Berens, C. (1994) Mechanisms underlying expression of TN10 encoded tetracycline resistance. *Annual Review of Microbiology* 48, 345–369.

Hughes, D. (2003). Exploiting genomics, genetics and chemistry to combat antibiotic resistance. *Nature Reviews Genetics* 4, 432–441.

Huovinen, P. (2001) Resistance to trimethoprim-sulfamethoxazole. *Clinical Infectious Diseases* 32, 1608–1614.

Jacobs, C., Frère, J.-M., and Normark, S. (1997) Cytosolic intermediates for cell wall biosynthesis and degradation control inducible β-lactam resistance in Gram-negative bacteria. *Cell* 88, 823–832.

Jacoby, G.A. (2005) Mechanisms of resistance to quinolones. *Clinical Infectious Diseases* 41(Suppl 2), S120–S126.

Jacoby, G.A. (2009) AmpC beta-lactamases. *Clinical Microbiology Reviews* 22, 161–182.

Jeannot, K., Sobel, M.L., El Garch, F., Poole, K., and Plésiat, P. (2005) Induction of the MexXY efflux pump in *Pseudomonas aeruginosa* is dependent on drug–ribosome interaction. *Journal of Bacteriology,* 187, 5341–5346.

Jones, R.N., Ross, J.E., Castanheira, M., and Mendes, R.E. (2008) United States resistance surveillance results for linezolid (LEADER Program for 2007). *Diagnostic Microbiology and Infectious Disease* 62, 416–426.

Kim, J., Lim, Y.M., Jeong, Y.S., and Seol, S.Y. (2005) Occurrence of CTX-M-3, CTX-M-15, CTX-M-14, and CTX-M-9 extended-spectrum beta-lactamases in *Enterobacteriaceae* clinical isolates in Korea. *Antimicrobial Agents and Chemotherapy* 49, 1572–1575.

Lartigue, M.F., Poirel, L., Decousser, J.W., and Nordmann, P. (2005) Multidrug-resistant *Shigella sonnei* and *Salmonella enterica* serotype *typhimurium* isolates producing CTX-M beta-lactamases as causes of community-acquired infection in France. *Clinical Infectious Diseases* 40, 1069–1070.

Leavis, H., Top, J., Shankar, N., Borgen, K., Bonten, M., van Embden, J., and Willems, R.J. (2004) A novel putative enterococcal pathogenicity island linked to the *esp* virulence gene of *Enterococcus faecium* and associated with epidemicity. *Journal of Bacteriology* 186, 672–682.

Leavis, H.L., Willems, R.J.L., van Wamel, W.J.B., Schuren, F.H., Caspers, M.P.M., and Bonten, M.J.M. (2007) Insertion sequence-driven diversification creates a globally dispersed emerging multiresistant subspecies of *E. faecium*. *PLoS Pathology* 3(1): e7.

Leiros, H.-K.S., Kozielski-Stuhrmann, S., Kapp, U., Terradot, L., Leonard, G.A., and McSweeney, S.M. (2004) Structural basis of 5-nitroimidazole antibiotic resistance: the crystal structure of NimA from *Deinococcus radiodurans*. *The Journal of Biological Chemistry* 279, 55840–55849.

Leitsch, D., Kolarich, D., Binder, M., Stadlmann, J., Altmann, F., and Duchene, M. (2009) *Trichomonas vaginalis*: metronidazole and other nitroimidazole drugs are reduced by the flavin enzyme thioredoxin reductase and disrupt the cellular redox system. Implications for nitroimidazole toxicity and resistance. *Molecular Microbiology* 72, 518–536.

Levine, D.P., Fromm, B.S., and Reddy, B.R. (1991) Slow response to vancomycin or vancomycin plus rifampin in methicillin-resistant *Staphylococcus aureus* endocarditis. *Annals of Internal Medicine* 115, 674–680.

Linares, J. (2001) The VISA/GISA problem: therapeutic implications. *Clinical Microbiology and Infection* 7, 8–15.

Liu, Y.Y., Wang, Y., Walsh, T.R., Yi, L.X., Zhang, R., Spencer, J., Doi, Y., Tian, G., Dong, B., Huang, X. *et al.* (2016) Emergence of plasmid-mediated colistin resistance mechanism MCR-1 in animals and human beings in China: a microbiological and molecular biological study. *The Lancet Infectious Diseases* 16, 161–8.

Magnet, S., Courvalin, P., and Lambert, T. (2001) Resistance-nodulation-cell division-type efflux pump involved in aminoglycoside resistance in *Acinetobacter baumannii* strain BM4454. *Antimicrobial Agents and Chemotherapy* 45, 3375–3380.

Mao, W., Warren, M.S., Lee, A., Mistry, A., and Lomovskaya, O. (2001) MexXY–OprM efflux pump is required for antagonism of aminoglycosides by divalent cations in *Pseudomonas aeruginosa*. *Antimicrobial Agents and Chemotherapy* 45, 2001–2007.

Marshall, S.H., Donskey, C.J., Hutton-Thomas, R., Salata, R.A., and Rice, L.B. (2002) Gene dosage and linezolid resistance in *Enterococcus faecium* and *Enterococcus faecalis*. *Antimicrobial Agents and Chemotherapy* 46, 3334–3336.

Matagne, A., Lamotte-Brasseur, J., and Frère, J.M. (1998) Catalytic properties of class A beta-lactamases: efficiency and diversity. *Biochemistry Journal* 330, 581–598.

Maus, C.E., Plikaytis, B.B., and Shinnick, T.M. (2005) Molecular analysis of cross-resistance to capreomycin, kanamycin, amikacin, and viomycin in *Mycobacterium tuberculosis*. *Antimicrobial Agents and Chemotherapy* 49, 3192–3197.

McOsker, C.C. and Fitzpatrick, P.M. (1994) Nitrofurantoin: mechanism of action and implications for resistance development in common uropathogens. *Journal of Antimicrobial Chemotherapy* 33(Suppl A), 23–30.

Mishra, N.N., Yang, S.J., Sawa, A., Rubio, A., Nast, C.C., Yeaman, M.R., and Bayer, A.S. (2009) Analysis of cell membrane characteristics of in vitro-selected daptomycin-resistant strains of methicillin-resistant *Staphylococcus aureus*. *Antimicrobial Agents and Chemotherapy* 53, 2312–2318.

Murray, I.A. and Shaw, W.V. (1997) *O*-Acetyltransferases for chloramphenicol and other natural products. *Antimicrobial Agents and Chemotherapy* 41, 1–6.

Murray, I.A., Martinez-Suarez, J.V., Close, T.J., and Shaw, W.V. (1990) Nucleotide sequences of genes encoding the type II chloramphenicol acetyltransferases of *Escherichia coli* and *Haemophilus influenzae*, which are sensitive to inhibition by thiol-reactive reagents. *Biochemical Journal* 272, 505–510.

Muthaiyan, A., Silverman, J.A., Jayaswal, R.K., and Wilkinson, B.J. (2008) Transcriptional profiling reveals that daptomycin induces the *Staphylococcus aureus* cell wall stress stimulon and genes responsive to membrane depolarization. *Antimicrobial Agents and Chemotherapy* 52, 980–990.

Naas, T. and Nordmann, P. (1999) OXA-type beta-lactamases. *Current Pharmaceutical Design* 5, 865–879.

Navia, M.M., Ruiz, J., and Vila, J. (2002) Characterization of an integron carrying a new class D beta-lactamase (OXA-37) in *Acinetobacter baumannii*. *Microbial Drug Resistance* 8, 261–265.

Nikaido, H. (1998) Multiple antibiotic resistance and efflux. *Current Opinion in Microbiology,* 1, 516–523.

Noble, W.C., Virani, Z., and Gee, R.G.A. (1992) Co-transfer of vancomycin and other resistance genes from *Enterococcus faecalis* NCTC 12201 to *Staphylococcus aureus*. *FEMS Microbiology Letters,* 93, 195–198.

Nordmann, P. and Poirel, L. (2005) Emergence of plasmid-mediated resistance to quinolones in *Enterobacteriaceae*. *Journal of Antimicrobial Chemotherapy* 56, 463–469.

Pan, X.S. and Fisher, L.M. (1999) *Streptococcus pneumoniae* DNA gyrase and topoisomerase IV: overexpression, purification, and differential inhibition by fluoroquinolones. *Antimicrobial Agents and Chemotherapy* 43, 1129–1136.

Paterson, D.L., Hujer, K.M., Hujer, A.M., Yeiser, B., Bonomo, M.D., Rice, L.B., and Bonomo, R.A. (2003) Extended-spectrum beta-lactamases in *Klebsiella pneumoniae* bloodstream isolates from seven countries: dominance and widespread prevalence of SHV- and CTX-M-type beta-lactamases. *Antimicrobial Agents and Chemotherapy* 47, 3554–3560.

Paton, R., Miles, R.S., Hood, J., and Amyes, S.G.B. (1993) ARI-1: β-lactamase-mediated imipenem resistance in *Acinetobacter baumannii*. *International Journal of Antimicrobial Agents* 2, 81–88.

Périchon, B. and Courvalin, P. (2009) VanA-type vancomycin-resistant *Staphylococcus aureus*. *Antimicrobial Agents and Chemotherapy* 53, 4580–4587.

Périchon, B., Goussard, S., Walewski, V., Krizova, L., Cerqueira, G., Murphy, C., Feldgarden, M., Wortman, J., Clermont, D., Nemec, A., and Courvalin, P. (2014) Identification of 50 class D beta-lactamases and 65 *Acinetobacter*-derived cephalosporinases in *Acinetobacter* spp. *Antimicrobial Agents and Chemotherapy* 58, 936–949.

Phillipon, A., Labia, R., and Jacoby, G.A. (1989) Extended-spectrum β-lactamases. *Antimicrobial Agents and Chemotherapy* 33, 1131–1136.

Piddock, L.J. (1999) Mechanisms of fluoroquinolone resistance: an update 1994–1998. *Drugs* 58, 11–18.

Pitout, J.D.D., Nordmann, P., and Poirel, L. (2015) Carbapenemase-producing *Klebsiella pneumonia*, a key pathogen set for global nosocomial dominance. *Antimicrobial Agents and Chemotherapy* 59, 5873–5884.

Poirel, L., Naas, T., Guibert, M., Chaibi, E.B., Labia, R., and Nordmann, P. (1999) Molecular and biochemical characterization of VEB-1, a novel class A extended-spectrum beta-lactamase encoded by an *Escherichia coli* integron gene. *Antimicrobial Agents and Chemotherapy* 043, 573–581.

Poirel, L., Collet, L., and Nordmann, P. (2000) Carbapenem-hydrolyzing metallo-beta-lactamase from a nosocomial isolate of *Pseudomonas aeruginosa* in France. *Emerging Infectious Diseases* 6, 84–85.

Poirel, L., Gerome, P., De Champs, C., Stephanazzi, J., Naas, T., and Nordmann, P. (2002) Integron-located *oxa-32* gene cassette encoding an extended-spectrum variant of OXA-2 beta-lactamase from *Pseudomonas aeruginosa*. *Antimicrobial Agents and Chemotherapy* 46, 566–569.

Poirel, L., Dortet, L., Bernabeu, S., and Nordmann, P. (2011) Genetic features of $bla_{\text{NDM-1}}$-positive *Enterobacteriaceae*. *Antimicrobial Agents and Chemotherapy* 55, 5403–5407.

Poole, K., Krebes, K., Mcnally, C., and Neshat, S. (1993) Multiple antibiotic resistance in *Pseudomonas aeruginosa*: evidence for involvement of an efflux operon. *Journal of Bacteriology* 175, 7363–7372.

Prystowsky, J., Siddiqui, F., Chosay, J., Shinabarger, D.L., Millichap, J., Peterson, L.R., and Noskin, G.A. (2001) Resistance to linezolid: characterization of mutations in rRNA and comparison of their occurrences in Vancomycin-resistant enterococci. *Antimicrobial Agents and Chemotherapy* 45, 2154–2156.

Race, P.R., Lovering, A.L., Green, R.M., Ossor, A., White, S.A., Searle, P.F., Wrighton, C.J., and Hyde, E.I. (2005) Structural and mechanistic studies of *Escherichia coli* nitroreductase with the antibiotic nitrofurazone. Reversed binding orientations in different redox states of the enzyme. *The Journal of Biological Chemistry* 280, 13256–13264.

Rice, L.B. (1998) Tn*916*-family conjugative transposons and dissemination of antimicrobial resistance determinants. *Antimicrobial Agents and Chemotherapy* 42, 1871–1877.

Rice, L.B., Carias, L.L., Hutton-Thomas, R., Sifaoui, F., Gutmann, L., and Rudin, S.D. (2001) Penicillin-binding protein 5 and expression of ampicillin resistance in *Enterococcus faecium*. *Antimicrobial Agents and Chemotherapy* 45, 1480–1486.

Rodriguez-Martinez, J.-M., Nordmann, P., Fortineau, N., and Poirel, L. (2010) VIM-19, a metallo-β-lactamase with increased carbapenemase activity from *Escherichia coli* and *Klebsiella pneumoniae*. *Antimicrobial Agents and Chemotherapy* 54, 471–476.

Ruzin, A., Keeney, D., and Bradford, P.A. (2005a) AcrAB efflux pump plays a role in decreased susceptibility to tigecycline in *Morganella morganii*. *Antimicrobial Agents and Chemotherapy* 49, 791–793.

Ruzin, A., Visalli, M.A., Keeney, D., and Bradford, P.A. (2005b) Influence of transcriptional activator RamA on expression of multidrug efflux pump AcrAB and tigecycline susceptibility in *Klebsiella pneumoniae*. *Antimicrobial Agents and Chemotherapy* 49, 1017–1022.

Rybkine, T., Mainardi, J.-L., Sougakoff, W., Collatz, E., and Gutmann, L. (1998) Penicillin-binding protein 5 sequence alterations in clinical isolates of *Enterococcus faecium* with different levels of β-lactam resistance. *The Journal of Infectious Diseases* 178, 159–163.

Sahm, D.F., Marsilio, M.K., and Piazza, G. (1999) Antimicrobial resistance in key bloodstream bacterial isolates: electronic surveillance with The Surveillance Network Database-USA. *Clinical Infectious Diseases* 29, 259–263.

Sanders, W.E.J. and Sanders, C.C. (1988) Inducible β-lactamases: clinical and epidemiologic implications for use of newer cephalosporins. *Reviews of Infectious Diseases* 10, 830–838.

Schnappinger, D. and Hillen, W. (1996) Tetracyclines: antibiotic action, uptake, and resistance mechanisms. *Archives of Microbiology,* 165, 359–369.

Schwarz, S., Kehrenberg, C., Doublet, B., and Cloeckaert, A. (2004) Molecular basis of bacterial resistance to chloramphenicol and florfenicol. *FEMS Microbiology Reviews* 28, 519–542.

Shea, M.E. and Hiasa, H. (1999) Interactions between DNA helicases and frozen topoisomerase IV- quinolone-DNA ternary complexes. *The Journal of Biological Chemistry,* 274, 22747–22754.

Skold, O. (2000) Sulfonamide resistance: mechanisms and trends. *Drug Resistance Updates* 3, 155–160.

Soltani, M., Beighton, D., Philpott-Howard, J., and Woodford, N. (2000) Mechanisms of resistance to quinupristin–dalfopristin among isolates of *Enterococcus faecium* from animals, raw meat, and hospital patients in Western Europe. *Antimicrobial Agents and Chemotherapy* 44, 433–436.

Stapleton, P.D., Shannon, K.P., and French, G.L. (1999) Carbapenem resistance in *Escherichia coli* associated with plasmid-determined CMY-4 β-lactamase production and loss of an outer membrane protein. *Antimicrobial Agents and Chemotherapy* 43, 1206–1210.

Su, Y.A., He, P., and Clewell, D.B. (1992) Characterization of the *tetM* determinant of Tn*916*: evidence for regulation by transcriptional attenuation. *Antimicrobial Agents and Chemotherapy,* 36, 769-778.

Swedberg, G., Ringertz, S., and Skold, O. (1998) Sulfonamide resistance in *Streptococcus pyogenes* is associated with differences in the amino acid sequence of its chromosomal dihydropteroate synthase. *Antimicrobial Agents and Chemotherapy* 42, 1062–1067.

Tam, V.H., Schilling, A.N., Vo, G., Kabbara, S., Kwa, A.L., Wiederhold, N.P., and Lewis, R.E. (2005) Pharmacodynamics of polymyxin B against *Pseudomonas aeruginosa*. *Antimicrobial Agents and Chemotherapy* 49, 3624–3630.

Then, R.L. (1982) Mechanisms of resistance to trimethoprim, the sulfonamides, and trimethoprim-sulfamethoxazole. *Reviews of Infectious Diseases* 4, 261–269.

Then, R.L. and Angehrn, P. (1979) Low trimethoprim susceptibility of anaerobic bacteria due to insensitive dihydrofolate reductases. *Antimicrobial Agents and Chemotherapy* 15, 1–6.

Tomasz, A. (1983) Murein hydrolases: enzymes in search of a physiologic function. In: Hackenbeck, R., Holtje, J., and Labischinski, H. (eds) *The Target of Penicillin.* Walter de Gruyter, Berlin, pp. 155–172.

Top, J., Willems, R., and Bonten, M. (2008) Emergence of CC17 *Enterococcus faecium*: from commensal to hospital-adapted pathogen. *FEMS Immunology and Medical Microbiology* 52, 297–308.

Tran, J.H. and Jacoby, G.A. (2002) Mechanism of plasmid-mediated quinolone resistance. *Proceedings of the National Academy of Sciences of the United States of America* 99, 5638–5642.

Tran, J.H., Jacoby, G.A., and Hooper, D.C. (2005) Interaction of the plasmid-encoded quinolone resistance protein Qnr with *Escherichia coli* DNA gyrase. *Antimicrobial Agents and Chemotherapy* 49, 118–125.

Tzeng, Y.L., Ambrose, K.D., Zughaier, S., Zhou, X., Miller, Y.K., Shafer, W.M., and Stephens, D.S. (2005) Cationic antimicrobial peptide resistance in *Neisseria meningitidis*. *Journal of Bacteriology* 187, 5387–5396.

Vakulenko, S.B. and Mobashery, S. (2003) Versatility of aminoglycosides and prospects for their future. *Clinical Microbiology Reviews* 16, 430–450.

Vila, J., Navia, M., Ruiz, J., and Casals, C. (1997) Cloning and nucleotide sequence analysis of a gene encoding an OXA-derived beta-lactamase in *Acinetobacter baumannii*. *Antimicrobial Agents and Chemotherapy* 41, 2757–2759.

Vincent, S., Minkler, P., Bincziewski, B., Etter, L., and Shlaes, D.M. (1992) Vancomycin resistance in *Enterococcus gallinarum*. *Antimicrobial Agents and Chemotherapy* 36, 1392–1399.

Walsh, T.R., Toleman, M.A., Poirel, L., and Nordmann, P. (2005) Metallo-beta-lactamases: the quiet before the storm? *Clinical Microbiology Reviews* 18, 306–325.

Wehrli, W. (1983) Rifampin: mechanisms of action and resistance. *Reviews of Infectious Diseases* 5(Suppl 3), S407–S411.

Weisblum, B. (1995) Erythromycin resistance by ribosome modification. *Antimicrobial Agents and Chemotherapy* 39, 577–585.

White, D.G., Hudson, C., Maurer, J.J., Ayers, S., Zhao, S., Lee, M.D., Bolton, L., Foley, T., and Sherwood, J. (2000) Characterization of chloramphenicol and florfenicol resistance in *Escherichia coli* associated with bovine diarrhea. *Journal of Clinical Microbiology* 38, 4593–4598.

Willmott, C.J., Critchlow, S.E., Eperon, I.C., and Maxwell, A. (1994) The complex of DNA gyrase and quinolone drugs with DNA forms a barrier to transcription by RNA polymerase. *Journal of Molecular Biology* 242, 351–363.

Wolter, N., Smith, A.M., Farrell, D.J., Schaffner, W., Moore, M., Whitney, C.G., Jorgensen, J. H., and Klugman, K.P. (2005) Novel mechanism of resistance to oxazolidinones, macrolides, and chloramphenicol in ribosomal protein L4 of the *Pneumococcus*. *Antimicrobial Agents and Chemotherapy* 49, 3554–3557.

Yamane, K., Wachino, J., Doi, Y., Kurokawa, H., and Arakawa, Y. (2005) Global spread of multiple aminoglycoside resistance genes. *Emerging Infectious Diseases* 11, 951–953.

Zervos, M.J. and Schaberg, D.R. (1985) Reversal of in vitro susceptibility of enterococci to trimethoprim-sulfamethoxazole by folinic acid. *Antimicrobial Agents and Chemotherapy* 28, 446–448.

Zhong, P. and Shortridge, V.D. (2000) The role of efflux in macrolide resistance. *Drug Resistance Updates* 3, 325–329.

Ziha-Zarifi, I., Llanes, C., Kohler, T., Pechere, J.C., and Plesiat, P. (1999) In vivo emergence of multidrug-resistant mutants of *Pseudomonas aeruginosa* overexpressing the active efflux system MexA–MexB–OprM. *Antimicrobial Agents and Chemotherapy* 43, 287–291.

6 Antimicrobial Resistance: Selection vs. Induction

Rafael Araos,[1] Jose M. Munita,[1,2] and Cesar A. Arias[2,3]*

[1]Facultad de Medicina Clínica Alemana—Universidad del Desarrollo, Santiago, Chile; [2]University of Texas Medical School at Houston, Texas, US; [3]Universidad El Bosque, Bogota, Colombia

Introduction

> I have called this principle, by which each slight variation, if useful, is preserved, by the term of natural selection.
>
> Charles Darwin, *The Origin of Species*, Chapter 3: Struggle for Existence

The emergence and dissemination of antimicrobial resistance among bacteria is a well-recognized worldwide public health threat. Infections caused by multidrug-resistant organisms (MDROs) are associated with higher mortality (up to fivefold) than those caused by susceptible organisms (Schwaber *et al.*, 2006), causing an enormous financial burden to healthcare systems (Howell, 2013). Furthermore, the emergence of MDROs in clinical settings is complicated by the lack of new antimicrobials. Indeed, the development of novel antimicrobial compounds has slowed down since major pharmaceutical companies made the decision to stop their anti-infective development programs. This worrisome situation results in clinicians having to deal with few—if any—alternatives to treat infections due to MDROs (Laxminarayan *et al.*, 2013).

Antimicrobial use (and misuse) is believed to be one of the most important factors implicated in the increase prevalence of MDROs, which is reaching epidemic proportions. The effect of antimicrobials on promoting resistance on bacterial communities is multifactorial, but it can be summarized from three aspects: (i) antimicrobial challenge increases the ability of bacteria to acquire resistance via mutations or the acquisition of resistance determinants via horizontal gene transfer (Alekshun and Levy, 2007); (ii) the presence of antimicrobials in specific biological niches (such as the gastrointestinal tract of humans) can profoundly disturb the normal microbiota and interfere with the host's immune mechanisms, which contribute to the eradication of MDROs, thereby further promoting conditions for overgrowth of resistant bacteria and the potential to disseminate their resistance determinants to other coresident bacteria (Rice, 2005; Taur and Pamer, 2013); and (iii) bacteria have evolved to respond to antimicrobial attack via altering their genome in such a way that evolution (as quoted above) becomes an important part of the bacterial life cycle in order to outcompete other organisms in difficult environments, such as the human host.

With the above rationale and in an effort to understand resistance as an inherent evolutionary characteristic of bacteria, we will attempt to answer the following question: is the emergence of resistance due to the "natural selection" (to quote Darwin) of bacterial subpopulations?; or is it the result of induction of genetic determinants already present (or acquired) in the bacterial population that are "activated" in the presence of the antibiotic molecule? Both phenomena are likely to play a role in the development of the resistance phenotype and, from the evolutionary point of view, selection and induction may represent adaptations to the environment.

Although several definitions of both phenomena (selection vs. induction) are possible, the clarification

*Corresponding author. E-mail: cesar.arias@uth.tmc.edu

of these terms has important evolutionary and biological connotations. Thus, we will approach the question from the genetic perspective, and in order to focus this discussion, we will propose that: (i) the *selection of antimicrobial resistance* refers to the phenomenon by which it is possible to select subpopulations of microorganisms due to the natural occurrence of random mutations or gene acquisition (preadaptation theory); and (ii) the *induction of antimicrobial resistance* involves the theoretical potential of some drugs to induce heritable adaptive responses in the exposed organisms (post-adaptation theory) (Fig. 6.1).

Seminal work in the 1950s (Luria and Delbrück, 1943; Lederberg and Lederberg, 1952) provided experimental evidence that strongly supported the selection theory of preadaptation. Indeed, the experimental approach showed that antimicrobial resistance among bacteria was genetically determined, and that its expression was secondary to the exogenous selection (e.g., antimicrobial effect) of naturally occurring mutants already present in the original bacterial community. Moreover, the selection hypothesis was further supported and complemented by the discovery of mobile genetic elements, which were shown to be able to transfer resistance determinants between microorganisms. In support of this concept, recent studies (D'Costa *et al.*, 2011; Wright and Poinar, 2012) have shown that many antimicrobial resistance determinants have been present in nature since ancient times, long before the introduction of antimicrobials for use in human health and farming (Perry and Wright, 2013).

These resistance genes are usually found in many ecosystems, where they have been acquired by environmental bacteria and represent the evolutionary consequences of the interaction between these microorganisms and their habitats. It has been hypothesized that a natural pool of resistance genes, or "resistome,"

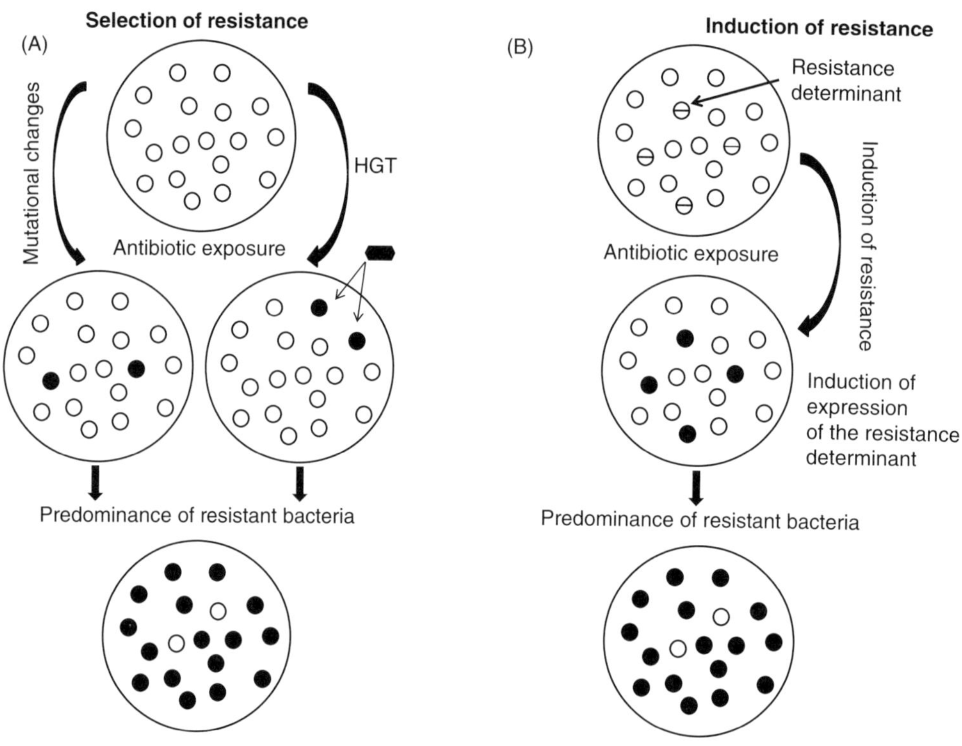

Fig. 6.1. The selection vs. induction paradigm—the preadaptation theory. (A) Selection. Genetic determinants of resistance arise through spontaneous mutations or the acquisition of resistant genes from other bacteria via horizontal gene transfer (HGT). After antibiotic exposure, susceptible bacteria are killed and resistant mutants predominate. (B) Induction—the post-adaptation theory. A small group of bacteria in the population harbor resistance determinants that are not expressed under normal conditions. Under antibiotic pressure, the expression of resistance genes is triggered. Bacteria carrying the resistance determinants predominate.

could be accessed by many human bacterial pathogens as the source of antimicrobial resistance genes through horizontal gene transfer (HGT), particularly under conditions of external stress (antibiotic attack). In such a context, the induction of resistance (or "post-adaptive" resistance) refers specifically to the transient and adaptive response of a microorganism that usually requires the presence of a particular antimicrobial in order to activate or repress regulatory systems that control the expression of already existing resistance determinants. These genes are normally under strict control because their expression poses an important risk to the bacterial cell due to an increased fitness cost to the microorganism jeopardizing an evolutionary advantage among competing microorganisms (Sanchez-Romero and Casadesus, 2014).

From a mechanistic perspective, the selection–induction conundrum is less evident. Conceptually, both terms are likely to be valid and the two phenomena can coexist. In general terms, MDROs selected during infections become clinically relevant only when the environment is exposed to the antimicrobial selective pressure, unless the resistant determinant is associated with a marked fitness advantage (which is rare). In the latter scenario, the resistant mutants may become more competitive and potentially take the place of the susceptible population. Therefore, from a mechanistic point of view, the selection paradigm does not necessarily exclude the possible occurrence of preadaptive or post-adaptive phenotypes in response to specific environmental challenges. In the following sections, and in order to further clarify the above concepts, we will discuss examples of clinically relevant resistance mechanisms that are either "selected" or "induced," and have become evident by the use of antimicrobials.

Selection of Antimicrobial Resistance

As mentioned above, and following the preadaptation theory, the selection of resistance is often generated by the occurrence of two important genetic events: (i) mutations of specific genes encoding a particular antimicrobial target or enzyme involved in a metabolic pathway that interferes with the mechanism of action of the antimicrobial; and (ii) the acquisition of genes via HGT, which could be promoted by antimicrobial molecules that may not necessarily be directly related to the mechanism of resistance (e.g., fluoroquinolone antimicrobials may select for cephalosporin resistance by promoting the exchange of mobile elements between bacteria that encode certain β-lactamases). Thus, in the following sections, we will discuss examples of these two scenarios separately.

The role of mutations

Antimicrobial resistance can be selected by specific mutations that can affect the cell homeostasis at three different levels in order to withstand the antibiotic attack: (i) modification of the antimicrobial target; (ii) disruptions of specific regulatory pathways; and (iii) alterations in drug uptake (Rice, 2010). Initially, we will discuss classical examples of resistance that emerge by mutational events, which are intended to show how these mechanisms vary in complexity from simple one-step mutations to highly sophisticated combinations of genomic changes.

Rifampin resistance

The antimicrobial activity of the rifamycin group of antimicrobials (rifampin, or RIF, is the most common example) relies on their ability to block transcription through the inhibition of bacterial RNA polymerase (RNAP, or Rpo). The interaction between the drug and the enzyme occurs at a specific site in the β-subunit of the enzyme, which is highly conserved among prokaryotes. The RIF binding pocket is located deeply within the RNAP main DNA/RNA channel, in close vicinity to the active site of the enzyme. At least 23 amino acid residues surround the binding pocket, and 12 of them directly interact with the RIF molecule, mainly through van der Waals' forces (Campbell *et al.*, 2001).

The emergence of resistance to RIF is an example of a relatively simple model of antimicrobial resistance and it is one of the few cases where point mutations are sufficient to confer high levels of resistance that are clinically relevant. Indeed, a single-step mutation in the *rpoB* gene (which encodes the β-subunit of RNAP) is capable of generating the resistance phenotype. These mutations result in amino acid substitutions at the RIF binding site, which decrease the affinity of the drug for its target. Of note is that the catalytic activity of RNAP is retained, so permitting transcription to occur in the resistant derivatives without major changes. Importantly, mutations may occur spontaneously and do not necessarily require selective antimicrobial pressure. However, exposure to RIF of a bacterial population that harbors a small proportion of cells with randomly acquired *rpoB* mutations results in the rapid killing of the susceptible bacteria, and in the selection of the RIF-resistant subpopulation, which will eventually predominate (Jin and Gross, 1988; Telenti and Imboden, 1993; Tupin *et al.*, 2010).

Fluoroquinolone resistance

The mechanism of action of fluoroquinolones involves the inhibition of DNA synthesis owing to the interaction of these agents with bacterial DNA and two structurally similar enzymes that participate in DNA replication: (i) DNA gyrase (also known as topoisomerase II); and (ii) topoisomerase IV. The affinity of a particular fluoroquinolone for either DNA gyrase or topoisomerase IV is variable, and the most sensitive enzyme is considered to be the primary target of the drug. Typically, DNA gyrase is the primary target of fluoroquinolones in Gram-negative bacteria and topoisomerase IV is the primary target in Gram-positive microorganisms (Hooper, 2001). Fluoroquinolone resistance frequently emerges as a result of chromosomal mutations that determine modifications of the target enzymes. Additionally, a decrease in intracellular drug accumulation and protection of the DNA target (mediated by the plasmid-borne quinolone resistance gene, *qnr*, which encodes a protein that protects DNA gyrase from quinolones) may also play a role in fluoroquinolone resistance (Hooper, 2001; Strahilevitz *et al.*, 2009). It is generally accepted that combinations of these mechanisms of resistance are usually found in clinical isolates (Rice, 2012).

The structural changes in DNA gyrase and/or topoisomerase IV derive from the occurrence of multiple mutations in the genes encoding the enzyme subunits of DNA gyrase (GyrA–GyrB) and topoisomerase IV (ParC–ParE). The mutational events leading to resistance accumulate over time, with the first mutation usually producing minor changes in susceptibility to the drugs; the elevation of the minimum inhibitory concentration (MIC) varies according to the effect of the mutation itself and the affinity of the fluoroquinolone to secondary drug targets.

In most cases, after the first-step mutation, the fluoroquinolones continue to exert some degree of inhibition of the alternate topoisomerase. Thus, a second mutational event, compromising the structure of the auxiliary target must occur in order to generate a fully resistant phenotype, although in some cases (e.g., *Staphylococcus aureus* and *Pseudomonas aeruginosa*), a single mutation may be enough to generate the resistance phenotype (Coronado *et al.*, 1995; Hooper, 2001). Usually though, a series of additive mutations affecting both targets are needed to obtain a clinically relevant increase in the MIC. Interestingly, it has been estimated that the rate of sequential mutational events affecting the target site of fluoroquinolones is relatively low (around 10^{-14} to 10^{-16}), which would hardly account for the high amount of fluoroquinolone resistance currently observed in clinical practice (Strahilevitz *et al.*, 2009). Therefore, it is thought that, as mentioned above, the co-occurrence of mutations with increased expression of genes involved in fluoroquinolone efflux or plasmid-mediated resistance determinants (*qnr* genes) may be responsible for high levels of quinolone resistance. Fluoroquinolone resistance is then an interesting model for the selection of resistance whereby an accumulation of sequential mutations that result in target modification are the initial events, and these are subsequently potentiated by other mechanisms, including those that affect the amount of drug that reaches the target and/or the production of additional proteins that interact with the target. This phenomenon could also occur in reverse order, in which target mutations emerge in a bacterial background already harboring *qnr* and/or overexpressing efflux pumps.

Linezolid resistance

Linezolid is a synthetic drug that belongs to the oxazolidinone family of antimicrobials. Its antibacterial spectrum of activity includes a wide range of Gram-positive microorganisms, including multidrug-resistant (MDR) pathogens such as methicillin-resistant *S. aureus* (MRSA), vancomycin-resistant enterococci (VRE), and penicillin-resistant *Streptococcus pneumoniae*. The mechanism of action of linezolid is considered to be unique, and stems from its ability to bind to the A site of the bacterial ribosome, thereby inhibiting important stages of bacterial protein synthesis. The initial characterization of linezolid resistance by serial-passage experiments indicated that mutations involving the ribosomal peptidyl-transferase center were associated with the resistance phenotype. Indeed, linezolid binds to the large ribosomal subunit through important interactions with the 23S rRNA of the 50S ribosomal subunit (Leach *et al.*, 2011). The existence of other classes of antimicrobials that also target the peptidyl transferase center (e.g., macrolides, lincosamides, streptogramins, and phenicols) raised theoretical concerns on the possibility of cross-resistance between these compounds and linezolid. However, interactions of linezolid and the ribosome have been shown to be highly specific (Toh *et al.*, 2007).

Although resistance to linezolid has been described in MRSA and VRE strains, its prevalence is low, and it generally occurs in patients that have been previously exposed to the agent, particularly during prolonged courses of treatment (Mendes *et al.*, 2014). The most common cause of linezolid resistance in MRSA and VRE is the presence of chromosomal mutations affecting the binding sites of the drug in the ribosome. The genes that normally encode for the bacterial 23S rRNA exist in multiple copies, and consequently, mutations have to accumulate in order to produce a clinically relevant resistant clone. This genetic redundancy or gene-dosage effect, together with the notion that a "linezolid resistome" may not be available in nature due to the synthetic characteristics of the antibiotic and its novel mechanism of action, may explain, in part, the protracted emergence of linezolid resistance in clinical practice (Eliopoulos *et al.*, 2004). Another mechanism of linezolid resistance is due to the acquisition of a gene usually carried on transposons (designated *cfr*, for chloramphenicol–florfenicol resistance) that encode a dimethylase enzyme that is capable of methylating a specific nucleotide (A2603, *Escherichia coli* numbering) of the 23S rRNA, which alters the interaction of the drug with its target. This mechanism was likely selected by other drugs that act on the peptidyl-transferase center and affect linezolid as well. Contrary to the apparent nontransferability of chromosomal resistance, the spreading potential of this plasmid-borne resistance determinant might be relevant in the evolution of linezolid and oxazolidinone resistance in the future.

The case of linezolid highlights the importance of antimicrobial selective pressure in the emergence of antimicrobial resistance. Even if theoretically improbable (in this case because the compound was not natural but rather synthetic) bacteria are likely to successfully evolve and succeed. If at any point, a mobile genetic element causing linezolid resistance emerges, the potential for dissemination will significantly increase.

Daptomycin (DAP) resistance in enterococci

Daptomycin (DAP) is an antimicrobial lipopeptide with a broad spectrum of activity against Gram-positive bacteria. Its mechanism of action involves the interaction of the antimicrobial with the bacterial cell membrane in a calcium-dependent manner. A crucial step in the mechanism of action of DAP is the ability of the molecule to oligomerize and translocate from the outer to the inner leaflet of the cell membrane. This event is likely regulated by the presence of negatively charged phospholipids such as phosphatidylglycerol and cardiolipin. Moreover, recent evidence suggests that DAP preferentially binds to the cell membrane at regions of active peptidoglycan synthesis (i.e., the bacterial division septum) (Pogliano *et al.*, 2012). Although DAP lacks approval by the US Food and Drug Administration (FED) for such purposes, many clinicians use it for deep-seated VRE infections because it is one of the few compounds that retains potent in vitro bactericidal activity against these organisms. Nevertheless, the emergence of DAP resistance during therapy in VRE is an important limitation in clinical settings, and this phenomenon is a perfect example of selection of in vivo resistance with major clinical implications.

Using whole genome analyses of vancomycin-resistant *Enterococcus faecalis* isolates recovered before and after DAP therapy, Arias *et al.* (2011) showed that in vivo the DAP resistance phenotype was associated with codon deletions of three genes: (i) a member (LiaF) of a three-component regulatory system (LiaFSR), previously involved in cell membrane/envelope homeostasis; (ii) an enzyme involved in phospholipid metabolism (GdpD, glycerophosphodiester phosphodiesterase); and (iii) a cardiolipin synthase (Cls). Using allelic replacements, it was shown that changes in these chromosomally encoded genes (part of the bacterium "core genome") were sufficient for the development of resistance. However, similar to the previously discussed issue with fluoroquinolones, a single mutation in one of the above genes (codon deletion of *liaF*) was associated with small changes in the DAP MIC (from 1 to 4 μg/ml—remaining within the susceptible range, as the breakpoint is 4 μg/ml), but it was sufficient to abolish the in vitro bactericidal activity of the antimicrobial agent (Munita *et al.*, 2013).

Also using an in vitro system, Miller *et al.* (2013) recently provided an evolutionary perspective to the DAP selection paradigm. The investigators were able to show that the selection of resistance to DAP in vancomycin-resistant *E. faecalis* followed an ordered and sequential mutational pathway. Interestingly, most of the initial successful mutations that led to resistance originated in genes encoding the LiaFSR system (the same system as identified in the in vivo study described above), and after changes in these genes, other substitutions (such as those in Cls) "completed" the initial phenotype and raised the DAP MIC above the breakpoint. These results provided compelling evidence that bacteria attempt a variety of genetic evolutionary trajectories in order to adapt to antimicrobial attack, with the "final aim of surviving." Only successful genetic pathways that are compatible with an efficient cell homeostasis are selected during the antimicrobial attack. These "trial and error" attempts may take several generations, but once an initial successful trajectory is identified, subsequent mutations are likely to maximize and compensate for the initial effect in order to gain a competitive advantage in a hostile environment. Of note, the main changes identified involved the deletion or insertions of codons in regions that contain repetitive sequences encoding the same amino acid, suggesting that such regions are initial targets for the antimicrobial driven evolution of protein function in certain bacterial species. Furthermore, certain strains may be prone to readily mutate and these "hypermutator" isolates are likely to be often found in clinical settings after exposures to multiple antibiotics.

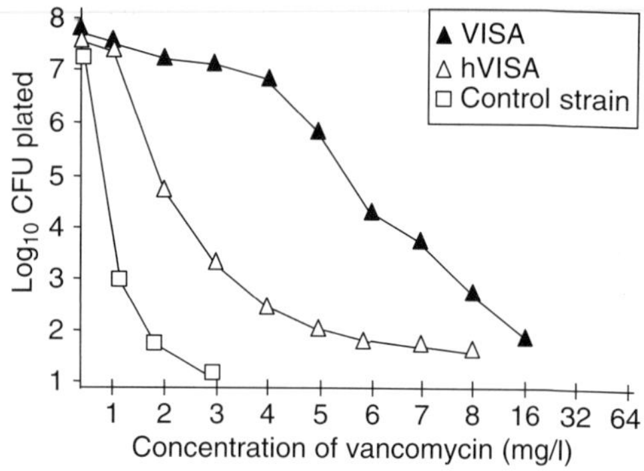

Fig. 6.2. Population analysis of the vancomycin-intermediate *Staphylococcus aureus* (VISA) Mu50 strain and the hetero-VISA (hVISA) Mu3 strain in the presence of vancomycin to illustrate the phenomenon of heteroresistance. The control strain, a vancomycin-susceptible methicillin-resistant *S. aureus* (MRSA), exhibits total inhibition of growth at low vancomycin concentrations. In contrast, Mu3, the hetero-VISA strain contained a subpopulation of bacteria that was able to grow at higher vancomycin concentrations (up to 9 mg/l). Finally, Mu50, the VISA strain, is able to maintain an important amount of cells in the population growing at concentrations of vancomycin >4 mg/l. Adapted from Hiramatsu *et al.*, (1997).

Heteroresistance and the selection phenomenon

Early studies on the response of *S. aureus* to the antimicrobial challenge (by methicillin) revealed that this organism had the ability to produce subpopulations of cells capable of withstanding the antimicrobial attack and that these cells were selected upon exposure to the antimicrobial molecule (Chambers, 1997). The most clinically relevant example of this phenomenon of heteroresistance is the emergence of *S. aureus* with intermediate susceptibility to vancomycin (VISA, vancomycin-intermediate *S. aureus*). From the clinical point of view, heteroresistance is commonly seen in VISA strains (designated hetero-VISA, or hVISA), and refers to subpopulations (ca. 10^2–10^3 CFU) that are capable of surviving at concentrations of vancomycin above 2 µg/ml (the vancomycin breakpoint) (Fig. 6.2). Because only a small proportion of cells within the population exhibit this phenotype, detection in the clinical laboratory becomes difficult and cumbersome, resulting in many of these strains being reported as "susceptible" to vancomycin by standard susceptibility testing. As a result, vancomycin is used in patients infected with these isolates, which likely predisposes to therapeutic failure in deep-seated infections. Indeed, the first VISA strain identified in Japan in 1997 (Mu50) was derived from an hVISA isolate (designated Mu3) (Hiramatsu *et al.*, 1997). The prevalence of the hVISA/VISA phenotype has been difficult to estimate owing to the lack of well-standardized operational definitions and to different detection methodologies. According to published studies, the prevalence varies significantly between different geographical locations, ranging from 0 to 50% of clinical MRSA isolates (Ike *et al.*, 2001; Arakawa *et al.*, 2004; Horne *et al.*, 2009). According to one review article that covered 14 original studies, the estimated prevalence of MRSA strains with the hVISA/VISA phenotype was approximately 2.16% (0–8.24%) (Liu and Chambers, 2003).

Similar to what was discussed above for DAP in enterococci, the emergence of the VISA phenotype appears to be the result of a sequential and ordered process that usually involves initial adaptive events derived from specific genetic changes in regulatory systems that control cell envelope homeostasis. Indeed, in a landmark in vivo study by Mwangi *et al.* (2007), selection of the VISA phenotype under vancomycin exposure in a patient with infective endocarditis caused by an MRSA strain was initiated by mutations in the VraTSR system (which in staphylococci is the homolog of LiaFSR system in enterococci), followed by mutations in other regulatory genes predicted to be involved in cell wall homeostasis. Notably, the initial mutation was sufficient to increase the MIC of vancomycin to 4 μg/ml with minimal effect on the daptomycin MIC; however, subsequent mutations selected upon continuous exposure to vancomycin also affected DAP susceptibility.

Likewise, using an in vitro selection system, selection of the VISA phenotype upon vancomycin exposure was the result of an initial mutation in the same VraTSR system (resulting in hVISA) and a subsequent mutation in the GraSR system (resulting in VISA), which has also been implicated in cell envelope homeostasis (Howden *et al.*, 2014). These findings support the notion that selection of the VISA phenotype involves genetic pathways that result in important remodeling of the cell envelope (as seen above with DAP) in order to survive antimicrobial attack and perhaps evade the innate immune system. In support of this concept, the selection of such strains usually results in similar changes in cell envelope metabolism and structure that include increased transport of cell wall components through the cytoplasmic membrane, an increased cell surface positive charge, thickening of the cell wall and decreased autolytic activity, among others (Neoh *et al.*, 2008).

Finally, a striking feature of some hVISA/VISA strains is their ability to revert from one phenotype to another (or even to a fully vancomycin-susceptible phenotype) in the absence of vancomycin exposure. Therefore, there seems to be a "price" to pay for developing resistance; this is a clear example of the ability of bacteria to adapt to the environment by means of their remarkable genetic plasticity. Another important issue that emerges from this discussion is that bacteria appear to develop common strategies to respond to alterations of vital structures (e.g., the cell envelope), regardless of the causative agent. As a result, the strategy taken to survive antimicrobial attack by a particular compound may also affect several other classes of antimicrobials whose targets involve the same bacterial structure. So the selection of resistance in the context of global adaptation may affect the susceptibility of a variety of different classes of antimicrobials (with different cell targets). This concept is important because targeting the cell adaptation machinery and producing "anti-adaptation" antibiotics may be a strategy to overcome or prevent the development of resistance in the future.

Horizontal gene transfer

HGT involves mechanisms by which bacteria are capable of transferring genetic material between members of the same or different species, and it has been considered the main mediator of the spread of antimicrobial resistance. Under natural circumstances, bacteria can acquire external genetic material by three mechanisms: (i) conjugation (bacterial sex); (ii) transformation (the acquisition of naked DNA); and (iii) transduction (phage mediated). These events can be markedly stimulated by the use of antimicrobials, and HGT has been implicated in the dissemination of resistance to aminoglycosides, macrolides, glycopeptides, tetracyclines, β-lactams, and fluoroquinolones, among others. Antimicrobials act as catalyzers of HGT by increasing the burden of resistance genes in specific environments. Thus, under the selective pressure exerted by antimicrobials, as the susceptible population is extinguished, the remaining resistant populations encounter more opportunities to exchange genetic material (Gogarten *et al.*, 2009).

Mobile elements such as transposons and integrons are important examples of genetic vehicles that bacteria use in order to respond to the antibiotic challenge. These elements have been shown to be a crucial part of bacterial evolution and key players in the ability of bacteria to adapt to unfavorable conditions. The evolution and spread of β-lactamases is a good example of how HGT can contribute to the selection of resistance. β-Lactams, one of the most commonly prescribed classes of antibiotics, are strongly associated with the selection of antimicrobial-resistant bacteria. It is accepted that owing to the enormous amount of β-lactam use in clinical settings, bacteria have responded by producing a wide range of β-lactamases that have adapted to the challenge. The malleability and versatility of the β-lactamase genes (several hundreds have been reported so far) (Jacoby, 2014) have put an immense pressure on current antimicrobials, making β-lactams almost obsolete for the treatment of an important number of Gram-negative organisms.

The evolution of β-lactamases has been driven mainly by their ability to mobilize from one cell to another between members of the same or different species. β-Lactams are naturally occurring antimicrobials and, therefore, bacteria have developed a versatile way to deal with the presence of these compounds. One of the most efficient mechanisms for transferring antimicrobial resistance genes is represented by integrons. These genetic elements are site-specific recombination systems that provide an efficient and rather simple mechanism for the addition of new genes into bacterial chromosomes. As such, this is a robust strategy of genetic interchange that drives bacterial evolution. Integrons are capable of recruiting open reading frames as gene cassettes and provide an appropriate platform for the expression of all these cassettes (Cambray *et al.*, 2010) (Fig. 6.3).

Among the most conspicuous examples of drug-driven bacterial evolution (and in our context of the selection of antimicrobial resistance) is the acquisition of genes encoding carbapenemase enzymes in Gram-negative organisms. Carbapenem resistance among the *Enterobacteriaceae* and non-fermenter Gram-negative bacilli represent major challenges for clinicians. Several mechanisms are involved in carbapenem resistance, and these include drug efflux, porin modifications, overexpression of AmpC (ampicillin C hydrolyzing) enzymes, and the production of carbapenemases. Among the latter enzymes, KPCs, *Klebsiella pneumoniae* carbapenemases, which belong to the Ambler class A β-lactamases) (Bush and Jacoby, 2010), have been recognized as one of the most important carbapenem resistance determinants, and their rapid dissemination throughout the globe has raised great concern for the potential consequences associated with their spread (Nordmann *et al.*, 2009).

The KPC genes are located on mobile genetic elements, particularly in plasmids that harbor a Tn*3*-like transposon, Tn*4401*. The physical associations of Tn*4401* with promoters, several insertion sequences, and other resistance genes define this mobile element as an integron that has successfully spread within and between species (Munoz-Price and Quinn, 2009). This genetic "shuttle" constitutes a clear example of the ability of bacteria to capture DNA through highly promiscuous mechanisms in order to respond to a strong antimicrobial selective pressure.

Another prime example of the influence of HGT on the selection of antimicrobial resistance is the dissemination of the CTX-M extended spectrum β-lactamases—so named for their greater activity against cefotaxime than other oxyimino-β-lactam substrates— among Gram-negative bacteria. In fact, the origin of CTX-M enzymes appears to be in members of the genus *Kluyvera*, which are environmental bacteria with no major pathogenic significance

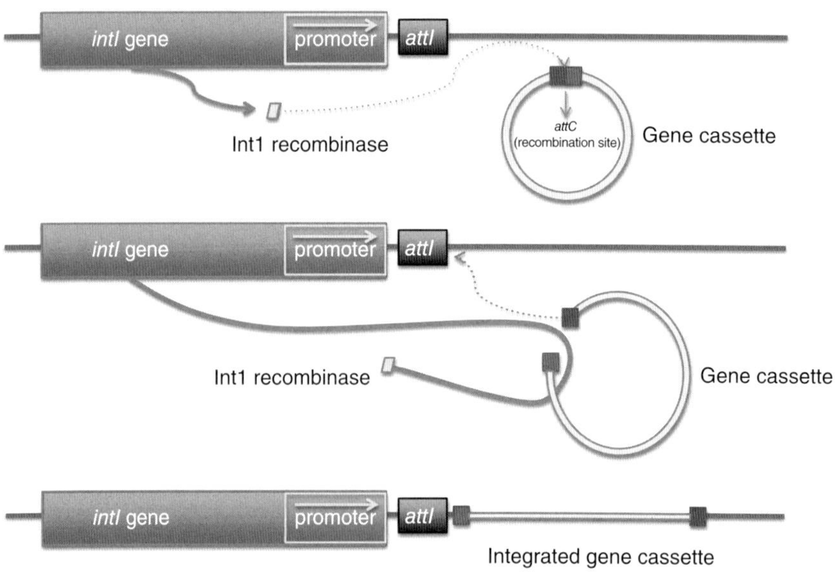

Fig. 6.3. Schematic representation of an integron. The basic components of an integron include an integrase gene (*intI*), a promoter and an *attI* site located downstream of the promoter. Additionally, *attC* is a repeat that flanks gene cassettes to be integrated at the *attI* site. Integrase-mediated acquisition of the gene cassette at the *attI* site is shown. Adapted from Cambray *et al.*, (2010).

for humans (Rossolini *et al.*, 2008). *Kluyvera ascobata* and *K. georgiana* have been found to be the sources of CTXM-1/2 and CTXM-8/9, respectively (Bonnet, 2004). Genes encoding CTX-M enzymes have been found in transposable elements that include insertion sequences (IS*Ecp1*) with the ability to comobilize using a one-ended transposition mechanism (Rossolini *et al.*, 2008). This genetic element has been shown to be able to capture chromosomal CTX-M genes from *Kluyvera* spp. and promote mobilization in *E. coli* (Poirel *et al.*, 2005; Lartigue *et al.*, 2006). Moreover, transposable elements (such as Tn*402*-like transposons) that can be captured by a broad range of conjugative plasmids or phage-like sequences can serve as the vehicle for the dissemination of CTX-M. Consequently, CTX-M enzymes became the most prevalent extended-spectrum β-lactamase enzymes worldwide, underscoring the notion that the hijacking of β-lactamase genes plus further dissemination of mobile elements carrying them is one of the most successful mechanisms of spreading antimicrobial resistance determinants.

Among Gram-positive bacteria, HGT has played an important role in the dissemination of vancomycin resistance. Indeed, most of the acquired *van* (vancomycin resistant) gene clusters (see below) have been disseminated in enterococci through the exchange of transposable elements (e.g., Tn*1546*, Tn*1547*, and others). Although mostly restricted to enterococci, such elements have been now captured by native conjugative plasmids of other Gram-positive species, including *S. aureus*, raising concerns about the dissemination of this resistance trait among MRSA (Rossi *et al.*, 2014).

Induction of Antimicrobial Resistance

As mentioned in the Introduction to the chapter, we will discuss the concept of the induction of antimicrobial resistance as a phenomenon that requires the existence of antimicrobial resistance determinants whose expression is modulated by the presence or absence of the antimicrobial molecule (or its effect therein). Such regulation of the expression of the antimicrobial resistance determinant may be performed at different levels but the most common examples observed in clinically relevant organisms involve protein and RNA-dependent regulation. Therefore, we will present examples of such phenomena that have important clinical relevance.

Induction of antimicrobial resistance modulated at the protein level

Penicillin resistance in Staphylococcus aureus

The regulation of β-lactamase expression in bacteria is perhaps one of the most common examples of induction of antimicrobial resistance in nature, in both Gram-positive and Gram-negative organisms. A good example of such phenomena is the production of β-lactamase in *S. aureus*. In this organism, β-lactamase production is mediated by the expression of the *blaZ* gene, which is part of a plasmid-borne gene complex that includes two additional genes, namely *blaR*1 and *blaI* that divergently encode for proteins with regulatory functions. BlaR1 acts as a membrane sensor and signal transducer, and the homodimer BlaI serves as a gene repressor that binds to the promoter region of the *blaZ–blaI–blaR*1 gene complex. In the absence of β-lactams, the expression of the β-lactamase is almost undetectable; however, if BlaR1 senses the extracellular presence of β-lactams, it initiates a signal cascade that results in the cleavage of BlaI from the promoter of *blaZ* and permits the initiation of transcription. Thus, *S. aureus* is able to increase the production of β-lactamase enzymes and fully express the penicillin resistance phenotype (Zhang *et al.*, 2001).

MRSA isolates emerged as a result of the acquisition of a mobile DNA fragment harboring the gene *mecA*, which encodes for the penicillin-binding protein (PBP) 2a that has low affinity for β-lactam antimicrobials. The expression of this resistant trait also involves the interaction of a gene complex that includes *mecA*, *mecl* (encoding for the repressor protein Mecl) and *mecR1* (that encodes for the signal transducer protein MecR1). All of these genes are present in a mobile DNA element called staphylococcal cassette chromosome *mec* (SCC*mec*). The SCC*mec* DNA also contains insertion sequences and genes coding for specific recombinases designated as cassette chromosome recombinases (*ccr*). It is currently accepted that the *mecA* gene was probably derived from a coagulase-negative staphylococcus (*S. fleurettii*) that, under the β-lactam selective pressure, was able to recombine with the SCC*mec* element of a methicillin-susceptible *S. aureus* (lacking *mecA*) (Moellering, 2012; Hiramatsu *et al.*, 2013).

In a manner similar to the expression of the *blaZ* gene complex, MecR1 function involves sensing the presence of β-lactams in the environment, which triggers a signal transduction cascade that removes the MecI repressor from its DNA binding site, thereby permitting transcription of both *mecA* and its regulatory genes. These events lead to the production of PBP2a, which is the hallmark of methicillin resistance in *S. aureus*. Of note, the plasmid-encoded *blaZ* gene complex seems to be very stable, but the SCC*mec* DNA can frequently be the subject of deletions or mutations affecting the *mecI* repressor, leading to the selection of constitutive mutants that express the antimicrobial resistance phenotype in the absence of the antimicrobial (Chambers, 1997; Katayama *et al.*, 2001; Deurenberg *et al.*, 2007). This illustrates how the phenomena of induction and selection can coexist to express antimicrobial resistance in vivo.

AmpC expression in Gram-negative organisms

The regulation of the AmpC β-lactamase is one of the most striking examples of the induction of antimicrobial resistance occurring during therapy. Many Gram-negative organisms harbor in their chromosome genes that encode for AmpC β-lactamase whose expression is induced by the presence of β-lactam antibiotics (Jacoby, 2009). The mechanism of AmpC induction is complex, but usually results from the disruption of cell wall biosynthesis caused by the β-lactam molecule. In general, the trigger for induction is the accumulation of oligopeptide derivatives of peptidoglycan precursors. The accumulation of 1,6-anhydro-*N*-muramic acid tri-, tetra- and pentapeptides in bacterial cells exposed to β-lactams competes with the "normal" UDP-N-acetylmuramic acid oligopeptides produced during cell wall synthesis for a binding site in AmpR, a LysR family transcriptional regulator (Jacoby, 2009). Displacement of UDP-*N*-acetylmuramic acid oligopeptides results in conformational changes in AmpR that activate the transcription of *ampC*. A second layer of regulation is provided by AmpD, an *N*-acetyl-muramyl-L-alanine amidase that prevents the accumulation of 1,6-anhydro-*N*-muramic acid derivatives by removing stem peptides from this moiety, so reducing their concentrations and subsequently the expression of AmpC.

In clinical isolates, high-level expression of the AmpC β-lactamase (usually designated as "stable derepression") is frequently due to a mutation in *ampD* (Schmidtke and Hanson, 2006). Additional events leading to hyperproduction of AmpC include mutations in *ampR* or *ampG*. The latter encodes a transmembrane permease that transports oligopeptides resulting from cell wall homeostasis from the periplasmic space to the cytosol (Lindquist *et al.*, 1993). Moreover, regulation of this complex system differs between species. For example, *E. coli* and *Acinetobacter baumannii* lack the *ampR* gene and the regulation of *ampC* is achieved (at least in *E. coli*) at the promoter level (Jaurin *et al.*, 1981). In *Serratia marcescens*, regulation is accomplished by AmpR and at the mRNA level by the formation of specific loops, similar to what is observed in resistance to macrolide antimicrobials (see below) (Mahlen *et al.*, 2003). Some strains of *P. aeruginosa* harbor more than a single copy of the *ampD* gene. For instance, *P. aeruginosa* PAO1 possesses three *ampD* genes and upregulation of *ampC* has been shown to occur after the inactivation of each gene in a stepwise fashion (Juan *et al.*, 2006). In conclusion, AmpC regulation is a complex phenomenon in Gram-negative bacteria and serves to illustrate that the induction of antimicrobial resistance is an important mechanism that has evolved several layers of complexity, suggesting that tight regulation of such systems is required for bacteria to gain an evolutionary advantage against antimicrobials and, perhaps, against competing bacteria.

Vancomycin resistance and dependence

The expression of vancomycin resistance is another important example of how resistance can be induced by the presence of an antimicrobial molecule. All gene clusters encoding the enzymes necessary for the synthesis of peptidoglycan precursors ending in D-alanine–D-lactate or D-alanine–D-serine (with destruction of the normal D-alanine–D-alanine ending precursors) are regulated by classical two-component regulatory systems. These systems encompass a transmembrane sensor that is capable of detecting the antimicrobial molecule or an alteration in cell wall synthesis upon antimicrobial exposure. As a result, the histidine kinase activity of the sensor is able to phosphorylate the response regulator and that, in turn, activates the transcription of the gene cluster. Several lines of evidence suggest that this relatively tight regulation of the expression of vancomycin resistance plays an important role in preserving bacterial fitness, including the ability of enterococci to disseminate in vivo (Foucault *et al.*, 2010). Indeed, constitutive expression of the *van* cluster appears to be deleterious to the cell by reducing its ability to efficiently synthesize and cross-link the cell wall components, leading to important effects in vivo (Hong *et al.*, 2005).

The malleability of the response to the presence of the antimicrobial could induce bacteria to evolutionary extremes of "drug addiction" in certain circumstances. Some strains of VRE develop mutations that make them dependent on the presence of the glycopeptide antimicrobial. This phenomenon is designated vancomycin dependence and is an example of a post-adaptive mechanism that could be considered as "reverse induction." In this case, the VRE develop mutations in the native D-Ala–D-Ala ligase, making the enzyme incapable of synthesizing the normal D-Ala–D-Ala-ending peptidoglycan precursors. Thus, these organisms rely entirely on the constant induction of their *van* clusters in order to synthesize their cell walls (Van Bambeke *et al.*, 1999). Therefore, the presence of the antimicrobial is vital to maintain a constant induction of the system and fulfill this dependence to the drug. This extreme phenotype seems to be of rare occurrence as it is readily reversible, but it underscores the fact that bacteria are capable of the ultimate sacrifice in order to gain an evolutionary advantage—even if that means paying a high biological price for survival.

Induction of antimicrobial resistance based on posttranscriptional modifications

The most studied example of this type of induction is the regulation of the expression of 23S rRNA methylases encoded by the *erm* (erythromycin ribosomal methylase) and *cfr* genes in response to macrolide and oxazolidinone (linezolid) challenge. As mentioned above, macrolides (and oxazolidinones) exert their antimicrobial effect by binding to the 50S subunit of the bacterial ribosome, where they specifically interact with nucleotides of the domain V of the 23S ribosomal RNA (rRNA) in the peptidyl-transferase center (PTC). The interaction of the antimicrobials with the PTC blocks the elongation of the polypeptide chain, arresting protein synthesis (Katz and Ashley, 2005). The Erm-mediated dimethylation of A2058 affects several classes of antimicrobials, including macrolides, lincosamides, and streptogramin B (MLS_B) compounds (the MLS_B phenotype). The presence of the methyl groups is thought to generate a structural change that prevents the macrolide from binding to its target. The expression of the MLS_B phenotype can be inducible or constitutive and, in both cases, an intact *erm* gene is required, illustrating an inducible mechanism using posttranscriptional gene control.

The best characterized inducible MLS_B phenotype is associated with the *ermC* gene in *S. aureus*, and involves modifications in an operon conformed by the *ermC* gene, an upstream gene that encodes a leader peptide, and an intergenic region between these genes. Under noninducing conditions (no antibiotic present), due to its conformation, the mRNA generated from this operon only allows leader peptide mRNA to be translated, hence preventing translation of the *ermC* mRNA (Fig. 6.4). In the presence of an inducer, most commonly erythromycin (but also clarithromycin and azithromycin), the antibiotic causes stalling of the ribosome translating the leader peptide. This stalling causes a conformational change in the *ermC* mRNA, unmasking the methylase ribosomal binding site and resulting in efficient translation (Fig. 6.4). Mutations occurring in the leader segment of the operon can switch the system from an inducible to a constitutive mode of expression of this mechanism of resistance (Katz and Ashley, 2005; Bailey *et al.*, 2008). So a sophisticated mRNA-based control has evolved to tightly regulate the methylation of nucleotides in the peptidyl transferase site in order to respond to the antibiotic challenge. This example of regulation ensures a high efficiency of action against the antibiotic, and at the same time reduces the possible fitness cost for the bacterial population.

Conclusion

Antimicrobial resistance is a complex and multidimensional phenomenon, and is the most remarkable example of evolution in nature. The natural occurrence of antimicrobials, which have been present since bacteria evolved on Earth, has been the principle driving force of adaptation in order for these organisms to outcompete and survive—the main purpose of natural evolution. Human beings have massively produced and used antimicrobials in a relatively short span of time compared with the billions of years that bacteria have populated this planet. Hence, antibiotic consumption appears to be the main driver of antibiotic resistance by "accelerating" the ability of bacteria to adapt, resulting in a direct impact on human health and a worldwide crisis. As an example of this phenomenon, genomic studies have determined that the evolution and dissemination of a hospital-associated lineage of multidrug-resistant *E. faecium* (one of the most difficult and recalcitrant bacteria to treat in clinical settings) occurred about 75 years

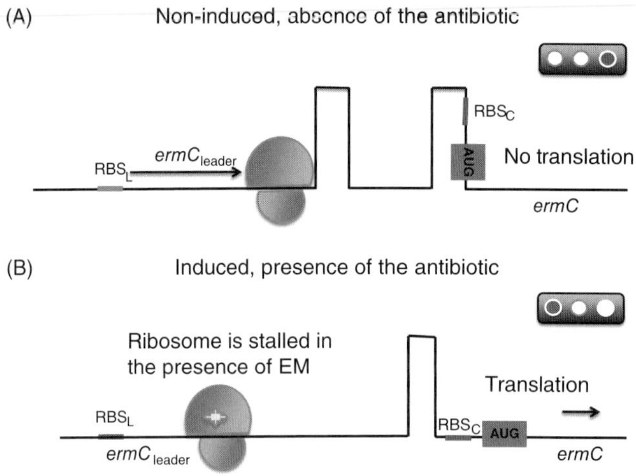

Fig. 6.4. Schematic representation of the posttranscriptional control of the *ermC* (erythromycin ribosomal methylase) gene. (A) Under non-inducing conditions, the ErmC leader peptide is produced and the *ermC* mRNA forms two hairpins, thus preventing the ribosome from recognizing the ribosomal binding site (RBS) of *ermC* (RBS_C). As a result, translation is inhibited. (B) After exposure to erythromycin (EM, the "star"-type symbol), the antibiotic interacts with the ribosome and binds tightly to the leader peptide, stalling progression of translation. This phenomenon releases the *ermC* RBS, RBS_C, and permits translation. Key: AUG, initiation codon; RBS_L ribosomal binding site of the leader; the ribosome is represented by the two incomplete circles on either side of the mRNA.

ago, coinciding with the introduction of antibiotics into clinical medicine (Lebreton *et al.*, 2013).

As specified above, differentiating the concepts of selection vs. induction of antibiotic resistance is not straightforward. Our approach has been by no means definitive, and different interpretations of the same theme may be appropriate. Our conceptual approach is based in a rather simplistic concept of gene acquisition vs. the expression of genetic determinants. However, a caveat is that the induction of a gene involved in resistance has a "selective" advantage, suggesting that the selection of antibiotic resistance has broader connotations. Moreover, the availability of a substantial pool of antibiotic resistance genes in the environment (the resistome) offers immense possibilities for horizontal gene transfer which play an important role in bacterial evolution. From a clinical perspective, differentiation of the concepts of selection vs. induction is more than academic. Indeed, the emergence of resistance in pathogenic bacteria under the pressure of antibiotics may have a profound impact on the population as a whole as bacteria have the ability to disseminate with particular ease. Therefore, an understanding of the factors that drive evolution may lead to a more judicious use of antimicrobials and, potentially, an improvement in the clinical approach to recalcitrant bacterial infections.

References

Alekshun, M.N. and Levy, S.B. (2007) Molecular mechanisms of antibacterial multidrug-resistance. *Cell* 128, 1037–1050.

Arakawa, Y., Ike, Y., and Nagasawa, M. (2004) Where has vancomycin-heterogeneously resistant *Staphylococcus aureus* gone? *The Lancet* 363, 1401.

Arias, C.A., Panesso, D., McGrath, D.M., Qin, X., Mojica, M.F., Miller, C., Diaz, L., Tran, T.T., Rincon, S., Barbu, E.M. *et al.* (2011) Genetic basis for in vivo daptomycin resistance in enterococci. *The New England Journal of Medicine* 365, 892–900.

Bailey, M., Chettiath, T., and Mankin, A.S. (2008) Induction of *erm*(C) expression by non-inducing antibiotics. *Antimicrobial Agents and Chemotherapy* 52, 866–874.

Bonnet, R. (2004) Growing group of extended-spectrum beta-lactamases: the CTX-M enzymes. *Antimicrobial Agents and Chemotherapy*. 48, 1–14.

Bush, K. and Jacoby, G.A. (2010) Updated functional classification of β-lactamases. *Antimicrobial Agents and Chemotherapy* 54, 969–976.

Cambray, G., Guerout, A.-M., and Mazel, D. (2010) Integrons. *Annual Review of Genetics* 44, 141–166.

Campbell, E.A., Korzheva, N., Mustaev, A., Murakami, K., Nair, S., Goldfarb, A., and Darst, S.A. (2001) Structural mechanism for rifampicin inhibition of bacterial RNA polymerase. *Cell* 104, 901–912.

Chambers, H.F. (1997) Methicillin resistance in staphylococci: molecular and biochemical basis and clinical implications. *Clinical Microbiology Reviews* 10, 781–791.

Coronado, V.G., Edwards, J.R., Culver, D.H., and Gaynes, R.P. (1995) Ciprofloxacin resistance among nosocomial *Pseudomonas aeruginosa* and *Staphylococcus aureus* in the United States. National Nosocomial Infections Surveillance (NNIS) System. *Infection Control and Hospital Epidemiology* 16, 71–75.

D'Costa, V.M., King, C.E., Kalan, L., Morar, M., Sung, W.W.L., Schwarz, C., Froese, D., Zazula, G., Calmels, F., Debruyne, R. *et al.* (2011) Antibiotic resistance is ancient. *Nature* 477, 457–461.

Deurenberg, R.H., Vink, C., Kalenic, S., Friedrich, A.W., Bruggeman, C.A., and Stobberingh, E.E. (2007) The molecular evolution of methicillin-resistant *Staphylococcus aureus*. *Clinical Microbiology and Infection* 13, 222–235.

Eliopoulos, G.M., Meka, V.G., and Gold, H.S. (2004) Antimicrobial resistance to linezolid. *Clinical Infectious Diseases* 39, 1010–1015.

Foucault, M.-L., Depardieu, F., Courvalin, P., and Grillot-Courvalin, C. (2010) Inducible expression eliminates the fitness cost of vancomycin resistance in enterococci. *Proceedings of the National Academy of Sciences of the United States of America* 107, 16964–16969.

Gogarten, M.B., Gogarten, J.P., and Olendzenski, L.C. (eds) (2009) *Horizontal Gene Transfer: Genomes in Flux*. Methods in Molecular Biology 532, Humana Press (Springer), Totowa, New Jersey.

Hiramatsu, K., Aritaka, N., Hanaki, H., Kawasaki, S., Hosoda, Y., Hori, S., Fukuchi, Y., and Kobayashi, I. (1997) Dissemination in Japanese hospitals of strains of *Staphylococcus aureus* heterogeneously resistant to vancomycin. *The Lancet* 350, 1670–1673.

Hiramatsu, K., Ito, T., Tsubakishita, S., Sasaki, T., Takeuchi, F., Morimoto, Y., Katayama, Y., Matsuo, M., Kuwahara-Arai, K., Hishinuma, T. and Baba, T. (2013) Genomic basis for methicillin resistance in *Staphylococcus aureus*. *Infection and Chemotherapy* 45, 117–136.

Hong, H.-J., Hutchings, M.I., Hill, L.M., and Buttner, M.J. (2005) The role of the novel Fem protein VanK in vancomycin resistance in *Streptomyces coelicolor*. *The Journal of Biological Chemistry* 280, 13055–13061.

Hooper, D.C. (2001) Emerging mechanisms of fluoroquinolone resistance. *Emerging Infectious Diseases* 7, 337–341.

Horne, K.C., Howden, B.P., Grabsch, E.A., Graham, M., Ward, P.B., Xie, S., Mayall, B.C., Johnson, P.D.R., and Grayson, M.L. (2009) Prospective comparison of the clinical impacts of heterogeneous vancomycin-intermediate methicillin-resistant *Staphylococcus aureus* (MRSA) and vancomycin-susceptible MRSA. *Antimicrobial Agents and Chemotherapy* 53, 3447–3452.

Howden, B.P., Peleg, A.Y., and Stinear, T.P. (2014) The evolution of vancomycin intermediate *Staphylococcus aureus* (VISA) and heterogenous-VISA. *Infection, Genetics and Evolution. Journal of Molecular Epidemiology and Evolutionary Genetics of Infectious Diseases* 21, 575–582.

Howell, L. (ed.) (2013) *World Economic Forum. Global Risks 2013*, 8th edn. Insight Report. An Initiative of the Risk Response Network. World Economic Forum, Cologny/Geneva, Switzerland.

Ike, Y., Arakawa, Y., Ma, X., Tatewaki, K., Nagasawa, M., Tomita, H., Tanimoto, K., and Fujimoto, S. (2001) Nationwide survey shows that methicillin-resistant *Staphylococcus aureus* strains heterogeneously and intermediately resistant to vancomycin are not disseminated throughout Japanese hospitals. *Journal of Clinical Microbiology* 39, 4445–4451.

Jacoby, G.A. (2009) AmpC beta-lactamases. *Clinical Microbiology Reviews* 22, 161–182.

Jacoby, G.A. (2014) >β-Lactamase Classification and Amino Acid Sequences for TEM, SHV and OXA Extended-Spectrum and Inhibitor Resistant Enzymes. Available at: http://www.lahey.org/Studies/ [in the process of transitioning, see http://www.ncbi.nlm.nih.gov/pathogens/submit_beta_lactamase/ for details] (accessed 3 May 2016).

Jaurin, B., Grundström, T., Edlund, T., and Normark, S. (1981) The *E. coli* beta-lactamase attenuator mediates growth rate-dependent regulation. *Nature* 290, 221–225.

Jin, D.J. and Gross, C.A. (1988) Mapping and sequencing of mutations in the *Escherichia coli rpoB* gene that lead to rifampicin resistance. *Journal of Molecular Biology* 202, 45–58.

Juan, C., Moyá, B., Pérez, J.L., and Oliver, A., (2006) Stepwise upregulation of the *Pseudomonas aeruginosa* chromosomal cephalosporinase conferring high-level beta-lactam resistance involves three AmpD homologues. *Antimicrobial Agents and Chemotherapy* 50, 1780–1787.

Katayama, Y., Ito, T., and Hiramatsu, K. (2001) Genetic organization of the chromosome region surrounding *mecA* in clinical staphylococcal strains: role of IS431-mediated *mecI* deletion in expression of resistance in *mecA*-carrying, low-level methicillin-resistant *Staphylococcus haemolyticus*. *Antimicrobial Agents and Chemotherapy* 45, 1955–1963.

Katz, L. and Ashley, G.W. (2005) Translation and protein synthesis: macrolides. *Chemical Reviews* 105, 499–528.

Lartigue, M.-F., Poirel, L., Aubert, D., and Nordmann, P. (2006) In vitro analysis of ISEcp1B-mediated mobilization of naturally occurring beta-lactamase gene *bla*$_{\text{CTX-M}}$ of *Kluyvera ascorbata*. *Antimicrobial Agents and Chemotherapy* 50, 1282–1286.

Laxminarayan, R., Duse, A., Wattal, C., Zaidi, A.K., Wertheim, H.F., Sumpradit, N., Vlieghe, E., Hara, G.L., Gould, I.M., and Goossens, H. (2013) Antibiotic resistance—the need for global solutions. *The Lancet Infectious Diseases* 13, 1057–1098.

Leach, K.L., Brickner, S.J., Noe, M.C., and Miller, P.F. (2011) Linezolid, the first oxazolidinone antibacterial agent. *Annals of the New York Academy of Sciences* 1222, 49–54.

Lebreton, F., van Schaik, W., Manson McGuire, A., Godfrey, P., Griggs, A., Mazumdar, V., Corander, J., Cheng, L., Saif, S., Young, S. *et al.* (2013) Emergence of epidemic multidrug-resistant *Enterococcus faecium* from animal and commensal strains. *mBio* 4(4), e00534–13.

Lederberg, J. and Lederberg, E.M. (1952) Replica plating and indirect selection of bacterial mutants. *Journal of Bacteriology* 63, 399–406.

Lindquist, S., Weston-Hafer, K., Schmidt, H., Pul, C., Korfmann, G., Erickson, J., Sanders, C., Martin, H.H., and Normark, S. (1993) AmpG, a signal transducer in chromosomal beta-lactamase induction. *Molecular Microbiology* 9, 703–715.

Liu, C. and Chambers, H.F. (2003) *Staphylococcus aureus* with heterogeneous resistance to vancomycin: epidemiology, clinical significance, and critical assessment of diagnostic methods. *Antimicrobial Agents and Chemotherapy* 47, 3040–3045.

Luria, S.E. and Delbrück, M. (1943) Mutations of bacteria from virus sensitivity to virus resistance. *Genetics* 28, 491–511.

Mahlen, S.D., Morrow, S.S., Abdalhamid, B., and Hanson N.D. (2003) Analyses of *ampC* gene expression in *Serratia marcescens* reveal new regulatory properties. *Journal of Antimicrobial Chemotherapy* 51, 791–802.

Mendes, R.E., Flamm, R.K., Hogan, P.A., Ross, J.E., and Jones, R.N. (2014) Summary of linezolid activity and resistance mechanisms detected during the 2012 LEADER Surveillance program for the United States. *Antimicrobial Agents and Chemotherapy* 58, 1243–1247.

Miller, C., Kong, J., Tran, T.T., Arias, C.A., Saxer, G., and Shamoo, Y. (2013) Adaptation of *Enterococcus faecalis* to daptomycin reveals an ordered progression to resistance. *Antimicrobial Agents and Chemotherapy* 57, 5373–5383.

Moellering, R.C. (2012) MRSA: the first half century. *Journal of Antimicrobial Chemotherapy.* 67, 4–11.

Munita, J.M., Tran, T.T., Diaz, L., Panesso, D., Reyes, J., Murray, B.E., and Arias, C.A. (2013) A *liaF* codon deletion abolishes daptomycin bactericidal activity against vancomycin-resistant *Enterococcus faecalis*. *Antimicrobial Agents and Chemotherapy* 57, 2831–2833.

Munoz-Price, L.S. and Quinn, J.P. (2009) The spread of *Klebsiella pneumoniae* carbapenemases: a tale of strains, plasmids and transposons. *Clinical Infectious Diseases* 49, 1739–1741.

Mwangi, M.M., Wu, S.W., Zhou, Y., Sieradzki, K., de Lencastre, H., Richardson, P., Bruce, D., Rubin, E., Myers, E., Siggia, E.D., and Tomasz, A. (2007) Tracking the in vivo evolution of multidrug resistance in *Staphylococcus aureus* by whole-genome sequencing. *Proceedings of the National Academy of Sciences of the United States of America* 104, 9451–9456.

Neoh, H., Cui, L., Yuzawa, H., Takeuchi, F., Matsuo, M., and Hiramatsu, K. (2008) Mutated response regulator *graR* is responsible for phenotypic conversion of *Staphylococcus aureus* from heterogeneous vancomycin-intermediate resistance to vancomycin-intermediate resistance. *Antimicrobial Agents and Chemotherapy* 52, 45–53.

Nordmann, P., Cuzon, G., and Naas, T. (2009) The real threat of *Klebsiella pneumoniae* carbapenemase-producing bacteria. *The Lancet Infectious Diseases* 9, 228–236.

Perry, J.A. and Wright, G.D. (2013) The antibiotic resistance "mobilome:" Searching for the link between environment and clinic. *Frontiers in Microbiology* 4: 138.

Pogliano, J., Pogliano, N., and Silverman, J.A. (2012) Daptomycin-mediated reorganization of membrane architecture causes mislocalization of essential cell division proteins. *Journal of Bacteriology* 194, 4494–4504.

Poirel, L., Lartigue, M.-F., Decousser, J.-W., and Nordmann, P. (2005) ISEcp1B-mediated transposition of $bla_{\text{CTX-M}}$ in *Escherichia coli*. *Antimicrobial Agents and Chemotherapy* 49, 447–450.

Rice, L.B. (2005) Antibiotics and gastrointestinal colonization by vancomycin-resistant enterococci. *European Journal of Clinical Microbiology and Infectious Diseases* 24, 804–814.

Rice, L.B. (2010) Genetics of resistance. In: Corvalin, P., Leclercrq, R. and Rice, L.B. (eds) *Antibiogram* [English version of the French *Antibiogramme* published in 1985]. Eska Publishing, Portland, Oregon/ASM Press, Washington, DC, p. 25–36.

Rice, L.B. (2012) Gastrointestinal bacteria will have its way. *The Journal of Infectious Diseases* 206, 1334–1335.

Rossi, F., Diaz, L., Wollam, A., Panesso, D., Zhou, Y., Rincon, S., Narechania, A., Xing, G., Di Gioia, T.S.R., Doi, A. *et al.* (2014) Transferable vancomycin resistance in a community-associated MRSA lineage. *The New England Journal of Medicine* 370, 1524–1531.

Rossolini, G.M., D'Andrea, M.M., and Mugnaioli, C. (2008) The spread of CTX-M-type extended-spectrum beta-lactamases. *Clinical Microbiology and Infection* 1, 33–41.

Sanchez-Romero, M.A. and Casadesus, J. (2014) Contribution of phenotypic heterogeneity to adaptive antibiotic resistance. *Proceedings of the National Academy of Sciences of the United States of America* 111, 355–360.

Schmidtke, A.J. and Hanson, N.D. (2006) Model system to evaluate the effect of *ampD* mutations on AmpC-mediated beta-lactam resistance. *Antimicrobial Agents and Chemotherapy* 50, 2030–2037.

Schwaber, M.J., Navon-Venezia, S., Kaye, K.S., Ben-Ami, R., Schwartz, D., and Carmeli, Y. (2006) Clinical and economic impact of bacteremia with extended-spectrum-β-lactamase-producing *Enterobacteriaceae*. *Antimicrobial Agents and Chemotherapy* 50, 1257–1262.

Strahilevitz, J., Jacoby, G.A., Hooper, D.C., and Robicsek, A. (2009) Plasmid-mediated quinolone resistance: a multifaceted threat. *Clinical Microbiology Reviews* 22, 664–689.

Taur, Y. and Pamer, E.G. (2013) The intestinal microbiota and susceptibility to infection in immunocompromised patients. *Current Opinion in Infectious Diseases* 26, 332–337.

Telenti, A. and Imboden, P. (1993) Detection of rifampicin-resistance mutations in *Mycobacterium tuberculosis*. *The Lancet* 341, 647–650.

Toh, S.-M., Xiong, L., Arias, C.A., Villegas, M.V., Lolans, K., Quinn, J., and Mankin, A.S. (2007) Acquisition of a natural resistance gene renders a clinical strain of methicillin [sic]-resistant *Staphylococcus aureus* resistant to the synthetic antibiotic linezolid. *Molecular Microbiology* 64, 1506–1514.

Tupin, A., Gualtieri, M., Roquet-Banères, F., Morichaud, Z., Brodolin, K., and Leonetti, J.-P. (2010) Resistance to rifampicin: at the crossroads between ecological, genomic and medical concerns. *International Journal of Antimicrobial Agents* 35, 519–523.

Van Bambeke, F., Chauvel, M., Reynolds, P.E., Fraimow, H.S., and Courvalin, P. (1999) Vancomycin-dependent *Enterococcus faecalis* clinical isolates and revertant mutants. *Antimicrobial Agents and Chemotherapy* 43, 41–47.

Wright, G.D. and Poinar, H. (2012) Antibiotic resistance is ancient: implications for drug discovery. *Trends in Microbiology* 20, 157–159.

Zhang, H.Z., Hackbarth, C.J., Chansky, K.M., and Chambers, H.F. (2001) A proteolytic transmembrane signaling pathway and resistance to beta-lactams in staphylococci. *Science* 291, 1962–1965.

7 Colonization and Its Importance for the Emergence of Clinical Resistance

Curtis J. Donskey*

Case Western Reserve University and Louis Stokes Cleveland Veterans Affairs Medical Center, Cleveland, Ohio, US

Introduction

Selective pressure exerted by antimicrobials plays a central role in the emergence and dissemination of antimicrobial-resistant pathogens. Systemic antimicrobials exert selective pressure not only on infecting microorganisms, but also on the normal microbiota of the host (e.g., the gastrointestinal tract, genitourinary tract, upper respiratory tract, and skin). Although a minority of those acquiring colonization with resistant pathogens develop infection, colonized individuals serve as a major reservoir for transmission. The intestinal tract is a particularly important source for the dissemination of resistant pathogens, including *Enterococcus* spp., Gram-negative bacilli, *Clostridium difficile*, *Candida* spp., and *Staphylococcus aureus* (Vollaard and Clasener, 1994; Cole *et al.*, 1996; Donskey *et al.*, 2000a,b; Bhalla *et al.*, 2003; Ray *et al.*, 2003; Donskey, 2004). The clinical manifestations of these pathogens are diverse, but a common pathogenesis is involved in their colonization of and dissemination from the intestinal tract (Donskey, 2004). Moreover, the colonization of the intestinal tract is often associated with the contamination of other sites, including the skin and the genitourinary tract.

An understanding of the role of colonization in the emergence and dissemination of resistant pathogens is essential for development of effective stewardship interventions. This chapter will examine common factors that facilitate the colonization and subsequent infection and transmission of antimicrobial-resistant healthcare-associated pathogens. The primary focus will be intestinal colonization and the critical role of antimicrobial selective pressure. Findings from studies involving animal models and healthy human volunteers will be used to illustrate general concepts on the effects of antibiotics on pathogen colonization. The applicability of these general concepts to clinical settings and implications for antimicrobial stewardship will be emphasized.

Colonization Resistance

The human colon contains as many as 10^{12} bacteria/g of contents and hundreds of bacterial species (Vollaard and Clasener, 1994). These indigenous bacteria provide a critical host defense, termed colonization resistance, by inhibiting colonization by exogenously introduced pathogens and preventing the overgrowth of resident bacteria, which are usually present in low numbers (e.g., *Escherichia coli*, enterococci) (Vollaard and Clasener, 1994). Multiple mechanisms may contribute to colonization resistance, including the depletion of nutrients, prevention of access to adherence sites or niches associated with the mucosa, and production of inhibitory substances or conditions (e.g., short-chain fatty acids, anaerobic conditions) (Vollaard and Clasener, 1994; Pultz *et al.*, 2005). Recent studies have identified some bacterial species or combinations of species that may contribute to colonization resistance. In antibiotic-treated mice, partial restoration of colonization resistance to *C. difficile* and

*E-mail: curtisd123@yahoo.com

vancomycin-resistant enterococci (VRE) was attained through the administration of isolates from the bacterial families *Lachnospiraceae* (phylum *Firmucutes*) and *Barnsiella* (phylum *Bacteroidetes*) (Reeves *et al.*, 2012; Ubeda *et al.*, 2013), respectively. A mixture of six phylogenetically diverse intestinal bacteria restored colonization resistance to *C. difficile* in mice, suggesting that the synergistic action of multiple organisms might be required (Lawley *et al.*, 2012).

Pathogenesis of Intestinal Colonization and Dissemination of Pathogens

A variety of factors may facilitate the intestinal colonization and dissemination of pathogens (Fig. 7.1) (Donskey, 2004). Some of these factors are intrinsic to the patient populations involved, whereas others are modifiable. Most studies focus on *C. difficile* and antimicrobial-resistant pathogens, but the same pathogenesis contributes to the overgrowth and transmission of susceptible organisms. Pathogens may be members of the indigenous microbiota present upon admission, or they may be acquired exogenously (Fig. 7.1). *Enterobacter*, *Candida albicans*, and ampicillin-susceptible *Enterococcus faecalis* strains often emerge from the indigenous microbiota, whereas VRE and *Candida glabrata* are usually acquired in healthcare settings (Donskey, 2004).

Exposure to pathogens

The hands of healthcare workers are the major source of transmission of pathogens from patient

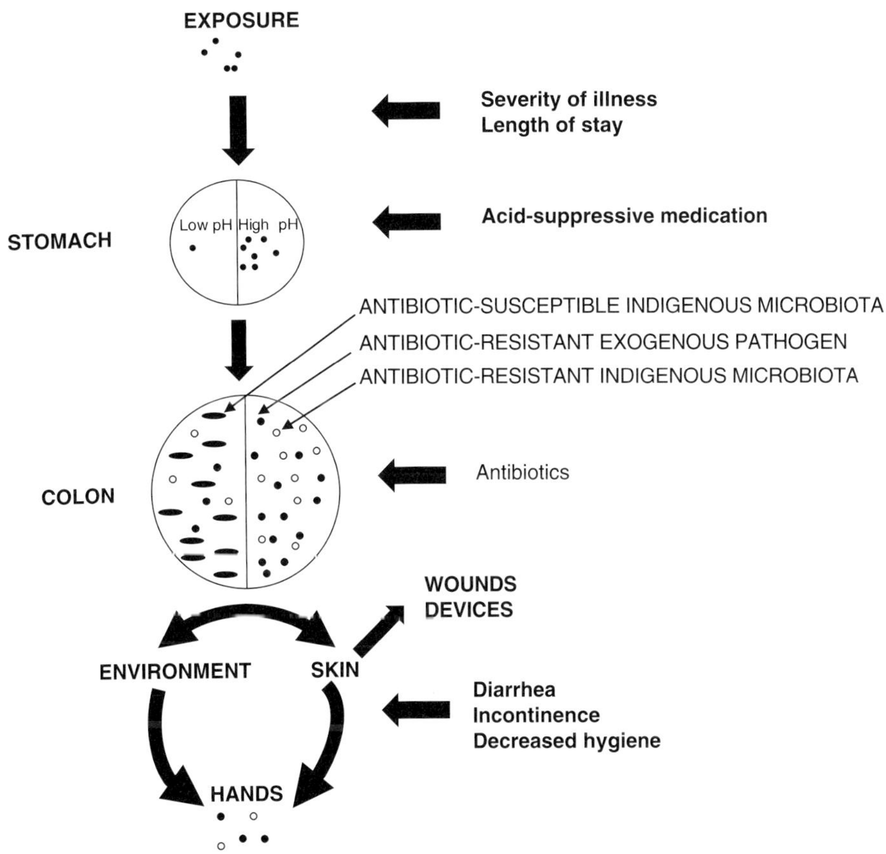

Fig. 7.1. Factors that facilitate the intestinal overgrowth and transmission of nosocomial pathogens. The left half of the circles illustrates the presence of normal acidity in the stomach and intact indigenous microbiota in the colon; the right half illustrates the effects of increased stomach pH and antibiotic-selective pressure in the colon. Adapted from Donskey (2004).

to patient (Bhalla, *et al.*, 2004). Hands may become contaminated through either contact with patients or contact with contaminated environmental surfaces (Bhalla *et al.*, 2004). Patients may also acquire pathogens directly from contaminated surfaces. Increased severity of illness and prolonged hospital stays place patients at increased risk of acquisition of pathogens, in part because these factors are associated with increased opportunity for interaction with healthcare workers and with contaminated surfaces and devices.

Reduction in gastric acidity

Gastric acid provides an important host defense by killing ingested pathogens (Rao *et al.*, 2006). Many studies have demonstrated an association between medications that inhibit the production of stomach acid (e.g., proton pump inhibitors or PPIs, and H_2 blockers) and healthcare-associated pathogens, including *C. difficile*, *C. albicans*, methicillin-resistant *S. aureus* (MRSA), VRE, and extended-spectrum β-lactamase (ESBL)-producing *Enterobacteriaceae* (Rao *et al.*, 2006). In mice, PPI treatment elevated gastric pH and facilitated colonization of the intestinal tract by VRE and ESBL-producing *Klebsiella pneumoniae* (Stiefel *et al.*, 2006).

The mechanism by which PPIs may promote *C. difficile* infection is unclear because acidic gastric contents and pH 1 or 2 buffers do not kill *C. difficile* spores (Rao *et al.*, 2006). Under acidic conditions, salivary nitrites are converted to reactive nitrogen compounds that could potentially kill *C. difficile* spores, but there is uncertainty on whether the usual physiological levels of nitrite are sufficient to provide sporicidal activity (Rao *et al.*, 2006; Cunningham *et al.*, 2014). An alternative hypothesis is that PPI therapy may promote *C. difficile* infection by alteration of the intestinal microbiota (Vesper *et al.*, 2009).

Alteration of the colonic microbiota

Antimicrobials that are excreted into the intestinal tract exert selective pressure on the microbiota. The magnitude of the effect of antimicrobials is determined by the concentrations achieved, the degree of inactivation that occurs, and the activity of the agents under in vivo conditions (Vollaard and Clasener, 1994; Donskey, 2004). Selective pressure results in the inhibition of susceptible members of the indigenous microbiota and facilitates overgrowth by the antibiotic-resistant members of the indigenous microbiota as well as by ingested antibiotic-resistant pathogens (Fig. 7.1). A more detailed discussion of antimicrobial selective pressure is provided in the next section.

Antibiotic treatment may facilitate pathogen colonization by disrupting several mechanisms by which the normal microbiota provides colonization resistance. First, the cecal microbiota of mice inhibit colonization by *C. difficile* in part through competition for nutrients, including carbohydrates present in mucin (e.g., sialic acids, *N*-acetylglucosamine) (Wilson and Perini, 1988). Recently, it has been demonstrated that antibiotic treatment increases the availability of sialic acids that can be used to support the growth of *C. difficile* and *Salmonella typhimurium* (Ng *et al.*, 2013). Others have demonstrated that antibiotic treatment increases the availability of nutrients that support colonization by *C. difficile* (e.g., sugar alcohols, fructose) (Theriot *et al.*, 2014). Secondly, antibiotic treatment may reduce the production of inhibitory substances or conditions (Pultz *et al.*, 2005). Finally, antibiotic treatment may promote *C. difficile* in part by inhibiting microorganisms that convert primary bile salts to secondary bile salts, thereby enhancing the germination of ingested spores (i.e., germination is stimulated by primary bile salts but inhibited by secondary bile salts) (Giel *et al.*, 2010).

Shedding of pathogens

The dissemination of pathogens from the intestinal tract occurs through fecal contamination, resulting in the contamination of environmental surfaces and the skin of patients (Donskey, 2004). Antimicrobial-induced overgrowth of pathogens contributes to their dissemination. For example, an increased density of VRE and *C. difficile* in stools has been associated with increased frequency of environmental and skin contamination (Sethi *et al.*, 2009, 2010). Fecal incontinence and diarrhea contribute to fecal contamination. Factors that reduce standards of hygiene (e.g., altered mental status, reduced mobility, indwelling devices) also contribute to the likelihood of contamination.

Immune suppression

The immune system plays a major role in the pathogenesis of *C. difficile* infection (Kyne *et al.*, 2001). A systemic antibody response to *C. difficile* toxins provides protection against the development of acute diarrhea and against recurrence (Kyne *et al.*, 2001). The immune system also plays a role in preventing clinical disease caused by other healthcare-associated pathogens. However, the impact of host immunity on colonization by these organisms is less clear.

Disruption of normal barriers leading to infection

Infection typically occurs in a minority of those colonized with resistant pathogens (Donskey, 2004). Frequently, such infection occurs as a result of the disruption of normal barriers that prevent access of colonizing organisms to sites where infection may occur. For example, many chemotherapy agents disrupt the integrity of the intestinal lining, thus facilitating the translocation of pathogenic bacteria or *Candida* spp. Similarly, devices, catheters, and surgical wounds provide a route for pathogens to reach normally sterile sites. In some cases, procedures result in the direct inoculation of pathogens into sterile sites (e.g., transrectal ultrasound-guided biopsy of the prostate introduces *E. coli* into the prostate and urinary tract) (Dumford *et al.*, 2013).

Antibiotics and Pathogen Colonization

Antibiotic selective pressure

Bacteria possess a remarkable ability to develop and acquire resistance to antibiotics. Common mechanisms of resistance include the modification of drug targets, production of inactivating enzymes such as β-lactamases and aminoglycoside-modifying enzymes, the action of efflux pumps, and alterations in outer membrane proteins that are associated with decreased permeability to antibiotics (Donskey, 2006). Antibiotic exposure does not directly induce these resistance mechanisms. Rather, antibiotic therapy exerts selective pressure, i.e., by the inhibition of competing microbiota but not resistant organisms. In individual patients, selective pressure may facilitate the emergence of new resistant mutants or of preexisting subpopulations of resistant organisms. For example, ceftazidime therapy may eliminate susceptible Gram-negative bacilli, while allowing expansion of the population of a new mutant of *K. pneumoniae* that harbors an ESBL, or of a preexisting subpopulation of *Enterobacter* spp. that constitutively hyperproduces chromosomal cephalosporinases. Numerous clinical studies have documented the emergence of resistant Gram-negative bacilli during antibiotic therapy (Donskey, 2006). Once resistant pathogens have emerged, antibiotics play a crucial role in their spread from patient to patient.

Because healthcare-associated pathogens are often multidrug-resistant, many different antibiotics may facilitate their colonization and dissemination. For example, the third-generation cephalosporins, trimethoprim/sulfamethoxazole, ciprofloxacin, and aminoglycosides have all been associated with ESBL-producing Gram-negative bacilli (Wiener *et al.*, 1999; Asensio *et al.*, 2000). In an outbreak of ESBL-producing Gram-negative bacilli in nursing homes, most patients had not received prior ceftazidime (Wiener *et al.*, 1999). Rather, the receipt of ciprofloxacin or trimethoprim/sulfamethoxazole was an independent risk factor for colonization (Wiener *et al.*, 1999). Similarly, many classes of antibiotics, including penicillins, cephalosporins, fluoroquinolones, lincosamides, metronidazole, and glycopeptides (e.g., vancomycin, teicoplanin), have been associated with VRE colonization and infection (Donskey *et al.*, 2000a,b).

Clinical studies

Many clinical studies, primarily with retrospective case-control design, have examined the association between antibiotics and colonization or infection with resistant pathogens and *C. difficile* infection. Although these studies have the advantage of providing information from clinical settings, some limitations should be acknowledged. First, patients often receive multiple antibiotics simultaneously or sequentially, making it difficult to determine the effect of individual agents. Secondly, classes of antibiotics are often grouped together for analysis despite the fact that individual agents within the class differ significantly. For example, cephalosporins include agents with potent anti-anaerobic activity that are excreted in high concentrations in bile (e.g., cefotetan) and agents with relatively little anti-anaerobic activity and minimal biliary excretion (e.g., cefepime) (Donskey, 2004). Finally, studies that evaluate the effect of formulary changes on rates of colonization or infection with nosocomial pathogens are subject to bias due to the difficulties in controlling for non-antimicrobial confounding factors.

Given these limitations of clinical studies, it is useful to consider whether associations between antibiotics and pathogens are consistent across multiple studies and are microbiologically plausible. For example, the association between third-generation cephalosporins and VRE is consistent across multiple studies and plausible because these agents alter the intestinal microbiota but have minimal activity against enterococci (Donskey *et al.*, 2000a,b). Similarly, third-generation cephalosporins have been associated with intestinal colonization by *Candida* spp., cephalosporin-resistant Gram-negative bacilli, and *C. difficile*, and they have minimal or no activity against these pathogens (Pultz *et al.*, 2005). Thus, it is plausible that the restriction of third-generation cephalosporins might be an effective strategy for reducing colonization by multiple pathogens.

Animal models

Mouse models provide a useful means to directly compare the effects of antibiotics on intestinal colonization with pathogens. Figure 7.2 provides a general framework for considering the effects of antibiotics on pathogen colonization based upon studies involving VRE, resistant Gram-negative bacilli, and *C. difficile* (Donskey *et al.*, 1999; Donskey *et al.*, 2000a; Stiefel *et al.*, 2004; Donskey, 2004, 2006; Adams *et al.*, 2007; Owens *et al.*, 2008; Perez *et al.*, 2011). Figure 7.2A shows an establishment model (i.e., pathogens administered orally during and/or after the completion of subcutaneous antibiotic treatment). Figure 7.2B shows a persistence model (i.e., high-density colonization established before antibiotic treatment to assess the effect of treatment on persistence). Antibiotics are classified into four categories, as described next.

Several factors determine the impact of an antibiotic on establishment of pathogen colonization. First, does the antibiotic alter the indigenous microbiota that provide colonization resistance? If an antibiotic alters colonization resistance but lacks activity against pathogens (category A; e.g., ceftriaxone alters the indigenous anaerobic microbiota but lacks significant activity against VRE, *C. difficile*, and ESBL-producing Gram-negative bacilli), vulnerability to pathogen colonization may persist during treatment and for a period of days to weeks after the discontinuation of treatment while the microbiota recovers. If an antibiotic does not alter the microbiota, it does not promote pathogen colonization at any time of exposure (category B; e.g., cefepime, aztreonam). Secondly, does the antibiotic inhibit pathogen colonization? For example, piperacillin/tazobactam (category C) achieves sufficient concentrations to inhibit the establishment of colonization by *C. difficile* and some strains of VRE and ESBL-producing *K. pneumoniae* during treatment. Finally, what is the timing of pathogen exposure relative to antibiotic treatment? Antibiotics such as piperacillin/tazobactam inhibit colonization by susceptible pathogens during treatment, but mice become vulnerable to colonization during the period of recovery of the microbiota.

For antibiotics with inhibitory activity against pathogens, an additional consideration is whether colonization is present prior to antibiotic exposure? As shown in Fig. 7.2B, antibiotics with modest inhibitory activity against pathogens may not have sufficient activity to eliminate preexisting high-density colonization. For example, piperacillin/tazobactam inhibits the establishment of colonization by VRE and ESBL-producing *K. pnuemoniae* (Fig. 7.2A), but promotes persistent high-density colonization by VRE and some strains of ESBL-producing *K. pneumoniae* if preexisting high-density colonization is present (Fig. 7.2B). Oral nonabsorbed antibiotics can be very effective for the decolonization of pathogens (category D; e.g., oral ramoplanin for VRE, polymyxin E for resistant Gram-negative bacilli). However, frequent relapses of colonization are common after treatment with agents that cause significant alteration of the indigenous microbiota (e.g., ramoplanin) (Stiefel *et al.*, 2004). In contrast, relapse (to colonization) of carbapenem-resistant *K. pneumoniae* did not occur after decolonization treatment with oral polymyxin E or gentamicin (Perez *et al.*, 2011).

Studies with healthy volunteers

Many studies have evaluated the effect of antibiotics on the intestinal microbiota of healthy human volunteers (Sullivan *et al.*, 2001). In general, the findings from these studies have been similar to those from mouse model studies. Antibiotics that inhibit anaerobes but lack anti-enterococcal activity (e.g., clindamycin, cephalosporins) promote overgrowth of enterococci (Donskey *et al.*, 2000b). Antibiotics that inhibit anaerobes without inhibiting *Enterobacteriaceae* (e.g., clindamycin, linezolid, oral vancomycin) promote overgrowth of indigenous *Enterobacteriaceae* and frequent emergence of new resistant Gram-negative bacilli. Antibiotics that cause relatively little disruption of the anaerobic microbiota (e.g., trimethoprim/

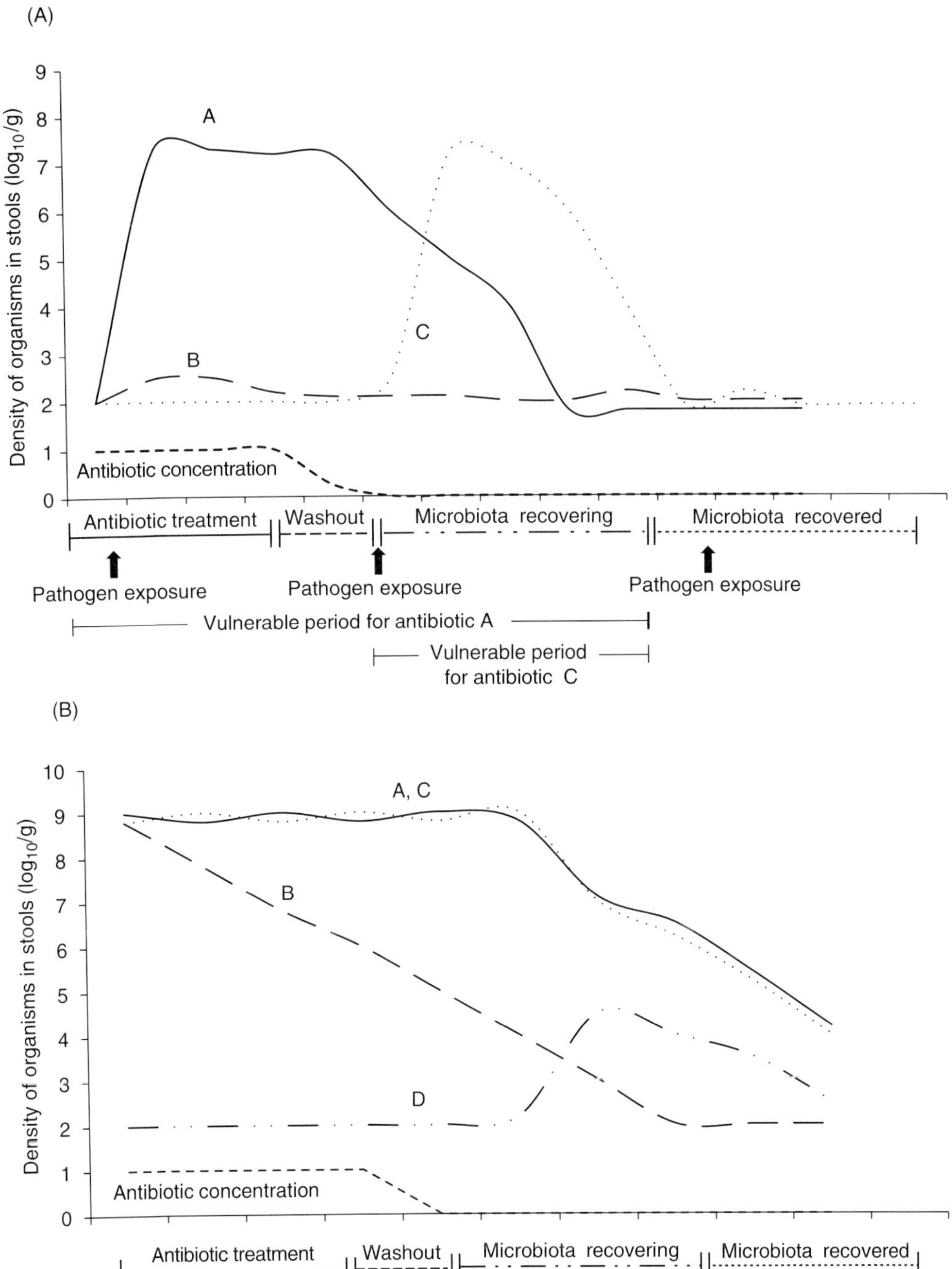

Fig. 7.2. General framework for considering the effect of antibiotics on pathogen colonization based upon several studies involving vancomycin-resistant enterococci (VRE), multidrug-resistant Gram-negative bacilli, and *Clostridium difficile*. (A) The effect of antibiotic treatment on the establishment of colonization by pathogens during and/or after completion of antibiotic treatment (an establishment model); (B) The effect of antibiotic treatment on the persistence of previously established colonization (a persistence model). Four potential effects of antibiotic treatment are shown in

Continued

sulfamethoxazole, cefadroxil, ciprofloxacin) may also promote the emergence of resistant Gram-negative bacilli, but promote less overgrowth of Gram-negative bacilli and enterococci.

Implications for Antimicrobial Stewardship and Infection Control

Infection control interventions that target multiple pathogens promoted or transmitted by the same mechanisms have been termed horizontal strategies (Edmond and Wenzel, 2013). In general, horizontal strategies may be preferred over vertical strategies that focus on the prevention of colonization and infection due to specific pathogens (Edmond and Wenzel, 2013). Because many healthcare-associated pathogens share similar risk factors and pathogenesis, there is an opportunity to implement control measures that limit the transmission of multiple pathogens. This section of the chapter will highlight several horizontal strategies related to antimicrobial stewardship and infection control. In addition, it will provide examples where targeted approaches may be advantageous.

Antibiotic formulary restriction or substitution

One strategy to control antibiotic-resistant pathogens and *C. difficile* is the implementation of formulary changes that involve restrictions in the use of antibiotics that have frequently been associated with particular pathogens, while substituting agents that may be less likely to promote those pathogens. Third-generation cephalosporins have frequently been targeted for restriction because they have been associated with intestinal colonization by multiple pathogens and with *C. difficile* infection. Substitution of piperacillin/tazobactam, cefepime, imipenem, or ticarcillin/clavulanate for third-generation cephalosporins has been associated with reductions of ESBL-producing organisms (Donskey, 2006). The substitution of piperacillin/tazobactam for third-generation cephalosporins has also been associated with reductions in *C. difficile* infection and VRE colonization or infection (Quale *et al.*, 1996; Owens *et al.*, 2008). As noted previously, the reductions associated with this formulary substitution are microbiologically plausible due to the fact that piperacillin/tazobactam, unlike cephalosporins, exhibits inhibitory activity against these pathogens and may suppress colonization in the intestinal tract or at other sites (Donskey *et al.*, 1999, 2000a).

Fluoroquinolones have also been associated with colonization by or infection with multiple pathogens, including resistant Gram-negative bacilli, VRE, and *C. difficile* (Donskey, 2004). The restriction of fluoroquinolones has been associated with reductions in fluoroquinolone-resistant Gram-negative bacilli and *C. difficile* infection (Donskey, 2004; Adams *et al.*, 2007). The current epidemic strain of *C. difficile*, termed North American PFGE type 1 or polymerase chain reaction ribotype 027, exhibits high-level resistance to fluoroquinolones (Adams *et al.*, 2007). In addition to reducing the incidence of *C. difficile* infection, the restriction of fluoroquinolones has been associated with a reduction in the proportion of infections due to the fluoroquinolone-resistant epidemic strain (Muto *et al.*, 2007). Similarly, the restriction of clindamycin has been effective in controlling outbreaks associated with clindamycin-resistant epidemic *C. difficile* strains, with concurrent reduction in the proportion of infecting isolates that were clindamycin resistant (Owens *et al.*, 2008). In mice, fluoroquinolones and clindamycin suppressed the growth of *C. difficile* strains susceptible to these agents, but promoted

Fig. 7.2. Continued.

the figures: A, antibiotics that alter the anaerobic microbiota and have minimal or no activity against pathogens (e.g., clindamycin or ceftriaxone do not inhibit VRE or ceftriaxone-resistant Gram-negative bacilli or clindamycin-resistant *C. difficile* strains); B, antibiotics that do not alter the anaerobic microbiota that provide colonization resistance (e.g., cefepime or aztreonam) do not promote colonization; C, antibiotics that alter the anaerobic microbiota but have modest or moderate activity against pathogens (e.g., piperacillin/tazobactam for VRE and extended-spectrum β-lactamase (ESBL)-producing Gram-negative bacilli; piperacillin/tazobactam and tigecycline for *C. difficile*); and D, nonabsorbed oral antibiotics that have potent activity against pathogens and achieve high concentrations in the colon (e.g., ramoplanin for VRE; polymyxin E for resistant Gram-negative bacilli). As noted in the text, relapses of VRE colonization are common after treatment with ramoplanin in mice, an agent that causes significant alteration of the indigenous microbiota, but relapse (to colonization) of carbapenem-resistant *Klebsiella pneumoniae* did not occur after decolonization treatment with oral polymyxin E or gentamicin. Adapted from Donskey, 2004.

that of resistant strains (Adams *et al.*, 2007). These data suggest that fluoroquinolones and clindamycin may exert selective pressure on *C. difficile* strains circulating in hospitals, favoring the proliferation of resistant epidemic strains.

Although formulary restriction or substitution is a promising stewardship strategy, it has limitations. The substitution of one agent for another may lead to the emergence of resistance to the new agent, a phenomenon that has been referred to as "squeezing the balloon" (Burke, 1998). Because multiple agents may promote pathogens, the restriction of agents such as third-generation cephalosporins might produce relatively little benefit unless concurrent efforts are made to restrict other high-risk agents.

Interventions to reduce overuse of all antibiotics or of multiple high-risk antibiotics

Antimicrobials are often used unnecessarily in hospitals and nursing homes (Hecker *et al.*, 2003). In two recent studies, reductions in the incidence of *C. difficile* infection have been achieved through stewardship interventions focusing on reducing the overuse of all antibiotics or of multiple high-risk agents. Reductions in the use of broad-spectrum antibiotics (i.e., cephalosporins and amoxicillin/clavulanate), while increasing the use of narrow-spectrum agents (i.e., benzyl penicillin, amoxicillin, trimethoprim) resulted in a 65% reduction in the incidence of *C. difficile* infection (Fowler *et al.*, 2007). Similarly, the implementation of a stewardship intervention to reduce use of high-risk antibiotics (e.g., cephalosporins, fluoroquinolones, macrolides) and total antibiotics resulted in a significant reduction in *C. difficile* infections (Valiquette *et al.*, 2007).

Reductions in the duration of therapy

Increased duration of antimicrobial therapy may be associated with increased risk for *C. difficile* infection and the acquisition of resistant pathogens (Donskey, 2004; Owens *et al.*, 2008). In addition, there is increasing evidence that the duration of therapy may be safely reduced for a variety of conditions, including complicated urinary tract infections (UTIs), pyelonephritis, and community-acquired pneumonia (Dunbar *et al.*, 2003; Peterson *et al.*, 2008; Rice, 2008). Moreover, two recent studies have also demonstrated that reducing the duration of antimicrobial therapy may be associated with reduced isolation of resistant organisms (Singh *et al.*, 2000; Chastre *et al.*, 2003).

Based on the available evidence and on practical considerations, efforts to better delineate optimal durations of antibiotic therapy to maximize efficacy whilst minimizing collateral damage have been advocated (Rice, 2008). One unmet need in this effort is a better understanding of the ecological impact of prolonged antibiotic administration. It is plausible that prolonged courses of broad-spectrum antibiotics may result in greater and more long-lasting disruption of the normal microbiota. However, limited data is available on the impact of different durations of therapy on the microbiota, particularly in hospitalized patients.

Decolonization with oral nonabsorbed antimicrobials

The goal of oral nonabsorbed antibiotic administration is to selectively inhibit pathogens in the gastrointestinal tract without disturbing the anaerobic microbiota (i.e., selective decontamination) (Vollaard and Clasener, 1994). Nonabsorbed oral antibiotics are administered with or without concurrent intravenous agents that are relatively narrow in spectrum. Selective decontamination has great potential because it addresses a major source of pathogen dissemination, although it may be associated with overgrowth and infections due to pathogens that are resistant to the agents being administered. In part, this reflects the fact that selective decontamination regimens are often not truly selective. For example, oral ramoplanin inhibits VRE and Gram-positive anaerobes in the colon, and therefore may facilitate the overgrowth of Gram-negative pathogens (Stiefel *et al.*, 2004).

Restoration or preservation of the indigenous intestinal microbiota

Based upon a recent randomized trial, the transplantation of fecal microbiota is now regarded as a standard therapy for patients with multiple recurrences of *C. difficile* infection (van Nood *et al.*, 2013). Modifications of the technique, such as the use of universal donors, simplified and defined collections of bacteria, and oral administration through capsules, will make fecal transplantation more feasible for widespread use. Studies are needed to determine the effectiveness of this approach as a means to eliminate other pathogens that colonize the intestinal tract.

While promising, the effectiveness of transplantation of the fecal microbiota is compromised in settings where patients require repeated courses of antibiotic therapy. Thus, there is a need for novel strategies to preserve the microbiota in the face of repeated antibiotic exposure. Because antibiotic activity within the lumen of the intestinal tract is not necessary for the treatment of most infections, we have postulated that the oral administration of β-lactamase enzymes could preserve colonization resistance during parenteral β-lactam antibiotic therapy by degrading the portion of antibiotic that is excreted into the intestinal tract. We demonstrated that the oral administration of recombinant β-lactamases to mice prevented piperacillin or piperacillin/tazobactam-induced changes in the indigenous microbiota and preserved colonization resistance against healthcare-associated pathogens (Stiefel *et al.*, 2005). This approach to the preservation of colonization resistance is currently being developed for testing in humans receiving β-lactam antibiotics.

Targeted decolonization or prophylaxis of resistant pathogens

Although horizontal approaches to infection prevention have advantages, there are situations where targeted decolonization of resistant pathogens may be preferable. The recent emergence of fluoroquinolone-resistant *E. coli* as an important cause of infection after transrectal ultrasound-guided biopsy of the prostate (TRUBP) provides one illustration. Two approaches have been successful in reducing infections due to fluoroquinolone-resistant *E. coli*. The first approach is to modify the prophylaxis for all patients undergoing TRUBP (e.g., the addition of parenteral aminoglycoside to ciprofloxacin, or the substitution of ceftriaxone for ciprofloxacin) (Dumford *et al.*, 2013). The second approach has been to screen for the presence of rectal colonization with fluoroquinolone-resistant *E. coli*, and to use targeted modification of prophylaxis only for those with such rectal colonization (Suwantarat *et al.*, 2013).

The second approach offers several advantages that are consistent with antimicrobial stewardship principles. First, only a minority of patients undergoing TRUBP are colonized with fluoroquinolone-resistant *E. coli*, and colonization is a strong predictor of infection risk (Dumford *et al.*, 2013). The targeted approach directs modification of prophylaxis only to the patients at risk, whereas the alternative approach results in the prescription of unnecessary or broader than necessary antibiotics for many low-risk patients. Such unnecessary therapy places patients at risk of adverse drug reactions and *C. difficile* infection. Secondly, routine addition of aminoglycosides may be a suboptimal approach in settings with high rates of aminoglycoside resistance in *E. coli* (Suwantarat *et al.*, 2013). Finally, nontargeted approaches may increase the risk of emergence of increasingly resistant pathogens.

Maintenance of stomach acidity

Many studies have demonstrated that PPIs are frequently prescribed in healthcare facilities without a clear indication (Sheikh-Taha *et al.*, 2012). Given the association between PPIs and multiple healthcare-associated pathogens, there is an urgent need for studies to evaluate the impact of reducing the unnecessary use of these agents on intestinal colonization by pathogens.

Conclusion

Antimicrobial selective pressure has contributed to steadily increasing rates of antimicrobial resistance among healthcare-associated pathogens. An improved understanding of the pathogenesis of colonization and of the impact of antibiotic treatment on the colonization and dissemination of resistant pathogens may contribute to more effective control strategies. Some promising antimicrobial stewardship approaches include formulary substitutions, interventions to reduce the overuse of all antibiotics or of multiple high-risk antibiotics, reductions in the duration of therapy, decolonization of pathogens with oral nonabsorbed antimicrobials, restoration or preservation of the indigenous intestinal microbiota, and targeted decolonization or prophylaxis of resistant pathogens. There is also a need for studies to explore the potential benefits of non-antimicrobial approaches, such as reductions in the unnecessary use of proton pump inhibitors.

Acknowledgment

This work was supported by the US Department of Veterans Affairs.

References

Adams, D.C., Riggs, M.M., and Donskey, C.J. (2007) Effect of fluoroquinolone treatment on growth of and toxin production by epidemic and non-epidemic *Clostridium difficile* in the cecal contents of mice. *Antimicrobial Agents and Chemotherapy* 51, 2674–2678.

Asensio, A., Oliver, A., González-Diego, P., Baquero, F., Pérez-Díaz, J.C., Ros, P., Cobo, J., Palacios, M., Lasheras, D., and Cantón, R. (2000) Outbreak of a multiresistant *Klebsiella pneumoniae* strain in an intensive care unit: antibiotic use as risk factor for colonization and infection. *Clinical Infectious Diseases* 30, 55–60.

Bhalla, A., Pultz, N.J., Ray, AJ., Hoyen, C.K., Eckstein, E.C., and Donskey, C.J. (2003) Antianaerobic antibiotic therapy promotes overgrowth of antibiotic-resistant Gram-negative *bacilli* and vancomycin-resistant enterococci in the stool of colonized patients. *Infection Control and Hospital Epidemiology* 24, 644–649.

Bhalla, A., Pultz, N.J., Gries, D.M., Ray, A.J., Eckstein, E.C., Aron, D.C., and Donskey, C.J. (2004) Acquisition of nosocomial pathogens on hands after contact with environmental surfaces near hospitalized patients. *Infection Control and Hospital Epidemiology* 25, 164–167.

Burke, J.P. (1998) Antibiotic resistance—squeezing the balloon? *JAMA (The Journal of the American Medical Association)* 280, 1270–1271.

Chastre, J., Wolff, M., Fagon, J.Y., Chevret, S., Thomas, F., Wermert, D., Clementi, E., Gonzalez, J., Jusserand, D., Asfar, P. *et al.* (2003) Comparison of 8 vs. 15 days of antibiotic therapy for ventilator-associated pneumonia in adults: a randomized trial. *JAMA (The Journal of the American Medical Association)* 290, 2588–2598.

Cole, G.T., Halawa, A.A., and Anaissie, E.J. (1996) The role of the gastrointestinal tract in hematogenous dissemination of candidiasis: from the laboratory to the bedside. *Clinical Infectious Diseases* 22(Suppl 2), S73–S88.

Cunningham, R., Mustoe, E., Spiller, L., Lewis, S., and Benjamin, N. (2014) Acidified nitrite: a host defence against colonization with *C. difficile* spores? *Journal of Hospital Infections* 86, 155–157.

Donskey, C.J. (2004) The role of the intestinal tract as a reservoir and source for transmission of nosocomial pathogens. *Clinical Infectious Diseases* 39, 219–226.

Donskey, C.J. (2006) Antibiotic regimens and intestinal colonization with antibiotic-resistant Gram-negative bacilli. *Clinical Infectious Diseases* 43(Suppl 2), S62–S69.

Donskey, C.J., Hanrahan, J.A., Hutton, R.A., and Rice, L.B. (1999) Effect of parenteral antibiotic administration on persistence of vancomycin-resistant *Enterococcus faecium* in the mouse gastrointestinal tract. *The Journal of Infectious Diseases* 180, 384–390.

Donskey, C.J., Hanrahan, J.A., Hutton, R.A., and Rice, L.B. (2000a) Effect of parenteral antibiotic administration on establishment of colonization with vancomycin-resistant *Enterococcus faecium* in the mouse gastrointestinal tract. *The Journal of Infectious Diseases* 18, 1830–1833.

Donskey, C.J., Chowdhry, T.K., Hecker, M.T., Hoyen, C.K., Hanrahan, J.A., Hujer, A.M., Hutton-Thomas, R.A., Whalen, C.C., Bonomo, R.A., and Rice, L.B. (2000b) Effect of antibiotic therapy on the density of vancomycin-resistant enterococci in the stool of colonized patients. *The New England Journal of Medicine* 343, 1925–1932.

Dumford, D., 3rd., Suwantarat, N., Bhasker, V., Kundrapu, S., Zabarsky, T.F., Draw, P., Zhu, H., and Donskey, C.J. (2013) Outbreak of fluoroquinolone-resistant *Escherichia coli* infections after transrectal ultrasound-guided biopsy of the prostate. *Infection Control and Hospital Epidemiology* 34, 269–273.

Dunbar, L.M., Wunderink, R.G., Habib, M.P., Smith, L.G., Tennenberg, A.M., Khashab, M.M., Wiesinger, B.A., Xiang, J.X., Zadeikis, N., and Kahn, J.B. (2003) High-dose, short-course levofloxacin for community-acquired pneumonia: a new treatment paradigm. *Clinical Infectious Diseases* 37, 752–760.

Edmond, M.B. and Wenzel, R.P. (2013) Screening inpatients for MRSA—case closed. *The New England Journal of Medicine* 368, 2314–2315.

Fowler, S., Webber, A., Cooper, B.S., Phimister, A., Price, K., Carter, Y., Kibbler, C.C., Simpson, A.J., and Stone, S.P. (2007) Successful use of feedback to improve antibiotic prescribing and reduce *Clostridium difficile* infection: a controlled interrupted time series. *Journal of Antimicrobial Chemotherapy* 59, 990–995.

Giel, J.L., Sorg, J.A., Sonenshein, A.L., and Zhu, J. (2010) Metabolism of bile salts in mice influences spore germination in *Clostridium difficile*. *PLoS ONE* 5(1): e8740.

Hecker, M.T., Aron, D.C., Patel, N.P., Lehman, M.K., and Donskey, C.J. (2003) Unnecessary use of antimicrobials in hospitalized patients: current patterns of misuse with an emphasis on the antianaerobic spectrum of activity. *Archives of Internal Medicine* 163, 972–978.

Kyne, L., Warny, M., Qamar, A., and Kelly, C.P. (2001) Association between antibody response to toxin A and protection against recurrent *Clostridium difficile* diarrhoea. *The Lancet* 357, 189–193.

Lawley, T.D., Clare, S., Walker, A.W., Stares, M.D., Connor, T.R., Raisen, C., Goulding, D., Rad, R., Schreiber, F., Brandt, C. *et al.* (2012) Targeted restoration of the intestinal microbiota with a simple, defined bacteriotherapy resolves relapsing *Clostridium difficile* disease in mice. *PLoS Pathogens* 8(10): e1002995.

Muto, C.A., Blank, M.K., Marsh, J.W., Vergis, E.N., O'Leary, M.M., Shutt, K.A., Pasculle, A.W., Pokrywka, M., Garcia, J.G., Posey, K. *et al.* (2007) Control of an outbreak of infection with the hypervirulent *Clostridium difficile* BI strain in a university hospital using a comprehensive 'bundle' approach. *Clinical Infectious Diseases* 45, 1266–1273.

Ng, K.M., Ferreyra, J.A., Higginbottom, S.K., Lynch, J.B., Kashyap, P.C., Gopinath, S., Naidu, N., Choudhury, B., Weimer, B.C., Monack, D.M., and Sonnenburg, J.L. (2013) Microbiota-liberated host sugars facilitate post-antibiotic expansion of enteric pathogens. *Nature* 502, 96–101.

Owens, R.C., Jr., Donskey, C.J., Gaynes, R.P., Loo, V.G., and Muto, C.A. (2008) Antimicrobial-associated risk factors for *Clostridium difficile* infection. *Clinical Infectious Diseases* 46(Suppl 1), S19–S31.

Perez, F., Pultz, M.J., Endimiani, A., Bonomo R.A., and Donskey, C.J. (2011) Effect of antibiotic treatment on establishment and elimination of intestinal colonization by KPC-producing *Klebsiella pneumoniae* in mice. *Antimicrobial Agents and Chemotherapy* 55, 2585–2589.

Peterson, J., Kaul, S., Khashab, M., Fisher, A.C., and Kahn, J.B. (2008) A double-blind, randomized comparison of levofloxacin 750 mg once-daily for five days with ciprofloxacin 400/500 mg twice-daily for 10 days for the treatment of complicated urinary tract infections and acute pyelonephritis. *Urology* 71, 17–22.

Pultz, N.J., Stiefel, U., Helfand, M.S., and Donskey, C.J. (2005) Mechanisms by which the anaerobic microflora of the colon inhibit establishment of colonization by vancomycin-resistant enterococci. *The Journal of Infectious Diseases* 191, 949–956.

Quale, J., Landman, D., Saurina, G., Atwood, E., DiTore, V., and Patel, K. (1996) Manipulation of a hospital antimicrobial formulary to control an outbreak of vancomycin-resistant *enterococci*. *Clinical Infectious Diseases* 23, 1020–1025.

Rao, A., Jump, R.L., Pultz, N.J., Pultz, M.J., and Donskey, C.J. (2006) In vitro killing of nosocomial pathogens, including *Clostridium difficile* spores, by acid and acidified nitrite. *Antimicrobial Agents and Chemotherapy* 50, 3901–3904.

Ray, A.J., Pultz, N.J., Bhalla, A., Aron, D.C., and Donskey, C.J. (2003) Coexistence of vancomycin-resistant *Enterococcus* and *Staphylococcus aureus* in the intestinal tracts of hospitalized patients. *Clinical Infectious Diseases* 37, 875–881

Reeves, A.E., Koenigsknecht, M.J., Bergin, I.L., and Young, V.B. (2012) Suppression of *Clostridium difficile* in the gastrointestinal tracts of germfree mice inoculated with a murine isolate from the family *Lachnospiraceae*. *Infection and Immunity* 80, 3786–3794.

Rice, L.B. (2008) The Maxwell Finland Lecture: for the duration—rational antibiotic administration in an era of antimicrobial resistance and *Clostridium difficile*. *Clinical Infectious Diseases* 46, 491–496.

Sethi, A.K., Al-Nassir, W., Nerandzic, M.M., and Donskey, C.J. (2009) Skin and environmental contamination with vancomycin-resistant enterococci in patients receiving oral metronidazole or oral vancomycin treatment for *Clostridium difficile*-associated disease. *Infection Control and Hospital Epidemiology* 30, 13–17.

Sethi, A.K., Al-Nassir, W.N., Nerandzic, M.M., Bobulsky, G.S., and Donskey, C.J. (2010) Persistence of skin contamination and environmental shedding of *Clostridium difficile* during and after treatment of *C. difficile* infection. *Infection Control and Hospital Epidemiology* 31, 21–27.

Sheikh-Taha, M., Alaeddine, S., and Nassif, J. (2012) Use of acid suppressive therapy in hospitalized non-critically ill patients. *The World Journal of Gastrointestinal Pharmacology and Therapeutics* 3, 93–96.

Singh, N., Rogers, P., Atwood, C.W., Wagener, M.M., and Yu, V.L. (2000) Short course empiric antibiotic therapy for patients with pulmonary infiltrates in the intensive care unit: a proposed solution for indiscriminate antibiotic prescription. *American Journal of Respiratory and Critical Care Medicine* 162, 505–511.

Stiefel, U., Pultz, N.J., Helfand, M.S., and Donskey, C.J. (2004) Efficacy of oral ramoplanin for inhibition of vancomycin-resistant *Enterococcus* intestinal colonization in mice. *Antimicrobial Agents and Chemotherapy* 48, 2144–2148.

Stiefel, U., Pultz, N.J., Harmoinen, J., Koski, P., Lindevall, K., Bonomo, A., Helfand, M.S., and Donskey, C.J. (2005) Orally administered recombinant metallo-β-lactamase preserves colonization resistance of piperacillin/tazobactam-treated mice. *Antimicrobial Agents and Chemotherapy* 49, 5190–5191.

Stiefel, U., Rao, A., Pultz, M.J., Jump, R.L., Aron, D.C., and Donskey, C.J. (2006) Suppression of gastric acid production by proton pump inhibitor treatment facilitates colonization of the large intestine by vancomycin-resistant *Enterococcus* spp. and *Klebsiella pneumoniae* in clindamycin-treated mice. *Antimicrobial Agents and Chemotherapy* 50, 3905–3907.

Sullivan, A., Edlund, C., and Nord, C.E. (2001) Effect of antimicrobial agents on the ecological balance of human microbiota. *The Lancet Infectious Diseases* 1, 101–114.

Suwantarat, N., Dumford, D.M., 3rd., Ponce-Terashima, R., Kundrapu S., Zabarsky, T.F., Zhu, H., and Donskey, C.J. (2013) Modification of antimicrobial prophylaxis based on rectal culture results to prevent fluoroquinolone-resistant *Escherichia coli* infections after prostate biopsy. *Infection Control and Hospital Epidemiology* 34, 973–976.

Theriot, C.M., Koenigsknecht, M.J., Carlson, P.E., Jr., Hatton, G.E., Nelson, A.M., Li, B., Huffnagle, G.B., Z, Li. J., and Young, V.B. (2014) Antibiotic-induced shifts in the mouse gut microbiome and metabolome increase susceptibility to *Clostridium difficile* infection. *Nature Communications* 5: 3114.

Ubeda, C., Bucci, V., Caballero, S., Djukovic, A., Toussaint, N.C., Equinda, M., Lipuma, L., Ling, L., Gobourne, A., No, D. *et al.* (2013) Intestinal microbiota containing *Barnesiella* species cures vancomycin-resistant *Enterococcus faecium* colonization. *Infection and Immunity* 81, 965–973.

Valiquette, L., Cossette, B., Garant, M.P., Diab, H., and Pépin, J. (2007) Impact of a reduction in the use of high-risk antibiotics on the course of an epidemic of *Clostridium difficile*-associated disease caused by the hypervirulent NAP1/027 strain. *Clinical Infectious Diseases* 45(Suppl 2), S112–S121.

van Nood, E., Vrieze, A., Nieuwdorp, M., Fuentes, S., Zoetendal, E.G., de Vos, W.M., Visser, C.E., Kuijper, E.J., Bartelsman, J.F., Tijssen, J.G. *et al.* (2013) Duodenal infusion of donor feces for recurrent *Clostridium difficile*. *The New England Journal of Medicine* 368, 407–415.

Vesper, B.J., Jawdi, A., Altman, K.W., Haines, G.K., 3rd., Tao, L., and Radosevich, J.A. (2009) The effect of proton pump inhibitors on the human microbiota. *Current Drug Metabolism* 10, 84–89.

Vollaard, E.J. and Clasener, H.A. (1994) Colonization resistance. *Antimicrobial Agents and Chemotherapy* 38, 409–414.

Wiener, J., Quinn, J.P., Bradford, P.A., Goering, R.V., Nathan, C., Bush, K., and Weinstein, R.A. (1999) Multiple antibiotic-resistant *Klebsiella* and *Escherichia coli* in nursing homes. *JAMA (The Journal of the American Medical Association)* 281, 517–523.

Wilson, K.H. and Perini, F. (1988) Role of competition for nutrients in suppression of *Clostridium difficile* by the colonic microflora. *Infection and Immunity* 56, 2610–2614.

8 Antibiotic Resistance: Associations and Implications for Antibiotic Usage Strategies to Control Multiresistant Bacteria

Louis B. Rice*

The Warren Alpert School of Medicine of Brown University, Rhode Island Hospital, Providence, Rhode Island, US

Introduction

Pathogenic bacteria employ a wide range of mechanisms to resist the toxic effects of the many antimicrobial agents that we use to treat infected patients. We now have extensive knowledge of the molecular and biochemical mechanisms of resistance for virtually all resistance phenotypes encountered in the clinical setting. In general terms, antimicrobial resistance may be intrinsic or acquired. Acquired resistance may be mutational or through the acquisition of foreign genes. Acquired genes may be single or multiple, and the expression of resistance may be due to the effects of one gene or several. In this chapter, I will discuss the emergence and spread of antimicrobial resistance in the context of antimicrobial use. I will express skepticism about our ability to control resistance through the clever use of antibiotics. I will promote the notion that the best way to reduce resistance is to reduce overall antibiotic exposure, and that the safest way to reduce antimicrobial exposure is to shorten courses of therapy as much as possible. By so doing, it is hoped that our antimicrobial stewardship strategies will be better able to counteract the root causes of resistance.

Antimicrobial usage strategies employed to try to prevent resistance fall into two general categories. One group of strategies favors antimicrobial manipulation, whereas the other group targets reductions in antibiotic use. Examples of the first group of strategies include the use of multiple agents to prevent the emergence of resistance (exemplified by the strategies for treating tuberculosis), the use of dosing strategies to prevent resistance (the mutant prevention concentration), and the avoidance of specific classes of antibiotics (such as cephalosporins or fluoroquinolones) with substitution by another class that presumably does not select for the resistance of interest. Examples of the second group of strategies include prior approval and restriction strategies, "narrowing down" therapy when a pathogen is identified, and reducing the lengths of therapy. In this chapter, I will argue that the safest and most effective strategy for controlling resistance is the last one.

Concepts and Misconceptions About Antimicrobial Resistance

Misconception 1: Induction of resistance is an important mechanism to consider in developing antimicrobial strategies

A major misconception among many who use antibiotics is that exposure to antibiotics commonly "induces" resistance to those antibiotics (for a more thorough discussion of induction and selection, see Chapter 6 by Rafael Araos *et al.*). While the induction of resistant strains does occur, and in some cases is

*E-mail: lrice@lifespan.org

of clinical importance, it is far less common than the "selection" of such strains. Induction refers to a regulatory phenomenon in which resistance traits are only expressed in response to the presence of the antibiotic. This expression may occur at high levels, but should become negligible once the antibiotic is removed from the medium. Examples of induction that are commonly cited include the induction of the chromosomal β-lactamase of *Enterobacter* spp. by cefoxitin or clavulanic acid (Sanders and Sanders, 1988), the induction of the *Pseudomonas aeruginosa* chromosomal β-lactamase by imipenem (Livermore, 1992), the induction of the *Staphylococcus aureus* β-lactamase by β-lactam antibiotics (Hackbarth and Chambers, 1993), the induction of enterococcal glycopeptide resistance by exposure to vancomycin (Arthur *et al.*, 1992), and the induction of MLS_B (macrolides, lincosamides, and streptogramin B)-type resistance in streptococci and staphylococci by exposure to the macrolide erythromycin (Horinouchi and Weisblum, 1980).

The recognition of inducible resistance mechanisms has, over the years, inspired the development of antimicrobial agents whose activity against resistant strains hinges on the fact that they are poor inducers of the resistance mechanism. Examples include the development of extended-spectrum cephalosporins, which are poor inducers of the *Enterobacter* β-lactamase, the use of teicoplanin, which is a poor inducer of the VanB vancomycin resistance determinant, or the use of clindamycin, a poor inducer of the MLS_B resistance determinant. Despite its intellectual appeal, the efficacy of strategies focused on the use of noninducing antibiotics against bacteria with inducible resistance has been disappointing. The problem with focusing on induction per se is that regulatory mechanisms can often be overcome by amino acid changes resulting from single-point mutations in affected genes. Consequently, the use of noninducing antibiotics in the clinical setting can be associated with the selection of constitutively expressing mutants that are now highly resistant to the noninducing antibiotics.

The most compelling example of this phenomenon is the development of extended-spectrum cephalosporins (cefotaxime, ceftizoxime, ceftriaxone). These agents were considered major advances at the time of their development, to a large extent because they were not inducers of the *Enterobacter* chromosomal β-lactamase (an AmpC, or ampicillin C-hydrolyzing β-lactamase), and hence were active in vitro and in vivo against these problematic pathogens. Unfortunately, they proved to be quite effective selectors of the rare (ca. 10^{-9}/CFU) mutants in which the regulatory mechanism had been bypassed by a single point mutation in the *ampD* gene, resulting in high-level constitutive expression (also referred to as stable derepression) of AmpC (Jacobs *et al.*, 1994, 1995, 1997). Because extended-spectrum cephalosporins are quite susceptible to hydrolysis by AmpC, the resulting mutants are highly resistant to most β-lactam antibiotics. Clinical use of extended-spectrum cephalosporins to treat infections caused by *Enterobacter* spp. was soon associated with the emergence of resistant *Enterobacter* strains (Chow *et al.*, 1991), leading to the present recommendation that these agents be avoided when treating high-inoculum infections caused by *Enterobacter* spp. Similar selection of constitutively expressing strains has been reported with the clinical use of teicoplanin to treat Van B-type vancomycin-resistant enterococci (Kaatz *et al.*, 1990) and the use of clindamycin to treat erythromycin-resistant staphylococci and streptococci (Siberry *et al.*, 2003).

Concept 1: Most resistance results from selection rather than induction

In most instances, the emergence and spread of antimicrobial resistance is a natural selection process. The *Merriam-Webster Collegiate Dictionary* defines natural selection as: "a natural process that results in the survival and reproductive success of individuals or groups best adjusted to their environment and that leads to the perpetuation of genetic qualities best suited to that particular environment." In the case of bacteria, antibiotic resistance offers a selective advantage for reproduction to those bacteria that are resistant to the inhibitory or bactericidal activity of an antibiotic-rich environment. The selection of naturally resistant species, of preexisting resistant mutants, or of strains that have acquired resistance genes are very common mechanisms by which resistance emerges in the clinical setting. For example, *Enterococcus faecium* is intrinsically resistant to cephalosporins (Williamson *et al.*, 1983). This is a known characteristic of the genus and, as such, cephalosporins are not commonly used to treat enterococcal infections. However, as cephalosporins are highly effective agents against a wide variety of pathogenic bacteria, they are frequently used in the clinical setting. Cephalosporin administration often yields an environment in the gastrointestinal (GI) tract that is toxic to many resident bacteria, and thus favorable for proliferation of enterococci. Because of this, clinical reports appear with some frequency linking the use of cephalosporins with the emergence of GI colonization or clinical infection with resistant enterococci (Bonten *et al.*, 1998a).

The selection of a preexisting subpopulation of resistant mutants is most starkly demonstrated by the example of rifampin, one of our most active and rapidly bactericidal antibiotics. Resistance to rifampin occurs through single-point mutations in the *rpoB* gene, which encodes bacterial RNA polymerase (Wehrli, 1983). In most species, functional mutations in *rpoB* that confer high levels of resistance to rifampin will be present at a rate of 10^{-9}/CFU. Consequently, populations that exceed 10^9 will predictably contain mutants resistant to rifampin. For example, an active tuberculosis cavity may contain as many as 10^{12} tubercle bacilli. Treatment with rifampin alone will, therefore, predictably select out the resistant subpopulation, which will soon become the dominant population. It is for this reason that single-agent therapy with rifampin is not recommended for the treatment of tuberculosis.

Another class of antimicrobial agent for whom point mutations in genes encoding targets are important is the fluoroquinolones. These agents act by interfering with cellular topoisomerases. Point mutations in the genes encoding these topoisomerases may result in amino acid changes in these enzymes that reduce their fluoroquinolone binding affinity and result in reduced susceptibility (Martinez *et al.*, 1998). In contrast to rifampin, single topoisomerase mutations confer only a modest level of resistance, depending on whether the mutated gene encoding the target enzyme is the primary target of the fluoroquinolone in question or not (Martinez *et al.*, 1998). Multiple mutations are generally required to confer high levels of resistance (Hooper, 1998).

Misconception 2: Targeting the "mutant prevention concentration" is a viable strategy for reducing the problem of resistance

The relatively modest increments in minimum inhibitory concentration (MIC) associated with single topoisomerase point mutations have led to the popularity of the concept of the "mutant prevention concentration" or MPC (Drlica and Zhao, 2007). The logic behind identifying the MPC is to administer a dose of fluoroquinolone that will result in local fluoroquinolone concentrations that exceed the highest MIC conferred by a first-order mutant with a single-point mutation. By doing this, the growth of mutant strains can be inhibited and thereby the emergence of resistance prevented. The MPC is a logical and compelling theory and has been shown to be quite effective in animal models of infection treated by fluoroquinolones. Yet fluoroquinolone resistance is now widespread. Why?

There are several reasons why the MPC concept has not prevented the emergence of resistance to fluoroquinolones. The first and most obvious is that the vast majority of fluoroquinolone prescriptions are not designed with the concept of suppressing resistance in mind. Standard doses are administered to a variety of individuals, with a variety of weights, renal functions, and metabolisms, and with infections in a variety of body sites. As a result, the target concentrations for suppressing mutants may not have been achieved in many cases. The second reason is that we now understand that there are more "auxiliary" mechanisms of resistance to fluoroquinolones than we appreciated when we first started using these agents (Strahilevitz *et al.*, 2009). These mechanisms include the expression of efflux pumps, some intrinsic and some acquired, the acquisition of genes that confer protection to the topoisomerases, and the acquisition of a modifying gene that inactivates ciprofloxacin and norfloxacin. The effect of these mechanisms is to increase the "baseline" MIC of the susceptible organism, a fact that will not be appreciated in the clinical setting because the information available to the clinician is not sufficiently detailed to make that judgment (Rice, 2012). In the presence of these auxiliary mechanisms, the impact of a single-point mutation may be to increase the MIC to beyond what had been identified as the MPC of a susceptible strain. The final reason is that the problem of resistance in the hospital setting is, in actuality, only uncommonly associated with the emergence of resistance in the infecting strain at the time of the infection. Resistance emerges and is acquired elsewhere, in places like the GI tract, where our ability to control the concentrations of antibiotic is limited, and our ability to predict or control the flora is minimal.

Concept 2a: Emergence of resistance frequently occurs at sites distant to and in bacteria unrelated to those causing the infection being treated

As we discuss the emergence and spread of antimicrobial resistance, we must first acknowledge that antibiotics given to patients do not target only the microorganisms at the site of infection. Antibiotics that achieve significant clinical use are selected for their broad volume of distribution, and they inhibit and kill in an indiscriminate manner. As a result, areas of the body colonized by microorganisms unrelated to any disease process will experience selective pressure similar to that experienced at the site of infection. These areas include the skin, the oral and nasal cavities, the upper respiratory tract, the upper and lower GI tract, the perineum, and the urinary tract, among others. Depending on the specific antimicrobial agent being used, there may be subinhibitory or superinhibitory concentrations found in many of these locations. Subinhibitory or barely inhibitory concentrations may select out resistant mutants of the resident bacteria, while superinhibitory concentrations will reduce the normal population and provide an environment where naturally resistant or newly resistant bacteria can thrive.

One salient example of such a phenomenon is the emergence of resistance to ciprofloxacin in staphylococci. When ciprofloxacin was first introduced into clinical use, it was truthfully marketed as the first effective oral therapy for the treatment of methicillin-resistant *S. aureus* infections (MRSA). However, within 1 year of its clinical use, medical centers were reporting dramatic increases in the rates of ciprofloxacin resistance in MRSA strains (Blumberg *et al.*, 1991). In most cases, this resistance was due to the accumulation of point mutations in the cellular topoisomerase genes, resulting in amino acid changes that lowered the affinity for the fluoroquinolone. The rapidity with which skin or nasal colonization with resistant staphylococci occurred in one clinical study was striking, and consistent with the rapid emergence observed in the clinical setting (Kotilainen *et al.*, 1990). Unlike many antibiotics, the fluoroquinolones permeate many regions of the body, including the skin and mucous membranes. Yet the concentrations of antibiotics achieved in these regions are difficult to predict, and they are very possibly at a level that could promote selection of first-order mutants. These areas of typical staphylococcal colonization were, therefore, turned into natural selection regions in patients treated by ciprofloxacin or other fluoroquinolones. Subsequent volunteer studies documented the rapid emergence of fluoroquinolone resistance in viridans streptococcal strains colonizing the human pharynx after the experimental administration of levofloxacin (Fantin *et al.*, 2009).

Another of the major locations for emergence of resistance is the human GI tract, which is routinely inhabited by large numbers of diverse types of bacteria. The bulk of these bacteria are anaerobes and are not (to our knowledge) pathogenic to humans. They are, in fact, beneficial to us in that they maintain the homeostasis of our digestive system, helping us digest different kinds of foods and to manufacture and absorb important vitamins. Increasing evidence also suggests that our microflora is critical to the normal development of the immune system (Purchiaroni *et al.*, 2013). Some antibiotics achieve very high concentrations in the human GI tract after intravenous administration. Ceftriaxone, for example, achieves concentrations as high as 5000 μg/ml in the bile after routine dosing (Hayton *et al.*, 1986). Such concentrations are more than sufficient to suppress the growth of many GI colonizers, but they do not suppress the growth of ampicillin-resistant *E. faecium*, whose ceftriaxone MICs can exceed 10,000 μg/ml (Donskey *et al.*, 1999b). Early studies associated ampicillin-resistant *E. faecium* GI colonization with the administration of extended-spectrum cephalosporins, and in some hospitals these pathogens became prevalent coincident with the increased use of these agents (Grayson *et al.*, 1991).

The emergence of vancomycin resistance in *E. faecium* presents an interesting counterpoint to the ampicillin resistance story. Vancomycin was introduced into clinical use in the late 1950s as a treatment for penicillin-resistant *S. aureus*, which was becoming an increasingly prevalent nosocomial problem. After nearly 30 years of use, no resistance to vancomycin had been reported in either staphylococci or enterococci, but in the late 1980s, reports emerged from Europe of the isolation of enterococcal strains expressing high-level resistance to vancomycin (Shlaes *et al.*, 1989a,b). Two major determinants (VanA and VanB) were described, both of which were transferable between enterococcal strains, suggesting that they were acquired determinants. In Europe, the emergence of these strains was attributed to the use of avoparcin, a vancomycin analogue, to promote growth in food animals (Woodford, 1998). In support of this epidemiology was the identification of vancomycin-resistant enterococci in the feces of food animals, in food items sold in grocery stores, and in the feces of community dwellers (Woodford, 1998). There was logic to the association: vancomycin administered intravenously achieved negligible concentrations in the GI tract, at least for the first several days, and so had not historically exerted significant selective pressure favoring the emergence of resistance in GI colonizers. The poor oral absorption of avoparcin administered to animals by mouth guaranteed very high concentrations in the GI tract, leading to the colonization of food animals by naturally resistant strains. Enterococci, which are tolerant to the action of glycopeptides such as avoparcin or vancomycin, were inhibited but not killed and, therefore, were available to acquire resistance determinants from naturally resistant bacteria. The fact that these determinants were present on transferable elements promoted the acquisition. Curiously, though, rates of infection by these bacteria in European hospitals were negligible in the early years (Woodford, 1998).

The emergence of vancomycin-resistant enterococci in the US followed a very different pattern. Avoparcin has never been licensed in the US, and when animals in the US were tested for colonization by vancomycin-resistant enterococci, none were found (Coque *et al.*, 1996). Similarly, vancomycin-resistant enterococci were not found colonizing community dwellers in the US (Coque *et al.*, 1996). However, in contrast to Europe, vancomycin-resistant enterococci quickly became important pathogens in US hospitals, particularly in immunocompromised patients (Vergis *et al.*, 2001). Also in contrast to Europe, US vancomycin-resistant enterococci expressed very high levels of resistance to ampicillin (Centers for Disease Control and Prevention, 1993; Descheemaeker *et al.*, 1999). Clinical studies began to associate colonization and infection by vancomycin-resistant enterococci with cephalosporin use (Moreno *et al.*, 1995; Bonten *et al.*, 1998b). Eventually, it was recognized that in the US, vancomycin-resistant enterococcal strains emerged from a clonal complex that had not only acquired high-level ampicillin resistance, but had also acquired putative virulence determinants that promoted infection in hospitalized patients (Top *et al.*, 2008). These strains had become prevalent in US hospitals in association with increased use of extended-spectrum cephalosporins, and so were available to acquire mobile vancomycin-resistance determinants when selective pressure was applied by oral vancomycin treatment of *Clostridium difficile* infections in the 1980s and 1990s (Bartlett *et al.*, 1978). The differences in these two epidemiologies reinforces two points: that the mammalian GI tract is an important site for the emergence of resistance in enterococci; and that the impact of antimicrobial administration on the emergence of resistant strains is often unpredictable.

The impact of cephalosporins on colonization by ampicillin- and vancomycin-resistant *E. faecium* strains was examined in a series of animal studies using a murine model of enterococcal colonization (Donskey *et al.*, 1999a, 2000b; Rice *et al.*, 2004; Lakticova *et al.*, 2006). These studies showed that cephalosporins such as ceftriaxone promoted the establishment of high-level colonization of the mouse GI tract by *E. faecium* after the introduction of minimal numbers of organisms to the stomach, but that ceftriaxone did not promote the persistence of high-level colonization over time. The persistence of high-level colonization was promoted by the administration of agents with potent activity against anaerobic bacteria. Agents with poor enterococcal activity in the upper GI tract and potent activity against anaerobic bacteria (such as cefotetan and clindamycin) promoted both the establishment and persistence of enterococcal colonization. Subsequent animal studies confirmed that it is the activity (or lack of activity) of agents against enterococci in the upper GI tract that is responsible for the establishment of colonization. Human studies confirmed the association of the administration of agents with potent anti-anaerobic activity with increases in the concentrations of enterococci in the feces (Donskey *et al.*, 2000a). As such, it is not possible to come up with a "perfect" antibiotic to minimize resistant enterococcal colonization, although the frequent association of extended-spectrum cephalosporins in clinical studies suggests that predisposing to the establishment of colonization is more important than persistent high-level colonization in perpetuating vancomycin-resistant enterococcal outbreaks in the hospital.

Concept 2b: There is always a last dose

In discussing antimicrobial dosage strategies designed to minimize the emergence of resistance, considerable attention has been focused on developing regimens designed to maintain inhibitory concentrations throughout the dosing interval. In this manner, a "mutant selection window" is avoided. As noted above, such strategies have been developed for dosing fluoroquinolones, and these strategies have proven quite effective in animal studies of infection. Unfortunately, the translation of these studies to the clinical setting is problematic, primarily because there will always be a last dose. After the last dose, local concentrations of antibiotic will decay at a defined rate. During that decay, the concentrations will pass through the mutant selection window, during which time first-order resistant mutants will have a selective advantage. This concept was elegantly demonstrated by Fantin *et al.* (2009), who followed the feces of volunteers administered ciprofloxacin for 42 days after a 14 day course of therapy. They did not observe the emergence of fluoroquinolone-resistant strains during the treatment interval (when the concentrations of ciprofloxacin in the feces were at suprainhibitory levels), but did observe the emergence of these strains in the post-therapy period, during which time the fecal concentrations of ciprofloxacin declined. Follow-up studies showed that the resistant strains that emerged were not detectable at the start of therapy, suggesting either that they were present in numbers too small to detect or that they were acquired during therapy (de Lastours *et al.*, 2012).

Misconception 3: Preclinical studies demonstrating a low frequency of resistance emergence are reliable predictors of what will emerge in the clinical setting

One of the requirements for drug licensure is to perform studies determining the rate of spontaneous resistance to the antibiotic under development. These studies generally consist of plating large inocula of bacteria onto fixed concentrations of the antibiotic and determining how many colonies grow. Studies may also involve exposure to sequentially higher concentrations of the antibiotic, in an effort to determine whether a stepwise emergence of resistance is likely to occur. These studies are very useful in that they are effective at predicting that resistance *will* emerge in the clinical setting, but they cannot be used to predict that resistance *will not* emerge. In other words, if resistance emerges at a moderate or high rate in such studies, it will almost certainly emerge in the clinical setting. However, if resistance does not emerge in vitro, one cannot safely predict that it will not emerge in the clinical setting because these experiments shed no light on whether preexisting determinants conferring resistance to the new antibiotic are already present in the environment.

Concept 3: Acquired resistance genes are readily available

There are many antibiotics for which neither alterations in the regulation of intrinsic genes nor point mutations can achieve meaningful levels of resistance. Resistance to β-lactam antibiotics in *S. aureus* or *Escherichia coli*, for example, is only very uncommonly the result of point mutations in the target penicillin-binding proteins. While small increments in vancomycin MICs can be achieved through chronic exposure of staphylococci to vancomycin, such resistance is clinically a rare event and still something of a mechanistic mystery (van Hal and Paterson, 2011). Resistance to such antibiotics can, therefore, be best achieved through the acquisition of exogenous DNA. The mechanisms by which DNA is acquired are many and varied, and depend, to some degree, on the species. *S. aureus*, for example, is known to be a target of a large number of bacteriophages. It has been supposed for years that transduction (the transfer of DNA through the action of bacteriophages) is a major mechanism of DNA transfer in staphylococci, yet only recently has the transduction of methicillin resistance been demonstrated experimentally (Scharn *et al.*, 2013). One recent study identified a large number of antibiotic resistance genes (including the penicillin resistance *mec* genes and a variety of β-lactamase genes) within fecal bacteriophages (Quiros *et al.*, 2014). *Enterobacteriaciae* such as *E. coli* employ a wide variety of conjugative plasmids (extrachromosomal, independently replicating DNA) to transfer a multitude of genes. Enterococci employ plasmids as well as conjugative transposons—members of a larger group of elements known as ICEs (integrating conjugative elements)—to transfer tetracycline resistance, vancomycin resistance, and other genetic material. Naturally transformable (meaning having the ability to absorb DNA directly from the environment) species such as *Streptococcus pneumoniae*, viridans streptococci, and *Neisseria* spp. achieve resistance to β-lactam antibiotics through the formation of mosaic genes created by splicing sections of PBP (penicillin-binding protein) genes from less susceptible species into their own PBP genes via homologous recombination (Dowson *et al.*, 1994). As most antimicrobial agents are natural products, it stands to reason that there will be determinants for resistance in the environment employed either by the antibiotic producers themselves or by species that need to live in close proximity to these producers. Pathogenic bacteria have proven quite adept at finding and assimilating these genes.

Precise Control Over Antimicrobial Exposures in the Clinical Setting Is an Illusion

If we cannot fine-tune the exposure to antibiotics in a controlled setting of a clinical study using volunteers (see the above description of studies by Fantin *et al.* (2009) under Concept 2b), what hope do we have of fine-tuning antimicrobial exposures in the clinical and community settings? After all, most antibiotics sold in the US are not even administered to humans, but are given to animals and spread elsewhere in the agricultural sector. Many antibiotic prescriptions are administered by patients themselves, in the community setting, where we have little if any control over the timing and quantity of dosing. Even in the hospital setting, our dosing rarely takes into account specific pharmacokinetics, in particular the increased clearance associated with many critical illnesses. Moreover, antibiotics are frequently switched with no consideration for how the timing of such switching will affect the emergence and spread of resistance. In many cases, multiple antibiotics are administered when a single agent would suffice, and courses of therapy are, in general, considerably longer than emerging evidence suggests that they need to be to treat infection (Rice, 2008).

Practical Implications for Stewardship Strategies Focusing on Antimicrobial Manipulation

Combination antibiotics to prevent resistance

The use of antituberculous agents in combination has been an effective tool to prevent the emergence of resistance during therapy for two simple reasons. The first is that *Mycobacterium tuberculosis*, as far as we know, does not exchange DNA with other species. As such, the only mechanism of achieving resistance to the commonly used antibiotics is the selection of strains with point mutations in genes conferring resistance. As the rate of these point mutations can be determined experimentally, it becomes a simple mathematics experiment to determine that the *M. tuberculosis* population in the typical lung cavity (ca. 10^{12} CFU) contains mutants resistant to either isoniazid (mutant rate ca. 10^{-8}) (David, 1970) or rifampin (mutant rate ca. 10^{-10}) (David, 1970), but will not (in the absence of a preexisting resistant strain) contain organisms expressing resistance to both agents ($10^{-8} \times 10^{-10} = 10^{-18}$). The second reason is that the common agents used to treat tuberculosis are not frequently used to treat other infections. Hence, we have no reason to monitor whether resistance is emerging in other bacteria elsewhere in the body. If one had to pick species–antimicrobial combinations for which this strategy will work, *M. tuberculosis* and isoniazid–rifampin would top the list.

Yet, multidrug-resistant *M. tuberculosis* is a major problem worldwide. This seeming contradiction results not from a flaw in the design of antituberculous therapy, but rather a flaw in the execution. Tuberculosis requires prolonged therapy with multiple antibiotics. In settings where patients are left to their own devices to administer the therapy, compliance with the regimen can be sporadic. It is for this reason that directly observed therapy has been recommended for patients that are being treated (Lauzardo and Peloquin, 2012). Problems become amplified if a resistant strain emerges, because the recognition of resistance is sometimes delayed, leading to what is essentially monotherapy, which will promote resistance to the active antimicrobial. Furthermore, a failure to adequately address infection control precautions (as happened in US hospitals early in the HIV era) can result in substantial transmission before the problem is recognized (Edlin *et al.*, 1992).

Given the complexity of executing a combination strategy against resistance in tuberculosis, it should come as no surprise that there are no good data to support the effectiveness of combination therapy in the prevention of resistance in bacteria that commonly infect hospitalized patients. The use of agents for multiple infections, caused by multiple pathogens, in multiple locations, administered by different routes, and the fact that the antimicrobial agents we use are broad in their spectrum and highly active against many commensal bacteria, virtually guarantee off-target effects. Moreover, the existence of mobile genetic elements that may confer resistance to a variety of agents means that combining antibiotics is as likely to augment selective pressure as it is to reduce it.

Pharmacodynamic approaches to preventing antimicrobial resistance

The science of antimicrobial pharmacodynamics has advanced substantially over the past two decades and has provided insights into optimal regimens for treating a variety of infections. These insights involve dosing strategies that take advantage of pharmacodynamic parameters that have been shown experimentally to optimize the killing of bacteria in vitro and in vivo. For example, understanding the concept of concentration-dependent killing resurrected the antibiotic daptomycin by suggesting that higher doses at less frequent intervals would improve killing while minimizing muscle toxicity (Tally *et al.*, 1999). The same concept allowed aminoglycosides to be administered more safely once a day with equal effectiveness (Nicolau *et al.*, 1995). The fact that β-lactam antibiotics exhibit time-dependent killing has helped to define optimal dosing regimens for these agents (Hope *et al.*, 2012). Monte Carlo simulations offer a relatively inexpensive way to determine the likelihood of target attainment in a population, given predetermined pharmacodynamic parameters, and so have helped pharmaceutical manufacturers determine the optimal dosing for clinical trials (Lee *et al.*, 2010).

Unfortunately, for reasons delineated earlier in this chapter, pharmacodynamic analysis has not been as useful in devising strategies for preventing resistance. Resistance occurs "around the edges" in both location and time, and it is not possible to control all of these antimicrobial exposures. Pharmacodynamic analysis has suggested a useful principle in one recent study: that prolonged courses of therapy offer a greater opportunity for the overgrowth of resistant subpopulations compared with susceptible populations (Drusano *et al.*, 2009). As such, shortening courses of therapy will be predicted to result in less resistance.

"Switching" strategies

The medical literature is replete with quasi-experimental studies documenting the alleviation of a resistance outbreak after switching the predominant antibiotic used to a different class. Among the more common of these stories is the reduction of prevalence of extended-spectrum β–lactamase-producing *Klebsiella pneumoniae* in hospitals, or in individual units, after reducing the use of extended-spectrum cephalosporins and replacing them with a non-cephalosporin, such as imipenem or piperacillin–tazobactam (Rice *et al.*, 1990; Meyer *et al.*, 1993). The logic behind the strategy is compelling—mutations in narrow-spectrum enzymes that confer resistance to extended-spectrum cephalosporins do not confer resistance to carbapenems or, in most cases, to β-lactamase inhibitors. The short-term results are often good—it appears that simply reducing the selective pressure exerted by extended-spectrum cephalosporins favors the reemergence of the narrow-spectrum enzymes. Unfortunately, the alternative antibiotics often come with their own disadvantages, particularly the carbapenems. The emergence and spread of carbapenem-resistant *K. pneumoniae* is now a significant problem in the US and around the world. Several different carbapenemases, of different classes, are now found worldwide in *K. pneumoniae* strains, portending a time when strains that are untreatable may become commonplace (Munoz-Price *et al.*, 2013). Perhaps the best depiction of the consequences associated with exchanging the overuse of one antibiotic with the overuse of another came from the late James Rahal, who described sequential outbreaks of resistant strains that followed the replacement of ceftazidime by cefotetan, then by imipenem (Rahal *et al.*, 2002). In each case, mechanisms of resistance emerged that compromised the efficacy of the "new" antibiotic, without reestablishing the efficacy of the former agent.

A systematic, prophylactic switching strategy that has been tested in several institutions has been antibiotic cycling. Cycling involves the systematic alternating of empiric antibiotic therapy between different agents in an effort to avoid emergence and spread or resistance to any one of them. Such strategies have generally been employed only within specific units within the hospital. For example, an intensive care unit (ICU) may designate ceftazidime as the empiric antibiotic of choice for a 3 month period, followed by a 3 month period in which imipenem is the preferred agent, followed by piperacillin–tazobactam and so on. There are several problems with such strategies. The first is a practical one—it is difficult to obtain physician buy-in, especially because many physicians have personal preferences for specific antimicrobial agents. Compliance with the recommended protocols has often been no better than 50% in most clinical studies. The deeper problem is that the selective pressure created by some broad-spectrum antibiotics results in the emergence of resistance to other antibiotics. As detailed above, cephalosporin use is associated with the emergence of vancomycin-resistant enterococci. Ciprofloxacin resistance in *P. aeruginosa* can be selected by exposure to imipenem. Reductions in the prevalence of resistance have not been universally seen with cycling protocols (Kollef, 2006), and in at least one instance a cycling protocol was associated with an outbreak of resistant *P. aeruginosa* (Hedrick *et al.*, 2008).

Toward a Coherent and Simple Antimicrobial Stewardship Strategy That Will Minimize Antimicrobial Resistance

There are a variety of reasons to promote the concept of antimicrobial stewardship. One is to control the costs of antibiotics. As payment systems for medical care change, US institutions will have a stronger motivation to control the costs of care. Pharmacy costs are among the fastest growing costs to medical institutions. Traditional programs designed to reduce the costs of antimicrobials have included prior approval programs for certain "expensive" antimicrobials, automatic substitutions of less expensive antibiotics, automatic stop orders, and limiting data reporting on expensive antibiotics. While some of these measures have the effect of lowering costs, they also effectively come between the physicians and his or her patient at the time of acute illness. As such, they risk alienating practitioners and compromising the care of sick patients.

A second reason to promote stewardship programs is to optimize the chance of therapeutic success for antimicrobial courses. This can best be done by developing rapid diagnostic tests to determine the causative organisms of the infection and their susceptibility to antimicrobials, and pharmacy involvement to insure optimal dosing. We are presently entering into an era of unprecedented advancement in the ability to rapidly identify the causative organisms in bacterial infections (Caliendo *et al.*, 2013). We are also arguably entering into an era of multiresistant pathogens and a dearth of therapeutic options. It appears likely that new therapeutics designed to address these multiresistant pathogens, if they can be developed, will be highly specific in the pathogens that they will treat (i.e., they may be able to treat a multiresistant *P. aeruginosa*, but not a multiresistant *Acinetobacter baumanni*) and will also much more expensive than traditional antibiotics. Therapeutic success will depend upon the rapid identification of the pathogen, while financial viability will depend upon using the new and more expensive antimicrobial agent only when it is necessary. Both of these factors argue for the need for reliable and accurate rapid diagnostic tests and, equally important, thorough understandings of what the tests are capable of and how best to use them. In addition, the availability of well-trained pharmacists to help with antimicrobial selection and optimize dosing regimens will be integral to success.

The third important reason to promote stewardship programs is to minimize the selective pressure favoring the emergence of resistance. As described in the earlier parts of this chapter, the emergence and spread of resistance is a complex matter. It is unlikely that we will be able to develop strategies in which the way in which we use a large amount of antibiotics will minimize resistance, because it is the use of antibiotics that promotes resistance. More antibiotic use is never the solution to resistance.

Hence, any strategy designed to reduce the emergence and spread of resistance must reduce the use of antibiotics. This cannot be accomplished by reducing antibiotic dosing, because that strategy is more likely to promote, rather than reduce, resistance. Reducing overall antimicrobial selective pressure while maintaining optimized dosing regimens can, therefore, be done in only one of three ways: we could administer antibiotics to only those patients that need them; we could avoid administering multiple antibiotics when a single antibiotic will suffice; or we could administer antibiotics only for as long as is absolutely necessary.

Withholding antibiotics from those who do not need them is a theoretically appealing strategy, but its implementation creates a significant practical problem. There are ample data in the medical literature to indicate that failure to institute effective antimicrobial therapy early in the course of serious infections is associated with worse outcomes (Alvarez-Lerma and ICU-Acquired Pneumonia Study Group, 1996; Luna *et al.*, 1997; Rello *et al.*, 1997; Leibovici *et al.*, 1998; Ibrahim *et al.*, 2000). Those data virtually mandate that patients be given antibiotics if infection is in the differential diagnosis (and infection is almost always in the differential diagnosis for patients admitted to the modern hospital). The same problem exists for strategies focused on not giving multiple antibiotics. Resistance is so prevalent, even in the community, that treatment with single agents often does not cover the range of suspected pathogens. In order to be certain that potential pathogens are covered by an empiric regimen, it is often required that more than one agent be used.

Pursuing a strategy based on administering antimicrobial agents for only as long as necessary avoids the problems associated with empirically withholding therapy, but it assumes several things. The first is that current antimicrobial regimens are longer than they need to be. There are compelling data on optimal lengths of therapy for only relatively few infections. It is well established that uncomplicated urinary tract infections in otherwise healthy young females can be treated effectively with a single dose of an active antibiotic in the majority of cases (Gupta *et al.*, 2011). Endocarditis due to highly susceptible viridans streptococci can be treated for 2 weeks if penicillin is combined with streptomycin (Wilson *et al.*, 1984). Several sexually transmitted diseases can be effectively treated with a single dose regimen (Workowski and Bolan, 2015). However, durations of therapy for the majority of infections treated in the hospital, where the problems of resistance are most acute, have never been established by credible trials. Our recommended regimens of 7 days, 10 days, 14 days, etc., are obviously based more on our traditional divisions of time than on empiric data. It is therefore safe to assume that therapy can be shortened in a substantial number of infections.

The second reason for pursuing a strategy based on administering antimicrobial agents only for as long as necessary is that prolonged courses of therapy are associated with increases in resistance. There are several levels of evidence that suggest this is the case. In vitro, resistance to specific antibiotics can often be selected with a single-step selection. When it cannot, a multistep selection (the in vitro correlate of prolonged dosing) will often select for resistance. In vitro pharmacodynamic studies indicate that shorter courses of therapy are less likely to be associated with the emergence of resistant subpopulations than are longer courses (Drusano *et al.*, 2009). Two clinical studies examining the effectiveness of shorter courses compared with longer courses implicated longer courses in the emergence of resistance. Singh *et al.* (2000) compared 3 day therapy with ciprofloxacin with the physician choice therapy (median 9 days) for the treatment of pneumonia in an ICU. They found that the regimens were equally effective, but that resistant strains were more likely to emerge in the longer therapy group. Chastre *et al.* (2003) studied the relative efficacy of 8 vs. 15 days of therapy for well-documented ventilator-associated pneumonia. They found that for most pathogens (nonfermenters being the possible exception), the shorter duration of therapy was equally effective and associated with reduced isolation of resistant strains.

The third assumption is that shortening durations of therapy will reduce antimicrobial resistance. Truthfully, there are few good data bearing on this question, but it stands to reason that reducing overall selective pressure by any means will have a positive effect on resistance. One remarkable study from the 1960s showed a dramatic reduction in aminoglycoside resistance after discontinuing the use of all antibiotics in a neurosurgical unit (Price and Sleigh, 1970). Another more recent study showed reductions in resistance in an ICU after efforts were made to reduce the duration of therapy to 14 days (Marra *et al.*, 2009). It is clear that we need carefully performed studies to identify not only the minimum durations of therapy for specific illnesses, but, more importantly, we need to define surrogate markers to indicate when antimicrobial therapy is no longer necessary. Procalcitonin (whose concentrations are raised in bacterial infections) has demonstrated its potential as a guide for reducing therapy durations without increasing adverse outcomes (Kopterides *et al.*, 2010). It seems likely that we should be able to identify one or more such surrogate markers that will offer us reliable information on the need for further antimicrobial therapy. If we can, then therapy can be individualized and we can be confident that we are prudently using our precious antimicrobial resources.

Conclusion

Our current difficulties with antimicrobial resistance represent the cumulative efforts of our microbiota to adapt to the selective pressure we began applying 75 years ago with the first clinical use of penicillin. Modern medicine has developed in significant part because of the availability of antibiotics. It is unquestionable that transplantation or aggressive chemotherapy would not be as effective and safe as they are without the availability of effective antibiotics to address the unavoidable infectious complications associated with these procedures. As it not possible to envision the modern medical world without effective antibiotics, it is not possible for us to cease their use in an effort to reverse resistance. Therefore, we likely must learn to live with the current levels of resistance.

Whether the resistance problem gets worse, or becomes completely unmanageable, will depend on our actions. Effective antimicrobial stewardship will be a critical element to forestalling a worsening of our resistance problem. It is important that stewardship strategies acknowledge the complexity of resistance and its mechanisms, and do not pursue convenient or clever strategies based on outmuscling or outwitting the bacteria. We must find ways to safely reduce the overall antimicrobial exposure in our patients. Reducing lengths of therapy is the logical place to start.

References

Alvarez-Lerma, F. and ICU-Acquired Pneumonia Study Group (1996) Modification of empiric antibiotic treatment in patients with pneumonia acquired in the intensive care unit. *Intensive Care Medicine* 22, 387–394.

Arthur, M., Molinas, C., and Courvalin, P. (1992) The VanS–VanR two-component regulatory system controls synthesis of depsipeptide peptidoglycan precursors in *Enterococcus faecium* 4147. *Journal of Bacteriology* 174, 2582–2591.

Bartlett, J.G., Chang, T.W., Gurwith, M., Gorbach, S.L., and Onderdonk, A.B. (1978) Antibiotic-associated pseudomembranous colitis due to toxin-producing clostridia. *The New England Journal of Medicine* 298, 531534.

Blumberg, H.M., Rimland, D., Carroll, D.J., Terry, P., and Wachsmuth, I.K. (1991) Rapid development of ciprofloxacin resistance in methicillin-susceptible and -resistant *Staphylococcus aureus*. *The Journal of Infectious Diseases* 163, 1279–1285.

Bonten, M.J., Slaughter, S., Ambergen, A.W., Hayden, M.K., van Voorhis, J., Nathan, C., and Weinstein, R.A. (1998) The role of "colonization pressure" in the spread of vancomycin-resistant enterococci: an important infection control variable. *Archives of Internal Medicine* 158, 1127–1132.

Caliendo, A.M., Gilbert, D.N., Ginocchio, C.C., Hanson, K.E., May, L., Quinn, T.C., Tenover, F.C., Alland, D., Blaschke, A.J., Bonomo, R.A. *et al.* (2013) Better tests, better care: improved diagnostics for infectious diseases. *Clinical Infectious Diseases* 57(Suppl 3), S139–S170.

Centers for Disease Control and Prevention (1993) Nosocomial enterococci resistant to vancomycin—United States, 1989–1993. *Morbidity and Mortality Weekly Report* 42, 597–599.

Chastre, J., Wolff, M., Fagon, J.Y., Chevret, S., Thomas, F., Wermert, D., Clementi, E., Gonzalez, J., Jusserand, D., Asfar, P. *et al.* (2003) Comparison of 8 vs. 15 days of antibiotic therapy for ventilator-associated pneumonia in adults: a randomized trial. *JAMA (The Journal of the American Medical Association)* 290, 2588–2598.

Chow, J.W., Fine, M.J., Shlaes, D.M., and Quinn, J.P. (1991) *Enterobacter* bacteremia: clinical features and emergence of antibiotic resistance during therapy. *Annals of Internal Medicine* 115, 585–590.

Coque, T.M., Tomayko, J.F., Ricke, S.C., Okhyusen, P.C., and Murray, B.E. (1996) Vancomycin-resistant enterococci from nosocomial, community, and animal sources. *Antimicrobial Agents and Chemotherapy* 40, 2605–2609.

David, H.L. (1970) Probability distribution of drug-resistant mutants in unselected populations of *Mycobacterium tuberculosis*. *Applied Microbiology* 20, 810–814.

de Lastours, V., Cambau, E., Guillard, T., Marcade, G., Chau, F., and Fantin, B. (2012) Diversity of individual dynamic patterns of emergence of resistance to quinolones in *Escherichia coli* from the fecal flora of healthy volunteers exposed to ciprofloxacin. *The Journal of Infectious Diseases* 206, 1399–1406.

Descheemaeker, P.R.M., Chapelle, S., Devriese, L.A., Butaye, P., Vandamme, P., and Goossens, H. (1999) Comparison of glycopeptide-resistant *Enterococcus faecium* isolates and glycopeptide resistance genes of human and animal origin. *Antimicrobial Agents and Chemotherapy* 43, 2032–2037.

Donskey, C.J., Hanrahan, J.A., Hutton, R.A., and Rice, L.B. (1999a) Effect of parenteral antibiotic administration on persistence of vancomycin-resistant *Enterococcus faecium* in the mouse gastrointestinal tract. *The Journal of Infectious Diseases* 180, 384–390.

Donskey, C.J., Schreiber, J.R., Jacobs, M.R., Shekar, R., Salata, R.A., Gordon, S., Whalen, C.C., Smith, F., Rice, L.B., and the Northeast Ohio Vancomycin-Resistant Enterococcus Surveillance Program (1999b) A polyclonal outbreak of predominantly VanB vancomycin-resistant enterococci in northeast Ohio.. *Clinical Infectious Diseases* 29, 573–579.

Donskey, C.J., Chowdhry, T.K., Hecker, M.T., Hoyen, C.K., Hanrahan, J.A., Hujer, A.M., Hutton-Thomas, R.A., Whalen, C.C., Bonomo, R.A., and Rice, L.B. (2000a) Effect of antibiotic therapy on the density of vancomycin-resistant enterococci in the stool of colonized patients. *The New England Journal of Medicine* 343, 1925–1932.

Donskey, C.J., Hanrahan, J.A., Hutton, R.A., and Rice, L.B. (2000b) Effect of parenteral antibiotic administration on establishment of colonization with vancomycin-resistant *Enterococcus faecium* in the mouse gastrointestinal tract. *The Journal of Infectious Diseases* 181, 1830–1833.

Dowson, C.G., Coffey, T.J., and Spratt, B.G. (1994) Origin and molecular epidemiology of penicillin-binding-protein-mediated resistance to beta-lactam antibiotics. *Trends in Microbiology* 2, 361–366.

Drlica, K. and Zhao, X. (2007) Mutant selection window hypothesis updated. *Clinical Infectious Diseases* 44, 681–688.

Drusano, G.L., Liu, W., Brown, D.L., Rice, L.B., and Louie, A. (2009) Impact of short-course quinolone therapy on susceptible and resistant populations of *Staphylococcus aureus*. *The Journal of Infectious Diseases* 199, 219–226.

Edlin, B.R., Tokars, J.I., Grieco, M.H., Crawford, J.T., Williams, J., Sordillo, E.M., Ong, K.R., Kilburn, J.O., Dooley, S.W., Castro, K.G. *et al.* (1992) An outbreak of multidrug-resistant tuberculosis among hospitalized patients with the acquired immunodeficiency syndrome. *The New England Journal of Medicine* 326, 1514–1521.

Fantin, B., Duval, X., Massias, L., Alavoine, L., Chau, F., Retout, S., Andremont, A., and Mentre, F. (2009) Ciprofloxacin dosage and emergence of resistance in human commensal bacteria. *The Journal of Infectious Diseases* 200, 390–398.

Grayson, M.L., Eliopoulos, G.M., Wennersten, C.B., Ruoff, K.L., Degirolami, P.C., Ferraro, M.-J., and Moellering, R.C., Jr. (1991) Increasing resistance to beta-lactam antibiotics among clinical isolates of *Enterococcus faecium*: a 22-year review at one institution. *Antimicrobial Agents and Chemotherapy* 35, 2180–2184.

Gupta, K., Hooton, T.M., Naber, K.G., Wullt, B., Colgan, R., Miller, L.G., Moran, G.J., Nicolle, L.E., Raz, R., Schaeffer, A.J., and Soper, D.E. (2011) International clinical practice guidelines for the treatment of acute uncomplicated cystitis and pyelonephritis in women: a 2010 update by the Infectious Diseases Society of America and the European Society for Microbiology and Infectious Diseases. *Clinical Infectious Diseases* 52, e103–e120.

Hackbarth, C.J. and Chambers, H.F. (1993) *blaI* and *blaR1* Regulate beta-lactamase and Pbp 2a production in methicillin-resistant *Staphylococcus aureus*. *Antimicrobial Agents and Chemotherapy* 37, 1144–1149.

Hayton, W.L., Schandlik, R., and Stoeckel, K. (1986) Biliary excretion and pharmacokinetics of ceftriaxone after cholecystectomy. *European Journal of Clinical Pharmacology* 30, 445–451.

Hedrick, T.L., Schulman, A.S., Mcelearney, S.T., Smith, R.L., Swenson, B.R., Evans, H.L., Truwit, J.D., Scheld, W.M., and Sawyer, R.G. (2008) Outbreak of resistant *Pseudomonas aeruginosa* infections during a quarterly cycling antibiotic regimen. *Surgical Infections* 9, 139–152.

Hooper, D.C. (1998) Bacterial topoisomerases, anti-topoisomerases, and anti-topoisomerase resistance. *Clinical Infectious Diseases* 27(Suppl 1), S54–S63.

Hope, W.W., Goodwin, J., Felton, T.W., Ellis, M., and Stevens, D.A. (2012) Population pharmacokinetics of conventional and intermittent dosing of liposomal amphotericin B in adults: a first critical step for rational design of innovative regimens. *Antimicrobial Agents and Chemotherapy* 56, 5303–5308.

Horinouchi, S. and Weisblum, B. (1980) Posttranscriptional modification of mRNA conformation: mechanism that regulates erythromycin-induced resistance. *Proceedings of the National Academy of Sciences of the United States of America* 77, 7079–7083.

Ibrahim, E.H., Sherman, G., Ward, S., Fraser, V.J., and Kollef, M.H. (2000) The influence of inadequate antimicrobial treatment of bloodstream infections on patient outcomes in the ICU setting. *Chest* 118, 146–155.

Jacobs, C., Huang, L.J., Bartowsky, E., Normark, S., and Park, J.T. (1994) Bacterial cell wall recycling provides cytosolic muropeptides as effectors for beta-lactamase induction. *EMBO Journal* 13, 4684–4694.

Jacobs, C., Joris, B., Jamin, M., Klarsov, K., van Beeumen, J., Mengin-Lecreuix, D., van Heijenoort, J., Park, J.T., Normark, S., and Frère, J.-M. (1995) AmpD, essential for both β-lactamase regulation and cell wall recycling, is a novel cytosolic *N*-acetylmuramyl-l-alanine amidase. *Molecular Microbiology* 15, 553–559.

Jacobs, C., Frère, J.-M., and Normark, S. (1997) Cytosolic intermediates for cell wall biosynthesis and degradation control inducible β-lactam resistance in Gram-negative bacteria. *Cell* 88, 823–832.

Kaatz, G.W., Seo, S.M., Dorman, N.J., and Lerner, S.A. (1990) Emergence of teicoplanin resistance during therapy of *Staphylococcus aureus* endocarditis. *The Journal of Infectious Diseases* 162, 103–108.

Kollef, M.H. (2006) Is antibiotic cycling the answer to preventing the emergence of bacterial resistance in the intensive care unit? *Clinical Infectious Diseases* 43(Suppl 2), S82–S88.

Kopterides, P., Siempos, I.I., Tsangaris, I., Tsantes, A., and Armaganidis, A. (2010) Procalcitonin-guided algorithms of antibiotic therapy in the intensive care unit: a systematic review and meta-analysis of randomized controlled trials. *Critical Care Medicine* 38, 2229–2241.

Kotilainen, P., Nikoskelainen, J., and Huovinen, P. (1990) Emergence of ciprofloxacin-resistant coagulase-negative staphylococcal skin flora in immunocompromised patients receiving ciprofloxacin. *The Journal of Infectious Diseases* 161, 41–44.

Lakticova, V., Hutton-Thomas, R., Meyer, M., Gurkan, E., and Rice, L.B. (2006) Antibiotic-induced enterococcal expansion in the mouse intestine occurs throughout the small bowel and correlates poorly with suppression of competing flora. *Antimicrobial Agents and Chemotherapy* 50, 3117–3123.

Lauzardo, M. and Peloquin, C.A. (2012) Antituberculosis therapy for 2012 and beyond. *Expert Opinion on Pharmacotherapy* 13, 511–526.

Lee, L.S., Kinzig-Schippers, M., Nafziger, A.N., Ma, L., Sorgel, F., Jones, R.N., Drusano, G.L., and Bertino, J.S., Jr. (2010) Comparison of 30-min and 3-h infusion regimens for imipenem/cilastatin and for meropenem evaluated by Monte Carlo simulation. *Diagnostic Microbiology and Infectious Disease* 68, 251-258.

Leibovici, L., Shraga, I., Drucker, M., Konigsberger, H., Samra, Z., and Pitlik, S.D. (1998) The benefit of appropriate empirical antibiotic treatment in patients with bloodstream infection. *Journal of Internal Medicine* 244, 379–386.

Livermore, D.M. (1992) Interplay of impermeability and chromosomal beta-lactamase activity in imipenem-resistant *Pseudomonas aeruginosa*. *Antimicrobial Agents and Chemotherapy,* 36, 2046–2048.

Luna, C.M., Vujacich, P., Niederman, M.S., Vay, C., Gherardi, C., Matera, J., and Jolly, E.C. (1997) Impact of BAL data on the therapy and outcome of ventilator-associated pneumonia. *Chest* 111, 676–685.

Marra, A.R., de Almeida, S.M., Correa, L., Silva, M., Jr., Martino, M.D., Silva, C.V., Cal, R.G., Edmond, M.B., and dos Santos, O.F. (2009) The effect of limiting antimicrobial therapy duration on antimicrobial resistance in the critical care setting. *American Journal of Infection Control* 37, 204–209.

Martinez, J.L., Alonso, A., Gomez-Gomez, J.M., and Baquero, F. (1998) Quinolone resistance by mutations in chromosomal gyrase genes. Just the tip of the iceberg? *Journal of Antimicrobial Chemotherapy,* 42, 683–688.

Meyer, K.S., Urban, C., Eagan, J.A., Berger, B.J., and Rahal, J.J. (1993) Nosocomial outbreak of *Klebsiella* infection resistant to late-generation cephalosporins. *Annals of Internal Medicine,* 119, 353–358.

Moreno, F., Grota, P., Crisp, C., Magnon, K., Melcher, G.P., Jorgensen, J.H., and Patterson, J.E. (1995) Clinical and molecular epidemiology of vancomycin-resistant *Enterococcus faecium* during its emergence in a city in southern Texas. *Clinical Infectious Diseases* 21, 1234–1237.

Munoz-Price, L.S., Poirel, L., Bonomo, R.A., Schwaber, M.J., Daikos, G.L., Cormican, M., Cornaglia, G., Garau, J., Gniadkowski, M., Hayden, M.K. *et al.* (2013) Clinical epidemiology of the global expansion of *Klebsiella pneumoniae* carbapenemases. *The Lancet Infectious Diseases* 13, 785–796.

Nicolau, D.P., Freeman, C.D., Belliveau, P.P., Nightingale, C.H., Ross, J.W., and Quintiliani, R. (1995) Experience with a once-daily aminoglycoside program administered to 2,184 adult patients. *Antimicrobial Agents and Chemotherapy* 39, 650–655.

Price, D.J. and Sleigh, J.D. (1970) Control of infection due to *Klebsiella aerogenes* in a neurosurgical unit by withdrawal of all antibiotics. *The Lancet* 2, 1213–1215.

Purchiaroni, F., Tortora, A., Gabrielli, M., Bertucci, F., Gigante, G., Ianiro, G., Ojetti, V., Scarpellini, E., and Gasbarrini, A. (2013) The role of intestinal microbiota and the immune system. *European Review for Medical and Pharmacological Sciences,* 17, 323–333.

Quiros, P., Colomer-Lluch, M., Martinez-Castillo, A., Miro, E., Argente, M., Jofre, J., Navarro, F., and Muniesa, M. (2014) Antibiotic resistance genes in the bacteriophage DNA fraction of human fecal samples. *Antimicrobial Agents and Chemotherapy* 58, 606–609.

Rahal, J.J., Urban, C., and Segal-Maurer, S. (2002) Nosocomial antibiotic resistance in multiple Gram-negative species: experience at one hospital with squeezing the resistance balloon at multiple sites. *Clinical Infectious Diseases* 34, 499–503.

Rello, J., Gallego, M., Mariscal, D., Sonora, R., and Valles, J. (1997) The value of routine microbial investigation in ventilator-associated pneumonia. *American Journal of Respiratory and Critical Care Medicine* 156, 196–200.

Rice, L.B. (2008) The Maxwell Finland Lecture: for the duration—rational antibiotic administration in an era of antimicrobial resistance and *Clostridium difficile*. *Clinical Infectious Diseases* 46, 491–496.

Rice, L.B. (2012) Mechanisms of resistance and clinical relevance of resistance to beta-lactams, glycopeptides, and fluoroquinolones. *Mayo Clinic Proceedings* 87, 198–208.

Rice, L.B., Willey, S.H., Papanicolaou, G.A., Medeiros, A.A., Eliopoulos, G.M., Moellering, R.C., Jr., and Jacoby, G.A. (1990) Outbreak of ceftazidime resistance caused by extended-spectrum beta-lactamases at a Massachusetts chronic-care facility. *Antimicrobial Agents and Chemotherapy* 34, 2193–2199.

Rice, L.B., Lakticova, V., Helfand, M.S., and Hutton-Thomas, R. (2004) *In vitro* antienterococcal activity explains associations between exposures to antimicrobial agents and risk of colonization by multiresistant enterococci. *The Journal of Infectious Diseases,* 190, 2162–2166.

Sanders, W.E.J. and Sanders, C.C. (1988) Inducible beta-lactamases: clinical and epidemiologic implications for use of newer cephalosporins. *Reviews of Infectious Diseases* 10, 830–838.

Scharn, C.R., Tenover, F.C., and Goering, R.V. (2013) Transduction of staphylococcal cassette chromosome *mec* elements between strains of *Staphylococcus aureus*. *Antimicrobial Agents and Chemotherapy* 57, 5233–5238.

Shlaes, D.M., Bouvet, A., Devine, C., Shlaes, J.H., Al-Obeid, S., and Williamson, R. (1989a) Inducible, transferable resistance to vancomycin in *Enterococcus faecalis* A256. *Antimicrobial Agents and Chemotherapy* 33, 198–203.

Shlaes, D.M., Al-Obeid, S., Shlaes, J.H., Boisivon, A., and Williamson, R. (1989b) Inducible, transferable resistance to vancomycin in *Enterococcus faecium*, D399. *Journal of Antimicrobial Chemotherapy* 23, 503–508.

Siberry, G.K., Tekle, T., Carroll, K., and Dick, J. (2003) Failure of clindamycin treatment of methicillin-resistant *Staphylococcus aureus* expressing inducible clindamycin resistance in vitro. *Clinical Infectious Diseases* 37, 1257–1260.

Singh, N., Rogers, P., Atwood, C.W., Wagener, M.M., and Yu, V.L. (2000) Short-course empiric antibiotic therapy for patients with pulmonary infiltrates in the intensive care unit. A proposed solution for indiscriminate antibiotic prescription. *American Journal of Respiratory and Critical Care Medicine* 162, 505–511.

Strahilevitz, J., Jacoby, G.A., Hooper, D.C., and Robicsek, A. (2009) Plasmid-mediated quinolone resistance: a multifaceted threat. *Clinical Microbiology Reviews* 22, 664–689.

Tally, F.P., Zeckel, M., Wasilewski, M.M., Carini, C., Berman, C.L., Drusano, G.L., and Oleson, F.B., Jr. (1999) Daptomycin: a novel agent for Gram-positive infections. *Expert Opinion on Investigational Drugs* 8, 1223–1238.

Top, J., Willems, R. and Bonten, M. (2008) Emergence of CC17 *Enterococcus faecium*: from commensal to hospital-adapted pathogen. *FEMS Immunology and Medical Microbiology* 52, 297–308.

van Hal, S.J. and Paterson, D.L. (2011) Systematic review and meta-analysis of the significance of heterogeneous vancomycin-intermediate *Staphylococcus aureus* isolates. *Antimicrobial Agents and Chemotherapy* 55, 405–410.

Vergis, E.N., Hayden, M.K., Chow, J.W., Snydman, D.R., Zervos, M.J., Linden, P.K., Wagener, M.M., Schmitt, B., and Muder, R.R. (2001) Determinants of vancomycin resistance and mortality rates in enterococcal bacteremia. A prospective multicenter study. *Annals of Internal Medicine* 135, 484–492.

Wehrli, W. (1983) Rifampin: mechanisms of action and resistance. *Reviews of Infectious Diseases* 5(Suppl 3), S407–S411.

Williamson, R., Calderwood, S.B., Moellering, R.C.J., and Tomasz, A. (1983) Studies on the mechanism of intrinsic resistance to beta-lactam antibiotic in Group D streptococci. *Journal of General Microbiology* 129, 813–822.

Wilson, W.R., Wilkowski, C.J., Wright, A.J., Sande, M.A., and Geraci, J.E. (1984) Treatment of streptomycin-susceptible and streptomycin-resistant enterococcal endocarditis. *Annals of Internal Medicine* 100, 816–823.

Woodford, N. (1998) Glycopeptide-resistant enterococci: a decade of experience. *Journal of Medical Microbiology* 47, 849–862.

Workowski, K.A. and Bolan, G.A. (2015) Sexually transmitted diseases treatment guidelines, 2015. *Morbidity and Mortality Weekly Report(MMWR)* 64(3), 1–138. Available at: http://www.cdc.gov/mmwr/pdf/rr/rr6403.pdf (accessed 5 May 2016).

9 The Role of Active Surveillance in the Prevention of Healthcare-acquired Infections and Antibiotic Stewardship

GERALD A. CAPRARO*

Carolinas Pathology Group, Carolinas HealthCare System, Charlotte, North Carolina, US

Introduction

Know thy self, know thy enemy.
Sun Tzu *The Art of War*, ca. 6th century BC

As one component of a formal infection control program, the goal of surveillance is to determine the endemic rates of infection within hospitals as a means to reduce the risk of healthcare-acquired infections (HAIs). On an annual basis, HAIs affect 1.7 million patients, account for over 100,000 deaths, and contribute an estimated $55 billion in healthcare costs (Klevens *et al*., 2007; Smith and Coast, 2013). The increased morbidity and mortality associated with HAIs underscore the importance of infection control within the healthcare facility.

Hospital-based surveillance should provide an understanding of the endemic rates of infection and the prevalence of antibiotic-resistant organisms (AROs) within the institution. Once baseline rates are known, it becomes easier to identify outbreaks when they arise. The US Centers for Disease Control and Prevention (CDC) recommends a systematic approach including: (i) prospective surveillance on a regular basis by trained Infection Preventionists (IPs) using standardized definitions; (ii) the analysis of infection rates using established epidemiologic and statistical methods (e.g., calculation of rates using appropriate denominators that reflect duration of exposure, and the use of statistical process control charts for trending rates); (iii) regular use of data in decision-making; and (iv) employment of an effective and trained healthcare epidemiologist who develops infection control strategies and policies, and serves as a liaison with the medical community and administration (Haley, 1995; Pottinger *et al*., 1997; Gaynes *et al*., 2001; Benneyan *et al*., 2003). The standardization of surveillance methodology has become particularly important as US states begin mandating the reporting of HAI rates to the public (McKibben *et al*., 2005).

Goals, Elements, and Methods of Surveillance

The basic goals and elements of surveillance include using standardized definitions of infection, finding and collecting cases of HAIs, tabulating data, using appropriate denominators that reflect the duration of risk, analyzing and interpreting the data, reporting important deviations from endemic rates (i.e., epidemics, outbreaks) to the patient caregivers and to administration, implementing appropriate control measures, auditing adherence rates for recommended measures, and assessing the efficacy of the control measures. Healthcare facilities can utilize a variety of methods for surveillance, but most experts agree that the use of more than one method, including some combination of clinical chart review and data

*E-mail: gerald.capraro@carolinashealthcare.org

mining, is important to enhance surveillance and the reliability of data (Haley, 1995; Pottinger *et al.*, 1997; Gaynes *et al.*, 2001; Benneyan *et al.*, 2003). The use of software specifically designed for infection control programs provides real-time tracking of trends and timely interventions when clusters are identified. The infection control team should participate in the development and updating of electronic medical record systems for the healthcare facility to be sure that the surveillance needs will be met.

The success of this approach was originally documented by the CDC in a "Study on the Efficacy of Nosocomial Infection Control," or SENIC project (Haley *et al.*, 1980), later updates on which found a 32% decrease in nosocomial infections in hospitals with active surveillance programs compared with hospitals without these programs (Haley *et al.*, 1985a,b). Aggregate data from hospitals participating in the "National Nosocomial Infection Surveillance System" (NNIS) demonstrated that in the 10 year period from 1990 to 1999, nosocomial bloodstream infections decreased by 44% in medical intensive care units (ICU), 32% in pediatric ICUs, and 31% in surgical ICUs (CDC, 2000). Recent data have demonstrated the continuation of this trend. The most recent report (Dudeck *et al.*, 2015), from the CDC's National Healthcare Safety Network (NHSN) showed a 23% decrease in the rates of central line-associated bloodstream infections (CLABSI) in critical care units compared with the prior year. Non-critical care areas showed a 19% reduction in CLABSI from the previous year. Catheter-associated urinary tract infections (CAUTI) showed only a slight decrease of 1.5% in critical care locations but a 12% decrease in non-critical care units compared with the previous report. These data, while reflecting a continuing decrease in the rate of CLABSI and CAUTI in hospitals, do not show the same significant decrease as in previous years. This slowdown may be due to the continued overall increase in the number of facilities reporting to this program, as more healthcare facilities begin to participate in federal quality reporting programs, which require participants to use the NHSN protocol for reporting data.

Use of the NHSN surveillance protocol

HAI surveillance using the NHSN methodology has become widely accepted as a means for hospitals to compare their institutional rates with those of other healthcare facilities across the country. This approach has generally targeted areas of the hospital with the highest rates of infection, highest impact of infection, and highest likelihood of antibiotic resistance. These areas have traditionally included ICUs, cardiothoracic surgery units, and hematology/oncology units. However, with an increasing number of states now mandating the reporting of HAI rates to the public, hospital-wide surveillance is becoming more prevalent. Not unexpectedly, this requires a massive dedication of both financial and personnel resources.

Use of electronic medical records

For hospitals that have employed electronic medical records (EMRs), the task of surveillance becomes somewhat less daunting from a resource perspective, as data collection becomes more automated. Computer-based algorithms can be written to extract the necessary information directly from the EMR. Denominator data, which previously had to be collected on a daily basis by tedious review of individual patient charts, can now be accomplished via extraction of data entered into the EMR as part of the daily nursing assessment. Likewise, information about bloodstream infections and infections caused by AROs can be retrieved from the EMR with relative ease by automated extraction from microbiology laboratory reports.

Use of active surveillance cultures

In some cases, active surveillance cultures are necessary. With HAIs placing such a heavy burden on the healthcare system, hospital-based infection control committees have the responsibility to engage the participation of clinical microbiology laboratories in rapidly and accurately detecting nosocomial pathogens, including their antimicrobial resistance patterns, and preventing their spread throughout the facility (Diekema and Saubolle, 2011). Although this adds additional burden to microbiology laboratories that are under their own cost-containment mandates, laboratories must be ready and willing to advise the infection control team as to the most appropriate and cost-effective methods for achieving surveillance goals. Whilst the importance of active surveillance cultures from all patients admitted to the hospital is unclear (Maki *et al.*, 1982; Sehulster and Chinn, 2003; Diekema and Pfaller, 2011, 2013), the role of the microbiology laboratory is vital within the framework of the infection control program. Laboratories can encourage and participate in active surveillance cultures in a targeted fashion in units where there is an indication of ongoing transmission of methicillin-resistant *Staphylococcus aureus* (MRSA) or other AROs (Siegel *et al.*, 2007).

MRSA surveillance

Jain *et al.* (2011), as part of the largest systematic MRSA surveillance program in a hospital setting, screened over 1.7 million patients over a 3 year period across the US as part of a comprehensive infection control program that also included contact isolation precautions for colonized or infected patients, and an emphasis on hand hygiene. This approach resulted in a 59% overall decrease in MRSA infection rates, from 1.5/1000 patient days to 0.6/1000 patient days. This decrease was observed in bacteremia, pneumonia, skin/soft tissue infections, and urinary tract infections due to MRSA. Additionally, as a by-product of this surveillance program, the authors identified an overall decrease in *Clostridium difficile* and healthcare-associated vancomycin-resistant *Enterococcus* (VRE) infections in non-ICU settings of 61 and 73%, respectively, which was attributed to decreased empiric vancomycin usage.

Some healthcare facilities have taken the approach of universal culture screening of patients at admission. In a three-hospital system in Illinois that initiated a universal screening and decolonization program for MRSA nasal carriage on admission, Hacek *et al.* (2009) reported a 20% decrease in the total number of clinical *S. aureus* isolates recovered by the microbiology laboratory following the implementation of this program. The decrease was specifically due to a 38% decrease in clinical MRSA isolates, while methicillin-susceptible *S. aureus* (MSSA) rates remained steady. In a separate study from the same three-hospital system, Robiscek *et al.* (2008) compared the rates of MRSA clinical disease following surveillance for ICU admission vs. universal surveillance for hospital admission. Compared with a baseline MRSA disease rate of 8.9/10,000 patient days, ICU surveillance decreased the MRSA rate to 7.4/10,000 patient days, and universal surveillance decreased the MRSA rate to 3.9/10,000 patient days. Additionally, universal surveillance also resulted in a decrease in MRSA disease in the 30 day period following discharge. However, although universal MRSA screening programs have been successfully implemented in some systems, the strength of evidence for this approach continues to be debated (Glick *et al.*, 2014).

VRE surveillance

Infections with Gram-positive organisms other than *S. aureus* also present a risk the transmission of HAIs in the hospital setting. There is evidence that infections caused by VRE can be reduced through routine surveillance for colonization in high-risk patients and the contact isolation of colonized or infected patients (Tucci *et al.*, 1997; Grayson *et al.*, 1999; Ostrowsky *et al.*, 2001). Because vancomycin resistance in *E. faecium* and *E. faecalis* is largely due to the presence of the vancomycin resistance genes *vanA* and *vanB*, which are present on mobile genetic elements, being able to detect and control these species is important for infection control. Grayson *et al.* (1999) showed that renal patients are at particular risk of VRE colonization. Some 574 high-risk patients from the renal and oncology units, and the ICU in a hospital in Victoria, Australia, were screened for rectal colonization with VRE. The authors detected 12 patients who were colonized with VRE. There was a high degree of clonality among these isolates, with 50% showing genetic similarity by pulsed-field gel electrophoresis (PFGE). Attempts to decolonize these patients with ampicillin, amoxicillin, or oral bacitracin were of limited success, with only two patients having had their VRE cleared upon follow-up rectal cultures after therapy.

Point-prevalence studies were also employed to determine the extent of VRE colonization in a consortium of 32 hospitals in the midwestern US (Ostrowsky *et al.*, 2001). During the 3 year study period, a total of 5708 patients were screened, of which 75 were colonized with VRE. There was an overall decrease in the prevalence of VRE at these facilities from 2.2% at the beginning of the study period to 0.5% at the end. The number of facilities that had at least one patient colonized with VRE also decreased from 15 at the beginning of the study to five at the conclusion of the study. These data were used to inform aggressive infection control practices, including obtaining surveillance cultures and the isolation of colonized or infected patients.

A study by Tucci *et al.* (1997) at Long Island Jewish Medical Center in New York showed the impact the microbiology laboratory could have on VRE surveillance. All cultures from adult and pediatric patients performed in the laboratory (including stool and urine cultures of low colony count) were screened for VRE, and a full antimicrobial susceptibility profile was obtained for all VRE isolates from these clinical specimens. Over a 9 month period, 138 cases of VRE were identified, with 80 patients (58%) determined to be colonized and 58 patients (42%) having infection with VRE, according to the National Nosocomial Infection Study definitions (CDC, 1981). Nosocomial acquisition of VRE was demonstrated to occur in 114 (83%) out of 138 patients. Several risk factors for VRE acquisition were identified, including prior use of vancomycin and prior stay in the ICU or PCU (progressive care unit), and the authors endorse a risk-based strategy for the isolation of high-risk or colonized patients, in line with current guidelines. Additionally, they noted that improved housekeeping practices and strict adherence to universal precautions may limit the spread of VRE and other nosocomially acquired organisms. It is important to underscore the contribution of the microbiology laboratory in this study—without its contribution 48% of VRE isolates would have gone undetected. This represents a significant increase in workload for the laboratory, and individual healthcare facilities will have to decide for themselves the level of support that they will require from their laboratories, as well as the levels of personnel and financial support the laboratory will receive to perform these, largely nonreimbursed, tasks.

Many studies assessing the role of surveillance for VRE have not included outcomes other than colonization prevalence and, in some cases, a cursory analysis of isolate clonality. However, in an elegant study that demonstrated the impact of VRE surveillance on patient care, Price *et al.* (2003) compared the rates of VRE bacteremia and the degree of VRE clonality, as determined by PFGE, between two (neighboring) hospitals. Hospital A, with 700 beds and 35,000 admissions a year, did not routinely perform VRE surveillance for rectal colonization. Hospital B, a 683 bed facility with 34,000 admissions a year actively performed VRE rectal colonization screening on high-risk patients. Both facilities were tertiary care, academic medical centers within the same metropolitan area (Chicago, Illinois). During the study period, hospital A reported a total of 218 patients with VRE bacteremia, for a calculated rate of 17.1 patients with VRE bloodstream isolates per 100,000 patient days. Hospital B identified 72 patients with VRE bacteremia during the study period, for a calculated rate of 8.2/100,000 patient days. Hospital B, with an active VRE surveillance program for high-risk patients, saw a 2.1-fold lower rate of VRE bacteremia than hospital A with no surveillance program. Additionally, the majority of VRE isolates that caused bloodstream infections at hospital A were clonally related, with the four most predominant clones being responsible for >75% of all patients with VRE bacteremia. VRE isolates were more polyclonal at hospital B, with the four most predominant clones being responsible for only 37% of VRE bacteremia. While previous studies have shown a correlation between vancomycin usage and the rate of VRE infections (Stosor *et al.*, 1998; Kim *et al.*, 1999; Bhavnani *et al.*, 2000), vancomycin use did not account for the differences observed in this study, as the mean defined daily dose of vancomycin per 1000 patient days a year was not significantly different between the two institutions.

Surveillance of Gram-negative organisms

Much of the work on surveillance cultures for HAIs has historically been focused on Gram-positive organisms, although Gram-negative bacteria, particularly multidrug-resistant Gram-negative rods (MDR-GNRs), may present a more serious risk for the acquisition of HAI. Diekema and Saubolle (2011) speculated that the gains made for MRSA surveillance might not be applicable to MDR-GNRs because the antibiotic resistance mechanisms for these are multiple and generally encoded on mobile genetic elements (Peleg and Hooper, 2010), and because decolonization practices are largely unavailable. Attempting to provide evidence for improved outcomes following the surveillance of MDR-GNRs, Ridgway *et al.* (2014) performed surveillance testing of ICU patients in four affiliated hospitals on Illinois and found an overall sensitivity of 58.8%. Of the organisms recovered in this study, *Enterobacter* spp., *Klebsiella* spp., and *Pseudomonas aeruginosa* showed the highest sensitivities at 80, 66.7, and 43.2%, respectively. Other organisms recovered included *Acinetobacter baumanii*, *Escherichia coli*, and *Proteus mirabilis*, with sensitivities of 50, 36.7, and 11.1%, respectively. The gold standard comparator in this study was the culture of clinical specimens within 30 days of the surveillance culture. A surveillance culture that was positive for at least one MDR-GNR that was also recovered in the clinical culture was defined as a true positive. The overall sensitivity of this approach may be higher than is observed, given that 60% of patients in this study had no previous positive clinical culture for MDR-GNR.

A recent study focused on the detection of *Klebsiella pneumoniae* carbapenemase (KPC)-producing *Enterobacteriaceae* in four long-term acute care hospitals in metropolitan Chicago (Hayden *et al.*, 2015). A bundled intervention included rectal surveillance on admission and every other week, contact isolation and geographic separation of KPC-positive patients, and daily bathing with chlorhexidine gluconate. The prevalence of KPC rectal colonization declined 45.8% in the preintervention period to 34.3% early in the intervention period, and then reached a plateau. Highlighting the effectiveness of this approach at preventing KPC transmission, the admission prevalence during the intervention period remained steady at 20.6%. Additionally, the intervention bundle resulted in a statistically significant decrease in KPC recovered from any clinical culture from a baseline rate of 3.7 per 1,000 patient days to 2.5 per 1,000 patient days, a 32% reduction. These data cannot be directly attributed to the use of surveillance for KPC-producing *Enterobacteriaceae*; nonetheless, a bundle that included a surveillance component resulted in a statistically significant decrease in this important pathogen. The scant evidence and paucity of outcomes data for MDR-GNR notwithstanding, several publications have highlighted successful surveillance programs for these bacteria.

The Role of Active Surveillance in MRSA Control

HAIs are a key concern to patients and hospitals because these infections are associated with high morbidity and mortality, and significantly increase the costs of healthcare in the US. Whilst there is a plethora of literature highlighting the successful implementation of infection control programs that included an active surveillance component in significantly reducing both the prevalence of MRSA and the rate of MRSA infections in the hospital, others have observed no reduction in these parameters following an active surveillance program (Harbarth *et al.*, 2008). In the largest controlled study of active surveillance, nearly 22,000 patients at a Swiss teaching hospital were split into two arms. In the control arm, standard infection control practices were used. These were defined as contact isolation for MRSA carriers, dedicated personal protective equipment (PPE) for caregivers, adjustment of perioperative antibiotic prophylaxis of MRSA carriers, a computerized MRSA alert system, and topical decolonization with nasal mupirocin and chlorhexidine bathing for 5 days. In the experimental arm, standard infection control practices were also used, but these patients were also screened for MRSA colonization using a rapid, multiplex polymerase chain reaction (PCR) test prior to or at admission to the participating surgical wards. Admission screening during the intervention period identified a total of 515 MRSA carriers among the 10,193 screened patients (5.1%). The total number of patients with nosocomial MRSA infections in the control and experimental groups was 76 and 93, respectively. The total incidence of nosocomial MRSA acquisition per 1000 patient days was not statistically different between the control and the experimental periods, at 1.59 and 1.69, respectively. Despite the identification of 337 previously unknown MRSA carriers, the incidence of nosocomial MRSA infections did not decrease during the intervention period. No differences were observed in patient characteristics between the two arms of the study.

In a similar study from a London teaching hospital, Jeyaratnam *et al.* (2008) compared the use of standard infection control practices plus either conventional culture-based methods for MRSA surveillance (the culture-based control arm of the study) or a rapid PCR test on admission (the intervention arm of the study). Of the 6888 patients included in this study, a total of 461 (6.7%) were positive for MRSA colonization. In the intervention arm, the rapid PCR assay identified carriers 24 h faster than those in the culture-based control arm. Rapid reporting of MRSA carrier status resulted in a reduction in the number of inappropriate preemptive isolation days from a total of 399 in the control arm to 277 in the intervention arm. This finding alone is important from an ethical standpoint, because patients under contact isolation precautions receive fewer visits from healthcare providers, have less contact time with clinicians during their hospital stay, and, as a consequence, may suffer both clinically and psychologically (Knowles, 1992; Kirkland and Weinstein, 1999; Evans *et al.*, 2003; Saint *et al.*, 2003; Stelfox *et al.*, 2003). The most significant finding from this study was the lack of difference in the nosocomial acquisition of MRSA infection between the control and intervention groups. Some 108 (3.2%) patients in the control arm acquired MRSA infections while 99 patients (2.8%) in the intervention arm acquired nosocomial MRSA, indicating no significant reduction of MRSA acquisition in the hospital.

To add to the debate, the topic of active MRSA surveillance was the subject of a recent "Point-Counterpoint" feature in the *Journal of Clinical Microbiology* (Peterson and Diekema, 2010). Dr Peterson presented the "Point" in favor of active surveillance programs for MRSA. Citing several studies demonstrating the successful reduction in MRSA infections, he outlined the key aspect of active surveillance programs: screen all patients on admission with isolation and decolonization for MRSA carriers. This seek and destroy approach has been successful in a multitude of facilities; however, the key "Counterpoint" to this principle, presented by Dr Diekema, is practices that do not include active surveillance (e.g., hand hygiene practices) (Harrington *et al.*, 2007), which target more than simply MRSA, have been similarly successful in reducing the rates of a variety of HAIs, including *C. difficile*, GNRs, and others. Ultimately, healthcare facilities will have to decide for themselves using a risk-assessment approach what is the best scenario to reduce the rate of MRSA infections.

The Role of the Clinical Microbiology Laboratory in Surveillance and Infection Control

The most important role of the clinical microbiology laboratory as part of the infection control team is to promptly and accurately detect nosocomial pathogens and their antimicrobial resistance patterns (Barenfanger *et al.*, 2009; Diekema and Pfaller, 2011). It is also important for the laboratory to work with both the infection control and information technology departments to determine how best to deliver microbiology laboratory results in a streamlined fashion, so that outbreaks and clustered increases in prevalence can be quickly identified and acted upon. The microbiology laboratory can provide online culture information about individual patients, outbreaks of infection, and antibiotic susceptibility patterns of pathogens in periodic antibiotic susceptibility summary reports (i.e., antibiograms). The laboratory also can assist with surveillance cultures and the facilitation of molecular typing of isolates during outbreak investigations. If microbiology laboratory work is not performed on-site, it is imperative to assure that the services needed to support an effective infection control program will be available.

Surveillance challenges, aside from those discussed earlier in this chapter, also include: novel and newly emerging pathogens such as the pandemic 2009 H1N1 influenza A virus, enterovirus D-68, and Middle East Respiratory Syndrome coronavirus (MERS CoV); novel antimicrobial-resistant pathogens such as vancomycin-intermediate and -resistant *S. aureus* (VISA, VRSA), and carbapenem-resistant *Enterobacteriaceae* (CRE); and new governmental and public health mandates as in those states in which active surveillance programs have been mandated by legislative action and public reporting of HAI rates. Finally, the microbiology laboratory is under increasing pressure to provide rapid test results. As hospital lengths of stay decrease, the window of clinical relevance and clinically actionable results becomes smaller and smaller (Peleg and Hooper, 2010). Rapid diagnostic testing of clinical specimens for the identification of respiratory, gastrointestinal tract, and central nervous system viruses, as well as *Bordetella pertussis*, is particularly important for facilities that care for pediatric patients.

The knowledge gained as part of an active hospital-based surveillance program can continue to inform better infection control and antimicrobial stewardship practices. With financial, technical, and philosophical support from hospital administrations, this approach can be successful in decreasing HAIs and lowering overall healthcare costs. It is over this key premise that laboratories find themselves at a rate-limiting step. Because of consistent reductions in reimbursement from governmental healthcare payers since the mid-1990s, most hospitals and medical centers are unwilling or unable to dedicate the necessary personnel and financial resources to establishing an aggressive infection control program that includes active surveillance of important nosocomial pathogens (Peterson *et al.*, 2001). Peterson and Noskin (2001), in a study that showed a 23% reduction in healthcare-associated infections using an integrated infection control program that included on-site molecular typing of nosocomial pathogens to determine clonality, suggested that hospital-based clinical microbiology laboratories should be offered easily obtainable annual federal grants to enhance their laboratory support for infection control purposes. At the time of this study, the authors reported a cost of $400,000 p.a. to offer this service in their laboratory at the Northwestern Memorial Hospital in Chicago to support infection control practices. Performing this testing in-house resulted in a cost avoidance of over $2.1 million annually.

Whilst specific laboratory methods are outside the scope of this chapter, there are general administrative decisions that must be made when the clinical microbiology laboratory is preparing to engage in active surveillance. Fundamental knowledge, usually from historical analysis of antimicrobial susceptibility testing results, is necessary to understand the endemicity of certain prevalent organisms in the healthcare facility. For example, if a historical report yields information that the laboratory has recovered only three CRE isolates over a 5 year time frame, then active surveillance for CRE may be considered unnecessary, and precious financial and personnel resources would be available for other laboratory priorities. Likewise, if the prevalence of CRE is demonstrated to be extremely high, once again active surveillance may be unnecessary, and a clinical assumption could be made to empirically utilize an antimicrobial regimen that would cover CRE until such an isolate was ruled out. However, many healthcare facilities likely fall into a middle ground where CRE isolates are recovered fairly routinely, yet the prevalence is not exceedingly high. These facilities may derive the most benefit from an active surveillance program targeted at determining CRE endemicity. In general, Diekema and Saubolle (2011) argue that the use of active surveillance cultures for MDR-GNR should be limited to new introductions of problematic MDR-GNRs such as CRE, or in the context of a potential outbreak. For all other scenarios, standard infection control practices should be the prevailing approach—those so-called "horizontal" approaches to infection prevention that apply to all at-risk patient populations and are designed to prevent infections due to all pathogens (e.g., hand hygiene and bundled practices for prevention of device-associated infections). A risk assessment is the most appropriate course of action in order to determine how best to utilize precious laboratory resources; and of course, these decisions should be made in collaboration with the infection control team.

Concluding Remarks

One school of thought for determining the feasibility of implementing an active surveillance program was put forth by Harris *et al.* (2006) specifically to inform decisions about MDR-GNR screening. These authors propose the need for understanding the following: (i) the performance of available screening methods for organisms of interest (e.g., sensitivity, specificity, and frequency of screening); (ii) the proportion of MDR-GNR infections that is due to in-hospital transmission; (iii) the "undetected ratio" of colonized patients (the proportion of colonized patients not detected by clinical cultures); and finally (iv) the effectiveness of contact precautions in preventing MDR-GNR transmission. Although this understanding is largely available for Gram-positive organisms like MRSA and VRE, for which there are highly sensitive assays to detect the single gold standard resistance mechanisms (i.e., the *mecA* penicillin resistance gene and the vancomycin resistance gene *vanA*), among the Gram-negative bacteria, more data are needed (Maragakis and Perl, 2010), though some studies have begun to address these questions (Weintrob *et al.*, 2010).

In the current era of healthcare epidemiology, which is characterized by increased scrutiny and government regulation, consumer demands for more transparency and accountability, and calls for rapid reductions in rates of HAI, coupled with the threat of non-reimbursement for the treatment of nosocomial infections, infection control programs are more important than ever. Hospitals must develop an infection control plan that outlines the scope of the infection control program, and defines the specific goals of the program as well as the metrics to be used for assessment of progress toward those goals. The infection control team should conduct an annual risk assessment, and the findings should be incorporated in the infection control plan for the upcoming year.

Increasing healthcare costs and concomitant decreases in reimbursement have created enormous financial pressures on healthcare facilities. It has become increasingly important to avoid complications from HAIs that prolong hospital stays. This truth introduces an economic incentive for hospitals to focus on quality medical care. In the US, Medicare recently introduced a new policy to provide no additional reimbursement for the treatment of HAIs. Commercial payers, in turn, have enacted similar policies. Coupled with the recent attention to HAIs by the media and consumer advocacy organizations, this policy has foist infection control programs into a new era of much greater scrutiny and a demand for higher accountability. Hospitals will have to demonstrate a commitment to HAI prevention in order to be reimbursed for the care they provide and to remain accredited by The Joint Commission or other accreditation agency.

The infection control team faces numerous challenges. These include: increasing severity of illness in hospitalized patients; a higher prevalence of immunocompromised patients (who present a significant risk not only for HAI but also for infections from their own microbial flora); fiscally responsible appropriation of scarce resources (especially in light of unfunded mandates by states requiring active surveillance as a component of infection control programs); and the need to demonstrate to administration and clinician-clients alike that the approach taken has resulted in better outcomes, savings in healthcare dollars, and improved customer satisfaction. Additionally, emerging infectious diseases and the threat of bioterrorism require a quick response from infection control programs in order to protect both patients and healthcare workers, and to prevent disease transmission. Lastly, but perhaps most importantly, infection control teams must demonstrate that their policies and practices are grounded in evidenced-based medicine. Critical evaluation of the literature is required of the infection control team in order that evidenced-based protocols are implemented.

The clinical microbiology laboratorian, as a central pillar of the infection control team, is well suited to providing guidance and consultation on each of these challenges. The clinical microbiologist, particularly a specialist certified as a Diplomate by the American Board of Medical Microbiology (ABMM), the American Board of Pathology (ABP), or the American Board of Medical Laboratory Immunology (ABMLI), or their equivalent certified by other organizations (Baron *et al.*, 2013), is one of the most valuable laboratory partners in infection control. In the current fiscal climate, in which microbiology laboratories are being asked to do more sophisticated testing, while still keeping their budgets in line, it is this laboratorian who can critically evaluate the literature to provide informed recommendations for testing based on in-depth knowledge of both medical and laboratory science, and make decisions, in conjunction with the entire infection control team, to provide the most appropriate course of action for his or her facility.

References

Barenfanger, J., Bente, J., and Havener, G. (2009) Optimal performance for clinical microbiologists and their interaction with infection control staff. *Clinical Microbiology Newsletter* 31, 9–15.

Baron, E.J., Miller, J.M., Weinstein, M.P., Richter, S.S., Gilligan, P.H., Thomson, R.B., Bourbeau, P., Carroll, K.C., Kehl, S.C., Dunne, W.M. *et al.* (2013) A guide to utilization of the microbiology laboratory for diagnosis of infectious diseases: 2013 recommendations by the Infectious Diseases Society of America (IDSA) and the American Society for Microbiology (ASM). *Clinical Infectious Diseases* 57, e22–e121.

Benneyan, J.C., Lloyd, R.C., and Plsek, P.E. (2003) Statistical process control as a tool for research and healthcare improvement. *Quality and Safety in Health Care* 12, 458–464.

Bhavnani, S.M., Drake, J.A., Forrest, A., Deinhart, J.A., Jones, R.N., Biedenbach, D.J., Ballow, C.H., and the National Nosocomial Resistance Surveillance Group (2000) A nationwide, multicenter, case-control study comparing risk factors, treatment and outcome for vancomycin-resistant and -susceptible enterococcal bacteremia. *Diagnostic Microbiology and Infectious Disease* 36, 145–158.

CDC (Centers for Disease Control) (1981) *Guidelines for the Prevention and Control of Nosocomial Infections 1981*. US Centers for Disease Control [now US Centers for Disease Control and Prevention], Atlanta, Georgia. Reprinted in *Infection Control* 2(2), 117–121 (1981), Available at: http://www.jstor.org/stable/pdf/25702743.pdf (accessed 6 May 2016).

CDC (Centers for Disease Control and Prevention) (2000) Monitoring hospital-acquired infections to promote patient safety—United States, 1990–1999. *Morbidity and Mortality Weekly Report (MMWR)* 49, 149–153.

Diekema, D.J. and Pfaller, M.A. (2011) Infection control epidemiology and clinical microbiology. In: Versalovic, J., Carroll, K., Funke, G., Jorgensen, J.H., Landry, M., and Warnock, D. (eds) *Manual of Clinical Microbiology*, 10th edn. ASM Press, Washington, DC, pp. 73–84.

Diekema, D.J. and Pfaller, M.A. (2013) Rapid detection of antibiotic-resistant organism carriage for infection prevention. *Clinical Infectious Diseases* 56, 1614–1620.

Diekema, D.J. and Saubolle, M.A. (2011) Clinical microbiology and infection prevention. *Journal of Clinical Microbiology* 49, S57–S60.

Dudeck, M.A., Edwards, J.R., Allen-Bridson, K., Gross, C., Malpiedi, P.J., Peterson, K.D., Pollock, D.A., Weiner, L.M., and Sievert, D.M. (2015) National healthcare safety network report, data summary for 2013, device-associated module. *American Journal of Infection Control* 43, 206–221.

Evans, H.L., Shaffer, M.M., Hughes, M.G., Smith, R.L., Chong, T.W., Raymond, D.P., Pelletier, S.J., Pruett, T.L., and Sawyer, R.G. (2003) Contact isolation in surgical patients: a barrier to care? *Surgery* 134, 180–188.

Gaynes, R., Richards, C., Edwards, J., Emori, T.G., Horan, T., Alonso-Echanove, J., Fridkin, S., Lawton, R., Peavy, G., Tolson, J., and the National Nosocomial Infections Surveillance (NNIS) System Hospitals

(2001) Feeding back surveillance data to prevent hospital-acquired infections. *Emerging Infectious Diseases* 7, 295–298.

Glick, S.B., Samson, D.J., Huang, E.S., Vats, V., Aronson, N., and Weber, S.G. (2014) Screening for methicillin-resistant *Staphylococcus aureus*: a comparative effectiveness review. *American Journal of Infection Control* 42, 148–155.

Grayson, M.L., Grabsch, E.A., Johnson, P.D., Olden, D., Aberline, M., Li, H.Y., Hogg, G., Abbott, M., and Kerr, P.G. (1999) Outcome of a screening program for vancomycin-resistant enterococci in a hospital in Victoria. *Medical Journal of Australia* 171, 133–136.

Hacek, D.M., Paule, S.M., Thomson, R.B., Robiscek, A., and Peterson, L.R. (2009) Implementation of a universal admission surveillance and decolonization program for methicillin-resistant *Staphylococcus aureus* (MRSA) reduces the number of MRSA and total number of *S. aureus* isolates reported by the clinical laboratory. *Journal of Clinical Microbiology* 47, 3749–3752.

Haley, R.W. (1995) The scientific basis for using surveillance and risk factor data to reduce nosocomial infection rates. *Journal of Hospital Infection* 30, 3–14.

Haley, R.W., Quade, D., Freeman, H.E., and Bennett, J.V. (1980) Study on the Efficacy of Nosocomial Infection Control (SENIC Project). Summary of study design. *American Journal of Epidemiology* 111, 472–485.

Haley, R.W., Culver, D.H., White, J.W., Morgan, W.M., Emori, T.G., Munn, V.P., and Hooton, T.M. (1985a) The efficacy of infection surveillance and control programs in preventing nosocomial infections in US hospitals. *American Journal of Epidemiology* 121, 182–205.

Haley, R.W., Morgan, W.M., Culver, D.H., White, J.W., Emori, T.G., Mosser, J., and Hughes, J.M. (1985b) Update from the SENIC project. Hospital infection control: recent progress and opportunities under prospective payment. *American Journal of Infection Control* 13, 97–108.

Harbarth, S., Fankhauser, C., Schrenzel, J., Christenson, J., Gervaz, P., Bandiera-Clerc, C., Renzi, G., Vernaz, N., Sax, H., and Pittet, .D. (2008) Universal screening for methicillin-resistant *Staphylococcus aureus* at hospital admission and nosocomial infection in surgical patients. *JAMA (Journal of the American Medical Association)* 299, 1149–1157.

Harrington, G., Watson, K., Bailey, M., Land, G., Borrell, S., Houston, L., Kehoe, R., Bass, P., Cockroft, E., Marshall, C. *et al.* (2007) Reduction in hospital wide incidence of infection or colonization with methicillin-resistant *Staphylococcus aureus* with use of antimicrobial hand hygiene and statistical process control chart. *Infection Control and Hospital Epidemiology* 28, 837–844.

Harris, D.P., McGregor, J.C., and Furuno, J.P. (2006) What infection control interventions should be undertaken to control multidrug-resistant Gram-negative bacteria. *Clinical Infectious Diseases* 43, S57–S61.

Hayden, M.K., Lin, M.Y., Lolans, K., Weiner, S., Blom, D., Moore, N.M., Fogg, L., Henry, D., Lyles, R., Thurlow, C. *et al.* for the CDC Epicenters Program (2015) Prevention of colonization and infection by *Klebsiella pneumoniae* carbapenemase-producing *Enterobacteriaceae* in long term acute care hospitals. *Clinical Infectious Diseases* 60, 1153–1161.

Jain, R., Kralovic, S.M., Evans, M.E., Ambrose, M., Simbartl, L.A., Obrosky, D.S., Render, M.L., Freyberg, R.W., Jernigan, J.A., Muder, R.R. *et al.* (2011) Veterans Affairs initiative to prevent methicillin-resistant *Staphylococcus aureus* infection. *The New England Journal of Medicine* 364, 1419–1430.

Jeyaratnam, D., Whitty, C.J., Phillips, K., Liu, D., Orezzi, C., Ajoku, U., and French, G.L. (2008) Impact of rapid screening tests on acquisition of methicillin-resistant *Staphylococcus aureus*: cluster randomized crossover trial. *British Medical Journal* 336, 927–934.

Kim, W.J., Weinstein, R.A., and Hayden, M.K. (1999) The changing molecular epidemiology and establishment of endemicity of vancomycin-resistance in enterococci at one hospital over a 6-year period. *The Journal of Infectious Diseases* 179, 163–171.

Kirkland, K.B. and Weinstein, J.M. (1999) Adverse effects of contact isolation. *The Lancet* 354, 1177–1178.

Klevens, R.M., Edwards, J.R., Richards, C.L., Horan, T.C., Gaynes, R.P., Pollock, D.A., and Cardo, D.M. (2007) Estimating health care-associated infections and deaths in US hospitals, 2002. *Public Health Reports* 122, 160–166.

Knowles, H.E. (1992) The experience of infectious patients in isolation. *Nursing Times* 89, 53–56.

Maki, D.G., Alvarado, C.J., Hassemer, C.A., and Zilz, M.A. (1982) Relation to the inanimate hospital environment to endemic nosocomial infection. *The New England Journal of Medicine* 25, 1562–1566.

Maragakis, L.L. and Perl, T.M. (2010) How can we stem the rising tide of multidrug-resistant Gram-negative bacilli? *Infection Control and Hospital Epidemiology* 31, 338–340.

McKibben, L., Horan, T.C., Tokars, J.I., Fowler, G., Cardo, D.M., Pearson, M.L., Brennan, P.J., and the Healthcare Infection Control Practices Advisory Committee (2005) Guidance on public reporting of healthcare-associated infections: recommendations of the Healthcare Infection Control Practices Advisory Committee. *Infection Control and Hospital Epidemiology* 26, 580–587.

Ostrowsky, B.E., Trick, W.E., Sohn, A.H., Quirk, S.B., Holt, S., Carson, L.A., Hill, B.C., Arduino, M.J., Kuehnert, M.J., and Jarvis, W.R. (2001) Control of vancomycin-resistant enterococcus in health care facilities in a region. *The New England Journal of Medicine* 344, 1427–1433.

Peleg, A.Y. and Hooper, D.C. (2010) Current concepts: hospital-acquired infections due to Gram-negative bacteria. *The New England Journal of Medicine* 309, 1804–1813.

Peterson, L.R. and Diekema, D.J. (2010) Point-counterpoint: to screen or not to screen for methicillin-resistant *Staphylococcus aureus*. *Journal of Clinical Microbiology* 48, 683–689.

Peterson, L.R. and Noskin, G.A. (2001) New technology for detecting multidrug-resistant pathogens in the clinical microbiology laboratory. *Emerging Infectious Diseases* 7, 306–311.

Peterson, L.R., Hamilton, J.D., Baron, E.J., Tompkins, L.S., Miller, J.M., Wilfert, C.M., Tenover, F.C., and Thomson, R.B. (2001) Role of clinical microbiology laboratories in the management and control of infectious diseases and the delivery of health care. *Clinical Infectious Diseases* 32, 605–611.

Pottinger, J.M., Herwaldt, L.A., and Perl, T.M. (1997) Basics of surveillance: an overview. *Infection Control and Hospital Epidemiology* 18, 513–527.

Price, C.S., Paule, S., Noskin, G.A., and Peterson, L.R. (2003) Active surveillance reduces the incidence of vancomycin-resistant enterococcal bacteremia. *Clinical Infectious Diseases* 37, 921–928.

Ridgway, J.P., Peterson, L.R., Thomson, R.B., Miller, B.A., Wright, M.O., Schora, D.M., and Robicsek, A. (2014) Sensitivity of surveillance testing for multidrug-resistant Gram negative bacteria in the intensive care unit. *Journal of Clinical Microbiology* 52, 4047–4048.

Robiscek, A., Beaumont, J.L., Paule, S.M., Hacek, D.M., Thomson, R.B., Kaul, K.L., King, P., and Peterson, L.R. (2008) Universal surveillance for methicillin-resistant *Staphylococcus aureus* in 3 affiliated hospitals. *Annals of Internal Medicine* 148, 409–418.

Saint, S., Higgins, L.A., Nallamothu, B.K., and Chenoweth, C. (2003) Do physicians examine patients in contact isolation less frequently? A brief report. *American Journal of Infection Control* 31, 354–356.

Sehulster, L. and Chinn, R.Y.W. (2003) Guidelines for environmental infection control in health-care facilities: recommendations of CDC and the Healthcare Infection Control Practices Advisory Committee (HICPAC). *Morbidity Mortality Weekly Report (MMWR)* 52(RR10), 1–42.

Siegel, J.D., Rhinehart, E., Jackson, M., Chiarello, L., and the Healthcare Infection Control Practices Advisory Committee (HICPAC) (2007) Management of multidrug-resistant organisms in healthcare settings, 2006. *American Journal of Infection Control* 35, S165–S193.

Smith, R. and Coast, J. (2013) The true cost of antimicrobial resistance. *British Medical Journal* 346, 1493–1497.

Stelfox, H.T., Bates, D.W., and Redelmeier, D.A. (2003) Safety of patients isolated for infection control. *JAMA (Journal of the American Medical Association)* 290, 1899–1905.

Stosor, V., Peterson, L.R., Postelnick, M., and Noskin, G.A. (1998) *Enterococcus faecium* bacteremia: does vancomycin resistance make a difference? *Archives of Internal Medicine* 158, 522–527.

Tucci, V., Haran, M.A., and Isenberg, H.D. (1997) Epidemiology and control of vancomycin-resistant enterococci in an adult and children's hospital. *American Journal of Infection Control* 25, 371–376.

Weintrob, A.C., Roediger, M.P., Barber, M., Summers, A., Fieberg, A.M., Dunn, J., Seldon, V., Leach, F., Huang, X.Z., Nikolich, M.P., and Wortmann, G.W. (2010) Natural history of colonization with Gram-negative multidrug-resistant organisms among hospitalized patients. *Infection Control and Hospital Epidemiology* 31, 310–337.

10 The Role of the Antibiogram in Antibiotic Stewardship

Gary V. Doern*

University of Iowa Carver College of Medicine, Iowa City, Iowa, US

Introduction

For the purposes of this discussion, the term antibiogram will be used to refer to a cumulative summary of in vitro antimicrobial susceptibility test results obtained with bacteria and/or fungi recovered from patients with infection over a defined period of time in a given healthcare setting. The Joint Commission on Accreditation of Healthcare Organizations (JCAHO) requires that a cumulative antibiogram is generated in all acute care hospitals in the US at least once annually. Guidelines for constructing cumulative antibiograms have been developed and promulgated by the Clinical and Laboratory Standards Institute (CLSI) (Hindler *et al.*, 2014). When crafted appropriately and distributed wisely, the information contained in cumulative antibiograms can serve as a valuable tool for optimizing antimicrobial therapy in patients with infection, an extremely useful information resource for active antimicrobial stewardship programs, and, ultimately, a vehicle for diminishing the burden of antimicrobial resistance in many different healthcare settings.

Creating Antibiograms

Restrict to isolates that are clinically significant

The antimicrobial susceptibility test information included in cumulative antibiograms should be restricted to organisms of known or at least highly suspect clinical significance. The inclusion of results obtained with nonclinically significant isolates can lead to erroneous and potentially clinically misleading information, and should be avoided. Recognizing that in the laboratory it is often difficult if not impossible to know with certainty that a given isolate is of clinical significance, rather than a commensal or a contaminant, every effort should be made to ascertain the significance of an isolate prior to including its antimicrobial susceptibility results in a cumulative antibiogram. The overarching importance of this was recently demonstrated by Bantar *et al.* (2007) in a study in Argentina. These authors examined the results of antibiograms crafted in the laboratory with and without prior careful clinical assessment of patients with infection. They noted substantial overestimation of susceptibility test results (i.e., an underestimation of resistance) in the non-assessed laboratory specimens tested in contrast to those that had been previously clinically validated—the error that one would most like to avoid.

Eliminate duplicate isolates

It is also very important when crafting cumulative antibiograms to avoid the inclusion of susceptibility test results that have been obtained from what are essentially multiple isolates of the same organism from the same patient (Shannon and French, 2002; Horvat *et al.*, 2003; Magee, 2004; Hindler *et al.*, 2014). Especially in care areas where patients reside for prolonged periods of time and where chronic infections often occur (e.g., oncology, medical and surgical intensive care settings, bone marrow and solid organ transplants), multiple isolates of the same organism are frequently recovered from a given patient during their hospitalization. For the purpose of crafting a cumulative antibiogram, these should only be counted once (Shannon and French, 2002; Horvat *et al.*,

*E-mail: gdoern@nc.rr.com

2003; Hindler and Stelling, 2007). A simple way of accomplishing this is to include only the first isolate of a given organism from an individual patient. Indeed, according to the dictates of the CLSI, it is only the first isolate from a given patient per year that should be included among the results compiled for a cumulative antibiogram, irrespective of how many distinct hospitalizations that patient should experience during that year, or the length and complexity of those hospitalizations (Hindler *et al.*, 2014).

While it is important to avoid the inclusion of duplicate isolates, this dictate of the CLSI may be overly restrictive, as it does not take into account the emergence of resistant organisms that often occurs in individual patients who experience prolonged periods of hospitalization with attendant exposure to multiple antibiotics. Capturing only the first isolate of a particular organism from each patient potentially violates one of the most basic precepts of antimicrobial susceptibility testing, namely, when in doubt, err on the side of being conservative. That is, it is better to include rather than to exclude organisms that are resistant.

A simple way to avoid missing important resistant organisms is to do the following. In addition to tabulating the first isolate of a given organism from an individual patient over a 1 year period (irrespective of the number of hospitalizations), also include any secondary isolates of that organism that may arise that are characterized by a more resistant susceptibility profile. Defining exactly what constitutes a definitive change with respect to resistance can be challenging. However, with careful consideration in consultation with infectious disease physicians and clinical pharmacology staff, guidelines can readily be developed that identify new organisms that have emerged as resistant. Having created such guidelines, rules-based algorithms can be programmed into laboratory information systems that allow for easy and reliable identification of these organisms for inclusion in the antibiogram.

Percent susceptible, percent resistant, or minimum inhibitory concentration (MIC) values?

For the purposes of a cumulative antibiogram, the percentage of isolates determined to be susceptible should be tallied. Information on the percentage of intermediate or resistant isolates or the results of quantitative MIC tests should be maintained in the laboratory and can be provided to healthcare professionals (HCPs) upon specific request.

How many isolates?

At least 30 unique patient isolates of a particular organism tested against a given antimicrobial agent within no more than a 1 year period are required before that drug–bug combination can reliably be reported on a cumulative antibiogram (Hindler *et al.*, 2014).

Which antimicrobial agents should be included?

Only agents that can reliably be tested in the laboratory against specific organisms are candidates for inclusion in cumulative antibiograms. With this as a starting point, additional criteria for selecting which antimicrobial test results should be provided in cumulative antibiograms include the demonstrated clinical value of agents in the treatment of infections due to a specific pathogen, practice patterns in individual institutions, and lastly, formulary composition. Final decisions as to which susceptibility test results should be included in antibiograms should always be made in consultation with infectious diseases physicians, as well as the clinical pharmacy.

The report format

Example of cumulative antibiograms are provided in Tables 10.1–10.4. Please note that for the purpose of facilitating use of the information, various types of Gram-positive and Gram-negative bacteria are separated. Some antibiograms may combine these individual reports into one card or document. Yeast and anaerobic organisms are often presented in separate antibiograms if those isolates routinely have their susceptibility testing performed within the institution.

Compiling susceptibility test results and composing antibiograms

Today, the requisite information for compiling antibiograms can typically be easily accessed either from laboratory information systems or directly from the instrumentation that is used to perform antimicrobial susceptibility tests.

Table 10.1. Example cumulative antibiogram: cumulative antimicrobial susceptibility of Gram-negative bacteria (January 1–December 31, 20XX).

	% Susceptibility to antimicrobial														
Organism (number tested)	Cefazolin	Ceftriaxone	Ampicillin	Ampicillin/ Sulbactam	Piperacillin/ Tazobactam	Cefepime	Ceftazidime	Gentamicin	Tobramycin	Meropenem	Ertapenem	Aztreonam	Trimethoprim/ Sulfamethoxazole	Ciprofloxacin	Nitrofurantoin (urine only)
Enterobacteriaceae															
Citrobacter freundii group (97)	1	84	0	19	88	100	84	95	96	100	99	85	85	89	33
C. koseri (37)	95	100	0	95	100	100	100	100	100	100	100	100	100	100	71
Enterobacter aerogenes (108)	0	90	0	0	88	99	87	99	98	100	100	89	97	94	31
E. cloacae complex (272)	0	82	0	0	84	99	83	98	98	100	98	85	88	97	24
Escherichia coli (2505)	89	96	59	62	97	97	96	93	92	100	100	97	78	82	98
Klebsiella oxytoca (137)	43	96	0	52	91	99	99	99	99	100	100	96	97	99	84
K. pneumoniae (633)	93	99	0	84	97	99	99	99	98	100	100	99	92	97	48
Morganella morganii (48)	0	96	0	15	88	100	90	88	88	100	100	94	73	71	–
Proteus mirabilis (238)	86	99	75	–	92	100	99	92	92	100	100	97	82	68	0
Serratia marcescens (94)	0	100	0	0	99	100	100	99	82	99	99	99	–	93	0
Non-*Enterobacteriaceae*															
Acinetobacter baumanii complex (35)	–	71	–	91	83	89	89	89	86	91	–	–	–	86	–
Pseudomonas aeruginosa (726)[a]	–	–	–	–	91	88	92	90	95	91	–	80	–	80	–
Stenotrophomonas maltophilia (103)[b]	–	–	–	–	–	–	47	–	–	–	–	–	96	–	–

[a]For serious infections caused by *P. aeruginosa,* the susceptible category for piperacillin/tazobactam implies the need for combination therapy with an aminoglycoside with in vitro activity against *P. aeruginosa*. High-dose or extended infusions of piperacillin/tazobactam and other β-lactam antibiotics should be considered. Clinical failure has been associated with monotherapy.

[b]*S. maltophilia* susceptibility rates for additional agents tested: 84% levofloxacin, 100% minocycline.

Table 10.2. Example cumulative antibiogram: cumulative antimicrobial susceptibility of staphylococci (January 1–December 31, 20XX).

Organism (number tested)	% Susceptibility to antimicrobial								
	Penicillin	Oxacillin[a]	Vancomycin	Linezolid	Daptomycin	Gentamicin	Clindamycin	Trimethoprim/ Sulfamethoxazole	Tetracycline[b]
Staphylococcus aureus (1610)	0	62	100	100	99	99	70	98	94
Oxacillin-resistant *S. aureus* (MRSA)[c] (616)	0	0	100	100	99	97	56	97	94
Oxacillin-susceptible *S. aureus* (MSSA) (994)	0	100	100	100	99	100	79	98	95
Coagulase negative staphylococci (517)	0	38	100	100	99.8	81[d]	–	–	84

[a]Oxacillin-susceptible staphylococci are susceptible to other penicillinase-stable penicillins (e.g., nafcillin, dicloxacillin), β-lactam/β-lactamase inhibitor combinations, relevant cephalosporins, and carbapenems.
[b]Staphylococci that are susceptible to tetracycline are susceptible to doxycycline and minocycline. Organisms resistant to tetracycline may still be susceptible to doxycycline or minocycline.
[c]MRSA (methicillin-resistant *S. aureus*) is a term that is also used for oxacillin-resistant *S. aureus* (or ORSA).
[d]14 isolates failed testing for gentamicin therefore resistance in these isolates cannot be ruled out.

Table 10.3. Example cumulative antibiogram: cumulative antimicrobial susceptibility of streptococci, enterococci, and Gram-positive bacteria (January 1–December 31, 20XX).

Organism (number tested)	% Susceptibility to antimicrobial												
	Penicillin	Ampicillin	Ceftriaxone	Vancomycin	Linezolid	Daptomycin	Gentamicin Synergy	Streptomycin Synergy	Erythromycin	Clindamycin	Nitrofurantoin (urine only)	Meropenem	Gentamicin
Viridans group streptococci (94)	72/22[a,b]	–	91	100	–	–	–	–	55	87	–	–	–
Group B strep (*Streptococcus agalactiae*) (83)	100	–	100	100	–	–	–	–	47	55	–	–	–
S. anginosus (171)	96	–	98	100		–	–	–	64	83	–	–	–
Enterococcus species (943)	–	78	–	82	98	99	85[c]	82[c]	–	–	84	–	–
Aerobic diphtheroids (26)	54	–	33	100	–	–	–	–	37	–	–	96	96

[a]% susceptible/% intermediate.
[b]Penicillin-intermediate viridans group strep isolates may require combination therapy with an aminoglycoside for bactericidal action.
[c]Predicts synergistic activity of the aminoglycoside with a cell-wall active agent (e.g., ampicillin, vancomycin) to which the *Enterococcus* isolate is also susceptible.

Additional information

It is instructive to append additional information to antibiogram reports, such as the cost and appropriate dosage regimens of the agents included in the report and contact information for the clinical microbiology laboratory director(s).

The Compilation and Distribution of Cumulative Antibiograms

How often should cumulative antibiograms be compiled?

Cumulative antibiograms should be compiled at least once annually. In large and extremely busy care settings,

Table 10.4. Example cumulative antibiogram: cumulative antimicrobial susceptibility of *Streptococcus pneumoniae* (January 1–December 31, 20XX).

Organism (number tested)	% Susceptibility to antimicrobial							
	Penicillin (IV)	Penicillin (oral)	Ceftriaxone	Vancomycin	Moxifloxacin	Erythromycin	Clindamycin	Trimethoprim/ Sulfamethoxazole
S. pneumoniae (100)	–[a]	43	–[a]	100	96	30	75	60
Meningitis (100)	43	–	79	–	–	–	–	–
Non-meningitis (100)	83	–	93	–	–	–	–	–

[a]Pneumococcal breakpoints for intravenous penicillin and ceftriaxone differ based on diagnosis.

biannual compilation of antibiograms can perhaps be justified, remembering that an underlying premise of accurate antibiograms is the requirement of having at least 30 unique patient isolates of a given organism tested against relevant antimicrobials. This requirement may be the ultimate dictate in how frequently a new antibiogram can and should be compiled.

Developing care area-specific antibiograms

The concept of developing antibiograms for specific care areas is highly desirable, given the caveat that sufficient numbers of unique patient isolates are tested. Assuming adequate numbers of isolates, antibiograms restricted to care areas such as the emergency department, medical, surgical, pediatric and neonatal intensive care units (ICUs), oncology, transplant and cystic fibrosis (CF) departments can be extremely valuable. One important consideration in crafting care area-specific antibiograms is to ensure that the organisms that are included actually originated in and are representative of that care area. In light of the frequency with which patients in acute care hospitals move from one place to another, this can be challenging. Another consideration that is germane to care area-specific antibiograms is to restrict the list of antimicrobial test results reported to those agents that are directly relevant to that care area. Stated another way, drugs that may be relevant to isolates of *Staphylococcus aureus* from patients in a CF clinical service may be very different from the drugs that are relevant to patients in the medical ICU with *S. aureus* infection.

Who should receive cumulative antibiograms?

All HCPs in a given institution who have anything at all to do with antimicrobial usage should be provided with cumulative antibiograms. It is essential, however, that the results of these cumulative antibiograms be retained in a given institution, accessible only to the relevant HCPs. They should not be made available to patients, the lay public, or individuals connected to the pharmaceutical industry.

How should cumulative antibiograms be conveyed?

In the past, hard copies of cumulative antibiograms were typically compiled, printed, and distributed in the form of small folded cardboard inserts that fit snuggly into either the breast or coat pockets of a white lab coat. There is probably still some value in this reporting approach. In addition, or perhaps as a replacement for this reporting mechanism, cumulative antibiogram information should always be conveniently accessible to HCPs on computer terminals in provider offices and in patient care areas. Further, the availability of this information on personal tablets, and other pertinent handheld information devices, is highly desirable.

One potential extension of the electronic conveyance of cumulative antibiogram information is the potential for instantaneous updates of antibiogram information as new susceptibility test results are forthcoming in the laboratory. Obviously, this would require rigorous control and oversight of what information gets passed along, but continuous updates of antibiograms are something that will certainly be possible in the near future.

Another innovative approach to reporting cumulative antibiogram information is to directly attach the cumulative results for a given organism–antimicrobial combination directly to an isolate of that organism that has been recovered from a patient with infection, but prior to the availability of susceptibility test results for that specific isolate. There is often a period of up to 16–24 h after an infecting pathogen has been identified in the laboratory before the results of susceptibility tests are forthcoming. In this case, the cumulative susceptibility information might be used to achieve some degree of focused therapy over and above whatever might otherwise have been started empirically. This has been referred to as 'predictive' reporting.

The Use of Cumulative Antibiogram Information

The information included in cumulative antibiograms can be used as a guide in initiating empiric antimicrobial therapy in individual patients, as a tool for assessing emerging antimicrobial resistance problems within a given healthcare setting, and as the basis for making decisions on the structure of antimicrobial formularies.

Concluding Remarks

In this chapter, the basic ingredients of cumulative antibiograms have been outlined. In addition, issues such as the distribution and application of antibiograms have been addressed. In sum, appropriately crafted cumulative antibiograms play a central role in any functional antimicrobial stewardship program.

References

Bantar, C., Alcazar, G., Franco, D., Salamone, E., Vesco, E., Stieben, T., Obaid, F., Fiorillo, A., Izaguirre, M., and Oliva, M.E. (2007) Are laboratory based antibograms reliable to guide the selection of empirical antimicrobial treatment in patients with hospital-acquired infections? *Journal of Antimicrobial Chemotherapy* 59, 140–143.

Hindler, J.F. and Stelling, J. (2007) Analysis and presentation of cumulative antibiograms: a new consensus guidline from the Clinical and Laboratory Standards Institute. *Clinical Infectious Diseases* 44, 867–873.

Hindler, J.A., Barton, M., Erdman, S.M., Evangelista, A.T., Jenkins, S.G., Johnston, J., Lewis, J.S., II, Luper, D., Master, R.N., Nimmo, N., and Stelling, J. (2014) *Analysis and Presentation of Cumulative Antimicrobial Test Data; Approved Guideline—Fourth Edition*. CLSI Document M39-A4, Vol. 34, No. 2, Replaces M39-A3, Vol. 29 No. 6, Clinical and Laboratory Standards Institute, Wayne, Pennsylvania.

Horvat, R.T., Klutman, N.E., Lacy, M.K., Grauer, D., and Wilson, M. (2003) Effect of duplicate isolates of methicillin-susceptible and methicillin-resistant isolates of *Staphylococcus aureus* on antibiogram data. *Journal of Clinical Microbiology* 41, 4611–4616.

Magee, J.T. (on behalf of the Welsh Antibiotic Study Group) (2004) Effects of duplicate and screening isolates on surveillance of community and hospital resistance. *Journal of Antimicrobial Chemotherapy* 54, 155–162.

Shannon, K.P. and French, C.L. (2002) Validation of the NCCLS proposal to use results only from the first isolates of a species per patient in the calculation of susceptibility frequencies. *Journal of Antimicrobial Chemotherapy* 50, 965–969.

11 Selective Reporting and Antimicrobial Stewardship

Christopher D. Doern[1] and Carey-Ann D. Burnham[2*]

[1]*Virginia Commonwealth University, Richmond, Virginia, US;* [2]*Washington University in St. Louis School of Medicine, St. Louis, Missouri, US*

Introduction

The widespread availability of anti-infective agents was a major step forward for clinical medicine, resulting in a significant improvement in human health. However, the interval between widespread use of antibiotics and the development of antibiotic resistance was not long, highlighting the importance of the judicious use of these compounds (Spink and Ferris, 1945; NNIS System, 2004). Within the last decade, there has been a growing emphasis on antimicrobial stewardship, the primary goals of which are to preserve optimal patient outcomes while minimizing the adverse events associated with antimicrobial use, such as toxicity, *Clostridium difficile* infection, and the selection of resistant microorganisms (Dellit *et al.*, 2007; Fishman, 2012; Hensgens *et al.*, 2012). A successful antimicrobial stewardship program requires an interprofessional effort (including, but not limited to, infectious diseases, pharmacy, and laboratories), a formal support structure, and a commitment from hospital leadership. Electronic decision support tools may be part of the stewardship program (Demonchy *et al.*, 2014; Forrest *et al.*, 2014; Hum *et al.*, 2014; Rhoads *et al.*, 2014).

The reports generated by the clinical microbiology laboratory can have a major impact on the selection of antimicrobial therapy, and the optimization of antimicrobial therapy is an important patient safety issue, as it has been well established that for serious infections, inappropriate or delayed antimicrobial therapy is associated with worse outcomes (Alvarez-Lerma, 1996; Ibrahim *et al.*, 2000; Lodise *et al.*, 2003). The clinical laboratory must work collaboratively with the antimicrobial stewardship team and produce reports that will support best practices for antibiotic use. This chapter will discuss approaches to the selective reporting of results from specimens submitted to the microbiology laboratory and considerations for microbiology laboratory reporting on antimicrobial stewardship.

Selective Reporting of Microorganisms Recovered in Culture

It is generally accepted that clinical microbiology laboratories can produce results that are analytically accurate, and measures such as quality control and proficiency testing are intended as checks in the analytical accuracy of results. Nonetheless, a truly effective microbiology lab will not be satisfied to merely produce accurate results, but rather to issue reports that transmit information that assists with the interpretation of culture findings, and, in some instances, provide consultative comments with the data that are presented.

An important consideration when interpreting the culture results from specimens submitted to the microbiology laboratory is the clinical relevance of the microorganisms growing in the culture. This requires a somewhat complex assessment of the body site from which the specimen is obtained, and takes into consideration whether the specimen was obtained intraoperatively, and if the body site is normally sterile or is normally colonized with resident microbiota. Additional considerations when judging the clinical relevance of microbes recovered in specimens include the status of the immune system of the patient, the absence or presence of

*Corresponding author. E-mail: cburnham@path.wustl.edu

polymorphonuclear or squamous epithelial cells in the specimen Gram stain, and the predominant bacterial morphotype(s) observed in the specimen Gram stain.

It can be argued that blood cultures are one of the most important specimen types evaluated by the clinical microbiology laboratory (Byl *et al.*, 1999; Pien *et al.*, 2010; Weinstein and Doern, 2011). However, the interpretation of blood culture results is not always straightforward (Riedel and Carroll, 2010). While the recovery of some organisms from blood cultures, such as *Enterobacteriaceae*, *Pseudomonas aeruginosa*, *Staphylococcus aureus*, and *Candida* spp., is almost always suggestive of a true infection, this is not the case for all organisms (Riedel and Carroll, 2010; Kirn and Weinstein, 2013). There exists a subset of organisms that are common culture contaminants, but can also represent true infection, such as the coryneform bacteria, coagulase-negative staphylococci, *Propionibacterium* spp., and viridians group streptococci (VGS). Thus, the laboratory must carefully consider the patient population as well as the number of positive cultures when reporting these commonly contaminant organisms (Weinstein *et al.*, 1998; Riedel and Carroll, 2010). In most cases, susceptibility testing should not be performed when a single blood culture is positive for a common contaminant, as reporting susceptibility results in this setting can send an inappropriate message that the isolate is clinically significant and requires treatment.

A guidepost of antimicrobial stewardship programs is to avoid treating colonizing or contaminating flora. This starts with the collection of a proper specimen by the healthcare provider—for example, surface swabs do not inform the microbiology of deep wound infections (Bowler *et al.*, 2001). Once a quality specimen is collected and submitted to the laboratory, the wording and format of microbiology laboratory reports must be considered carefully and can have a tremendous impact on the subsequent use of antibiotics (Ackerman *et al.*, 1979, 1980). For body sites where the recovery of resident microbiota is expected, such as from mucosal surfaces, it is important that the laboratory report indicates this. That is, the isolates representing normal flora should not be routinely identified to species level and susceptibility testing should not be performed on such isolates. Laboratories may wish to develop standard reporting comments for use in these situations, and it is incumbent upon the microbiology laboratory to send a clear message when the organisms isolated simply represent normal flora rather than pathogenic bacteria. For instance, listing out the names of bacteria from body sites from which they would be expected to be found has been demonstrated to contribute to the inappropriate use of antibiotics (Lee and McLean, 1977; Ackerman *et al.*, 1979).

In some circumstances, culture results reveal that a specimen is grossly mixed with multiple taxa that have been isolated. In these cases, it may be unnecessary and inefficient to perform identification and antimicrobial susceptibility testing on all of the organisms present. Rather, the laboratory should indicate that the specimen is grossly mixed with different morphotypes of bacteria, and specifically seek out organisms of pathogenic significance and/or organisms that are likely to be resistant to antimicrobial agents (Bowler *et al.*, 2001). Knowledge of the absence or presence of these specific species can aid in the streamlining of antimicrobial therapy, especially in chronic and frequently polymicrobial infections. Some key organisms that should always be sought in this situation include *S. aureus*, vancomycin resistant *Enterococcus* (VRE), *P. aeruginosa*; also, depending upon the context, *Enterobacteriaceae* may be identified and queried for the presence of unusual or extended-spectrum resistance (such as carbapenem resistance).

Urinary tract infections (UTIs) are usually considered to be the most common type of bacterial infection (Foxman, 2002, 2003). It is estimated that UTIs account for more than 7 million office visits and more than 1 million encounters in the emergency department (ED) annually (Foxman, 2002, 2003). In addition, UTIs are a common nosocomial infection and are estimated to represent nearly one third of reported hospital-acquired infections (HAIs) every year (Gould *et al.*, 2010; Kwon *et al.*, 2012). Urine specimens usually represent a very large portion of the workload of the clinical microbiology laboratory. The work-up of urine cultures can be challenging for the microbiology lab, and interpreting the results of these cultures can be equally challenging for the physician receiving the laboratory report. In many cases, the recovery of microorganisms in urine specimens represents asymptomatic bacteriuria rather than true infection.

Numerous experts and studies have suggested that urine cultures are rarely necessary for the clinical presentation of acute, uncomplicated cystitis in a young female with the first episode of such symptoms (Burd and Kehl, 2011; Gupta *et al.*, 2011). Nevertheless, for hospitalized or immunosuppressed individuals, those with recurrent disease, or those at risk of infection with resistant microorganisms, urine cultures may be necessary to guide antimicrobial therapy when the suspicion for UTI is high. That said, these are also individuals at risk for asymptomatic bacteriuria. Urine dipstick findings and/or the presence of pyuria do not include or exclude the diagnosis of UTI (Koken *et al.*, 2002; De Rosa *et al.*, 2010; Burd and Kehl, 2011), so it is important that the laboratory educates physicians on the utility of urine cultures, and sends a clear message that urine cultures should only be ordered from patients who have signs and symptoms of a UTI (Walker *et al.*, 2000; Linares *et al.*, 2011; Leis *et al.*, 2013, 2014a).

Urine cultures are typically inoculated to solid medium using a calibrated loop, so that an approximation of the colony count in the specimen can be performed; this colony count is frequently used to guide both the laboratory work-up and the interpretation of the clinical significance of the culture. The classical definition of UTI has existed for over half a century—that is, a urine culture with ≥100,000 colony forming units (CFU)/ml of a single bacterial isolate is highly suggestive of a true infection. However, there is great debate over this number; whereas some studies suggest that it lacks sensitivity, other studies suggest that a mixture of numerous bacterial species does not rule out infection (Orenstein and Wong, 1999; Langley, 2005; McCarter *et al.*, 2009; Burd and Kehl, 2011; Kwon *et al.*, 2012; Hilt *et al.*, 2014). Thus, laboratories must establish cutoffs for reporting culture findings from urine, including standards for what colony count is clinically significant, and how to report specimens that represent grossly mixed bacterial flora. Antimicrobial susceptibility testing should not be performed on urine specimens that do not meet laboratory criteria for significance, as this would contribute to unnecessary administration of antimicrobial therapy (Kwon *et al.*, 2012).

For a completely different approach to stewardship and reducing unnecessary antimicrobial therapy in the setting of asymptomatic bacteriuria, Leis *et al.* (2014b) did not report the results of urine cultures from noncatheterized patients in the medical or surgical wards of the hospital. Rather, these cultures were reported as a message stating that the majority of positive urine cultures represent asymptomatic bacteriuria, but that if urinary tract infection was suspected to call the laboratory to trigger the release of culture results. This approach resulted in a significant reduction in the use of antibiotics for asymptomatic bacteriuria (Leis *et al.*, 2014b), and while it may not be broadly suitable for all laboratories, the findings do highlight the fact that it is difficult not to act upon culture reports if they exist in the medical record.

Culture results of respiratory specimens must clearly convey which organism(s) might be pathogenic. This analysis should start with analysis of the Gram stain. If the Gram stain of a sputum specimen reveals an abundance of squamous epithelial cells—indicative of gross contamination with oral secretions and flora—the culture should not be evaluated further. If the specimen is suitable for analysis, the Gram stain report should present a clear view of the most likely pathogen. A report of numerous bacterial morphotypes may either result in excessive empiric therapy, or it may be ignored, whereas a report stating the predominant pathogen (for example: "Abundant polymorphonuclear cells", "Moderate Gram-positive cocci in pairs suggestive of *Streptococcus pneumoniae*", and "Mixed microorganisms suggestive of normal respiratory flora") helps to solidify the message to be conveyed by the Gram stain.

In throat cultures, it is imperative that laboratories only report the absence or presence of pathogens at this body site, such as *Streptococcus pyogenes* or *Arcanobacterium haemolyticum*, and do not report the presence of bacteria that commonly colonize this body site, such as *S. aureus* or *Neisseria meningitidis*. Reporting the presence of these colonizers can lead to undue anxiety, trigger unnecessary medical procedures, and result in initiation of antibiotic therapy that is not needed. This would be contradictory to the efforts of a stewardship program.

Selective Reporting of Antimicrobial Agents Based on Specimen Type

Susceptibility data should only be reported to the provider if the isolate recovered, source or site of the specimen, and clinical details suggest that therapy is likely to be medically indicated. Once this has been established, the laboratory must carefully select which drugs to report based on the specimen type, from the viewpoints of both antimicrobial stewardship and for patient safety.

The recovery of a pathogenic bacterium from cerebrospinal fluid (CSF) is a medical emergency. In almost all cases, susceptibility data are needed to direct appropriate therapy. However, not all antimicrobial agents are able to cross the blood–brain barrier and gain access to the CSF, so the laboratory must only report agents with activity in the setting of meningitis. Some agents may appear susceptible in vitro, but would not be clinically active—for example, the Clinical and Laboratory Standards Institute (CLSI), in its M100 document, *Performance Standards for Antimicrobial Susceptibility Testing*, directs laboratories not to routinely report the following agents from bacteria isolated from the CSF: agents administered by an oral route only, and first- and second-generation cephalosporins, cephamycins, macrolides, lincosamides, tetracyclines, and fluoroquinolones (CLSI, 2014).

Conversely, there are antimicrobial agents for which achievable concentrations in the bloodstream are low, but the agent is concentrated efficiently in the urine. These agents might be reported only from urine cultures, but not from other specimen types; examples include nitrofurantoin and fosfomycin. In addition, uncomplicated UTIs are frequently treated using oral antimicrobial therapy, so it is imperative that clinical laboratories test and report these agents on isolates from urinary specimens. To take this one step further, some agents concentrate so effectively in the urinary system that testing and reporting antimicrobial susceptibly results, where interpretive criteria are based on the concentration of the agent that can be achieved systemically, may be irrelevant in the setting of UTI. For example, the CLSI states that susceptibility testing is not typically advised when *Staphylococcus saprophyticus* is isolated from urine specimens, because these infections are effectively treated by the agents that are usually employed in the treatment of uncomplicated UTI, such as trimethoprim–sulfamethoxazole, nitrofurantoin, or a fluoroquinolone (CLSI, 2014).

There are circumstances where the recovery of an organism from a particular specimen type is highly suggestive of infection, but where antimicrobial therapy is considered unnecessary or may even be contraindicated. For example, in an otherwise healthy host, antimicrobial therapy is usually unnecessary for infection with nontyphoidal *Salmonella* spp., as this is usually a relatively mild, self-limiting illness. Antimicrobial therapy does not reduce the duration of illness, but may prolong the duration of convalescent carriage with the organism, which contributes to ongoing spread in the community (Yen *et al.*, 2002; Stoycheva and Murdjeva, 2006; Chen *et al.*, 2013) Thus, laboratories should not routinely report susceptibility results from fecal isolates of nontyphoidal *Salmonella* spp. but rather include a comment indicating that these can be performed on request if needed (for example, if the patient is immunosuppressed).

There are also circumstances in which reporting antimicrobial susceptibility results may have a negative impact on patient outcome. In the setting of diarrhea due to *Escherichia coli* O157, antibiotics do not ameliorate the course of illness and may actually accelerate disease progression. Antimicrobial therapy in the setting of hemolytic uremic syndrome due to *E. coli* O157 is associated with worse outcomes (Tarr *et al.*, 2002, 2005; Dundas *et al.*, 2005; Wong *et al.*, 2012).

Finally, the reporting of specific minimum inhibitory concentration (MIC) results is often advocated with good intentions, but these data may be misinterpreted. The end user of the data may not be aware of the relative concentration of the various antimicrobials in different body compartments in order that the data may be placed in a clinical context. A common point of confusion is to simply select the antimicrobial agents with the lowest MIC, which frequently does not correlate with the most appropriate choice of therapy for a given infection. So, regardless of the body site, specific MIC data should be reported sparingly.

Selective Reporting of Antimicrobial Agents—Cascade Reporting

Once it has been established that a culture result is clinically relevant and that the reporting of antimicrobial susceptibility results is indicated, cascade reporting is one option that laboratories may choose for the selective reporting of antimicrobial agents. With cascade reporting, a limited subset of antibiotics (generally narrow spectrum) is initially reported. This subset can be expanded to include broader spectrum antibiotics if the first-tier options test as resistant. This strategy of reporting is based on the premise that providers will not prescribe antibiotics that are not listed in the susceptibility report. While this seems intuitive, McNulty *et al.* (2011) provide objective evidence that this is indeed true. In a study where the investigators changed their reporting scheme from routinely reporting (listing) amoxicillin–clavulanate (amox–clav) in the susceptibility report to listing cephalexin, they reduced the prescription of amox–clav by 70%, with cephalexin use concomitantly increasing ninefold (McNulty *et al.*, 2011). Cunney *et al.* (2000) provided further support for this strategy by demonstrating that providers were less likely to prescribe antibiotics if a susceptibility report was not provided. Thus, selective cascade reporting can be a powerful tool in driving antibiotic usage (both the type of agents prescribed and overall use) as a component of an antimicrobial stewardship program.

The CLSI M100 document mentioned above includes a table (Table 1A, pp. 38–42) with suggested groupings of antimicrobial agents that should be considered for routine testing and reporting by clinical laboratories (CLSI, 2014). The suggestions in this table take into account the identity of the organism, US Food and Drug Administration (FDA) clinical indications for antimicrobial agents, and spectra of activity for these antimicrobial agents. The categories within this table include "primary test and report," "primary test/selectively report," "supplemental/report selectively," and "supplemental for urine only." There are many different strategies that could be employed in cascade reporting, but the CLSI recommends one that is common amongst clinical microbiology laboratories, which is to cascade by increasing potency and spectrum. As an example, the CLSI document has ceftazidime, gentamicin, tobramycin, and piperacillin listed as primary agents to be tested and reported for *P. aeruginosa*. Included in the category of "primary test but selectively report" are several additional options, such as amikacin, cefepime, ciprofloxacin, piperacillin–tazobactam, and the carbapenems (excluding ertapenem).

The strategy behind a laboratory's antibiotic cascade reporting should always be developed in collaboration with infectious disease (ID) physicians, the pharmacy, and other interested stakeholders. A few of the factors that will need to be considered in developing a reporting cascade are: the cost of an antibiotic, the formulary, spectrum of activity, and patient population, and a local antibiogram.

Laboratories utilizing automated susceptibility testing systems with large antibiotic panels can suppress or release results within their cascade. This is a benefit of such panels as the information for an entire cascade may be provided in one test. As a result, technologists will rarely be required to perform secondary testing to satisfy the reporting cascade. In addition, this can improve turnaround time because secondary results can be reported instantaneously based on the results of the primary agents in the cascade. Laboratories using non-automated systems such as disk or gradient diffusion may be required to develop a primary and a secondary panel to facilitate cascade reporting. According to the CLSI M02-A11 document, *Performance Standards for Antimicrobial Disk Susceptibility Tests*, up to 12 disks can be placed on a 150 mm agar plate for susceptibility testing (CLSI, 2012). This provides enough antibiotic options to cover most organisms, but laboratories may wish to have a secondary panel for highly resistant organisms.

One consideration for laboratories is the decision to always report antimicrobial agents that are determined to be resistant during the course of routine testing, regardless of the typical reporting cascade. From a patient safety perspective, it may be hazardous to withhold this information from the provider given the chance that the agent may have been used empirically. This scenario is an important discussion for the stewardship team to have prior to the implementation of a cascade reporting strategy.

The rules that govern cascade reporting can be complex and is oftentimes too detailed (and/or used too rarely) to rely on individuals to enact; reliably implementing an antibiotic cascade requires that reporting algorithms be programmed into the laboratory information system or automated susceptibility testing software. That said, it is important to have the laboratory algorithm for cascade reporting documented in formal procedures so that technologists can refer to them and ensure that the cascade program is performing properly. It is not uncommon for cascade algorithms to fire erroneously as a result of seemingly unrelated programming changes. Therefore, it is critical that laboratory staff be familiar with the cascade reports and monitor their accuracy.

Restricted Reporting of Antibiotics to Infectious Disease Specialists

ID specialists should have an important role in antimicrobial stewardship and setting the criteria that govern the program. Selective reporting strategies should consider these individuals and ensure that they have access to all the information needed to manage their patients. If possible, laboratories may consider granting IDs specialist access to restricted antibiotic reports that are hidden from other users. This kind of access can allow ID specialists to answer treatment questions without having to consult the laboratory. The downside of this approach is that the ID specialist may see results that were suppressed for reasons unrelated to antimicrobial stewardship. Results are frequently suppressed because confirmatory testing is necessary, because an agent is tested only as a surrogate to report other agents (e.g., cefoxitin and *Staphylococcus*), or because the results have not been validated. ID specialists will need to have an understanding of these limitations if they are to be granted special access to laboratory reports.

In addition, ID specialists can have an active role in the enforcement of the program's restrictions. In these types of programs, a provider must obtain approval from an ID specialist in order to use a restricted antibiotic. This practice can serve several important purposes. First, ID specialist intervention improves the likelihood that antibiotics will be used appropriately and within the guidelines set forth by the antimicrobial stewardship program. In a study in a Turkish teaching hospital, Beović *et al.* (2009) evaluated the impact of ID specialists on antibiotic consumption and cost. In their study, they compared three units: unit A used restricted antibiotics with the approval of the head of the unit who was not an ID specialist; unit B used restricted antibiotics with the approval of ID specialists; and unit C had ID specialists prescribing all antibiotics. The authors found a significant decrease in the defined daily doses (DDD) of antibiotics in both units B and C, but not in unit A. Although just a single center study, this illustrates the impact that ID specialist approval can have on antibiotic use. In a second Turkish study by Ozkurt *et al.* (2005) antibiotic use was compared before and after the implementation of an ID specialist-based intervention program, and similar discoveries were made. Each evaluation period was 1 year and the authors found that ID intervention reduced the rate of antibiotic use from 52.7 to 36.7% ($P < 0.001$). In addition, the appropriate use of antibiotics was increased from 55.5 to 66.4% ($P < 0.05$) and the rate of culture-based treatment regimens also increased. Lastly, Ozkurt *et al.* found that for restricted antibiotics the rate of appropriate use was 88.4%, but that for unrestricted antibiotics it was only 58.2% ($P < 0.001$). This shows a clear benefit to intervention-based stewardship programs and suggests that all prescriptions would benefit from ID specialist review. Of course, this approach is not practical in most institutions, but it does show that ID specialists can play an important role in an antimicrobial stewardship program.

Oncology patients frequently require complex and broad-spectrum antibiotic management. Even though oncology physicians are accustomed to managing these patients, Granwehr and Kontoyiannis (2013) found that ID specialist consultation could still play an important role in the proper use of antibiotics. In their review, they found that ID consultation was particularly useful in managing *S. aureus* bacteremia, candidemia, and Hepatitis C virus (HCV) infection. They state that the benefits derived included decreased cost, decreased complications in care, and, most importantly, a reduction in mortality (Pragman *et al.*, 2012). This further illustrates the point that ID specialist involvement in antimicrobial stewardship is broadly beneficial, even when it involves non-ID services that are familiar with prescribing antibiotics.

Selective Reporting of Antibiotics in the Outpatient Setting

Most antimicrobial stewardship efforts are focused on inpatient prescribing practices. However, a significant amount of antibiotics is used in the outpatient setting, and this is, therefore, an area of interest both for stewardship programs and to the general medical community. The unnecessary use of broad-spectrum antimicrobials in the outpatient setting is most certainly a contributor to the ongoing emergence of antibiotic resistance. There are several challenges confronting programs that want to manage outpatient prescribing. From the perspective of selective reporting, the clinical microbiology laboratory may not be aware of the status of the patient, i.e., whether he/she is an inpatient or an outpatient. Further complicating the issue is that the time required to complete a microbiology culture may span the period of time from which a patient goes from being seen in an ED, to being admitted, to being discharged. This makes it nearly impossible for laboratories to selectively report antibiotics based on patient location. Szymczak *et al.* (2014) identified another significant challenge to implementing outpatient antimicrobial stewardship programs. Their survey of pediatricians from primary care practices who were involved in an antimicrobial stewardship intervention revealed deep skepticism among providers, and many respondents flatly ignored the recommendations. An additional barrier to program adherence was patient/parent pressure for antibiotic prescription. Interestingly, respondents agreed that overprescription in the outpatient setting was a problem, but they believed that it was driven primarily by the behaviors of non-pediatric physicians.

Although Szymczak *et al.* (2014) identified some of the challenges in outpatient antimicrobial stewardship, a cluster randomized trial by Gerber *et al.* (2013) showed that education, combined with audit and feedback, can improve adherence to prescribing guidelines in the outpatient setting. Education has a clear role in outpatient stewardship, but the role of the microbiology laboratory is less clear and less proven for the reasons discussed above. A Cochrane review evaluating the impact of professional interventions on outpatient antimicrobial stewardship identified no studies meeting their criteria that showed any benefit to selective reporting in the outpatient setting (Arnold and Straus, 2005). Nonetheless, the same principles that govern inpatient reporting can apply to outpatient reporting.

Laboratories should provide susceptibility testing results for the narrowest spectrum antibiotics needed to manage a given infection. Outpatients are usually treated with oral medications, and laboratories should ensure that results for these medications are provided. A common dilemma for physicians is transitioning patients from intravenous therapy (given while admitted) to oral therapy upon discharge, and laboratories may want to consider this when designing their reporting cascades.

Reporting for Specific Patient Populations

Clinical microbiology laboratories provide testing for a wide variety of patients that have different antibiotic treatment needs. It can be challenging for the laboratory to reliably identify what group a patient belongs to, but, where possible, reporting should be tailored to specific patient populations. Examples of patients that may have unique stewardship needs include neonates, children, hematology/oncology patients, solid organ transplant (SOT) patients, and cystic fibrosis (CF) patients. These patient populations are frequently immunocompromised and treated empirically with broad-spectrum antibiotics; they are at risk of developing infection with highly resistant organisms. As a result, the antibiotic management of these patients is different than for other patients and the laboratory reporting should accommodate those differences. For example, a common empiric antibiotic choice in a hematology/oncology patient who presents with fever and neutropenia might be piperacillin–tazobactam out of concern for infection with *P. aeruginosa*. Many laboratories will restrict the reporting of antibiotics like piperacillin–tazobactam in other patient populations, but may want to consider making it available in cases where it is commonly used in empiric treatment. This strategy may apply to any patient population in which the specific antibiotic needs of the patients are different from those of the overall population.

Similarly, laboratories often employ patient-specific work-up strategies. Hematology/oncology patients often develop mucositis and, as a result, become bacteremic with mouth flora organisms such as VGS (Blijlevens *et al.*, 2001). Laboratories typically provide susceptibility results for these isolates in hematology/oncology patients, but limit work-up in other patient populations where VGS is a common contaminant. By not providing susceptibility results for those isolates likely to be contaminants, laboratories hope to discourage unnecessary antibiotic use.

Antibiotic cycling is a controversial stewardship practice that is a form of controlling antibiotic prescribing according to patient population. This approach requires that patients are given empiric antibiotics based on location (usually intensive care units, ICUs), and then, over time, patients in that location are cycled to different empiric antibiotics. The hypothesis is that this strategy will reduce long-term exposure to any one broad-spectrum antibiotic and therefore reduce the selective pressure for resistance. Unfortunately, clinical studies have yielded mixed results on this approach, and it remains unclear whether it is an effective strategy for controlling the emergence of resistance (Bergstrom *et al.*, 2004; Brown and Nathwani, 2005). From the laboratory perspective, it may be difficult to provide unit-specific susceptibility reports to accommodate antibiotic cycling changes. Most studies suggest that antibiotic cycles should last between 3 and 4 months, and laboratories may struggle to keep their reporting relevant to each unit. One of the hurdles can be the laboratory information system (LIS) that is needed and/or electronic medical record (EMR) updates to operationalize this differential reporting. Thus, it may be difficult to have laboratory-selective reporting drive cycling programs.

Interpretive Comments

EMRs are frequently limited in space and/or flexibility for the reporting of laboratory results. In addition, they are typically optimized for reporting discrete values. In light of this, clinical microbiology laboratories must present complex information in concise ways that may not allow for detailed explanations in the way that surgical pathology reports do. If the LIS and EMR are capable, brief interpretive comments amended to reports can be a valuable way of enhancing provider understanding. These reports should be used judiciously as "alert fatigue" is a well-documented phenomenon in which users ignore extraneous information that is presented too frequently (Baysari *et al.*, 2014).

In susceptibility reporting, there are a variety of ways in which laboratories can enhance understanding and guide therapeutic decision making with the addition of interpretive comments. There has been widespread adoption of matrix-assisted laser desorption ionization time-of-flight mass spectrometry (MALDI-TOF MS) for organism identification. One of the primary benefits of MALDI-TOF MS is that it provides more rapid results than growth-based identification systems. A consequence of this is the prolonged interval of the availability of organism identification prior to susceptibility results. This is somewhat analogous to the results provided from molecular testing performed directly from positive blood culture media. In both circumstances, laboratories are providing actionable information, but without the complement of definitive susceptibility results. These are situations in which comments could be used to help providers to interpret the results. Organism-specific comments pertaining to the treatment of methicillin-resistant *S. aureus* (MRSA), methicillin-susceptible *S. aureus* (MSSA), *E. faecium*, *E. faecalis*, beta-hemolytic streptococci, and *P. aeruginosa* may help providers to select appropriate antibiotics before the definitive susceptibility results become available, especially for tailoring therapy for organisms that are predictably susceptible to narrow-spectrum agents. Therapeutic guidelines vary widely, but Table 11.1 presents a few examples of situations and comments that are widely accepted and may be considered.

Interpretive comments can also be used to help explain culture results for different clinical situations. Cunney *et al.* (2000) included interpretive comments for 169 positive cultures from sputum, urine, and wound specimens. In the study, they included comments for all 169 cultures but only released susceptibility results on 17% of those cultures. The comments provided guidance in situations such as fecal contamination of wound specimens, multidrug resistant organisms (MDROs), and superficial abscesses that could be managed without antibiotic therapy (Table 11.1). The goal of the program was to limit susceptibility reporting, advise providers in culture interpretation, and avoid unnecessary antibiotic utilization. It is difficult to know how effective the interpretive comments were as there was no control group in the study. However, the authors did find that withholding susceptibility results was an effective tool in limiting inappropriate antibiotic use.

Table 11.1. Examples of interpretive comments for microbiology reports. Adapted from Cunney *et al.* (2000) and HICSIG Wiki (2014).

Clinical situation	Interpretive comment example
Culture yielding an organism that may represent colonization or contamination	This isolate most likely represents contamination or colonization and is of questionable clinical significance.
Stool culture positive for *Campylobacter*	*Campylobacter* gastroenteritis is typically self-resolving and does not require antibiotic treatment. In severe or prolonged infection and during pregnancy, erythromycin treatment is recommended.
Mixed urine culture result	This urine culture is not indicative of urinary tract infection and reflects urethral contamination of the specimen. It is suggested that a repeat specimen be obtained using proper collection technique.
Abscess drainage sufficient, antibiotics not warranted	Drainage of superficial abscesses may be sufficient to cure infection. Consider withholding antibiotic therapy if patient likely to respond to drainage alone.
Multiply resistant isolate	Multidrug resistant organism isolated. Infection control precautions required to minimize transmission to other patients.
Susceptibility results not provided but may be necessary if clinically indicated	Contact the laboratory if susceptibility testing is clinically indicated.

Mechanisms of Reporting and Decision Support

An important function of a clinical microbiology laboratory is to perform susceptibility testing on organisms causing infection. A key, but often overlooked, component of susceptibility testing is the manner in which the information is conveyed or reported. In the modern hospital, the vast majority of microbiological information is communicated through a LIS into an EMR. This information is then used by a treating physician to determine whether his or her therapeutic choices are likely to be effective. In terms of antimicrobial stewardship, the reporting of susceptibility results is the primary mechanism by which the microbiology laboratory impacts therapeutic choices. Therefore, it is vital that this information be conveyed in accurate and understandable ways.

The literature suggests that when treating potentially septic patients, most antibiotic decisions are made before any microbiological information is available (Munson *et al.*, 2003). Because the causative pathogen is not known at the time of antibiotic selection, physicians commonly start with broad-spectrum antibiotics that cover a wide variety of pathogens. The use of these broad-spectrum antibiotics is a focus of stewardship programs for three reasons: (i) broad-spectrum antibiotics are also some of the most potent antibiotics used to treat difficult infections; (ii) broad-spectrum antibiotics have activity against some of the most resistant pathogens and are often the only treatment option; and (iii) most importantly, the use of an antibiotic drives resistance to that antibiotic (Goettsch *et al.*, 2000; Kahlmeter, 2003). For these reasons, the goal of a microbiology laboratory is to quickly identify causative pathogens and report susceptibility results that provide options for de-escalating therapy to narrower spectrum antibiotics.

Mechanisms of reporting

The success of any selective reporting program depends on how effectively information is communicated to and utilized by the provider. Laboratories have a number of selective reporting strategies that can be used to guide antibiotic utilization, such as limiting the release of susceptibility results according to organism, specimen type, patient population, and resistance profile. An alternative to these strategies is selectively identifying clinical situations in which unique reporting mechanisms can be employed (i.e., the selective utilization of a reporting mechanism). The rest of this section will discuss various mechanisms that can be used to selectively report susceptibility results as well as their respective strengths and weaknesses.

Susceptibility results are most frequently presented through the EMR. A strength of this approach is that information transfer from the LIS to the EMR occurs quickly and accurately. In addition, the results can be stored indefinitely and can generally be accessed at any time and from any location with a secure Internet connection.

Although EMRs quickly transfer information, there is generally no notification that a result is ready and therefore a lag exists between the time of reporting and the time at which a physician sees and acts upon the result. It is perhaps for this reason that Bruins *et al.* (2005), in a study in a hospital in the Netherlands, found that some physicians actually prefer hard copy reporting. They also found that the time of reporting was improved with the use of an EMR, but that ultimately this had no impact on provider decision making. The reasons for these findings are unclear—it may be that they are an artifact of the single center study that was done, or it may be that they provide insight into the way that microbiology data are used. In an Australian study conducted in an ED, Callen *et al.* (2010) obtained evidence of the latter. They found that 32% of microbiology results were not viewed on the day they were released. In an earlier study, Kilpatrick and Holding (2001) found that 45% of the emergency (biochemistry) test results available electronically for patients seen in the accident and emergency department and the acute medical admissions ward of a UK hospital were never accessed at all.

Another contributing factor to physicians seeing and acting upon microbiology results may be that their timing can be unpredictable, making it difficult for providers to know when to look for a result. In the absence of direct notification, a lag between result reporting and viewing is not surprising. This point is illustrated nicely by Munson *et al.* (2003), who analyzed the impact of Gram stain results from positive blood cultures versus definitive susceptibility testing on antibiotic management. Gram stain results were telephoned, while susceptibility results were passively reported into the EMR. Not only did the authors find that the telephoned result had a greater impact on patient care, they also found that when a change was warranted, it occurred much more quickly when the result was telephoned rather than passively reported.

Even though these data suggest that telephoning all results is the best way to ensure that they are used in a timely manner, this is not a practical solution for the laboratory. Laboratories do not have the staffing required to operationalize this form of communication on a routine basis, and neither would physicians want to be interrupted with a high volume of phone calls. One alternative is an automated pager system that could automatically generate a page that would alert providers to the availability of actionable results. The clinical impact of an automated paging approach to susceptibility reporting has not been evaluated. However, Kuperman *et al.* (1999) conducted a randomized controlled trial in a hospital in Massachusetts that analyzed the impact of automatic alerts on the reporting of critical chemistry values. In this system, rule-based logic was used to notify physicians of critical results. The outcome was a significant reduction in the time it took to act upon the result and get patients on to appropriate therapy.

Clinical decision support systems

Although ID physicians receive specialty training in the interpretation and utilization of susceptibility testing results, they comprise a very small percentage of the providers that act upon this information. Because of this, it is critical that these results be communicated in a clear and concise manner. Coupat *et al.* (2013) conducted a survey of residents training in general practice and found that 124 of 325 (38%) claimed to be uncomfortable with interpreting antibiotic susceptibility data. One tool that laboratories can use to facilitate proper antibiotic selection is a clinical decision support system (CDSS). These systems can be useful in driving adherence to clinical practice guidelines as well as for improving antibiotic prescribing practices. CDSSs exist in several forms, but generally share some of the same features. Patient data can be monitored in real time, and antibiotic organism mismatches can be identified and brought to the attention of providers. In addition, CDSSs are capable of making real-time or point-of-care recommendations for empiric antibiotic choices. Criteria such as patient age and presenting symptoms can be considered by a CDSS and result in a specific antibiotic recommendation. Litvin *et al.* (2013) designed their CDSS to incorporate the "Get Smart" antibiotics prescribing program of the US Centers for Disease Control and Prevention (CDC) as well as to aid in the proper diagnosis of acute respiratory infection. This system operated in real time as providers wrote their notes. As information was entered into predefined fields, the CDSS was able to provide patient-specific recommendations. Other CDSSs require a more personal interaction by providing alerts to antimicrobial stewardship teams who then go on to intervene, usually through a direct conversation. These types of interventions can be triggered by any criteria the stewardship program desires. Common examples are identifying organism/antibiotic mismatches, inappropriately broad-spectrum antibiotics, or unnecessarily long courses of treatment in the setting of negative cultures.

There is a wide variety of CDSSs available, but they are all dependent on the ability to reliably collect data. This can be a challenge as LISs and EMRs vary in their interfacing capabilities. In addition, there are often significant costs associated with building and maintaining these interfaces. Once the interfaces are built, it is important (and required by regulatory agencies) to periodically validate the data being provided, as seemingly unrelated programming changes can have unintended consequences for CDSS data capture.

A recent development that may be leveraged to improve antimicrobial stewardship is the widespread adoption of the "smartphone." Most providers are now expected to have access to these devices because they function as pagers as well as granting access to e-mail, hence making them ideal targets for a CDSS. A study by May *et al.* (2014) surveyed ED providers at eight sites in three cities in the US and found that 89% used their smartphones to aid in antibiotic decision making. In a pilot study, Payne *et al.* (2014) introduced a smartphone application to a group of junior doctors in a UK hospital. The application provided information about disease management and antibiotic dosing guidelines. Interestingly, smartphone use increased the likelihood of referring to hospital guidelines, but was viewed by healthcare staff as being unprofessional. In contrast, Charani *et al.* (2013) in a survey of five teaching hospitals in London found that most clinicians liked having antibiotic information on their smartphones and that 71% felt that it improved their antibiotic knowledge.

Numerous studies have documented the positive impact that decision support tools can have on antibiotic decision making (Patel *et al.*, 2012; Gonzales *et al.*, 2013; Litvin *et al.*, 2013; Mainous

Table 11.2. Key points in selective reporting and antimicrobial stewardship.

Reports issued by the clinical microbiology laboratory influence antimicrobial prescription practices.
Reporting clinically insignificant organisms encourages inappropriate antimicrobial usage. Reports should only include identification and susceptibility results for clinically significant organisms.
Selective reporting should consider the site of infection, as antimicrobial concentration is variable by body site and thus an antimicrobial may not be clinically effective despite in vitro susceptibility.
Antibiotic cascades can selectively report antibiotics based on the susceptibility profile of individual isolates.
Intervention by Infectious diseases (ID) specialists improves adherence to antimicrobial stewardship guidelines.
Selective reporting should consider the management of outpatient infections (which frequently require oral antimicrobial therapy).
Selective reporting should account for the specific requirements of patients, with special consideration for populations such as children and oncology patients.
Interpretive comments can be used to facilitate understanding of susceptibility reports and stewardship program guidelines.
Clinical decision support systems (CDSSs) can improve compliance with stewardship program guidelines and alert ID specialists to situations requiring intervention.

et al., 2013). Mainous *et al.* (2013) showed that a CDSS decreased inappropriate prescribing practices of broad-spectrum antibiotics in acute respiratory infections. Patel *et al.* (2012) studied the impact of a CDSS on deescalating from piperacillin–tazobactam (Pip/Tazo) once the culture results became available. Through an alert and intervention system they observed a decrease in the mean doses per patient from 16 to ten, resulting in an overall mean decrease of 1800 doses/month. Litvin *et al.* (2013) used a progress note template to implement a CDSS that targeted antibiotics used for inappropriate indications, the use of inappropriately broad antibiotics, and the use of antibiotics for sinusitis and bronchitis. In total the CDSS was used over 38,000 times during the study. The results showed that the CDSS did not change the use of antibiotics for those diagnoses that rarely require antibiotics. However, they did show that the use of broad-spectrum antibiotics for acute respiratory infections decreased by over 16%. These results were sustained over a 27 month period. While there are many different platforms and strategies for employing a CDSS, it is clear from these studies and others, that this is an effective means by which a stewardship program can influence antibiotic utilization.

Summary

The clinical microbiology laboratory is an important component of any antimicrobial stewardship program, and the reports generated by the clinical microbiology laboratory impact the selection of antimicrobial therapy. The clinical laboratory must work collaboratively with the antimicrobial stewardship team to produce guidelines regarding selective reporting, resulting in laboratory reports that will support best practices for antibiotic use. Table 11.2 summarizes some key points in selective reporting and antimicrobial stewardship.

References

Ackerman, V.P., Pritchard, R.C., Obbink, D.J., Bradbury, R., and Lee, A. (1979) Consumer survey on microbiology reports. *The Lancet* 1, 199–202.

Ackerman, V.P., Pritchard, R.C., Groot Obbink, D.J., Bradbury, R., and Lee, A. (1980) Reporting practices of microbiology laboratories. *Journal of Clinical Pathology* 33, 830–835.

Alvarez-Lerma, F. and ICU-Acquired Pneumonia Study Group (1996) Modification of empiric antibiotic treatment in patients with pneumonia acquired in the intensive care unit. *Intensive Care Medicine* 22, 387–394.

Arnold, S.R. and Straus, S.E. (2005) Interventions to improve antibiotic prescribing practices in ambulatory care. *Cochrane Database of Systematic Reviews* 2005(4): CD003539.

Baysari, M.T., Westbrook, J.I., Richardson, K., and Day, R.O. (2014) Optimising computerised alerts within electronic medication management systems: a synthesis of four years of research. *Studies in Health Technology and Informatics* 204, 1–6.

Beović, B., Kreft, S., Seme, K., and Čižman, M. (2009) The impact of total control of antibiotic prescribing by infectious disease specialist on antibiotic consumption and cost. *Journal of Chemotherapy* 21, 46–51.

Bergstrom, C.T., Lo, M., and Lipsitch, M. (2004) Ecological theory suggests that antimicrobial cycling will not reduce antimicrobial resistance in hospitals. *Proceedings of the National Academy of Sciences of the United States of America* 101, 13285–13290.

Blijlevens, N.M., Donnelly, J.P., and de Pauw, B.E. (2001) Empirical therapy of febrile neutropenic patients with mucositis: Challenge of risk-based therapy. *Clinical Microbiology and Infection* 7(Suppl 4), 47–52.

Bowler, P.G., Duerden, B.I., and Armstrong, D.G. (2001) Wound microbiology and associated approaches to wound management. *Clinical Microbiology Reviews* 14, 244–269.

Brown, E.M. and Nathwani, D. (2005) Antibiotic cycling or rotation: a systematic review of the evidence of efficacy. *Journal of Antimicrobial Chemotherapy* 55, 6–9.

Bruins, M., Oord, H., Bloembergen, P., Wolfhagen, M., Casparie, A., Degener, J., and Ruijs, G. (2005) Lack of effect of shorter turnaround time of microbiological procedures on clinical outcomes: a randomised controlled trial among hospitalised patients in the Netherlands. *European Journal of Clinical Microbiology and Infectious Diseases* 24, 305–313.

Burd, E.M. and Kehl, K.S. (2011) A critical appraisal of the role of the clinical microbiology laboratory in the diagnosis of urinary tract infections. *Journal of Clinical Microbiology* 49, S34–S38.

Byl, B., Clevenbergh, P., Jacobs, F., Struelens, M.J., Zech, F., Kentos, A., and Thys, J.-P. (1999) Impact of infectious diseases specialists and microbiological data on the appropriateness of antimicrobial therapy for bacteremia. *Clinical Infectious Diseases* 29, 60–68.

Callen, J., Paoloni, R., Georgiou, A., Prgomet, M., and Westbrook, J. (2010) The rate of missed test results in an emergency department: an evaluation using an electronic test order and results viewing system. *Methods of Information in Medicine* 49, 37–43.

Charani, E., Kyratsis, Y., Lawson, W., Wickens, H., Brannigan, E.T., Moore, L.S.P., and Holmes, A.H. (2013) An analysis of the development and implementation of a smartphone application for the delivery of antimicrobial prescribing policy: lessons learnt. *Journal of Antimicrobial Chemotherapy* 68, 960–967.

Chen, H.M., Wang, Y., Su, L.H., and Chiu, C.H. (2013) Nontyphoid salmonella infection: microbiology, clinical features, and antimicrobial therapy. *Pediatrics and Neonatology* 54, 147–152

CLSI (2012) *M02-A11: Performance Standards for Antimicrobial Disk Susceptibility Tests; Approved Standard—Eleventh Edition*. Clinical and Laboratory Standards Institute, Wayne, Pennsylvania.

CLSI (2014) *M100-S24: Performance Standards for Antimicrobial Susceptibility Testing; Twenty-Fourth Informational Supplement*. Clinical and Laboratory Standards Institute, Wayne, Pennsylvania.

Coupat, C., Pradier, C., Degand, N., Hofliger, P., and Pulcini, C. (2013) Selective reporting of antibiotic susceptibility data improves the appropriateness of intended antibiotic prescriptions in urinary tract infections: a case-vignette randomised study. *European Journal of Clinical Microbiology and Infectious Diseases* 32, 627–636.

Cunney, R., Aziz, H.A., Schubert, D., McNamara, E., and Smyth, E. (2000) Interpretative reporting and selective antimicrobial susceptibility release in non-critical microbiology results. *Journal of Antimicrobial Chemotherapy* 45, 705–708.

De Rosa, R., Grosso, S., Bruschetta, G., Avolio, M., Stano, P., Modolo, M.L., and Camporese, A. (2010) Evaluation of the Sysmex UF1000i flow cytometer for ruling out bacterial urinary tract infection. *Clinica Chimica Acta* 411, 1137–1142.

Dellit, T.H., Owens, R.C., McGowan, J.E., Jr., Gerding, D.N., Weinstein, R.A., Burke, J.P., Huskins, W.C., Paterson, D.L., Fishman, N.O., Carpenter, C.F. *et al.* (2007) Infectious Diseases Society of America and the Society for Healthcare Epidemiology of America guidelines for developing an institutional program to enhance antimicrobial stewardship. *Clinical Infectious Diseases* 44, 159–177.

Demonchy, E., Dufour, J.C., Gaudart, J., Cervetti, E., Michelet, P., Poussard, N., Levraut, J., and Pulcini, C. (2014) Impact of a computerized decision support system on compliance with guidelines on antibiotics prescribed for urinary tract infections in emergency departments: a multicentre prospective before-and-after controlled interventional study. *Journal of Antimicrobial Chemotherapy* 69, 2857–2863.

Dundas, S., Todd, W.T., Neill, M.A., and Tarr, P.I. (2005) Using antibiotics in suspected haemolytic–uraemic syndrome: antibiotics should not be used in *Escherichia coli* O157:H7 infection. *BMJ* 330, 1209.

Fishman, N. (2012) Policy statement on antimicrobial stewardship by the Society for Healthcare Epidemiology of America (SHEA), the Infectious Diseases Society of America (IDSA), and the Pediatric Infectious Diseases Society (PIDS). *Infection Control and Hospital Epidemiology* 33, 322–327.

Forrest, G.N., Van Schooneveld, T.C., Kullar, R., Schulz, L.T., Duong, P., and Postelnick, M. (2014) Use of electronic health records and clinical decision support systems for antimicrobial stewardship. *Clinical Infectious Diseases* 59(Suppl 3), S122–S133.

Foxman, B. (2002) Epidemiology of urinary tract infections: incidence, morbidity, and economic costs. *The American Journal of Medicine* 113(Suppl 1A), 5S–13S.

Foxman, B. (2003) Epidemiology of urinary tract infections: incidence, morbidity, and economic costs. *Disease-a-Month* 49, 53–70.

Gerber, J.S., Prasad, P.A., Fiks, A.G., Localio, A.R., Grundmeier, R.W., Bell, L.M., Wasserman, R.C., Keren, R., and Zaoutis, T.E. (2013) Effect of an outpatient antimicrobial stewardship intervention on broad-spectrum antibiotic prescribing by primary care pediatricians: a randomized trial. *JAMA (The Journal of the American Medical Association)* 309, 2345–2352.

Goettsch, W., van Pelt, W., Nagelkerke, N., Hendrix, M.G.R., Buiting, A.G.M., Petit, P.L., Sabbe, L.J.M., van Griethuysen, A.J.A., and de Neeling, A.J. (2000) Increasing resistance to fluoroquinolones in *Escherichia coli* from urinary tract infections in the Netherlands. *Journal of Antimicrobial Chemotherapy* 46, 223–228.

Gonzales, R., Anderer, T., McCulloch, C.E., Maselli, J.H., Bloom, F.J., Jr., Graf, T.R., Stahl, M., Yefko, M., Molecavage, J., and Metlay, J.P. (2013) A cluster randomized trial of decision support strategies for reducing antibiotic use in acute bronchitis. *JAMA Internal Medicine* 173, 267–273.

Gould, C.V., Umscheid, C.A., Agarwal, R.K., Kuntz, G., Pegues, D.A., and the Healthcare Infection Control Practices Advisory Committee (2010) Guideline for prevention of catheter-associated urinary tract infections 2009. *Infection Control and Hospital Epidemiology* 31, 319–326.

Granwehr, B.P. and Kontoyiannis, D.P. (2013) The impact of infectious diseases consultation on oncology practice. *Current Opinion in Oncology* 25, 353–359.

Gupta, K., Hooton, T.M., Naber, K.G., Wullt, B., Colgan, R., Miller, L.G., Moran, G.J., Nicolle, L.E., Raz, R., Schaeffer, A.J. *et al.* (2011) International clinical practice guidelines for the treatment of acute uncomplicated cystitis and pyelonephritis in women: a 2010 update by the Infectious Diseases Society of America and the European Society for Microbiology and Infectious Diseases. *Clinical Infectious Diseases* 52: e103–e120.

Hensgens, M.P., Goorhuis, A., Dekkers, O.M., and Kuijper, E.J. (2012) Time interval of increased risk for *Clostridium difficile* infection after exposure to antibiotics. *Journal of Antimicrobial Chemotherapy* 67, 742–748.

HICSIG Wiki (2014) Microbiology reporting practice. Healthcare Infection Control Special Interest Group, Australasian Society for Infectious Diseases, Surry Hills, New South Wales, Australia. Available at: http://hicsigwiki.asid.net.au/index.php?title=Microbiology_reporting_practice (accessed 28 October 2014).

Hilt, E.E., McKinley, K., Pearce, M.M., Rosenfeld, A.B., Zilliox, M.J., Mueller, E.R., Brubaker, L., Gai, X., Wolfe, A.J., and Schreckenberger, P.C. (2014) Urine is not sterile: use of enhanced urine culture techniques to detect resident bacterial flora in the adult female bladder. *Journal of Clinical Microbiology* 52, 871–876.

Hum, R.S., Cato, K., Sheehan, B., Patel, S., Duchon, J., DeLaMora, P., Ferng, Y.H., Graham, P., Vawdrey, D.K., Perlman, J. *et al.* (2014) Developing clinical decision support within a commercial electronic health record system to improve antimicrobial prescribing in the neonatal ICU. *Applied Clinical Informatics* 5, 368–387.

Ibrahim, E.H., Sherman, G., Ward, S., Fraser, V.J., and Kollef, M.H. (2000) The influence of inadequate antimicrobial treatment of bloodstream infections on patient outcomes in the ICU setting. *Chest* 118, 146–155.

Kahlmeter, G. (2003) An international survey of the antimicrobial susceptibility of pathogens from uncomplicated urinary tract infections: the ECO.SENS Project. *Journal of Antimicrobial Chemotherapy* 51, 69–76.

Kilpatrick, E.S. and Holding, S. (2001) Use of computer terminals on wards to access emergency test results: a retrospective audit. *BMJ* 322, 1101–1103.

Kirn, T.J. and Weinstein, M.P. (2013) Update on blood cultures: how to obtain, process, report, and interpret. *Clinical Microbiology and Infection* 19, 513–520.

Koken, T., Aktepe, O.C., Serteser, M., Samli, M., Kahraman, A., and Dogan, N. (2002) Determination of cut-off values for leucocytes and bacteria for urine flow cytometer (UF-100) in urinary tract infections. *International Urology and Nephrology* 34, 175–178.

Kuperman, G.J., Teich, J.M., Tanasijevic, M.J., Ma'Luf, N., Rittenberg, E., Jha, A., Fiskio, J., Winkelman, J., and Bates, D.W. (1999) Improving response to critical laboratory results with automation: results of a randomized controlled trial. *JAMIA (Journal of the American Medical Informatics Association)* 6, 512–522.

Kwon, J.H., Fausone, M.K., Du, H., Robicsek, A., and Peterson, L.R. (2012) Impact of laboratory-reported urine culture colony counts on the diagnosis and treatment of urinary tract infection for hospitalized patients. *American Journal of Clinical Pathology* 137, 778–784.

Langley, J.M. (2005) Defining urinary tract infection in the critically ill child. *Pediatric Critical Care Medicine* 6(3 Suppl), S25–S29.

Lee, A. and McLean, S. (1977) The laboratory report: a problem in communication between clinician and microbiologist? *The Medical Journal of Australia* 2, 858–860.

Leis, J.A., Gold, W.L., Daneman, N., Shojania, K., and McGeer, A. (2013) Downstream impact of urine cultures ordered without indication at two acute care teaching hospitals. *Infection Control and Hospital Epidemiology* 34, 1113–1114.

Leis, J.A., Palmay, L., Elligsen, M., Walker, S.A., Lee, C., and Daneman, N. (2014a) Lessons from audit and feedback of hospitalized patients with bacteriuria. *American Journal of Infection Control* 42, 1136–1137.

Leis, J.A., Rebick, G.W., Daneman, N., Gold, W.L., Poutanen, S.M., Lo, P., Larocque, M., Shojania, K.G., and McGeer, A. (2014b) Reducing antimicrobial therapy for asymptomatic bacteriuria among noncatheterized inpatients: a proof-of-concept study. *Clinical Infectious Diseases* 58, 980–983.

Linares, L.A., Thornton, D.J., Strymish, J., Baker, E., and Gupta, K. (2011) Electronic memorandum decreases unnecessary antimicrobial use for asymptomatic bacteriuria and culture-negative pyuria. *Infection Control and Hospital Epidemiology* 32, 644–648.

Litvin, C.B., Ornstein, S.M., Wessell, A.M., Nemeth, L.S., and Nietert, P.J. (2013) Use of an electronic health record clinical decision support tool to improve antibiotic prescribing for acute respiratory infections: the ABX-TRIP study. *Journal of General Internal Medicine* 28, 810–816.

Lodise, T.P., McKinnon, P.S., Swiderski, L., and Rybak, M.J. (2003) Outcomes analysis of delayed antibiotic

treatment for hospital-acquired *Staphylococcus aureus* bacteremia. *Clinical Infectious Diseases* 36, 1418–1423.
Mainous, A.G., III, Lambourne, C.A., and Nietert, P.J. (2013) Impact of a clinical decision support system on antibiotic prescribing for acute respiratory infections in primary care: quasi-experimental trial. *JAMIA (Journal of the American Medical Informatics Association)* 20, 317–324.
May, L., Gudger, G., Armstrong, P., Brooks, G., Hinds, P., Bhat, R., Moran, G.J., Schwartz, L., Cosgrove, S.E., Klein, E.Y. *et al.* (2014) Multisite exploration of clinical decision making for antibiotic use by emergency medicine providers using quantitative and qualitative methods. *Infection Control and Hospital Epidemiology* 35, 1114–1125.
McCarter, Y.S., Burd, E.M., Hall, G.S., Zervos, M., and Sharp, S.E. (ed.) (2009) *Laboratory Diagnosis of Urinary Tract Infections*. Cumitech (Cumulative Techniques and Procedures in Clinical Microbiology) 2C, American Society of Microbiology, Washington, DC.
McNulty, C.A., Lasseter, G.M., Charlett, A., Lovering, A., Howell-Jones, R., Macgowan, A., and Thomas, M. (2011) Does laboratory antibiotic susceptibility reporting influence primary care prescribing in urinary tract infection and other infections? *Journal of Antimicrobial Chemotherapy* 66, 1396–1404.
Munson, E.L., Diekema, D.J., Beekmann, S.E., Chapin, K.C., and Doern, G.V. (2003) Detection and treatment of bloodstream infection: laboratory reporting and antimicrobial management. *Journal of Clinical Microbiology* 41, 495–497.
NNIS System (2004) National Nosocomial Infections Surveillance (NNIS) System report, data summary from January 1992 through June 2004, issued October 2004. *American Journal of Infection Control* 32, 470–485.
Orenstein, R. and Wong, E.S. (1999) Urinary tract infections in adults. *American Family Physician* 59, 1225–1234, 1237.
Ozkurt, Z., Erol, S., Kadanali, A., Ertek, M., Ozden, K., and Tasyaran, M.A. (2005) Changes in antibiotic use, cost and consumption after an antibiotic restriction policy applied by infectious disease specialists. *Japanese Journal of Infectious Diseases* 58, 338–343.
Patel, J., Esterly, J.S., Scheetz, M.H., Bolon, M.K., and Postelnick, M.J. (2012) Effective use of a clinical decision-support system to advance antimicrobial stewardship. *AJHP (American Journal of Health-System Pharmacy)* 69, 1543–1544.
Payne, K.F., Weeks, L., and Dunning, P. (2014) A mixed methods pilot study to investigate the impact of a hospital-specific iPhone application (iTreat) within a British junior doctor cohort. *Health Informatics Journal* 20, 59–73.
Pien, B.C., Sundaram, P., Raoof, N., Costa, S.F., Mirrett, S., Woods, C.W., Reller, L.B., and Weinstein, M.P. (2010) The clinical and prognostic importance of positive blood cultures in adults. *The American Journal of Medicine* 123, 819–828.
Pragman, A.A., Kuskowski, M.A., Abraham, J.M., and Filice, G.A. (2012) Infectious disease consultation for *Staphylococcus aureus* bacteremia improves patient management and outcomes. *Infectious Diseases in Clinical Practice* 20, 261–267.
Rhoads, D.D., Sintchenko, V., Rauch, C.A., and Pantanowitz, L. (2014) Clinical microbiology informatics. *Clinical Microbiology Reviews* 27, 1025–1047.
Riedel, S. and Carroll, K.C. (2010) Blood cultures: key elements for best practices and future directions. *Journal of Infection and Chemotherapy* 16, 301–316.
Spink, W.W. and Ferris, V. (1945) Quantitative action of penicillin inhibitor from penicillin-resistant strains of staphylococci. *Science* 102, 221–223.
Stoycheva, M.V. and Murdjeva, M.A. (2006) Antimicrobial therapy of salmonellosis—current state and perspectives. *Folia Medica* 48, 5–10.
Szymczak, J.E., Feemster, K.A., Zaoutis, T.E., and Gerber, J.S. (2014) Pediatrician perceptions of an outpatient antimicrobial stewardship intervention. *Infection Control and Hospital Epidemiology* 35(Suppl 3), S69–S78.
Tarr, P.I., Watkins, S.L., and Neill, M.A. (2002) Risk of hemolytic uremic syndrome from antibiotic treatment of *Escherichia coli* O157:H7 colitis. *JAMA (The Journal of the American Medical Association)* 288, 3111–3112.
Tarr, P.I., Gordon, C.A., and Chandler, W.L. (2005) Shiga-toxin-producing *Escherichia coli* and haemolytic uraemic syndrome. *The Lancet* 365(9464), 1073–1086.
Walker, S., McGeer, A., Simor, A.E., Armstrong-Evans, M., and Loeb, M. (2000) Why are antibiotics prescribed for asymptomatic bacteriuria in institutionalized elderly people? A qualitative study of physicians' and nurses' perceptions. *CMAJ (Canadian Medical Association Journal)* 163, 273–277.
Weinstein, M.P. and Doern, G.V. (2011) A critical appraisal of the role of the clinical microbiology laboratory in the diagnosis of bloodstream infections. *Journal of Clinical Microbiology* 49, S26–S29.
Weinstein, M.P., Mirrett, S., Van Pelt, L., McKinnon, M., Zimmer, B.L., Kloos, W., and Reller, LB. (1998) Clinical importance of identifying coagulase-negative staphylococci isolated from blood cultures: evaluation of MicroScan Rapid and Dried Overnight Gram-Positive panels versus a conventional reference method. *Journal of Clinical Microbiology* 36, 2089–2092.
Wong, C.S., Mooney, J.C., Brandt, J.R., Staples, A.O., Jelacic, S., Boster, D.R., Watkins, S.L., and Tarr P.I. (2012) Risk factors for the hemolytic uremic syndrome in children infected with *Escherichia coli* O157:H7: a multivariable analysis. *Clinical Infectious Diseases* 55(1), 33–41.
Yen, M.H., Huang, Y.C., Chiu, C.H., and Lin, T.Y. (2002) Duration of antimicrobial therapy for non-typhoid *Salmonella bacteremia* in healthy children. *Journal of Microbiology, Immunology, and Infection (Wei Mian yu Gan Ran za Zhi)* 35(2), 94–98.

12 The Role of New Diagnostics to Enhance Antibiotic Stewardship Efforts

Kimberle C. Chapin and April M. Bobenchik*

The Warren Alpert School of Medicine of Brown University, Providence, Rhode Island, US

Introduction

Clinical diagnostics performed in the clinical microbiology laboratory for the purposes of identifying infectious diseases or antibiotic susceptibility has changed significantly in the past 20 years. The transitions have come on three major fronts: less culture-based to more rapid technologies and molecular based assays; less subjective interpretive reporting to an increase in automation and objective results; and finally, how implementation of new technologies into the laboratory and clinical practice achieves success. The drivers for rapid diagnostic implementation are as likely to be initiated from outside the laboratory as from within the lab itself, and often stem from initiatives to support other hospital or healthcare goals, including: rapid PCR testing for *Clostridium difficile* toxin to clarify infection control issues; a more rapid and sensitive methicillin-resistant *Staphylococcus aureus*/*S. aureus* (MRSA/SA) nasal preadmission screening for surgery patients to optimize preoperative and postoperative antibiotic choices; and rapid tests for sepsis or syndromic conditions to clarify anti-infective treatment. The potential impact that newer diagnostic technologies can play is critical in antibiotic stewardship efforts as labs can often provide rapid actionable results from which an immediate clinical decision can be made of the appropriate treatment.

This chapter will: highlight specific categories of diagnostics that have shown clear outcome benefits, or the potential for benefits in conjunction with antibiotic stewardship efforts; identify components to consider when implementing a rapid technology, and trends in diagnostics for infectious diseases and susceptibility data; and, finally, discuss the role of rapid diagnostics and antibiotic stewardship as part of evidence-based care practice models to enhance quality for the healthcare system.

Traditional versus Rapid Microbiology

The goal of the microbiology laboratory has always been to provide the most relevant interpretation so that the correct diagnosis of the infectious disease being sought and the appropriate treatment could be rendered for the patient. Traditionally, the majority of results were obtained from the performance of stains, with morphological interpretation followed by culture growth or cytopathic effect(s) and/or biochemical tests, all in the context of an experienced subjective interpretation and reported days later. The most rapid test to an actionable result included a Gram stain, performed on spinal fluid, tissues or a positive blood culture, and rapid antigen tests (Beekmann *et al.*, 2003; Bonner *et al.*, 2003; Barenfanger *et al.*, 2008; Lean *et al.*, 2014). These simple but clinically significant tests led the way in documenting how microbiology results performed in conjunction with reporting to the appropriate healthcare provider could result in timely alterations in the course of a patient's treatment, resulting in more appropriate antibiotic therapy and even decreased mortality, and in cost savings to healthcare systems.

However, this menu was limited and microbiology was never considered to be a true rapid response laboratory with these (earlier) technologies. Jump forward to the current microbiology laboratory

*Corresponding author. E-mail: april_bobenchik@brown.edu

and the introduction of rapid final results (1 h to 1 day), including near-patient care diagnostics, tests with greatly increased sensitivity and/or specificity compared with traditional methodologies, an expanded menu of pathogens for syndromic conditions, antibiotic resistance targets and/or biomarkers, and, suddenly, the *product* from the microbiology laboratory is a real-time game changer for clinical care (Bartlett, 2011). The choice of the technologies implemented in a laboratory is no longer simply *if* it will diagnose the infectious disease but *how* those results will positively impact subsequent patient care as well as resulting in cost avoidance downstream. Therefore, importantly, as technological advances become more costly to implement from a laboratory standpoint, the decision to implement certain diagnostics is necessarily done in conjunction with the goals and needs of the particular healthcare system, and by the partnering of specialty areas with the laboratory, and with infectious diseases and infection control.

Emergency medicine, etc., and the antibiotic stewardship team, can help to clarify to administration the potential added value and critical role that these new technologies can play in providing rapid actionable results from which immediate clinical and anti-infective decisions can be made. Because the implementation of rapid diagnostics has been so successful in stewardship, and antibiotic resistance has become a worldwide threat, the US Centers for Disease Control and Prevention (CDC), as part of its "Get Smart for Healthcare Program", has recently highlighted the use of diagnostic tools in patient care as an area of great interest, and recommended further research to determine how they can best be applied to stewardship efforts (http://www.cdc.gov/getsmart/healthcare/implementation/core-elements.html).

Considerations Prior to Implementation of Rapid Diagnostics

Many factors need to be in place and considered prior to the implementation of diagnostics to optimize the effect of the rapid result from the laboratory. It is not as simple as buying the "box" on which the assay will be run. How this process should work is ideally developed between the stewardship and the laboratory teams in conjunction with other stakeholders to prioritize which assays will benefit the institution the best. Key components include: medical staff education and clear expectations for the use and interpretation of the test(s); adequate laboratory information and electronic medical record (EMR) resources; and administrative support for the action plan subsequent to implementation, especially if the follow-up based on the results of a test will include a physician-specific report card on how well recommendations are being followed.

Consider the following:

1. Test preimplementation data gathering:
   a. What are the major pathogen/antibiotic issues that need assessment in the healthcare setting?
   b. What parameters of a test are most desired for the situation or goal of initiating the test? E.g., is de-escalation or discontinuation of an anti-infective the goal and so is a test with a high negative predictive value needed? Or is sensitivity and a directed antibiotic the goal?
   c. What is typically done in the context of a patient with the disease currently? This allows assessment of measurable goals and parameters to consider:
      i. How often is Infectious Diseases (ID) consulted?
      ii. What are the empiric antibiotics and duration of use?
      iii. What ancillary testing is used to define the illness?
      iv. What are the length of stay (LOS) and readmissions for the same international classification of diseases (ICD-10) code?
   d. How do different populations affect technology choice, reporting and antibiotic stewardship follow-up? E.g.,
      i. Pediatric versus adult?

For instance, in a setting with both pediatric and adult patients, a result of Gram-positive cocci (GPC) in clusters suggestive of *Staphylococcus* spp. from a blood culture is less likely to be treated empirically in pediatric patients, but is almost uniformly to be treated in adults. Thus, the parameters assessed by stewardship might be the time to the appropriate *initiation* of an antibiotic in pediatrics but the *de-escalation* of an unnecessary antibiotic in adults.

2. Laboratory factors:
   a. What are the resources and time line needed by the laboratory to implement the testing?
   b. Is capital equipment and budget justification to administration necessary?
   c. Will the testing replace or be complementary to current methods, and in the latter case, will it require more resources such as personnel and/or space?

d. Will the test be restricted to specific specialties?
e. What is the schedule for reporting testing results, personnel needs, and training to perform?

3. Reporting and interpretation:
   a. How and to whom will the result be reported? Is there the need for an interpretation comment?
   b. How will the result be transmitted to the stewardship team and what is the expectation on the time line for follow-up on the result by the stewardship team?
      i. This may alter the lab testing schedule.
   c. What are the recommendations from the stewardship team to the medical staff on how to treat?
      i. Antibiotic choices and duration for the organism/syndrome?
      ii. Discontinuation of antibiotics if the test is negative or other etiology is identified, such as viral?
      iii. Duration of treatment?
4. Post-implementation tabulation and performance assessment:
   a. What will be tabulated once the technology is in place to make sure it is working correctly, and who will monitor it?
      i. Time to add, de-escalate to an effective antibiotic or discontinue an antibiotic, or add an antiviral?
      ii. Before patient discharge, do cultures need to be obtained and results assessed?
      iii. LOS, morbidity, mortality, readmission, cost savings per patient?
      iv. How will feedback be presented to the medical staff and to the laboratory and stewardship team on performance?
      v. How to adjust the process if goals are not what are expected?

Essentially, the greater the effort and thought put into the process of test implementation, reporting, and goals prior to the "go-live" date, the more likely will be the result in optimizing the overall benefits of the test system.

The Rapid Diagnostic Test Menu

An astounding array of rapid advanced technologies has emerged for the diagnosis of infectious disease in the midst of other traditional simpler but rapid and informative tests that are still clinically important. The Gram stain and rapid antigen tests used in the appropriate context are such traditional examples that continue to provide both clinically relevant information and the ability to tailor therapy within minutes. The technologies presented in Tables 12.1 and 12.2 are more recent developments. They include a range of pathogens, toxins, and resistance markers that vary in laboratory complexity and cost, and in turnaround time (TAT) to report a result, as well as assays that rely on initial bacterial growth or colony development, and those that are direct sample detection methods. Specifics on the technological components, the targets they identify, and the clinical outcome benefits are referred to in several publications and reviews that are referenced (Wolk and Dunne, 2011; Bauer *et al.*, 2014; Kothari *et al.*, 2014). In addition, diagnostic company websites often provide excellent videos of their technology unique diagnostic features and details of peer-reviewed publications. Examples of emerging less well-known technologies include GeneWEAVE's Smarticles molecular diagnostics technology (details available at: https://vimeo.com/126622853) and Accelerate Diagnostic's in vitro diagnostics (details available at: http://acceleratediagnostics.com/), both which address rapid susceptibility and antibiotic resistance, as well as BacterioScan's rapid urinary tract infection (UTI) identification, which also addresses antibiotic resistance (details available at: http://www.bacterioscan.com).

While most of new and rapid tests involve added costs to the microbiology laboratory, many have shown clinical benefits and cost savings for the healthcare system that justify support for the additional laboratory expenditure.

Sepsis

The success of rapid diagnostics and antibiotic stewardship efforts have been most commonly reported in the diagnosis and management of sepsis, with benefits identified as the appropriate initiation or de-escalation of antibiotics, and decreases in LOS and hospital costs, as well as morbidity and mortality.

The majority of these technologies involve not direct detection from the patient's blood, but growth from a positive blood culture, which subsequently yields enough organisms to perform a Gram stain and then specific rapid testing from the broth or from a single colony from a culture plate on the following day. Thus, an initial Gram stain interpretation helps to guide what test platform or multiplex test follows. The Protein Nucleic Acid

Table 12.1. Microbiology techniques for rapid detection of pathogens associated with sepsis.

Specimen source	Technology	Name (and manufacturer)	Target(s)	Technical		Published antibiotic stewardship outcomes
				Benefits	Limitations	
Blood culture (BC) broth or colony	Immunochromatographic qualitative assay	PBP2a Culture Colony Test (Alere™)	Penicillin binding protein 2a (detects MRSA[a])	No equipment Rapid identification of MRSA from colony	Sensitivity of test varies in published literature	No studies to date
	Protein Nucleic Acid Fluorescent in situ hybridization (PNA FISH®)	Quick FISH™ PNA FISH™ Traffic Light PNA FISH™ *Xpress* FISH™ (AdvanDx)	4 Gram-positive organisms 3 Gram-negative organisms 5 *Candida* spp. 1 resistance marker	TAT[b] 30–90 min Fluorescent scope with Texas red filter already in most labs High specificity	Not automated Cannot differentiate between different species of *Candida*	Yes, for Gram-positive organisms and *Candida*
	Biplex PCR	Xpert MRSA/SA[a] BC (Cepheid)	MRSA *Staphylococcus aureus*	Fully automated TAT 1 h	Equipment costly	Yes
	Multiplex PCR	Verigene® Gram Negative Blood Culture Test (BC-GN)[c] (Nanosphere)	8 Gram-negative organisms 6 resistance markers	Fully automated Broad array of targets TAT 2.5 h	Multiple steps Equipment costly	No studies to date
		Verigene® Gram-Positive Blood Culture Test (BC-GP)[d] (Nanosphere)	12 Gram-positive organisms 3 resistance markers	Fully automated Broad array of targets TAT 2.5 h	Multiple steps, Equipment costly	Yes
		BioFire FilmArray® Blood Culture Identification (BCID) (bioMérieux)	11 Gram-negative organisms 8 Gram-positive organisms 5 *Candida* spp. 3 resistance markers	Fully automated Simple setup Broad panel with potential to eliminate Gram stain TAT 1 h	Equipment costly Only one test per instrument	Yes

Continued

Table 12.1. Continued.

Specimen source	Technology	Name (and manufacturer)	Target(s)	Technical		Published antibiotic stewardship outcomes
				Benefits	Limitations	
	Matrix-assisted laser desorption/ionization time-of-flight (MALDI-TOF)	MALDI biotyper MS (Bruker Corp) Vitek MS (bioMérieux)	From colony growth: bacteria, yeast, molds, mycobacteria	Identification 15 min Low reagent costs Colony from any specimen type	Occupies large space Costly equipment US FDA[e]-cleared database limitations	Yes (additional positive studies in syndromes other than sepsis)
Direct detection from blood drawn sample	RT (real-time)-PCR	LightCycler® SeptiFast (Roche Molecular Systems)	10 Gram-negative organisms 8 Gram-positive organisms 5 *Candida* spp. *Aspergillus fumigatus*	Broad array, including *Aspergillus*	Not approved in US	Yes
	PCR with sequencing	SepsiTest™ (Molzym)	Bacteria and fungi	Broad array	Not approved in US Time consuming	Yes
	Nanoparticle-based PCR	T2 Candida (T2Biosystems®)	5 *Candida* spp.	Fully automated, Direct from blood High NPV[f]	Costly equipment	No studies to date

[a]MRSA/SA, methicillin resistant *Staphylococcus aureus*/*S. aureus*; [b]TAT, turnaround time; [c]BC-GN, Gram-negative blood culture; [d]BC-GP, Gram-positive blood culture; [e]Food and Drug Administration; [f]NPV, negative predictive value.

(PNA) Fluorescent in situ hybridization (FISH) assays require only a fluorescent microscope, and are very specific, and so they allow almost any lab to introduce this technology. When PNA is performed in conjunction with antibiotic stewardship in place, positive outcomes are common. Similarly, while there is an initial outlay of equipment to perform the BioFire Blood Culture Identification (BCID) assay, which detects 24 of the most common Gram-negative (GN), Gram-positive (GP), and yeast targets, the extensive menu makes the performance of the Gram stain potentially unnecessary. Coupling the Sepsityper™ extraction kit along with Matrix-assisted laser desorption/ionization time-of-flight (MALDI-TOF MS) allows for the direct detection of a number of different organisms, including aerobes, anaerobes, and yeast from positive blood cultures (Schieffer *et al.*, 2014). This then opens up significant flexibility for ease of testing in many lab settings. Assays for the direct or rapid detection of pathogens in blood and/or of susceptibilities are becoming more common, but the data are still unclear and/or contradictory on the clinical performance and outcome benefits as far as stewardship practices are concerned, and further assessment is needed. Details of these technologies are presented in Table 12.1.

Rapid testing for infections other than sepsis

Hospital-acquired infections

Hospital-acquired infections (HAIs, also known as "healthcare-acquired infections") are a considerable burden to healthcare systems and have significant implications for reimbursement. Thus, with the goal of decreased HAI transmission, rapid and sensitive tests, along with targeted antibiotic treatment and isolation precautions, are major patient safety and quality measurement goals for hospitals (CDC, 2000).

The most commonly used rapid test for HAIs is a PCR for *C. difficile* toxins, with the requirement to address appropriate treatment, infection control precautions, and bed placement. The PCR technologies are all statistically equivalent, but differ on the equipment needed, their TATs, and their ability to perform random, small batch, or large batch testing (see Table 12.2). While there is sparse literature on antibiotic stewardship practices per se, there is a significant body of data showing that a concerted HAI team approach among infection control, microbiology, nursing, pharmacy, and administration can result in decreased transmission, morbidity, and mortality (Mermel *et al.*, 2013).

Active, point prevalence or targeted high-risk screening for MRSA/SA in presurgical patients and for MRSA in medical patients is a more controversial issue, and has shown varied results for decreased HAI transmission (Hacek *et al.*, 2009; Peterson and Diekema, 2010). However, for many sites, especially the presurgical screening for MRSA/SA in high-risk groups, the identification of carriers, with directed treatment, has commonly resulted in a subsequent decrease in transmission rates, postsurgical infections, and decreased overall costs (Kim *et al.*, 2010). For laboratories, it is necessary to clarify whether the increased sensitivity of a molecular assay, faster TAT than culture, and automation versus traditional and/or chromogenic media justify the fourfold increase in cost/test (Peterson and Diekema, 2010).

Syndromic panels

Another growing area of rapid testing and one with emerging possibilities for antibiotic stewardship is the use of multiplex PCR assays for syndromic conditions (see Table 12.2).

RESPIRATORY DISEASES Acute respiratory illness is the most common reason that patients come into the healthcare system, and many of them leave with an antibiotic when they have a viral etiology (Chan *et al.*, 2011). Thus, the potential to de-escalate antibiotics in cases when a viral etiology is diagnosed, or to correctly use antiviral medication depending on whether influenza is detected or not, justify a concerted effort for antibiotic stewardship to address (Jennings *et al.*, 2009; Byington *et al.*, 2012; McCulloh *et al.*, 2014). Data is beginning to present as to the value of the respiratory viral panel as well as the rapid PCR influenza tests in addressing these changes in use of anti-infectives and infection control issues for admitted patients. In addition, several other antivirals (other than for influenza) are in development, and with the specific identification of a respiratory virus possible, targeted therapy as we currently use for bacterial infections may be possible. Table 12.2 presents details of both influenza A and B assays, as well as of the respiratory multiplex assays. The available panels of these assays differ in the presence of several bacterial targets in some, whereas others include only viral targets. In addition, the recent arrival of a rapid TB (tuberculosis)

Table 12.2. Microbiology techniques for rapid detection of pathogens associated with conditions other than sepsis.

Syndrome	Specimen source	Technology	Name (and manufacturer)	Target(s)	Technical benefits	Technical limitations	Published antibiotic stewardship outcomes
Meningitis	Cerebral spinal fluid (CSF)	SinglePlex PCR	Xpert EV (Cepheid)	*Enterovirus* (EV)	Ability to discontinue antibiotics TAT 2.5 h	Single meningitis pathogen	Yes
			Simplexa™ HSV-1 & 2 (Focus Diagnostics)	HSV-1 and HIV-2 (herpes simplex virus types 1 and 2)	Rapid TAT 1 h		No studies to date
		Multiplex PCR	FilmArray™ Meningitis/ encephalitis (ME) panel (bioMérieux)	22 Bacterial, viral, and fungal targets	Comprehensive syndromic panel TAT 1 h	Equipment cost, single test per instrument	No studies to date
Acute respiratory illness	Nasopharyngeal swab	Isothermal PCR	Alere™ i Influenza A and B test (Alere™)	Influenza A and B	Waived point of care test (POCT) TAT 15 min	No subtyping, clinical performance not as good as other PCR assays	No studies to date
		Biplex PCR	Xpert Flu (Cepheid)	Influenza A and B	Can be used for real-time isolation and treatment TAT 1 h	First-generation test with decreased sensitivity of Flu A detection	Yes
		Triplex PCR	Prodesse Pro Flu+ Prodesse ProFast+ (Hologic)	Influenza A and B RSV[a] Influenza A (including subtypes) Additional targets: parainfluenza, *Metapneumovirus*, Adenovirus	Can customize menu TAT 2.5 h	Multiple manipulations	No studies to date
			Xpert Flu/RSV XC (Cepheid)	Influenza A and B, RSV	May allow discrimination of 2 major viruses in pediatric settings TAT 1 h	Initial data 2nd generation test with improved detection for Influenza A/B	No studies to date
			Simplexa™ FluA/B RSV (Focus Diagnostics)	Influenza A and B, RSV	TAT 2 h	Small batch testing (1–6)	No studies to date

		Multiplex PCR	FilmArray® Respiratory Panel (bioMérieux)	17 Viruses (including Influenza A subtypes)	TAT 1 h	Costly equipment, one test per instrument	Yes (pediatrics only)
			eSensor® Respiratory Viral Panel (RVP) (GenMark)	14 Viruses (including Influenza A subtypes)	Test up to 24 samples at a time TAT 6h		No studies to date
			Verigene® Respiratory Virus plus test (RV+) (Nanosphere)	Influenza A and B, RSV			No studies to date
			xTAG® Respiratory Viral Panel (RVP) (Luminex)	10 Viruses (including Influenza A subtypes)			Yes (pediatrics > adults)
Diarrhea consistent with *Clostridium difficile* infection	Stool	SinglePlex PCR	Multiple manufacturers Performance parameters not statistically significantly different between manufacturers	tcdB and/ or tcdA toxin gene	Aids in patient isolation TAT 1–2.5 h	May detect carriers, limit testing to appropriate patients and restrictions on retesting	Yes
Acute gastroenteritis	Stool in Cary–Blair medium or fresh specimen	Multiplex	BD Max™ Enteric Panel (BD Biosciences)	*Salmonella*, *Shigella*, *Campylobacter*, shiga toxins (sxt-1,sxt-2)	Streamlines lab workflow and increased sensitivity and detection of additional pathogens compared with culture TAT 2.5 h	Equipment costs, batch testing	No studies to date
			BioFire FilmArray® Gastrointestinal (GI) Panel (bioMérieux)	22 Bacterial, virus, parasite, and toxin targets	Streamlines lab workflow, increased sensitivity and detection of additional pathogens compared with culture TAT 1 h	Equipment costs, one test per instrument	No studies to date

Continued

Table 12.2. Continued.

Syndrome	Specimen source	Technology	Name (and manufacturer)	Target(s)	Technical benefits	Technical limitations	Published antibiotic stewardship/ outcomes
			xTAG® Gastrointestinal Pathogen Panel (GPP) (Luminex)	15 Bacterial, virus, parasite, and toxin targets	Increased sensitivity and detection of additional pathogens compared with culture TAT 6 h	Equipment costs, labor intensive, batch testing	Yes
			Verigene® Enteric Pathogens (EP) (Nanosphere)	9 Bacterial, virus, and toxin targets	Increased sensitivity and detection of additional pathogens compared with culture TAT 3 h	Equipment costs, labor intensive, batch testing	No studies to date
Mycobacterium tuberculosis respiratory disease	Respiratory specimens	PCR	Xpert MTB/RIF (Cepheid)	*M. tuberculosis* and rifampin resistance mutations	Both smear positive and negative specimens, allows removal from isolation in suspected cases of TB after 2 negative tests TAT 1 h	Equipment costs	Yes

[a]Respiratory syncytial virus.

PCR test (Xpert® MTB/RIF) has allowed more timely identification, treatment, and isolation protocols, importantly with a high enough negative predictive value to be able to remove patients from isolation and provide cost savings to the hospital (Lippincott *et al.*, 2014).

ACUTE GASTROENTERITIS Currently, there are multiple US Food and Drug Administration (FDA)-cleared multiplex assays for the detection of stool pathogens, with more in development. Every panel has a slightly different menu of pathogens, with all containing the major reportable or culturable pathogens of significance to public health—*Salmonella* spp., *Shigella* spp., *Campylobacter* spp., and shiga toxins sxt-1 and sxt-2. Because it is not uncommon for adults to be prescribed metronidazole and/or ciprofloxacin empirically, a specific diagnosis would certainly limit antibiotic use, and in some studies, direct treatment to the pathogen actually identified. In both adults and pediatric patients, what has been shown to date in most studies is that providers cannot accurately diagnose gastrointestinal (GI) pathogens readily based on symptoms alone, and secondary to overlapping clinical presentations, and that a significant increase in the number of pathogens detected is found with multiplex testing (Dunbar *et al.*, 2013; Buss *et al.*, 2014; Stockmann *et al.*, 2015). There is a great deal to learn about these assays as far as the interpretation of results goes, as it has not been possible to identify many of these pathogens previously with routine methods. However, because the diagnosis in the laboratory will be more streamlined compared with culture and other techniques, laboratories will probably implement this technology, and there are likely to be infection control and stewardship benefits downstream from diagnosing specific infections rapidly (Buss *et al.*, 2014; Goldenberg *et al.*, 2015).

MENINGITIS/ENCEPHALITIS Currently, there is only one multiplex PCR diagnostic in this group. The BioFire Meningitis/Encephalitis Panel has now been cleared by the FDA and the results are in press. No outcome studies have yet been published.

In contrast, Enterovirus (EV) PCR has been used successfully to de-escalate antibiotics in EV positive patients as well as reduce LOS in febrile infants, and has clear value as a single-plex test because it is the most common meningitis etiology in pediatrics (Dewan *et al.*, 2010).

Deficiencies in Current Rapid Assays

While rapid diagnostics has been a great step forward in infectious disease diagnosis and antibiotic stewardship efforts, these assays present some diagnostic and overall healthcare issues. With the molecular multiplex assays that have multiple targets, including resistance markers, the debate is whether the target detected is actually representative of disease or colonization and/or if resistance will be expressed. This is of greatest concern for the detection of *C. difficile* toxin, for antibiotic resistance targets for the purpose of treatment, and for the interpretation of multiple targets in a single sample with multiplex assays, where it is not uncommon to find 10–20% of specimens with more than one pathogen. Thus, for the general provider, interpretation and treatment become more complicated, and education on and the reporting of results from the laboratory of a critical nature (Baron *et al.*, 2013). Importantly, laboratory personnel and those treating the infections cannot presume to know all of the correct answers at this stage. As these assays become more common, it will be exciting to watch the growing understanding of the pathology of diseases and identify how better guidance on treatment can be established. For molecular assays in general, there is the obvious potential advantages of a more sensitive and specific result, and a faster TAT, but the cost benefit for some of that performance will have to be carefully weighed to justify the increased costs overall in the healthcare picture.

For simpler routine technologies, such as MALDI-TOF, the issue is a rapid identification, but with no susceptibility data. This can be addressed in part by using "predictive" reporting from an institution's antibiogram, or by performing direct susceptibility testing from positive broth cultures, but currently the lack of susceptibility data inhibits antibiotic stewardship from providing clear guidance. New diagnostics in development direct from clinical samples could potentially address some of the deficiencies seen here. These include unique systems that rely on a component of phenotypic analysis, along with genomic components, to address the problem of nucleic acid alone "detected" by a molecular assay. The process occurs in a sped-up fashion of 1–4 h, which is acceptable for clinical care, along with sophisticated algorithms to determine identification and susceptibility.

Point-of-care Testing (POCT)

Waived POCT devices exist for molecular based rapid influenza A/B testing and other tests are in development (Table 12.2). This testing potentially brings the era of molecular diagnostics literally to the bedside, and directed antibiotics/antivirals even more rapidly available to the patient with a test that may be better than the current technology. As the assay menu expands, so will the ability to have these instruments become handheld devices with direct connectivity to the EMR. It would not be surprising if these test systems became more commonplace to support timely antibiotic stewardship goals. The cost–benefit compared with rapid response labs, the potential oversight and comparison with other much less expensive methods will have to be assessed per analyte so as not to make them cost prohibitive for the patient and the healthcare system.

Partnering of Diagnostic Companies with Pharmaceuticals

Not surprisingly, diagnostics companies are partnering with pharmaceutical companies to expand the development of rapid resistance detection tests. Given the issue of worldwide resistance issues, this seems to be a logical pursuit and one that will likely enhance development on both the anti-infective and diagnostic side, as well as addressing problems from a different viewpoint as to identifying what is critically needed and how to get it done in a more effective manner.

Evidence-based Care Process Models

Evidence-based care process models (EB-CPMs) are now becoming more commonly implemented within a healthcare system with the ultimate goal of quality improvement for patients. Both diagnostic testing and clear antibiotic stewardship processes are key to the success of these models. For example, in one large pediatric system assessing febrile infants that appear well, the EB-CPM included a broader menu of diagnostic testing to start with, including rapid *Enterovirus* testing and respiratory viral panel testing, as well as clear antibiotic empiric choices and duration. While initial diagnostic testing costs were increased, the benefits were statistically significant for numerous other quality measures, including earlier discharge without antibiotics in patients diagnosed with a viral pathogen, no missed meningitis cases, no readmissions for screening and brief intervention (SBI), patients more likely to receive only the antibiotics recommended, overall cost savings realized in decreased LOS, and reductions in antibiotic prescribing and ancillary testing not recommend by the EB-CPM. With the advent of the EMR, these process models are even easier to implement and will become a standard of care with rapid diagnostic testing and antibiotic stewardship as the main components (Byington *et al.*, 2012).

Conclusions

Increased diagnostic capabilities for rapid detection of infectious disease pathogens, toxins, and resistance markers have allowed the making of major improvements in anti-infective treatment and patient care. In addition, many of these rapid technologies have become simple enough to be placed in most clinical laboratories to enhance antibiotic stewardship efforts in almost any healthcare setting. Often, the results generated are more sensitive than those from traditional culture, more specific for the desired pathogen than biochemical identifications, and allow the diagnosis of an array of different pathogens simultaneously with syndromic panels. Importantly, rapid infectious disease diagnostics as part of an overall care process that includes antibiotic stewardship, coordination with many departments, and optimization for the patient population served, allows favorable clinical outcomes, targeted therapy, improvements in patient safety, and overall cost savings. As healthcare systems change, expand, and become both multi-site and multiple provider practices, the coordination of choices of technology implementation will continue to play a critical role in antibiotic stewardship goals.

References

Barenfanger, J., Graham, D.R., Kolluri, L., Sangwan, G., Lawhorn, J., Drake, C.A., Verhulst, S.J., Peterson, R., Moja, L.B., Ertmoed, M.M. *et al.* (2008) Decreased mortality associated with prompt Gram staining of blood cultures. *American Journal of Clinical Pathology* 130, 870–876.

Baron, E.J., Miller, J.M., Weinstein, M.P., Richter, S.S., Gilligan, P.H., Thomson, R.B., Bourbeau, P., Carroll, K.C., Kehl, S.C., Dunne, W.M. *et al.* (2013) A guide to utilization of the microbiology laboratory for diagnosis of infectious diseases: 2013 recommendations by the Infectious Diseases Society of America (IDSA) and the American Society for Microbiology (ASM). *Clinical Infectious Diseases* 57, e22–e121.

Bartlett, J.G. (2011) John Bartlett's Game Changers in Infectious Diseases: 2011. Available at: http://www.medscape.com/viewarticle/751959 (accessed 10 May 2016).

Bauer, K.A., Perez, K.K., Forrest, G.N., and Goff, D.A. (2014) Review of rapid diagnostic tests used by antimicrobial stewardship programs. *Clinical Infectious Diseases* 59(Suppl 3), S134–S145.

Beekmann, S.E., Diekema, D.J., Chapin, K.C., and Doern, G.V. (2003) Effects of rapid detection of bloodstream infections on length of hospitalization and hospital charges. *Journal of Clinical Microbiology* 41, 3119–3125.

Bonner, A.B., Monroe, K.W., Talley, L.I., Klasner, A.E., and Kimberlin, D.W. (2003) Impact of the rapid diagnosis of influenza on physician decision-making and patient management in the pediatric emergency department: results of a randomized, prospective, controlled trial. *Pediatrics* 112, 363–367.

Buss, S.N., Leber, A., Chapin, K.C., Fey, F.D., Bankowski, J.K., Rogatcheva, M., Kanack, K.J., and Bourzac, K.M. (2014) Multicenter evaluation of the BioFire FilmArray gastrointestinal panel for etiologic diagnosis of infectious gastroenteritis. *Journal of Clinical Microbiology* 53, 915–925.

Byington, C.L., Reynolds, C.C., Korgenski, K., Sheng, X., Valentine, K.J., Nelson, R.E., Daly, J.A., Osguthorpe, R.J., James, B., Savitz, L. *et al.* (2012) Costs and infant outcomes after implementation of a care process model for febrile infants. *Pediatrics* 130(1): e16–e24.

CDC (Centers for Disease Control and Prevention) (2000) Monitoring hospital-acquired infections to promote patient safety—United States, 1990–1999. *Morbidity and Mortality Weekly Report (MMWR)* 49, 149–153.

Chan, P.A., Mermel, L.A., Andrea, S.B., McCulloh, R., Mills, J.P., Echenique, I., Levoen, F., Rybak, N., Cunha, C., Machan, J.T. *et al.* (2011) Distinguishing characteristics between pandemic 2009–2010 influenza A (H1N1) and other viruses in patients hospitalized with respiratory illness. *PLoS ONE* 6(9): e24734.

Dewan, M., Zorc, J.J., Hodinka, R.L., and Shah, S.S. (2010) Cerebrospinal fluid enterovirus testing in infants 56 days or younger. *Archives of Pediatrics and Adolescent Medicine* 164, 824–830.

Dunbar, S.A., Zhang, H., and Tang, Y.-W. (2013) Advanced techniques for detection and identification of microbial agents of gastroenteritis. *Clinics in Laboratory Medicine* 33, 527–552.

Goldenberg, S.D., Bacelar, M., Brazier, P., Bisnauthsing, K., and Edgeworth, J.D. (2015) A cost benefit analysis of the Luminex xTag gastrointestinal pathogen panel for detection of infectious gastroenteritis in hospitalised patients. *Journal of Infection* 70, 504–511.

Hacek, D.M., Paule, S.M., Thomson, R.B., Robiscek, A., and Peterson, L.R. (2009) Implementation of a universal admission surveillance and decolonization program for methicillin-resistant *Staphylococcus aureus* (MRSA) reduces the number of MRSA and total number of *S. aureus* isolates reported by the clinical laboratory. *Journal of Clinical Microbiology* 47, 3749–3752.

Jennings, L.C., Skopnik, H., Burckhardt, I., Hribar, I., Del Piero, L., and Deichmann, K.A. (2009) Effect of rapid influenza testing on the clinical management of paediatric influenza. *Influenza and Other Respiratory Viruses* 3, 91–98.

Kim, D.H., Spencer, M., Davidson, S.M., Li, L., Shaw, J.D., Gulczynski, D., Hunter, D.J., Martha, J.F., Miley, G.B., Parazin, S.J. *et al.* (2010) Institutional prescreening for detection and eradication of methicillin-resistant *Staphylococcus aureus* in patients undergoing elective orthopaedic surgery. *The Journal of Bone and Joint Surgery* 92, 1820–1826.

Kothari, A., Morgan, M., and Haake, D.A. (2014) Emerging technologies for rapid identification of bloodstream pathogens. *Clinical Infectious Diseases* 59, 272–278.

Lean, W.L., Arnup, S., Danchin, M., and Steere, A.C. (2014) Rapid diagnostic tests for group A *Streptococcal pharyngitis*: a meta-analysis. *Pediatrics* 134, 771–781.

Lippincott, C.K., Miller, M.B., Popowitch, E.B., Hanrahan, C.F., and Van Rie, A. (2014) Xpert MTB/RIF assay shortens airborne isolation for hospitalized patients with presumptive tuberculosis in the United States. *Clinical Infectious Diseases* 59, 186–192.

McCulloh, R.J., Andrea, S., Reinert, S., and Chapin, K. (2014) Potential utility of multiplex amplification respiratory viral panel testing in the management of acute respiratory infection in children: a retrospective analysis. *Journal of the Pediatric Infectious Diseases Society* 3, 146–153.

Mermel, L.A., Jefferson, J., Blanchard, K., Parenteau, S., Mathis, B., Chapin, K., and Machan, J.T. (2013) Reducing *Clostridium difficile* incidence, colectomies, and mortality in the hospital setting: a successful multidisciplinary approach. *The Joint Commission Journal on Quality and Patient Safety* 39, 298–305.

Peterson, L.R. and Diekema, D.J. (2010) Point–counterpoint. To screen or not to screen for methicillin-resistant *Staphylococcus aureus*. *Journal of Clinical Microbiology* 48, 683–689.

Schieffer, K.M., Tan, K.E., Stamper, P.D., Somogyi, A., Andrea, S.B., Wakefield, T., Romagnoli, M., Chapin, K.C., Wolk, D.M., and Carroll, K.C. (2014) Multicenter evaluation of the Sepsityper (TM) extraction kit and MALDI-TOF MS for direct identification of positive blood culture isolates using the BD BACTEC (TM)

FX and VersaTREK((R)) diagnostic blood culture systems. *Journal of Applied Microbiology* 116, 934–941.
Stockmann, C., Rogatcheva, M., Harrel, B., Vaughn, M., Crisp, R., Poritz, M., Thatcher, S., Korgenski, E.K., Barney, T., Daly, J., and Pavia, A.T. (2015) How well does physician selection of microbiologic tests identify *Clostridium difficile* and other pathogens in paediatric diarrhoea? Insights using multiplex PCR-based detection. *Clinical Microbiology and Infection* 21, 179.e9–179.e15.
Wolk, D.M. and Dunne, W., Jr. (2011) New technologies in clinical microbiology. *Journal of Clinical Microbiology* 49(9 Supplement), S62–S67.

13 Epidemiology of *Staphylococcus aureus* and Enterococci and an Overview of Antimicrobial Resistance

ALLISON H. BARTLETT AND ROBERT S. DAUM*

The University of Chicago Medicine, Chicago, Illinois, US

Introduction

Gram-positive bacteria are common causes of bloodstream and other infections among hospitalized patients in the US. Methicillin-resistant *Staphylococcus aureus* (MRSA) or *S. aureus* isolates resistant to nearly all β-lactam antimicrobials, and vancomycin-resistant enterococci (VRE) are of particular concern. Often, patients with an infecting isolate, particularly MRSA, are hospitalized, frequently in the intensive care unit (ICU), because of the infection and the community origin of the antibiotic-resistant isolate is not apparent. MRSA colonization and infections, particularly of the skin, are common in the community in many locales in the US and elsewhere. Indeed, an epidemic of skin and soft tissue infections (SSTIs) occurred in the US in the 1990s and in the first decade of the 21st century, although it may be abating now. Community-origin SSTIs continue to be a problem in many parts of the US and elsewhere in the world. VRE are also a problem in patients requiring intensive care, but as described below, they are found among healthy individuals in the community and in animals as well. Colonization with MRSA and VRE is associated with an increased risk of developing invasive infections. Rates of colonization vary, as do institutional policies on isolation and state laws on screening.

Staphylococcus aureus

S. aureus is a major cause of morbidity and mortality. It is the most virulent member of the genus *Staphylococcus*. Unlike other members of the genus, *S. aureus* is well endowed with a variety of virulence factors.

Superficial and invasive *S. aureus* infections occur in previously healthy persons. Infections of the skin range from impetigo to abscess formation, cellulitis, or lymphadenitis, particularly of the cervical lymph nodes in children or adjacent to an infectious focus. *S. aureus* also may cause several important ocular infections, including conjunctivitis, preseptal cellulitis, and endophthalmitis. *S. aureus* is an important cause of endocarditis and is the leading cause of this in many regions of the world. (Bashore *et al.*, 2006). The clinical manifestations may be particularly severe, and the infected heart valve may have been previously normal, especially when the mitral or aortic valves are involved. Pericarditis may be an isolated syndrome or may accompany endocarditis.

Hematogenous seeding of a bone or joint may result in osteomyelitis, septic arthritis, or even bursitis. *S. aureus* is a cause of several respiratory tract infections, including an occasional otitis media and pneumonia. The latter may be a severe, necrotizing process with high mortality. Central nervous system (CNS) infections are infrequent and usually involve a surgically introduced portal of entry for the organism, such as extension from an infected sinus, a dermal sinus, or a meningomyelocele. CNS infectious syndromes also include subdural and epidural empyemas. A spinal epidural abscess adjacent to the dura may be caused by *S. aureus* too, especially in adults. Rarely, *S. aureus* may infect the urinary tract or be recovered from the urine of a patient with high-grade bacteremia or a renal abscess.

*Corresponding author. E-mail: rdaum@peds.bsd.uchicago.edu

S. aureus frequently complicates surgical procedures in which the integument was breached and medical interventions in which indwelling foreign bodies are used in management. The insertion of a plastic, metal, or Gore-Tex device provides an opportunity for *S. aureus* to persistently adhere. Success in removing the organism(s) from the inserted device has posed a great challenge, even when the isolate is susceptible to the antibiotic used. Thus, certain patients, such as those needing hemodialysis, those with indwelling venous catheterization, indwelling intravascular Gore-Tex patches, artificial prostheses, or cerebrospinal fluid flow diversionary devices, are at high risk of *S. aureus* infections.

A variety of veterinary infections are caused by *S. aureus* as well. For example, mastitis in cows is an economically important disease in dairy ruminants. Horses, chickens, dogs, and cats are among the animals known to be colonized or infected with *S. aureus*, with documented transmission to humans from horses (Weese *et al.*, 2006). Mounting evidence suggests that pigs may represent a new important reservoir for CA-MRSA (community-acquired MRSA) strains. ST398 is the most commonly reported MRSA sequence type among livestock in Europe and the US. Transmission to pig farm workers has occurred too, although invasive disease has rarely been documented. Because of its extensive propensity to cause disease in animals, *S. aureus* has been a subject for investigations related to veterinary vaccine development.

S. aureus is also the cause of a number of toxinoses with clinical manifestations that reflect the effects of one or more elaborated and released exotoxins. Examples include toxic shock syndrome, scalded skin syndrome, and food poisoning.

In addition, attention has been called to septic illnesses caused by *S. aureus* with multiorgan failure that likely reflect the activity of one or more *S. aureus* toxins. Examples include so-called severe sepsis, (Mongkolrattanothai and Daum, 2005) an illness with a sometimes fulminant course, purpura fulminans (Kravitz *et al.*, 2005), and the Waterhouse–Friderichsen syndrome (Adem *et al.*, 2005).

S. aureus is the leading cause of hospital-acquired infections (HAIs, also used to mean healthcare-acquired infections) in the US, accounting for 16% of HAIs overall. More than half of healthcare-associated *S. aureus* infections (range 44–59% depending on HAI type) were due to MRSA in data reported to the National Healthcare Safety Network at the Centers for Disease Control and Prevention (CDC) in 2009–2010 (Sievert *et al.*, 2013).

Antibiotic resistance in *S. aureus*

Resistance to penicillin was described shortly after its introduction into clinical practice in the 1940s and is mediated by a transposon-borne β-lactamase. The development of β-lactamase-resistant semi-synthetic penicillins (classically methicillin) solved this problem until resistance to these new agents was first described (Jevons, 1961). Methicillin resistance requires the presence of the *mecA* gene, which produces penicillin binding protein (PBP) 2a, a peptidoglycan-synthesizing enzyme that has a low affinity for β-lactam antibiotics compared with native PBPs (Brown and Reynolds, 1980). The *mecA* gene is located on a mobile genetic element, the staphylococcal chromosome cassette *mec* (SCC*mec*) (Ito *et al.*, 1999); to date, 12 SCC*mec* elements have been described.

Hospital-acquired MRSA (HA-MRSA)

MRSA was first described in healthcare settings, especially in the hospital, and isolates were noted to be resistant to additional antibiotic classes as well as penicillin. The two most common SCC*mec* types in MRSA (until the emergence of CA-MRSA in the 1990s) in the US, SCC*mec* II and III, also contain genes that confer resistance to antimicrobials of other classes, including aminoglycosides, macrolides, and tetracyclines (Hiramatsu *et al.*, 2002).

Community-acquired MRSA (CA-MRSA)

Initially described in the late 1990s, MRSA infections in people with no healthcare exposure became widespread in the following decades (Herold *et al.*, 1998; Los Angeles County Jail *et al.*, 2003; Kaplan *et al.*, 2005), The epidemiologic definition of CA-MRSA put forth by the CDC in 2000 is an MRSA infection in an outpatient or within 48 h of hospitalization in a patient without healthcare-associated risk factors for MRSA infection (hemodialysis, surgery, residence in a long-term care facility, hospitalization during the past year, presence of an indwelling catheter or percutaneous device at the time of culture, or previous history of MRSA) (Morrison *et al.*, 2006).

As the CA-MRSA epidemic evolved, and understanding of the pathogenic mechanisms deepened, definitions become increasingly difficult. CA-MRSA strains in the US are predominantly USA300, which corresponds to sequence type (ST) 8 (Enright *et al.*, 2000; McDougal *et al.*, 2003). The former is a pulsed field gel electrophoretic (PFGE) strain classification and the latter grouping is derived from multilocus sequence classification (MLST). In general, CA-MRSA isolates, unlike HA-MRSA isolates, are susceptible to multiple antibiotics (clindamycin, tetracyclines), and carry the genes for Panton–Valentine leukocidin (a pore-forming, two-component toxin) (Naimi *et al.*, 2003). Unlike HA-MRSA isolates, CA-MRSA isolates contain the smaller, more streamlined SCC*mec* IV, which does not contain additional antibiotic resistance genes (Hiramatsu *et al.*, 2002).

Definitive studies are lacking, but the available information suggests that invasive infection with CA-MRSA is associated with a lower risk of mortality than infection with HA-MRSA strains (Nair *et al.*, 2014).

Vancomycin-intermediate/resistant S. aureus *(VISA/VRSA)*

Vancomycin low-level or intermediate resistant *S. aureus* (VISA), first described in 1997, are isolates with decreased susceptibility to vancomycin (minimum inhibitory concentration or MIC >2–8 μg/ml) and can be isolated during vancomycin therapy (Gardete and Tomasz, 2014). Therapeutic failure is common among patients with VISA strains (Fridkin *et al.*, 2003; Majcherczyk *et al.*, 2008). High-level vancomycin-resistant *S. aureus* (VRSA) strains (MIC ≥16 μg/ml; many with a MIC ≥ 100 μg/ml)) were reported in the US in 2002 (Sievert *et al.*, 2008). Resistance in these strains occurs by a mechanism different from VISA strains— the *vanA* gene cluster that mediates vancomycin resistance in *Enterococcus* was spontaneously transferred to *S. aureus*, presumably at a site of co-colonization or coinfection (Sievert *et al.*, 2008).

Epidemiology/prevalence of MRSA

Rates of colonization

There is significant geographic variation in the prevalence of MRSA colonization. There is also variation in the results obtained depending on the method of assessment: nasal sampling alone underestimates colonization by as much as half when compared with methods that also screen other sites (the throat, axilla, perineum, wounds) (Matheson *et al.*, 2012; McKinnell *et al.*, 2013; Kumar *et al.*, 2015). In addition, longitudinal studies suggest that colonization is fluid with frequent loss, acquisition of new strains, and high rates of polyclonal colonization (Kumar *et al.*, 2015). The most recent estimates (from 2004) suggest that 1.5% of noninstitutionalized, healthy US adult residents have nasal colonization with MRSA (Gorwitz *et al.*, 2008). Certain populations have much higher rates of colonization. The prevalence of MRSA colonization among adult patients admitted to the ICU is approximately 11% (Nair *et al.*, 2011), and prevalence rates over 50% were reported in skilled nursing facilities (Fisch *et al.*, 2012). High rates have been reported in other populations as well.

The detection of MRSA colonization is important because it is associated with an increased risk of developing a MRSA infection. Hospitalized adults with newly identified MRSA colonization had about a 30% rate of developing a subsequent MRSA infection in the 18 months following the detection of MRSA colonization (Huang and Platt, 2003), and half of these infections required readmission to the hospital (Huang *et al.*, 2011). According to a retrospective study of more than 350,000 inpatients in the US Department of Veterans Affairs (VA) system in 2007–2010, patients who acquired MRSA colonization in the hospital were 42% more likely to die in the year following admission than patients who were not colonized by MRSA; patients who developed an MRSA HAI were 49% more likely to die in the year following admission to hospital than patients who were not colonized by MRSA (Nelson *et al.*, 2015).

Epidemiology of MRSA by infection type

Invasive infections

The incidence of invasive MRSA increased from 1999 to 2005, with annual MRSA-related hospitalizations more than doubling from ~ 125,000 to ~ 280,000 during that time (Klein *et al.*, 2007). The estimated incidence of all MRSA infections in US academic medical centers increased from 20.9/1000 hospital discharges to 41.7/1000 hospital discharges in 2003–2008 (David *et al.*, 2012). The trend then appeared to be reversing, with incidence rates of healthcare-associated MRSA infections falling approximately 9% a year from 2005 to 2012 (to a low of 18.74 infections/100,000 population in 2012) (Kallen *et al.*, 2010; Johnson *et al.*, 2014). Rates of MRSA central line-associated bloodstream infections in ICUs in the US decreased by 49.6% from 1997 to 2007 (Burton *et al.*, 2009).

Bacteremia

S. aureus is the most frequent pathogen associated with bloodstream infections in North America (Diekema *et al.*, 2002). After increasing steadily since the 1980s (Karchmer, 2000), rates of MRSA central line-associated bloodstream infections in ICUs in the US decreased by 49.6% from 1997 to 2007 (Burton *et al.*, 2009). While the rates of line-associated bacteremia are decreasing, MRSA bacteremia remains a serious problem (David *et al.*, 2014). Meta-analysis shows that the mortality rate of MRSA bacteremia is nearly twice as high as mortality for MSSA (methicillin-susceptible *S. aureus*) bacteremia (Cosgrove *et al.*, 2003), and is associated with a longer hospital stay (Abramson and Sexton, 1999), and a higher cost (Cosgrove *et al.*, 2005).

Patients with MRSA colonization at ICU admission were significantly more likely to have bacteremia due to MRSA compared with patients without nasal MRSA colonization at the time of ICU admission (PR 5.64, 95% CI, 3.86–8.26; P<.001) (Nair *et al.*, 2011).

Pneumonia

As with many other types of invasive MRSA infections, the incidence of MRSA pneumonia increased in the 1990s and 2000s owing mostly to CA-MRSA (Rubinstein *et al.*, 2008). While other invasive MRSA infections are decreasing, the incidence of healthcare-related MRSA pneumonia is increasing, at least in community hospitals (Lewis *et al.*, 2014). As with other invasive infections, there is excess morbidity and mortality with MRSA pneumonia: in one case series, nearly a third of adult patients with MRSA hospital-acquired or healthcare-associated pneumonia died within 30 days (Tadros *et al.*, 2013). Pneumonia caused by MRSA resistant to vancomycin was associated with an odds ratio (OR) of mortality of 2.5 (Tadros *et al.*, 2013).

MRSA pneumonia has a seasonality, with a peak in the winter months (December–February), aided by an association between influenza infection and MRSA pneumonia (Kallen *et al.*, 2009; Klein *et al.*, 2013; Lewis *et al.*, 2014). Nasal screening for MRSA colonization in patients with pneumonia may be used for antibiotic de-escalation. In adults with pneumonia, the negative predictive value of a MRSA nasal swab was 98.5%, suggesting that patients with a negative nasal swab are unlikely to have MRSA pneumonia (Tilahun *et al.*, 2015).

Epidemiology of MRSA by patient population

ICU

The prevalence of MRSA colonization among adult patients admitted to the ICU is about 11% (Nair *et al.*, 2011). Some 14% of the colonizing strains were CA-MRSA (USA300, USA400 or USA100). Patients colonized with CA-MRSA strains were less likely to have been hospitalized in the previous 12 months (prevalence ratio (PR) 0.39, 95% confidence interval (CI), 0.21–0.73) and less likely to be older (PR 0.97, 95% CI, 0.95–0.98 a year) than patients colonized with HA-MRSA strains (Nair *et al.*, 2011). Patients with MRSA colonization at ICU admission were significantly more likely to have bacteremia due to MRSA compared with patients without nasal MRSA colonization at the time of ICU admission (PR 5.64, 95% CI 3.86–8.26; $P < 0.001$), but there was no difference in overall mortality (Nair *et al.*, 2011).

Neonatal intensive care unit (NICU)

S. aureus is the second-most common pathogen causing hospital-acquired infections in the NICU, after coagulase-negative staphylococci (Hocevar *et al.*, 2012). The incidence of late-onset MRSA infections tripled between 1995 and 2004 (Lessa *et al.*, 2009), but there has been a decrease of ~10% p.a. from 2005 to 2010 (Iwamoto *et al.*, 2013). The prevalence of MRSA colonization in infants in NICUs (and pediatric ICUs, or PICUs) is higher for infants born out of hospital (3.6–5.8%) than in hospital (0.2–1.6%) infants (Myers *et al.*, 2011; Zervou *et al.*, 2014). The acquisition of MRSA during the stay in the NICU or PICU is as high as 6.1% (Zervou *et al.*, 2014). Similar to what has been shown in adults, the relative risk of MRSA infection for neonates colonized with MRSA at admission is 24.2 (95% CI, 8.9–66.0) (Zervou *et al.*, 2014).

Dialysis

MRSA colonization is more prevalent in hemodialysis patients (7.2%; 95% CI, 4.9–9.9%) compared with peritoneal dialysis patients (1.3%; 95% CI, 0.5–2.4%) (Zacharioudakis *et al.*, 2014). Colonization is also more prevalent in hospitalized patients compared with outpatients (14.2%; 95% CI, 8.0–21.8% vs. 5.4%; 95% CI, 3.5–7.7%) (Zacharioudakis *et al.*, 2014). The probability of developing MRSA infection in 6–20 months is higher in MRSA-colonized patients (19%) than in noncolonized patients (2%) (Zacharioudakis *et al.*, 2014).

Liver transplantation

Prospective evaluation of liver transplantation candidates and recipients with and without MRSA colonization revealed a significant difference in the probability of developing MRSA infection. MRSA-colonized patients had a 50% probability of developing an MRSA infection within 1 year. Noncolonized patients had a 5% probability (Russell *et al.*, 2008). MRSA colonization was associated with a decreased 1 year survival (35 vs. 45% for noncolonized patients), but not with increased overall risk of death (Russell *et al.*, 2008). MRSA infection was associated with an increased risk of death (adjusted odds ratio (OR) 2.29, 95% CI, 1.78–4.46) (Russell *et al.*, 2008).

Skilled nursing facilities and nursing homes

Point prevalence studies in skilled nursing facilities have shown a range of MRSA colonization from 10 to 50% (Fisch *et al.*, 2012). Rates of colonization with antibiotic-resistant organisms including MRSA, and rates of infection are significantly higher in patients with indwelling devices (Wang *et al.*, 2012). Similarly, the results of a prospective observational study of nursing home residents with indwelling feeding tubes and/or urinary catheters demonstrated that 50/120 (42%) developed MRSA colonization, mostly (78%) transiently (Gibson *et al.*, 2014). Importantly, screening the nares alone would have failed to ascertain 45.5% of persistent carriers and 74.4% of transient carriers at the time of acquisition; inclusion of sampling of the groin, device insertion sites, and perianal areas identified all carriers (Gibson *et al.*, 2014).

Surveillance

Reliance on clinical microbiological cultures to identify patients colonized with MRSA fails to identify 85% of MRSA-colonized patients (Cassandra *et al.*, 2016). Active surveillance has been implemented in many settings in the hope of decreasing the amount of MRSA transmission and subsequent MRSA infections, morbidity, and mortality. A recent review of the evidence found that universal screening of all hospitalized patients for MRSA carriage decreases the rate of healthcare associated MRSA infection, but the strength of evidence to support this effect is low (Glick *et al.*, 2014).

Discontinuation of contact precautions

A randomized, controlled trial in hospitalized patients with MRSA colonization demonstrated that active screening is superior to passive screening for the discontinuation of MRSA contact precautions and leads to cost savings (Shenoy *et al.*, 2013). A study of long-term inpatients (>30 days) in a facility with an active screening protocol for MRSA and VRE, and a baseline MRSA carriage rate of 46%, demonstrated that 11% of MRSA carriers cleared MRSA colonization at a median of 23 days (interquartile range (IQR) 14–39 days) and only 20% of those became recolonized (Ghosh *et al.*, 2014). MRSA clearance led to discontinuation of contact isolation and was associated with a significant cost saving ($289,112); however, the impact on transmission control was not discussed (Ghosh *et al.*, 2014).

Decolonization

In some patient populations, decolonization has been shown to be effective in preventing subsequent infections. Methods for decolonization include the intranasal application of 2% mupirocin ointment, bathing with 2% chlorhexidine or triclosan, or a combination of these. Meta-analysis of *S aureus* screening and decolonization prior to elective orthopedic procedures is effective at preventing wound complications and is cost saving (Chen *et al.*, 2013). Decision analysis modeling that compared three methods of MRSA prevention in the ICU suggested that a policy of "universal decolonization without screening" (5 days of 2% intranasal mupirocin and daily 2% chlorhexidine baths) is more cost effective than either nasal screening and contact precautions for MRSA-positive patients or "targeted decolonization" (nasal screening, contact precautions, and decolonization of MRSA carriers) (Ziakas *et al.*, 2015). Decolonization with intranasal mupirocin and chlorhexidine bathing has also been effective in decreasing MRSA transmission in long-term care facilities (Schora *et al.*, 2014). Increased rates of mupirocin resistance have not been reported in large-scale trials evaluating its effectiveness (Bode *et al.*, 2013), although decolonization is less effective in patients who have a mupirocin-resistant MRSA strain (Robicsek *et al.*, 2009; Fritz *et al.*, 2013). To date, there does not appear to be a relationship between extended use of chlorhexidine baths and the prevalence of chlorhexidine-resistant MRSA isolates (Schlett *et al.*, 2014).

Vancomycin Resistant Enterococci (VRE)

Enterococcal infections

Enterococci can be components of the normal intestinal flora. The recognition of VRE is a relatively recent development, with the first clinical cases reported in 1988 in France and the UK (Leclercq *et al.*, 1988; Uttley *et al.*, 1988). In the subsequent two decades, VRE has become an important nosocomial pathogen in the US and worldwide (Frieden *et al.*, 1993; Bonten *et al.*, 2001). The initial reports of VRE were among isolates of *E. faecium*, and this remains the predominant species of VRE today.

Enterococcus spp.

E. faecalis

E. faecalis is the clinically predominant species of enterococcus in the US and throughout the world, and rates of vancomycin resistance vary based on geographic location, patient population, and comorbid condition (Biedenbach *et al.*, 2004; Putnam *et al.*, 2010). Surveillance of bloodstream infection isolates from US medical centers in 2010 revealed a 4% rate of vancomycin resistance in *E. faecalis*, with uniform susceptibility to ampicillin and 99.8% susceptibility to linezolid (Arias *et al.*, 2012). Rates of vancomycin resistance in all clinically significant isolates are higher, with up 13.8% resistance reported in a multi-institution survey (Edelsberg *et al.*, 2014).

E. faecium

Rates of vancomycin resistance and antibiotic resistance in *E. faecium* are in general much higher than in *E. faecalis*, and have continued to rise. In 1997–2002, rates of vancomycin resistance among bloodstream infections (BSIs) increased from 40 to 61% (Biedenbach *et al.*, 2004). By 2010, rates of resistance in hospital-acquired infection was as high as 80–89% (Arias *et al.*, 2012; Edelsberg *et al.*, 2014). Nearly all of the *E faecium* isolates were susceptible to daptomycin, linezolid, and quinupristin/dalfopristin (99.7, 98.6, and 94.3%, respectively) (Arias *et al.*, 2012). The rate of ampicillin resistance from bloodstream infection isolates in 2010 was 94.6% (Arias *et al.*, 2012).

Other enterococcal species

E. gallinarum and *E. casseliflavus* are uncommonly isolated from clinical sources and generally have intrinsic low-level resistance to vancomycin.

Epidemiology/prevalence of VRE

There is considerable geographic variability in rates of VRE, with high rates of colonization detected in healthy people and livestock in Europe (Bonten *et al.*, 2001). The high colonization rates in Europe have been attributed in part to the use of the antibiotic avoparicin, a glycopeptic antibiotic that was in wide use in livestock in Europe from the 1970s until it was banned in the late 1990s (Bonten *et al.*, 2001). Clinical isolates collected from sites across the world reveal disparate rates of vancomycin resistance among *E. faecium* isolates: 14.1% in the Asia/Pacific; 31.5% in Europe; 48.1% in Latin America; and 76% in North America (Putnam *et al.*, 2010; Cattoir and Leclercq, 2013). The rate of vancomycin resistance among *E. faecalis* isolates was much lower: 0.01% in the Asia/Pacific; 1.5% in Europe; 3% in Latin America; and 5.6% in North America) (Putnam *et al.*, 2010; Cattoir and Leclercq, 2013).

In 2006–2007, 12% of HAIs were caused by *Enterococcus* sp. (the third most common pathogen), and VRE was the second most common antibiotic-resistant pathogen causing HAI (Hidron *et al.*, 2008). Some 33% of *Enterococcus* isolates causing device-associated HAI (central line-associated bloodstream infections, CLABSI; catheter-associated urinary tract infections, CAUTI; and ventilator-associated pneumonia, VAP) were resistant to vancomycin (Hidron *et al.*, 2008).

Identified risk factors associated with VRE colonization and infection include: previous receipt of vancomycin and/or other broad-spectrum antibiotics, prolonged hospital stay, immunosuppression, severe underlying disease, or intra-abdominal surgery (Hidron *et al.*, 2008).

Epidemiology of VRE by infection type

Bacteremia

Enterococcus spp. are the fourth most frequent pathogens associated with bloodstream infections in North America (Diekema *et al.*, 2002). Rates of VRE causing nosocomial bloodstream infection are as high as 15–30%, with much lower rates of resistance seen in community-acquired BSIs (Biedenbach *et al.*, 2004).

Endocarditis

Hemodialysis and liver transplantation have been identified as risk factors for the acquisition of VRE-infective endocarditis (IE) (Forrest *et al.*, 2011).

Urinary tract infections (UTIs)

In hospitalized patients, *Enterococcus* spp. are the third most common cause of UTIs (Foxman, 2010). A large, single center study comparing VSE (vancomycin-susceptible enterococci) and VRE bacteriuria found no difference in severity of infection (cystitis or pyelonephritis compared with asymptomatic bacteriuria) and no difference in patient outcome (Khair *et al.*, 2013). VRE infection was more common among African–American patients and among patients with renal insufficiency—the latter may be a marker for frequent exposure to the healthcare system (Khair *et al.*, 2013).

VRE occurrence stratified by patient population

ICU

A meta-analysis of the burden of VRE in ICUs showed that about 10% patients admitted to an ICU are colonized at admission and an additional 10% will become colonized during their ICU stay (Ziakas *et al.*, 2013). The rate of development of invasive infection ranged widely among colonized patients—from 0 to 45% (Ziakas *et al.*, 2013). The risk of developing invasive infection by VRE in patients not colonized with VRE was low (<2%) (Ziakas *et al.*, 2013).

Hematology/oncology

Colonization with VRE is associated with an increased risk of developing invasive VRE infection. Hematopoietic stem cell transplant (HSCT) patients are at high risk of becoming colonized or infected with VRE because of the prolonged periods of immunosuppression during conditioning and in preventing graft-versus-host disease (GVHD).

Risk factors for developing VRE BSIs in cancer patients known to be colonized with VRE include the use of vancomycin and gastrointestinal procedures (Zaas *et al.*, 2002). It has been reported that as many as 13.4–29% of VRE-colonized cancer patients subsequently develop VRE bacteremia (Zaas *et al.*, 2002; Matar *et al.*, 2006).

Liver transplantation

Prospective evaluation of liver transplantation candidates and recipients with and without VRE colonization revealed a significant difference in the probability of developing VRE infection. VRE-colonized patients had a 30% probability of developing a VRE infection within 1 year after recognition; VRE-noncolonized patients had a 15% probability (Russell *et al.*, 2008). VRE colonization was associated with an increased risk of death (adjusted OR 2.12, 95% CI, 1.27–3.54) (Russell *et al.*, 2008). VRE infection was associated with an increased risk of death (adjusted OR 2.65, 95% CI, 1.53–4.58) (Russell *et al.*, 2008). Only 2.6% of patients were co-colonized with both VRE and MRSA (Russell *et al.*, 2008).

NICU

Outbreaks of VRE have been reported in NICUs, with colonization reported in up to 39.9% of neonates screened (Iosifidis *et al.*, 2013). The rate of asymptomatic colonization by VRE of infants transferred to NICUs was 1.7% (Macnow *et al.*, 2013). Factors associated with colonization include older postnatal age (age of transfer to the NICU), lower gestational age, and lower birthweight (Macnow *et al.*, 2013). Of the 1751 infants studied, only six had a positive culture for VRE (0.3%), and none were positive for VRE colonization at the time of transfer (Macnow *et al.*, 2013).

Transmission

Nosocomial transmission of VRE is often due to proximity to VRE-positive patients who have not been isolated and is an important risk factor for the acquisition of VRE (Byers *et al.*, 2001). Evaluation of the roommates of patients identified to have VRE colonization or infection (index patients) revealed several factors associated with an increased the risk of VRE acquisition: older age (of the roommate); the presence of a urinary catheter during the time of exposure; and longer exposure to the index patient (Zhou *et al.*, 2008). Some 21% of exposed roommates developed VRE colonization, and all of those who acquired VRE had shared a room with an index patient for more than 3 days (Zhou *et al.*, 2008).

Patient factors influencing transmission include the site of colonization (intestinal tract vs. groin vs. skin) (Bonten *et al.*, 2001). Antibiotic use, especially vancomycin, third-generation cephalosporins, and anti-anaerobic agents, has also been associated with an increased prevalence of VRE (Edmond *et al.*, 1995; Fridkin *et al.*, 2001).

Contamination of medical equipment (blood pressure cuffs, stethoscopes, thermometers) and surfaces (bedside rails, medical charts) also contribute to nosocomial spread (Bonten *et al.*, 2001). Colonization pressure is another key contributor to nosocomial spread of VRE: the more patients on a unit that are colonized with VRE, the higher the chances of spreading VRE to an uncolonized patient (Bonten *et al.*, 1998).

Transmission control

Undetected fecal carriers of VRE represent a large reservoir that may contribute to the transmission of VRE. In addition to humans, many animals have been shown to be colonized with VRE. Other factors associated with transmission include environmental contamination and antibiotic pressure (Bonten *et al.*, 2001). Initiatives focusing on decreasing the inappropriate vancomycin use in the ICU are associated with decreased prevalence of VRE (Fridkin *et al.*, 2002). Switching from detergent-based room cleaning to bleach cleaning and disinfection significantly decreases environmental contamination with VRE, new VRE colonization of inpatients, and VRE bacteremia, while not affecting the rates of patients colonized with VRE at admission (Grabsch *et al.*, 2012).

Persistence

There are limited data on the duration of carriage of VRE, but the proportion of colonized patients appears to decrease over time, with no patients remaining positive for over 4 years (Karki *et al.*, 2013). Also in this study, within the limited number of paired isolates available for evaluation, five of 11 patients had different strain or species types on follow-up sampling, suggesting recolonization rather than persistence of VRE. In the absence of risk factors such as antibiotic exposure and hospitalization, a patient with a remote history (i.e., >4 years) of VRE may not require contact isolation (Karki *et al.*, 2013). A South Korean study evaluating the persistence of VRE colonization after hospital discharge found that 90% of patients cleared VRE colonization within 30 weeks of discharge; factors associated with prolonged VRE carriage included antibiotic use during admission (specifically fluoroquinolone use in subgroup analysis), surgery during admission, transfer to another institution, and dialysis (Sohn *et al.*, 2013).

Duration of isolation

There remains a gap in knowledge of the appropriate duration of contact isolation for a patient with VRE colonization. The Healthcare Infection Control Practices Advisory Committee (HICPAC) suggested in 2006 that contact precautions could be discontinued when three of more surveillance cultures over 1–2 weeks are negative, in the absence of other risk factors (no antibiotic therapy for several weeks, no draining wound, no VRE epidemic) (Siegel *et al.*, 2007).

Surveillance

Guidelines written by the Society for Healthcare Epidemiology of America (SHEA) in 2003 advocated the use of active surveillance cultures (ASC) in conjunction with other measures, such as contact precautions, to control the spread of VRE and MRSA (Muto *et al.*, 2003). Subsequent guidelines from HICPAC suggest that ASC might be added to control programs if routine practices do not decrease the incidence of prevalence of multidrug resistant organisms (MDROs), or if an MDRO is identified in a unit that has not been previously infected (Siegel *et al.*, 2007). Routine surveillance results in the early initiation of contact precautions and decreased transmission, although the cost of surveillance and contact precautions must be considered when making an institution-specific infection control plan (Huang *et al.*, 2007).

A study of long-term inpatients (>30 days) in a facility with an active screening protocol and a baseline VRE carriage rate of 64% demonstrated that 18% of VRE carriers cleared VRE colonization at a median of 26.5 days (IQR 13–45.5 days) and that only 7% became recolonized (Ghosh *et al.*, 2014).

Mechanism of antibiotic resistance in enterococci

Ampicillin

The rate and mechanism of ampicillin resistance differs between *E. faecalis* and *E. faecium*. Ampicillin resistance is uncommon in *E. faecalis*, and the mechanism by which it occurs is uncertain, although β-lactamase production has been reported rarely. A change in the amino acid sequence of PBP 5 has been found in the majority of *E. faecium* isolates and results in significantly decreased affinity for ampicillin and, therefore, high-level ampicillin resistance (MIC > 64 μg/ml) (Arias *et al.*, 2012).

Vancomycin

Vancomycin is a glycopeptide antibiotic that works by inhibiting the synthesis of peptidoglycan, the major component of bacterial cell walls. Specifically, vancomycin binds to the D-Ala–D-Ala terminus of the nascent peptidoglycan chain and causes steric hindrance that blocks subsequent peptidoglycan synthesis. Glycopeptide resistance occurs when bacteria produce alternate pentapeptide precursors ending in D-Ala–D-Lac or D-Ala–D-Ser (reviewed in Courvalin, 2006). There are eight types of acquired resistance that are classified on a phenotypic and genotypic basis: VanA, VanB, VanD, VanE, VanG, VanL, VanM, and VanN). A ninth resistance mechanism, VanC, is intrinsic to *E. gallinarum* and *E. casseliflavus* (Courvalin, 2006). The origin of the *van* operons in *Enterococcus* spp. remains uncertain, although related operons have been identified in glycopeptide-producing organisms (Marshall *et al.*, 1998) and in the vancomycin-resistant biopesticide *Paenibacillus popilliae* (Patel *et al.*, 2000).

VANA The VanA phenotype is predominant among VRE isolates in North America (76%), and less commonly in Europe (40%) (Deshpande *et al.*, 2007). VanA was the first type of glycopeptide resistance identified, and results in inducible high-level resistance to vancomycin and teicoplanin (MIC > 64 mg/l), another glycopeptide not licensed in the US. Resistance is medicated by transposon Tn*1546*, which encodes three important enzymes: VanH, a dehydrogenase that generates D-Lac from pyruvate; VanA, a ligase that links D-Ala and D-Lac, and VanX, which hydrolyzes the wild type D-Ala–D-Ala terminal dipeptide (Cattoir and Leclercq, 2013)—the resulting D-Ala–D-Lac peptidoglycan stem has significantly lower affinity for vancomycin than the standard D-Ala–D-Ala terminus (Bugg *et al.*, 1991).

The *vanA* gene has been found in *S aureus*, and renders it vancomycin resistant (VRSA) (Sievert *et al.*, 2008).

VANB The VanB phenotype results from a similar transposon-borne operon, which leads to the terminal D-Ala–D-Lac dipeptide. Unlike VanA, however, the inducible resistance of VanB results in a variable, but generally lower level of resistance to vancomycin (MIC 16–32 mg/l) with continued susceptibility to teicoplanin (Cattoir and Leclercq, 2013).

VANC VanC, unlike other types of glycopeptide resistance, is intrinsic to enterococci, specifically *E. gallinarum* and *E. casseliflavus*. VanC leads to low-level resistance in these species.

VanC-containing organisms are more common in Europe than in the US, and appear to be associated with a decreased risk of progression from colonization to invasive infection (0.4%) (Sutter *et al.*, 2010). In addition, because epidemics of VanC have not been reported, and because the resistance genes are intrinsic (rather than potentially transmittable via a transposon or plasmid), some institutions advocate using standard precautions (rather than contact isolation) for patients colonized with VanC-containing organisms (Sutter *et al.*, 2010).

OTHER VAN TYPES Resistance types other than VanA, B, and C are very infrequent. VanM and VanD are similar to VanA and VanB, with a D-Ala–D Lac dipeptide and moderate-to-high level vancomycin and teicoplanin resistance. VanE, VanG, VanL, and VanN have a D-Ala–D-Ser terminal dipeptide and low-level vancomycin resistance with maintained teicoplanin susceptibility.

Linezolid and tedizolid

Linezolid and tedizolid have US Food and Drug Administration (FDA) approval for the treatment of VRE infections. They inhibit ribosomal protein synthetic and are bacteriostatic. There are concerns with toxicity during long-term use.

Transferrable resistance to linezolid occurs via the action of Cfr (for chloramphenicol–florfenicol resistance) (Long *et al.*, 2006), an rRNA methyltransferase that leads to methylation of the ribosome, blocking access to 23S rRNAs, a site where multiple antibiotics (chloramphenicol, clindamycin, streptogramin A) work (Long *et al.*, 2006). The largest retrospective study to determine risk factors for VRE with decreased susceptibility found that HSCT or solid organ transplant, and the receipt of immunosuppressive medications were associated on univariate analysis, but that on multivariate analysis, only receipt of linezolid in the prior year remained an independent risk factor for infection with VRE with decreased susceptibility to linezolid (Santayana *et al.*, 2012). The majority (30/48, 63%) of cases had no documented exposure to linezolid in the year prior to infection (Santayana *et al.*, 2012).

Quinupristin/dalfopristin

Quinupristin/dalfopristin is a synergistic combination of two streptogramins that interfere with protein synthesis (Hershberger *et al.*, 2004). It is the second FDA-approved compound for the treatment of vancomycin-resistant *E. faecium* infections; *E. faecalis* is inherently resistant to quinupristin/dalfopristin because these bacteria have the gene *lsa* (for lincosamide and streptogramin A resistance), which codes a species-specific ATP-binding protein (Singh *et al.*, 2002). *E. faecium* can develop resistance via several mechanisms, including active transport, enzymatic modification of the antibiotic, or alteration of the target site (Hershberger *et al.*, 2004).

Virginiamycin is a streptogramin used as a feed additive to promote weight gain in chickens, turkeys, swine, and cattle. The use of virginiamycin in animals can lead to quinupristin/dalfopristin resistance in *E faecium*, thus preventing the use of quinupristin/dalfopristin for the treatment of human enterococcal infection. Enterococci can serve as the host for the spread of multiple resistance genes on a single element from one species to another (Hershberger *et al.*, 2004).

Daptomycin

Daptomycin is approved by the FDA for the treatment of susceptible Gram-positive organisms and has been used widely to treat VRE infections (Cantón *et al.*, 2010). The antibiotic acts by altering the membrane potential of Gram-positive organisms without causing cell lysis via interaction with the cytoplasmic membrane (Silverman *et al.*, 2003; Cotroneo *et al.*, 2008). In *E. faecium*, a survey of antibiotic resistance in admissions to 19 US hospitals showed that daptomycin resistance had increased to about 4% of clinical isolates, higher than had been reported up to then; *E. faecalis* resistance rates remained very low, at 0.2% (Edelsberg *et al.*, 2014). The genetic basis for resistance is incompletely understood, but appears to involve multiple mutations that result in diversion of the antibiotic molecule from the division septum and "trapping" of it in a separate membrane region (Tran *et al.*, 2013).

References

Abramson, M.A. and Sexton, D.J. (1999) Nosocomial methicillin-resistant and methicillin-susceptible *Staphylococcus aureus* primary bacteremia: at what costs? *Infection Control and Hospital Epidemiology* 20, 408–411.

Adem, P.V., Montgomery, C.P., Husain, A.N., Koogler, T.K., Arangelovich, V., Humilier, M., Boyle-Vavra, S., and Daum, R.S. (2005) *Staphylococcus aureus* sepsis and the Waterhouse–Friderichsen syndrome in children. *The New England Journal of Medicine* 353, 1245–1251.

Arias, C.A., Mendes, R.E., Stilwell, M.G., Jones, R.N., and Murray, B.E. (2012) Unmet needs and prospects for oritavancin in the management of vancomycin-resistant enterococcal infections. *Clinical Infectious Diseases* 54, S233–S238.

Bashore, T.M., Cabell, C. and Fowler, V., Jr. (2006) Update on infective endocarditis. *Current Problems in Cardiology* 31, 274–352.

Biedenbach, D.J., Moet, G.J., and Jones, R.N. (2004) Occurrence and antimicrobial resistance pattern comparisons among bloodstream infection isolates from the SENTRY Antimicrobial Surveillance Program (1997–2002). *Diagnostic Microbiology and Infectious Disease* 50, 59–69.

Bode, L.G.M., Kluytmans, J.A., Wertheim, H.F., Bogaers, D., Vandenbroucke-Grauls, C.M., Roosendaal, R., Troelstra, A., Box, A.T., Voss, A., van der Tweel, I. *et al.* (2013) Preventing surgical-site infections in nasal carriers of *Staphylococcus aureus*. *The New England Journal of Medicine* 362, 9–17.

Bonten, M.J.M., Slaughter, S., Ambergen, A.W., Hayden, M.K., van Voorhis, J., Nathan, C., and Weinstein, R.A. (1998) The role of "colonization pressure" in the spread of vancomycin-resistant enterococci: an important infection control variable. *Archives of Internal Medicine* 158, 1127–1132.

Bonten, M.J.M., Willems, R., and Weinstein, R.A. (2001) Vancomycin-resistant enterococci: why are they here, and where do they come from? *The Lancet Infectious Diseases* 1, 314–325.

Brown, D.F.J. and Reynolds, P.E. (1980) Intrinsic resistance to β-lactam antibiotics in *Staphylococcus aureus*. *FEBS Letters* 122, 275–278.

Bugg, T.D., Wright, G.D., Dutka-Malen, S., Arthur, M., Courvalin, P., and Walsh, C.T. (1991) Molecular basis for vancomycin resistance in *Enterococcus faecium* BM4147: biosynthesis of a depsipeptide peptidoglycan precursor by vancomycin resistance proteins VanH and VanA. *Biochemistry* 30, 10408–10415.

Burton, D.C., Edwards, J.R., Horan, T.C., Jernigan, J.A., and Fridkin, S.K. (2009) Methicillin-resistant *Staphylococcus aureus* central line-associated bloodstream infections in US intensive care units, 1997–2007. *JAMA (The Journal of the American Medical Association)* 301, 727–736.

Byers, K.E., Anglim, A.M., Anneski, C.J., Germanson, T.P., Gold, H.S., Durbin, L.J., Simonton, B.M., and Farr, B.M. (2001) A hospital epidemic of vancomycin-resistant *Enterococcus*: risk factors and control. *Infection Control and Hospital Epidemiology* 22, 140–147.

Cantón, R., Ruiz-Garbajosa, P., Chaves, R.L., and Johnson, A.P. (2010) A potential role for daptomycin in enterococcal infections: What is the evidence? *Journal of Antimicrobial Chemotherapy* 65, 1126–1136.

Cassandra, D., Salgado, M.M., and Farr, B.M. (2016) What proportion of hospital patients colonized with methicillin-resistant *Staphylococcus aureus* are identified by clinical microbiological cultures? *Infection Control and Hospital Epidemiology* 27, 116–121.

Cattoir, V. and Leclercq, R. (2013) Twenty-five years of shared life with vancomycin-resistant enterococci: is it time to divorce? *Journal of Antimicrobial Chemotherapy* 68, 731–742.

Chen, A., Wessel, C., and Rao, N. (2013) *Staphylococcus aureus* screening and decolonization in orthopaedic surgery and reduction of surgical site infections. *Clinical Orthopaedics and Related Research* 471, 2383–2399.

Cosgrove, S.E., Sakoulas, G., Perencevich, E.N., Schwaber, M.J., Karchmer, A.W., and Carmeli, Y. (2003) Comparison of mortality associated with methicillin-resistant and methicillin-susceptible *Staphylococcus aureus* bacteremia: a meta-analysis. *Clinical Infectious Diseases* 36, 53–59.

Cosgrove, S.E., Qi, Y., Kaye, K.S., Harbarth, S., Karchmer, A.W., and Carmeli, Y. (2005) The impact of methicillin resistance in *Staphylococcus aureus* bacteremia on patient outcomes: mortality, length of stay, and hospital charges. *Infection Control and Hospital Epidemiology* 26, 166–174.

Cotroneo, N., Harris, R., Perlmutter, N., Beveridge, T., and Silverman, J.A. (2008) Daptomycin exerts bactericidal activity without lysis of *Staphylococcus aureus*. *Antimicrobial Agents and Chemotherapy* 52, 2223–2225.

Courvalin, P. (2006) Vancomycin resistance in Gram-positive cocci. *Clinical Infectious Diseases* 42, S25–S34.

David, M.Z., Medvedev, S., Hohmann, S.F., Ewigman, B., and Daum, R.S. (2012) Increasing burden of methicillin-resistant *Staphylococcus aureus* hospitalizations at US Academic Medical Centers, 2003–2008. *Infection Control and Hospital Epidemiology* 33(8), 782–789.

David, M.Z., Daum, R.S., Bayer, A.S., Chambers, H.F., Fowler, V.G., Jr., Miller, L.G., Ostrowsky, B., Baesa, A., Boyle-Vavra, S., Eells, S.J. *et al.* (2014) *Staphylococcus aureus* bacteremia at 5 US academic medical centers, 2008–2011: significant geographic variation in community-onset infections. *Clinical Infectious Diseases* 59, 798–807.

Deshpande, L.M., Fritsche, T.R., Moet, G.J., Biedenbach, D.J., and Jones, R.N. (2007) Antimicrobial resistance and molecular epidemiology of vancomycin-resistant enterococci from North America and Europe: a report from the SENTRY antimicrobial surveillance program. *Diagnostic Microbiology and Infectious Disease* 58, 163–170.

Diekema, D.J., Pfaller, M.A., and Jones, R.N. (2002) Age-related trends in pathogen frequency and antimicrobial susceptibility of bloodstream isolates in North America: SENTRY Antimicrobial Surveillance Program, 1997–2000. *International Journal of Antimicrobial Agents* 20, 412–418.

Edelsberg, J., Weycker, D., Barron, R., Li, X., Wu, H., Oster, G., Badre, S., Langeberg, W.J., and Weber, D.J. (2014) Prevalence of antibiotic resistance in US hospitals. *Diagnostic Microbiology and Infectious Disease* 78, 255–262.

Edmond, M.M., Ober, J.F., Weinbaum, D.L., Pfaller, M.A., Hwang, T., Sanford, M.D., and Wenzel RP. (1995) Vancomycin-resistant *Enterococcus faecium* bacteremia: risk factors for infection. *Clinical Infectious Diseases* 20, 1126–1133.

Enright, M.C., Day, N.P., Davies, C.E., Peacock, S.J., and Spratt, B.G. (2000) Multilocus sequence typing for characterization of methicillin-resistant and methicillin-susceptible clones of *Staphylococcus aureus*. *Journal of Clinical Microbiology* 38, 1008–1015.

Fisch, J., Lansing, B., Wang, L., Symons, K., Cherian, K., McNamara, S., and Mody, L. (2012) New acquisition of antibiotic-resistant organisms in skilled nursing facilities. *Journal of Clinical Microbiology* 50, 1698–1703.

Forrest, G.N., Arnold, R.S., Gammie, J.S., and Gilliam, B.L. (2011) Single center experience of a vancomycin resistant enterococcal endocarditis cohort. *Journal of Infection* 63, 420–428.

Foxman, B. (2010) The epidemiology of urinary tract infection. *Nature Reviews Urology* 7, 653–660.

Fridkin, S.K., Edwards, J.R., Courval, J.M., Hill, H., Tenover, F.C., Lawton, R., Gaynes, R.P., McGowan, J.E., Jr., Intensive Care Antimicrobial Resistance Epidemiology

(ICARE) Project; and the National Nosocomial Infections Surveillance (NNIS) System Hospitals (2001) The effect of vancomycin and third-generation cephalosporins on prevalence of vancomycin-resistant enterococci in 126 U.S. adult intensive care units. *Annals of Internal Medicine* 135, 175–183.

Fridkin, S.K., Lawton, R., Edwards, J.R., Tenover, F.C., McGowan, J.E., Jr., Gaynes, R.P., Intensive Care Antimicrobial Resistance Epidemiology (ICARE) Project, and the National Nosocomial Infections Surveillance (NNIS) System Hospitals (2002) Monitoring antimicrobial use and resistance: comparison with a national benchmark on reducing vancomycin use and vancomycin-resistant enterococci. *Emerging Infectious Diseases Journal* 8, 702–707.

Fridkin, S.K., Hageman, J., McDougal, L.K., Mohammed, J., Jarvis, W.R., Perl, T.M., Tenover, F.C., and the Vancomycin-Intermediate *Staphylococcus aureus* Epidemiology Study Group (2003) Epidemiological and microbiological characterization of infections caused by *Staphylococcus aureus* with reduced susceptibility to vancomycin, United States, 1997–2001. *Clinical Infectious Diseases* 36, 429–439.

Frieden, T.R., Munsiff, S.S., Low, D.E., Willey, B.M., Williams, G., Faur, Y., Eisner, W., Warren, S., and Kreiswirth, B. (1993) Emergence of vancomycin-resistant enterococci in New York City. *The Lancet* 342, 76–79.

Fritz, S.A., Hogan, P.G., Camins, B.C., Ainsworth, A.J., Patrick, C., Martin, M.S., Krauss, M.J., Rodriguez, M., and Burnham, C.A. (2013) Mupirocin and chlorhexidine resistance in *Staphylococcus aureus* in patients with community-onset skin and soft tissue infections. *Antimicrobial Agents and Chemotherapy* 57, 559–568.

Gardete, S. and Tomasz, A. (2014) Mechanisms of vancomycin resistance in *Staphylococcus aureus. JCI (The Journal of Clinical Investigation)* 124, 2836–2840.

Ghosh, A., Jiao, L., Al-Mutawa, F., O'Neill, C., Mertz, D., and Hamilton Health Sciences Infection Prevention and Control Team (2014) Value of an active surveillance policy to document clearance of meticillin-resistant *Staphylococcus aureus* and vancomycin-resistant enterococci amongst inpatients with prolonged admissions. *Journal of Hospital Infection* 88, 230–233.

Gibson, K.E., McNamara, S.E., Cassone, M., Perri, M.B., Zervos, M., Mody, L., and Targeted Infection Prevention Study Team (2014) Methicillin-resistant *Staphylococcus aureus*: site of acquisition and strain variation in high-risk nursing home residents with indwelling devices. *Infection Control and Hospital Epidemiology* 35, 1458–1465.

Glick, S.B., Samson, D.J., Huang, E.S., Vats, V., Aronson, N., and Weber, S.G. (2014) Screening for methicillin-resistant *Staphylococcus aureus:* a comparative effectiveness review. *American Journal of Infection Control* 42, 148–155.

Gorwitz, R.J., Kruszon-Moran, D., McAllister, S.K., McQuillan, G., McDougal, L.K., Fosheim, G.E., Jensen, B.J., Killgore, G., Tenover, F.C., and Kuehnert, M.J. (2008) Changes in the prevalence of nasal colonization with *Staphylococcus aureus* in the United States, 2001–2004. *The Journal of Infectious Diseases* 197, 1226–1234.

Grabsch, E.A., Mahony, A.A., Cameron, D.R., Martin, R.D., Heland, M., Davey, P., Petty, M., Xie, S., and Grayson, M.L. (2012) Significant reduction in vancomycin-resistant enterococcus colonization and bacteraemia after introduction of a bleach-based cleaning–disinfection programme. *Journal of Hospital Infection* 82, 234–242.

Herold, B.C., Immergluck, L.C., Maranan, M.C., Lauderdale, D.S., Gaskin, R.E, Boyle-Vavra, S., and Leit, C.D. (1998) Community-acquired methicillin-resistant *Staphylococcus aureus* in children with no identified predisposing risk. *JAMA (The Journal of the American Medical Association)* 279, 593–598.

Hershberger, E., Donabedian, S., Konstantinou, K., and Zervos, M.J. (2004) Quinupristin–dalfopristin resistance in Gram-positive bacteria: mechanism of resistance and epidemiology. *Clinical Infectious Diseases* 38, 92–98.

Hidron, A.I., Edwards, J.R., Patel, J., Horan, T.C., Sievert, D.M., Pollock, D.A., Fridkin, S.K. National Healthcare Safety Network Team, and Participating National Healthcare Safety Network Facilities (2008) NHSN Annual Update. Antimicrobial-resistant pathogens associated with healthcare-associated infections: annual summary of data reported to the National Healthcare Safety Network at the Centers for Disease Control and Prevention, 2006–2007. *Infection Control and Hospital Epidemiology* 29, 996–1011.

Hiramatsu, K., Katayama, Y., Yuzawa, H., and Ito, T. (2002) Molecular genetics of methicillin-resistant *Staphylococcus aureus. International Journal of Medical Microbiology* 292, 67–74.

Hocevar, S.N., Edwards, J.R., Horan, T.C., Morrell, G.C., Iwamoto, M., and Lessa, F.C. (2012) Device-associated infections among neonatal intensive care unit patients: incidence and associated pathogens reported to the National Healthcare Safety Network, 2006–2008. *Infection Control and Hospital Epidemiology* 33, 1200–1206.

Huang, S.S. and Platt, R. (2003) Risk of methicillin-resistant *Staphylococcus aureus* infection after previous infection or colonization. *Clinical Infectious Diseases* 36, 281–285.

Huang, S.S., Rifas-Shiman, S.L., Pottinger, J.M., Herwaldt, L.A., Zembower, T.R., Noskin, G.A., Cosgrove, S.E., Perl, T.M., Curtis, A.B., Tokars, J.L. *et al.* (2007) Improving the assessment of vancomycin-resistant enterococci by routine screening. *The Journal of Infectious Diseases* 195, 339–346.

Huang, S.S., Hinrichsen, V.L., Datta, R., Spurchise, L., Miroshnik, I., Nelson, K., and Platt, R. (2011) Methicillin-resistant *Staphylococcus aureus* infection and hospitalization in high-risk patients in the year following detection. *PLoS ONE* 6(9): e24340.

Iosifidis, E., Evdoridou, I., Agakidou, E., Chochliourou, E., Protonotariou, E., Karakoula, K., Stathis, I., Sofianou, D., Drossou-Agakidou, V., Pournaras, S., and Roilides, E. (2013) Vancomycin-resistant *Enterococcus* outbreak in a neonatal intensive care unit: epidemiology, molecular analysis and risk factors. *American Journal of Infection Control* 41, 857–861.

Ito, T., Katayama, Y., and Hiramatsu, K. (1999) Cloning and nucleotide sequence determination of the entire *mec* DNA of pre-methicillin-resistant *Staphylococcus aureus* N315. *Antimicrobial Agents and Chemotherapy* 43, 1449–1458.

Iwamoto, M., Mu, Y., Lynfield, R., Bulens, S.N., Nadle, J., Aragon, D., Petit, S., Ray, S.M., Harrison, L.H., Dumyati, G. *et al*. (2013) Trends in invasive methicillin-resistant *Staphylococcus aureus* infections. *Pediatrics* 132(4): e817–e824.

Jevons, M. (1961) "Celbenin"-resistant staphylococci. *BMJ* 1, 124–125.

Johnson, N., Hayes, L.D., Brown, K., Hoo, E.C., and Ethier, K.A. (2014) CDC National Health Report: leading causes of morbidity and mortality and associated behavioral risk and protective factors—United States, 2005–2013. *Morbidity and Mortality Weekly Report (MMWR)* 63(Suppl 04), 3–27.

Kallen, A.J., Brunkard, J., Moore, Z., Budge, P., Arnold, K.E., Fosheim, G., Finelli, L., Beekmann, S.E., Polgreen, P.M., Gorwitz, R., and Hageman, J. (2009) *Staphylococcus aureus* community-acquired pneumonia during the 2006 to 2007 influenza season. *Annals of Emergency Medicine* 53, 358–365.

Kallen, A.J., Mu, Y., Bulens, S., Reingold, A., Petit, S., Gershman, K., Ray, S.M., Harrison, L.H., Lynfield, R., Dumyati, G. *et al*. (2010) Health care–associated invasive MRSA infections, 2005–2008. *JAMA (The Journal of the American Medical Association)* 304, 641–647.

Kaplan, S.L., Hulten, K.G., Gonzalez, B.E., Hammerman, W.A., Lamberth, L., Versalovic, J., and Mason, E.O., Jr. (2005) Three-year surveillance of community-acquired *Staphylococcus aureus* infections in children. *Clinical Infectious Diseases* 40, 1785–1791.

Karchmer, A.W. (2000) Nosocomial bloodstream infections: organisms, risk factors, and implications. *Clinical Infectious Diseases* 31(Supplement 4), S139–S143.

Karki, S., Land, G., Aitchison, S., Kennon, J., Johnson, P.D.R., Ballard, S.A., Leder, K., and Cheng, A.C. (2013) Long-term carriage of vancomycin-resistant enterococci in patients discharged from hospitals: a 12-year retrospective cohort study. *Journal of Clinical Microbiology* 51, 3374–3379.

Khair, H.N., VanTassell, P., Henderson, J.P., Warren, D.K., Marschall, J., and the CDC Prevention Epicenters Program (2013) Vancomycin resistance has no influence on outcomes of enterococcal bacteriuria. *Journal of Hospital Infection* 85, 183–188.

Klein, E., Smith, D.L., and Laxminarayan, R. (2007) Hospitalizations and deaths caused by methicillin-resistant *Staphylococcus aureus*, United States, 1999–2005. *Emerging Infectious Disease Journal* 13, 1840.

Klein, E.Y., Sun, L., Smith, D.L., and Laxminarayan, R. (2013) The changing epidemiology of methicillin-resistant *Staphylococcus aureus* in the United States: a national observational study. *American Journal of Epidemiology* 177, 666–674.

Kravitz, G.R., Dries, D.J., Peterson, M.L., and Schlievert, P.M. (2005) Purpura fulminans due to *Staphylococcus aureus*. *Clinical Infectious Diseases* 40, 941–947.

Kumar, N., David, M.Z., Boyle-Vavra, S., Sieth, J., and Daum, R.S. (2015) High *Staphylococcus aureus* colonization prevalence among patients with skin and soft tissue infections and controls in an urban emergency department. *Journal of Clinical Microbiology* 53, 810–815.

Leclercq, R., Derlot, E., Duval, J., Courvalin, P. (1988) Plasmid-mediated resistance to vancomycin and teicoplanin in *Enterococcus faecium*. *The New England Journal of Medicine* 319, 157–161.

Lessa, F.C., Edwards, J.R., Fridkin, S.K., Tenover, F.C., Horan, T.C., and Gorwitz, R.J. (2009) Trends in incidence of late-onset methicillin-resistant *Staphylococus aureus* infection in neonatal intensive care units: data from the National Nosocomial Infections Surveillance System, 1995–2004. *The Pediatric Infectious Disease Journal* 28, 577–581.

Lewis, S.S., Walker, V.J., Lee, M.S., Chen, L., Moehring, R.W., Cox, C.E., Sexton, D.J., and Anderson, D.J. (2014) Epidemiology of methicillin-resistant *Staphylococcus aureus* pneumonia in community hospitals. *Infection Control and Hospital Epidemiology* 35, 1452–1457.

Long, K.S., Poehlsgaard, J., Kehrenberg, C., Schwarz, S., and Vester, B. (2006) The Cfr rRNA methyltransferase confers resistance to phenicols, lincosamides, oxazolidinones, pleuromutilins, and streptogramin A antibiotics. *Antimicrobial Agents and Chemotherapy* 50, 2500–2505.

Los Angeles County Jail, Los Angeles County Department of Health Services, and CDC (Centers for Disease Control and Prevention) (2003) Public health dispatch: outbreaks of community-associated methicillin-resistant *Staphylococcus aureus* skin infections—Los Angeles County, 2002–2003. *Morbidity and Mortality Weekly Report (MMWR)* 52, 88.

Macnow, T., O'Toole, D., DeLaMora, P., Murray, M., Rivera, K., Whittier, S., Ross, B., Jenkins, S., Saiman, L., and Duchon, J. (2013) Utility of surveillance cultures for antimicrobial resistant organisms in infants transferred to the neonatal intensive care unit. *The Pediatric Infectious Disease Journal* 32(12): e443–e450.

Majcherczyk, P.A., Barblan, J.L., Moreillon, P., and Entenza, J.M. (2008) Development of glycopeptide-intermediate resistance by *Staphylococcus aureus* leads to attenuated infectivity in a rat model of endocarditis. *Microbial Pathogenesis* 45, 408–414.

Marshall, C.G., Lessard, I.A., Park, I., and Wright, G.D. (1998) Glycopeptide antibiotic resistance genes in glycopeptide-producing organisms. *Antimicrobial Agents and Chemotherapy* 42, 2215–2220.

Matar, M.J., Safdar, A., and Rolston, K.V.I. (2006) Relationship of colonization with vancomycin-resistant enterococci and risk of systemic infection in patients with cancer. *Clinical Infectious Diseases* 42, 1506–1507.

Matheson, A., Christie, P., Stari, T., Kavanagh, K., Gould, I.M., Masterton, R., and Reilly, J.S. (2012) Nasal swab screening for methicillin-resistant *Staphylococcus aureus*—how well does it perform? A cross-sectional study. *Infection Control and Hospital Epidemiology* 33, 803–808.

McDougal, L.K., Steward, C.D., Killgore, G.E., Chaitram, J.M., McAllister, S.K., Tenover, F.C. (2003) Pulsed-field gel electrophoresis typing of oxacillin-resistant *Staphylococcus aureus* isolates from the United States: establishing a national database. *Journal of Clinical Microbiology* 41, 5113–5120.

McKinnell, J.A., Huang, S.S., Eells, S.J., Cui, E., and Miller, L.G. (2013) Quantifying the impact of extranasal testing of body sites for methicillin-resistant *Staphylococcus aureus* colonization at the time of hospital or intensive care unit admission. *Infection Control and Hospital Epidemiology* 34, 161–170.

Mongkolrattanothai, K. and Daum, R.S. (2005) Impact of community-associated methicillin-resistant *Staphylococcus aureus* on management of the skin and soft tissue infections in children. *Current Infectious Disease Reports* 7, 381–389.

Morrison, M.A., Hageman, J.C., and Klevens, R.M. (2006) Case definition for community-associated methicillin-resistant *Staphylococcus aureus*. *The Journal of Hospital Infection* 62(2), 241.

Muto, C.A., Jernigan, J.A., Ostrowsky, B.E., Richet, H.M., Jarvis, W.R., Boyce, J.M., Farr, B.M., and SHEA. (2003) SHEA guideline for preventing nosocomial transmission of multidrug-resistant strains of *Staphylococcus aureus* and Enterococcus. *Infection Control and Hospital Epidemiology* 24, 362–386.

Myers, P.J., Marcinak, J., David, M.Z., Zychowski, D.L., Boyle-Vavra, S., Weber, S., and Daum, R.S. (2011) Universal admission screening for methicillin-resistant *Staphylococcus aureus* in a level IIID neonatal intensive care unit: the first 9 months. *Infection Control and Hospital Epidemiology* 32, 398–400.

Naimi, T.S., LeDell, K.H., Como-Sabetti, K., Borchardt, S.M., Boxrud, D.J., Etienne, J., Johnson, S.K., Vandenesch, F., Fridkin, S., O'Boyle, C. *et al.* (2003) Comparison of community- and health care-associated methicillin-resistant *Staphylococcus aureus* infection. *JAMA (The Journal of the American Medical Association)* 290, 2976–2984.

Nair, N., Kourbatova, E., Poole, K., Huckabee, C.M., Murray, P., Huskins, W.C., and Blumberg, H.M. (2011) Molecular epidemiology of methicillin-resistant *Staphylococcus aureus* (MRSA) among patients admitted to adult intensive care units: the STAR*ICU trial. *Infection Control and Hospital Epidemiology* 32, 1057–1063.

Nair, R., Ammann, E., Rysavy, M., and Schweizer, M.L. (2014) Mortality among patients with methicillin-resistant *Staphylococcus aureus* USA300 versus non-USA300 invasive infections: a meta-analysis. *Infection Control and Hospital Epidemiology* 35, 31–41.

Nelson, R.E., Stevens, V.W., Jones, M., Samore, M.H., and Rubin, M.A. (2015) Health care-associated methicillin-resistant *Staphylococcus aureus* infections increases the risk of postdischarge mortality. *American Journal of Infection Control* 43, 38–43.

Patel, R., Piper, K., Cockerill, F.R., 3rd, Steckelberg, J.M., and Yousten, A.A. (2000) The biopesticide *Paenibacillus popilliae* has a vancomycin resistance gene cluster homologous to the enterococcal VanA vancomycin resistance gene cluster. *Antimicrobial Agents and Chemotherapy* 44, 705–709.

Putnam, S.D., Sader, H.S., Moet, G.J., Mendes, R.E., and Jones, R.N. (2010) Worldwide summary of telavancin spectrum and potency against Gram-positive pathogens: 2007 to 2008 surveillance results. *Diagnostic Microbiology and Infectious Disease* 67, 359–368.

Robicsek, A., Beaumont, J.L., Thomson, R.B., Jr., Govindarajan, G., and Peterson, L.R. (2009) Topical therapy for methicillin-resistant *Staphylococcus aureus* colonization: impact on infection risk. *Infection Control and Hospital Epidemiology* 30, 623–632.

Rubinstein, E., Kollef, M.H., and Nathwani, D. (2008) Pneumonia caused by methicillin-resistant *Staphylococcus aureus*. *Clinical Infectious Diseases* 46(Supplement 5), S378–S385.

Russell, D.L., Flood, A., Zaroda, T.E., Acosta, C., Riley, M.M.S., Busuttil, R.W., and Pegues, D.A. (2008) Outcomes of colonization with MRSA and VRE among liver transplant candidates and recipients. *American Journal of Transplantation* 8, 1737–1743.

Santayana, E.M., Grim, S.A., Janda, W.M., Layden, J.E., Lee, T.A., and Clark, N.M. (2012) Risk factors and outcomes associated with vancomycin-resistant *Enterococcus* infections with reduced susceptibilities to linezolid. *Diagnostic Microbiology and Infectious Disease* 74, 39–42.

Schlett, C.D., Millar, E.V., Crawford, K.B., Cui, T., Lanier, J.B., Tribble, D.R., and Ellis, M.W. (2014) Prevalence of chlorhexidine-resistant methicillin-resistant *Staphylococcus aureus* following prolonged exposure. *Antimicrobial Agents and Chemotherapy* 58, 4404–4410.

Schora, D.M., Boehm, S., Das, S., Patel, P.A., O'Brien, J., Hines, C., Burdsall, D., Beaumont, J., Peterson, K., Fausone, M., and Peterson, L.R.. (2014) Impact of Detection, Education, Research and Decolonization without Isolation in Long-term care (DERAIL) on methicillin-resistant *Staphylococcus aureus* colonization and transmission at 3 long-term care facilities. *American Journal of Infection Control* 42(10, Supplement), S269–S273.

Shenoy, E.S., Kim, J., Rosenberg, E.S., Cotter, J.A., Lee H., Walensky, R.P., and Hooper, D.C. (2013) Discontinuation of contact precautions for methicillin-resistant *Staphylococcus aureus*: a randomized controlled trial comparing passive and active screening with culture and polymerase chain reaction. *Clinical Infectious Diseases* 57, 176–184.

Siegel, J.D., Rhinehart, E., Jackson, M., Chiarello, L., and Healthcare Infection Control Practices Advisory Committee (2007) Management of multidrug-resistant organisms in health care settings, 2006. *American Journal of Infection Control* 35(10, Suppl 2), S165–S193.

Sievert, D.M., Rudrik, J.T., Patel, J.B., McDonald, L.C., Wilkins, M.J. and Hageman, J.C. (2008) Vancomycin-resistant *Staphylococcus aureus* in the United States, 2002–2006. *Clinical Infectious Diseases* 46, 668–674.

Sievert, D.M.P., Ricks, R., Edwards, J.R., Schneider, A., Patel, J., Srinivasan, A., Kallen, A., Limbago, B., Fridkin, S., and the National Healthcare Safety Network (NHSN) Team and Participating NHSN Facilities (2013) Antimicrobial-resistant pathogens associated with healthcare-associated infections: summary of data reported to the National Healthcare Safety Network at the Centers for Disease Control and Prevention, 2009–2010. *Infection Control and Hospital Epidemiology* 34, 1–14.

Silverman, J.A., Perlmutter, N.G., and Shapiro, H.M. (2003) Correlation of daptomycin bactericidal activity and membrane depolarization in *Staphylococcus aureus*. *Antimicrobial Agents and Chemotherapy* 47, 2538–2544.

Singh, K.V., Weinstock, G.M., and Murray, B.E. (2002) An *Enterococcus faecalis* ABC homologue (Lsa) is required for the resistance of this species to clindamycin and quinupristin–dalfopristin. *Antimicrobial Agents and Chemotherapy* 46, 1845–1850.

Sohn, K.M., Peck, K.R., Joo, E.J., Ha, Y.E., Kang, C.I., Chung, D.R., Lee, N.Y., and Song, J.H. (2013) Duration of colonization and risk factors for prolonged carriage of vancomycin-resistant enterococci after discharge from the hospital. *International Journal of Infectious Diseases* 17(4): e240–e246.

Sutter, S.T., Frei, R., Dangel, M., Gratwohl, A., Bonten, M. and Widmer, A.F. (2010) Not all patients with vancomycin-resistant Enterococci need to be isolated. *Clinical Infectious Diseases* 51, 678–683.

Tadros, M., Williams, V., Coleman, B.L., McGeer, A.J., Haider, S., Lee, C., Iacovides, H., Rubinstein, E., John, M., Johnston, L. *et al*. (2013) Epidemiology and outcome of pneumonia caused by methicillin-resistant *Staphylococcus aureus* (MRSA) in Canadian hospitals. *PLoS ONE* 8(9): e75171.

Tilahun, B., Faust, A.C., McCorstin, P., and Ortegon, A. (2015) Nasal colonization and lower respiratory tract infections with methicillin-resistant *Staphylococcus aureus*. *American Journal of Critical Care* 24, 8–12.

Tran, T.T., Panesso, D., Mishra, N.N., Mileykovskaya, E., Guan, Z., Munita, J.M., Reyes, J., Diaz, L., Weinstock, G.M., Murray, B.E. *et al*. (2013) Daptomycin-resistant *Enterococcus faecalis* diverts the antibiotic molecule from the division septum and remodels cell membrane phospholipids. *MBio* 4(4): pii: e00281–13.

Uttley, A.H., Collins, C.H., Naidoo, J., and George, R.C. (1988) Vancomycin-resistant enterococci. *The Lancet* 1, 57–58.

Wang, L., Lansing, B., Symons, K., Flannery, E.L., Fisch, J., Cherian, K., McNamara, S.E. and Mody, L. (2012) Infection rate and colonization with antibiotic-resistant organisms in skilled nursing facility residents with indwelling devices. *European Journal of Clinical Microbiology and Infectious Diseases* 31, 1797–1804.

Weese, J.S., Rousseau, J., Willey, B.M., Archambault, M., McGeer, A., and Low, D.E. (2006) Methicillin-resistant *Staphylococcus aureus* in horses at a veterinary teaching hospital: frequency, characterization, and association with clinical diseases. *Journal of Veterinary Internal Medicine* 20, 182–186.

Zaas, A.K., Song, X., Tucker, P., and Perl, T.M. (2002) Risk factors for development of vancomycin-resistant enterococcal bloodstream infection in patients with cancer who are colonized with vancomycin-resistant enterococci. *Clinical Infectious Diseases* 35, 1139–1146.

Zacharioudakis, I.M., Zervou, F.N., Ziakas, P.D., and Mylonakis, E. (2014) Meta-analysis of methicillin-resistant *Staphylococcus aureus* colonization and risk of infection in dialysis patients. *Journal of the American Society of Nephrology* 25, 2131–2141.

Zervou, F.N., Zacharioudakis, I.M., Ziakas, P.D., and Mylonakis, E. (2014) MRSA colonization and risk of infection in the neonatal and pediatric ICU: a meta-analysis. *Pediatrics* 133(4): e1015–e1023.

Zhou, Q., Moore, C., Eden, S., Tong, A., McGeer, A., and Mount Sinai Hospital Infection Control Team (2008) Factors associated with acquisition of vancomycin resistant Enterococci (VRE) in roommate contacts of patients colonized or infected with VRE in a tertiary care hospital. *Infection Control and Hospital Epidemiology* 29, 398–403.

Ziakas, P.D., Thapa, R., Rice, L.B., and Mylonakis, E. (2013) Trends and significance of VRE colonization in the ICU: a meta-analysis of published studies. *PLoS ONE* 8(9): e75658.

Ziakas, P.D., Zacharioudakis, I.M., Zervou, F.N., and Mylonakis, E. (2015) Methicillin-resistant *Staphylococcus aureus* prevention strategies in the ICU: a clinical decision analysis. *Critical Care Medicine* 43, 382–393.

14 Epidemiology of Multidrug-resistant Gram-negative Organisms

Evangelia-Theophano Piperaki,[1] Antonis Markogiannakis,[2] Leonidas Tzouvelekis,[1] and George L. Daikos[1]*

[1]*Medical School, National and Kapodistrian University of Athens, Greece;* [2]*Laiko General Hospital, Athens, Greece*

Introduction

Hospital infections caused by multidrug-resistant (MDR, resistant to ≥3 antibiotic classes), extremely drug-resistant (XDR, resistant to all but two classes, such as polymyxins and glycylcyclines), and pandrug-resistant (PDR, resistant to all commercially available antibiotics) bacteria are a global phenomenon, and of these, Gram-negative organisms are of particular importance (Magiorakos *et al*., 2012). Nevertheless, the isolation frequencies of such organisms differ widely depending on the setting. As a rule of thumb, in settings where infection control practices are inadequate and the incidence of healthcare-associated (HCA) infections is high, the phenomenon of antimicrobial resistance is more prominent. The consequences are dire; infections caused by Gram-negative multidrug-resistant organisms (MDROs) dramatically increase morbidity and mortality. In addition, they are costly and have, on many occasions, literally led to severe limitations of healthcare resources. MDROs are not new in healthcare facilities; ever since the 1980s, strains of enterobacterial species have exhibited MDR phenotypes that were resistant to newly introduced extended-spectrum cephalosporins (ESCs), as well as aminoglycosides. Since then, there has been a clear propensity for an increasing frequency of MDROs, and for the escalation of antimicrobial resistance phenotypes. However, a direct comparison with the current situation is not feasible, as the patient population and clinical practices have changed (e.g., there is a higher proportion of immunocompromised and debilitated patients, there are more invasive procedures and aggressive therapies, and novel antimicrobials have been introduced).

According to estimates in a recent report from the US Centers for Disease Control and Prevention (CDC, 2013), more than 2 million people in the US fall sick every year with antibiotic-resistant infections, with at least 23,000 dying as a result. In the same report, infections caused by carbapenem-resistant *Enterobacteriaceae*, MDR *Acinetobacter*, and MDR *Pseudomonas* are characterized as "urgent" and "serious" threats.

Overview of the Main Resistance Mechanisms

Antibiotic-inactivating enzymes

β-Lactamases

β-Lactams are the most widely used antibiotics, mainly because of their excellent selective toxicity and bactericidal action. Moreover, the industry has overcome many antimicrobial chemotherapy-related problems by developing β-lactam molecules with improved properties (e.g., more favorable kinetics, enhanced activity against specific pathogens, etc.). Conversely, bacterial populations have developed, and continue to develop, a variety of resistance mechanisms, the most important being the production of β-lactamases.

*Corresponding author. E-mail: gldaikos@gmail.com

The most widespread β-lactamases, although of lesser clinical significance, are the plasmid-mediated penicillinases of the TEM (named after the patient from whom it was isolated—Temoniera), SHV (named so because sulfhydryl reagents had a variable effect on their substrate specificity) and OXA (oxacillin hydrolyzing) types. These are encountered in *Enterobacteriaceae* more frequently than in *Pseudomonas* and *Acinetobacter*. TEM and SHV penicillinases were the progenitors of extended-spectrum β-lactamases (ESBLs), which evolved through point mutations that resulted in an enlarged active site capable of accommodating the bulky ESCs. These ESBLs have predominated among MDR *Enterobacteriaceae* since the 1990s. A limited spread to non-fermenters has also occurred. Of note, virtually all ESBLs are transferred by plasmids that also mediate resistance to wide variety of other antimicrobials, such as aminoglycosides, tetracyclines, and co-trimoxazole. The prevalence of TEM and SHV ESBL producers has drastically reduced following the "invasion" of CTX-M ESBLs (named for their greater activity against cefotaxime than other oxyimino-β-lactam substrates). The extended-spectrum properties of these enzymes are inherent. The $bla_{\text{CTX-M}}$ β-lactamase genes originated from the genus *Kluyvera*. They have been mobilized by the insertion sequence (IS) element IS*Ecp1*, which was subsequently captured by plasmids circulating among *Enterobacteriaceae*, mainly *Escherichia coli* and *Klebsiella pneumoniae*. For reasons that remain unclear, the majority of CTX-M producers are also resistant to fluoroquinolones.

Another example of the mobilization of *bla* genes from chromosomes to plasmids via IS elements is those encoding for the AmpC enzymes (cephalosporinases). The most important of these are those of the CMY-2 type derived from the *Citrobacter freundii* chromosome. Their preferable β-lactam substrates are the early cephalosporins, such as cephalothin, whereas the newer cephalosporins are poorly hydrolyzed. Nonetheless, cephalosporinases are usually produced in large quantities and thus are able to confer clinically significant resistance to ESCs. It is of interest that cephalosporinase-encoding plasmids found in *Enterobacteriaceae* often coexist in the same organism with plasmids encoding other potent β-lactamases.

The dissemination of organisms producing ESBLs led to the increased use of carbapenem antimicrobials, which partly explains the global spread of Gram-negative MDROs producing carbapenemases. Acquired carbapenemases encountered in MDROs comprise a heterogeneous group of β-lactamases exhibiting significant structural and functional differences. It should be noted that carbapenems are not the most preferable substrates for many of these enzymes. Therefore, the term 'carbapenemase' mainly reflects the clinical impact of inactivation of carbapenems.

The most widespread carbapenemase among the *Enterobacteriaceae* is the *K. pneumoniae* carbapenemase (KPC)-2 and its variants (KPC-3 to KPC-13). Apart from carbapenems, KPCs are also capable of hydrolyzing penicillins, newer cephalosporins, and aztreonam. Consequently, KPC producers exhibit either decreased susceptibility or resistance to virtually all available β-lactams. The minimum inhibitory concentrations (MICs) of carbapenems may vary significantly. However, according to the current breakpoints adopted by the Clinical and Laboratory Standards Institute (CLSI), the majority of KPC producers are classified as carbapenem resistant. The bla_{KPC} genes in enterobacterial species are carried by different plasmids, suggesting the operation of mobilization mechanisms. Indeed, bla_{KPC} genes are associated with the Tn*4401* transposon, a fact that explains their acquisition by distinct genetic units (Tzouvelekis *et al*., 2012; Diene and Rolain, 2014). The carriage of bla_{KPC} by various plasmids has apparently facilitated the dissemination of the gene to various strains of enterobacteria and, to a lesser extent, to strains of *Pseudomonas aeruginosa*. KPCs are also occasionally detected in *Acinetobacter baumannii*.

As important as the KPCs are the metallo-β-lactamases (MβLs or MBLs) of the NDM (New Delhi MβL), VIM (Verona integron-encoded MβP), and IMP (imipenem) types. MβLs characteristically possess divalent cations (usually Zn^{2+}) in the active site that are essential for β-lactam hydrolysis. Their distribution among MDROs is largely similar to that of KPCs. Despite their structural differences, MβLs exhibit a broad hydrolysis spectrum that includes all β-lactams except aztreonam. Consequently, the MβL-producing MDROs display either decreased susceptibility or (more often) resistance to penicillins, penicillin–inhibitor combinations, cephamycins, newer cephalosporins, and carbapenems, though the MICs, especially of the latter drugs, may vary considerably. The bla_{MBL} genes are usually carried by multiresistant transmissible plasmids that have facilitated their dissemination.

One of the latest developments is the emergence and rapid spread of OXA-48 carbapenemase and a certain few of its derivatives, such as OXA-163 and OXA-181. These are encoded by a limited number of plasmid types and are found in MDR enterobacteria but their role in β-lactam resistance in pseudomonads and acinetobacters is, if anything, negligible. The substrate spectrum of OXA-48-type β-lactamases is rather unusual in including penicillins and carbapenems, but not the newer cephalosporins. Unfortunately, the majority of OXA-48-positive isolates coproduce ESBLs with activity against ESCs and, therefore, their phenotypes resemble those of the KPC producers, even though the MICs of carbapenems are comparatively lower. Finally, a phylogenetically distinct type of potent OXA carbapenemases (OXA-23, OXA-40, and OXA-58) is increasingly found, almost exclusively, in MDR *Acinetobacter* spp. (Diene and Rolain, 2014).

Aminoglycoside-modifying enzymes

The production of aminoglycoside-modifying enzymes (AMG-MEs) is the primary mechanism conferring resistance to aminoglycosides in MDROs. AMG-MEs are classified in three families: (i) AMG acetyltransferases (AACs), which catalyze the acetylation of amino groups in the aminoglycoside molecule; (ii) AMG nucleotidyltranferases (ANTs), which are capable of adenylating hydroxyl-groups; and (iii) AMG phosphotransferases (APHs), which phosphorylate hydroxyl groups. These chemical modifications effectively reduce the binding of AMGs to the small ribosomal subunits and, consequently, their ability to interfere with protein synthesis.

Numerous AMG-MEs with various substrate patterns have been identified so far. Evidently, the most important are those capable of inactivating one or more of the most clinically used AMGs, i.e., gentamicin (GM), tobramycin (TM), netilmicin (NET), and amikacin (AN) and are also widely disseminated among MDROs (Becker and Cooper, 2013). An example of such an AMG-ME is the AAC(6′)-I conferring resistance to TM, NET, and AN, and encountered frequently among enterobacteria and non-fermenters, as well as AAC(3)-II (resistance pattern GM, TM, NET), which is more commonly encountered in *Enterobacteriaceae*. Specific mention is merited for AAC(6′)-Ib-cr, a variant of AAC(6′)-Ib that is able to also inactivate fluoroquinolones through acetylation.

The genes encoding AMG-MEs are usually carried by multiresistant, transmissible plasmids that encode potent β-lactamases as well.

Decreased outer membrane permeability

Decreased expression of outer membrane porins

Decreased outer membrane permeability is a common, but rather complementary, antibiotic resistance mechanism in MDR enterobacterial species (Fernández and Hancock, 2012). It is due to decreased expression or even complete loss of the main outer membrane porins (OMPs). The latter form pores allowing the passive diffusion of nutrients and also of hydrophilic antibiotics. The mechanism affects mainly the β-lactams and, to a lesser extent, other antimicrobials such as fluoroquinolones. The mechanism usually operates in a synergic fashion with the production of β-lactamases, resulting in high-level resistance to β-lactams.

Loss of OMPs is not included among the important resistance mechanisms in non-fermenters, though there are two notable exceptions, both involving carbapenems. *P. aeruginosa* expresses an OMP (OprD) which is the preferred channel for the entrance of imipenem; loss of this porin—a common event during imipenem monotherapy—results in high-level resistance to the drug. Decreased expression of CarO, an OMP of *A. baumannii* is also associated with carbapenem resistance.

Operation of efflux pumps

Virtually all bacterial species possess efflux pumps, which are membrane-located, energy-dependent systems that are able to expel a wide variety of harmful substances, including antibiotics. The operation of efflux pumps seems to be more important in *P. aeruginosa* and *Acinetobacter* spp. than in *Enterobacteriaceae* (Fernández and Hancock, 2012).

The efflux pumps involved in antibiotic resistance in MDROs are classified into four families: SMR (small multidrug resistant), MF (major facilitator), MATE (multidrug and toxic extrusion), and RND (resistance, nodulation, division). The latter family is the most significant, comprising pumps capable of causing resistance to virtually all important antibiotics (β-lactams, aminoglycosides, fluoroquinolones, and, probably, tigecycline). RND pumps are inherent in *P. aeruginosa* and *A. baumannii*. Their structure comprises a cytoplasmic membrane pump protein, a periplasmic protein (membrane fusion protein), and a channel-forming outer membrane protein (i.e., an OMP).

REPRESENTATIVE EFFLUX PUMPS

AbeS AbeS is an SMR pump encountered in *A. baumannii*. The main component of the pump is a protein spanning the cytoplasmic membrane. The energy to carry out extrusion is derived from the membrane's electrochemical potential. The pump preferentially expels fluoroquinolones, macrolides, and chloramphenicol.

TetA The TetA pump is probably one of the most common plasmid-mediated resistance mechanisms. A drug/proton antiporter, as is typical for the MF family, this pump extrudes tetracyclines through a transmembrane protein.

MdfA The MdfA pump is an MF pump encountered mainly in MDR *E. coli*. MdfA exhibits a wide substrate spectrum including aminoglycosides, tetracyclines, and fluoroquinolones.

AbeM The AbeM pump is an efflux system of the MATE type and shares a structural similarity with MF pumps. Yet, to carry out the extrusion of antibiotics, these pumps use the sodium gradient of the membrane. AbeM is found in MDR *A. baumannii*, and its main substrates are fluoroquinolones and aminoglycosides.

AcrAB–TolC The AcrAB–TolC pump is often encountered in MDR enterobacteria, and less frequently in *A. baumannii*. Similar to all pumps of the RND family, it operates as a drug/proton antiporter. Its activity spectrum includes fluoroquinoles and tetracyclines.

AbeABC, AbeFGH, and AbeIJK These three pumps are *A. baumanni* specific. The substrates of the AbeABC pump are the fluoroquinolones, aminoglycosides, and tetracyclines. The AbeFGH and AbeIJK pumps confer resistance to fluoroquinolones, β-lactams, and tetracyclines.

MexAB, MexCD, MexEF, and MexXY These four pumps are *P. aeruginosa* specific. The substrate spectrum of each pump is as follows: MexAB for fluoroquinolones, tetracyclines, and β-lactams (except for imipenem); MexCD for fluoroquinolones and tetracyclines; MexEF for fluoroquinolones and carbapenems; and MexXY for fluoroquinolones, tetracyclines, aminoglycosides, and β-lactams (except for meropenem and cefepime).

RND pumps and tigecycline resistance Tigecycline acts as an inhibitor of protein synthesis in a fashion similar to tetracyclines, but it is not affected by the tetracycline resistance mechanisms. Due to its good in vitro activity, it is heavily used against MDROs. Yet there is a clear trend toward increasing MICs for tigecycline. The examination of a limited number of MDR *K. pneumoniae* isolates indicated overexpression of the AcrB–TolC pump as the possible mechanism of tigecycline resistance.

Antibiotic target alterations

This functional group includes a wide variety of quite different resistance mechanisms affecting the activity of several antimicrobial classes. Here, the most clinically relevant of these mechanisms will be briefly discussed.

Point mutations conferring resistance to fluoroquinolones

Levels of fluoroquinolone resistance have been rising worldwide since the 1990s. High-level resistance to fluoroquinolones in *Enterobacteriaceae* is primarily mediated by specific mutations in the chromosomally encoded enzymes DNA gyrase and topoisomerase IV, resulting in a drastic decrease in drug–enzyme affinity (Aldred *et al*., 2014). Similar mutations conferring fluoroquinolone resistance are also common among MDR *P. aeruginosa* and *Acinetobacter* spp.

Ribosomal modification and resistance to aminoglycosides

High-level resistance to all available AMGs may arise by the specific methylation of the target of the drugs, 16S rRNA. This process is mediated by the plasmid-encoded aminoglycoside resistance methyltransferases Arm and Rmt (Galimand *et al*., 2005; Becker and Cooper, 2013). Arm and Rmt have been identified in both *Enterobacteriaceae* and non-fermenters, albeit at low frequencies. Some of the Arm-encoding plasmids also mediate the production of NDM carbapenemases, thereby conferring extensive resistance patterns.

Ribosomal protection and resistance to tetracyclines

A large number of *tet* genes conferring resistance to tetracyclines have been identified. Approximately one third of these genes encode proteins protecting the ribosomes from the action of tetracyclines. The most characteristic of these proteins is TetM, and this is also the most widespread among MDROs owing to its association with mobile elements.

Dihydrofolate reductases (DHFRs) and resistance to trimethoprim

Trimethoprim binds to bacterial DHFR and inhibits the reduction of dihydrofolic to tetrahydrofolic acid, which is essential for DNA synthesis. Trimethoprim resistance in enterobacteria (*P. aeruginosa* and *Acinetobacter* spp. are inherently resistant to the drug) is commonly mediated by multiresistant plasmids encoding DHFR variants that are not inhibited by trimethoprim while retaining their enzymatic activity.

Colistin resistance and lipopolysaccharide alterations

Recent studies indicate an increasing isolation frequency of polymyxin-resistant MDROs. Polymyxins (Polymyxin B, and Polymyxin E, also known as colistin) interact with and destabilize lipopolysaccharides (LPSs) through the replacement of calcium and magnesium. Resistance can be developed mainly through LPS modifications, such as substitution of the phosphate groups in lipid A by phosphoethanolamine (PEtN) and 4-amino-4-deoxy-L-arabinose (LAra4N) (lipid A is the inner part of the LPS outer membrane of Gram-negative bacteria and is responsible for their toxicity). These substitutions drastically reduce the negative charge of lipid A, thereby decreasing its affinity for the positively charged polymyxins. These mechanisms are common in *P. aeruginosa*, *K. pneumoniae* and *E. coli*. *A. baumannii* does not possess the LAra4N biosynthesis-related genes, indicating that Ara4N modification of lipid A is not required for polymyxin resistance, although it does have the system that results in the addition of PEtN to lipid A. Another polymyxin resistance mechanism unique to *A. baumannii* is the complete loss of LPS resulting from the inactivation of the lipid A biosynthesis genes (Olaita *et al.*, 2014).

Until now, the aforementioned colistin resistance mechanisms were known to be chromosomal. However, recently, a transmissible plasmid-borne gene, *mcr-1*, that confers resistance to colistin, has been described and has already been reported from multiple geographic regions. This gene encodes the production of a PEtN transferase that results in the addition of PEtN to lipid A (Liu *et al.*, 2016).

Global Epidemiology and International Clones

It has become evident that within a typical pathogenic species, only a few clones, clusters or lineages are excessively represented among the clinical isolates of the species from particular types of infections. Some of the most "successful" bacterial clones serve as stable hosts and starting points for the spread of antimicrobial resistance genetic elements, such as genes, transposons, integrons, and plasmids. The reasons for the exceptional success of certain high-risk clones and their association with particular epidemic resistance elements have not been elucidated.

It is generally believed that an international MDR high-risk clone must be characterized by global distribution, association with a variety of antimicrobial resistance determinants, and the ability to be effectively transmitted and persist in colonized hosts for prolonged time intervals (>6 months); it must also exhibit increased virulence and fitness, as well as the potential to cause severe and/or recurrent infections.

Klebsiella pneumoniae

A prototype high-risk clone is *K. pneumoniae* sequence type (ST) 258, the predominant member of clonal complex (CC) 258, which also contains its single locus variants ST11, ST340, and ST512. Members of this clonal complex that contain the bla_{KPC} gene have been dominating the pandemic spread of KPC-producing *K. pneumoniae*. Initially identified in the US in the late 1990s, they were later also reported from Israel, Greece, Italy, Poland, Norway, Sweden, Brazil, Canada, and South Korea in the mid-2000s. Of the ST258 single locus variants: ST11 predominates in Asia, and has also been reported from South America and European countries; ST512 has been reported from Italy, Israel, and Colombia; and ST340 is present in Brazil as well as in Greece.

Other significant international *K. pneumoniae* clones are the ST147 clone, which is present in North America, Europe, and Asia, and produces carbapenemases of the KPC, NDM, and OXA types, as well as the ST15 clone, which harbors OXA-48, being found both in Europe and in North America.

Most resistance factors in clone ST258 are plasmid borne and this clone often harbors more than one plasmid, each encoding for multiple resistance determinants to various antibiotic groups, including aminoglycosides, β-lactams, fluoroquinolones, sulfonamides, trimethoprim, and chloramphenicol. Antibiotic choice in clinical practice is further limited by the often observed decrease in outer membrane permeability.

Studies from different countries indicate that hospitalized populations, including neonatal patients, are rapidly and effectively colonized with *K. pneumoniae* ST258, and that a significant proportion remains so, up to a year after exposure. Nevertheless, in contrast to *E. coli*, which is readily detectable in the community, clone ST258 seems to remain confined mostly to healthcare-associated environments (Lopez-Cerero and Almirante, 2014; Mathers *et al.*, 2015).

Virulence in *K. pneumoniae* is mainly determined by the K antigens of its capsular polysaccharide, which render it resistant to phagocytosis. Clone ST258 lacks well-established virulence factors such as the genes for the capsular antigens K1, K2, and K5, and for aerobactin. Indeed, it has recently been shown to be highly vulnerable to complement lysis and phagocytosis. It has, however, been associated with the integrative conjugative element ICEKp258.2, which carries gene clusters for a type IV pilus and a type III restriction-modification system; it is believed that these determinants may have contributed to the ecological success of clone ST258 by facilitating the movement of resistance genes, enhancing transmissibility and survival, and modulating exchanges of mobile elements.

Escherichia coli

The quintessential international multiresistant high-risk clone, later to be known as ST131, was initially identified in the UK and Canada in the mid-2000s as a $bla_{CTX\text{-}M\text{-}15}$-encoding *E. coli*. The same clone was subsequently reported from several countries, including Spain, France, Portugal, Switzerland, Italy, Croatia, Lebanon, Turkey, India, Kuwait, South Korea, Japan, the US, South Africa, and Brazil. Further studies in the late 2000s revealed epidemic spread of ST131, predominantly in the community of all inhabited continents, apparently within the same time frame, and without an immediately obvious link between the patients. The most prevalent lineage within the ST131 clone is the H30 (gene *fimH30*) subclone, which comprises the majority of fluoroquinolone-resistant isolates that also harbor CTX-M-15.

The total prevalence of ST131 among *E. coli* isolates, as determined by surveillance studies, varies by geographic location from 10% to approximately 30%, and encompasses up to 80% of fluoroquinolone-resistant isolates and 50% of ESBL producers. It is most commonly associated with community onset extraintestinal infections, especially urinary tract infections (UTIs) and bloodstream infections (BSIs), particularly in patients with previous contact with healthcare settings. ST131 is frequently found in elderly inhabitants of nursing homes and long-term healthcare facilities, but has also been detected in animals, food, and the environment. Although isolated in hospital settings, it has not been linked to nosocomial and intensive care unit (ICU) outbreaks. History of travel, particularly to the Indian subcontinent, Africa, and the Middle East was associated with community UTIs in returning travellers, though follow-up studies that demonstrated similar rates of rectal colonization in travellers and non-travellers indicate that international travel is not essential for the pandemic spread of the clone. Studies have indicated that ST131 is easily transmissible from person to person, particularly between family members, but whether this transmission is more efficient than in other *E. coli* clones remains to be determined. Rectal colonization rates range from 0 to 44%, depending on geographical area and population characteristics, but whether the colonizing abilities of this clone differ from those of other isolates also remains unclear.

Most of the virulence factors possessed by ST131 are more likely related to its ability to colonize the human host rather than cause infection. No differences in cure or mortality rates were noted between patients with infections due to ST131 and other extraintestinal *E. coli* isolates. Notwithstanding, ST131 is more prevalent among invasive than non-invasive *E. coli* isolates, while H30-Rx, the *bla*CTX-M-15 encoding sublineage of H30, has been epidemiologically associated with BSI and sepsis (Mathers *et al.*, 2015).

Based on the resistance determinants that have been characterized, isolates belonging to the ST131 clone are frequently resistant to fluoroquinolones, cephalosporins, monobactams, aminoglycosides, trimethoprim–sulfamethoxazole, and, possibly, also to tetracyclines and chloramphenicol.

The evolution of carbapenem-resistant community-acquired pathogens like *E. coli* is a daunting possibility. Currently, the most commonly encountered carbapenemases in *E. coli* isolates are NDM and OXA-48, and infections with these are usually associated with travel to geographical areas of endemicity, such as the Indian subcontinent and North Africa or Turkey, respectively. The bla_{NDM} carbapenemase gene was initially detected in *E. coli* ST131 in 2010 in travellers returning from the Indian subcontinent, followed by bla_{VIM} from Italy, bla_{KPC} from Ireland, Italy, France, the US and China, bla_{OXA-48} from the UK, Ireland, Algeria, and Spain, and bla_{IMP} from Taiwan. Given its universal dissemination, a carbapenemase-producing *E. coli* ST131 clone would present a serious emerging public health threat.

Acinetobacter baumannii

A. baumannii is primarily an opportunistic nosocomial pathogen that commonly causes ventilator-associated pneumonia (VAP) and BSIs. Three international clonal lineages known as I, II, and III (ICLI, ICLII, and ICLIII) currently dominate the *A. baumannii* clinical isolate population; these correspond to CC1/CC109, CC2/CC92, and CC3/CC187 (there are two multilocus sequence typing (MLST) schemes for *A. baumannii* typing). The majority of outbreak-associated isolates fall within the first two clonal lineages; initially described in numerous European hospitals, these are currently recognized in more than 30 countries in Europe, Asia, Africa, Australia, and the Americas. More specifically, ICLII, the most widespread *A. baumannii* clonal lineage, comprises isolates from 34 countries from all continents, containing genes encoding OXA-23 OXA-24/40, OXA-58, VIM-1, and NDM-1. The second largest group, ICLI, so far comprises isolates from 30 countries, and hosts the genes for OXA-23, OXA-58, VIM-4, and NDM-1 (Karah *et al.*, 2012).

Pseudomonas aeruginosa

P. aeruginosa is a ubiquitous environmental species, and one of the commonest nosocomial pathogens. It is responsible for a wide range of opportunistic acute and chronic infections in compromised hosts. Carbapenem resistance is common in this MDR species, both through horizontal transfer, chiefly of metallo-carbapenemases, and through diminished drug permeability due to porin loss and/or increased drug efflux. *P. aeruginosa* exhibits great diversity of STs, with substantial overlap between isolates of environmental and clinical origin. Two MDR clones, ST235 and ST111, currently comprise most of the isolates that harbor MBL genes, mostly VIM and IMP, but occasionally NDM. ST235 has been reported from numerous European countries, including the UK and Russia, and also from Asia, South America, and Africa (Wright *et al.*, 2015). ExoU is a cytotoxic exotoxin transported by the type III secretion system of *P. aeruginosa*, and its presence has been associated with unfavorable clinical outcomes and higher mortality rates in patients with *P. aeruginosa* pneumonia and BSI. ExoU occurs more frequently in carbapenem-resistant *P. aeruginosa* isolates and is disproportionately more prevalent among CC235 isolates.

P. aeruginosa is present in soil, water, plants, and many animals, although it is rarely a member of the normal human microbial flora. Within the hospital, it is recovered from a variety of moist environments and is known to readily colonize the gastrointestinal and respiratory tracts of hospitalized patients, particularly those with a breach of the skin or mucosal integrity, immunosuppression, and disruption of normal microbiota. It has been detected in sludge and water samples collected upstream and downstream from wastewater treatment plants, and also in hospital wastewater. Most of the MDR isolates from these aquatic sources belong to the ST235 and ST111 high-risk clones.

The main characteristics of the international high-risk clones of all four species that have just been described are presented in Table 14.1.

Hospital epidemiology

The Gram-negative MDROs are mainly encountered in acute- and long-term-care facilities. These organisms can colonize not only debilitated, immunocompromised, or critically ill patients, but also previously healthy patients who are hospitalized in healthcare settings with inadequate infection control practices. By the time the first case of MDRO infection is recognized in a healthcare facility, these organisms may have already spread widely and colonized a substantial number of patients (a typical iceberg effect). Colonization may be extensive and pass largely unnoticed in institutions located in endemic regions, as has been evident in several studies. For instance, during an outbreak of carbapenemase-producing *K. pneumoniae* in Israel, a

Table 14.1. Characteristics of international high-risk clones. Data extracted from Galimond *et al.* (2005), Diene and Rolain (2014), and Fernández and Hancock (2012).

High-risk clone	Geographic distribution	Predominant resistance mechanism[a]	Extent of resistance[b]
Escherichia coli			
ST131	Global	CTX-M-15	MDR
Klebseilla pneumoniae			
ST258	Global	KPC-2, KPC-3	XDR, PDR
ST11	South America, East Asia, South Asia, Europe	KPC-2, NDM-1, VIM, OXA-48	XDR, PDR
ST340	West Asia, North America, Europe	NDM, KPC, KPC-2, NDM-1	XDR, PDR
ST512	Europe, West Asia	KPC-3	XDR, PDR
ST147	North America, Europe, East Asia, West Asia, South Asia	KPC, KPC-2, NDM-1, OXA-48, OXA-48-like, VIM	XDR, PDR
ST15	North America, Europe	OXA-48	XDR, PDR
Acinetobacter baumannii			
ICLI CC1/CC109	Global	NDM, OXA-23, OXA-58, VIM	XDR, PDR
ICLII CC2/CC92	Global	NDM, OXA-23, OXA-24/40, OXA-58, VIM	XDR, PDR
Pseudomanas aeruginosa			
ST235	Europe, Asia, South America, Africa	IMP, VIM	XDR, PDR
ST111	Europe	IMP, VIM	XDR, PDR

[a]See text for details of these resistance mechanisms.
[b]MDR, multidrug resistant; XDR, extremely drug resistant; PDR, pandrug resistant.

point prevalence survey demonstrated that 16 (5.4%) of 298 patients screened were colonized with carbapenemase producers, and notably, 11 (69%) of these carriers would have remained undetected without the performance of active surveillance cultures. MDROs seem to have a high potential for spread not only from patient to patient within a healthcare facility but also through the "cycling" of patients between institutions in the same region and/or across borders from high- to low-prevalence countries.

The gastrointestinal tract is the main colonization site of *Enterobacteriaceae*, although other anatomic sites including the nasopharynx, surgical wounds, ulcers, and the respiratory and urinary tracts, may also become colonized by MDR *Enterobacteriaceae* and other MDROs. Several organisms can also be found in various environmental niches. For instance, *P. aeruginosa* is commonly associated with moist environmental sources, and *A. baumannii* with both dry and moist areas. Environmental contamination may play a key role in the dissemination of these organisms and cause nosocomial outbreaks. Nevertheless, it appears that contamination of the inanimate environment with fermenters is probably less important for intrahospital dissemination than is contamination with non-fermenters.

Several variables have been recognized as risk factors for MDRO acquisition. Those most frequently reported are prolonged hospitalization, stay in the same room with a colonized patient, stay in an ICU, poor functional status, previous use of antibiotics, malignancies, solid organ or stem cell transplantation, and the use of multiple invasive devices, among others. After colonization with MDROs has occurred, it may persist for days to months, or even years. During the carriage of MDROs, shedding and transmission to other patients may take place. Moreover, the transmission of mobile genetic elements containing resistance genes to other strains colonizing the gastrointestinal tract may occur, resulting in new MDRO clones and species, which, in turn, may become the source of new outbreaks.

The usual sequence of events in microbe cross-transmission from patient to patient in healthcare facilities is the following: (i) presence of microbes on the patient's skin and/or in the patient's environment; (ii) transfer of these organisms to the hands of healthcare workers (HCWs); (iii) microbe survival on the hands of HCWs; (iv) incorrect hand cleansing by HCWs; and (v) cross-transmission to other patients. In certain regions, the spread of ESBLs (CTX-M) and carbapenemases (NDM-1 and OXA-48) occurs primarily in the community via the fecal–oral route, either by food-borne or waterborne transmission. As with other enteric bacteria, waterborne outbreaks are often of a much larger scale than food-borne epidemics. In a globalized world, where travel, migration, medical tourism, and trade between countries with different levels of sanitary conditions is common, repeated importation of carbapenemase producers is to be expected from countries where these bacteria are prevalent in the community.

In a subset of patients, MDROs will migrate to an anatomic site(s), such as the respiratory tract, the urinary tract, surgical sites, as well as medical devices, and infection will occur. The size of the subset that develops infection probably varies with patient and pathogen factors, as well as the characteristics of the competing normal flora. A variety of risk factors, such as stay in an ICU, invasive procedures, exposure to antibiotics, and various underlying diseases have been strongly associated with life-threatening MDRO infections. In general, the ICU setting appears to promote the progression from colonization to infection. However, there also seems to be a strong link between developing MDRO infection and factors directly related to the host (e.g., solid organ or stem cell transplantation, solid tumors, diabetes mellitus, and renal failure). A wide spectrum of clinical infections is caused by MDROs, including primary or catheter-related bacteremia, nosocomial pneumonia, surgical site and wound infections, intraabdominal infections, and UTIs. These infections are associated with increased mortality ranging from 30 to 70%. Older age, severity of underlying disease, severe sepsis and/or septic shock, high APACHE (Acute Physiologic and Chronic Health Evaluation) score, and inappropriate antibiotic therapy are among the factors that have been associated with adverse outcomes (Akova *et al.*, 2012; Tzouvelekis *et al.*, 2012).

Preventing infections will reduce the burden of MDROs in healthcare settings. The implementation of a bundle of infection control measures for the optimal management of vascular and urinary catheters, prevention of lower respiratory tract infection in intubated patients, accurate diagnosis of infectious etiologies, and judicious antimicrobial use are all clinical practices that should be incorporated into routine patient care in order to decrease the burden of MDRO infections. Guidance for these preventive practices is given by the *CDC Campaign to Reduce Antimicrobial Resistance in Healthcare Settings* (CDC, 2004), now manifested in the CDC's "Get Smart for Healthcare Campaign" (see www.cdc.gov/drugresistance/healthcare/). This is a multifaceted, evidence-based approach with four parallel strategies: infection prevention; accurate and prompt diagnosis and treatment; prudent use of antimicrobials; and prevention of transmission (Akova *et al.*, 2012; Tzouvelekis *et al.*, 2012).

Several guidance documents on the control of Gram-negative MDROs have been published over the last few years. Of special note are two documents from the CDC (Siegel *et al.*, 2006; CDC, 2015), a technical report from the European Centre for Disease Control and Prevention Technical Report (ECDC, 2014), and the European Society of Clinical Microbiology and Infectious Diseases (ESCMID) guidelines (Tacconelli *et al.*, 2014). Although there are some differences between these guidance documents, many similarities and common themes are the rule. Thus, these documents should be read as complementing each other. There has been consistent evidence for the effectiveness of a bundle of control measures, such as the early detection of colonized patients by active surveillance cultures, contact precautions, and cohort nursing care for colonized patients with carbapenemase-producing Gram-negatives. Other measures have been employed too, such as antibiotic restriction, environmental surface decontamination, patient decolonization with antiseptic bathing, and healthcare staff education; however, their effectiveness is unclear. Environmental decontamination may play a more important role in eradicating non-fermenters such as *A. baumannii* and *P. aeruginosa*, which have the ability to survive on surfaces.

A summary of prevention strategies for containing the spread of carbapenemase-producing *Enterobacteriaceae* in acute- and long-term care facilities, as proposed by the CDC, is presented in Table 14.2. Ideally, in endemic settings, all interventions should be coordinated at a regional or even national level, and across the healthcare system, as was recently supported by experience from Israel. Guidelines for effective intervention must be prepared before Gram-negative MDROs have entered the region, and these should be implemented imme-

Table 14.2. Summary of prevention strategies for containing carbapenemase-producing *Enterobacteriaceae* (CPE) in acute- and long-term care facilities. From CDC (2015).

Core measures

1. Hand hygiene.
2. Contact precautions:
   - Preemptive contact precautions might be used for high-risk patients.
3. Patient and staff cohorting:
   - When available, cohort CPE colonized or infected patients and the staff that care for them, even if patients are housed in single rooms;
   - If the number of single patient rooms is limited, reserve these rooms for patients with highest risk for transmission (e.g. incontinence, diarrhea).
4. Minimize use of invasive devices.
5. Promote antimicrobial stewardship.
6. Screening:
   - Screen patients with epidemiologic links to unrecognized CPE colonized or infected patients and/or conduct point prevalence surveys of units containing unrecognized CPE patients.

Supplemental measures for healthcare facilities with CPE transmission

1. Conduct active surveillance testing:
   - Screen high-risk patients for CPE at admission, or at admission and periodically during their hospital stay;
   - Preemptive contact precaution can be used while results of admission surveillance testing are pending;
   - Consider screening patients transferred from facilities known to have CPE at admission.
2. Chlorhexidine bathing:
   - Bathe patients with 2% chlorhexidine.

diately upon the detection of MDROs by clinical culture. Communication channels at the local, regional, and national levels should be established in advance, in order to facilitate rapid notification and feedback.

Conclusions

During the last decade, a rather unexpected development appears to have taken place: a literally worldwide spread of distinct multidrug resistant Gram-negative clones. There can be no doubt that increased population mobility, ranging from uncontrolled massive immigration to medical tourism, has substantially contributed to this phenomenon. While it is unlikely that such extensive dissemination of resistant pathogens did not occur in the past (albeit to a lesser extent), it probably went unnoticed or was overlooked, partly owing to the lack of the currently available highly discriminating molecular methods. In any case, it is clear that we are now facing a consequence of what most people call "globalization."

Judging the current situation, it is evident that we have been caught unprepared. Indeed, several virtually untreatable strains have achieved global dissemination. The problem is largely under control in developed countries. However, in other vast geographic areas, we cannot even roughly estimate its extent. The only reasonable course of action for the international community is a massive mobilization of human and financial resources in order to combat this major threat to public health.

References

Akova, M., Daikos, G.L., Tzouvelekis, L., and Carmeli, Y. (2012) Interventional strategies and current clinical experience with carbapenemase-producing Gram-negative bacteria. *Clinical Microbiology and Infection* 18, 439–448.

Aldred, K.J., Kerns, R.J., and Osheroff, N. (2014) Mechanism of quinolone action and resistance. *Biochemistry* 53, 1565–1574.

Becker, B. and Cooper, M.A. (2013) Aminoglycoside antibiotics in the 21st century. *ACS Chemical Biology* 8, 105–115.

CDC (2004) *CDC Campaign to Reduce Antimicrobial Resistance in Healthcare Settings*. Available at: http://www.kliinikum.ee/infektsioonikontrolliteenistus/doc/oppematerjalid/longterm.pdf (accessed 12 May 2016).

CDC (2013) *Antibiotic Resistance Threats in the United States, 2013*. Centers for Disease Control and Prevention, Atlanta, Georgia. Available at: http://www.cdc.gov/drugresistance/pdf/ar-threats-2013-508.pdf (accessed 12 May 2016).

CDC (2015) *Facility Guidance for Control of Carbapenem-resistant Enterobacteriaceae (CRE): November 2015. Update CRE Toolkit*. Centers for Disease Control and Prevention, Atlanta, Georgia. Available at: http://www.

cdc.gov/hai/pdfs/cre/CRE-guidance-508.pdf (accessed 12 May 2016).
Diene, S.M. and Rolain, J.M. (2014) Carbapenemase genes and genetic platforms in Gram-negative bacilli: *Enterobacteriaceae*, *Pseudomonas* and *Acinetobacter* species. *Clinical Microbiology and Infection* 20, 831–838.
ECDC (2014) *Systemic Review of the Effectiveness of Infection Control Measures to Prevent the Transmission of Carbapenemase-producing* Enterobacteriaceae *through Cross-border Transfer of Patients*. ECDC Technical Report, European Centre for Disease Control and Prevention, Stockholm. Available at: http://ecdc.europa.eu/en/publications/Publications/CPE-systematic-review-effectiveness-infection-control-measures-to-prevent-transmission-2014.pdf (accessed 12 May 2016).
Fernández, L. and Hancock, R.E.W. (2012) Adaptive and mutational resistance: the role of porins and efflux pumps in the development of drug resistance. *Clinical Microbiology Reviews* 25, 661–681.
Galimand, M., Sabtcheva, S., Courvalin, P., and Lambert, T. (2005) Worldwide disseminated *armA* aminoglycoside resistance methylase gene is borne by composite transposon Tn*1548*. *Antimicrobial Agents and Chemotherapy* 49, 2949–2953.
Karah, N., Sundsfjord, A., Towner, K., and Samuelsen, Ø. (2012) Insights into the global molecular epidemiology of carbapenem non-susceptible clones of *Acinetobacter baumannii*. *Drug Resistance Updates* 15, 237–247.
Liu, Y.Y., Wang, Y., Walsh, T.R., Yi, L.-X., Zhang, R., Spencer, J., Doi, Y., Tian, G., Dong, B., Huang, X. *et al*. (2016) Emergence of plasmid-mediated colistin resistance mechanism MCR-1 in animals and human beings in China: a microbiological and molecular biological study. *The Lancet Infectious Diseases* 16, 161–168.
Lopez-Cerero, L. and Almirante, B. (2014) Epidemiology of infections caused by carbapenem-resistant *Enterobacteriaceae*: reservoirs and transmission mechanisms. *Enfermedades Infecciosas y Microbiología Clínica* 32(Suppl 4), 10–16.
Magiorakos, A.-P., Srinivasan, A., Carey, R.B., Carmeli, Y., Falagas, M.E., Giske, C.G., Harbarth, S., Hindler, J.F., Kahlmeter, G., Olsson-Liljequist, B. *et al*. (2012) Multidrug-resistant, extensively drug-resistant and pandrug-resistant bacteria: an international expert proposal for interim standard definitions for acquired resistance. *Clinical Microbiology and Infection* 18, 268–281.
Mathers, A.J., Peirano, G., and Pitout, J.D. (2015) The role of epidemic resistance plasmids and international high-risk clones in the spread of multidrug-resistant *Enterobacteriaceae*. *Clinical Microbiology Reviews* 28, 565–591.
Olaita, A.O., Morand, S., and Rolain, J.M. (2014) Mechanisms of polymyxin resistance: acquired and intrinsic resistance in bacteria. *Frontiers in Microbiology* 5: 643.
Siegel, J.D., Rhinehart, E., Jackson, M., Chiarello, L., and HICPAC (Healthcare Infection Control Practices Advisory Committee) (2006) *Management of Multidrug-Resistant Organisms In Healthcare Settings, 2006*. Centers for Disease Control and Prevention, Atlanta, Georgia. Available at: http://www.cdc.gov/hicpac/pdf/MDRO/MDROGuideline2006.pdf (accessed 12 May 2016).
Tacconelli, E., Cataldo, M.A., Dancer, S.J., De Angelis, G., Falcone, M., Frank, U., Kahlmeter, G., Pan, A., Petrosillo, N., Rodríguez-Baño, J. *et al*. (2014) ESCMID guidelines for the management of the infection control measures to reduce transmission of multidrug-resistant Gram-negative bacteria in hospitalized patients. *Clinical Microbiology and Infection* 20(Suppl 1), 1–55. Available at: http://www.clinicalmicrobiologyandinfection.com/article/S1198-743X%2814%2960007-0/pdf (accessed 12 May 2016).
Tzouvelekis, L.S., Markogiannakis, A., Psichogiou, M., Tassios, P.T., and Daikos, G.L. (2012) Carbapenemases in *Klebsiella pneumoniae* and other *Enterobacteriaceae*: an evolving crisis of global dimensions. *Clinical Microbiology Reviews* 25, 682–707.
Wright, L.L., Turton, J.F., Livermore, D.M., Hopkins, K.L., and Woodford, N. (2015) Dominance of international 'high-risk clones' among metallo-β-lactamase-producing *Pseudomonas aeruginosa* in the UK. *Journal of Antimicrobial Chemotherapy* 70, 103–110.

15 Pathogenesis and Epidemiology of *Clostridium difficile* Infection: Implications for Antibiotic Stewardship

BLANCA E. GONZALEZ[1] AND PHILIP TOLTZIS[2]

[1]*Cleveland Clinic Children's Hospital and Cleveland Clinic Lerner College of Medicine of Case Western Reserve University, Cleveland, Ohio;* [2]*Rainbow Babies and Children's Hospital and Case Western Reserve University School of Medicine, Cleveland, Ohio*

Introduction

Clostridium difficile, a spore-forming, obligate-anaerobic Gram-positive bacillus, is the most common cause of healthcare-associated diarrhea in North America. The impact of the organism has increased progressively since it was discovered. *C. difficile* was initially identified in the 1930s as a gastrointestinal commensal in young children. In the mid-1970s, accumulating clinical and laboratory evidence indicated that toxigenic *C. difficile* was the cause of clindamycin-associated diarrhea and pseudomembranous colitis (Bartlett *et al.*, 1977; Bartlett, 1981, 2004). By the 1980s, it was established that approximately 20% of all antibiotic-associated diarrhea was caused by *C. difficile*; however, it was no longer clindamycin but rather cephalosporins and penicillin that were associated with the greatest number of cases of *C. difficile* infections (CDIs), owing to the widespread use of these antimicrobials. The importance of *C. difficile* rose to even higher levels during the first decade of the new millennium owing to a dramatic increase in the frequency of CDI in North America and Europe. Several population-based studies documented that the annual incidence of CDI more than doubled from 2002 to 2008 (Kim *et al.*, 2008; Zilberberg *et al.*, 2008a,b; Nylund *et al.*, 2011; Lessa *et al.*, 2012). Recent studies suggest that *C. difficile* has surpassed methicillin-resistant *Staphylococcus aureus* (MRSA) as the most frequently encountered hospital-acquired organism in the US (Lessa *et al.*, 2012), resulting in \$1–3 billion of excess cost to the American healthcare system each year (O'Brien *et al.*, 2007; Dubberke and Wertheimer, 2009).

The strong association between CDI and antecedent antibiotic exposure recommends *C. difficile*-associated diarrhea as an ideal candidate for antibiotic stewardship intervention. The next section will review the clinical manifestations, pathogenesis, and epidemiology of CDI, providing the groundwork and rationale for antibiotic stewardship for this condition; it is followed by a description of representative CDI-specific antimicrobial (antibiotic) stewardship programs (ASPs).

Background: Manifestations, Pathogenesis, Epidemiology

Clinical manifestations

In humans, CDI causes gastrointestinal symptoms with a wide range of clinical severity. Approximately 2% of the healthy adult population, and nearly 20% of adult inpatients, excrete the organism asymptomatically (Samore *et al.*, 1994). When symptoms

*Corresponding author. E-mail: Philip.Toltzis@UHHospitals.org

do occur, they vary from mild, self-limited diarrhea to life-threatening disease. The typical patient has multiple days of fever, abdominal cramping, and profuse diarrhea. Severe disease, which occurs most frequently in the elderly, is complicated by toxic megacolon with septic and hypovolemic shock. Severe manifestations typically are associated with prominent leukocytosis, elevation of serum creatinine, and low albumin (Zar *et al.*, 2007). Depending on the strain, mortality can exceed 5% (Morrison *et al.*, 2011; Walker *et al.*, 2012).

In mild cases, CDI can be managed by cessation of the offending antibiotics and supportive therapy. In more severe cases, therapies specific for *C. difficile* are required. The mainstay therapies for CDI for over 40 years have been oral metronidazole and oral vancomycin. For many years, these agents were considered therapeutically equivalent, with cure rates exceeding 90%, but metronidazole was preferred due to its lower cost and to concerns raised during the 1990s that the administration of oral vancomycin could select populations of vancomycin-resistant enterococci (VRE). More recent studies have demonstrated that oral vancomycin is the superior drug for severe CDI (Zar *et al.*, 2007), and the preference for vancomycin in patients with severe disease has been included in current consensus guidelines (Cohen *et al.*, 2010).

Approximately 20–30% of persons with CDI suffer from recurrent disease within 2 months of an apparent response to the first course of therapy. Genetic comparisons of the initial strain versus isolates from the recurrence indicate that the majority of cases are due to relapse with the original organism, and the remainder are the result of infection with a newly acquired strain (Kamboj *et al.*, 2011; Eyre *et al.*, 2012; Marsh *et al.*, 2012). To date, recurrence has not been due to the emergence of metronidazole- or vancomycin-resistant organisms. Rather, both metronidazole and vancomycin have broad activity against resident colonic flora as well as *C. difficile* and therefore perpetuate the dysbiosis (also called dysbacteriosis, or microbial imbalance) of the colonic bacterial populations that predisposed the patient to CDI in the first place. The intraluminal environment thus remains inviting either for retained spores, which then may germinate into toxin-producing organisms after the cessation of antibiotics (accounting for relapse), or for the re-acquisition of organisms resulting from repeated exposure to the healthcare environment (accounting for reinfection). Fidaxomincin is a recently introduced macrocyclic antibiotic that possesses activity against *C. difficile* but has relatively mild effects on the resident colonic flora. This observation has been cited to explain the relatively low rates of recurrent disease after fidaxomcin administration (Louie *et al.*, 2011; Cornely *et al.*, 2012).

Pathogenesis

CDI results from the inadvertent ingestion of *C. difficile* spores acquired from the hands of caregivers or others with close contact with the patient, or from contaminated inanimate surfaces. Early animal studies indicated that ingestion of <10 spores could result in clinical disease if the animal had been exposed to antibiotics. The sporulated organisms convert to the metabolically active vegetative state in the small intestines after exposure to bile acids (Seekatz and Young, 2014), but the principal site of *C. difficile*-associated disease in humans is the colon.

The pathogenicity of *C. difficile* depends on the coincidence of two factors, namely, perturbation of the normal colonic flora and the elaboration of cytotoxins by the infecting organisms. In most instances, the normal colonic microbiome renders the intraluminal environment inhospitable to colonization by *C. difficile*, a phenomenon termed "colonization resistance." For successful infection to occur, the patient must experience some event that results in disturbance of the healthy colonic bacterial populations. The most common such event is the administration of antibiotics. Some antibiotic families are particularly prone to result in *C. difficile* infection, most notably clindamycin and the broad-spectrum β-lactam agents, especially the cephalosporins (Owens *et al.*, 2008; Brown *et al.*, 2013; de Blank *et al.*, 2013), but exposure to virtually any antibiotic class may predispose a person to CDI. Recently, the emergence of fluoroquinolone-resistant strains has highlighted the association between exposure to this family of drugs and acquisition of CDI as well (Owens *et al.*, 2008). Investigators have demonstrated that even a short course of antibiotics may have a very prolonged effect on gut microbial diversity and that the reconstitution of this environment after antibiotic cessation may not result in repopulation with the beneficial pre-antibiotic flora (Dethlefsen and Relman, 2011).

Other conditions may also result in colonic dysbiosis, and thereby predispose to CDI; these include colonic surgery, dysmotility conditions, inflammatory bowel disease, chemotherapy, and intrinsic immunosuppression (Seekatz and Young, 2014). Likewise, the change in gastric pH conferred by histamine 2 receptor antagonists and protein pump inhibitors (PPIs) may affect the composition of bacterial populations in more distal gut locations, possibly accounting for the association of these classes of drugs with CDI (Morrison *et al.*, 2011; Mezoff and Cohen, 2013; Tleyjeh *et al.*, 2013; Seekatz and Young, 2014). Advanced age, also known to predispose to *C. difficile* diarrhea, is similarly associated with alterations of the normal microbiome (Rea *et al.*, 2012; Seekatz and Young, 2014).

The mechanism of colonization resistance is uncertain. The human microbiome is extraordinarily complex and varies from person to person, and it has been difficult to identify specific bacterial genera that are necessary to prevent stable colonization by *C. difficile*. Indeed, there is some evidence that it is the metabolic by-products of the normal colonic flora, rather than the bacteria themselves, that may be most important in inhibiting colonization (Seekatz and Young, 2014). Additional data suggest that colonization resistance is associated with the diversity of gut microbial populations as well as the presence of specific bacterial classes. Infants less than 1–2 years of age, with a developing microbiome considerably less diverse than that recorded in older children and adults, have a high incidence of colonization with toxigenic *C. difficile* (Jangi and Lamont, 2010; Ferraris *et al.*, 2012). Conversely, fecal transplants, in which stool flora from healthy patients are transferred to the colon of patients with multiply recurrent CDI, have been effective in producing a clinical cure (Russell *et al.*, 2010; Rubin *et al.*, 2013; van Nood *et al.*, 2013). Recent studies have demonstrated that patients successfully treated by fecal transplantation establish a microbiome of similar diversity to that of the donor and considerably more complex than that resident in the host during the episode of CDI (van Nood *et al.*, 2013; Walia *et al.*, 2014).

Once it successfully colonizes, *C. difficile* produces disease through the expression of two cytotoxins, called Toxin A and B (Voth and Ballard, 2005), and these are responsible for the intense colonic inflammation characteristic of this infection. Organisms that do not elaborate these toxins do not cause disease. Historically, Toxin A was felt to constitute the principal toxin, but recently, clinical isolates that are Toxin A–/Toxin B+ have been identified that are disease producing (O'Connor *et al.*, 2009). The genes encoding these toxins, *tcdA* and *tcdB*, respectively, lie within a 19.6 kB pathogenicity locus, which additionally encodes a positive regulator (*tcdD*) and a negative regulator (*tcdC*) for toxin production. Both toxins are largely homologous glucosyltransferases. The toxins enter the eukaryotic cell through surface binding and endocytosis. They then glycosylate small eukaryotic GTP hydrolases (GTPases), which results in the disruption of host intracellular signaling functions, the most notable downstream manifestation of which is the loss of actin, disturbance of the cellular cytoskeleton, and cell death (Voth and Ballard, 2005). A third toxin, called binary toxin, has been identified in some populations of *C. difficile*; while this toxin also has in vitro activity against actin, its role in the pathogenicity of *C. difficile* is uncertain, as strains that produce binary toxin in the absence of Toxins A and B typically are non-pathogenic (Kelly and LaMont, 2008).

Recent data additionally suggest that host factors may contribute to symptoms in CDI. Both adults and children with *C. difficile*-associated diarrheal disease elaborate cytokines (IL-8) and chemokines (CXCL-5) in the gastrointestinal tract (El Feghaly *et al.*, 2012, 2013). Stool mRNA quantities encoding both of these products correlate with disease severity and persistence; by contrast, the burden of excreted *C. difficile* has no such apparent clinical correlation (El Feghaly *et al.*, 2012; El Feghaly *et al.*, 2013).

Common to other toxigenic infectious diseases, recovery from *C. difficile* infection is at least partially dependent on the appearance of antibodies that neutralize the toxins (Kyne *et al.*, 2000). The majority of patients who become symptomatic do so shortly after acquisition of the organism rather than after prolonged colonization, suggesting that in the former case, symptoms begin before antitoxin antibody production or boosting can occur (Samore *et al.*, 1994; McDonald *et al.*, 2007). Poor systemic antibody response to the enterotoxins is associated with recurrent disease (Kelly and LaMont, 2008), and the infusion of both polyclonal and monoclonal antitoxin antibodies (Kelly and LaMont, 2008; Lowy *et al.*, 2010) has been used successfully in reducing the rate of recurrence.

Epidemiology

Until recently, *C. difficile* was acquired almost exclusively in the hospital setting. *C. difficile* spores are particularly resistant to hospital disinfectants, except for those containing peroxide or bleach (Wilcox *et al.*, 2003), and they remain viable on inanimate surfaces for weeks (Samore *et al.*, 1996; Mayfield *et al.*, 2000; Shaughnessy *et al.*, 2011). These observations have led to the adaption of "terminal disinfection" protocols in most hospitals, in which rooms that have been recently vacated are thoroughly hand cleaned with bleach-containing compounds or by apparatuses which spray the room with hydrogen peroxide (Kaatz *et al.*, 1988; Otter *et al.*, 2009). As the organisms are resistant to many hand hygiene agents, particularly those that are alcohol based, the use of hand gels is discouraged in caring for the patient with CDI; vigorous scrubbing with conventional hand-disinfection agents, which leads to the physical removal of the organism from the upper layers of the dermis, is felt to be more effective in ridding the hands of spores (Cohen *et al.*, 2010).

In the early 2000s, a new strain of *C. difficile* was identified in Quebec that soon became epidemic throughout central Canada (Loo *et al.*, 2005; Hubert *et al.*, 2007), and then rapidly spread throughout the US (McDonald *et al.*, 2005), into western Europe and further east, accounting at least in part for the worldwide explosive increase in CDI during the ensuing decade. This strain, with the ribotype classification 027, became known as the North American Pulsed Field Type 1 (NAP1) strain. The NAP1 strain is hypervirulent, characteristically producing disease more severe than that associated with non-epidemic strains, especially in the elderly (Redelings *et al.*, 2007; Labbé *et al.*, 2008; Walker *et al.*, 2012; Miller *et al.*, 2013). In vitro, the NAP1 strain sporulates more readily than other clinical strains (Merrigan *et al.*, 2010), thus accounting for its rapid spread, and produces more Toxins A and B than conventional strains (Warny *et al.*, 2005; Curry *et al.*, 2007; Merrigan *et al.*, 2010), accounting for its heightened virulence. The latter phenomenon has been ascribed to deletions in the *tcdC* inhibitor gene within the pathogenicity locus. Virtually all NAP1 isolates contain binary toxin. They are typically resistant to the fluoroquinolones, and the frequent use of these agents among hospitalized patients has been hypothesized as a key factor in promoting the horizontal spread of NAP1 in inpatient populations. Other hypervirulent strains have recently been identified, particularly a ribotype 078 strain in northern Europe (Goorhuis *et al.*, 2008; Walker *et al.*, 2012), but the epidemiological importance of these have not yet equaled that associated with NAP1.

In the middle of the first decade of the 2000s, American reports documented CDI in otherwise healthy individuals who had not had recent exposure to antibiotics or to inpatient healthcare facilities (Benson *et al.*, 2007; Baker *et al.*, 2010; Kutty *et al.*, 2010; Kuntz *et al.*, 2011, 2012; Khanna *et al.*, 2012, 2013; Chitnis *et al.*, 2013; Le Saux *et al.*, 2014; Wendt *et al.*, 2014). These included children and young adults, particularly pregnant women, some of whom exhibited severe symptoms. These reports constituted the first evidence that *C. difficile* had escaped the confines of inpatient hospital wards and had become a community-acquired pathogen. With expanding numbers of such cases, a committee advisory to the US Centers for Disease Control and Prevention (CDC) developed a case definition for community-associated *C. difficile* infection (CA-CDI), namely, the development of symptoms <48 h after hospital admission in a patient with no contact with a healthcare facility within the prior 12 weeks (McDonald *et al.*, 2007). By 2014, several population-based surveys, mostly in the US, documented the prominence of CA-CDI as frequently constituting 40% or more of all CDI cases (Kutty *et al.*, 2010; Kuntz *et al.*, 2011; Khanna *et al.*, 2012). Compared with persons with hospital-acquired infection, patients with CA-CDI tend to be younger, have less antecedent exposure to antibiotics, and possess fewer comorbidities. Conversely, however, approximately 70–75% of CA-CDI patients have, in fact, received antibiotics within the past 90 days (Hirshon *et al.*, 2011; Naggie *et al.*, 2011; Khanna *et al.*, 2012; Kuntz *et al.*, 2012; Brown *et al.*, 2013; Tschudin-Sutter *et al.*, 2013; Wendt *et al.*, 2014), and a substantial majority have comorbidities requiring frequent visits to an outpatient medical facility (Chitnis *et al.*, 2013). This latter observation has raised the possibility that outpatient settings may be a source for *C. difficile* acquisition, and some investigators have subsequently documented the contamination of inanimate surfaces by *C. difficile* in outpatient clinics (Jury *et al.*, 2013a). Other potential sources of *C. difficile* in CA-CDI that have been surmised include infants who are asymptomatically excreting toxigenic organisms (Stoesser *et al.*, 2011; Rousseau *et al.*, 2012).

Conversely, studies have repeatedly failed to identify *C. difficile* as a cause of community acquired diarrhea in otherwise healthy children (Vernacchio *et al.*, 2006; Denno *et al.*, 2012; Valentini *et al.*, 2013). Although the organism can be isolated in small proportions of children presenting to an outpatient facility, or admitted to the hospital with community-onset diarrhea, controlled studies have documented that *C. difficile* can be isolated among similar patients without diarrhea at least as frequently (Denno *et al.*, 2012).

Clostridium difficile Infections and Antimicrobial Stewardship

The prominence of antecedent antibiotic exposure as a risk factor in CDI, and the observation that most CDI is acquired in inpatient settings, prima facie renders the disease an ideal target for antibiotic stewardship as a strategy to reduce its incidence and spread. With government agencies requiring mandatory reporting of CDI rates, and with a potential effect on reimbursement, there are now compelling financial as well as medical reasons to consider ASPs to reduce the incidence of CDI. The results of several single-center studies indicate that ASPs are effective in this regard (Hsu *et al.*, 2010). In June 2014, the Society for Healthcare Epidemiology of America (SHEA) and the Infectious Diseases Society of America (IDSA) collaboratively released practice recommendations on strategies to prevent CDI in acute hospital settings (Dubberke *et al.*, 2014). The initiation of an ASP program was included in these guidelines, with quality of evidence level II.

A number of approaches have been employed in designing ASPs to reduce the incidence of CDI. These include the following:

- formulary restriction and preauthorization of selected antibiotics;
- prospective audit of antibiotic use with intervention and feedback;
- establishing stewardship as one component of a global infection control program to reduce hospital-acquired *C. difficile* infection; and
- establishing stewardship to guide the diagnosis and treatment of CDI.

Published experience describing experience with each of these strategies is summarized below, followed by a brief account of systematic reviews and meta-analyses of the role of ASPs in reducing CDI.

Formulary restriction and preauthorization

The restriction of antibiotics has been utilized as an infection prevention measure to contain *C. difficile* outbreaks. In 1994, an outbreak of *C. difficile* in a hospital in Tucson, Arizona was linked to the increased use of clindamycin, with a fivefold increase in the number of cases of CDI. The restriction of this antibiotic from the formulary was effective in controlling the epidemic (Pear *et al.*, 1994). A recent study from the UK examined the effect of antibiotic restriction on hospital CDI incidence (Aldeyab *et al.*, 2012). Specifically, the program restricted "high-risk antibiotics" according to their propensity to result in CDI, namely, second- and third-generation cephalosporins, fluoroquinolones, and clindamycin. The impact of the policy on CDI rates was studied from 2004 to 2010 (pre- and postintervention) using interrupted time sequence. The study found both a reduction in the trend ($P = 0.008$) of high-risk antibiotic use and a reduction in the incidence of CDI after the intervention. Conversely, Climo *et al.* (1998), from a Veterans Affairs (VA) Medical Center in Richmond, Virginia, reported that the restriction of clindamycin, instituted in response to a *C. difficile* outbreak at the VA Center, led to the increased use of other antibiotics with anaerobic activity, such as imipenem, ticarcillin/clavulanate, metronidazole, cefotetan, and piperacillin. While the intervention resulted in a significant reduction in new cases of CDI, this observation illustrates how single antimicrobial restriction may be effective for a short period of time, but that without the monitoring of other antibiotics, the cycle of overuse of antibiotics is likely to occur.

Prospective audit with intervention and feedback

Fowler *et al.* (2007), using a combination of narrow-spectrum antibiotic usage policies reinforced by feedback, achieved a significant reduction of CDI along with a significant impact on antimicrobial prescription patterns. The study took place on three wards that admitted elderly patients in a London teaching hospital, where a cephalosporin restriction policy was already in place; audit and feedback methods were then added to prevent a rise of CDI incidence. The antibiotic recommendations, provided on laminated cards given to all physicians, preferred the use of the narrow-spectrum antibiotics benzylpenicillin, trimethoprim, and amoxicillin, and the avoidance of cephalosporins and amoxicillin/clavulanate, with feedback provided by an infectious diseases physician or a clinical pharmacologist. The primary outcome of the study was a change in the level and linear trend of prescription of targeted antibiotics, and the secondary outcome was monthly CDI rates. The intervention resulted in a significant reduction in amoxicillin/clavulanate and cephalosporin use in the periods after the intervention and a decrease in incidence of CDI (incidence rate ratio (IRR) 0.35, [95% CI] 0.17–0.73, $P = 0.009$).

Carling *et al.* (2003) published the first study to measure the long-term outcomes of a multiple-antibiotic control-based initiative. The program consisted of prospective auditing of several target antibiotics (cephalosporins, aztreonam, quinolones, and imipenem) with individualized recommendations by a team consisting of an infectious disease specialist and pharmacist in a hospital in Boston, Massachusetts. Other interventions included pharmacy restrictions, automatic 7 day stop orders of all antibiotics, education to staff, and limiting the release of susceptibilities of bacteria to certain antibiotics. The program resulted in a decrease in incidence of CDI from 2.2 cases per 1000 patient days to 1.4 cases per 1000 patient days ($P = 0.002$) in the 3 years following the implementation of the program.

There currently are limited data on which approach—formulary restriction versus prospective audit and feedback—is superior in attaining the desired outcomes, and it is likely that a combination of these strategies yields the best results. The Cochrane Review by Davey *et al.* (2013) concluded that restrictive policies are better when rapid changes are needed, but that restrictive and prospective audit and feedback programs had similar effects on antimicrobial prescription and microbiological outcomes after 6 months. A single center study from the University of Pennsylvania utilizing a quasi-experimental design compared both strategies (Mehta *et al.*, 2014). The investigators looked at a 24 month period before and after implementing an ASP that evolved from the restriction of selected antimicrobials to prospective audit and feedback. The change to a prospective audit and feedback strategy actually resulted in an increase in the use of the targeted antibiotics compared with use during the period when antibiotic restriction was employed. Clearly, more studies are needed to compare the two strategies. Until such evidence is available, each hospital should evaluate the extent of its antimicrobial overuse/misuse problem, identify antibiotics that may be driving CDI rates, and select the approach that would be most acceptable within the culture of that institution.

Antimicrobial stewardship as one component of a global infection control program to reduce hospital-acquired *C. difficile* infection

ASPs have been employed as an important component of an infection control bundle to prevent or contain *C difficile* outbreaks. Valiquette *et al.* (2007) reported their experience using antimicrobial stewardship as an infection control tool. In 2003, a surge in nosocomial CDI at their 683-bed referral hospital in Quebec prompted the adoption of strict infection control measures. In the initial phase, staff education, strict hand washing and isolation policies were adopted in addition to environmental disinfection with sodium hypochlorite. These measures were further strengthened a few months later, and terminal room disinfection with hydrogen peroxide was instituted. Despite these measures, the number of nosocomial cases of *C. difficile* did not significantly decline; hence, a third phase of measures was implemented, consisting of interventions to optimize antimicrobial prescribing in the hospital. Guidelines for decreasing the use of antibiotics associated with *C. difficile* infections were created by a team of infectious disease physicians and pharmacists. Educational materials for physicians and pharmacists were distributed. Target antibiotics (those associated with CDI in the institution concerned) were not restricted, but feedback was given directly to physicians suggesting alternative antibiotics for treating conditions. Following this intervention, CDI incidence decreased by 60% between 2003–2004 and 2005–2006, and total and targeted antibiotic consumption decreased by 23% and 54% respectively.

Muto *et al.* (2007) implemented a bundle in a stepwise fashion to contain an outbreak of the epidemic NAP1 *C. difficile* strain in a university hospital in Pittsburgh, Pennsylvania. A component of the bundle included the restriction of antibiotics such as clindamycin, levofloxacin, and ceftriaxone, which were associated with CDI at the facility. The implementation of the stewardship program led to a composite decline of the use of these antibiotics by 54%. The authors speculated that the ASP may have contributed to the sustained low incidence of CDI, but they noted that fluoroquinolone use increased toward the end of the study period, while the incidence of CDI continued to decrease; the contribution of antibiotic management within the bundle could not, therefore, be confidently assessed.

Antimicrobial stewardship to guide the diagnosis and treatment of *C. difficile* infection

To improve the outcomes of patients with CDI, treatment must be instituted promptly and adequately. Despite published guidelines for the diagnosis and treatment of CDI (Cohen *et al.*, 2010), the management of these infections remains suboptimal. A survey of primary care physicians performed at a Michigan health system revealed that 59% of the physicians were aware of the existence of the guidelines but only 31% had read them (Byker *et al.*, 2009). Others have reported delays in the implementation of empirical therapy for severe cases or prescription of empirical therapy for mild cases (Kundrapu *et al.*, 2013). Hence, a stewardship program focused on the treatment of *C. difficile* could improve adherence to guidelines and avoid delay in the treatment of severe cases.

To date, only one study has been published testing whether such guideline initiatives are effective (Jury *et al.*, 2013b). The goals of the program were prompt initiation of therapy for documented cases of CDI, increased adherence to practice guidelines, and the reduction of empiric therapy for cases with low suspicion of CDI or mild-to-moderate symptoms. The implementation of the initiative involved education of physicians on the diagnosis and treatment of CDI and an order set in the electronic medical record (EMR). The microbiology laboratory alerted stewardship program personnel with every positive *C. difficile* assay, which prompted a prospectively conducted review of the patient's chart and generated initial treatment recommendations, or feedback if the therapy initiated was not in accordance with guidelines. The program resulted in a statistically significant reduction in the delay of treatment of CDI, reduction in the use of metronidazole for severe infections, and reduced empirical treatment of mild cases (Jury *et al.*, 2013b).

Systematic reviews and meta-analyses of ASPs in reducing *C. difficile* infection

Single center experiences have demonstrated that ASPs have an important influence on CDI rates (Dubberke *et al.*, 2014; File *et al.*, 2014), but when systematic reviews have been performed, the effect is less clear (Davey *et al.*, 2013; Dodds Ashley *et al.*, 2014; McGregor and Furuno, 2014; Wagner *et al.*, 2014). The various methodologies and study designs, and the varying strength of evidence for patient outcomes make it difficult to ascertain the overall, generalizable impact of antibiotic stewardship on rates of CDI. Many studies have been done in the midst of a *C. difficile* outbreak with multiple interventions implemented at the same time, with confounding factors obscuring the effect of the ASP. A Cochrane Review of inpatient interventions to improve antimicrobial prescribing practices in general reviewed 507 full text articles of which only 89 met the inclusion criteria, and 45% of the included studies had a high risk of bias (Davey *et al.*, 2013). Five studies had CDI as a microbiological outcome and found that interventions aimed at changing antibiotic prescribing practices were associated with a decline in *C difficile*. However, few studies included reliable data on changes in antibiotic prescribing and only two of the studies were low bias. Notably, these two studies showed the least intervention effect on CDI. A second systematic review (Wagner *et al.*, 2014) found a paucity of well-designed studies to detect improvement in patient outcomes.

In 2012, the Greater New York Hospital Association, in collaboration with the US Agency for Healthcare Research and Quality (AHRQ) launched the ERASE *C. difficile* (**E**valuation and **R**esearch on **A**ntimicrobial **S**tewardship's **E**ffect on *Clostridium difficile*) Project, and the results of the experience were published recently (Ostrowsky *et al.*, 2014). Ten hospitals participated, and each had previously participated in a CDI initiative focusing on improved infection control and environmental cleaning practices. Six had additionally recently developed ASPs to control the incidence of *C. difficile*, employing methodologies included in a toolkit specifically designed for this purpose by the AHRQ (AHRQ and CDC, 2012); the remaining four institutions without ASPs served as non-intervention controls. The ASP hospitals included audit and feedback on the use of high-risk antibiotics as part of their interventions. The intervention hospitals recorded a decline in the use of target antibiotics, although this did not translate into a significant reduction in hospital-onset CDI, either when the intervention hospitals were compared with non-intervention hospitals or when the incidence of CDI in intervention hospitals was assessed over time. One of the principal problems with the project was that it took longer than anticipated to fully implement the ASP interventions and the time to detect a change in rates was likely insufficient.

Conclusions

There is a strong association between *C. difficile*-associated diarrhea and both antecedent antibiotic exposure and inpatient hospitalization. These observations highly recommend antibiotic stewardship as a key element in the control of CDI. Other factors remains important drivers of CDI rates, however, including adherence to hand hygiene, patient isolation procedures, the effectiveness of environmental disinfection policies, patient mix, local incidence of CA-CDI, and the relative local prevalence of NAP1 and other hypervirulent strains of *C. difficile*. It is likely, therefore, that ASPs are a necessary but not sufficient component of CDI containment. Future research should define the stewardship strategies most effective in reducing the frequency of CDI as well as the other vital components of a multipronged effort to reduce this preventable life-threatening condition.

References

Aldeyab, M.A., Kearney, M.P., Scott, M.G., Aldiab, M.A., Alahmadi, Y.M., Darwish Elhajji, F.W., Magee, F.A., and McElnay, J.C. (2012) An evaluation of the impact of antibiotic stewardship on reducing the use of high-risk antibiotics and its effect on the incidence of *Clostridium difficile* infection in hospital settings. *Journal of Antimicrobial Chemotherapy* 67, 2988–2996.

AHRQ and CDC (2012) *Evaluation and Research on Antimicrobial Stewardship's Effect on* Clostridium difficile *(ERASE* C. difficile*) Project: Toolkit for Reduction of* Clostridium difficile *Through Antimicrobial Stewardship*. AHRQ Publication No. 12-0082-EF. Prepared by Boston University School of Public Health, Montefiore Medical Center, and Greater New York Hospital Association/ United Hospital Fund for the Agency for Healthcare Research and Quality (AHRQ), Rockville, Maryland and the Centers for Disease Control and Prevention, Atlanta, Georgia. Available at: http://www.ahrq.gov/sites/default/files/publications/files/cdifftoolkit.pdf (accessed 13 May 2016).

Baker, S.S., Faden, H., Sayej, W., Patel, R., and Baker, R.D. (2010) Increasing incidence of community-associated atypical *Clostridium difficile* disease in children. *Clinical Pediatrics (Philadelphia)* 49, 644–647.

Bartlett, J.G. (1981) Antimicrobial agents implicated in *Clostridium difficile* toxin-associated diarrhea of colitis. *The Johns Hopkins Medical Journal* 149, 6–9.

Bartlett, J.G. (2004) Commentary: Bartlett, JG, Onderdonk, AB, Cisneros, RL, Kasper, D.L. Clindamycin-associated colitis due to a toxin-producing species of *Clostridium* in hamsters. *J. Infect Dis* 1977;136:701. *The Journal of Infectious Diseases* 190, 202–209.

Bartlett, J.G., Onderdonk, A.B., Cisneros, R.L., and Kasper, D.L. (1977) Clindamycin-associated colitis due to a toxin-producing species of *Clostridium* in hamsters. *The Journal of Infectious Diseases* 136, 701–705.

Benson, L., Song, X., Campos, J., and Singh, N. (2007) Changing epidemiology of *Clostridium difficile*-associated disease in children. *Infection Control and Hospital Epidemiology* 28, 1233–1235.

Brown, K.A., Khanafer, N., Daneman, N., and Fisman, D.N. (2013) Meta-analysis of antibiotics and the risk of community-associated *Clostridium difficile* infection. *Antimicrobial Agents and Chemotherapy* 57, 2326–2332.

Byker, G.L., Dinh, M.T., Gunaratnam, N.T., Robinson, E.A., Shehab, T.M., and Malani, A.N. (2009) Management of *Clostridium difficile* infection: survey of practices and compliance with national guidelines among primary care physicians. *Infection Control and Hospital Epidemiology* 30, 397–399.

Carling, P., Fung, T., Killion, A., Terrin, N., and Barza, M. (2003) Favorable impact of a multidisciplinary antibiotic management program conducted during 7 years. *Infection Control and Hospital Epidemiology* 24, 699–706.

Chitnis, A.S., Holzbauer, S.M., Belflower, R.M., Winston, L.G., Bamberg, W.M., Lyons, C., Farley, M.M., Dumyati, G.K., Wilson, L.E., Beldavs, Z.G. *et al.* (2013) Epidemiology of community-associated *Clostridium difficile* infection, 2009 through 2011. *JAMA Internal Medicine* 173, 1359–1367.

Climo, M.W., Israel, D.S., Wong, E.S., Williams, D., Coudron, P., and Markowitz, S.M. (1998) Hospital-wide restriction of clindamycin: effect on the incidence of *Clostridium difficile*-associated diarrhea and cost. *Annals of Internal Medicine* 128, 989–995.

Cohen, S.H., Gerding, D.N., Johnson, S., Kelly, C.P., Loo, V.G., McDonald, L.C., Pepin, J., Wilcox, M.H., Society for Healthcare Epidemiology of America, and Infectious Diseases Society of America (2010) Clinical practice guidelines for *Clostridium difficile* infection in adults: 2010 update by the Society for Healthcare Epidemiology of America (SHEA) and the Infectious Diseases Society of America (IDSA). *Infection Control and Hospital Epidemiology* 31, 431–455.

Cornely, O.A., Miller, M.A., Louie, T.J., Crook, D.W., and Gorbach, S.L. (2012) Treatment of first recurrence of *Clostridium difficile* infection: fidaxomicin versus vancomycin. *Clinical Infectious Diseases* 55(Suppl 2), S154–S161.

Curry, S.R., Marsh, J.W., Muto, C.A., O'Leary, M.M., Pasculle, A.W., and Harrison, L.H. (2007) *tcdC* Genotypes associated with severe TcdC truncation in an epidemic clone and other strains of *Clostridium difficile*. *Journal of Clinical Microbiology* 45, 215–221.

Davey, P., Brown, E., Charani, E., Fenelon, L., Gould, I.M., Holmes, A., Ramsay, C.R., Wiffen, P.J., and Wilcox, M. (2013) Interventions to improve antibiotic prescribing practices for hospital inpatients. *The Cochrane Database of Systematic Reviews* 4: CD003543.

de Blank, P., Zaoutis, T., Fisher, B., Troxel, A., Kim, J., and Aplenc, R. (2013) Trends in *Clostridium difficile* infection and risk factors for hospital acquisition of *Clostridium difficile* among children with cancer. *Journal of Pediatrics* 63, 699–705.e1.

Denno, D.M., Shaikh, N., Stapp, J.R., Qin, X., Hutter, C.M., Hoffman, V., Mooney, J.C., Wood, K.M., Stevens, H.J., Jones, R., Tarr, P.I., and Klein, E.J. (2012) Diarrhea etiology in a pediatric emergency department: a case control study. *Clinical Infectious Diseases* 55, 897–904.

Dethlefsen, L. and Relman, D.A. (2011) Incomplete recovery and individualized responses of the human distal gut microbiota to repeated antibiotic perturbation. *Proceedings of the National Academy of Sciences of the United States of America* 108(Suppl 1), 4554–4561.

Dodds Ashley, E.S., Kaye, K.S., DePestel, D.D., and Hermsen, E.D. (2014) Antimicrobial stewardship: philosophy versus practice. *Clinical Infectious Diseases* 59(Suppl 3), S112–S121.

Dubberke, E.R. and Wertheimer, A.I. (2009) Review of current literature on the economic burden of *Clostridium difficile* infection. *Infection Control and Hospital Epidemiology* 30, 57–66.

Dubberke, E.R., Carling, P., Carrico, R., Donskey, C.J., Loo, V.G., McDonald, L.C., Maragakis, L.L., Sandora, T.J., Weber, D.J., Yokoe, D.S., and Gerding, D.N. (2014) Strategies to prevent *Clostridium difficile* infections in acute care hospitals: 2014 update. *Infection Control and Hospital Epidemiology* 35, 628–645.

El Feghaly, R.E., Stauber, J.L., Deych, E., Gonzalez, C., Tarr, P.I., and Haslam, D.B. (2012) Markers of intestinal inflammation, not bacterial burden, correlate with clinical outcomes in *Clostridium difficile* infection. *Clinical Infectious Diseases* 56, 1713–1721.

El Feghaly, R.E., Stauber, J.L., Tarr, P.I., and Haslam, D.B. (2013) Intestinal inflammatory biomarkers and outcome in pediatric *Clostridium difficile* infections. *Journal of Pediatrics* 163, 1697–1704.e2.

Eyre, D.W., Walker, A.S., Wyllie, D., Dingle K.E., Griffiths, D., Finney, J., O'Connor, L., Vaughan, A., Crook, D.W., Wilcox, M.H., and Peto, T.E.A. (2012) Predictors of first recurrence of *Clostridium difficile* infection: implications for initial management. *Clinical Infectious Diseases* 55(Suppl 2), S77–S87.

Ferraris, L., Butel, M.J., Campeotto, F., Vodovar, M., Roze, J.C., and Aires, J. (2012) Clostridia in premature neonates' gut: Incidence, antibiotic susceptibility, and perinatal determinants influencing colonization. *PLoS ONE* 7(1): e30594.

File, T.M., Jr., Srinivasan, A., and Bartlett, J.G. (2014) Antimicrobial stewardship: importance for patient and public health. *Clinical Infectious Diseases* 59(Suppl 3), S93–S96.

Fowler, S., Webber, A., Cooper, B.S., Phimister, A., Price, K., Carter, Y., Kibbler, C.C., Simpson, A.J.H., and Stone, S.P. (2007) Successful use of feedback to improve antibiotic prescribing and reduce *Clostridium difficile* infection: a controlled interrupted time series. *Journal of Antimicrobial Chemotherapy* 59, 990–995.

Goorhuis, A., Bakker, D., Corver, J., Debast, S.B., Harmanus, C., Notermans, D.W., Bergwerff, A.A., Dekker, F.W., and Kuijper, E.J. (2008) Emergence of *Clostridium difficile* infection due to a new hypervirulent strain, polymerase chain reaction ribotype 078. *Clinical Infectious Diseases* 47, 1162–1170.

Hirshon, J.M., Thompson, A.D., Limbago, B., McDonald, L.C., Bonkosky, M., Heimer, R., Meek, J., Mai, V. and Braden, V. (2011) *Clostridium difficile* infection in outpatients, Maryland and Connecticut, USA, 2002-2007. *Emerging Infectious Diseases Journal* 17, 1946–1949.

Hsu, J., Abad, C., Dinh, M., and Safdar, N. (2010) Prevention of endemic healthcare-associated *Clostridium difficile* infection: reviewing the evidence. *The American Journal of Gastroenterology* 105, 2327–2339; quiz 40.

Hubert, B., Loo, V.G., Bourgault, A.M., Poirier, L., Dascal, A., Fortin, E., Dionne, M., and Lorange, M. (2007) A portrait of the geographic dissemination of the *Clostridium difficile* North American pulsed-field type 1 strain and the epidemiology of *C. difficile*-associated disease in Quebec. *Clinical Infectious Diseases* 44, 238–244.

Jangi, S. and Lamont, J.T. (2010) Asymptomatic colonization by *Clostridium difficile* in infants: implications for disease in later life. *Journal of Pediatric Gastroenterology and Nutrition* 51, 2–7.

Jury, L.A., Sitzlar, B., Kundrapu, S., Cadnum, J.L., Summers, K.M., Muganda, C.P., Deshpande, A., Sethi, A.J., and Donskey, C.J. (2013a) Outpatient healthcare settings and transmission of *Clostridium difficile*. *PLoS ONE* 8(7): e70175.

Jury, L.A., Tomas, M., Kundrapu, S., Sitzlar, B., and Donskey, C.J. (2013b) A *Clostridium difficile* infection (CDI) stewardship initiative improves adherence to practice guidelines for management of CDI. *Infection Control and Hospital Epidemiology* 34, 1222–1224.

Kaatz, G.W., Gitlin, S.D., Schaberg, D.R., Wilson, K.H., Kauffman, C.A., Seo, S.M., and Fekety, R. (1988) Acquisition of *Clostridium difficile* from the hospital environment. *American Journal of Epidemiology* 127, 1289–1294.

Kamboj, M., Khosa, P., Kaltsas, A., Babady, N.E., Son, C., and Sepkowitz, K.A. (2011) Relapse versus reinfection: surveillance of *Clostridium difficile* infection. *Clinical Infectious Diseases* 53, 1003–1006.

Kelly, C.P. and LaMont, J.T. (2008) *Clostridium difficile*—more difficult than ever. *The New England Journal of Medicine* 359, 1932–1940.

Khanna, S., Pardi, D.S., Aronson, S.L., Kammer, P.P., Orenstein, R., St Sauver, J.L., Harmsen, W.S., and Zinsmeister, A.R. (2012) The epidemiology of community-acquired *Clostridium difficile* infection: a population-based study. *The American Journal of Gastroenterology* 107, 89–95.

Khanna, S., Baddour, L.M., Huskins, W.C., Kammer, P.P., Faubion, W.A., Zinsmeister, A.R., Harmsen, W.S., and Pardi, D.S. (2013) The epidemiology of *Clostridium difficile* infection in children: a population-based study. *Clinical Infectious Diseases* 56, 1401–1406.

Kim, J., Smathers, S.A., Prasad, P., Leckerman, K.H., Coffin, S., and Zaoutis, T. (2008) Epidemiological features of *Clostridium difficile*-associated disease among inpatients at children's hospitals in the United States, 2001–2006. *Pediatrics* 122, 1266–1270.

Kundrapu, S., Jury, L.A., Sitzlar, B., Sunkesula, V.C., Sethi, A.K., and Donskey, C.J. (2013) Easily modified factors contribute to delays in diagnosis of *Clostridium difficile* infection: a cohort study and intervention. *Journal of Clinical Microbiology* 51, 2365–2370.

Kuntz, J.L., Chrischilles, E.A., Pendergast, J.F., Herwaldt, L.A., and Polgreen, P.M. (2011) Incidence of and risk factors for community-associated *Clostridium difficile* infection: a nested case-control study. *BMC Infectious Diseases* 11: 194.

Kuntz, J.L., Johnson, E.S., Raebel, M.A., Petrik, A.F., Yang, X., Thorp, M.L., Spindel, S.J., Neil, N., and Smith, D.H. (2012) Epidemiology and healthcare costs of incident *Clostridium difficile* infections identified in the outpatient healthcare setting. *Infection Control and Hospital Epidemiology* 33, 1031–1038.

Kutty, P.K., Woods, C.W., Sena, A.C., Benoit, S.R., Naggie, S., Frederick, J., Evans, S., Engel, J., and McDonald, L.C. (2010) Risk factors for and estimated incidence of community-associated *Clostridium difficile* infection, North Carolina, USA. *Emerging Infectious Diseases Journal* 16, 197–204.

Kyne, L., Warny, M., Qamar, A., and Kelly, C.P. (2000) Asymptomatic carriage of *Clostridium difficile* and serum levels of IgG antibody against toxin A. *The New England Journal of Medicine* 342, 390–397.

Labbé, A.C., Poirier, L., Maccannell, D., Louie, T., Savoie, M., Béliveau, C., Laverdière, M., and Pépin, J. (2008) *Clostridium difficile* infections in a Canadian tertiary care hospital before and during a regional epidemic associated with the BI/NAP1/027 strain. *Antimicrobial Agents and Chemotherapy* 52, 3180–3187.

Le Saux, N., Gravel, D., Mulvey, M.R., Dennis, J., Yasseen, S.S., III, Barrowman, N., and Bowes, J. (2014) Pediatric *Clostridium difficile* infection: 6-year active surveillance in a defined patient population. *Infection Control and Hospital Epidemiology* 35, 904–906.

Lessa, F.C., Gould, C.V., and McDonald, L.C. (2012) Current status of *Clostridium difficile* infection epidemiology. *Clinical Infectious Diseases* 55(Suppl 2), S65–S70.

Loo, V.G., Poirier, L., Miller, M.A., Oughton, M., Libman, M.D., Michaud, S., Bourgault, A.M., Nguyen, T., Frenette, C., Kelly, M. *et al.* (2005) A predominantly clonal multi-institutional outbreak of *Clostridium difficile*-associated diarrhea with high morbidity and mortality. *The New England Journal of Medicine* 353, 2442–2449.

Louie, T.J., Miller, M.A., Mullane, K.M., Weiss, K., Lentnek, A., Golan, Y., Gorbach, S., Sears, P., Shue, Y.K., and OPT-80-003 Clinical Study Group (2011) Fidaxomicin versus vancomycin for *Clostridium difficile* infection. *The New England Journal of Medicine* 364, 422–431.

Lowy, I., Molrine, D.C., Leav, B.A., Blair, B.M., Baxter, R., Gerding, D.N., Nichol, G., Thomas, W.D., Jr., Leney, M., Sloan, S. *et al.* (2010) Treatment with monoclonal antibodies against *Clostridium difficile* toxins. *The New England Journal of Medicine* 362, 197–205.

Marsh, J.W., Arora, R., Schlackman, J.L., Shutt, K.A., Curry, S.R., and Harrison, L.H. (2012) Association of relapse of *Clostridium difficile* disease with BI/NAP1/027. *Journal of Clinical Microbiology* 50, 4078–4082.

Mayfield, J.L., Leet, T., Miller, J., and Mundy, L.M. (2000) Environmental control to reduce transmission of *Clostridium difficile*. *Clinical Infectious Diseases* 31, 995–1000.

McDonald, L.C., Killgore, G.E., Thompson, A., Owens, R.C., Jr., Kazakova, S.V., Sambol, S.P., Johnson, S., and Gerding, D.N. (2005) An epidemic, toxin gene-variant strain of *Clostridium difficile*. *The New England Journal of Medicine* 353, 2433–2441.

McDonald, L.C., Coignard, B., Dubberke, E., Song, X., Horan, T., and Kutty, P.K. (2007) Recommendations for surveillance of *Clostridium difficile*-associated disease. *Infection Control and Hospital Epidemiology* 28, 140–145.

McGregor, J.C. and Furuno, J.P. (2014) Optimizing research methods used for the evaluation of antimicrobial stewardship programs. *Clinical Infectious Diseases* 59(Suppl 3), S185–S192.

Mehta, J.M., Haynes, K., Wileyto, E.P., Gerber, J.S., Timko, D.R., Morgan, S.C., Binkley, S., Fishman, N.O., Lautenbach, E., Zaoutis, T., and Centers for Disease Control and Prevention Epicenter Program (2014) Comparison of prior authorization and prospective audit with feedback for antimicrobial stewardship. *Infection Control and Hospital Epidemiology* 35, 1092–1099.

Merrigan, M., Venugopal, A., Mallozzi, M., Roxas, B., Viswanathan, V.K., Johnson, S., Gerding, D.N., and Vedantam, G. (2010) Human hypervirulent *Clostridium difficile* strains exhibit increased sporulation as well as robust toxin production. *Journal of Bacteriology* 192, 4904–4911.

Mezoff, E.A. and Cohen, M.B. (2013) Acid suppression and the risk of *Clostridium difficile* infection. *Journal of Pediatrics* 163, 627–630.

Miller, M., Gravel, D., Mulvey, M., Taylor, G., Boyd, D., Simor, A., Gardam, M., McGeer, A., Hutchinson, J., Moore, D., and Kelly, S. (2013) Health care-associated *Clostridium difficile* infection in Canada: patient age and infecting strain type are highly predictive of severe outcome and mortality. *Clinical Infectious Diseases* 50, 194–201.

Morrison, R.H., Hall, N.S., Said, M., Rice, T., Groff, H., Brodine, S.K., Slymen, D., and Lederman, E.R. (2011) Risk factors associated with complications and mortality in patients with *Clostridium difficile* infection. *Clinical Infectious Diseases* 53, 1173–1178.

Muto, C.A., Blank, M.K., Marsh, J.W., Vergis, E.N., O'Leary, M.M., Shutt, K.A., Pasculle, A.W., Pokrywka, M., Garcia, J.G., Posey, K. *et al.* (2007) Control of an outbreak of infection with the hypervirulent *Clostridium difficile* BI strain in a university hospital using a comprehensive "bundle" approach. *Clinical Infectious Diseases* 45, 1266–1273.

Naggie, S., Miller, B.A., Zuzak, K.B., Pence, B.W., Mayo, A.J., Nicholson, B.P., Kutty, P.K., McDonald, L.C., and Woods, C.W. (2011) A case-control study of community-associated *Clostridium difficile* infection: no role for proton pump inhibitors. *The American Journal of Medicine* 124, 276.e1–276.e7.

Nylund, C.M., Goudie, A., Garza, J.M., Fairbrother, G., and Cohen, M.B. (2011) *Clostridium difficile* infection in hospitalized children in the United States. *Archives of Pediatrics and Adolescent Medicine* 165, 451–457.

O'Brien, J.A., Lahue, B.J., Caro, J.J., and Davidson, D.M. (2007) The emerging infectious challenge of *Clostridium difficile*-associated disease in Massachusetts hospitals: clinical and economic consequences. *Infection Control and Hospital Epidemiology* 28, 1219–1227.

O'Connor, J.R., Johnson, S., and Gerding, D.N. (2009) *Clostridium difficile* infection caused by the epidemic BI/NAP1/027 strain. *Gastroenterology* 136, 1913–1924.

Ostrowsky, B., Ruiz, R., Brown, S., Chung, P., Koppelman, E., van Deusen Lukas, C., Guo, Y., Jalon, H., Sumer, Z., Araujo, C. *et al.* (2014) Lessons learned from implementing *Clostridium difficile*-focused antibiotic stewardship interventions. *Infection Control and Hospital Epidemiology* 35(Suppl 3), S86–S95.

Otter, J.A., Puchowicz, M., Ryan, D., Salkeld, J.A.G., Cooper, T.A., Havill, N.L., Tuozzo, K., and Boyce, J.M. (2009) Feasibility of routinely using hydrogen peroxide vapor to decontaminate rooms in a busy United States hospital. *Infection Control and Hospital Epidemiology* 30, 574–577.

Owens, R.C., Jr., Donskey, C.J., Gaynes, R.P., Loo, V.G., and Muto, C.A. (2008) Antimicrobial-associated risk factors for *Clostridium difficile* infection. *Clinical Infectious Diseases* 46(Suppl 1), S19–S31.

Pear, S.M., Williamson, T.H., Bettin, K.M., Gerding, D.N., and Galgiani, J.N. (1994) Decrease in nosocomial *Clostridium difficile*-associated diarrhea by restricting clindamycin use. *Annals of Internal Medicine* 120, 272–277.

Rea, M.C., O'Sullivan, O., Shanahan, F., O'Toole, P.W., Stanton, C., Ross, R.P., and Hill, C. (2012) *Clostridium difficile* carriage in elderly subjects and associated changes in the intestinal microbiota. *Journal of Clinical Microbiology* 50, 867–875.

Redelings, M.D., Sorvillo, F., and Mascola, L. (2007) Increase in *Clostridium difficile*-related mortality rates, United States, 1999–2004. *Emerging Infectious Diseases Journal* 13, 1417–1419.

Rousseau, C., Poilane, I., De Pontual, L., Maherault, A.C., Le Monnier, A., and Collignon, A. (2012) *Clostridium difficile* carriage in healthy infants in the community: a potential reservoir for pathogenic strains. *Clinical Infectious Diseases* 55, 1209–1215.

Rubin, T.A., Gessert, C.E., Aas, J., and Bakken, J.S. (2013) Fecal microbiome transplantation for recurrent *Clostridium difficile* infection: report on a case series. *Anaerobe* 19, 22–26.

Russell, G., Kaplan, J., Ferraro, M., and Michelow, I.C. (2010) Fecal bacteriotherapy for relapsing *Clostridium difficile* infection in a child: a proposed treatment protocol. *Pediatrics* 126, e239–e242.

Samore, M.H., DeGirolami, P.C., Tlucko, A., Lichtenberg, D.A., Melvin, Z.A., and Karchmer, A.W. (1994) *Clostridium difficile* colonization and diarrhea at a tertiary care hospital. *Clinical Infectious Diseases* 18, 181–187.

Samore, M.H., Venkataraman, L., DeGirolami, P.C., Arbeit, R.D., and Karchmer, A.W. (1996) Clinical and molecular epidemiology of sporadic and clustered cases of nosocomial *Clostridium difficile* diarrhea. *The American Journal of Medicine* 100, 32–40.

Seekatz, A.M. and Young, V.B. (2014) *Clostridium difficile* and the microbiota. *The Journal of Clinical Investigation* 124, 4182–4189.

Shaughnessy, M.K., Micielli, R.L., DePestel, D.D., Arndt, J., Strachan, C.L., Welch, K.B., and Chenoweth, C.E. (2011) Evaluation of hospital room assignment and acquisition of *Clostridium difficile* infection. *Infection Control and Hospital Epidemiology* 32, 201–206.

Stoesser, N., Crook, D.W., Fung, R., Griffiths, D., Harding, R.M., Kachrimanidou, M., Keshav, S., Peto, T.E., Vaughan, A., Walker, A.S., and Dingle, K.E. (2011) Molecular epidemiology of *Clostridium difficile* strains in children compared with that of strains circulating in adults with *Clostridium difficile*-associated infection. *Journal of Clinical Microbiology* 49, 3994–3996.

Tleyjeh, I.M., Abdulhak, A.B., Riaz, M., Garbati, M.A., Al-Tannir, M., Alasmari, F.A., Alghamdi, M., Khan, A.R., Erwin, P.J., Sutton, A.J., and Baddour, L.M. (2013) The association between histamine 2 receptor antagonist use and *Clostridium difficile* infection: a systematic review and meta-analysis. *PLoS ONE* 8(3): e56498.

Tschudin-Sutter, S., Tamma, P.D., Naegeli, A.N., Speck, K.A., Milstone, A.M., and Perl, T.M. (2013) Distinguishing community-associated from hospital-associated *Clostridium difficile* infections in children: implications for public health surveillance. *Clinical Infectious Diseases* 57, 1665–1672.

Valentini, D., Vittucci, A.C., Grandin, A., Tozzi, A.E., Russo, C., Onori, M., Menichella, D., Bartuli, A., and Villani, A. (2013) Coinfection in acute gastroenteritis predicts a more severe clinical course in children. *European Journal of Clinical Microbiology* 32, 909–915.

Valiquette, L., Cossette, B., Garant, M.P., Diab, H., and Pepin, J. (2007) Impact of a reduction in the use of high-risk antibiotics on the course of an epidemic of *Clostridium difficile*-associated disease caused by the hypervirulent NAP1/027 strain. *Clinical Infectious Diseases* 45(Suppl 2), S112–S121.

van Nood, E., Vrieze, A., Nieuwdorp, M., Fuentes, S., Zoetendal, E.G., de Vos, W.M., Visser, C.E., Kuijper, E.J., Bartelsman, J.F.W.M. (2013) Duodenal infusion of donor feces for recurrent *Clostridium difficile*. *The New England Journal of Medicine* 368, 407–415.

Vernacchio, L., Vezina, R.M., Mitchell, A.A., Lesko, S.M., Plaut, A.G., and Acheson, D.W. (2006) Diarrhea in American infants and young children in the community setting: incidence, clinical presentation and microbiology. *The Pediatric Infectious Disease Journal* 25, 2–7.

Voth, D.E. and Ballard, J.D. (2005) *Clostridium difficile* toxins: mechanism of action and role in disease. *Clinical Microbiology Reviews* 18, 247–263.

Wagner, B., Filice, G.A., Drekonja, D., Greer, N., MacDonald, R., Rutks, I., Butler, M., and Wilt, T.J. (2014) Antimicrobial stewardship programs in inpatient hospital settings: a systematic review. *Infection Control and Hospital Epidemiology* 35, 1209–1228.

Walia, R., Garg, S., Song, Y., Girotra, M., Cuffari, C., Fricke, W.F., and Dutta, S.K. (2014) Efficacy of fecal microbiota transplantation in 2 children with recurrent *Clostridium difficile* infection and its impact on their growth and gut microbiome. *Journal of Pediatric Gastroenterology and Nutrition* 59, 565–570.

Walker, A.S., Eyre, D.W., Wyllie, D.H., Dingle, K.E., Griffiths, D., Shine, B., Oakley, S., O'Connor, L., Finney, J., Vaughan, A. *et al.* (2012) Relationship between bacterial strain type, host biomarkers, and mortality in *Clostridium difficile* infection. *Clinical Infectious Diseases* 56, 1589–1600.

Warny, M., Pepin, J., Fang, A., Killgore, G., Thompson, A., Brazier, J., Frost, E., and McDonald, L.C. (2005) Toxin production by an emerging strain of *Clostridium difficile* associated with outbreaks of severe disease in North America and Europe. *The Lancet* 366, 1079–1084.

Wendt, J.M., Cohen, J.A., Mu, Y., Dumyati, G.K., Dunn, J.R., Holzbauer, S.M., Winston, L.G., Johnston, H.L., Meek, J.I., Farley, M.M. *et al.* (2014) *Clostridium difficile* infection among children across diverse US geographic locations. *Pediatrics* 133, 651–658.

Wilcox, M.H., Fawley, W.N., Wigglesworth, N., Parnell, P., Verity, P., and Freeman, J. (2003) Comparison of the effect of detergent versus hypochlorite cleaning on environmental contamination and incidence of *Clostridium difficile* infection. *Journal of Hospital Infection* 54, 109–114.

Zar, F.A., Bakkanagari, S.R., Moorthi, K.M., and Davis, M.B. (2007) A comparison of vancomycin and metronidazole for the treatment of *Clostridium difficile*-associated diarrhea, stratified by disease severity. *Clinical Infectious Diseases* 45, 302–307.

Zilberberg, M.D., Shorr, A.F., and Kollef, M.H. (2008a) Increase in *Clostridium difficile*-related hospitalizations among infants in the United States, 2000–2005. *The Pediatric Infectious Disease Journal* 27, 1111–1113.

Zilberberg, M.D., Shorr, A.F., and Kollef, M.H. (2008b) Increase in adult *Clostridium difficile*-related hospitalizations and case-fatality rate, United States, 2000–2005. *Emerging Infectious Diseases Journal* 14, 929–931.

16 The Role of the Hospital Epidemiologist in Supporting Antimicrobial Stewardship

Michael S. Calderwood*

Dartmouth-Hitchcock Medical Center, Lebanon, New Hampshire, US

Introduction

Joint guidelines by the Infectious Diseases Society of America (IDSA), the Society for Healthcare Epidemiology of America (SHEA), and the Association for Professionals in Infection Control and Epidemiology (APIC) highlight the importance of engaging experts across multiple disciplines to ensure the success of an antimicrobial stewardship program (Dellit *et al.*, 2007; Moody *et al.*, 2012). The "core members" of an antimicrobial stewardship team include an infectious diseases physician, a clinical pharmacist with infectious diseases training, a clinical microbiologist, a specialist in information systems, and an infection prevention and control team under the direction of a hospital epidemiologist (Dellit *et al.*, 2007). These individuals work together to improve patient outcomes through reductions in multidrug-resistant organisms (MDROs) and *C. difficile* infection (CDI).

The Hospital Epidemiologist and Antimicrobial Stewardship

The emergence and transmission of MDROs and CDI are a key patient safety concern. Healthcare-associated infections (HAIs) caused by these organisms have been linked to increases in hospital length of stay, cost, and mortality (Roberts *et al.*, 2010). Efforts to improve patient outcomes must be focused on reducing the acquisition of and infection by MDROs and CDI. This involves collaboration between antimicrobial stewardship programs and a number of key partners. The clinical microbiology laboratory must be able to rapidly identify MDROs and CDI, information system professionals must develop alerts to monitor and report individual cases and trends suggestive of possible outbreaks, and infection control practitioners (also known as infection preventionists) must institute and improve adherence to rigorous hand hygiene, transmission-based precautions, and environmental cleaning aimed at reducing further transmission (Goldmann *et al.*, 1996; Siegel *et al.*, 2006; Dellit *et al.*, 2007; Moody *et al.*, 2012).

The hospital epidemiologist and the antimicrobial stewardship team collaborate closely with the clinical microbiology lab. Given a recent increase in the use of rapid diagnostic assays (Bauer *et al.*, 2014), it is possible to both intervene earlier to halt the transmission of MDROs and CDI to other patients, and to tailor therapy and limit unnecessary antibiotic exposure. Collaboration with the clinical microbiology lab is also important for developing local antibiograms, which can be used to develop guidelines for a rationale for empiric antibiotic prescribing for common clinical conditions. Reductions in unnecessary antibiotic exposure have been linked to reduced colonization and infection by methicillin-resistant *Staphylococcus aureus* (MRSA), vancomycin-resistant *Enterococcus* (VRE), and multidrug-resistant (MDR) Gram-negative bacteria, along with reductions in CDI (Dancer *et al.*, 2013; Davey *et al.*, 2013; Feazel *et al.*, 2014).

*E-mail: michael.s.calderwood@hitchcock.org

In monitoring MDRO and CDI trends, automated surveillance using electronic data has become increasingly utilized to improve real-time tracking and cluster detection (Wright *et al.*, 2004; Huang *et al.*, 2010), in addition to its use for identifying specific HAIs (Klompas and Yokoe, 2009; Woeltje *et al.*, 2014). Local development, maintenance, and evolution of these automated tools requires close collaboration with information systems (IS or IT) specialists. Due to the increasing need to track and publicly report hospital data on multiple HAIs, infection prevention and control teams have a great deal of experience in working closely with IS specialists to support surveillance activities. This surveillance and reporting experience will be increasingly important as antimicrobial stewardship teams are called upon to publicly report antimicrobial usage and resistance data to the US Centers for Disease Control and Prevention (CDC). One expert panel recommended the public reporting both of days of therapy/1000 patient days and numbers of drug-resistant organisms (Morris *et al.*, 2012). This is in line with the data being requested by the CDC's Antimicrobial Use and Resistance (AUR) module (CDC, 2015), and is in addition to the public reporting of all CDI cases already mandated in US hospitals (CDC, 2016).

Once an increase in MDRO or CDI rates is identified within a given population, the persistence of these HAIs is driven by four factors (Siegel *et al.*, 2006):

1. The availability of vulnerable patients, such as those with a compromised immune system, those who have undergone recent surgery with healing wounds, and those with an indwelling device (i.e., urinary catheter, central venous catheter, endotracheal tube).
2. Overall antimicrobial use in the affected healthcare setting (selective pressure).
3. Number of patients either colonized or infected with a given pathogen in the affected healthcare setting (colonization pressure).
4. Adherence to infection prevention and control policies.

Hospital epidemiologists and infection control practitioners are involved in the implementation of prevention bundles targeting reductions in specific HAIs, such as central line-associated bloodstream infections, catheter-associated urinary tract infections, ventilator-associated infections, and surgical site infections (Yokoe *et al.*, 2014). As interventions by infection prevention and control teams reduce the number of device-related and procedure-related infections, the number of MDROs in a given healthcare setting declines (Moody *et al.*, 2012). This reduction in the prevalence of MDROs then leads to declines in empiric broad-spectrum antimicrobial use and reduction in the incidence of CDI (Siegel *et al.*, 2006). Antibiotic prescribing on a hospital ward impacts CDI risk, even in those patients not receiving antibiotics (Brown *et al.*, 2015). Similar causal links have been discussed between ward-level antimicrobial use and the overall prevalence of antimicrobial resistance (Dellit *et al.*, 2007). This is part of the reason for current policies promoting universal skin decolonization with chlorhexidine bathing in intensive care unit (ICU) settings. Such policies have shown reductions in the acquisition of MDROs (Climo *et al.*, 2013; Huang *et al.*, 2013).

Adherence to rigorous hand hygiene, transmission-based precautions, and environmental cleaning are also necessary to prevent the further transmission of MDROs and CDI within healthcare settings. Improvements in hand hygiene driven by education, feedback, and reminders have been shown to significantly reduce the prevalence of HAIs (Kirkland *et al.*, 2012; Schweizer *et al.*, 2014). In addition, transmission-based precautions are recommended to help prevent the spread of MDROs and CDI between patients within a healthcare facility (Siegel *et al.*, 2006, 2007). Finally, hospital surfaces have been shown to be involved in patient-to-patient transmission and have thus become a focus in efforts to halt the spread of MDROs and CDI. Patients admitted to a room where the prior room occupant had MRSA, VRE, and MDR Gram-negative bacteria, or *C. difficile* are at an increased risk of infection with the same organism (Weber *et al.*, 2013). However, adequate cleaning of the environment and equipment in a room can remove the higher risk of infection in the next room occupant (Datta *et al.*, 2011). Environmental services staff should have the appropriate education, written room-cleaning guidelines, checklists, and periodic observations to ensure good practice (Carling *et al.*, 2006; Weber *et al.*, 2013).

Conclusion

Working together, a healthcare facility's antimicrobial stewardship and infection prevention and control teams can jointly improve patient outcomes through a reduction in MDRO and CDI prevalence and transmission. Antimicrobial stewardship efforts target the selective pressure driving the emergence of MDROs and CDI (McGowan, 2012; Nowak *et al*., 2012), but interventions under the direction of a hospital epidemiologist are necessary to control the secondary spread of these infections once they occur, while simultaneously instituting best-practice bundles aimed at prevention.

References

Bauer, K.A., Perez, K.K., Forrest, G.N., and Goff, D.A. (2014) Review of rapid diagnostic tests used by antimicrobial stewardship programs. *Clinical Infectious Diseases* 59(Suppl 3), S134–S145.

Brown, K., Valenta, K., Fisman, D., Simor, A., and Daneman, N. (2015) Hospital ward prescribing and the risks of *Clostridium difficile* infection. *JAMA Internal Medicine* 175, 626–633.

Carling, P.C., Briggs, J.L., Perkins, J., and Highlander, D. (2006) Improved cleaning of patient rooms using a new targeting method. *Clinical Infectious Diseases* 42, 385–388.

CDC (2015) *Updated Operational Guidance for Acute Care Hospitals for 2015. Operational Guidance for Acute Care Hospitals to Report Facility-Wide Inpatient (FacWideIN)* Clostridium difficile *Infection (CDI) Laboratory-Identified (LabID) Event Data to CDC's NHSN for the Purpose of Fulfilling CMS's Hospital Inpatient Quality Reporting (IQR) Requirements*. Centers for Disease Control and Prevention, Atlanta, Georgia. Available at: http://www.cdc.gov/nhsn/PDFs/mrsa-cdi/FINAL-ACH-CDI-Guidance.pdf (accessed 27 April 2015).

CDC (2016) Surveillance for Antimicrobial Use and Antimicrobial Resistance Options. Resources for NHSN Users Already Enrolled. National Healthcare Safety Network, Centers for Disease Control and Prevention, Atlanta, Georgia. Available at: http://www.cdc.gov/nhsn/acute-care-hospital/aur/ (accessed 16 May 2016).

Climo, M.W., Yokoe, D.S., Warren, D.K., Perl, T.M., Bolon, M., Herwaldt, L.A., Weinstein, R.A., Sepkowitz, K.A., Jernigan, J.A., Sanogo, K., and Wong, E.S. (2013) Effect of daily chlorhexidine bathing on hospital-acquired infection. *The New England Journal of Medicine* 368, 533–542.

Dancer, S.J., Kirkpatrick, P., Corcoran, D.S., Christison, F., Farmer, D., and Robertson, C. (2013) Approaching zero: temporal effects of a restrictive antibiotic policy on hospital-acquired *Clostridium difficile*, extended-spectrum β-lactamase-producing coliforms and methicillin-resistant *Staphylococcus aureus*. *International Journal of Antimicrobial Agents* 41, 137–142.

Datta, R., Platt, R., Yokoe, D.S., and Huang, S.S. (2011) Environmental cleaning intervention and risk of acquiring multidrug-resistant organisms from prior room occupants. *Archives of Internal Medicine* 171, 491–494.

Davey, P., Brown, E., Charani, E., Fenelon, L., Gould, I.M., Holmes, A., Ramsay, C.R., Wiffen, P.J., and Wilcox, M. (2013) Interventions to improve antibiotic prescribing practices for hospital inpatients. *Cochrane Database of Systematic Reviews* 4: CD003543.

Dellit, T.H., Owens, R.C., McGowan, J.E., Jr., Gerding, D.N., Weinstein, R.A., Burke, J.P., Huskins, W.C., Paterson, D.L., Fishman, N.O., Carpenter, C.F. *et al*. (2007) Infectious Diseases Society of America and the Society for Healthcare Epidemiology of America Guidelines for developing an institutional program to enhance antimicrobial stewardship. *Clinical Infectious Diseases* 44, 159–177.

Feazel, L.M., Malhotra, A., Perencevich, E.N., Kaboli, P., Diekema, D.J., and Schweizer, M.L. (2014) Effect of antibiotic stewardship programmes on *Clostridium difficile* incidence: a systematic review and meta-analysis. *Journal of Antimicrobial Chemotherapy* 69, 1748–1754.

Goldmann, D.A., Weinstein, R.A., Wenzel, R.P., Tablan, O.C., Duma, R.J., Gaynes, R.P., Schlosser, J., and Martone, W.J. (1996) Strategies to prevent and control the emergence and spread of antimicrobial-resistant microorganisms in hospitals. A challenge to hospital leadership. *JAMA (The Journal of the American Medical Association)* 275, 234–240.

Huang, S.S., Yokoe, D.S., Stelling, J., Placzek, H., Kulldorff, M., Kleinman, K., O'Brien, T.F., Calderwood, M.S., Vostok, J., Dunn, J., and Platt, R. (2010) Automated detection of infectious disease outbreaks in hospitals: a retrospective cohort study. *PLoS Medicine* 7(2): e1000238.

Huang, S.S., Septimus, E., Kleinman, K., Moody, J., Hickok, J., Avery, T.R., Lankiewicz, J., Gombosev, A., Terpstra, L., Hartford, F. *et al*. (2013) Targeted versus universal decolonization to prevent ICU infection. *The New England Journal of Medicine* 368, 2255–2265.

Kirkland, K.B., Homa, K.A., Lasky, R.A., Ptak, J.A., Taylor, E.A., and Splaine, M.E. (2012) Impact of a hospital-wide hand hygiene initiative on healthcare-associated infections: results of an interrupted time series. *BMJ Quality and Safety* 21, 1019–1026.

Klompas, M. and Yokoe, D.S. (2009) Automated surveillance of healthcare-associated infections. *Clinical Infectious Diseases* 48, 1268–1275.

McGowan, J.E., Jr. (2012) Antimicrobial stewardship—the state of the art in 2011: focus on outcome and methods. *Infection Control and Hospital Epidemiology* 33, 331–337.

Moody, J., Cosgrove, S.E., Olmsted, R., Septimus, E., Aureden, K., Oriola, S., Patel, G.W., and Trivedi, K.K. (2012) Antimicrobial stewardship: a collaborative partnership between infection preventionists and health care epidemiologists. *American Journal of Infection Control* 40, 94–95.

Morris, A.M., Brener, S., Dresser, L., Daneman, N., Dellit, T.H., Avdic, E., and Bell, C.M. (2012) Use of a structured panel process to define quality metrics for antimicrobial stewardship programs. *Infection Control and Hospital Epidemiology* 33, 500–506.

Nowak, M.A., Nelson, R.E., Breidenbach, J.L., Thompson, P.A., and Carson, P.J. (2012) Clinical and economic outcomes of a prospective antimicrobial stewardship program. *American Journal of Health-System Pharmacy* 69, 1500–1508.

Roberts, R.R., Scott, R.D., II, Hota, B., Kampe, L.M., Abbasi, F., Schabowski, S., Ahmad, I., Ciavarella, G.G., Cordell, R., Solomon, S.L. *et al.* (2010) Costs attributable to healthcare-associated infection in hospitalized adults and a comparison of economic methods. *Medical Care* 48, 1026–1035.

Schweizer, M.L., Reisinger, H.S., Ohl, M., Formanek, M.B., Blevins, A., Ward, M.A., and Perencevich, E.N. (2014) Searching for an optimal hand hygiene bundle: a meta-analysis. *Clinical Infectious Diseases* 58, 248–259.

Siegel, J.D., Rhinehart, E., Jackson, M., Chiarello, L., and HICPAC (Healthcare Infection Control Practices Advisory Committee) (2006) *Management of Multidrug-Resistant Organisms in Healthcare Settings, 2006*. Centers for Disease Control and Prevention, Atlanta, Georgia. Available at: http://www.cdc.gov/hicpac/pdf/MDRO/MDROGuideline2006.pdf (accessed 27 April 2015).

Siegel, J.D., Rhinehart, E., Jackson, M., Chiarello, L., and HICPAC (Healthcare Infection Control Practices Advisory Committee) (2007) *2007 Guideline for Isolation Precautions: Preventing Transmission of Infectious Agents in Healthcare Settings*. Available at: http://www.cdc.gov/hicpac/pdf/isolation/isolation2007.pdf (accessed 27 April 2015).

Weber, D.J., Anderson, D., and Rutala, W.A. (2013) The role of the environment in healthcare-associated infections. *Current Opinion in Infectious Diseases* 26, 338–344.

Woeltje, K.F., Lin, M.Y., Klompas, M., Wright, M.O., Zuccotti, G., and Trick, W.E. (2014) Data requirements for electronic surveillance of healthcare-associated infections. *Clinical Infectious Diseases* 35, 1083–1091.

Wright, M.O., Perencevich, E.N., Novak, C., Hebden, J.N., Standiford, H.C., and Harris, A.D. (2004) Preliminary assessment of an automated surveillance system for infection control. *Infection Control and Hospital Epidemiology* 25, 325–332.

Yokoe, D.S., Anderson, D.J., Berenholtz, S.M., Calfee, D.P., Dubberke, E.R., Ellingson, K.D., Gerding, D.N., Haas, J.P., Kaye, K.S., Klompas, M. *et al.* (2014) A compendium of strategies to prevent healthcare-associated infections in acute care hospitals: 2014 updates. *Infection Control and Hospital Epidemiology* 35, 967–977.

17 Principles of Pharmacokinetic/ Pharmacodynamic Optimization for Antibiotic Dosing

Islam M. Ghazi, Joseph L. Kuti, and David P. Nicolau*

Center for Anti-Infective Research and Development, Hartford Hospital, Hartford, Connecticut, US

Introduction

The use of antimicrobial therapy presents infectious diseases clinicians with an increasing number of challenges. In particular, the rapid development of multidrug-resistant organisms (MDROs) is overwhelming the spectra of available antibiotics, and new agents are needed. Despite this paucity of antibiotics, regulatory, financial and technical issues have hindered the development of novel antimicrobial agents. Dose optimization through the application of pharmacokinetic/pharmacodynamic (PK/PD) principles is a pivotal strategy for maintaining the efficacy and utility of the current antibiotic armamentarium. In this chapter, we review PK/PD concepts and provide examples of how these concepts can be used to modify and optimize antimicrobial drug regimens in the clinical setting.

Pharmacokinetics and Pharmacodynamics

Pharmacokinetics (PK) describes the influences exerted on a drug by the body from its administration to its removal: namely absorption, distribution, metabolism, and excretion. As a result of the path of an antibiotic through the body, the concentrations of that antibiotic take on a specific profile in the blood and tissues after administration. This profile is a result of the pharmacokinetic parameters that allow the characterization of a drug's PK profile. Common pharmacokinetic parameters utilized to describe the PK profile of an antibiotic include the maximum concentration achieved after a dose (Cmax), the minimum concentration achieved after a dose (Cmin), the area under the concentration vs. time curve (AUC), and the antibiotic clearance (CL) or half-life ($t_{1/2}$), which are used to estimate the exposure profile of the compound. Also of importance is the protein binding capacity of an antibiotic, because only free drug concentrations are responsible for microbiological activity. As such, the above pharmacokinetic parameters are often corrected for the fraction of the drug that is unbound and noted by '*f*' in front (e.g., *f*AUC, *f*Cmax) (Levison, 2004). Notably, several physiologic and disease-specific variables can affect antibiotic PK parameters and thus increase or decrease drug exposure. Aspects of PK in special populations will be discussed briefly later in this chapter.

Pharmacodynamics (PD) describes the effect of the drug on the bacteria. Of note, infectious diseases is the only field where the drug target is not specifically the human host but another living organism residing within the host and causing the infection. As a result, the pharmacokinetic parameters described above are linked to a surrogate measure of activity, which is typically described as the minimum inhibitory concentration (MIC) of the antibiotic. The MIC is the lowest concentration of antibiotic that inhibits visible growth in vitro. The lower the MIC, the greater potency the antibiotic has against the organism; vice versa, more resistant organisms have higher antibiotic MICs,

*Corresponding author. E-mail: david.nicolau@hhchealth.org

thereby requiring higher concentrations of the antibiotic concerned to inhibit growth. Therefore, common pharmacodynamic indices include the *f*Cmax/MIC ratio, the *f*AUC/MIC ratio, and the time (T) that free drug concentrations are maintained above the MIC (*f*T > MIC), the latter of which is often depicted as a percentage of the dosing interval (% *f*T > MIC) (see Fig. 17.1) (Ambrose *et al.*, 2007).

Antibiotics can manifest different killing or inhibitory patterns on bacteria. When higher concentrations are able to achieve greater or more rapid killing, then the antibiotic is said to exert concentration-dependent killing. In contrast, when increasing the concentration no longer achieves greater bacterial killing, the antibiotic is said to exert concentration-independent killing. This latter profile is more commonly called time-dependent killing because there is a stronger correlation between the *f*T > MIC and bacterial killing compared with the *f*Cmax/MIC ratio. Additionally, there are antibiotics that exert both concentration- and time-dependent killing; these compounds often have *f*AUC/MIC as the strongest correlate with bacterial killing. Common antibiotic classes that display concentration-dependent killing at standard clinical dosing regimens include the aminoglycoside, fluoroquinolone, and polymyxin antibiotics (Table 17.1). In contrast, β-lactams, macrolides, and tetracyclines often exhibit time-dependent killing profiles. Finally, other antibiotics (those that characteristically often have a prolonged $T_{1/2}$) fall in the middle of these two classes and are best classified by *f*AUC/MIC exposure; examples include vancomycin, daptomycin, linezolid, tigecycline, and many of the newer approved antibiotics (lipoglycopeptides) (Drusano, 2007).

In the next few sections, we will review specific classes of popular antibiotics, provide data to support PK/PD profiles, and provide clinical examples of how the modification of dosing regimens can lead to optimal PK/PD exposure.

Aminoglycosides

Since the introduction of streptomycin in the 1940s, aminoglycosides have been valuable agents in the treatment of many infections, notably those caused by Gram-negative bacteria. Traditional dosing of gentamicin and tobramycin in the 1970s and 1980s was in the form of low mg/kg doses (e.g., 60–100 mg) administered every 8–12 h. Early clinical studies suggested that high peak serum levels of aminoglycosides were correlated with successful clinical response (Moore *et al.*, 1987). In patients with Gram-negative pneumonia, a gentamicin or tobramycin peak of 7 μg/ml or higher, and an amikacin peak of 28 μg/ml or higher were associated

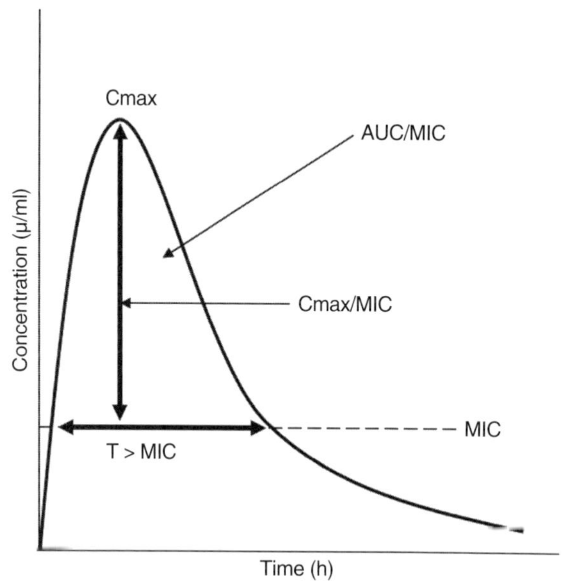

Fig. 17.1. Pharmacokinetic/pharmacodynamic indices that describe antimicrobial effect of an antibiotic. Key: Cmax = maximum concentration; AUC/MIC = area under curve/minimum inhibitory concentration; T > MIC = time above the MIC.

Table 17.1. Summary of pharmacodynamics (PD) classification, exposure threshold, and strategy to optimize exposure for different antibiotic classes.

Antibiotic class	PD profile	PD indices	Clinical optimization strategy
Aminoglycosides (gentamicin, tobramycin, amikacin)	Concentration dependent	*f*Cmax/MIC ≥ 10–12; total AUC/MIC ≥ 156	High dose once daily, can be individualized using therapeutic drug monitoring and MIC
β-Lactams:			Prolonged infusion (extended or continuous), greater doses can be used for higher MIC
Penicillins	Time dependent	*f*T/MIC ≥ 50	
Cephalosporins		*f*T/MIC ≥ 60–70	
Carbapenems		*f*T/MIC ≥ 30–40	
Fluoroquinolones	Concentration dependent	*f*Cmax ≥ 10–12; total AUC/MIC > 125 for Gram-negative bacteria; *f*AUC/MIC > 30–50 for Gram-positive bacteria	Use the most potent dose to maximize AUC/MIC ratio while tolerable
Glycolipopeptides:			
Vancomycin	Time dependent	Total AUC/MIC > 400	Maximize daily dose:MIC; target trough concentration of 15–20 μg/ml
Daptomycin	Concentration dependent	*f*AUC/MIC; *f*Cmax/MIC[a]	Maximize dose relative to MIC
Telavancin	Concentration dependent for new lipopeptides	AUC/MIC	Dose optimized during development
Dalbavancin		AUC/MIC	
Oritavancin		AUC/MIC, Cmax/MIC	
Polymyxins	Concentration dependent	*f*AUC/MIC > 12–15; total AUC/MIC > 60	Maximize overall daily dose:MIC as tolerable; consider dosing algorithm
Oxazolidinones:			
Linezolid	Time dependent	Total AUC/MIC > 110	Maximize dose in relation to MIC; approved dose is optimized for bacteria with MIC 2 μg/ml
Tedizolid		*f*AUC/MIC[a]	
Glycylcyclines			
Tigecycline	Time dependent	*f*AUC/MIC	Approved dosage is optimized for most susceptible bacteria; can increase daily dose to 200 mg for serious infections
Macrolides/azalides	Time dependent	AUC/MIC[a]	Not applicable

[a]Clinically relevant AUC/MIC targets for these antibiotics have not been well established.

with better clinical outcomes (Moore *et al.*, 1984). In an attempt to link exposure to outcomes, it was demonstrated that high Cmax/MIC and AUC/MIC ratios predicted bacterial killing better than T > MIC (Noone and Pattison, 1974; Craig *et al.*, 1991).

The attainment of Cmax/MIC ratios in the range of 10–12 was established by various studies to maximize clinical benefits. In patients with pneumonia, it was observed that the first Cmax/MIC predicted length of time to becoming afebrile and the second Cmax/MIC predicted time to leukocyte count resolution; when the Cmax/MIC ≥10 in the first 48 h of treatment, there was a 90% probability of both temperature and leukocyte resolution by day 7 (Kashuba *et al.*, 1999). In order to achieve this exposure, an aminoglycoside dose typically has to be increased above the standard 1–2 mg/kg to obtain a higher Cmax concentration, as well as have the dosing interval extended to allow drug concentrations to fall below a threshold that would provide safe administration.

The administration of aminoglycosides by this method has become known clinically as "high-dose, extended-interval" or "once daily" aminoglycoside dosing. The Hartford Nomogram was developed at our institution in an effort to quickly and accurately select the optimal dosing interval to administer high-dose gentamicin or tobramycin (Nicolau *et al.*, 1995). A fixed dose of 7 mg/kg was employed based on computer simulation to target a Cmax of approximately 20 µg/ml; this concentration targeted a Cmax/MIC ratio of at least ten against *Pseudomonas aeruginosa*, which had a gentamicin MIC_{90} of 2 µg/ml at the time. After initially treating approximately 2200 patients, the data demonstrated no decline in clinical response, while a reduction in nephrotoxicity was observed. Notably, when the gentamicin MIC_{90} increased above 2 µg/ml in later years, a change to tobramycin 7 mg/kg, which still had a MIC_{90} of 2 µg/ml, was made in order to maintain achievement of a Cmax/MIC of at least ten. Many other once-daily or extended-interval aminoglycoside dosing programs, including 5 mg/kg once daily up to 10–14 mg/kg daily, as well as individualized dosing with aggressive therapeutic drug monitoring to achieve the first Cmax/MIC of at least ten have been described in the literature. No single high-dose, extended-interval dosing strategy has proven superior to another; therefore, specific protocols should be based on the target pathogens at the institution concerned, their aminoglycoside MICs, individual patient pharmacokinetics, and the likelihood of prescriber compliance.

As briefly highlighted above, high-dose extended-interval aminoglycoside dosing also appears to diminish the rate of nephrotoxicity and ototoxicity of these antibiotics, which were largely responsible for their declined use in the clinical setting. Mechanistically, the uptake of aminoglycosides into the proximal convoluted tubular cells of the kidney follows a low-affinity, high-capacity mechanism that is linear and saturable (Guiliano *et al.*, 1986). The primary factor limiting toxicity is saturation, which does not occur when low doses are repeated often in a 24 h period (i.e., 1–2 mg/kg every 8–12 h, or q8–12h). In contrast, high-dose, extended-interval administration leads to less accumulation in the kidneys due to a saturation of this influx mechanism. In a prospective, double-blind study of high-dose, extended interval versus standard dosing of aminoglycosides, AUC_{0-24} was the primary predictor of the probability of nephrotoxicity, but that standard dosing had a much steeper probability curve, thereby suggesting earlier and greater nephrotoxicity (Rybak *et al.*, 1999).

In line with animal and human studies demonstrating PK/PD optimization of aminoglycosides with high-dose, extended intervals, Monte Carlo simulation has provided further support. In a recent study, simulations for regimens of 5mg/kg dose administered once daily (i.e., 400 mg q24h assuming an 80 kg patient) vs. 2.5 mg/kg twice daily (i.e., 200 mg q12h for the 80 kg patient) were conducted and the data explored (Drusano *et al.*, 2007). The probability of efficacy as well as toxicity with each dosing regimen at MICs of 0.25, 0.5 and 1 µg/ml was tested. The end points queried by the investigators were a scenario in which the dosing regimen had a less than 10% chance of toxicity and a greater than 90% chance of clinical success. With respect to nephrotoxicity, 61% of the simulated patients receiving the 200 mg every 12 h dosing regimen had less than a 10% chance of toxicity. In contrast, the likelihood of low toxicity with the once daily regimen was nearly 100%. As the authors used an AUC/MIC ratio as the predictor of efficacy, the AUC_{0-24} values for these simulated dosing regimens were identical and, therefore, both regimens achieved the same likelihoods of efficacy. These likelihoods decreased as the MIC increased from 0.25 to 1 µg/ml. The authors concluded that the window of opportunity between obtaining efficacy and avoiding toxicity is wider with the once daily regimen.

In a later paper by the same authors (Drusano and Louie, 2011), an optimization function metric "Δ" was defined as the difference between probability of good clinical response and probability of toxicity with the purpose of identifying an optimal dosing regimen that has the largest difference between those two probabilities (i.e., high efficacy with low toxicity). Simulated dosing regimens included 5, 7 and 10 mg/kg once and twice daily against hypothetical pathogens with MICs of 0.25, 0.5, 1.0, 2.0, and 4.0 µg/ml. Among twice daily regimens, a dose of 5 mg/kg q12h produced probability of target attainments (PTAs) of 100, 95, and 80% at MICs of 0.25, 0.5 and 1.0 µg/ml, respectively; however, the PTAs declined substantially at MICs of 2.0 and 4.0 µg/ml, to 65 and 50%, respectively. Increasing doses to 7 and 10 mg/kg q12h resulted in modest increases in probability (7 mg/kg yielded 70 and 60% with MICs of 2 and 4 mg/l; moreover, 10 mg/kg yielded 80 and 65%), but that came at the expense of increased toxicity (50 and 80% toxicity, respectively) The investigators concluded that a 5 mg/kg q12h dose would not provide optimal likelihood for efficacy if gentamicin or tobramycin MICs were >1 µg/ml; furthermore, 7 and 10 mg/kg q12h dosing regimens produced unacceptable toxicity rates. In contrast, simulating a 10 mg/kg dose administered once daily against an MIC of 4 µg/ml resulted in an 80% probability of efficacy and a near zero probability of toxicity.

A myriad of clinical studies have evaluated high-dose, extended interval aminoglycoside dosing in various patient populations, including pneumonia, urinary tract infections (UTIs), cystic fibrosis pulmonary exacerbations, post-partum endometritis, pelvic inflammatory disease, and febrile neutropenia. Collectively, these studies have demonstrated that compared with standard dosing, high-dose, extended-interval aminoglycoside dosing achieves similar if not better efficacy, particularly at higher MICs, while at the same time minimizing nephrotoxicity.

β-Lactams

The most abundantly used class of antibiotics clinically is the β-lactams, owing to their relative safety, wide spectrum of activity, and numerous clinical indications. Since the discovery of penicillin early in the 20th century, these agents have been the cornerstone in the treatment of various Gram-positive and Gram-negative infections. The inaugural work by Harry Eagle in the late 1940s paved the way to understanding the pharmacodynamic concepts that relate to the efficacy of β-lactams. Using in vitro and in vivo animal models, Eagle observed the time-dependent killing pattern of β-lactams, as lower, more frequent doses of β-lactams produced better response rates than larger doses administered fewer times per day (Eagle *et al.*, 1950). As a result, the *f*T > MIC is the PK/PD parameter for this class of antibiotics that predicts efficacy.

The specific *f*T/MIC threshold varies in the different β-lactam classes. In animal infection models, penicillins, cephalosporins, and carbapenems require 30, 30–40, and 20% *f*T > MIC, respectively, to achieve bacteriostatic activity. Additionally, 50, 60–70, and 40% *f*T > MIC thresholds are linked to bactericidal activity, respectively (Drusano, 2004). Finally, Gram-positive bacteria are typically inhibited at lower *f*T > MIC exposures than are Gram-negative bacteria. PK/PD studies in humans are challenging, but those done to date have demonstrated similar PK/PD targets to those of animal studies. For cephalosporins, studies in pneumonia have observed an *f*T > MIC of approximately 50–60% to be predictive of clinical or microbiological outcomes (MacVane *et al.*, 2014b; Muller *et al.*, 2014). A study in patients receiving the carbapenem meropenem for lower respiratory tract infections observed microbiological success to be correlated with an *f*T > MIC of at least 54% (Li *et al.*, 2007). In patients with neutropenic fever, a meropenem *f*T > MIC of 80% was required for clinical response (Ariano *et al.*, 2005). Notably, these studies use clinical or microbiological end points (i.e., qualitative metrics), while in vivo animal infection models use the number of bacteria killed or changes in colony forming units (CFUs) (i.e., quantitative metrics). Given these differences, it is quite remarkable how close the thresholds are between the study designs. Importantly, it is therefore unlikely that 100% *f*T > MIC is needed to optimize killing with β-lactams.

There are several methods to increase the *f*T > MIC for β-lactams and thus optimize their dosing. This can be accomplished by increasing the dose, decreasing the dosing interval or prolonging the infusion time (see Fig. 17.2). Intuitively, increasing the dose would increase the time that the drug concentration is above the MIC during the dosing interval; however, this is inefficient as doubling the dose can only prolong the *f*T/MIC by approximately one half-life (i.e., usually only 30–60 min for most β-lactams) (Lodise *et al.*, 2006). Increasing the number of doses given over a 24 h period (e.g., three instead of two doses) significantly increases *f*T > MIC, but at the expense of dosing convenience and, potentially, cost. Extending the infusion time from traditionally 0.5 h to 3–4 h or continuously over the entire day is an efficient way to increase *f*T > MIC, thereby creating a more desirable pharmacodynamic profile and higher probability of attaining targets without the need to change the dose or dosing frequency.

The use of Monte Carlo simulation is a valuable tool for characterizing optimal antibiotic regimens with a high probability of target attainment, and has been widely applied to demonstrate the improvements in *f*T > MIC exposure with the use of prolonged or continuous infusion β-lactam dosing regimens. Briefly, Monte Carlo simulation takes into consideration patient variability in pharmacokinetics and then allows the simulation of thousands of concentration vs. time profiles for a specific dosing regimen. These profiles can then be further explored to determine the likelihood that a drug regimen will achieve a specific PK/PD threshold (e.g., an *f*T > MIC of 50%). This likelihood is referred to as a PTA when queried at a specific MIC and a cumulative fraction of response (CFR)

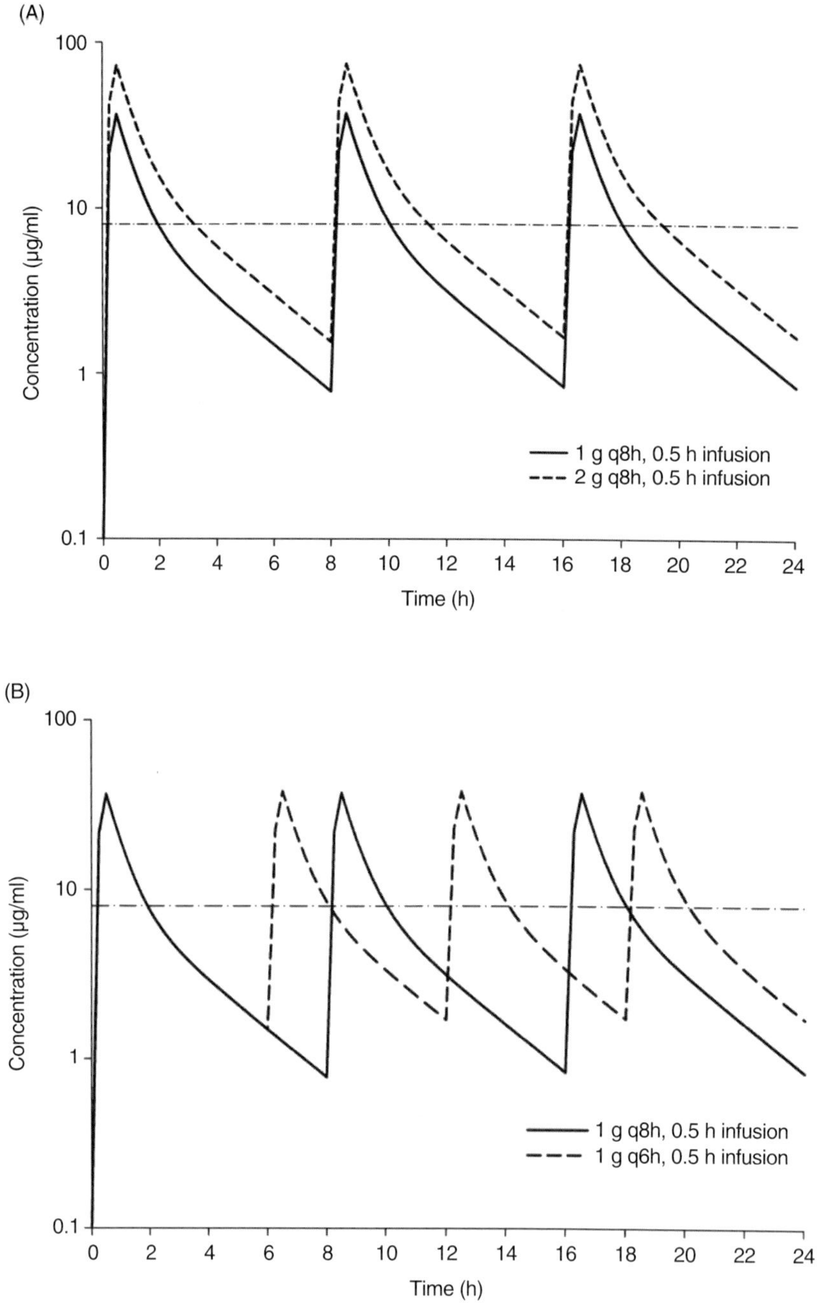

Fig. 17.2. Concentration vs. time profile for a typical intravenous β-lactam demonstrating the effect of: (A) doubling the dose from 1 g q8h (22% *f*T > MIC) to 2 g q8h (38% *f*T > MIC) as 0.5 h infusions; (B) decreasing the dosing interval from 1 g q8h (22% *f*T > MIC) to 1 g q6h (29% *f*T > MIC) as 0.5 h infusions; and (C) increasing the infusion time from 0.5 h (22% *f*T > MIC) to 3 h (38% *f*T > MIC) at a MIC of 8 μg/ml. Key: MIC, minimum inhibitory concentration; *f*T > MIC, time free drug concentrations are maintained above the MIC; q6h, q8h, every 6 h, every 8 h.

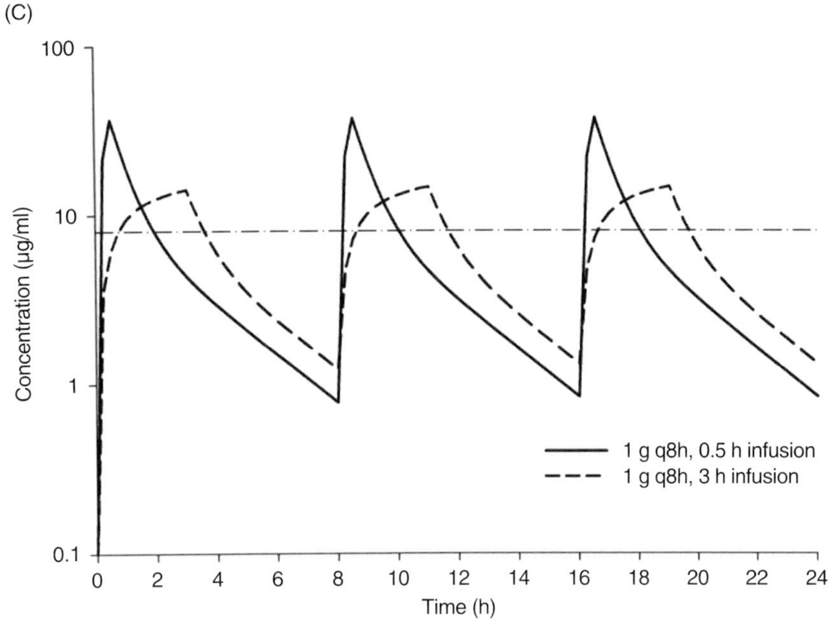

Fig. 17.2. Continued.

when queried against a distribution of organisms with varying MICs. A comprehensive review of Monte Carlo simulation studies assessing prolonged infusion has been published by George *et al.* (2012).

In brief, most simulation studies demonstrate a 5–20% improvement in CFR for cephalosporins and carbapenems when the same dose is simulated as a 3–4 h infusion compared with a standard 0.5 h infusion. Even larger increases in CFR can be observed for piperacillin–tazobactam (a β-lactam (penicillin)/β-lactamase inhibitor combination) when simulated as prolonged or continuous infusion. For example, using MIC data from the TRUST 12 surveillance study (conducted in 2008 across 56 US hospitals), meropenem simulated as 1 g q8h (0.5 h infusion) achieved 88% CFR against *P. aeruginosa* isolates (Koomanachai *et al.*, 2010). Simulating the same dose as a 3 h infusion improved the CFR to 94%. These probabilities were much lower at 64 and 69%, respectively, for the same dosing regimens against *Acinetobacter baumannii*. Piperacillin/tazobactam, simulated as 4.5 g q6h (0.5 h infusion), obtained a CFR of 77% against *P. aeruginosa*. The same dosage simulated as a 3 h infusion achieved 86% CFR. In general, increasing the dose and administration as a prolonged infusion can result in substantial increases in CFR. For instance, meropenem 2 g q8h as a 3 h infusion increased the CFR to 97 and 75%, respectively, against *P. aeruginosa* and *A. baumannii* in that study. Importantly, the changes in CFR are dependent on the percentage of isolates at each MIC, so if an institution has a substantial frequency of isolates residing just above the susceptibility breakpoint, prolonged and continuous infusion regimens will be very efficient in capturing those organisms.

A plethora of animal studies has been conducted to strengthen the finding that *f*T/MIC is the parameter of interest with respect to β-lactams and suggest a role for prolonged or continuous infusion; a few specific examples are provided here. In a series of studies pursuing the effect of continuous infusion vs. intermittent infusion of the cephalosporin ceftazidime in the rat lung infection model, the investigators sought to determine the total daily dose that kept 50% animals alive for 15 days after 4 days of ceftazidime treatment (50% protective dose or $PD_{50\%}$) (Roosendaal *et al.*, 1985, 1987). The investigators observed that the $PD_{50\%}$ of a continuous infusion regimen was 8–25% of that of the intermittent infusion. Using a humanized exposure (i.e., doses in animals that result in a concentration vs. time profile resembling that achieved in humans), the efficacy of a ceftazidime intermittent infusion was compared with that of a continuous infusion with or without the aminoglycoside tobramycin in a rabbit pneumonia rat model. The *f*T/MIC values for intermittent and continuous infusions of ceftazidime were 62 and 99%, respectively, and the animals receiving the continuous infusion regimen had greater sterilization of their septicemia (Croisier *et al.*, 2008).

Clinical data further support the findings of Monte Carlo simulation and *in vivo* animal infection experiments on prolonged and continuous infusion. In a phase III randomized controlled trial of a continuous vs. intermittent infusion of piperacillin–tazobactam in patients with complicated intra-abdominal infection, similar clinical and microbiological efficacy was observed between the infusion strategies; however, the organisms isolated during the study were largely *Enterobacteriaceae* and had piperacillin–tazobactam MICs well below the susceptibility breakpoint (Lau *et al.*, 2006). A retrospective study using a continuous infusion for various infection types (i.e., pneumonia, and intra-abdominal, skin and soft tissue infections) also identified similar clinical and microbiological outcomes, but these were achieved with a significantly lower piperacillin–tazobactam dose when administered as a continuous infusion (Grant *et al.*, 2002). Lodise *et al.* (2007) implemented a prolonged infusion piperacillin–tazobactam regimen (3.375 g q8h, 4 h infusion) in their hospital and observed a significant reduction in mortality among sicker patients infected with *P. aeruginosa*. These observations were replicated in a multicenter study across 14 hospitals and 359 patients with Gram-negative infections (Yost and Cappelletty, 2011).

For other β-lactams, a continuous infusion regimen of the cephalosporin cefepime was found to enhance the antibacterial effect and reduce the treatment duration in neurosurgical patients (Huang *et al.*, 2014). Bauer *et al.* (2013) utilized their local institution MIC data for *P. aeruginosa* to identify the cefepime regimen that would provide the highest target attainment; they then implemented a 2 g q8h, 4 h infusion regimen and observed a significant reduction in mortality compared with patients treated with the standard 30 min infusion. Meropenem infused over 4 h significantly reduced mortality in a cohort of Japanese pneumonia patients compared with standard infusion (Itabashi, 2007). A double-blind, randomized controlled trial of continuous vs. intermittent infusion β-lactams (piperacillin–tazobactam, meropenem, and the β-lactam (penicillin)/β-lactamase inhibitor combination ticarcillin–clavulanate) in patients with severe sepsis identified a significant improvement in rate of clinical cure with the continuous infusion group (76 vs. 50%, $P = 0.032$), but hospital survival was not different (90 vs. 80%, $P = 0.47$) (Dulhunty *et al.*, 2013).

A recent meta-analysis was conducted to compare prolonged infusion versus intermittent infusion of β-lactams; it included 18 randomized controlled trials, three prospective and eight retrospective studies of 2206 patients (Teo *et al.*, 2014). Generally, patients were admitted to intensive care units (ICUs) for all or part of their hospital stay. The treatments encompassed several cephalosporins, β-lactam/β-lactamase inhibitor combinations, penicillins, and carbapenems against a variety of Gram-positive and Gram-negative organisms. The analysis of the pooled data demonstrated a significant mortality advantage for the prolonged infusion group. Potential benefits in terms of clinical success were also detected, with no increase in toxicity. Another meta-analysis was performed on 14 randomized controlled trials (RCTs), retrospective, and prospective designed studies (Falagas *et al.*, 2013). The studies reported the use of meropenem, imipenem–cilastatin (a carbapenem/β-lactamase inhibitor combination) and piperacillin–tazobactam against both Gram-positive and Gram-negative organisms causing UTIs, sepsis, intra-abdominal infections, and pneumonia. Among the sicker patients, a mortality benefit was observed for those receiving their antibiotic as a prolonged infusion.

The surmounting evidence collected from different types of models, mathematical simulations, and clinical trials clearly points to the time-dependent killing pattern of β-lactams and highlights the importance of their administration mode on dose optimization. In summary, for sicker patients who are likely to be infected with higher MIC organisms, a combination of administering higher doses as prolonged or continuous infusions is an effective strategy for improving the PK/PD of these antibiotics.

Vancomycin

Vancomycin is one of the oldest antibiotics still in use today and remains the most frequently utilized agent for treatment of methicillin-resistant *Staphylococcus aureus* (MRSA) infections. In vitro model experiments found no increase in killing by vancomycin of *S. aureus* and *S. epidermidis* (both methicillin resistant and methicillin susceptible) with increasing concentrations 2–64 times above the MIC, thereby suggesting a time-dependent killing pattern (Löwdin *et al.*, 1998). Even with the long history of and extensive experience with vancomycin use, optimal dosing is yet to be defined. A consensus statement published in 2009 stated that AUC/MIC is the PK/PD parameter that drives the efficacy of vancomycin; furthermore, numerous studies have suggested that a target total drug AUC/MIC of approximately 400 is needed for successful vancomycin outcomes (Rybak *et al.*, 2009; Brown *et al.*, 2012; Holmes *et al.*, 2013). Notably, it is challenging to achieve these AUC/MIC ratios with standard vancomycin dosages in patients with normal renal function when the organism MIC is greater than 1 μg/ml; the susceptibility breakpoint for vancomycin is 2 μg/ml, meaning that some susceptible organisms will not be 'treatable' with standard vancomycin dosing regimens (Kuti *et al.*, 2008). This has led to clinical programs that monitor vancomycin MIC for serious MRSA infections, and substitute therapy based on MICs > 1 μg/ml (Murray *et al.*, 2013).

In order to optimize the PK/PD of vancomycin, the MIC is required along with individual vancomycin concentrations in the patient; the latter are easily accessible owing to the availability of a clinical assay in most hospitals worldwide. Despite AUC/MIC being the best predictor of clinical success (or mortality in some studies), therapeutic drug monitoring targeting troughs (i.e., the lowest drug concentration before the next dose) of 15–20 μg/ml are still advocated and widely practiced in hospitals across the world. It should be stressed that trough concentrations are not predictive of vancomycin AUC, and that multiple (at least two) vancomycin concentrations during the dosing interval are needed to accurately estimate AUC via the trapezoidal rule (Neely *et al.*, 2014). Alternatively, a single vancomycin sample can be incorporated into Bayesian pharmacokinetic software and the AUC estimated by the program (Pai *et al.*, 2014).

Due to its time-dependent killing profile, vancomycin has also been studied when administered as a continuous infusion (Wysocki *et al.*, 1995). Typically, continuous infusion requires the administration of a loading dose to reach the target concentration rapidly (20–25 mg/l in most studies), and then the daily vancomycin dose can be adjusted to maintain this concentration when infused continuously. A large meta-analysis has also suggested that continuous infusion vancomycin results in less nephrotoxicity than intermittent infusion vancomycin (Cataldo *et al.*, 2012). At the time of writing, the optimal method to administer vancomycin has yet to be determined. Intermittent infusions aiming for trough concentrations of 15–20 mg/l produce numerically higher 24 h AUC, which would benefit efficacy from an AUC/MIC perspective, but they also risk increased nephrotoxicity (DiMondi and Rafferty, 2013).

Influence of PK/PD on New Antimicrobial Development

The appreciable strides in the identification of relationships between exposure to antibiotics and bactericidal effect have led to the earlier incorporation of the principles of pharmacodynamics into drug development with the intent that approved dosing regimens will already have been optimized for use in the infections for which they are being studied.

Tigecycline is a broad-spectrum glycylcycline antibiotic with activity against resistant organisms including MRSA and multidrug-resistant Gram-negative bacteria. Tigecycline was approved for complicated skin and skin structure infections, and intra-abdominal infections at a dosage of 50 mg q12h after a 100 mg loading dose. However, this dosing regimen performed poorly in studies for the treatment of nosocomial pneumonia (Freire *et al.*, 2010). On analyzing tigecycline PK, the investigators observed lower AUCs in the subset of patients with ventilator-associated pneumonia (VAP), which was largely responsible for the decreased efficacy. As a result of these observations, investigators increased the tigecycline dose to 100 mg q12h and conducted a second study in patients with nosocomial pneumonia (Ramirez *et al.*, 2013). In this subsequent study, the higher tigecycline dosing regimen was non-inferior to imipenem–cilastatin, thereby demonstrating that PK/PD principles can be utilized to develop the optimal dosing regimen, but that different dosing regimens may be needed for different types of infections.

Another example of PK/PD principles being utilized in drug development includes the new lipopeptides and lipoglycopeptides, which are potent additions to vancomycin to treat MRSA infection

(Candiani *et al.*, 1999; Biedenbach *et al.*, 2007). There are currently three antibiotics that belong to this new class: telavancin, dalbavancin, and oritavancin. Telavancin was approved in 2009 and is administered at a dosage of 10 mg/kg daily (Laohavaleeson *et al.*, 2007). During preclinical development, concentration-dependent activity was observed in murine thigh infection experiments and AUC/MIC was demonstrated as the most suitable predictor of efficacy (Hegde *et al.*, 2004). The study identified that an AUC/MIC target of at least 219 was needed to achieve a 1 log reduction in colony counts against MRSA. This AUC/MIC target was applied in Monte Carlo simulations to support the dose of 10 mg/kg daily in patients with normal renal functions, and simulated patients achieved a 100% likelihood of obtaining this target at telavancin MICs up to 1 µg/ml and a 93.8% probability at 2 µg/ml (Lodise *et al.*, 2012). Probabilities were similarly high for dosage adjustments in patients with renal dysfunction (7.5 mg/kg daily for a creatinine clearance (CrCL) of 30–50 ml/min, and 10 mg/kg every 48 h for CrCL < 30 ml/min).

Dalbavancin was approved in 2014 for the treatment of acute bacterial skin and skin structure infections. Owing to a prolonged half-life (~346 h) and high protein binding (~93%), murine infection models have identified that the antibacterial activity best correlates with the AUC/MIC and that higher doses of dalbavancin over longer intervals produce a higher rate of bacterial eradication, indicating a concentration-dependent killing component. AUC/MIC ratios of 214–331 were required for bactericidal activity, whereas ratios of 31–100 culminated in bacteriostatic activity. As a result, dalbavancin was approved as a two-dose regimen, 1000 mg on day 1, followed by a single dose of 500 mg 1 week later. The optimization of this atypical dosing regimen is supported by two landmark clinical trials (Boucher *et al.*, 2014).

Oritavancin was approved in 2014 for acute bacterial skin and skin structure infections at as a single 1200 mg dose. Like dalbavancin, it is highly protein bound (~85%) and has a prolonged half-life (~245 h). In vitro studies depicted concentration-dependent killing against various strains of staphylococci and enterococci with differing resistance levels to vancomycin (McKay *et al.*, 2009). An animal model for oritavancin observed concentration-dependent killing as higher, infrequent doses resulted in more pronounced bactericidal activity (Patti *et al.*, 2009). The Phase II clinical study also observed clinical success to correlate with higher exposure (i.e., higher AUC/MIC and Cmax/MIC) (Bhavnani *et al.*, 2006). Of note, the official drug labeling designates AUC/MIC as the parameter most correlated with efficacy.

Ceftazidime–avibactam is an example of a new combination antibiotic with improved activity against Gram-negative bacteria producing broad-spectrum hydrolyzing enzymes. Ceftazidime is β-lactam, thereby displaying time-dependent bactericidal activity in which *f*T/MIC correlates with efficacy. It is combined with a β-lactamase inhibitor, avibactam, to effectively lower the MIC of the combination product compared with ceftazidime alone (Nichols *et al.*, 2012). Like other β-lactams, the ceftazidime–avibactam combination can be optimized by a prolonged infusion dosing strategy. In a study of humanized dose in a neutropenic murine thigh model, ceftazidime–avibactam 2000/500 mg q8h administered as a 2 h infusion was simulated against clinical *Enterobacteriaceae* isolates with MICs ≥8 µg/ml. The combination antibiotic elicited enhanced efficacy against isolates with MIC ≤ 16 µg/ml (*f*T/MIC ≥ 62%) and variable activity against isolates with MIC 32 µg/ml (*f*T/MIC ≥ 34%), results that are consistent with established targets for cephalosporins (MacVane *et al.*, 2014a). Notably, at the time of writing, a very few *Enterobacteriaceae* and *P. aeruginosa* have ceftazidime–avibactam MICs greater than or equal to 32 µg/ml, indicating that the selected dosing regimen has been optimized for targeted pathogens. It will be interesting to observe the efficacy of this compound in clinical studies, particularly against pathogens with MICs of 8–16 µg/ml.

PK/PD Considerations in Special Populations

Obesity

Obesity can in some instances affect antibiotic pharmacokinetic profiles, and in doing so, may result in drug exposures that are too low to treat the infection. Alternately, for some antibiotics that have weight-based dosing, the administration of very high doses when calculated using actual body weight leads to no improvement in efficacy, but an increase in toxicity. In general, calculations of drug dosing follow one of three approaches: fixed dosing, weight-based dosing, or body surface area-based dosing (Pai, 2012). Notably, studies in early drug development only incorporate healthy adults with a limited weight range; therefore, once a drug is approved for clinical use, the question of optimal dosing in obese patients becomes paramount. The alteration in body weight and composition can affect the attainment of PK/PD targets for antibiotics. Drug distribution depends on regional blood flow, drug lipophilicity, and protein binding, so increases in body fat are expected to augment the volume of distribution of lipophilic compounds, while hydrophilic compounds should be minimally affected (Rosell and Belfrage, 1979). Additionally, obesity has been shown to increase glomerular filtration rate and, consequently, increase drug clearance (Rea *et al.*, 2006).

Ideally, PK/PD studies with individual antibiotics are needed to identify changes in exposure in the obese population. For example, ertapenem PK was studied in healthy volunteers with normal weight (body mass index (BMI) 18.5–24.9 kg/m (Ambrose *et al.*, 2007), class I–II obesity (BMI 30–39.9 kg/m (Ambrose *et al.*, 2007), and class III obesity (BMI ≥40 kg/m (Chen *et al.*, 2006; Ambrose *et al.*, 2007). All patients received the standard dose of 1 g q24h. Despite receiving the same dose, obese patients displayed significantly lower AUC secondary to a larger volume of the central compartment. When PK profiles were used to simulate exposure by Monte Carlo simulation, normal weight subjects were able to achieve 20% *f*T > MIC with 90% probability at a MIC ≤ 0.5 µg/ml, while class III obese subjects were only able to achieve that target at MIC ≤ 0.25 µg/ml, suggesting that a higher ertapenem dosage would be required in these patients or that an alternative strategy to increase the T > MIC (e.g., prolonged infusion) is warranted. In contrast, when linezolid PK was studied in moderately and morbidly obese bariatric patients, the exposure was found to be similar to that of non-obese patients (Bhalodi *et al.*, 2013). As a result, the standard linezolid dosing regimen of 600 mg q12h should produce comparable AUCs in patients with weights up to 150 kg, implying that dose adjustment based on BMI alone is not warranted. Further studies conducted on individual drugs are required to identify the optimal weight-based dosing strategies.

Critically ill patients

Dosage optimization in the critically ill is of importance given the severity of infections in this population, their underlying diseases states and compromised situation, as well as the increased likelihood of infection with pathogens that have higher antibiotic MICs. Critically ill patients may demonstrate rapidly changing pathophysiologic phenomena as a result of multiple organ dysfunctions (Blot *et al.*, 2014). Briefly, volume of distribution can be increased in sepsis and septic shock as a consequence of vasodilatation, increased vascular permeability, and the contribution of intravenous fluid resuscitation (Hosein *et al.*, 2011). This is particularly important in cases of hydrophilic drugs (e.g., β-lactams, aminoglycosides, glycopeptides, polymyxin B), and leads to diminished plasma concentrations and the need for higher loading doses (Sandri *et al.*, 2013). Hypoalbuminemia frequently occurs in critically ill patients, instigating higher concentrations of unbound drug, which can therefore be eliminated more rapidly through glomerular filtration, thereby necessitating increased frequency of dosing (Roberts *et al.*, 2013). Augmented renal clearance (ARC) (Udy *et al.*, 2011) resulting from glomerular hyperfiltration may lead to decreased drug concentrations of antibiotics eliminated predominantly through the kidneys. ARC affects time-dependent antibiotics, and continuous infusion is a reasonable solution to maintain efficacy (Itabashi, 2007), while concentration-dependent antibiotics are less impacted by ARC. In general, knowledge of local resistance rates, MICs, and the use of the highest doses, combined with prolonged/continuous dosing regimens, can lead to the optimization of PK/PD in critically ill patients (Nicasio *et al.*, 2010).

Pediatrics

There is a paucity of literature to elucidate the application of PK/PD in the pediatric population. The common understanding is that the PK/PD principles of adults still apply to children. In one retrospective analysis, investigators inspected the relationship between *f*T/MIC and efficacy in sinusitis and otitis media (Craig and Andes, 1996). They concluded that cure rates for β-lactams and macrolides were 80–90% when *f*T/MIC exceeded 40–50%, targets that are similar to those described in adults. Notably, drug resistance is less common in pediatric patients, but when higher MIC organisms are suspected, similar dosage optimization strategies such as high-dose, extended interval aminoglycosides and prolonged infusion β-lactams can be employed, as evidenced from a study investigating the piperacillin–tazobactam PK in febrile neutropenic children (Cies *et al.*, 2014). The standard dose infused over 0.5 h was unlikely to reach PK/PD targets, and extended infusion of 100 mg/kg given q6h over 3 h or continuous infusion of 400 mg/kg were required to optimize dosing in that population. Similar findings have been identified with other β-lactams in the pediatric population (Courter *et al.*, 2009). In conclusion, the optimization of PK/PD indices in children requires the same modifications as in adults.

Renal replacement therapy

Acute renal failure is a common occurrence in critically ill patients that requires renal replacement therapy (RRT). Continuous renal replacement therapy (CRRT) modalities are all effective in removing hydrophilic drugs with high protein binding and renal clearance (Pea *et al.*, 2007). However, drugs with large volumes of distribution are minimally affected by CRRT and no incremental doses are mandated (Roberts and Lipman, 2006). Drug clearance is dependent on flow rate and sieving coefficient (SC) of the agent and, typically, higher doses are required as the SC increases, and maximal or near maximal doses are warranted for agents with an SC of one to maintain adequate exposure (Gilbert *et al.*, 2011). In contrast, intermittent hemodialysis (IHD) does not typically remove as much antibiotic as CRRT modalities, so substantial dosage reductions are generally required to avoid overexposure and toxicity, while the administration of extra doses at the end of the dialysis session or giving higher intra-dialytic doses are recommended to reduce the clearance of antibiotics during hemodialysis sessions and avoid subtherapeutic exposure (Barth and DeVincenzo, 1996; Marx *et al.*, 1998).

Conversion from intravenous to oral therapy

Optimizing dose and adhering to wise stewardship practice is not precluded to parenteral administration. Patients admitted to hospitals for severe illness are usually initiated on intravenous therapy. The availability of antimicrobial agents with improved oral bioavailability, such as oxazolidinones, fluoroquinolones, sulfamethoxazole-trimethorprim, metronidazole, and clindamycin, permits the transition of therapy when defined clinical criteria are met, so that PK/PD thresholds are maintained. There are three strategies to convert antibiotics from the intravenous to oral form: sequential therapy (Walters *et al.*, 1999), which is defined as replacing the parenteral therapy with the oral formulation of the same drug at a dose that obtains similar drug concentrations; switch therapy (Paladino *et al.*, 1991), which refers to changing parenteral therapy to an equally potent different drug; and step-down therapy (Paladino *et al.*, 2002), in which the parenteral therapy is modified to use a less potent oral agent. Ideally, to maintain PK/PD optimization, an oral antibiotic with near 100% bioavailability should be selected, or an antibiotic with lower bioavailability but dosage modifications used to counter this loss in PK/PD exposure. For example, oral respiratory fluoroquinolones (e.g., levofloxacin, moxifloxacin) will achieve similar PK/PD thresholds as the same dosage administered intravenously and thus would be considered sequential therapy. Another example is that due to lower bioavailability, oral dosages of amoxicillin should be increased to 1000 mg three times a day (TID) to maintain an adequate *f*T > MIC against contemporary *Streptococcus pneumoniae*.

Future Considerations for PK/PD

As demonstrated throughout this chapter, many opportunities exist to optimize the PK/PD of available antibiotics, particularly for older agents for which PK/PD principles were not originally considered when approved dosage regimens were selected. To optimize PK/PD for any antibiotic, the prescriber needs to understand the target drug exposure required for efficacy, the MIC of the infecting organism, and the concentrations achieved in the individual patient. When one considers these items though, several challenges quickly become apparent.

First, the MIC of the infecting organism is rarely identified by the clinical microbiology lab, which instead substitutes S (for susceptible), I (for intermediate), or R (for resistant); and when a MIC is possible, it is often not available until several days into the antimicrobial treatment. Rapid diagnostic identification and susceptibility testing is an emerging field that takes advantage of advances in technology. For the future, it is not too far a stretch to imagine the test that will identify the organism and provide the antibiotic MIC within the first 24 h of therapy, which will permit rapid selection of the appropriate drug and dosing regimen.

Secondly, with the exception of vancomycin and aminoglycosides, few other antibiotics have clinical assays available to determine drug concentrations in the blood. This makes individualization of the dosing regimen, and as a result PK/PD optimization, challenging. Currently, we "assume" that an individual patient has a PK profile that resembles the mean or the median of the population from drug-specific PK studies. More advanced population modeling permits investigators to identify relationships between patient CrCL, sex, or body weight, for example, and likely estimates for CL, Vc (the central volume of distribution, a hypothetical volume into which a drug initially distributes upon administration), etc., so that a concentration vs. time profile can be 'guessed' for a specific dosing regimen. The use of Bayesian PK software, which has become more user friendly, and even has become available on smartphones, will allow a better job to be done at dosage individualization (Roberts *et al.*, 2014). Nonetheless, antibiotic assays would still need to be developed and made clinically available to the prescriber, so that, for example, a 'cefepime level', could be ordered on a patient with a high MIC organism. Until this technology is widely available, it may be prudent and safe to incorporate these strategies into clinical practice based on a disease state focus so that optimal outcomes can be achieved for a range of pathogen and MIC profiles (Craig and Andes, 1996; Itabashi, 2007). Using our local MIC distribution data for patients with VAP, dosing regimens were optimized through the empiric use of once daily aminoglycosides, prolonged and continuous infusion β-lactams, and aggressive vancomycin dosing (Nicasio *et al.*, 2010).

Lastly, the exposure threshold needed for efficacy is one typically identified early in drug development, and particularly in animal infection models. While PK/PD evaluations are being incorporated into clinical trial design, the homogeneity of the patients, doses, and the MIC profile of the pathogens often does not lend itself to adequate PD target profile. As such, retrospective evaluations of trial data sets may better define the required PD exposure. Recently, this has been done for novel β-lactams (Ariano *et al.*, 2005; Li *et al.*, 2007; Crandon *et al.*, 2010; Muller *et al.*, 2013, 2014; MacVane *et al.*, 2014b). These targets can be used to design or support the prolonged dosing approach.

While the transformation of PK/PD principles into clinical practice is an evolving paradigm, what is clear is that enhanced exposure profiles lead to an increased probability of clinical and microbiological response. As a result of the current uncertainty in the early empiric period due to poorly defined PK, causative pathogen, and potency of the agents utilized, efforts should be made to employ PD-optimized regimens in the first several days of treatment. When data are available and a clinical response has been achieved, the antibiotic regimen can then be refined in accordance with good stewardship practices.

References

Ambrose, P.G., Bhavnani, S.M., Rubino, C.M., Louie, A., Gumbo, T., Forrest, A., and Drusano, G.L. (2007) Pharmacokinetics–pharmacodynamics of antimicrobial therapy: it's not just for mice any more. *Clinical Infectious Diseases* 44, 79–86.

Ariano, R.E., Nyhlén, A., Donnelly, J.P., Sitar, D.S., Harding, G.K., and Zelenitsky, S.A. (2005) Pharmacokinetics and pharmacodynamics of meropenem in febrile neutropenic patients with bacteremia. *Annals of Pharmacotherapy* 39, 32–38.

Barth, R.H. and DeVincenzo, N. (1996) Use of vancomycin in high-flux hemodialysis: experience with 130 courses of therapy. *Kidney International* 50, 929–936.

Bauer, K.A., West, J.E., O'Brien, J.M., and Goff, D.A. (2013) Extended-infusion cefepime reduces mortality in patients with *Pseudomonas aeruginosa* infections. *Antimicrobial Agents and Chemotherapy* 57, 2907–2912.

Bhalodi, A.A., Papasavas, P.K., Tishler, D.S., Nicolau, D.P., and Kuti, J.P. (2013) Pharmacokinetics of intravenous linezolid in moderately to morbidly obese adults. *Antimicrobial Agents and Chemotherapy* 57, 1144–1149.

Bhavnani, S.M., Passarell, J.A., Owen, J.S., Loutit, J.S., Porter, S.B., and Ambrose, P.G. (2006) Pharmacokinetic-pharmacodynamic relationships describing the efficacy of oritavancin in patients with *Staphylococcus aureus* bacteremia. *Antimicrobial Agents and Chemotherapy* 50, 994–1000.

Biedenbach, D.J., Ross, J.E., Fritsche, T.R., Sader, H.S., and Jones, R.N. (2007) Activity of dalbavancin tested against *Staphylococcus* spp. and β-hemolytic *Streptococcus* spp. isolated from 52 geographically diverse medical centers in the United States. *Journal of Clinical Microbiology* 45, 998–1004.

Blot, S.I., Pea, F., and Lipman, J. (2014) The effect of pathophysiology on pharmacokinetics in the critically ill patient—concepts appraised by the example of antimicrobial agents. *Advanced Drug Delivery Reviews* 77, 3–11.

Boucher, H.W., Wilcox, M., Talbot, G.H., Puttagunta, S., Das, A.F., and Dunne, M.W. (2014) Once-weekly dalbavancin versus daily conventional therapy for skin infection. *The New England Journal of Medicine* 370,

2169–2179. For correspondence, see *The New England Journal of Medicine* 371, 1160–1163.

Brown, J., Brown, K., and Forrest, A. (2012) Vancomycin AUC24/MIC ratio in patients with complicated bacteremia and infective endocarditis due to methicillin-resistant *Staphylococcus aureus* and its association with attributable mortality during hospitalization. *Antimicrobial Agents and Chemotherapy* 56, 634–638.

Candiani, G., Abbondi, M., Borgonovi, M., Romano, G., and Parenti, F. (1999) In-vitro and in-vivo antibacterial activity of BI 397, a new semi-synthetic glycopeptide antibiotic. *Journal of Antimicrobial Chemotherapy* 44, 179–192.

Cataldo, M.A., Tacconelli, E., Grilli, E., Pea, F., and Petrosillo, N. (2012) Continuous versus intermittent infusion of vancomycin for the treatment of Gram-positive infections: systematic review and meta-analysis. *Journal of Antimicrobial Chemotherapy* 67, 17–24.

Chen, M., Nafziger, A.N., Drusano, G.L., Ma, L., and Bertino, J.S. Jr (2006) Comparative pharmacokinetics and pharmacodynamic target attainment of ertapenem in normal-weight, obese and extremely obese adults. *Antimicrobial Agents and Chemotherapy* 50, 1222–1227.

Cies, J.J., Jain, J., and Kuti, J.L. (2014) Population pharmacokinetics of the piperacillin component of piperacillin/tazobactam in pediatric oncology patients with fever and neutropenia. *Pediatric Blood and Cancer* 62, 477–482.

Courter, J.D., Kuti, J.L., Girotto, J.E., and Nicolau, D.P. (2009) Optimizing bactericidal exposure for beta-lactams using prolonged and continuous infusions in the pediatric population. *Pediatric Blood and Cancer* 53, 379–385.

Craig, W.A. and Andes, D. (1996) Pharmacokinetics and pharmacodynamics of antibiotics in otitis media. *The Pediatric Infectious Disease Journal* 15, 255–259.

Craig, W.A., Redington, J., and Ebert, S.C. (1991) Pharmacodynamics of amikacin in vitro and in mouse thigh and lung infections. *Journal of Antimicrobial Chemotherapy* 27(Suppl C), 29–40.

Crandon, J.L., Bulik, C.C., Kuti, J.L., and Nicolau, D.P. (2010) Clinical pharmacodynamics of cefepime in patients with *Pseudomonas aeruginosa*. *Antimicrobial Agents and Chemotherapy* 54, 1111–1116.

Croisier, D., Martha, B., Piroth, L., and Chavanet, P. (2008) In vivo efficacy of humanized intermittent versus continuous ceftazidime in combination with tobramycin in an experimental model of pseudomonal pneumonia. *International Journal of Antimicrobial Agents* 32, 494–498.

DiMondi, V.P. and Rafferty, K. (2013) Review of continuous-infusion vancomycin. *Annals of Pharmacotherapy* 47, 219–227.

Drusano, G.L. (2004) Antimicrobial pharmacodynamics: critical interactions of 'bug and drug'. *Nature Reviews Microbiology* 2, 289–300.

Drusano, G.L. (2007) Pharmacokinetics and pharmacodynamics of antimicrobials. *Clinical Infectious Diseases* 45(Suppl 1), S89–S95.

Drusano, G.L. and Louie, A. (2011) Optimization of aminoglycoside therapy. *Antimicrobial Agents and Chemotherapy* 55, 2528–2531.

Drusano, G.L., Ambrose, P.G., Bhavnani, S.M., Bertino, J.S., Nafziger, A.N., and Louie, A. (2007) Back to the future: using aminoglycosides again and how to dose them optimally. *Clinical Infectious Diseases* 45, 753–760.

Dulhunty, J.M., Roberts, J.A., Davis, J.S., Webb, S.A.R., Bellomo, R., Gomersall, C., Shirwadkar, C., Eastwook, G.M., Myburgh, J., Paterson, D.L., and Lipman, J. (2013) Continuous infusion of beta-lactam antibiotics in severe sepsis: a multicenter double-blind, randomized controlled trial. *Clinical Infectious Diseases* 56, 236–244.

Eagle, H., Fleischman, R., and Musselman, A.D. (1950) Effect of schedule of administration on the therapeutic efficacy of penicillin; importance of the aggregate time penicillin remains at effectively bactericidal levels. *The American Journal of Medicine* 9, 280–299.

Falagas, M.E., Tansarli, G.S., Ikawa, K., and Vardakas, K.Z. (2013) Clinical outcomes with extended or continuous versus short-term intravenous infusion of carbapenems and piperacillin/tazobactam: a systematic review and meta-analysis. *Clinical Infectious Diseases* 56, 272–282.

Freire, A.T., Melnyk, V., Kim, M.J., Datsenko, O., Dzyublik, O., Glumcher, F., Chuang, Y.C., Maroko, R.T., Dukart, G., Cooper, C.A. *et al.* (2010) Comparison of tigecycline with imipenem/cilastatin for the treatment of hospital-acquired pneumonia. *Diagnostic Microbiology and Infectious Disease* 68, 140–151.

George, J.M., Towne, T.G., and Rodvold, K.A. (2012) Prolonged infusions of β-lactam antibiotics: implication for antimicrobial stewardship. *Pharmacotherapy* 32, 707–721.

Gilbert, B., Robbins, P., and Livornese, L.L., Jr. (2011) Use of antibacterial agents in renal failure. *Medical Clinics of North America* 95, 677–702.

Grant, E.M., Kuti, J.L., Nicolau, D.P., Nightingale, C., and Quintiliani, R. (2002) Clinical efficacy and pharmacoeconomics of a continuous-infusion piperacillin–tazobactam program in a large community teaching hospital. *Pharmacotherapy* 22, 471–483.

Guiliano, R.A., Verpooten, G.A., Verbist, L., Weeden, R., and DeBroe, M.E. (1986) In vivo uptake kinetics of aminoglycosides in the kidney cortex of rats. *Journal of Pharmacology and Experimental Therapeutics* 236, 470–475.

Hegde, S.S., Reyes, N., Wiens, T., Vanasse, N., Skinner, R., McCullough, J., Kaniga, K., Pace, J., Thomas, R., Shaw, J.P. *et al.* (2004) Pharmacodynamics of telavancin (TD-6424), a novel bactericidal agent, against Gram-positive bacteria. *Antimicrobial Agents and Chemotherapy* 48, 3043–3050.

Holmes, N.E., Turnidge, J.D., Munckhof, W.J., Robinson, J.O., Korman, T.M., O'Sullivan, M.V.N., Anderson, T.L., Roberts, S.A., Warren, S.J.C., Gao, W. *et al.* (2013) Vancomycin AUC/MIC ratio and 30-day mortality in patients with *Staphylococcus aureus* bacteremia. *Antimicrobial Agents and Chemotherapy* 57, 1654–1663.

Hosein, S., Udy, A.A., and Lipman, J. (2011) Physiological changes in the critically ill patient with sepsis. *Current Pharmaceutical Biotechnology* 12, 1991–1995.

Huang, H., Huang, S., Zhu, P., and Xi, X. (2014) Continuous versus intermittent infusion of cefepime in neurosurgical patients with post-operative intracranial infections. *International Journal of Antimicrobial Agents* 43, 68–72.

Itabashi, S. (2007) Clinical efficacy of prolonged (4 hour) drip infusion of meropenem against severe pneumonia. *The Japanese Journal of Antibiotics* 60, 161–170.

Kashuba, A.D., Nafziger, A.N., Drusano, G.L., and Bertino, J.S., Jr. (1999) Optimizing aminoglycoside therapy for nosocomial pneumonia caused by Gram-negative bacteria. *Antimicrobial Agents and Chemotherapy* 43, 623–629.

Koomanachai, P., Bulik, C.C., Kuti, J.L., and Nicolau, D.P. (2010) Pharmacodynamic modeling of intravenous antibiotics against Gram-negative bacteria collected in the United States. *Clinical Therapeutics* 32, 766–779.

Kuti, J.L., Kiffer, C.R.V., Mendes, C.M.F., and Nicolau, D.P. (2008) Pharmacodynamic comparison of linezolid, teicoplanin, and vancomycin against clinical isolates of *Staphylococcus aureus* and coagulase-negative staphylococci collected from hospitals in Brazil. *Clinical Microbiology and Infection* 14, 116–123.

Laohavaleeson, S., Kuti, J.L., and Nicolau, D.P. (2007) Telavancin: a novel lipoglycopeptide for serious Gram-positive infections. *Expert Opinion on Investigational Drugs* 16, 347–357.

Lau, W.K., Mercer, D., Itani, K.M., Nicolau, D.P., Kuti, J.L., Mansfield, D., and Dana, A. (2006) Randomized, open-label, comparative study of piperacillin–tazobactam administered by continuous infusion versus intermittent infusion for treatment of hospitalized patients with complicated intra-abdominal infection. *Antimicrobial Agents and Chemotherapy* 50, 3556–3561.

Levison, M.E. (2004) Pharmacodynamics of antimicrobial drugs. *Infectious Disease Clinics of North America* 18, 451–465.

Li, C., Du, X., Kuti, J., and Nicolau, D. (2007) Clinical pharmacodynamics of meropenem in patients with lower respiratory tract Infections. *Antimicrobial Agents and Chemotherapy* 51, 1725–1730.

Lodise, T.P., Lomaestro, B.M., and Drusano, G.L. (2006) Application of antimicrobial pharmacodynamic concepts into clinical practice: focus on beta-lactam antibiotics. Insights from the Society of Infectious Diseases Pharmacists. *Pharmacotherapy* 26, 1320–1332.

Lodise, T.P., Jr., Lomaestro, B., and Drusano, G.L. (2007) Piperacillin–tazobactam for *Pseudomonas aeruginosa* infection: clinical implications of an extended-infusion dosing strategy. *Clinical Infectious Diseases* 44, 357–363.

Lodise, T.P., Butterfield, J.M., Hegde, S.S., Samara, E., and Barriere, S.L. (2012) Telavancin pharmacokinetics and pharmacodynamics in patients with complicated skin and skin structure infections and various degrees of renal function. *Antimicrobial Agents and Chemotherapy* 56, 2062–2066.

Löwdin, E., Odenholt, I., and Cars, O. (1998) In vitro studies of pharmacodynamics properties of vancomycin against *Staphylococcus aureus* and *Staphylococcus epidermidis*. *Antimicrobial Agents and Chemotherapy* 42, 2739–2744.

MacVane, S.H., Kuti, J.L., and Nicolau, D.P. (2014a) Clinical pharmacodynamics of antipseudomonal cephalosporins in patients with ventilator associated pneumonia. *Antimicrobial Agents and Chemotherapy* 58, 1359–1364.

MacVane, S.H., Crandon, J.L., Nichols, W.W., and Nicolau, D.P. (2014b) In vivo efficacy of humanized exposures of ceftazidime–avibactam in comparison with ceftazidime against contemporary *Enterobacteriaceae* isolates. *Antimicrobial Agents and Chemotherapy* 58, 6913–6919.

Marx, M.A., Frye, R.F., Matzke, G.R., and Golper, T.A. (1998) Cefazolin as empiric therapy in hemodialysis-related infections: efficacy and blood concentrations. *American Journal of Kidney Diseases* 32, 410–414.

McKay, G.A., Beaulieu, S., Arhin, F.F., Parr, T.R., Jr., and Moeck, G. (2009) Time-kill kinetics of oritavancin and comparator agents against *Staphylococcus aureus*, *Enterococcus faecalis* and *Enterococcus faecium*. *Journal of Antimicrobial Chemotherapy* 63, 1191–1199.

Moore, R.D., Smith, C.R., and Lietman, P.S. (1984) Association of aminoglycoside plasma levels with therapeutic outcome in Gram-negative pneumonia. *The American Journal of Medicine* 77, 657–662.

Moore, R.D., Lietman, P.S., and Smith, C.R. (1987) Clinical response to aminoglycoside therapy: importance of the ratio of the peak concentration to minimal inhibitory concentration. *The Journal of Infectious Diseases* 155, 93–99.

Muller, A.E., Punt, N., and Mouton, J.W. (2013) Optimal exposures of ceftazidime predict the probability of microbiological and clinical outcomes in treatment of nosocomial pneumonia. *Journal of Antimicrobial Chemotherapy* 68, 900–906.

Muller, A.E., Punt, N., and Mouton, J.W. (2014) Exposure to ceftobiprole is associated with microbiological eradication and clinical cure in patients with nosocomial pneumonia. *Antimicrobial Agents and Chemotherapy* 58, 2512–2519.

Murray, K.P., Zhao, J.J., Davis, S.L., Kullar, R., Kaye, K.S., Lephart, P., and Rybak, M.J. (2013) Early use of daptomycin versus vancomycin for methicillin-resistant *Staphylococcus aureus* bacteremia with vancomycin minimum inhibitory concentration >1 mg/L: a matched cohort study. *Clinical Infectious Diseases* 56, 1562–1569.

Neely, M.N., Youn, G., Jones, B., Jelliffe, R.W., Drusano, G.L., Rodvold, K.A., and Lodise, T.P. (2014) Are vancomycin trough concentrations adequate for optimal dosing? *Antimicrobial Agents and Chemotherapy* 58, 309–316.

Nicasio, A.M., Eagye, K.J., Nicolau, D.P., Shore, E., Palter, M., Pepe, J., and Kuti, J.L. (2010) Pharmacodynamic-based clinical pathway for empiric antibiotic choice in patients with ventilator-associated pneumonia. *Journal of Critical Care* 25, 69–77.

Nichols, P., Levasseur, W., Li, J., and Das, S. (2012) A threshold concentration of avibactam (AVI) during the pharmacokinetic decline phase, below which lactamase inhibition in *Enterobacteriaceae* becomes ineffective. In: *ICAAC Abstracts: 52nd Interscience Conference on Antimicrobial Agents and Chemotherapy, September 9-12, 2012, San Francisco*. American Society for Microbiology, Washington, DC, AbstrA-1760.

Nicolau, D.P., Freeman, C.D., Belliveau, P.P., Nightingale, C.H., Ross, J.W., and Quintiliani, R. (1995) Experience with a once-daily aminoglycoside program administered to 2,184 adult patients. *Antimicrobial Agents and Chemotherapy* 39, 650–655.

Noone, P. and Pattison, J.R. (1974) The effective use of gentamicin in life threatening sepsis. *Postgraduate Medicine* 50(Suppl 4), 9–16.

Pai, M.P. (2012) Drug dosing based on weight and body surface area: mathematical assumptions and limitations in obese adults. *Pharmacotherapy* 32, 856–868.

Pai, M.P., Neely, M., Rodvold, K.A., and Lodise, T.P. (2014) Innovative approaches to optimizing the delivery of vancomycin in individual patients. *Advanced Drug Delivery Reviews* 77, 50–57.

Paladino, J.A., Sperry, H.E., Backes, J.M., Gelber, J.A., Serrianne, D.J., Cumbo, T.J., and Schentag, J.J. (1991) Clinical and economic evaluation of oral ciprofloxacin after an abbreviated course of intravenous antibiotics. *The American Journal of Medicine* 91, 462–470.

Paladino, J.A., Gudgel, L.D., Forrest, A., and Niederman, M.S. (2002) Cost-effectiveness of IV-to-oral switch therapy: azithromycin vs cefuroxime with or without erythromycin for the treatment of community-acquired pneumonia. *Chest* 122, 1271–1279.

Patti, G.J., Kim, S.J., Yu, T.Y., Dietrich, E., Tanaka, K.S., Parr, T.R., Jr., Far, A.R., and Schaefer, J. (2009) Vancomycin and oritavancin have different modes of action in *Enterococcus faecium*. *Journal of Molecular Biology* 392, 1178–1191.

Pea, F., Viale, P., Pavan, F., and Furlanut, M. (2007) Pharmacokinetic considerations for antimicrobial therapy in patients receiving renal replacement therapy. *Clinical Pharmacokinetics* 46, 997–1038.

Ramirez, J., Dartois, N., Gandjini, H., Yan, J.L., Korth-Bradley, J., and McGovern, P.C. (2013) Randomized Phase 2 trial to evaluate the clinical efficacy of two high-dosage tigecycline regimens versus imipenem–cilastatin for treatment of hospital-acquired pneumonia. *Antimicrobial Agents and Chemotherapy* 57, 1756–1762.

Rea, D.J., Heimbach, J.K., Grande, J.P., Textor, S.C., Taler, S.J., Prieto, M., Larson, T.S., Cosio, F.G., and Stegall, M.D. (2006) Glomerular volume and renal histology in obese and non-obese living kidney donors. *Kidney International* 70, 1636–1641.

Roberts, J.A. and Lipman, J. (2006) Antibacterial dosing in intensive care: pharmacokinetics, degree of disease, and pharmacodynamics of sepsis. *Clinical Pharmacokinetics* 45, 755–773.

Roberts, J.A., Pea, F., and Lipman, J. (2013) The clinical relevance of plasma protein binding changes. *Clinical Pharmacokinetics* 52, 1–8.

Roberts, J.A., Aziz, M.H.A., Lipman, J., Mouton, J.W., Vinks, A.A., Felton, T.W., Hope, W.W., Farkas, A., Neely, M.N., Schentag, J.J. *et al.* (2014) Individualized antibiotic dosing for patients who are critically ill: challenges and potential solutions. *The Lancet Infectious Diseases* 14, 498–509.

Roosendaal, R., Bakker-Woudenberg, I.A., van den Berg, J.C., and Michel, M.F. (1985) Therapeutic efficacy of continuous versus intermittent administration of ceftazidime in an experimental *Klebsiella pneumoniae* pneumonia in rats. *The Journal of Infectious Diseases* 152, 373–378.

Roosendaal, R., Bakker-Woudenberg, I.A., van den Berghe-van Raffe, M., Vink-van den Berg, J.C., and Michel, M.F. (1987) Comparative activities of ciprofloxacin and ceftazidime against *Klebsiella pneumoniae* in vitro and in experimental pneumonia in leukopenic rats. *Antimicrobial Agents and Chemotherapy* 31, 1809–1815.

Rosell, S. and Belfrage, E. (1979) Blood circulation in adipose tissue. *Physiological Reviews* 59, 1078–1104.

Rybak, M.J., Abate, B.J., Kang, S.L., Ruffing, M.J., Lerner, S.A., and Drusano, G.L. (1999) Prospective evaluation of the effect of an aminoglycoside dosing regimen on rates of observed nephrotoxicity and ototoxicity. *Antimicrobial Agents and Chemotherapy* 43, 1549–1555.

Rybak, M., Lomaestro, B., Rotschafer, J.C., Moellering, R., Craig, W., Billeter, M., Dalovisio, J.R., and Levine, D.P. (2009) Therapeutic monitoring of vancomycin in adult patients: a consensus review of the American Society of Health-System Pharmacists, the Infectious Diseases Society of America and the Society of

Infectious Diseases Pharmacists. *American Journal of Health-System Pharmacy* 66, 82–98.

Sandri, A.M., Landersdorfer, C.B., Jacob, J., Boniatti, M.M., Dalarosa, M.G., Falci, D.R., Behle, T.F., Bordinhao, R.C., Wang, J., Forrest, A. *et al.* (2013) Population pharmacokinetics of intravenous polymyxin B in critically ill patients: implications for selection of dosage regimens. *Clinical Infectious Diseases* 57, 524–531.

Teo, J., Liew, Y., Lee, W., and Kwa, A.L. (2014) Prolonged infusion versus intermittent boluses of β-lactam antibiotics for treatment of acute infections: a meta-analysis. *International Journal of Antimicrobial Agents* 43, 403–411.

Udy, A.A., Roberts, J.A., and Lipman, J. (2011) Implications of augmented renal clearance in critically ill patients. *Nature Reviews Nephrology* 7, 539–543.

Walters, D.J., Solomkin, J.S., and Paladino, J.A. (1999) Cost effectiveness of ciprofloxacin plus metronidazole versus imipenem–cilastatin in the treatment of intra-abdominal infections. *Pharmacoeconomics* 16, 551–561.

Wysocki, M., Thomas, F., Wolff, M.A., Pean, Y., Ravaud, Y., and Herman, B. (1995) Comparison of continuous with discontinuous intravenous infusion of vancomycin in severe MRSA infections. *Journal of Antimicrobial Chemotherapy* 35, 352–354.

Yost, R.J. and Cappelletty, D.M. (2011) The retrospective cohort of extended-infusion piperacillin–tazobactam (RECEIPT) study: a multicenter study. *Pharmacotherapy* 31, 767–775.

18 Optimal Use of Gram-negative Antibiotics in the Real World: Providing Effective Therapy while Minimizing Resistance

Jason M. Pogue,[1] Jessica K. Ortwine,[2] and Keith S. Kaye[3*]

[1]*Sinai-Grace Hospital, Detroit Medical Center, Wayne State University School of Medicine, Detroit, Michigan, US;* [2]*Parkland Hospital and Health System, University of Texas Southwestern Medical School, Dallas, Texas, US;* [3]*Detroit Medical Center and Wayne State University Health Center, Detroit, Michigan, US*

Introduction

Over the past 10 years, there has been an increasing awareness of and concern about multidrug-resistant (MDR) bacteria and the role that these organisms play in increasing patient morbidity and mortality. This concern is heightened in Gram-negative bacilli. The Infectious Diseases Society of America (IDSA) first brought this problem to light through their 2004 report *Bad Bugs, No Drugs: As Antibiotic Discovery Stagnates … A Public Health Crisis Brews* (IDSA, 2004) This report acknowledged that antibiotic resistance, while not a new phenomenon, had been increasing more rapidly than historically documented. Among the organisms that frequently display multidrug resistance, certain Gram-negative pathogens are especially worrisome. The Centers for Disease Control and Prevention (CDC, 2013), in its threat report, categorized carbapenem-resistant *Enterobacteriaceae* (CRE) as an urgent threat (the highest threat level) and MDR *Acinetobacter*, *Pseudomonas aeruginosa*, and extended-spectrum β-lactamase (ESBL)-producing *Enterobacteriaceae* as serious threats to domestic healthcare. The justification for such a high level of concern over these pathogens is multifactorial and is related, in part, to the rising incidence of infections due to these resistant pathogens, the burden that these infections have on the healthcare community in terms of morbidity and mortality, and the limited currently available treatment options for infections caused by these pathogens, combined with a lack of new antibiotics being developed to combat the increasing resistance.

Antimicrobial stewardship programs play a crucial role in both curtailing the development and spread of these pathogens, but, importantly, also in developing strategies to optimally manage infections caused by them. This chapter will discuss strategies that stewardship programs can employ to limit the emergence and spread of resistant Gram-negative bacilli, and optimal management strategies for the treatment of infections due to these problematic pathogens. These topics will be discussed by addressing clinically important stewardship questions surrounding resistant Gram-negative pathogens and the antibiotics used to treat them, and by providing evidence-based solutions.

*Corresponding author. E-mail: KKaye@dmc.org

Can Stewardship Programs Reduce the Incidence of Drug-resistant Gram-negative Bacilli?

The use of antimicrobials is known to be the most important, modifiable risk factor for the development of drug resistance, and therefore it is reasonable to assume that limiting their unnecessary use can limit the development of resistant organisms. However, do the data support this? Furthermore, and perhaps more importantly as we now live in an age where these resistant organisms are more prevalent, can we reverse the trends—or at least stop the progression of further resistance?

Among the non-fermenters, there are ample data on the impact of stewardship on resistance in *P. aeruginosa*. Del Arco *et al.* (2014) describe the impact that their stewardship program, which targeted broad-spectrum antibiotic usage, had on the resistance rates of *P. aeruginosa* in their institution. As part of a comprehensive stewardship effort addressing all broad-spectrum antimicrobials, the authors were able to demonstrate a 9% reduction in imipenem usage in year 1 of their program that remained consistent for the next 2 years. The investigators showed a stepwise decrease in the incidence of imipenem-resistant *P. aeruginosa* at their institution. Resistance rates were 14% in 2008 (the year before the program began), 8% in 2009, 6% in 2010, and finally 4% at the end of the study period in 2011. While limiting carbapenem usage is desirable, it is important to understand that β-lactam resistance in *P. aeruginosa* is known to be selected by many antimicrobial classes beyond the β-lactams due to the selection of non-specific multidrug efflux pumps. So strategies to limit use of structurally unrelated classes can also be implemented to stabilize or improve susceptibility profiles of the β-lactams.

Lewis *et al.* (2012) showed that through the restriction of ciprofloxacin and despite a significant increase in the usage of carbapenems during their study period, they were able to decrease the rate of both carbapenem and cefepime resistance, as well as ciprofloxacin resistance, in *P. aeruginosa*, with no impact seen on piperacillin–tazobactam susceptibilities. These results highlight the importance of limiting all antimicrobial exposures (rather than limiting a specific agent) in order to decrease the incidence of resistant pathogens. Unfortunately, for *Acinetobacter baumannii*, data pertaining to the association between stewardship and decreased resistance are lacking.

Evidence from the *Enterobacteriaceae* suggests that a similar impact to that described above for *P. aeruginosa* can be seen on both the incidence of patients infected with ESBL-producing organisms and on overall rates of infection by such organisms (henceforward designated "ESBL infections"). In an analysis of hospital infections caused by ESBL- (and AmpC β-lactamase)-producing Gram-negative bacteria by Knudsen *et al.* (2014), due to high rates of ESBL infection, the investigators shifted from empiric use of cefuroxime to the use of piperacillin–tazobactam. This approach decreased rates of infection by ESBL-producing *Klebsiella pneumoniae* from roughly 40% at the beginning of the study period, to ~10% by the end. Importantly though, in this analysis, there was a subsequent increase in piperacillin–tazobactam-resistant *P. aeruginosa*. This shows that there are potential trade-offs associated with limiting resistance to a particular antimicrobial and/or a particular pathogen. Programs directed at de-escalation and duration of therapy that focus on minimizing the unnecessary utilization of all antimicrobials (particularly broad-spectrum antimicrobials) are imperative and can have a favorable impact on resistance across pathogens. Therefore, it is prudent for institutions with high (or increasing) rates of resistant pathogens to develop robust programs targeting de-escalation and minimizing the durations of therapy for all antimicrobials, with an emphasis on carbapenems, fluoroquinolones, and other anti-pseudomonal β-lactams.

Can I Avoid a Carbapenem for the Treatment of Invasive ESBL Infections?

As previously described, the use of specific agents has associated risks for the development of resistant organisms but, undoubtedly, of greatest concern in the Gram-negative world is development of resistance to the carbapenems, our last "first-line" therapeutic option for isolates resistant to other β-lactams. Consequently, all efforts to conserve this important class of antimicrobials should be undertaken. Thus, the clinical question arises: can I use non-carbapenem options—notably cefepime and or β-lactam/β-lactamase inhibitor combinations (BLBLI)—for the treatment of ESBL infections if the pathogens have *in vitro* susceptibility to these agents? Potential reasons for avoiding these non-carbapenem agents are: (i) the long-standing track record of successful treatment with carbapenems for ESBL infections; (ii) the failure of third-generation cephalosporins in the treatment of ESBL infections despite apparent *in vitro* activity; (iii) the "inoculum effect," which is notably present with cephalosporins that show increased (and non-susceptible) minimum inhibitory concentrations (MICs) in the presence of a high inoculum of organisms that are similar to those seen in deep-seated infections; and (iv) a lack of convincing clinical data with non-carbapenems in the treatment of serious ESBL-producing infections.

Unfortunately, data are limited when looking at the efficacy of BLBLI combinations for ESBL infections, as the majority of publications comparing BLBLIs with other therapies are limited to small sample sizes (Vardakas *et al*., 2012). However, in 2012, Rodríguez-Baño *et al*. published data from a retrospective analysis of the largest cohort of patients yet studied that compared both empiric and definitive therapy with a BLBLI, and a carbapenem therapy for ESBL bloodstream infections. The Rodríguez-Baño *et al*. study serves as the best evidence to date that has looked at this question. The study included 103 patients in the empiric therapy cohort (BLBLI 72, carbapenem 31) and 174 in the definitive therapy cohort (BLBLI 54, carbapenem 120). After controlling for differences between the groups, the authors found no association of either the empiric therapy (hazard ratio (HR) 1.14, 95% confidence interval (CI) 0.29–4.40) or the definitive therapy (HR 0.76, 95% CI 0.28–2.07) with a BLBLI and increased mortality.

While these findings from the Rodríguez-Baño *et al*. (2012) study are on the surface encouraging, it is important for the reader to be aware that "sicker" patients received both empiric and definitive therapy with carbapenems. Although these numbers failed to reach statistical significance owing to the low number of patients in the study, patients who received carbapenems were more likely to be immunosuppressed, neutropenic, have a non-urinary or gastrointestinal infection source, and present in severe sepsis or septic shock. Additionally, 44% of patients in the analysis who had an empiric BLBLI received definitive therapy with a carbapenem, whereas over 80% of patients who received definitive therapy with a BLBLI started on therapy with a BLBLI. Only 25% of patients in the carbapenem definitive therapy cohort were empirically started on a carbapenem. When taken together, these findings suggest that patients empirically started on a BLBLI who responded to therapy were kept on a BLBLI, whereas those who were started on a BLBLI or other non-carbapenem antibiotic and were not responding were switched to a carbapenem. These findings inherently bias the results against the carbapenems, and despite attempts to control for differences; sicker patients were treated with carbapenems. Therefore, true interpretation of these data would suggest that BLBLIs could be appropriate therapy for an ESBL bloodstream infection in patients who were empirically started on BLBLIs, who had isolates susceptible to this agent, and who responded to empiric therapy.

The Rodríguez-Baño *et al*. (2012) study and another analysis by Retamar *et al*. (2013) looking at the impact of the MIC of the BLBLI on outcome in ESBL bloodstream infection are interesting, but difficult to interpret, as they only look at outcome as a function of MIC in patients empirically treated with a BLBLI without accounting for definitive therapy. The impact of MIC (and dosing) on outcome is an important area for future study in this area.

The issue of the appropriate MIC and dosing is further illustrated when looking at the role of cefepime for the treatment of ESBL bloodstream infections. As with BLBLIs, there is a lack of comprehensive, well-controlled evidence to support the use of cefepime when isolates retain in vitro susceptibility. While the numbers are small, the best evidence and clues on the use of cefepime for ESBL bloodstream infections was published in late 2012 (Lee *et al*., 2013). When the authors compared 17 patients who received definitive therapy with cefepime (dose 1–2 g every 8 h) with either 161 total patients or 17 propensity score-matched patients receiving carbapenems, they found that definitive therapy with cefepime (displaying in vitro susceptibility at an MIC $\leq$ 8 μg/ml) was associated with increased 30 day mortality (59% for cefepime, 17% for carbapenems, 12% for propensity score-matched carbapenem patients). In the multivariate model, definitive therapy with cefepime was associated with a 5.4-fold increased risk (95% CI 1.4–20.9) for 30 day mortality. Interestingly, mortality rates with cefepime varied as a function of MIC; they were of 17% (1/6) with MICs $\leq$ 1 μg/ml, 46% (5/11) for MICs of 2–8 μg/ml, and 100% (4/4) for MICs $\geq$ 16. These findings are particularly interesting because pharmacokinetic/pharmacodynamic (PK/PD) targets with cefepime dosing as used in this study should be obtainable up to MICs of 8 μg/ml.

So is there any role for treatment with non-carbapenem β-lactams with in vitro susceptibility for ESBL infections? Undoubtedly, the answer is yes. One important caveat to keep in mind is that the above data are for the treatment of bloodstream infections in more severely ill patients. A large proportion of ESBL infections are in the urinary tract, where high antimicrobial concentrations would outweigh any "hidden resistance" and the patients are often only moderately ill. BLBLIs and cefepime are certainly options in urinary tract infection due to ESBL producers when in vitro susceptibility is seen. In addition, urinary tract agents, notably nitrofurantoin and fosfomycin, also have a role in lower urinary tract infections. The situation is more complex with bloodstream and other systemic infections. While the exact role of dose optimization strategies (extended infusion, maximum dose) and MIC remains unsettled, it is important to note that in clinical practice many patients are started empirically on cefepime or piperacillin–tazobactam as broad-spectrum coverage prior to the identification of an ESBL. It usually takes 48–72 h from the time that a culture is obtained to identify to a pathogen. Once an ESBL producer is identified, clinicians should incorporate clinical response into antimicrobial decision making. If the patient is improving and in vitro susceptibility is seen to a BLBLI or to cefepime, it would be reasonable to continue the agent, but if the patient is not clinically improving, it would be prudent to switch to a carbapenem, the gold standard. In the empiric setting, carbapenems would be preferred if targeting ESBL producers, and recent evidence supports this recommendation (Tamma *et al.*, 2015).

What About AmpC Hyperproducing *Enterobacter*?

Treatment with a third-generation cephalosporin for invasive *Enterobacter* spp. has been associated with the selection of AmpC β-lactamase hyperproducing strains ("derepressed mutants") and clinical failure (Kaye *et al.*, 2001). However, cefepime, which is more stable to hydrolysis by this AmpC enzyme, has been proposed as a carbapenem-sparing alternative for *Enterobacter* infections. Similar to the ESBL debate, there is concern about an inoculum effect, and the clinical literature supporting cefepime treatment for deep-seated *Enterobacter* infections has been scarce. Encouragingly, the past 2 years have seen an increase in data, and while the numbers are limited, the results have been encouraging.

Hilty *et al.* (2013) looked at outcomes in 52 patients with either inducible or derepressed AmpC *Enterobacter* bloodstream infections. Clinical success was seen in 16/18 (89%) of patients receiving cefepime compared with 12/13 (92%) receiving carbapenems. Tamma *et al.* (2013) built on these findings with a well-controlled propensity score-matched analysis of 64 patients receiving cefepime or meropenem for proven AmpC derepressed *Enterobacteriaceae* infections (94% were *Enterobacter* spp.). A particular strength of this study is that the authors explicitly included only patients who received one of the two agents as both empiric and definitive therapy for the entire course. In addition, the dose was optimized at 1–2 g q8h (with renal dosing adjustments) for both agents. In the propensity score-matched cohorts, 30 day mortality rates were similar (31% cefepime vs. 34% meropenem; $P = 0.99$), and multivariate modeling for factors associated with mortality failed to identify either agent as a predictor. Similarly, Siedner *et al.* (2014) showed similar mortality rates in *Enterobacter* bloodstream infections treated with cefepime (7/42, 17%) and carbapenems (5/19, 26%) in patients receiving monotherapy.

Both the Hilty *et al.* (2013) and Tamma *et al.* (2013) analyses attempted to look at the impact of MIC on outcomes, but due to the high success rates and overall low MIC distributions (the majority of isolates had cefepime MICs of ≤ 2 μg/ml), meaningful conclusions cannot be drawn. These data, when taken together, suggest a potential role for cefepime in the treatment of AmpC hyperproducing organisms, particularly when MICs are low and patients are clinically responding to cefepime therapy.

Are All Carbapenems Created Equally?

As stated above, invasive infections due to ESBL producers and *Enterobacter* spp., particularly those in critically ill patients, will often require therapy with carbapenems. As previously discussed, the concern with increased usage of carbapenems is the development of carbapenem resistance in *P. aeruginosa*, *A. baumannii*, and *Enterobacteriaceae* due to selective pressure. With this in mind, ertapenem, the lone "group 1 carbapenem" offers a theoretical advantage in some scenarios over meropenem, imipenem–cilistatin, and doripenem, the "group 2 carbapenems." Ertapenem is unique in that, unlike the group 2 carbapenems, it lacks appreciable in vitro activity against *P. aeruginosa*, *A. baumannii*, and penicillin-susceptible enterococci. Therefore, in theory, preferential usage of ertapenem, which has proven efficacious in ESBL bloodstream infections (Collins *et al.*, 2012; Lee *et al.*, 2012) for the treatment of Enterobacteriaceae that warrant carbapenem therapy, could decrease the selective antimicrobial pressure on other problematic Gram-negative bacilli.

While it is encouraging is that ertapenem has been shown to be an effective agent for ESBL infections, the question that still warrants investigation is whether or not this theoretical selective pressure benefit holds true in patient-level data. In general, there are much more data looking at this association with *P. aeruginosa* than with *A. baumannii*, and for *Pseudomonas*, the data have been encouraging. Most data looking at the impact of the introduction of ertapenem on to the formulary for the susceptibility of *P. aeruginosa* to group 2 carbapenems have shown no worsening of susceptibility with increasing use of ertapenem. These findings are reassuring as they suggest that there is no selection of group 2 carbapenem resistance (Nicolau *et al.*, 2012). Perhaps most promising though were the findings from Goldstein *et al.* (2009). In this investigation, the authors showed that with the introduction of ertapenem on to the formulary they were able to significantly decrease the use of imipenem by ~30%, and saw a subsequent improvement in imipenem susceptibility in *P. aeruginosa* from a 65–75% range to 88% by the end of the study period. The authors were able to show that for every unit decrease in the monthly defined daily dose (DDD) of imipenem, there was an increase of 0.38% (P = 0.008) in the susceptibility of *P. aeruginosa* to imipenem in the same month. These findings suggest that if ertapenem introduction can lead to a reduction in group 2 carbapenem usage (by using it to target *Enterobacteriaceae* warranting carbapenem therapy), it can lead to improved group 2 carbapenem susceptibility in *P. aeruginosa*.

The data on *A. baumannii* are much more sparse and conflicting, consisting of only two reports. Yoon *et al.* (2014) studied the impact of a stewardship program's preferential use of ertapenem for *Enterobacteriaceae* on the incidence of carbapenem-resistant *A. baumannii*. Over the study period, the authors noted a stable amount of total carbapenem usage with a significant increase in ertapenem usage (from 2.7 DDD/1000 patient days to 7.2 DDD/1000 patient days; P = <0.0001), offset by a significant reduction in group 2 carbapenem usage (from 20.7 DDD/1000 patient days to 15.5 DDD/1000 patient days; P = 0.028). They noted that over the study period there was a significant increase in the percentage of carbapenem-resistant *A. baumannii* (from 52.2 to 69.9%; P <0.0001). However, when looking at the individual impact by linear regression, the authors only noted an association between total carbapenem (P = 0.0493) and group 2 carbapenems (P = 0.03) but not ertapenem (P = 0.94) with the proportion of carbapenem-resistant isolates. While these findings suggest that there was not a correlation between ertapenem usage and the isolation of carbapenem-resistant *A. baumannii*, it is clear that preferential usage of ertapenem and decreased usage of group 2 carbapenems did not stop the continued rise in proportion of carbapenem-resistant isolates. Additionally, Tan *et al.* (2014) found the strongest antibiotic risk factor association between ertapenem (r^2 = 0.76) and carbapenem resistance in *A. baumannii*—higher than that of piperacillin–tazobactam, cefepime, and group 2 carbapenems. Based on the findings of both of these analyses, there certainly are not clear data that the same selective pressure benefit seen in *P. aeruginosa* is also seen in *A. baumannii*. Based on current evidence, ertapenem should be considered as preferential for the treatment of infections due to *Enterobacteriaceae* requiring a carbapenem.

If My Patient Does Develop an Infection Due to a Carbapenem-resistant Organism, What is the Optimal Treatment?

Treatment options for carbapenem-resistant Gram-negative bacilli are sparse and each has potential issues, highlighting the need to limit the development of these pathogens. This section will discuss these options, highlighting the strengths and limitations of each. The following section will then focus on the data investigating the role of combination therapy vs. monotherapy with these agents.

Polymyxins

The polymyxins are the mainstay of therapy for carbapenem-resistant Gram-negative bacilli and all of the target organisms—*P. aeruginosa*, *A. baumannii*, and carbapenem-resistant *Enterobacteriaceae* (CRE)—retain high degrees of susceptibility to the polymyxins. There are two commercially available polymyxins; polymyxin E, or colistin, formulated as its inactive prodrug colistin methanesulfonate (CMS), and polymyxin B. The polymyxins have a unique mechanism of action in which the highly positively charged molecules interact electrostatically with the negatively charged phosphate groups of the lipid A of lipopolysaccharide (Nation *et al.*, 2014). The electrostatic interaction leads to the displacement of the stabilizing magnesium and calcium, and subsequent disruption of the outer membrane. This starts a bactericidal process that ultimately leads to cell death. This interaction is particularly important because it serves as the basis for the excellent synergy that is seen between polymyxins and other agents, as the lack of an intact outer membrane allows other antimicrobial agents to enter the cell and achieve higher concentrations at their target sites of action.

The polymyxins have a unique and interesting history. Originally available from the 1950s to the 1960s, these agents fell out of favor owing to their high rates of nephrotoxicity and neurotoxicity (approaching 50%), and the subsequent availability of less toxic alternatives (notably the aminoglycosides, followed by extended-spectrum cephalosporins). They reemerged in the 1990s with the rise of MDR and XDR (extremely drug resistant) organisms resistant to the agents that had replaced the polymyxins earlier. Because of the 20–30 year period of relative non-use, the polymyxins reemerged in spite of having little PK/PD evidence to support their labeled doses. Optimal dosing remains controversial and unclear.

With the reemergence of the polymyxins, colistin arose as the preferred polymyxin, largely because of an inaccurate belief that it was less toxic. Significant advancements have been achieved in the understanding of colistin in the past 10 years, and many limitations of the agent have been identified. Many, but not all, of these limitations stem from the fact that colistin is administered in the form of its inactive prodrug CMS, as mentioned above, and requires hydrolysis to the active moiety, colistin. It has been determined that CMS hydrolysis to active colistin is a slow (it can take 7 h or longer to reach maximum concentrations of colistin), inefficient (only about 20% of CMS gets converted prior to elimination), and unpredictable (even for patients with the same dose and creatinine clearance, concentrations of colistin can vary ~tenfold amongst them) (Nation *et al.*, 2014). While the slow conversion can be of particular concern in critically ill patients, where time to appropriate therapy is an important driver of outcomes, the unpredictability of the concentrations attained *in vivo* is perhaps the most alarming aspect of the use of colistin, due to the extremely narrow (if even existent) therapeutic index of polymyxins. For these reasons, increasing attention has recently been paid to polymyxin B, which is administered as the active drug, even though very little is known about the safety and efficacy of this drug.

In addition to the aforementioned issues and concerns with safely reaching therapeutic targets, polymyxins exhibit a high degree of heteroresistance in all three target organisms, and significant concern exists, supported by in vitro data, that monotherapy with these agents can rapidly lead to the selection of polymyxin-resistant strains. Because the optimal dose remains unclear, and because of heteroresistance concerns, and the questionable ability of polymyxins to reach PK/PD targets safely, most experts recommend that these agents should not be used as monotherapy.

Tigecycline

Tigecycline, which is a derivative of minocycline, is a glycylcycline antibiotic that exerts bacteriostatic effects via the inhibition of protein synthesis through interactions with the 30S ribosomal subunit. It has broad-spectrum Gram-positive and Gram-negative activity, including excellent in vitro activity against *A. baumannii* and *Enterobacteriaceae*, including carbapenem-resistant strains. Of note, like other tetracyclines, it lacks activity against *P. aeruginosa*.

Enthusiasm surrounding tigecycline for the treatment of CRE and CRAB (carbapenem-resistant *A. baumannii*) has been tempered by early reports that showed mixed results as well as the pharmacokinetic limitations of the agent. Tigecycline has an extremely large volume of distribution (7 9 l/kg), which leads to high concentrations in certain sites, such as the hepatobiliary tract. However, it leads to low concentrations in the serum (Cmax in serum < 1.0 µg/ml), and subsequently to low concentrations in urine due to a lack of renal elimination (Stein and Babinchak, 2013). Breakthrough bloodstream infections while on tigecycline have been reported with *A. baumannii*, and tigecycline has proven inferior to other options for treating CRE urinary tract infections. Additionally, in spite of the high volume of distribution seen with tigecycline, the epithelial lining fluid penetration ratio of tigecycline is 1.15 (Rubino *et al.*, 2007), suggesting similar pharmacokinetic limitations in the lung as are seen in bloodstream infections. In fact, tigecycline failed to demonstrate non-inferiority to imipenem–cilistatin for the treatment of hospital-acquired pneumonia (Freire *et al.*, 2010), and levels achieved with the current dosing strategies are considered to be the predominant reason why this occurred. Furthermore, multiple reports have demonstrated the development of resistance to tigecycline monotherapy when used for the treatment of *A. baumannii* infections. When the risk of development of resistance on therapy is taken into account in conjunction with the fact that CRAB and CRE cause infections in the lungs, blood, and urine, the role of tigecycline monotherapy remains questionable.

Minocycline

The pharmacokinetic limitations of tigecycline have led to a reevaluation of minocycline as a potential therapeutic option. While in vitro activity against CRE is limited, individual isolates may remain susceptible. The larger interest in minocycline lies in CRAB, where susceptibilities often exceed 75% (Pogue *et al.*, 2014). The most notable pharmacokinetic advantage of minocycline comes in the form of higher blood concentrations with maximum concentrations following a high dose (400 mg/day) approaching 5 μg/ml. In addition, minocycline has excellent bioavailability and can be given orally while still achieving excellent systemic levels. Moreover, minocycline has indications for the treatment of infections due to *Acinetobacter* spp. as well as respiratory tract and urinary tract infections due to susceptible *Klebsiella* spp. Clinical experience for carbapenem-resistant Gram-negative infections has been limited and exists almost exclusively against CRAB. While these initial reports have been encouraging, minocycline has largely been used as part of a combination regimen, most commonly with colistin, and more data are necessary to elucidate the exact role of this agent.

Aminoglycosides

Aminoglycosides, notably gentamicin, tobramycin, and amikacin can have activity against *P. aeruginosa*, *A. baumannii*, and *Enterobacteriaceae*, including XDR and carbapenem-resistant strains. The role of aminoglycosides as definitive therapeutic options, particularly as a monotherapy, is poorly defined and controversial. Aminoglycosides inhibit protein synthesis by binding to the 30S ribosomal subunit, but their bactericidal activity against many pathogens is thought to be mediated by poorly described secondary mechanisms. Aminoglycosides have been used as monotherapy and as part of combination therapy for the treatment of carbapenem-resistant organisms, although their utility as a monotherapy is questionable. Meta-analyses have shown that there is insufficient evidence to support aminoglycosides as first-line therapy for infections outside the urinary tract (Vidal *et al.*, 2007), and significant rates of nephrotoxicity remain a clinical concern. The poor volume of distribution of these agents is often used as the rationale for their clinical inferiority. However, there are no well-controlled data looking at the efficacy of monotherapy with extended interval or "once daily" dosing of aminoglycosides, a strategy that is known to optimize the activity of these agents, while minimizing their toxicity. Due in part to all of these issues, the role of aminoglycosides in the treatment of carbapenem-resistant organisms remains unclear.

Sulbactam

Sulbactam is a commercially available β-lactamase inhibitor of many class A β-lactamases, most commonly formulated with ampicillin, but it also has antibacterial activity itself via the inhibition of penicillin-binding proteins of a few organisms, including *A. baumannii*. Importantly, sulbactam activity does not appear to be compromised by some of the more common class D oxacillinases (Penwell *et al.*, 2015), and can retain activity in carbapenem-resistant strains when other β-lactamases, to which sulbactam is labile, are not coproduced. Unfortunately, little is known about the optimal strategy for administration of sulbactam, as the current dosing for ampicillin/sulbactam is driven by the ampicillin component, which lacks activity against *A. baumannii*. While the available retrospective data generally show similar outcomes with sulbactam-based regimens and comparators (Chu *et al.*, 2013), the inconsistent combinations of antibiotics in sulbactam-based regimens and the significantly different (or not described) dosing regimens used make interpretation difficult.

Recent PK/PD data suggest that exposures to what would be considered high doses of sulbactam (6–8 g sulbactam a day) would not achieve bactericidal activity for organisms with MICs greater than 1–2 µg/ml (Yokoyama *et al.*, 2015), which are below the susceptibility breakpoint of 4 µg/ml. It is important to note that this susceptibility breakpoint is based on the ampicillin/sulbactam breakpoint (driven by ampicillin breakpoints), which is not relevant to sulbactam and *A. baumannii*. Therefore, the optimal dosing and usage of sulbactam remains unclear.

Combination Therapy in the Treatment of Carbapenem-resistant Gram-negative Bacilli

All therapies for the treatment of carbapenem-resistant Gram-negative bacilli have significant limitations, which underscores the importance of previously described strategies for decreasing the emergence of these organisms. Because of these limitations, many experts recommend combination therapy. However, the recommended combinations are controversial. Whether or not multiple in vitro active agents (if available) or highly synergistic combinations (such as a polymyxin + a carbapenem regardless of in vitro susceptibility to the carbapenem) should be preferred, is an area of intense debate. Unfortunately, but not surprisingly, the clinical literature on this issue is sparse, of poor methodology, and often retrospective in nature, thus making definitive conclusions difficult. In this section, we will look at the recent advances in the area of combination therapy for the treatment of carbapenem-resistant Gram-negative bacilli. As there are no data comparing combination therapy vs. monotherapy for carbapenem-resistant *P. aeruginosa*, these data will focus on CRAB and CRE.

Carbapenem-resistant *A. baumannii*

All analyses assessing combination therapy vs. monotherapy for CRAB infections compare colistin monotherapy and combination therapy with colistin plus either a synergistic or active agent. There are three small-randomized controlled trials in this area and these will be the focus of this discussion. Two of these analyses, one comparing colistin vs. colistin + rifampin (Aydemir *et al.*, 2013), and the other comparing colistin vs. colistin + fosfomycin (Sirijatuphat and Thamlikitkul, 2014), have very limited numbers, so preventing definitive conclusions to be drawn, and therefore will not be the focus here. The focus will be on the largest analysis, which was published by Durante-Mangoni *et al.* in 2013.

The Durante-Mangoni *et al.* (2013) study was a multicenter, randomized, open-label Phase III trial comparing colistin + rifampin with colistin monotherapy for the treatment of XDR *A. baumannii* infections (~80% of patients had pneumonia). Some 209 patients were included in the final analysis, made up of 104 and 105 in the combination and monotherapy arms, respectively. There was no difference in the primary outcome of 30 day mortality (43% in both arms), infection-related mortality (26.6% in the monotherapy vs. 21.1% in the combination therapy; $P = 0.29$), or length of stay between the two groups, but patients in the combination therapy had a higher rate of microbiological cure (60.6 vs. 44.8%; $P = 0.034$). Multivariate modeling confirmed a lack of an association between therapy and mortality.

While the above data suggest a lack of improved outcome with combination therapy, the open-label nature of the study introduces significant confounders. As the treating physicians knew whether the patient was on combination or monotherapy and they were allowed to add any agent other than rifampin, the possibility exists that monotherapy patients were more likely to be given concomitant off-study antibiotics that could bias the impact of combination therapy. This potential limitation is important, because 21% of the patients in the monotherapy arm also received off-study meropenem (16%) or tigecycline (5%). Additionally, the choice of rifampin as a second drug warrants mention, as a higher percentage of patients in the combination arm developed hepatotoxicity (21 vs. 12%; $P = 0.13$) with ten patients in the rifampin arm having to discontinue the agent as a result of hepatotoxicity. How these patients were managed is not clearly described in the paper and as a result the impact that this hepatotoxicity had on outcomes is unclear.

Interestingly, the two other small studies mentioned above (Aydemir *et al.*, 2013; Sirijatuphat and Thamlikitkul, 2014) also showed an association between combination therapy and improved microbiological clearance, and, importantly, the fosfomycin trial showed no association between combination therapy and hepatotoxicity (Sirijatuphat and Thamlikitkul, 2014).

Carbapenem-resistant *Enterobacteriaceae*

There are no randomized controlled data comparing combination therapy and monotherapy for CRE.

However, over the past few years, four retrospective cohort studies have analyzed the impact of combination therapy (defined as the use of multiple agents with in vitro activity) vs. monotherapy in patients with CRE (largely *K. pneumoniae* producing carbapenememase, KPC) bacteremia. The first two involve small numbers and do not describe dose and/or MICs, and therefore will not be described in further detail (Zarkotou *et al.*, 2011; Qureshi *et al.*, 2012).

Tumbarello *et al.* (2012) published the first large retrospective analysis looking at the impact of combination therapy (defined as multiple agents with in vitro activity) vs. monotherapy in patients with KPC bloodstream infections. The most common source of infection was primary bloodstream infection (60%), followed by lung (22%), urinary tract (14%), and central-line infections (10%). Importantly, in this analysis, doses were optimized and included colistin at 6–9 million international units (IU)/day, tigecycline at 100–200 mg/day, meropenem at 6 g/day (q8h and administered as a 3 h infusion), and high-dose, extended-interval gentamicin.

The combination regimens used by Tumbarello *et al.* (2012) consisted of two (n = 56) or three (n = 23) drug combinations of these agents. The most common two drug regimen was colistin + tigecycline (n = 23), and the most common three drug regimen was colistin + tigecycline + meropenem (n = 16). The 46 monotherapy regimens largely consisted of colistin (n = 22) or tigecycline (n = 19). Mortality was seen in 27/79 (34%) patients receiving combination regimens compared with 25/46 (53%) patients receiving monotherapy (P = 0.02). Interestingly, mortality was seen in 23/56 (41%) of patients receiving the two-drug combinations compared with 4/23 (17%) patients receiving triple therapy. In multivariate analysis, the triple combination of colistin + tigecycline + meropenem was highly protective against death (odds ratio (OR) 0.11, CI 0.02–0.69). As meropenem MICs can vary significantly in KPC isolates, the authors went on to examine the association between meropenem MIC and death in patients receiving meropenem as part of their combination regimen. Mortality was seen in 0/5 (0 %) patients with meropenem MICs ≤ 2 µg/ml, 2/10 (20%) patients with an MIC of 4 µg/ml, 1/4 (25%) patients with MICs of 8 µg/ml, and 6/17 (35%) patients with MICs ≥ 16 µg/ml, suggesting an association between in vitro susceptibility (and increasing MIC) and outcome.

Daikos *et al.* (2014) published the largest retrospective study studying combination vs. monotherapy for CRE bloodstream infections. The most common source of bloodstream infection was primary bacteremia (40%), followed by lung (21%), gastrointestinal (14%), central-line (10%), and urinary tract bacteremia (10%). Combination therapy was defined as two or more active drugs. Similar to the doses described above for the Tumbarello *et al.* (2012) study, doses were optimized for each agent: 9 million IU/day for colistin, 100–200 mg/day for tigecycline, high-dose extended-interval dosing for aminoglycosides, and 3 g (imipenem or meropenem) to 6 g (meropenem)/day divided every 8 h for the carbapenems.

The analysis in the Daikos *et al.* (2014) study consisted of 175 patients who received at least one active drug, 72 of whom received monotherapy, most commonly with tigecycline (n = 27) or colistin (n = 22), and 103 who received two or three drug combination regimens that were carbapenem containing (n = 31) or sparing (n = 72). A carbapenem was considered active in isolates with MIC values of up to 8 µg/ml, because in vitro pharmacodynamic targets are considered attainable with the dosing that was used in this study. Mortality was seen in 28/103 (27%) patients receiving combination therapy vs. 32/72 (44%) patients receiving monotherapy (P = 0.018). Monotherapy was associated with an increased risk of death in multivariate modeling (HR 2.08, CI 1.23–3.51). The mortality benefit with combination therapy was most pronounced in patients whose combination regimens contained a carbapenem (6/31, 19%) compared with those who did not (22/72, 31%). Additionally, patients treated with a carbapenem in the setting of an isolate with higher, and pharmacodynamically unachievable MICs (MIC >8 µg/ml) had similar mortality rates 11/31 (36%), as did those patients who received carbapenem-sparing regimens, once again suggesting an association between dosing, treatment with carbapenems, and mortality.

So, do we use combination regimens for carbapenem-resistant Gram-negative bacilli?

Overall, the early evidence comparing combination therapy and monotherapy for the treatment of carbapenem-resistant organisms has been encouraging. However, in general, the quality of the evidence is low. For carbapenem-resistant *A. baumannii*, despite the higher quality of evidence, the small sample sizes combined with the open-label nature of the trials limit the conclusions that can be drawn. There are three published trials, all looking at colistin-based regimens combined with a synergistic second drug (rifampin or fosfomycin) compared with monotherapy. None of these analyze the addition of a second active agent, such as tigecycline, minocycline, or an aminoglycoside. In each of these studies, the investigators failed to find a difference in all cause- or infection-related mortality, but did find a consistent association between combination therapy and microbiologic cure. Conversely, in patients with CRE (largely KPC) bloodstream infections, a consistent mortality benefit was reported in three retrospective analyses studying the impact of combination therapy with two or more *active* agents (Qureshi *et al.*, 2012; Tumbarello *et al.*, 2012; Daikos *et al.*, 2014). While the retrospective nature of these analyses limits the conclusions that can be drawn, they do suggest that patients with invasive CRE infections would be best served by receiving combination regimens.

These findings, when taken together, often lead to more questions than answers. For example, why the survival advantage with CRE, but not with CRAB? One potential answer relates to the nature of the infections studied in these analyses. The CRE patients had bacteremia, largely from either a central line or non-identifiable source (primary bacteremia). These patients in general are the ones in whom clinicians feel confident that, if appropriately treated with optimal doses of active antibiotics, outcomes can be improved. In contrast, patients in the CRAB trials often had pneumonia, which is consistent with what is seen in clinical practice. These patients are often much more complex, and the exact role of *A. baumannii* in their ultimate clinical outcomes is unclear (e.g., does the patient have infection vs. colonization, and do they often die or have a prolonged stay in an intensive care unit (ICU) owing to multiple competing risks). When considering the findings from this perspective, the consistent improvement in microbiological cure seems even more interesting, as this might be the only outcome that can be reliably measured in terms of the impact of antimicrobial therapy. The selection of rifampin as the combination agent in the largest analysis also complicates matters even more, as this is a pharmacologically problematic agent, with many associated drug interactions and toxicity issues. Furthermore, the open-label nature of the study undoubtedly led clinicians to add other potential synergistic or active agents, which complicates the interpretation of the study results.

Other questions raised by these data include the following: If I choose to treat with a combination regimen, which should I choose? Do I choose multiple active agents or synergistic ones? Does my regimen need a polymyxin backbone for mechanistic synergy? Do I need two drugs or three drugs? If I choose a polymyxin, is polymyxin B superior to colistin? If my patient has a bloodstream infection, would minocycline offer an advantage over tigecycline? Do I need to include a carbapenem in my regimen? What if the MIC to carbapenems is > 8 μg/ml? Should I consider CRE and CRAB the same way when I am choosing a treatment regimen? Unfortunately, definitive answers do not yet exist to these questions.

As is evident from the preceding paragraph, there are many unknowns related to the optimal regimen for these patients, and well-controlled randomized trials are needed to provide answers to these questions, as exposure to other antimicrobial agents, notably carbapenems, without clear benefit is not without the significant ecological risks that have been described in detail throughout this chapter. In the interim, the best available evidence supports combination regimens with a polymyxin backbone, preferably combined with a second agent with in vitro activity if possible, or a highly synergistic one. Additionally, carbapenems appear to be an important component of the optimal regimen, although their exact role remains undefined. Finally, in a critically ill patient with CRE bacteremia, triple therapy with a polymyxin + dose-optimized meropenem + tigecycline is reasonable based on the available published data.

Summary and Conclusions

Infections due to problematic Gram-negative bacilli are increasing in incidence, are becoming more resistant, and are associated with high mortality rates. Stewardship programs should have three main targets for these organisms. First and foremost, infections due to these organisms, owing to their extensive resistance profiles, are associated with significant delays in time to initiation of appropriate antibiotics, a known driver of poor outcomes. Therefore, stewardship programs should have strategies in place to rapidly identify patients with these infections, as quickly as possible, so that the appropriate therapy can be initiated. Rapid identifications of resistant pathogens might incorporate strategies such as rapid diagnostic techniques, combination antibiograms in problematic ICUs, and being aware of recent antimicrobial exposures and any history of drug-resistant pathogens in an individual patient. A second strategy that stewardship programs can use is avoidance of the overuse of antimicrobials. As resistant Gram-negative infections are associated with significantly worse outcomes when compared with their susceptible counterparts, limiting the use of broad-spectrum agents, where possible, and optimizing (i.e., shortening) durations of therapy become important in decreasing the spread and development of resistant organisms. Finally, for patients who develop infections caused by these organisms, stewardship programs should be aggressive in placing patients on the most optimal antimicrobial regimen.

As the data presented in this chapter clearly show, there is no "one size fits all" regimen for the optimal management of patients with multidrug resistant Gram-negative infections. Patient-specific and organism-specific factors, such as site of infection, type of infecting organism, MICs, in vitro susceptibility of the study pathogen to different antimicrobials, clinical status of the patient, and clinical response should all be taken in to account when selecting the optimal antimicrobial regimen for a given patient with an infection due to a resistant Gram-negative pathogen.

References

Aydemir, H., Akduman, D., Piskin, N., Comert, F., Horuz, E., Terzi, A., Kokturk, F., Ornek, T., and Celebi, G. (2013) Colistin vs. the combination of colistin and rifampicin for the treatment of carbapenem-resistant *Acinetobacter baumannii* ventilator-associated pneumonia. *Epidemiology and Infection* 141, 1214–1222.

CDC (2013) *Antibiotic Resistance Threats in the United States, 2013*. Centers for Disease Control and Prevention, Atlanta, Georgia.

Chu, H., Zhao, L., Wang, M., Liu, Y., Gui, T., and Zhang, J. (2013) Sulbactam based therapy for *Acinetobacter baumannii* infection: a systematic review and meta-analysis. *Brazilian Journal of Infectious Diseases* 17, 389–394.

Collins, V.L., Marchaim, D., Pogue, J.M., Moshos, J., Bheemreddy, S., Sunkara, B., Shallal, A., Chugh, N., Eiseler, S., Bhargava, P. *et al.* (2012) Efficacy of ertapenem for treatment of bloodstream infections caused by extended-spectrum-β-lactamase-producing *Enterobacteriaceae*. *Antimicrobial Agents and Chemotherapy* 56, 2173–2177.

Daikos, G.L., Tsaousi, S., Tzouvelekis, L.S., Anyfantis, I., Psichogiou, M., Argyropoulou, A., Stefanou, I., Sypsa, V., Miriagou, V., Nepka, M. *et al.* (2014) Carbapenemase-producing *Klebsiella pneumoniae* bloodstream infections: lowering mortality by antibiotic combination schemes and the role of the carbapenems. *Antimicrobial Agents and Chemotherapy* 5, 32–38.

Del Arco, A., Tortajada, B., de la Torre, J., Olalla, J., Prada, J.L., Fernández, F., Rivas, F., García-Alegría, J., Faus, V., and Montiel, N. (2014) The impact of an antimicrobial stewardship programme on the use of antimicrobials and the evolution of drug resistance. *European Journal of Clinical Microbiology* 34, 247–251.

Durante-Mangoni, E., Signoriello, G., Andini, R., Mattei, A., De Cristoforo, M., Murino, P., Bassetti, M., Malacarne, P., Petrosillo, N., Galdieri, N. *et al.* (2013) Colistin and rifampicin compared with colistin alone for the treatment of serious infections due to extensively drug-resistant *Acinetobacter baumannii*: a multicenter, randomized clinical trial. *Clinical Infectious Diseases* 57, 349–358.

Freire, A.T., Melnyk, V., Kim, M.J., Datsenko, O., Dzyublik, O., Glumcher, F., Chuang, Y.-C., Maroko, R.T., Dukart, G., Cooper, C.A. *et al.* (2010) Comparison of tigecycline with imipenem/cilistatin for the treatment of hospital acquired pneumonia. *Diagnostic Microbiology and Infectious Disease* 68, 140–151.

Goldstein, E.J., Citron, D.M., Peraino, V., Elgourt, T., Meibohm, A.R., and Lu, S. (2009) Introduction of ertapenem into a hospital formulary: effect on antimicrobial usage and improved in vitro susceptibility of *Pseudomonas aeruginosa*. *Antimicrobial Agents and Chemotherapy* 53, 5122–5126.

Hilty, M., Sendi, P., Seiffert, S.N., Droz, S., Perreten, V., Hujer, A.M., Bonomo, R.A., Mühlemann, K., and Endimiani, A. (2013) Characterisation and clinical features of *Enterobacter cloacae* bloodstream infections occurring at a tertiary care university hospital in Switzerland: is cefepime adequate therapy? *International Journal of Antimicrobial Agents* 41, 236–249.

IDSA (2004) *Bad Bugs, No Drugs: As Antibiotic Discovery Stagnates … A Public Health Crisis Brews*. Infectious Diseases Society of America, Alexandria, Virginia. Available at: http://www.idsociety.org/uploadedFiles/IDSA/Policy_and_Advocacy/Current_Topics_and_Issues/Advancing_Product_Research_and_Development/Bad_Bugs_No_Drugs/Statements/As%20Antibiotic%20Discovery%20Stagnates%20A%20Public%20Health%20Crisis%20Brews.pdf (accessed 29 April 2016).

Kaye, K.S., Cosgrove, S., Harris, A., Eliopoulos, G.M., and Carmeli, Y. (2001) Risk factors for emergence of resistance to broad-spectrum cephalosporins among *Enterobacter* spp. *Antimicrobial Agents and Chemotherapy* 45, 2628–2630.

Knudsen, J.D., Andersen, S.E., and Bispebjerg Intervention Group (2014) A multidisciplinary intervention to reduce infections of ESBL- and AmpC-producing, Gram-negative bacteria at a University Hospital. *PLoS ONE* 9(1): e86457.

Lee, N.Y., Lee, C.C., Huang, W.H., Tsui, K.C., Hsueh, P.R., and Ko, W.C. (2012) Carbapenem therapy for bacteremia due to extended-spectrum-β-lactamase-producing *Escherichia coli* or *Klebsiella pneumoniae*: implications of ertapenem susceptibility. *Antimicrobial Agents and Chemotherapy* 56, 2888–2893.

Lee, N.Y., Lee, C.C., Huang, W.H., Tsui, K.C., Hsueh, P.R., and Ko, W.C. (2013) Cefepime therapy for monomicrobial bacteremia caused by cefepime-susceptible extended-spectrum beta-lactamase-producing *Enterobacteriaceae*: MIC matters. *Clinical Infectious Diseases* 56, 88–95.

Lewis, G.J., Fang, X., Gooch, M., and Cook, P.P. (2012) Decreased resistance of *Pseudomonas aeruginosa* with restriction of ciprofloxacin in a large teaching hospital's intensive care and intermediate care units. *Infection Control and Hospital Epidemiology* 33, 368–373.

Nation, R.L., Velkov, T., and Li, J. (2014) Colistin and polymyxin B: peas in a pod, or chalk and cheese? *Clinical Infectious Diseases* 9, 88–94.

Nicolau, D.P., Carmeli, Y., Crank, C.W., Goff, D.A., Graber, C.J., Lima, A.L., and Goldstein, E.J. (2012) Carbapenem stewardship: does ertapenem affect *Pseudomonas* susceptibility to other carbapenems? A review of the evidence. *International Journal of Antimicrobial Agents* 39, 11–15.

Penwell, W.F., Shapiro, A.B., Giacobbe, R.A., Gu, R.F., Gao, N., Thresher, J., McLaughlin, R.E., Huband, M.D., DeJonge, B.L., Ehmann, D.E., and Miller, A.A. (2015) Molecular mechanisms of sulbactam antibacterial activity and resistance determinants in *Acinetobacter baumannii*. *Antimicrobial Agents and Chemotherapy* 59, 1680–1689.

Pogue, J.M., Neelakanta, A., Mynatt, R.P., Sharma, S., Lephart, P., and Kaye, K.S. (2014) Carbapenem-resistance in Gram-negative bacilli and intravenous minocycline: an antimicrobial stewardship approach at the Detroit Medical Center. *Clinical Infectious Diseases* 9(Suppl 6), S388–S393.

Qureshi, Z.A., Paterson, D.L., Potoski, B.A., Kilayko, M.C., Sandovsky, G., Sordillo, E., Polsky, B., Adams-Haduch, J.M., and Doi, Y. (2012) Treatment outcome of bacteremia due to KPC-producing *Klebsiella pneumoniae*: superiority of combination antimicrobial regimens. *Antimicrobial Agents and Chemotherapy* 5, 10–13.

Retamar, P., López-Cerero, L., Muniain, M.A, Pascual, Á., Rodríguez-Baño, J., and ESBL-REIPI/GEIH Group (2013) Impact of the MIC of piperacillin-tazobactam on the outcome of patients with bacteremia due to extended-spectrum-β-lactamase-producing *Escherichia coli*. *Antimicrobial Agents and Chemotherapy* 57(7), 3402–3404.

Rodríguez-Baño, J., Navarro, M.D., Retamar, P., Picó1, E., Pascual, Á., and Extended-Spectrum Beta-Lactamases–Red Española de Investigación en Patología Infecciosa/Grupo de Estudio de Infección Hospitalaria (2012) β-Lactam/β-lactam inhibitor combinations for the treatment of bacteremia due to extended-spectrum β-lactamase-producing *Escherichia coli*: a post hoc analysis of prospective cohorts. *Clinical Infectious Diseases* 54, 167–174.

Rubino, C.M., Ma, L., Bhavnani, S.M., Korth-Bradley, J., Speth, J., Ellis-Grosse, E., Rodvold, K.R., Ambrose, P.G., and Drusano, G.L. (2007) Evaluation of tigecycline penetration into colon wall tissue and epithelial lining fluid using a population pharmacokinetic model and Monte Carlo simulation. *Antimicrobial Agents and Chemotherapy* 51(11), 4085–4089.

Siedner, M.J., Galar, A., Guzmán-Suarez, B.B., Kubiak, D.W., Baghdady, N., Ferraro, M.J., Hooper, D.C., O'Brien, T.F., and Marty, F.M. (2014) Cefepime vs other antibacterial agents for the treatment of *Enterobacter* species bacteremia. *Clinical Infectious Diseases* 58, 1554–1563.

Sirijatuphat, R. and Thamlikitkul, V. (2014) Preliminary study of colistin versus colistin plus fosfomycin for treatment of carbapenem-resistant *Acinetobacter baumannii* infections. *Antimicrobial Agents and Chemotherapy* 58, 5598–5601.

Stein, G.E. and Babinchak, T. (2013) Tigecycline: an update. *Diagnostic Microbiology and Infectious Disease* 75, 331–336.

Tamma, P.D., Girdwood, S.C.T., Gopaul, R., Tekle, T., Roberts, A.A., Harris, A.D., Cosgrove, S.E., and Carroll, K.C. (2013) The use of cefepime for treating Amp C-β-lactamase producing *Enterobacteriaceae*. *Clinical Infectious Diseases* 57, 781–788.

Tamma, P.D., Han, J.H., Rock, C., Harris, A.D., Lautenbach, E., Hsu, A.J., Avdic, E., Cosgrove, S.E., and Antibacterial Resistance Leadership Group (2015) Carbapenem therapy is associated with improved survival compared with piperacillin–tazobactam for patients with extended-spectrum β-lactamase bacteremia. *Clinical Infectious Diseases* 60, 319–25.

Tan, M.W., Lye, D.C., Ng, T.M., Nikolaou, M., and Tam, V.H. (2014) Mathematical model to quantify the effects of risk factors on carbapenem-resistant *Acinetobacter baumannii*. *Antimicrobial Agents and Chemotherapy* 58, 5239–5244.

Tumbarello, M., Viale, P., Viscoli, C., Trecarichi, E.M., Tumietto, F., Marchese, A., Spanu, T., Ambretti, S., Ginocchio, F., Cristini, F. *et al.* (2012) Predictors of mortality in blood stream infections caused by KPC producing *Klebsiella pneumoniae*: importance of combination therapy. *Clinical Infectious Diseases* 5, 4–50.

Vardakas, K.Z., Tansarli, G.S., Rafailidas, P.I., and Falagas, M.E. (2012) Carbapenems versus alternative antibiotics for the treatment of bacteraemia due to *Enterobacteriaceae* producing extended-spectrum β-lactamases: a systematic review and meta-analysis. *Journal of Antimicrobial Chemotherapy* 67, 2793–2803.

Vidal, L., Gafter-Gvili, A., Borok, S., Fraser, A., Leibovici, L., and Paul, M. (2007) Efficacy and safety of aminoglycoside monotherapy: systematic review and meta analysis of randomized controlled trials. *Journal of Antimicrobial Chemotherapy* 6, 4–57.

Yokoyama, Y., Matsumoto, K., Ikawa, K., Watanabe, E., Morikawa, N., and Takeda, Y. (2015) Population pharmacokinetic-pharmacodynamic target attainment analysis of sulbactam in patients with impaired renal function: dosing considerations for *Acinetobacter baumannii* infections. *Journal of Infection and Chemotherapy* 21, 284–289.

Yoon, Y.K., Yang, K.S., Lee, S.E., Kim, H.J., Sohn, J.W., and Kim, M.J. (2014) Effects of Group 1 versus Group 2 carbapenems on the susceptibility of *Acinetobacter baumannii* to carbapenems: a before and after intervention study of carbapenem-use stewardship. *PLoS ONE* 9(6): e99101.

Zarkotou, O., Pournaras, S., Tselioti, P., Dragoumanos, V., Pitiriga, V., Ranellou, K., Prekates, A., Themeli-Digalaki, K., and Tsakris, A. (2011) Predictors of mortality in patients with bloodstream infections caused by KPC-producing *Klebsiella pneumoniae* and impact of appropriate antimicrobial treatment. *Clinical Microbiology and Infection* 1, 79–803.

19 Optimal Use of Fluoroquinolones

Thomas J. Dilworth,[1] Ramy H. Elshaboury,[2] and John C. Rotschafer[3]*

[1]*Aurora St. Luke's Medical Center, Milwaukee, Wisconsin, US;* [2]*Massachusetts General Hospital, Boston, Massachusetts, US;* [3]*University of Minnesota College of Pharmacy, Minneapolis, Minnesota, US*

Introduction

In 2011, oral and injectable fluoroquinolones were prescribed to 23 million and 3.8 million patients, respectively, in the US (FDA, 2013). These compounds offer a broad antimicrobial spectrum with excellent efficacy and a fairly modest adverse drug reaction profile (Wright *et al.*, 2000). Responsible antibiotic utilization is a challenge for all antibiotics, but especially for the fluoroquinolones, as they are given orally or parenterally, and are used to treat a broad range of routine clinic-based to life-threatening hospital-based infections (Rotschafer *et al.*, 2011). The US Food and Drug Administration (FDA)-approved indications for fluoroquinolones include skin/soft tissue, urinary tract, and upper respiratory tract infections, three areas that are prone to antibiotic overuse and misuse.

Over the years, the pool of prescribers has also increased with the addition of nurse practitioners and physician assistants to the customary physician prescriber. Fluoroquinolones are particularly attractive to prescribers owing to their perceived safety profile, antimicrobial spectrum, and broad range of clinical indications, as well as the ability to use oral or injectable administration, or to switch administration routes. In many Third World countries, fluoroquinolones can be obtained over the counter without a prescription, and usage is totally unsupervised.

The overuse of antibiotics, including the fluoroquinolones, has permeated our entire society and the problem is worldwide. These agents are used as growth promoters in massive swine and poultry feedlots (Singer *et al.*, 2003; Wegener, 2003). Veterinarians also enjoy access to a wide spectrum of antibiotics, including fluoroquinolones, which, in some cases, can be bought over the counter or prescribed at their office or clinic. As a result of these agricultural practices, the local environment and the food chain are constantly exposed to low-level residual concentrations of antibiotics and the resultant resistant organisms that are generated.

Healthcare professionals have been regularly advised as to the consequences associated with the overuse and misuse of antibiotics, including the fluoroquinolones (Hecker *et al.*, 2003; Costelloe *et al.*, 2010; CDC, 2013; Fridkin *et al.*, 2014). The appropriate duration of use and dosing of antibiotics are major tenets of antimicrobial stewardship. Fluoroquinolones are perhaps the only class of antibiotics for which the pharmaceutical manufacturer has actually sought FDA indications to reduce the duration of therapy for community-acquired pneumonia and uncomplicated urinary tract infections. While short-course therapy can be used in specific situations, durations of 7 to 14 days remain common with other infections. Our ability to prospectively and reliably identify patients who can have their antibiotics stopped early, and those who will not do well unless the course of antibiotics is extended, is an ultimate goal in responsible antibiotic therapy.

Herein we describe the many challenges associated with the stewardship of fluoroquinolones and offer some suggestion as to how to optimally approach this task.

Fluoroquinolone Pharmacokinetics

Fluoroquinolones are generally well absorbed when given orally, and exhibit excellent oral bioavailability,

*Corresponding author. E-mail: rotsc001@umn.edu

approaching 99% with levofloxacin (Lubasch *et al.*, 2000; Stass *et al.*, 2001). Time to peak serum concentrations (Tmax) following oral administration is reached within 60–90 min in healthy volunteers. At comparable oral doses, levofloxacin and moxifloxacin produce higher peak serum concentrations (Cmax) than ciprofloxacin (3–6 mg/l vs. 2–2.5 mg/l, respectively) (Lubasch *et al.*, 2000; Stass *et al.*, 2001). The oral absorption of fluoroquinolones is negatively affected when they are coadministered with divalent cations and iron as a result of the significant chelation effects that occur (Brighty and Gootz, 2000; Stass and Kubitza, 2001). Once absorbed, fluoroquinolones exhibit significant tissue penetration owing to their relatively large volumes of distribution (ranging 1–5 l/kg) despite their moderate plasma protein binding (Brighty and Gootz, 2000; Stass *et al.*, 2001). Consequently, orally and intravenously administered fluoroquinolones achieve therapeutic concentrations in various tissues, including the central nervous system (CNS), lungs, prostate, and skin/soft tissue, and these often exceed the simultaneous serum concentrations. For example, the ratios of ciprofloxacin simultaneous tissue-to-serum concentrations can reach 0.3, 1.6, 2.1, 1.2, 13.3 and 30 in the CNS, in bronchial, lung and blister fluids, and in the kidneys and bile, respectively (Vance-Bryan *et al.*, 1990; Aminimanizani *et al.*, 2001; Nau *et al.*, 2010). The value of the 24 h free drug area under the serum concentration time curve ($fAUC_{24}$) generated by common oral and intravenous doses of fluoroquinolones ranges from 20 to 70 mg*h/l according to the agent used, with levofloxacin generating higher values approaching 70 when higher doses (750 mg/day) are used (Brighty and Gootz, 2000; Lubasch *et al.*, 2000; Wright *et al.*, 2000; Stass *et al.*, 2001; Wispelwey, 2005).

Ciprofloxacin and levofloxacin are largely eliminated via the kidneys, while moxifloxacin is metabolized via phase-II sulfate and glucuronide-conjugation to inactive metabolites, without appreciable renal elimination (Lubasch *et al.*, 2000; Barth *et al.*, 2008). While dosing adjustments for ciprofloxacin and levofloxacin are necessary for patients with renal impairment, no moxifloxacin dosing adjustments are recommended for those with significant hepatic dysfunction.

Fluoroquinolone Pharmacodynamics

Remarkable agreement between in vitro models, animal models, and clinical data all pointed to the concentration-dependent killing properties of fluoroquinolones against Gram-negative organisms. Initially, values of the ratio Cmax/MIC (minimum inhibitory concentration) were shown to correlate with outcomes using in vitro and animal Gram-negative infection models (Blaser *et al.*, 1987; Drusano *et al.*, 1993). Later, the focus shifted to AUC/MIC ratios as the focal predictor of microbiologic outcomes when studying ciprofloxacin and oflaxacin against strains of *Pseudomonas aeruginosa*, with AUC/MIC values of ≥100 emerging as the proposed breakpoints (Madaras-Kelly *et al.*, 1996; Odenholt and Cars, 2006). Furthermore, the ability of fluoroquinolones to suppress the emergence of resistant strains was shown to be concentration dependent (Louie *et al.*, 2009). The results of clinical trials described similar concentration-dependent killing against Gram-negative infections in human subjects: f-AUC_{24}/MIC breakpoints of above 100–125 correlated with successful outcomes, while values below 100 correlated with increased likelihood of mortality and the emergence of fluoroquinolone resistance (Forrest *et al.*, 1993; Preston *et al.*, 1998; Thomas *et al.*, 1998; Zelenitsky and Ariano, 2010). Over the years, proposed AUC/MIC breakpoints against Gram-negative organisms withstood the heterogeneity amongst studies in the models used, organism(s) tested, inoculum size, susceptibility testing, fluoroquinolone study agent, observed duration of effect, and growth media (Wright *et al.*, 2000).

Resistance to Fluoroquinolones

Gram-negative resistance

One of the most pressing reasons for implementing antimicrobial stewardship interventions for fluoroquinolones is the rising rate of resistance among Gram-negative pathogens. Increasing resistance rates among *Escherichia coli* have resulted in an Infectious Diseases Society of America (IDSA) recommendation to use an intravenous third-generation cephalosporin or an aminoglycoside in combination with a fluoroquinolone for pyelonephritis when local fluoroquinolone resistance is greater than 10% (Gupta *et al.*, 2011). The use of local susceptibility data is crucial to making effective antimicrobial stewardship decisions, but these data are often presented in aggregate for a hospital or a (regional) health(care) system. While informative on a broad level, data should be examined more closely in order to better inform treatment decisions. For example, patient-level factors may drive fluoroquinolone susceptibility patterns among Gram-negative pathogens within an antibiogram. Many facilities commonly develop antibiograms that are hospital unit specific, for example, they may apply to the intensive care unit (ICU). Yet, this level of granularity still does not capture the full influence of patient-level factors on fluoroquinolone resistance. The influence of infection site on antibiotic susceptibility among *E. coli* and *Klebsiella pneumoniae* isolates was examined over an 18-year period. Soriano *et al.* (2014) reported that decreases in antibiotic susceptibility were greater among non-urinary than urinary isolates in cases of urinary tract infections (UTIs), and make the case for creating source-specific antibiograms rather than relying solely on a composite antibiogram.

Prior fluoroquinolone use was a risk factor for multidrug resistant (MDR) *Enterobacteriaceae* UTIs in one study in an emergency department (Khawcharoenporn *et al.*, 2013). Of patients in the study with MDR UTIs, 73% were infected with a levofloxacin-resistant organism. In another study in a tertiary care hospital, an extended-spectrum β-lactamase (ESBL-) producing phenotype among nosocomially acquired *K. pneumoniae* isolates was linked with fluoroquinolone resistance in a retrospective case-control study (Wener *et al.*, 2010). In that study, 95% of isolates with an ESBL-producing phenotype were resistant to fluoroquinolones compared with only 18% of non ESBL-producing isolates (Wener *et al.*, 2010). Fluoroquinolone exposure was also significantly associated with ESBL-producing *K. pneumoniae* spp. in a propensity score-adjusted multivariable model (Wener *et al.*, 2010).

Kriengkauykiat *et al.* (2005) found that 80% of fluoroquinolone-resistant *P. aeruginosa* isolates were clonally distinct and that 72% of the fluoroquinolone-resistant strains had a phenotype in which the efflux pump was overexpressed. Antibiotic resistance in *P. aeruginosa* is mediated by gyrase and/or topoisomerase mutations as well as efflux pumps, and fluoroquinolones are substrates for all characterized *P. aeruginosa* efflux pumps (Hancock, 1998; Poole, 2000). As such, fluoroquinolone resistance in *P. aeruginosa* can lead to the overexpression of efflux pumps, which can confer resistance to other antibiotics. Empiric fluoroquinolone use among patients with fluoroquinolone-resistant *P. aeruginosa* was associated with poor clinical outcomes (Hsu *et al.*, 2005). While this result is intuitive, it is important for clinicians to analyze local *P. aeruginosa* susceptibility to fluoroquinolones in order to guide empiric antibiotic therapy for patients potentially infected with *P. aeruginosa*. Given the far-reaching impact that fluoroquinolone use can have on Gram-negative resistance within an institution, it seems reasonable to tailor antimicrobial stewardship efforts toward this class of antibiotics.

Changing institutional empiric antibiotic recommendations is one successful strategy that can be implemented (Nguyen *et al.*, 2008). The removal of fluoroquinolone therapy from local empiric antibiotic recommendations led to a 30% decrease in empiric fluoroquinolone use for *P. aeruginosa* infections and a 10% increase in *P. aeruginosa* susceptibility rates among ciprofloxacin and other anti-pseudomonal antibiotics at one institution (Nguyen *et al.*, 2008). A similar increase in *P. aeruginosa* susceptibility was observed following ciprofloxacin restriction in an ICU (Medina Presentado *et al.*, 2011).

Appropriate combination therapy with fluoroquinolones

An area in which fluoroquinolones are potentially overused is their routine use in combination regimens for empiric and, to a much lesser extent, definitive treatment of serious infections due to Gram-negative organisms. In daily practice, clinicians are faced with the realities of MDR Gram-negative pathogens. Consequently, combination antibiotic therapy is often used to ensure that there is at least one active agent against the presumed pathogen(s), especially in patients who are critically ill. Clinical data for combination therapies remain largely conflicting and without clear evidence for improved outcomes when compared with the appropriate empiric monotherapy (Boyd and Nailor, 2011; Tamma *et al.*, 2012). While most favorable clinical data supporting this practice have used aminoglycosides in combination with a β-lactam, fluoroquinolones in combination with β-lactams yielded comparable outcomes in a meta-analysis of eight randomized clinical trials of patients with febrile neutropenia (Bliziotis *et al.*, 2005). Fundamental assumptions to justify combination therapy include broadening the empiric Gram-negative spectrum, inducing potential synergy between different antibiotics, and reducing the risk of developing resistance during treatment (Boyd and Nailor 2011; Johnson *et al.*, 2011; Tamma *et al.*, 2012).

Combination therapy antibiograms have become valuable tools in determining optimal combination regimens based on local susceptibility patterns. The current literature shows combination regimens that include aminoglycosides and broad-spectrum anti-pseudomonal β-lactams provide better empiric coverage when compared with fluoroquinolones in combination with the same anti-pseudomonal β-lactams (Beardsley *et al.*, 2006; Mizuta *et al.* 2006; Christoff *et al.*, 2010; Johnson *et al.*, 2011). In one study of Gram-negative organisms resistant to piperacillin–tazobactam or cefepime, ciprofloxacin was active against <10% of isolates, while amikacin was active against >80% (Beardsley *et al.*, 2006). Additionally, data from clinical trials have not substantiated synergistic effects when fluoroquinolones are combined with a β-lactam or a β-lactam/β-lactamase inhibitor against *P. aeruginosa*, despite promising in vitro data (Boyd and Nailor, 2011; Tamma *et al.*, 2012). A small in vitro study of *P. aeruginosa* strains treated with levofloxacin and imipenem showed that combination therapy delayed the emergence of resistance (Tamma *et al.*, 2012). However, clinical data have to date shown no clear evidence that combination therapy prevents, or delays, the emergence of Gram-negative resistance while on treatment (Boyd and Nailor, 2011; Tamma *et al.* 2012). Nevertheless, combination regimens continue to be a mainstay of empiric treatments for critically ill patients, as is the case in patients with hospital-acquired and ventilator-associated lower respiratory tract infections, until final microbiologic data become available (ATS and IDSA, 2005).

With local susceptibility data in mind, antimicrobial stewardship providers are primed to develop appropriate combination antibiograms to evaluate the utility of fluoroquinolones in empiric combination therapy. Providers may be able to advocate against the routine use of fluoroquinolones if local susceptibility data show that they are inferior to aminoglycosides as a means of broadening empiric Gram-negative coverage. This can be achieved through steady stewardship efforts that may include establishing appropriate criteria for use, prospective audits, and feedback, and routine reporting of fluoroquinolone usage. Patient-specific risk factors for MDR Gram-negative organisms remain of great importance, and individual assessments should be made prior to starting combination regimens. Until further conclusive clinical data become available, antimicrobial stewardship initiatives will be necessary to ensure the appropriate use of combination regimens in critically ill patients.

Preserving fluoroquinolone activity against *Streptococcus pneumoniae*

Preserving fluoroquinolone activity against *S. pneumoniae* is also of utmost importance given the rising rate of its resistance to β-lactams. Invasive pneumococcal disease is primarily community acquired, so efforts to preserve fluoroquinolone efficacy against *S. pneumoniae* should be instituted in both the inpatient and outpatient settings. Hicks *et al.* (2011) found an association between outpatient antibiotic prescribing and drug-resistant invasive pneumococcal disease. Given their high bioavailability, fluoroquinolones are uniquely suited for outpatient therapy for respiratory and urinary tract infections, although it should be noted that unnecessary fluoroquinolone use in the outpatient setting is common (Shapiro *et al.*, 2014). Also, it has been shown that approximately 30% of *S. pneumoniae* isolates possess first-step fluoroquinolone mutations, most commonly mutations in the *gyrA* or *parC* genes (Hooper, 2002; Dalhoff, 2012), which encode the A units of DNA gyrase (topoisomerase II) and topoisomerase IV, respectively. Unfortunately, first-step mutations are not detected by traditional susceptibility testing, and pathogens with a first-step mutation are likely to develop additional mutations conferring fluoroquinolone-resistance if exposed to subtherapeutic dosing (Dalhoff, 2012). Increased fluoroquinolone use in Canada led to increased rates of fluoroquinolone resistance in *S. pneumoniae* (Adam *et al.*, 2009). Moreover, a recent study found an inverse relationship between the time of a patient's last antibiotic exposure and the risk of invasive, drug-resistant pneumococcal disease (Kuster *et al.*, 2014). Reports of fluoroquinolone treatment failure among patients with pneumococcal pneumonia have also been reported (Empey *et al.*, 2001; Urban *et al.*, 2001; Kays *et al.*, 2002; Davidson *et al.*, 2002; de Cueto *et al.*, 2008).

Side Effects of Fluoroquinolone Overuse

Fluoroquinolones and *Clostridium difficile*

C. difficile is arguably one of the worst human pathogens. The US Centers for Disease Control and Prevention (CDC) has listed it as a top priority when it comes to antimicrobial resistance and the necessary antibiotic development (CDC, 2013). *C. difficile*-associated diarrhea (CDAD) is responsible for 250,000 infections a year and 14,000 deaths (CDC, 2013). Many studies have investigated links between antibiotic use and CDAD, and earlier studies found an association between fluoroquinolone use and CDAD. However, more recent literature suggests that overall antibiotic exposure and not specific antibiotic exposure(s) is the leading cause of CDAD. Additionally, most studies were performed at a single center, which may limit their external validity.

Fluoroquinolone use, specifically levofloxacin use, was associated with CDAD in a case-control study (odds ratio (OR) 12.7, 95% confidence interval (CI) 2.6–61.6) (McCusker *et al.*, 2003). While the CI surrounding this association was large, similar findings have been observed in other studies (Muto *et al.*, 2005; Biller *et al.*, 2007; Labbé *et al.*, 2008). Fluoroquinolone use was also significantly associated with CDAD in a retrospective analysis of cases during an 18 month period at a Canadian hospital affected by the North American pulsed-field gel electrophoresis type 1 hypervirulent strain (NAP1) of *C. difficile* (Pepin *et al.*, 2005). Notably, cephalosporins, clindamycin, intravenous β-lactam/β-lactamase inhibitors and macrolides were also associated with CDAD (Pépin *et al.*, 2005). In a meta-analysis of studies examining the impact of specific antibiotic use, fluoroquinolones were less likely to be associated with CDAD than cephalosporins, clindamycin, carbapenems, and trimethoprim–sulfamethoxazole (Slimings and Riley, 2014). Also, a review by Weiss (2009) found that the link between fluoroquinolone exposure and CDAD was not conclusive. Furthermore, community-acquired CDAD is increasingly common, and the overuse of antibiotics in outpatient settings is a likely contributor (Depestel and Aronoff, 2013; Deshpande *et al.*, 2013; Marwick *et al.*, 2013; Gupta and Khanna, 2014); fluoroquinolone use was strongly associated with community-acquired CDAD in two of these studies (Deshpande *et al.*, 2013; Marwick *et al.*, 2013). Given the overuse of fluoroquinolones in the outpatient setting, it is imperative that clinicians institute stewardship strategies aimed at decreasing outpatient fluoroquinolone use.

Because of the potential link between fluoroquinolone use and CDAD, it is reasonable for antimicrobial stewardship clinicians to limit the use of fluoroquinolones in their institution. Fluoroquinolone restriction has been used effectively, along with changes to environmental cleaning services, to control a nosocomial CDAD outbreak at a community hospital (Kallen *et al.*, 2009). Dancer *et al.* (2013) also found that prohibiting the routine use of ciprofloxacin reduced its use by 72% and the rate of nosocomial CDAD by 77%. An interrupted time-series analysis examining the impact of high-risk antibiotic restrictions—those of second- and third-generation cephalosporins, clindamycin, and fluoroquinolones—on CDAD incidence rates found the intervention significantly decreased high-risk antibiotic use and CDAD incidence rate (Aldeyab *et al.*, 2012). These data suggest a potential link between fluoroquinolone use and CDAD. Therefore, antimicrobial stewardship clinicians should work to limit unnecessary fluoroquinolone use, which may decrease the rate of CDAD. Clinicians should also examine the possible role of fluoroquinolone use in local outbreaks of CDAD.

Fluoroquinolones and methicillin-resistant *Staphylococcus aureus*

Fluoroquinolone exposure has been shown to induce methicillin resistance in *S. aureus* in vitro, and clinical studies have also shown a correlation between fluoroquinolone use and methicillin-resistant *S. aureus* (MRSA) (Paterson, 2004). The mechanism by which fluoroquinolones contribute to methicillin resistance in *S. aureus* is not fully understood. Fluoroquinolones may exhibit activity against methicillin-susceptible *S. aureus* (MSSA), but they often lack activity against MRSA (Dalhoff, 2012). Consequently, it is possible that fluoroquinolone use may select for MRSA within a heterogeneous *S. aureus* population (Acar and Goldstein, 1997; Knight *et al.*, 2012). It has also been hypothesized that fluoroquinolone resistance may be essential for MRSA survival in the hospital setting. Knight *et al.* (2012) demonstrated that nosocomial MRSA clones only infrequently lost fluoroquinolone-resistance genes over a 10 year period, despite losing resistance genes for other antibiotics (as well as acquiring resistances).

Fluoroquinolone use has been associated with MRSA colonization and acquisition in hospitals (Dziekan *et al.*, 2000; Harbarth *et al.*, 2000; Salangsang *et al.*, 2010). Fluoroquinolone exposure has also been linked to MRSA infection and acquisition (Graffunder and Venezia, 2002; Weber *et al.*, 2003; Ernst *et al.*, 2005; Tacconelli *et al.*, 2008). MacDougall *et al.* (2005) found a significant association between levofloxacin use and MRSA among a subset of US hospitals from 1999 to 2003. Also, a recent prospective nested case-control study found previous fluoroquinolone use to be significantly associated with MRSA acquisition among long-term care facility patients (Couderc *et al.*, 2014). This was the first prospective study linking fluoroquinolone exposure to MRSA acquisition, but the generalizability of these findings may be limited because the study was performed in long-term care facilities. However, antibiotic use among patients at long-term care facilities impacts the bacterial ecology of (regional) health(care) systems because patients from long-term care facilities are frequently admitted to the associated hospital(s).

Tracking Fluoroquinolone Usage and Adverse Drug Events

As antimicrobial stewardship continues to become a focal part of best practices across a wide range of healthcare settings, the need for accurate estimates of antimicrobial utilization has been of great importance. Unfortunately, the validity of standardized, risk-adjusted tracking metrics has not been fully established (Ibrahim and Polk, 2012). Standardized metrics allow for the reporting of clinical and organizational outcomes of antimicrobial stewardship initiatives on a facility, local, and national level. They also create a framework for comparing the overall or agent/class-specific utilization with notable trends in antimicrobial resistance over time, and the rates of adverse effects related to different antimicrobials. Lastly, the availability of validated metrics affords clinicians the opportunity to benchmark antimicrobial utilization across a continuum of acute- and long-term care practice settings.

A number of metrics have been used at the individual-facility level, but lack of standardization hinders their widespread application for benchmarking across health systems at the local and national levels (Ibrahim and Polk, 2014). Two particular metrics have emerged as the most widely accepted for tracking and trending antimicrobial usage: days of therapy (DOT) and defined daily dose (DDD), both normalized per 1000 patient days or days-present denominators (Polk *et al.*, 2007). In an effort to facilitate the reporting of, participation in, and interpretation of antimicrobial utilization at a national level in the US, the CDC National Healthcare Safety Network (NHSN) launched the Antimicrobial Use and Resistance (AUR) modules in 2011 and 2014, respectively (CDC, 2016a,b). Currently, the AUR modules utilize antimicrobial DOT/1000 days present to collect and disseminate utilization data. Initial data from 19 participating hospitals in the US for which data were validated were available in March 2014, and showed a high degree of variability in fluoroquinolone usage rates (Fridkin *et al.*, 2014). The report also estimated that 15–16% of all patients admitted to 323 hospitals in the US in 2010 received at least one dose of a fluoroquinolone agent during their hospitalization. While in their infancy, AUR modules will undoubtedly enhance antimicrobial stewardship efforts at participating facilities and across the US, now and in the future. Additionally, regardless of the metric used (DOT or DDD), certain patient populations, such as pediatric patients and those with end-stage renal disease, can present unique challenges for the interpretation of institution-level data. With that in mind, antimicrobial stewardship clinicians should interpret available data with caution, as more in-depth analyses may be needed to fully explain trends in antimicrobial utilization, including fluoroquinolone utilization, over time.

Antimicrobial Stewardship Strategies for Fluoroquinolones

Formulary exclusions and formulary restrictions

One of the key strategies for managing fluoroquinolone use is to place appropriate restrictions on their use or by eliminating one or more agents from the formulary. Formulary management policies for the fluoroquinolones have been shown to decrease the rate of fluoroquinolone resistance among Gram-negative pathogens, decrease the incidence of MRSA and CDAD, and increase the susceptibility of Gram negatives to group 2 carbapenems, as well as producing other beneficial results (Aubert *et al.*, 2005; Bosso and Mauldin, 2006; Aspinall *et al.*, 2007; Ntagiopoulos *et al.*, 2007; Kallen *et al.*, 2009; Medina Presentado *et al.*, 2011; Lewis *et al.*, 2012; Dancer *et al.*, 2013; Jose *et al.*, 2014). Clinicians should continue to advocate for formulary restrictions on fluoroquinolone use, especially in the ICU setting, as this has been shown to decrease the rate of Gram-negative resistance (Aubert *et al.*, 2005; Ntagiopoulos *et al.*, 2007; Medina Presentado *et al.*, 2011; Lewis *et al.*, 2012). Clinicians practicing in subacute and/or long-term care facilities should also develop specific criteria for fluoroquinolone use, especially for UTIs and catheter-associated UTIs. Similar restrictions have been implemented on a national level in several countries. Furthermore, the use of alternative agents, such as β-lactams, should be encouraged and incorporated into institutional treatment protocols whenever possible, while reserving fluoroquinolones only for cases of a true β-lactam allergy or proven clinical superiority over alternative agents. Jaso *et al.* (2014) evaluated the impact of a pharmacist-led allergy assessment performed for all patients who were prescribed moxifloxacin, and this intervention greatly reduced moxifloxacin use in their institution.

Utilizing local susceptibility data

It is imperative that any antimicrobial stewardship efforts targeting fluoroquinolone use be supported by local antibiotic use and susceptibility data. The use of available data should focus on understanding various aspects of fluoroquinolone use: providers prescribing fluoroquinolones; how these agents are prescribed; to what patient population(s) fluoroquinolones are prescribed; fluoroquinolone treatment duration and any concurrent antibiotics used. Equally important is an in-depth understanding of local susceptibility data on: patients infected with fluoroquinolone-resistant organisms; sites where these organisms are recovered; and the antibiotic susceptibility profiles of these organisms—including susceptibility to combination antibiotic therapy. The literature is clear that patients in the emergency department and in long-term care facilities, those with upper and lower respiratory infections and/or UTIs, including outpatients, and those over the age of 65 years are the high users of fluoroquinolone antibiotics. It would be prudent to tailor fluoroquinolone-related stewardship interventions toward one or more of these patient populations and/or healthcare settings. Collaboration with the microbiology laboratory is also needed to optimize fluoroquinolone use. A marked reduction in fluoroquinolone use was demonstrated following provider education via a pharmacist-led antimicrobial stewardship program that focused on fluoroquinolone use and resistance among Gram-negative bacteria (Wong-Beringer *et al.*, 2009). The program also suppressed fluoroquinolone susceptibility results when isolates were susceptible to alternative agents, such as β-lactams—if providers cannot see a fluoroquinolone susceptibility result, they may be less likely to prescribe a fluoroquinolone.

Modifying institutional empiric antibiotic recommendations

The optimal use of fluoroquinolones in an institution should be based upon each agent's in vitro activity against various pathogens, as well as on pharmacokinetic/pharmacodynamic data (Lode, 2014). When appropriate, incorporating non-fluoroquinolone antibiotics into institutional empiric antibiotic guidelines as preferred therapy will support efforts to decrease fluoroquinolone use. However, fluoroquinolones may be used if local susceptibility data dictate the use of a fluoroquinolone for a certain disease state(s) and/or organism(s). One group of investigators was able to successfully eliminate fluoroquinolones from the empiric antibiotic recommendations at their institution, which in turn drastically reduced fluoroquinolone use and improved fluoroquinolone susceptibility rates among Gram-negative pathogens (Nguyen *et al.*, 2008; Wong-Beringer *et al.*, 2009). Clinicians are advised to keep local susceptibility data in mind when interpreting studies that examine the impact of changing institutional antibiotic recommendations, as geographical variability in susceptibility rates can influence empiric treatment recommendations.

Intravenous to oral interchanges

Switching from intravenous (IV) fluoroquinolone to oral therapy is appropriate for clinically stable patients, and is a recommended strategy for antimicrobial stewardship programs (SHEA, IDSA and PIDS, 2012). Given their high oral bioavailability, fluoroquinolones are an ideal antimicrobial class to implement IV to oral switch programs (Goff *et al.*, 2012; Jones *et al.*, 2012). Jones *et al.* (2012) examined avoidable IV fluoroquinolone use (defined as an IV dose given when the patient received at least one other medication enterally) at 128 Veterans Administration (VA) hospitals in the US. They found that 46.8% of IV fluoroquinolone use was avoidable and that it was more common in the ICU (Jones *et al.*, 2012). These findings demonstrate that IV to oral dosing programs must work to avoid oral antibiotic choices being unnecessarily broad in spectrum.

Provider education, audit and feedback

Provider education on appropriate fluoroquinolone use, along with prescription audits and feedback, can effectively optimize the use of fluoroquinolones (Nguyen *et al.*, 2008; Borde *et al.*, 2014a,b). Central to any feedback is a robust system with which the stewardship clinician can track fluoroquinolone usage and adverse events. This system should be informed by both the literature and the limitations of the electronic medical record and data collection capabilities. Infectious diseases consultations are also helpful in reducing fluoroquinolone use (Jump *et al.*, 2012; Borde *et al.*, 2014a,b). Provider education and feedback, including local susceptibility data, was also shown to drastically reduce fluoroquinolone use (Nguyen *et al.*, 2008; Wong-Beringer *et al.*, 2009). However, a recent study found that requiring prior authorization for antibiotics was more effective in reducing antibiotic overuse than provider education, audit, and feedback (Mehta *et al.*, 2014). That being said, many studies have shown that provider education, prescription audit, and feedback can reduce fluoroquinolone use, and clinicians are encouraged to adopt these practices for fluoroquinolone use at their facilities.

Clinical decision support and the electronic medical record

The data supporting clinical decision support within the electronic medical record as a means to enforce antimicrobial stewardship initiatives is well documented in the literature (Kullar *et al.*, 2013; Forrest *et al.*, 2014; Kullar and Goff, 2014). Because of the multiple settings in which fluoroquinolones are used, incorporating electronic clinical decision support is a conceivable strategy for extending the reach of antimicrobial stewardship initiatives beyond the areas that are directly overseen by antimicrobial stewardship clinicians, such as outpatient clinics. Hecker *et al.* (2014) implemented an electronic order set coupled with audit and feedback in order to enforce adherence to guideline recommendations for the treatment of uncomplicated cystitis in the emergency department. These authors demonstrated increased adherence to uncomplicated cystitis guidelines and reductions in both unnecessary antibiotic therapy and fluoroquinolone therapy for patients with cystitis. Finally, Rattinger *et al.* (2012) found that clinical decision support aimed at improving antibiotic prescribing for upper respiratory infections decreased unwarranted antibiotic use from 22 to 2% ($P < 0.0001$).

Emergency department and outpatient strategies

Fluoroquinolone use in the emergency department and in outpatient settings is high and often inappropriate and/or unnecessary (Lautenbach *et al.*, 2003; Linder *et al.*, 2005; Aspinall *et al.*, 2007; Caterino *et al.*, 2009; Donnelly *et al.*, 2014; Shapiro *et al.*, 2014). The CDC has issued a directive imploring clinicians to practice antimicrobial stewardship in the outpatient setting in order to reduce rates of CDAD (McDonald, 2012). Antimicrobial stewardship efforts aimed at reducing fluoroquinolone use should not disregard the potential impact of interventions in these clinical settings (Hecker *et al.*, 2014). These are relatively new areas of focus for antimicrobial stewardship, and limited literature is available describing outpatient interventions compared with interventions in the inpatient setting. However, there are recommendations for instituting antimicrobial stewardship in the emergency department (May *et al.*, 2013), where the following strategies have been shown to reduce fluoroquinolone use: electronic order sets with clinical decision support; prescription audit and feedback; formulary restriction; provider education; and modifying empiric antibiotic recommendations (Aspinall *et al.*, 2007; Borde *et al.*, 2014b; Hecker *et al.*, 2014). In the outpatient setting, limiting the use of fluoroquinolones for viral upper respiratory infections is of utmost importance. One outpatient stewardship strategy is to encourage prescribers to write delayed antibiotic prescriptions for respiratory infections, with explicit instructions to only dispense the prescription if symptoms do not improve within a few days' time (Arroll *et al.*, 2003; Little *et al.*, 2014). Restricting fluoroquinolone use in the outpatient setting is another strategy. An interrupted time-series analysis evaluated the impact of an outpatient fluoroquinolone restriction on MRSA incidence in the community, and showed that intervention was associated with a reduction in MRSA incidence in the community (Aldeyab *et al.*, 2014).

Conclusion

Without question, the use of fluoroquinolones ought to be part of every institution's antibiotic stewardship program. Nevertheless, the magnitude and the volume of surveillance required for this one antibiotic class would likely overwhelm most stewardship programs, which are also trying simultaneously to promote other stewardship efforts. Antibiotic stewardship efforts directed at fluoroquinolones might best be accomplished by the identification of specific programs/targets or goals (Table 19.1), and the use of pharmacy and therapeutics committee policies to drive desired prescriber behavior.

Table 19.1. Fluoroquinolone (FQ) antimicrobial stewardship checklist.

Minimum action	Optimal action
Automatic pharmacy dose optimization for FQs (intravenous to oral conversions, renal dosing, pharmacokinetic/ pharmacodynamic (PK/PD) optimization)	Detailed monitoring of local FQ susceptibility (e.g., by unit, infection site, provider, inpatient, outpatient, etc.)
Formulary restrictions on FQ use	Detailed tracking of FQ use (e.g., by unit, provider, inpatient, outpatient, etc.)
Appropriate empiric antibiotic recommendations (including combination antibiograms for Gram-negative pathogens)	'Internal benchmarking of FQ use by medical discipline' to be added to this list in between the currently listed second and third points
Tracking aggregate FQ use (using either days of therapy, DOT, or defined daily dose, DDD)	Stewardship interventions in the emergency department and the outpatient setting
Monitoring local FQ susceptibility	External benchmarking of internal FQ use
Provider education, audit and feedback on appropriate FQ use	'Track indications for FQ use' to be added as the final point in this column
Utilizing clinical decision support within the electronic medical record to guide appropriate FQ use	
Tracking *Clostridium difficile* ESBL and MRSA rates	
'Discourage FQ use for upper respiratory tract infections and skin/skin structure infections' to be added to the list for minimum action	

Annual usage of fluoroquinolones should be tracked within each institution and reported as defined daily doses or fluoroquinolone antibiotic days. Additionally, possible benchmarking of fluoroquinolone use by hospital ward, clinic, and prescriber might help to draw attention to areas or prescribers overusing these antibiotics. Regular monitoring of the institution antibiogram for changing patterns of fluoroquinolone susceptibility should be considered quarterly to semi-annually, and changes in patterns of *C. difficile* infection should be carefully addressed for a possible fluoroquinolone link.

Stewardship of fluoroquinolone antibiotics also provides some shared but unique challenges to hospital programs. The bacterial flora represented in the hospital antibiogram embodies a melting pot of the community. Patients might be in an ICU with a Gram-negative pathogen that is multidrug resistant, but the pathogen may have been created in and originated from a long-term care facility or the broader community itself, areas that are not impacted by the hospital's stewardship efforts or control. While other antibiotic classes are also used in the "community," in the case of fluoroquinolones, the exact same products are used in both environments at similar doses with ~100% bioavailability. Because fluoroquinolones have a variety of FDA-approved indications for some of the most common infectious diseases, the unique antibiotic pressure exerted on the microbial flora is extreme in both the hospital setting and in the community at large. Also, the "community" is very mobile. Regardless of the quality or control of the stewardship program in the hospital, the next MDR pathogen is only a ride away across town, the country, or the world.

Unfortunately, the use of fluoroquinolones and the overall misuse of antibiotics remains as an ongoing societal problem on a variety of levels. While stewardship programs are laudable, they can be expensive to run and clearly have boundaries. Attempts to rein in fluoroquinolone use throughout our society and throughout the world will clearly test our resources and resolve.

References

Acar, J.F. and Goldstein, F.W. (1997) Trends in bacterial resistance to fluoroquinolones. *Clinical Infectious Diseases* 24 (Suppl 1), S67–73.

Adam, H.J., Hoban, D.J., Gin, A.S., and Zhanel, G.G. (2009) Association between fluoroquinolone usage and a dramatic rise in ciprofloxacin-resistant *Streptococcus pneumoniae* in Canada, 1997–2006. *International Journal of Antimicrobial Agents* 34, 82–85.

Aldeyab, M.A., Kearney, M.P., Scott, M.G., Aldiab, M.A., Alahmadi, Y.M., Darwish Elhajji, F.W., Magee, F.A., and McElnay, J.C. (2012) An evaluation of the impact of antibiotic stewardship on reducing the use of high-risk antibiotics and its effect on the incidence of *Clostridium difficile* infection in hospital settings. *Journal of Antimicrobial Chemotherapy* 67, 2988–2996.

Aldeyab, M.A., Scott, M.G., Kearney, M.P., Alahmadi, Y.M., Magee, F.A., Conlon, G., and McElnay, J.C. (2014) Impact of an enhanced antibiotic stewardship on reducing methicillin-resistant *Staphylococcus aureus* in primary and secondary healthcare settings. *Epidemiology and Infection* 142, 494–500.

Aminimanizani, A., Beringer, P., and Jelliffe, R. (2001) Comparative pharmacokinetics and pharmacodynamics of the newer fluoroquinolone antibacterials. *Clinical Pharmacokinetics* 40, 169–187.

Arroll, B., Kenealy, T., and Kerse, N. (2003) Do delayed prescriptions reduce antibiotic use in respiratory tract infections? A systematic review. *British Journal of General Practice* 53, 871–877.

Aspinall, S.L., Metlay, J.P., Maselli, J.H., and Gonzales, R. (2007) Impact of hospital formularies on fluoroquinolone prescribing in emergency departments. *The American Journal of Managed Care* 13, 241–248.

ATS (American Thoracic Society) and IDSA (Infectious Diseases Society of America) (2005) Guidelines for the management of adults with hospital-acquired, ventilator-associated, and healthcare-associated pneumonia. *American Journal of Respiratory and Critical Care Medicine* 171, 388–416.

Aubert, G., Carricajo, A., Vautrin, A.C., Guyomarc'h, S., Fonsale, N., Page, D., Brunel, P., Rusch, P., and Zéni, F. (2005) Impact of restricting fluoroquinolone prescription on bacterial resistance in an intensive care unit. *The Journal of Hospital Infection* 59, 83–89.

Barth, J., Jager, D., Mundkowski, R., Drewelow, B., Welte, T., and Burkhardt, O. (2008) Single- and multiple-dose pharmacokinetics of intravenous moxifloxacin in patients with severe hepatic impairment. *Journal of Antimicrobial Chemotherapy* 62, 575–578.

Beardsley, J.R., Williamson, J.C., Johnson, J.W., Ohl, C.A., Karchmer, T.B., and Bowton, D.L. (2006) Using local microbiologic data to develop institution-specific guidelines for the treatment of hospital-acquired pneumonia. *Chest* 130, 787–793.

Biller, P., Shank, B., Lind, L., Brennan, M., Tkatch, L., Killgore, G., Thompson, A., and McDonald, L.C. (2007) Moxifloxacin therapy as a risk factor for *Clostridium difficile*-associated disease during an outbreak: attempts to control a new epidemic strain. *Infection Control and Hospital Epidemiology* 28, 198–201.

Blaser, J., Stone, B.B., Groner, M.C., and Zinner, S.H. (1987) Comparative study with enoxacin and netilmicin in a pharmacodynamic model to determine importance of ratio of antibiotic peak concentration to MIC for bactericidal activity and emergence of resistance. *Antimicrobial Agents and Chemotherapy* 31, 1054–1060.

Bliziotis, I.A., Michalopoulos, A., Kasiakou, S.K., Samonis, G., Christodoulou, C., Chrysanthopoulou, S., and Falagas, M.E. (2005) Ciprofloxacin vs an aminoglycoside in combination with a β-lactam for the treatment of febrile neutropenia: a meta-analysis of randomized controlled trials. *Mayo Clinic Proceedings* 80, 1146–1156.

Borde, J.P., Kaier, K., Steib-Bauert, M., Vach, W., Geibel-Zehender, A., Busch, H., Bertz, H., Hug, M., de With, K., and Kern, W.V. (2014a) Feasibility and impact of an intensified antibiotic stewardship programme targeting cephalosporin and fluoroquinolone use in a tertiary care university medical center. *BMC Infectious Diseases* 14: 201.

Borde, J.P., Kern, W.V., Hug, M., Steib-Bauert, M., de With, K., Busch, H.J., and Kaier, K. (2014b) Implementation of an intensified antibiotic stewardship programme targeting third-generation cephalosporin and fluoroquinolone use in an emergency medicine department. *Emergency Medicine Journal* 32, 509–515.

Bosso, J.A. and Mauldin, P.D. (2006) Using interrupted time series analysis to assess associations of fluoroquinolone formulary changes with susceptibility of gram-negative pathogens and isolation rates of methicillin-resistant *Staphylococcus aureus*. *Antimicrobial Agents and Chemotherapy* 50, 2106–2112.

Boyd, N. and Nailor, M.D. (2011) Combination antibiotic therapy for empiric and definitive treatment of Gram-negative infections: insights from the Society of Infectious Diseases Pharmacists. *Pharmacotherapy* 31, 1073–1084.

Brighty, K. and Gootz, T. (2000) Chemistry and mechanism of action of the quinolone antibacterials. In: Andriole, V.T. (ed.) *The Quinolones*, 3rd edn. Academic Press, San Diego, California, pp. 33–97.

Caterino, J.M., Weed, S.G., Espinola, J.A., and Camargo, C.A., Jr. (2009) National trends in emergency department antibiotic prescribing for elders with urinary tract infection, 1996–2005. *Academic Emergency Medicine* 16, 500–507.

CDC (2013) *Antibiotic Resistance Threats in the United States, 2013*. Centers for Disease Control and Prevention, Atlanta, Georgia. Available at: http://www.cdc.gov/drugresistance/threat-report-2013/pdf/ar-threats-2013-508.pdf (accessed 28 October 2014).

CDC (2016a) *Antimicrobial Use and Resistance (AUR) Module*. Centers for Disease Control and Prevention, Atlanta, Georgia. Available at: http://www.cdc.gov/nhsn/pdfs/training/aur/aur-training.pdf (accessed 18 May 2016).

CDC (2016b) *Antimicrobial Use and Resistance (AUR) Module*. Centers for Disease Control and Prevention, Atlanta, Georgia. Available at: http://www.cdc.gov/nhsn/PDFs/pscManual/11pscAURcurrent.pdf (accessed 18 May 2016).

Christoff, J., Tolentino, J., Mawdsley, E., Matushek, S., Pitrak, D., and Weber, S.G. (2010) Optimizing empirical antimicrobial therapy for infection due to Gram-negative pathogens in the intensive care unit: utility of a combination antibiogram. *Infection Control and Hospital Epidemiology* 31, 256–261.

Costelloe, C., Metcalfe, C., Lovering, A., Mant, D., and Hay, A.D. (2010) Effect of antibiotic prescribing in primary care on antimicrobial resistance in individual patients: systematic review and meta-analysis. *BMJ* 340: c2096.

Couderc, C., Jolivet, S., Thiébaut, A.C., Ligier, C., Remy, L., Alvarez, A.S., Lawrence, C., Salomon, J., Herrmann, J.L., and Guillemot, D. (2014) Fluoroquinolone use is a risk factor for methicillin-resistant *Staphylococcus aureus* acquisition in long-term care facilities: a nested case-case-control study. *Clinical Infectious Diseases* 59, 206–215.

Dalhoff, A. (2012) Global fluoroquinolone resistance epidemiology and implictions for clinical use. *Interdisciplinary Perspectives on Infectious Diseases* 2012: 976273.

Dancer, S.J., Kirkpatrick, P., Corcoran, D.S., Christison, F., Farmer, D., and Robertson, C. (2013) Approaching zero: temporal effects of a restrictive antibiotic policy on hospital-acquired *Clostridium difficile*, extended-spectrum β-lactamase-producing coliforms and meticillin-resistant *Staphylococcus aureus*. *International Journal of Antimicrobial Agents* 41, 137–142.

Davidson, R., Cavalcanti, R., Brunton, J.L., Bast, D.J., de Azavedo, J.C., Kibsey, P., Fleming, C., and Low, D.E. (2002) Resistance to levofloxacin and failure of treatment of pneumococcal pneumonia. *The New England Journal of Medicine* 346, 747–750.

de Cueto, M., Rodriguez, J.M., Soriano, M.J., Lopez-Cerero, L., Venero, J. and Pascual, A. (2008) Fatal levofloxacin failure in treatment of a bacteremic patient infected with *Streptococcus pneumoniae* with a preexisting *parC* mutation. *Journal of Clinical Microbiology* 46, 1558–1560.

Depestel, D.D. and Aronoff, D.M. (2013) Epidemiology of *Clostridium difficile* infection. *Journal of Pharmacy Practice* 26, 464–475.

Deshpande, A., Pasupuleti, V., Thota, P., Pant, C., Rolston, D.D., Sferra, T.J., Hernandez, A.V., and Donskey, C.J. (2013) Community-associated *Clostridium difficile* infection and antibiotics: a meta-analysis. *Journal of Antimicrobial Chemotherapy* 68, 1951–1961.

Donnelly, J.P., Baddley, J.W., and Wang, H.E. (2014) Antibiotic utilization for acute respiratory tract infections in U.S. emergency departments. *Antimicrobial Agents and Chemotherapy* 58, 1451–1457.

Drusano, G.L., Johnson, D.E., Rosen, M., and Standiford, H.C. (1993) Pharmacodynamics of a fluoroquinolone antimicrobial agent in a neutropenic rat model of *Pseudomonas* sepsis. *Antimicrobial Agents and Chemotherapy* 37, 483–490.

Dziekan, G., Hahn, A., Thüne, K., Schwarzer, G., Schäfer, K., Daschner, F.D., and Grundmann, H. (2000) Methicillin-resistant *Staphylococcus aureus* in a teaching hospital: investigation of nosocomial transmission using a matched case-control study. *The Journal of Hospital Infection* 46, 263–270.

Empey, P.E., Jennings, H.R., Thornton, A.C., Rapp, R.P., and Evans, M.E. (2001) Levofloxacin failure in a patient with pneumococcal pneumonia. *Annals of Pharmacotherapy* 35, 687–690.

Ernst, E.J., Raley, G., Herwaldt, L.A., and Diekema, D.J. (2005) Importance of control group selection for evaluating antimicrobial use as a risk factor for methicillin-resistant *Staphylococcus aureus* bacteremia. *Infection Control and Hospital Epidemiology* 26, 634–637.

FDA (2013) *FDA Requires Label Changes to Warn of Risk for Possibly Permanent Nerve Damage from Antibacterial Fluoroquinolone Drugs Taken by Mouth or by Injection*. FDA Drug Safety Communication, US Food and Drug Administration, Washington, DC. Available at: http://www.fda.gov/downloads/Drugs/DrugSafety/UCM365078.pdf (accessed 5 December 2014).

Forrest, A., Nix, D.E., Ballow, C.H., Goss, T.F., Birmingham, M.C., and Schentag, J.J. (1993) Pharmacodynamics of intravenous ciprofloxacin in seriously ill patients. *Antimicrobial Agents and Chemotherapy* 37, 1073–1081.

Forrest, G.N., Van Schooneveld, T.C., Kullar, R., Schulz, L.T., Duong, P., and Postelnick, M. (2014) Use of electronic health records and clinical decision support systems for antimicrobial stewardship. *Clinical Infectious Diseases* 59 (Suppl 3), S122–S133.

Fridkin, S., Baggs, J., Fagan, R., Magill, S., Pollack, L.A., Malpiedi, P., Slayton, R., Khader, K., Rubin, M.A., Jones, M. *et al*. (2014) Vital signs: improving antibiotic use among hospitalized patients. *Morbidity and Mortality Weekly Report (MMWR)* 63,194–200.

Goff, D.A., Bauer, K.A., Reed, E.E., Stevenson, K.B., Taylor, J.J., and West, J.E. (2012) Is the "low-hanging fruit" worth picking for antimicrobial stewardship programs? *Clinical Infectious Diseases* 55, 587–592.

Graffunder, E.M. and Venezia, R.A. (2002) Risk factors associated with nosocomial methicillin-resistant *Staphylococcus aureus* (MRSA) infection including previous use of antimicrobials. *Journal of Antimicrobial Chemotherapy* 49, 999–1005.

Gupta, A and Khanna, S. (2014) Community-acquired *Clostridium difficile* infection: an increasing public health threat. *Infection and Drug Resistance* 7, 63–72.

Gupta, K., Hooton, T.M., Naber, K.G., Wullt, B., Colgan, R., Miller, L.G., Moran, G.J., Nicolle, L.E., Raz, P., Schaeffer, A.J., and Soper, D.E. (2011) International clinical practice guidelines for the treatment of acute uncomplicated cystitis and pyelonephritis in women: a 2010 update by the Infectious Diseases Society of America and the European Society for Microbiology and Infectious Diseases. *Clinical Infectious Diseases* 52, e103–e120.

Hancock, R.E. (1998) Resistance mechanisms in *Pseudomonas aeruginosa* and other nonfermentative

Gram-negative bacteria. *Clinical Infectious Diseases* 27 (Suppl 1), S93–99.

Harbarth, S., Liassine, N., Dharan, S., Herrault, P., Auckenthaler, R., and Pittet, D. (2000) Risk factors for persistent carriage of methicillin-resistant *Staphylococcus aureus*. *Clinical Infectious Diseases* 31, 1380–1385.

Hecker, M.T., Aron, D.C., Patel, N.P., Lehmann, M.K., and Donskey, C.J. (2003) Unnecessary use of antimicrobials in hospitalized patients: current patterns of misuse with an emphasis on the antianaerobic spectrum of activity. *Archives of Internal Medicine* 163, 972–978.

Hecker, M.T., Fox, C.J., Son, A.H. Cydulka, R.K., Siff, J.E., Emerman, C.L., Sethi, A.K., Muganda, C.P., and Donskey, C.J. (2014) Effect of a stewardship intervention on adherence to uncomplicated cystitis and pyelonephritis guidelines in an emergency department setting. *PloS ONE* 9(2): e87899.

Hicks, L.A., Chien, Y.W., Taylor, T.H., Jr., Haber, M., and Klugman, K.P. (2011) Outpatient antibiotic prescribing and nonsusceptible *Streptococcus pneumoniae* in the United States, 1996–2003. *Clinical Infectious Diseases* 53, 631–639.

Hooper, D.C. (2002) Fluoroquinolone resistance among Gram-positive cocci. *The Lancet Infectious Diseases* 2, 530–538.

Hsu, D.I., Okamoto, M.P., Murthy, R., and Wong-Beringer, A. (2005) Fluoroquinolone-resistant *Pseudomonas aeruginosa*: risk factors for acquisition and impact on outcomes. *Journal of Antimicrobial Chemotherapy* 55, 535–541.

Ibrahim, O.M. and Polk, R.E. (2012) Benchmarking antimicrobial drug use in hospitals. *Expert Review of Anti-Infective Therapy* 10, 445–457.

Ibrahim, O.M. and Polk, R.E (2014) Antimicrobial use metrics and benchmarking to improve stewardship outcomes: methodology, opportunities, and challenges. *Infectious Disease Clinics of North America* 28, 195–214.

Jaso, T., Shea, K., Daley, M., Bissett, J., and Douglass, E. (2014) Antimicrobial stewardship strategy to decrease respiratory fluoroquinolone utilization in a large, urban, healthcare system. Poster presentation 230 at: *ID Week 2014, October 8–12, Philadelphia, PA*. IDWeek, Arlington, Virginia. Available at: https://idsa.confex.com/idsa/2014/webprogram/Paper45967.html (accessed 18 May 2016).

Johnson, S.J., Ernst, E.J., and Moores, K.G. (2011) Is double coverage of Gram-negative organisms necessary? *American Journal of Health-System Pharmacy* 68, 119–124.

Jones, M., Huttner, B., Madaras-Kelly, K., Nechodom, K., Nielson, C., Bidwell Goetz, M., Neuhauser, M.M., Samore, M.H., and Rubin, M.A. (2012) Parenteral to oral conversion of fluoroquinolones: low-hanging fruit for antimicrobial stewardship programs? *Infection Control and Hospital Epidemiology* 33, 362–367.

Jose, K., Wallace, M.R., and Jacobson, S. (2014) Effect of a fluoroquinolone restriction program on antibiotic utilization. Poster presentation K-342 at: *54th Interscience Conference on Antimicrobial Agents and Chemotherapy (ICAAC 2014), Washington, DC, 5–9 September 2014*. American Society for Microbiology, Washington, DC. Available at: http://www.abstractsonline.com/Plan/ViewAbstract.aspx?mID=3529&sKey=91d31c03-fc88-4244-9b67-9667c18838a0&cKey=d8e9d598-9fb1-4016-9ef9-78bcc3258356&mKey=5d6b1802-e453-486b-bcbb-b11d1182d8bb (accessed 18 May 2016).

Jump, R.L., Olds, D.M., Seifi, N., Kypriotakis, G., Jury, L.A., Peron, E.P., Hirsch, A.A., Drawz, P.E., Watts, B., Bonomo, R.A., and Donskey, C.J. (2012) Effective antimicrobial stewardship in a long-term care facility through an infectious disease consultation service: keeping a LID on antibiotic use. *Infection Control and Hospital Epidemiology* 33, 1185–1192.

Kallen, A.J., Thompson, A., Ristaino, P., Chapman, L., Nicholson, A., Sim, B.T., Lessa, F., Sharapov, U., Fadden, E., Boehler, R. *et al.* (2009) Complete restriction of fluoroquinolone use to control an outbreak of *Clostridium difficile* infection at a community hospital. *Infection Control and Hospital Epidemiology* 30, 264–272.

Kays, M.B., Smith, D.W., Wack, M.E., and Denys, G.A. (2002) Levofloxacin treatment failure in a patient with fluoroquinolone-resistant *Streptococcus pneumoniae* pneumonia. *Pharmacotherapy* 22, 395–399.

Khawcharoenporn, T., Vasoo, S., and Singh, K. (2013) Urinary tract infections due to multidrug-resistant *Enterobacteriaceae*: prevalence and risk factors in a Chicago emergency department. *Emergency Medicine International* 2013: 258517.

Knight, G.M., Budd, E.L., Whitney, L., Thornley, A., Al-Ghusein, H., Planche, T., and Lindsay, J.A. (2012) Shift in dominant hospital-associated methicillin-resistant *Staphylococcus aureus* (HA-MRSA) clones over time. *Journal of Antimicrobial Chemotherapy* 67, 2514–2522.

Kriengkauykiat, J., Porter, E., Lomovskaya, O., and Wong-Beringer, A. (2005) Use of an efflux pump inhibitor to determine the prevalence of efflux pump-mediated fluoroquinolone resistance and multidrug resistance in *Pseudomonas aeruginosa*. *Antimicrobial Agents and Chemotherapy* 49, 565–570.

Kullar, R. and Goff, D.A. (2014) Transformation of antimicrobial stewardship programs through technology and informatics. *Infectious Disease Clinics of North America* 28, 291–300.

Kullar, R., Goff, D.A., Schulz, L.T., Fox, B.C., and Rose, W.E. (2013) The "epic" challenge of optimizing antimicrobial stewardship: the role of electronic medical records and technology. *Clinical Infectious Diseases* 57, 1005–1013.

Kuster, S.P., Rudnick, W., Shigayeva, A., Green, K., Baqi, M., Gold, W.L., Lovinsky, R., Muller, M.P., Powis, J.E., Rau, N. *et al.* (2014) Previous antibiotic exposure and antimicrobial resistance in invasive pneumococcal disease: results from prospective surveillance. *Clinical Infectious Diseases* 59, 944–952.

Labbé, A.C., Poirier, L., Maccannell, D., Louie, T., Savoie, M., Béliveau, C., Laverdière, M., and Pépin, J. (2008) *Clostridium difficile* infections in a Canadian tertiary care hospital before and during a regional epidemic associated with the BI/NAP1/027 strain. *Antimicrobial Agents and Chemotherapy* 52, 3180–3187.

Lautenbach, E., Larosa, L.A., Kasbekar, N., Peng, H.P., Maniglia, R.J., and Fishman, N.O. (2003) Fluoroquinolone utilization in the emergency departments of academic medical centers: prevalence of, and risk factors for, inappropriate use. *Archives of Internal Medicine* 163, 601–605.

Lewis, G.J., Fang, X., Gooch, M., and Cook, P.P. (2012) Decreased resistance of *Pseudomonas aeruginosa* with restriction of ciprofloxacin in a large teaching hospital's intensive care and intermediate care units. *Infection Control and Hospital Epidemiology* 33, 368–373.

Linder, J.A., Huang, E.S., Steinman, M.A., Gonzales, R. and Stafford, R,S. (2005) Fluoroquinolone prescribing in the United States: 1995 to 2002. *The American Journal of Medicine* 118, 259–268.

Little, P., Moore, M., Kelly, J., Williamson, I., Leydon, G., McDermott, L., Mullee, M., and Stuart, B. (2014) Delayed antibiotic prescribing strategies for respiratory tract infections in primary care: pragmatic, factorial, randomised controlled trial. *BMJ* 348: g1606.

Lode, H.M. (2014) Preserving the efficacy of front-line fluoroquinolones through selective use to optimise clinical outcomes. *International Journal of Antimicrobial Agents* 43, 497–507.

Louie, A., Fregeau, C., Liu, W., Kulawy, R., and Drusano, G.L. (2009) Pharmacodynamics of levofloxacin in a murine pneumonia model of *Pseudomonas aeruginosa* infection: determination of epithelial lining fluid targets. *Antimicrobial Agents and Chemotherapy* 53, 3325–3330.

Lubasch, A., Keller, I., Borner, K., Koeppe, P., and Lode, H. (2000) Comparative pharmacokinetics of ciprofloxacin, gatifloxacin, grepafloxacin, levofloxacin, trovafloxacin, and moxifloxacin after single oral administration in healthy volunteers. *Antimicrobial Agents and Chemotherapy* 44, 2600–2603.

MacDougall, C., Harpe, S.E., Powell, J.P., Johnson, C.K., Edmond, M.B., and Polk, R.E. (2005) *Pseudomonas aeruginosa, Staphylococcus aureus*, and fluoroquinolone use. *Emerging Infectious Diseases* 11, 1197–1204.

Madaras-Kelly, K.J., Ostergaard, B.E., Hovde, L.B., and Rotschafer, J.C. (1996) Twenty-four-hour area under the concentration-time curve/MIC ratio as a generic predictor of fluoroquinolone antimicrobial effect by using three strains of *Pseudomonas aeruginosa* and an in vitro pharmacodynamic model. *Antimicrobial Agents and Chemotherapy* 40, 627–632.

Marwick, C.A., Yu, N., Lockhart, M.C., McGuigan, C.C., Wiuff, C., Davey, P.G., and Donnan, P.T. (2013) Community-associated *Clostridium difficile* infection among older people in Tayside, Scotland, is associated with antibiotic exposure and care home residence: cohort study with nested case-control. *Journal of Antimicrobial Chemotherapy* 68, 2927–2933.

May, L., Cosgrove, S., L'Archeveque, M., Talan, D.A., Payne, P., Jordan, J., and Rothman, R.E. (2013) A call to action for antimicrobial stewardship in the emergency department: approaches and strategies. *Annals of Emergency Medicine* 62, 69–77.e2.

McCusker, M.E., Harris, A.D., Perencevich, E., and Roghmann, M.C. (2003) Fluoroquinolone use and *Clostridium difficile*-associated diarrhea. *Emerging Infectious Diseases* 9, 730–733.

McDonald, L.C., Lessa, F., Sievert, D., Wise, M., Herrera, R., Gould, C., Malpiedi, P., Dudeck, M., Srinivasan, A., Fridkin, S. *et al.* (2012) Vital signs: preventing *Clostridium difficile* infections (2012) *Morbidity and Mortality Weekly Report (MMWR)* 61, 157–162.

Medina Presentado, J.C., Paciel Lopez, D., Berro Castiglioni, M., and Gerez, J. (2011) Ceftriaxone and ciprofloxacin restriction in an intensive care unit: less incidence of *Acinetobacter* spp. and improved susceptibility of *Pseudomonas aeruginosa*. *Revista Panamericana de Salud Pública* 30, 603–609.

Mehta, J.M., Haynes, K., Wileyto, E.P., Gerber, J.S., Timko, D.R., Morgan, S.C., Binkley, S., Fishman, N.O., Lautenbach, E., and Zaoutis, T. (2014) Comparison of prior authorization and prospective audit with feedback for antimicrobial stewardship. *Infection Control and Hospital Epidemiology* 35, 1092–1099.

Mizuta, M., Linkin, D.R., Nachamkin, I., Fishman, N.O., Weiner, M.G., Sheridan, A., and Lautenbach, E. (2006) Identification of optimal combinations for empirical dual antimicrobial therapy of *Pseudomonas aeruginosa* infection: potential role of a combination antibiogram. *Infection Control and Hospital Epidemiology* 27, 413–415.

Muto, C.A., Pokrywka, M., Shutt, K., Mendelsohn, A.B., Nouri, K., Posey, K., Roberts, T., Croyle, K., Krystofiak, S., Patel-Brown, S. *et al.* (2005) A large outbreak of *Clostridium difficile*-associated disease with an unexpected proportion of deaths and colectomies at a teaching hospital following increased fluoroquinolone use. *Infection Control and Hospital Epidemiology* 26, 273–280.

Nau, R., Sorgel, F., and Eiffert, H. (2010) Penetration of drugs through the blood-cerebrospinal fluid/blood–brain barrier for treatment of central nervous system infections. *Clinical Microbiology Reviews* 23, 858–883.

Nguyen, L.H., Hsu, D.I., Ganapathy, V., Shriner, K., and Wong-Beringer, A. (2008) Reducing empirical use of fluoroquinolones for *Pseudomonas aeruginosa* infections improves outcome. *Journal of Antimicrobial Chemotherapy* 61, 714–720.

Ntagiopoulos, P.G., Paramythiotou, E., Antoniadou, A., Giamarellou, H., and Karabinis, A. (2007) Impact of an antibiotic restriction policy on the antibiotic resistance patterns of Gram-negative microorganisms in an intensive care unit in Greece. *International Journal of Antimicrobial Agents* 30, 360–365.

Odenholt, I. and Cars, O. (2006) Pharmacodynamics of moxifloxacin and levofloxacin against *Streptococcus pneumoniae*, *Staphylococcus aureus*, *Klebsiella pneumoniae* and *Escherichia coli*: simulation of human plasma concentrations after intravenous dosage in an in vitro kinetic model. *Journal of Antimicrobial Chemotherapy* 58, 960–965.

Paterson, D.L. (2004) "Collateral damage" from cephalosporin or quinolone antibiotic therapy. *Clinical Infectious Diseases* 38(Suppl 4), S341–S345.

Pépin, J., Saheb, N., Coulombe, M.A., Alary, M.E., Corriveau, M.P., Authier, S., Leblanc, M., Rivard, G., Bettez, M., Primeau, V. *et al.* (2005) Emergence of fluoroquinolones as the predominant risk factor for *Clostridium difficile*-associated diarrhea: a cohort study during an epidemic in Quebec. *Clinical Infectious Diseases* 41, 1254–1260.

Polk, R.E., Fox, C., Mahoney, A., Letcavage, J., and MacDougall, C. (2007) Measurement of adult antibacterial drug use in 130 US hospitals: comparison of defined daily dose and days of therapy. *Clinical Infectious Diseases* 44, 664–670.

Poole, K. (2000) Efflux-mediated resistance to fluoroquinolones in Gram-negative bacteria. *Antimicrobial Agents and Chemotherapy* 44, 2233–2241.

Preston, S.L., Drusano, G.L., Berman, A.L., Fowler, C.L., Chow, A.T., Dornseif, B., Reichl, V., Natarajan, J., and Corrado, M. (1998) Pharmacodynamics of levofloxacin: A new paradigm for early clinical trials. *Journal of the American Medical Association* 279, 125–129.

Rattinger, G.B., Mullins, C.D., Zuckerman, I.H., Onukwugha, E., Walker, L.D., Gundlapalli, A., Samore, M. and DeLisle, S. (2012) A sustainable strategy to prevent misuse of antibiotics for acute respiratory infections. *PloS One* 7(12): e51147.

Rotschafer, J.C., Ullman, M.A., and Sullivan, C.J. (2011) Optimal use of fluoroquinolones in the intensive care unit setting. *Critical Care Clinics* 27, 95–106.

Salangsang, J.A., Harrison, L.H., Brooks, M.M., Shutt, K.A., Saul, M.I., and Muto, C.A. (2010) Patient-associated risk factors for acquisition of methicillin-resistant *Staphylococcus aureus* in a tertiary care hospital. *Infection Control and Hospital Epidemiology* 31, 1139–1147.

Shapiro, D.J., Hicks, L.A., Pavia, A.T., and Hersh, A.L. (2014) Antibiotic prescribing for adults in ambulatory care in the USA, 2007–2009. *Journal of Antimicrobial Chemotherapy* 69, 234–240.

SHEA, IDSA and PIDS (2012) Policy statement on antimicrobial stewardship by the Society for Healthcare Epidemiology of America (SHEA), the Infectious Diseases Society of America (IDSA), and the Pediatric Infectious Diseases Society (PIDS) *Infection Control and Hospital Epidemiology* 33, 322–327.

Singer, R.S., Finch, R., Wegener, H.C., Bywater, R., Walters, J., and Lipsitch, M. (2003) Antibiotic resistance—the interplay between antibiotic use in animals and human beings. *The Lancet Infectious Diseases* 3, 47–51.

Slimings, C. and Riley, T.V. (2014) Antibiotics and hospital-acquired *Clostridium difficile* infection: update of systematic review and meta-analysis. *Journal of Antimicrobial Chemotherapy* 69, 881–891.

Soriano, M.M., McCongehy, M.W., Harrington, A.T., and Danziger, L.H. (2014) Are urinary isolates more resistant than non-urinary isolates? In: *24th European Congress of Clinical Microbiology and Infectious Diseases, Barcelona, Spain, 10–13 May 2014*. European Society of Clinical Microbiology and Infectious Diseases (ESCMID), Basel, Switzerland, ECCMID 2014 Poster presentation P1060. Available at: https://www.escmid.org/escmid_publications/escmid_elibrary/material/?mid=16770 (accessed 18 May 2016).

Stass, H. and Kubitza, D. (2001) Effects of iron supplements on the oral bioavailability of moxifloxacin, a novel 8-methoxyfluoroquinolone, in humans. *Clinical Pharmacokinetics* 40(Suppl 1), 57–62.

Stass, H., Kubitza, D., and Schuhly, U. (2001) Pharmacokinetics, safety and tolerability of moxifloxacin, a novel 8-methoxyfluoroquinolone, after repeated oral administration. *Clinical Pharmacokinetics* 40(Suppl 1), 1–9.

Tacconelli, E., De Angelis, G., Cataldo, M.A., Pozzi, E., and Cauda, R. (2008) Does antibiotic exposure increase the risk of methicillin-resistant *Staphylococcus aureus* (MRSA) isolation? A systematic review and meta-analysis. *Journal of Antimicrobial Chemotherapy* 61, 26–38.

Tamma, P.D., Cosgrove, S.E., and Maragakis, L.L. (2012) Combination therapy for treatment of infections with Gram-negative bacteria. *Clinical Microbiology Reviews* 25, 450–470.

Thomas, J.K., Forrest, A., Bhavnani, S.M., Hyatt, J.M., Cheng, A., Ballow, C.H., and Schentag, J.J. (1998) Pharmacodynamic evaluation of factors associated with the development of bacterial resistance in acutely ill patients during therapy. *Antimicrobial Agents and Chemotherapy* 42, 521–527.

Urban, C., Rahman, N., Zhao, X., Mariano, N., Segal-Maurer, S., Drlica, K., and Rahal, J.J. (2001) Fluoroquinolone-resistant *Streptococcus pneumoniae* associated with levofloxacin therapy. *The Journal of Infectious Diseases* 184, 794–798.

Vance-Bryan, K., Guay, D.R., and Rotschafer, J.C. (1990) Clinical pharmacokinetics of ciprofloxacin. *Clinical Pharmacokinetics* 19, 434–461.

Weber, S.G., Gold, H.S., Hooper, D.C., Karchmer, A.W., and Carmeli, Y. (2003) Fluoroquinolones and the risk for methicillin-resistant *Staphylococcus aureus* in hospitalized patients. *Emerging Infectious Diseases* 9, 1415–1422.

Wegener, H.C. (2003) Antibiotics in animal feed and their role in resistance development. *Current Opinion in Microbiology* 6, 439–445.

Weiss, K. (2009) *Clostridium difficile* and fluoroquinolones: is there a link? *International Journal of Antimicrobial Agents* 33 (Suppl 1), S29–32.

Wener, K.M., Schechner, V., Gold, H.S., Wright, S.B., and Carmeli, Y. (2010) Treatment with fluoroquinolones or with β-lactam-β-lactamase inhibitor combinations is a risk factor for isolation of extended-spectrum-β-lactamase-producing *Klebsiella* species in hospitalized patients. *Antimicrobial Agents and Chemotherapy* 54, 2010–2016.

Wispelwey, B. (2005) Clinical implications of pharmacokinetics and pharmacodynamics of fluoroquinolones. *Clinical Infectious Diseases* 41(Suppl 2), S127–S135.

Wong-Beringer, A., Nguyen, L.H., Lee, M., Shriner, K.A., and Pallares, J. (2009) An antimicrobial stewardship program with a focus on reducing fluoroquinolone overuse. *Pharmacotherapy* 29, 736–743.

Wright, D.H., Brown, G.H., Peterson, M.L., and Rotschafer J.C. (2000) Application of fluoroquinolone pharmacodynamics. *Journal of Antimicrobial Chemotherapy* 46, 669–683.

Zelenitsky, S.A. and Ariano, R.E. (2010) Support for higher ciprofloxacin AUC 24/MIC targets in treating Enterobacteriaceae bloodstream infection. *Journal of Antimicrobial Chemotherapy* 65, 1725–1732.

20 Optimal Use of β-Lactam Antibiotics

Warren Rose* and Andrew Berti

School of Pharmacy University of Wisconsin, Madison, Wisconsin, US

Introduction

Since the beginning of the antibiotic era, β-lactam antibiotics have been the cornerstone of anti-infective therapy. The supreme clinical effectiveness of β-lactams can be attributed to many attractive traits of this antimicrobial class: a rapidly bactericidal mechanism of action combined with an excellent safety profile that allows patients to tolerate a relatively large amount of drug exposure. The foundation of antimicrobial pharmacodynamics can be largely attributed to investigations into the dose–activity relationships of β-lactam antibiotics to optimize in vivo activity (Craig, 1998). β-Lactam antibiotics are arguably the most diverse class of antimicrobials, with multiple agents in the subclasses of penicillin, cephalosporin, carbapenem, and monobactam antibiotics. This diversity also extends to the spectrum of coverage of this class, which ranges from narrow spectrum therapy with antistaphylococcal penicillins such as nafcillin, to carbapenem antibiotics possessing some of the broadest antimicrobial coverage. These factors have led to β-lactams being a primary target of antimicrobial stewardship programs so that the effectiveness of this critical class of antimicrobials can be preserved and the unintended consequence of overuse be limited.

The problem of antimicrobial resistance is exemplified by the evolution of resistance mechanisms to β-lactam antibiotics. The US Centers for Disease Control and Prevention (CDC) has identified public health threat levels of antibiotic resistance in the US, and β-lactam resistance plays a prominent role. Among the three resistant organisms listed as the highest level threats, β-lactam resistance contributes to two, the carbapenem-resistant *Enterobacteriaceae* and the cephalosporin-resistant *Neisseria gonorrhea*, while the third of these highest threats, *Clostridium difficile*, has arisen a consequence of the use of broad-spectrum antibiotics that include the β-lactams (CDC, 2013). The CDC report also describes other significant threats that compromise the effectiveness of β-lactams, such as methicillin-resistant *Staphylococcus aureus* (MRSA), extended-spectrum β-lactamase (ESBL)-producing *Enterobacteriaceae*, and multidrug resistant (MDR) *Pseudomonas aeruginosa*.

This chapter focuses on the optimal use of β-lactam antibiotics in light of the current climate of antibiotic resistance and the need for antibiotic stewardship. To provide a foundation for the optimal use of these diverse agents, the first part of the chapter provides an overview of the principles of β-lactam activity and pharmacology; the second part presents the evidence for optimizing their use in antimicrobial stewardship initiatives.

Principles of β-Lactam Antibiotics

Antimicrobial stewardship with β-lactam antibiotics first requires a detailed understanding of the classification, mechanism of action and of resistance, antimicrobial spectrum, pharmacokinetics/pharmacodynamics (PK/PD) of common clinical uses, and adverse events profile for each subclass, and further by individual agent when applicable. This knowledge is essential to designing stewardship protocols that will most effectively optimize outcomes in patients.

Chemistry

β-Lactam antibiotics all share a common basic structural feature, the β-lactam ring, which is essential for their antibiotic activity. Subclasses of β-lactams differ in the nucleus connected to the β-lactam ring,

*Corresponding author. E-mail: warren.rose@wisc.edu

as well in the side chain, and such differences may increase (or decrease) the binding affinity responsible for the spectrum of activity and/or drug delivery aspects of the molecule. Both penicillins and cephalosporins possess sulfur-containing rings connected to the β-lactam, with the penicillins containing a five-member thiazolidine ring and the cephalosporins a six-member dihydrothiazine ring (Craig and Andes, 2014; Doi and Chambers, 2014a). Carbapenem antibiotics differ by having a double bond between C2 and C3 and the sulfur in the five member thiazolidine ring replaced with a carbon atom (Doi and Chambers, 2014b). The monobactam antibiotic, aztreonam, is different than other β-lactams in that it contains only a single β-lactam ring (i.e., it is monocyclic) with a 1-sulfonic acid group that activates the β-lactam (Doi and Chambers, 2014b). β-Lactamase inhibitors are also β-lactam compounds but provide weak antibiotic activity. Instead, their chemical structures allow them to have greater affinity for β-lactamases and inactivate these enzymes, thereby allowing the primary β-lactam to remain active (Doi and Chambers, 2014a).

Mechanism of action

β-Lactams possess one of the most potent mechanisms for antibiotic activity. While the chemistry among the various β-lactams may differ, the mechanism of action for all agents in this class is conferred through its inhibition of bacterial cell wall synthesis, resulting in rapid bactericidal activity. Bacteria contain penicillin-binding proteins (PBPs) that catalyze a transpeptidase reaction that is responsible for cell wall cross-linking, formation, and integrity. β-Lactams covalently bind to PBPs and therefore halt cell wall formation, resulting in cell death. However, the relative affinity for different PBPs can vary with each β-lactam (Sauvage *et al.*, 2008). Likewise, PBP function and numbers can vary between bacterial genera. For example, Gram-positive cocci typically have three to five PBPs, whereas Gram-negative bacilli may contain up to ten PBPs. To further cloud the issue, between organisms, the PBPs found may have similar functions but different nomenclature, such as PBP1 of *S. aureus*, which is homologous to PBP3 of *Escherichia coli* (Sauvage *et al.*, 2008). The β-lactams that bind to a variety of PBPs across different organisms as well as those that target higher molecular weight PBPs typically have the highest spectrum of antimicrobial activity (Sauvage *et al.*, 2008).

Spectrum of activity

The spectrum of activity of β-lactams is highly diverse and is one of the factors responsible for their high utility and with significant implication for antimicrobial stewardship. The spectra of activity of each subclass are outlined below.

The penicillins consist of the natural penicillins, penicillinase-resistant (semi-synthetic) penicillins, aminopenicillins, carboxypenicillins, and acyl ureidopenicillins. Natural penicillins, such as penicillin V and penicillin G, are active against only non-β-lactamase-producing Gram-positive organisms, anaerobes, and Gram-negative organisms (cocci) (Doi and Chambers, 2014a). These agents are highly potent and effective against susceptible isolates and often remain the treatment of choice in these cases. Semi-synthetic penicillins, such as nafcillin, oxacillin, and dicloxacillin, remain active against strains that produce penicillinase, a trait that is commonly found among strains of *S. aureus* and *S. epidermidis*, as well as some *Streptococcus* spp. (Doi and Chambers, 2014a).

The aminopenicillins ampicillin and amoxicillin are not active against strains that produce β-lactamase, but otherwise they remain active against a broad range of Gram-positive, Gram-negative, and anaerobic bacteria, including *Enterococcus* spp. (Doi and Chambers, 2014a). The two agents in the carboxypenicillin subclass, carbenicillin and ticarcillin, are no longer manufactured for clinical use in the US. Lastly, the ureidopenicillins are commonly referred to as the anti-pseudomonal penicillins owing to their activity against *P. aeruginosa* and other resistant Gram-negative rods, but they also retain activity against β-lactamase-producing Gram-positive organisms (Doi and Chambers, 2014a). Piperacillin is currently the only agent in the ureidopenicillin subclass that is used, when it is co-formulated with tazobactam. The β-lactamase inhibitors, clavulanic acid (clavulanate), sulbactam, and tazobactam have been historically formulated with penicillins to increase their potency against β-lactamase producing strains (Doi and Chambers, 2014a). Recently, these agents along with newer β-lactamase inhibitors such as avibactam have been co-formulated with cephalosporins (Cho *et al.*, 2015), while carbapenem-β-lactamase inhibitor combinations are in development.

The cephalosporins are a diverse subgroup of β-lactams that are highly used in both inpatient and outpatient settings. Shortly after their discovery, a number of cephalosporin derivatives were developed by the pharmaceutical industry, and these resulted in a large number of clinical options being available. The six-member ring of cephalosporins is generally stable against basic β-lactamase enzymes, and so they remain active against a variety of organisms (Craig and Andes, 2014). Cephalosporins have historically been classified into four subclasses—first through fourth generation—based on development and targeted activity. This scheme has also been used to identify trends in activity among the different agents, but newer agents, such as ceftaroline and ceftolozane, may not fit into these classifications. The description of activity here will use the classic nomenclature, and the newer, uncharacteristic agents will be described separately. Also, other cephalosporins in the generation classes may be clinically available, but information on the most commonly used agents is presented as these are typical considerations for antimicrobial stewardship.

First-generation cephalosporins, most prominently cephalexin and cefazolin, are active against β-lactamase producing Gram-positive organisms of the staphylococci and streptococci, and have no appreciable activity against Gram-negative organisms. The second-generation cephalosporins, cefuroxime, cefaclor, and cefprozil, maintain activity against Gram-positive cocci but have a broadened spectrum that includes activity against Gram-negative bacilli and coccobacilli. The cephamycins are a subgroup of second-generation cephalosporins that includes cefoxitin and cefotetan, which are similar in activity to other second-generation cephalosporins, but also are active against Gram-negative anaerobes, including *Bacteroides* spp.

Third-generation cephalosporins are generally thought to have more pronounced Gram-negative activity but are not as potent against Gram-positive organisms compared with other cephalosporins. In this group, cefotaxime, ceftazidime, and ceftriaxone are commonly used intravenous options, while cefdinir, cefditoren, cefixime, cefpodoxime, and ceftibuten can be administered by the oral route. Of these agents, only ceftazidime is reliably active against *P. aeruginosa* (Craig and Andes, 2014). The β-lactamase inhibitor avibactam has been recently formulated with ceftazidime to target highly resistant β-lactamases produced by *P.* aeruginosa and *Enterobacteriaceae*, most notably the clinical threats imposed by *Klebsiella pneumoniae* carbapenemase (KPC)-producing strains and by *Enterobacteriaceae* producing OXA-48-type carbapenemases (Liscio *et al.*, 2015; Nicolau, 2015). Ceftriaxone is also a notable exception because it remains highly active against many gram-positive organisms. The only member of the fourth-generation cephalosporin subgroup, cefepime, is renowned for its broad-spectrum activity. It is active against many Gram-positive organisms and Gram-negative rods, including *P. aeruginosa*, and it retains some anaerobic activity (Craig and Andes, 2014). Although cefepime is a broad-spectrum agent, it may not necessarily be the most active against a given organism.

The most recently developed cephalosporins offer some distinction to the prior generational classes. Ceftaroline is a novel cephalosporin that is similar in activity to ceftriaxone, but it is able to bind to PBP2a. Therefore, it is the only cephalosporin with activity against MRSA (Rodvold and McConeghy, 2014), though a similar product, ceftobiprole, is available in many other developed countries (Dauner *et al.*, 2010). Despite their novel mechanism of action, resistance to these two cephalosporins in the treatment of MRSA has already been described (Chan *et al.*, 2015). Ceftolozane is the most recently developed cephalosporin, and is similar in structure to ceftazidime, but it is unique in spectrum because of its activity against AmpC (ampicillin C-hydrolyzing) β-lactamase and ESBL-producing *Enterobacteriaceae*, *P. aeruginosa*, and anaerobes when combined with tazobactam (Cho *et al.*, 2015).

Carbapenem antibiotics provide the widest spectrum of activity among the β-lactam class. The activity is derived from their ability to bind to specific PBPs in a variety of organisms, as well as their relative stability to degradation by β-lactamase enzymes. The available agents in this subclass include imipenem (formulated with cilastin to improve imipenem pharmacokinetics), meropenem, ertapenem, and doripenem. The antibiotic activity of these agents are similar overall, with some notable exceptions. Carbapenems are active against Gram-positive, Gram-negative, and anaerobic organisms. They maintain activity against many types of β-lactamases, including ESBLs. Imipenem, meropenem, and doripenem are active against non-fermenting Gram-negative rods, but meropenem and doripenem tend to be more active against *P. aeruginosa* and *Acinetobacter baumanni* and therefore tend to be used more for empiric treatment (Sader *et al.*, 2014). Ertapenem is unique in the carbapenem class because it can be considered a pseudomonad-sparing carbapenem as it does not have activity against non-fermenting Gram-negative rods. Despite their broad-spectrum activity, carbapenems are inactive against MRSA and *Enterococcus* spp., and their widespread use has resulted in carbapenemase-producing organisms, primarily in *K. pneumoniae* (Munoz-Price *et al.*, 2013). Carbapenem combinations with β-lactamase inhibitors are in development to target these carbapenemase-producing strains as well as other clinically challenging, MDR Gram-negative organisms (Lapuebla *et al.*, 2015).

Aztreonam is a narrow spectrum monobactam antibiotic due to its affinity for PBP3 in Gram-negative organisms. Its activity is mostly limited to enteric Gram-negative rods; it is not active against Gram-negative organisms that produce ESBLs or other advanced β-lactamases. It is also not active against Gram-positive organisms or anaerobes. *P. aeruginosa* has been shown to be susceptible to aztreonam, but its use is directed for this organism only after sensitivities have been confirmed. Because of its comparatively weak antibiotic spectrum and activity, aztreonam is often reserved for patients with allergic reactions to β-lactams, as it presents no IgE antibody-mediated cross-reactivity risk (Doi and Chambers, 2014b).

Mechanisms of resistance

The use of antibiotics selects for subpopulations of bacteria that are able to survive the exposure. Clonal expansion of these subpopulations or horizontal transfer of their resistance determinants can lead to resistance and treatment failure. These subpopulations typically rely upon one of three general strategies to survive antibiotic exposure. The bacteria can: (i) restrict access to the antibiotic target; (ii) modify the target to become less sensitive to the antibiotic; and/or (iii) produce enzymes that inactivate the antibiotic before it can reach its target. All three of these strategies have been employed by different bacteria to confer resistance to β-lactams.

The outer membrane that is present in all Gram-negative bacteria acts as a natural barrier that prevents β-lactam antibiotics from reaching their target in the inner membrane. However, most Gram-negative bacteria contain porin proteins in the outer membrane that allow hydrophilic compounds such as nutrients and signaling molecules to cross the otherwise hydrophobic barrier and enter the periplasmic space. β-Lactam antibiotics maintain activity against Gram-negative bacteria because they can use these porins to access the periplasm. Bacteria that restrict the number or size of these porins can reduce the flux across the outer membrane and increase their resistance to a broad range of β-lactams. *P. aeruginosa* is a common organism that employs this strategy to restrict access to the target in order to resist β-lactam activity.

As described previously, β-lactams exert their antibiotic effect by inactivating the PBPs that are essential for the enzymatic cross-linking of bacterial cell walls. Exposure to β-lactams has selected for bacteria that modify the antibiotic target by producing variants of the PBP enzymes that maintain cross-linking activity but have reduced affinity for β-lactams. The most well-known example of these alternative PBPs is the PBP2a encoded by the SCC*mec* (staphylococcal cassette chromosome *mec*) mobile genetic element of MRSA, but several other bacteria, including *Streptococcus pneumoniae*, *N. gonorrhoeae*, *Enterococcus faecium*, and *Haemophilus influenzae* have independently evolved their own PBPs with reduced affinity for β-lactam substrates. These changes typically result in reduced susceptibility to all classes of β-lactam except for the advanced-generation cephalosporins, such as ceftaroline and ceftobiprole (van Hal and Paterson, 2011).

Several bacteria are capable of producing enzymes that are able to reduce the active pharmacophore of β-lactam antibiotics and render them pharmacologically inert before they can reach their PBP target. These enzymes can be classified according to which β-lactam structure functions as an enzyme substrate. Penicillinases preferentially hydrolyze the penicillin class of β-lactams, whereas cephalosporinases preferentially hydrolyze cephalosporins. β-Lactamases are capable of hydrolyzing both penicillins and cephalosporins. These enzymes can be encoded on mobile genetic elements such as plasmids and transferred between organisms or they can be coded in the chromosome and stably maintained among all isolates of phylogenetically related bacteria. In a complication to sensitivity testing, β-lactamases may be constitutively expressed or induced by exposure to β-lactams, thus potentially leading to the misidentification of inducible-resistant bacteria as "sensitive." Alternatively, enzymes with β-lactamase activity may be produced constitutively but at levels too low to confer resistance. In this scenario, strains may be classified as "resistant" using rapid testing methods but remain clinically susceptible (Sader *et al.*, 2006).

The widespread use of β-lactams has resulted in a variety of β-lactamase enzymes, some of which are able to resist the effects of even the most potent antibiotics in this class. According to the Ambler classification system, β-lactamase enzymes are categorized into four classes (A, B, C, and D) based on their amino acid sequences (Bush and Jacoby, 2010). Classes A, C, and D use serine residues for β-lactam hydrolysis, while class B enzymes utilize divalent zinc ions. Class A enzymes can be either chromosomally encoded or plasmid mediated and include broad-spectrum β-lactamases (e.g., TEM-1, SHV-1), extended-spectrum β-lactamases (CTX-M-15), and carbapenemases (KPC-2). The class B enzymes, such as the NDM-1, IMP, and VIM enzymes are metallo-β-lactamases. Class C enzymes are chromosomally encoded (e.g. AmpC) or plasmid mediated (CMY-2), can be inducible, are resistant to commercially available β-lactamase inhibitors, and inactivate penicillins and most cephalosporins (with the exception of cefepime). Class D enzymes are referred to as oxacillinases as they favor oxacillin as a substrate, and are typically produced by non-fermenters such as *Pseudomonas* and *Acinetobacter* spp. Like the class A enzymes, they include broad-spectrum and carbapenemase producers (OXA-48), but typically are less potent and therefore rely on multiple resistance mechanisms to inactivate β-lactams (Bush and Jacoby, 2010).

Pharmacokinetics/pharmacodynamics

The diversity of β-lactams has allowed for extensive study into the PK/PD of each subclass and each representative agent. Although it seems inherent based on the current approaches with β-lactam treatment, the idea of changing PK/PD exposure to optimize β-lactam activity was a different approach to antimicrobial management. Using an in vivo mouse thigh or lung infection model, Dr. William Craig at the University of Wisconsin-Madison investigated different exposure parameters of a number of β-lactam antibiotics. Using the approach of providing the same amount of drug exposure (i.e., area under the concentration vs. time curve, or AUC), but altering the dose frequency, a significant correlation was noted between the antibacterial effect and the percentage of time (T) that the antibiotic levels remained above the minimum inhibitory concentration (MIC) for the organism concerned (Craig, 1998). The PK/PD concept of time above the MIC for β-lactams has been consistently found for all β-lactam antibiotics. However, the magnitude of time above the MIC has been shown to differ among the β-lactam classes for maximum target attainment needed for antibiotic activity.

As shown in Table 20.1, carbapenems require the lowest percentage of time above the organism MIC, with a higher time above the MIC for penicillins and the highest time for cephalosporins (Craig, 1998). This in vitro and in vivo discovery has been translated into the clinical setting with great success as described in the second part of the chapter on antimicrobial stewardship initiatives. Advances in computer modeling allow individual antimicrobial stewardship programs to identify the ability to treat resistant pathogen at their institutions. Software programs that perform Monte Carlo simulations—a technique for determining situational probability—can determine the probability of target attainment for a β-lactam regimen against a population of organisms with a given susceptibility profile (Patel *et al.*, 2010).

Adverse reactions

Overall, β-lactam antibiotics remain one of the safest types of antibiotics, given the relatively high doses administered and their widespread use in the general patient population. Even so, the potential for specific adverse events should be monitored as part of routine antimicrobial stewardship measures. Alleged allergic reactions to β-lactam antibiotics are one of the most commonly reported drug allergies, but true allergic reactions are much less common. Frequently, these reactions are misdiagnosed, resulting in unnecessary use of broad-spectrum or suboptimal therapies. Overall, β-lactams are safe drugs, with β-lactam allergy documented in less than 10% of individuals (Macy, 2014). For those patients who do experience a true adverse reaction to the antibiotic, clinical presentation can vary widely from a mild, transitory skin reaction resulting in minor patient discomfort to life-threatening anaphylaxis.

Type I hypersensitivities are medical emergencies involving IgE release, mast cell degranulation, vasodilation and smooth muscle contraction. Presentation may include urticaria, pruritus, flushing, and/or laryngeal edema. Extensive involvement, or lack of intervention with epinephrine, antihistamines, and corticosteroids may result in progression to airway

Table 20.1. Pharmacokinetic/pharmacodynamic target attainments for β-lactam antibiotics expressed as time required above minimum inhibitory concentration (MIC).

	Organism type	
β-Lactam class	Gram positive	Gram negative
Carbapenems	20–30%	40–50%
Penicillins	30–40%	50–60%
Cephalosporins	40–50%	60–70%

compromise. Type I hypersensitivity to penicillin can occur early (within 4 h of the first dose) or late (several days into therapy). The estimated prevalence of type I hypersensitivity to penicillin in the general population is 10–50 cases/100,000 individuals (Macy, 2014). A patient who experiences a type I hypersensitivity to a β-lactam agent should not receive additional doses of the same agent without first receiving desensitization therapy.

Because of the clinical effectiveness of β-lactam antibiotics, these agents may be preferred antimicrobials even in patients with documented allergic reactions. Therefore, many stewardship programs have a β-lactam desensitization protocols which are initiated with a dilute concentration of drug followed by a doubling of the dose every 15–30 mins until a full therapeutic dose is reached. This approach is usually successful in desensitizing the patient to β-lactams, although mild reactions, such as pruritus and urticaria, have been documented in up to 30% of patients during and after desensitization (Yates, 2008). Another widely used option is to treat patients with a β-lactam from a different subclass. A common example is to use cephalosporins in patients with a documented allergy to penicillins. Overall, the data surrounding the cross-reactivity among the β-lactam subclasses suggests that the risk is low (Gonzalez-Estrada and Radojicic, 2015), but studies to evaluate this effect have been poorly designed. It is generally accepted that the cross-reactivity risk is up to 10% among the β-lactam subclasses, but with actual use in the clinical setting, the cross-reactivity risk is lower than previously appreciated (Gonzalez-Estrada and Radojicic, 2015). Another approach for patients with a questionable history of β-lactam allergy is to order a penicillin skin test and graded dose challenge.

Serum sickness is a distinct immune-mediated hypersensitivity that can occur late in therapy (after 4–10 days). Presentation may include urticaria, pruritus, arthralgia, fever, and lymphadenopathy. Incidence of serum sickness due to β-lactams is low, occurring in 7–40 individuals/100,000 prescriptions administered. Discontinuation of the β-lactam typically results in symptom resolution, though when clinically indicated, it may be reasonable to continue β-lactam therapy with the offending agent for the duration of treatment with or without supportive corticosteroid therapy for symptom management.

Central nervous system (CNS) toxicity in the form of seizures has been connected to high-dose β-lactam therapy. The most common agents that have been associated with this adverse effect are penicillins, cefepime, and carbapenems. However, any β-lactam may have this potential when used at high doses (Asbel and Levison, 2000). The seizure potential of carbapenems has been of considerable debate since the introduction of imipenem, and data demonstrating that carbapenems may displace the inhibitory neurotransmitter γ-aminobutyric acid (GABA) in the CNS, which could result in convulsions (Miller *et al.*, 2011). Of the carbapenems, imipenem appears to have the highest risk, whereas the seizure risk with meropenem and doripenem is much lower (Zhanel *et al.*, 2009). Overall, the risk of CNS toxicity, including seizures, is increased with any β-lactam use in patients with renal dysfunction, as this could result in excessive levels in the absence of antibiotic stewardship intervention to adjust the renal dose (Chow *et al.*, 2003).

β-Lactam-induced renal dysfunction is another adverse event that can occur in patients. There are various mechanisms through which this can occur in the kidney. Interstitial nephritis induced by methicillin is the classic example, but this drug is no longer in clinical use. The related agents oxacillin and nafcillin have a much lower propensity for this effect, but interstitial nephritis with the use of these agents has been reported (Murray and Keane, 1992). This effect generally presents as acute renal failure, with hematuria, proteinuria, and pyuria. Patients may also have other signs of hypersensitivity during this acute phase, including fever, eosinophilia, and rash (Murray and Keane, 1992). Removal of the β-lactam usually reverses the toxicity. Cross-reactivity with other β-lactams has been reported, so avoiding the entire class is advisable in patients with β-lactam-induced interstitial nephritis. Other renal toxicity mechanisms have been described with multiple β-lactams, including glomerulonephritis and aminoglycoside-induced nephrotoxicity when they are used in combination.

One of the most concerning unintended consequences of broad-spectrum antibiotic therapy is the emergence of *C. difficile* infection (CDI), which has been associated with multiple β-lactams. In a systemic review of hospital-acquired *C. difficile* associated with antibiotics, third-generation cephalosporins had the highest association among all antibiotics with over threefold greater odds of developing CDI (Slimings and Riley, 2014). In the same study, the risk of developing CDI was higher than that associated with clindamycin, which has typically been associated with high-level risk. Of further concern, second- and fourth-generation cephalosporins were each associated with over twofold greater odds and penicillins with almost 1.5-fold greater odds of developing CDI (Slimings and Riley, 2014). This evidence provides notice for antimicrobial stewardship programs to reduce unnecessary broad-spectrum β-lactam exposure to limit the potential for CDI infection (Dellit *et al.*, 2007).

Antimicrobial Stewardship Initiatives for β-Lactam Antibiotics

Antimicrobial stewardship programs are designed to ensure that antibiotics are administered appropriately and in such a way as to derive maximum benefit and minimize toxicity. β-Lactams are the cornerstone of antimicrobial therapy in the hospital and community setting, so it is essential that prudent antimicrobial stewardship practices target this class of antibiotics to ensure their effectiveness. As has previously been discussed in greater detail in this chapter, the PK/PD parameter most associated with β-lactam effectiveness is time (T) above MIC of the organisms. Multiple strategies have been employed successfully by antimicrobial stewardship programs in order to optimize time above MIC and improve the effectiveness of β-lactams. Many mature stewardship programs have developed prolonged infusion administration as a strategy to optimize the activity β-lactams and prevent resistance development. Based on Monte Carlo modeling, extending the infusion time of β-lactams can significantly improve the likelihood of treating Gram-negative organisms (Lodise *et al.*, 2006).

Figure 20.1 shows the results of a Monte Carlo simulation of the probability of target attainment for the prolonged infusion of β-lactams compared with standard infusion. The probability of target attainment increases with prolonged infusion, particularly as the MIC of the organism increases, therefore the ability to treat organisms with reduced susceptibilities can still be achieved (Lee *et al.*, 2010). Piperacillin–tazobactam, meropenem, and cefepime are common β-lactams that are utilized as prolonged infusion. Administering β-lactams by prolonged infusion can result in fewer doses administered per day, resulting in lower total drug exposure without compromising antimicrobial effectiveness.

A retrospective study of 121 patients conducted at a surgical/medical intensive care unit (ICU) after implementing a prolonged infusion protocol for piperacillin–tazobactam and meropenem showed significant decreases in ventilator days (from 16.8 to 9.6 days, 95% confidence interval (CI) −12.4 to −2.4), ICU length of stay (from 15.3 to 10.7 days, 95% CI −8.3 to −1.4), and hospital length of stay (from 30.9 to 22.4 days, 95% CI −18.7 to −1.2) between the intermittent infusion and the prolonged infusion group. There was also a decrease in mortality in the prolonged infusion group (from 20.7 to 12.4%, odds ratio (OR) 0.54, 95% CI 0.18–1.66) that did not reach statistical significance. Use of the prolonged infusion was also associated with fewer doses of antibiotic and an estimated $10,000 of drug cost savings for the 54 patients included in the prolonged infusion group (Dow *et al.*, 2011). Several other examples of improved patient outcomes with this approach can be found in the literature, and prolonged β-lactam infusion has been accepted as a practice to improve outcome in patients with Gram-negative infections.

Considerations that may limit use of prolonged infusion include clinical scenarios such as meningitis, where there is a need for high maximum concentrations, or in a patient who requires significant intravenous access for other medications, particularly those with known drug–drug incompatibilities with β-lactams. Compatibility studies have recently been designed to evaluate vancomycin and piperacillin–tazobactam or cefepime compatibility at concentrations typically achieved during prolonged infusion. These studies have found no compatibility issues with these prolonged infusion β-lactams and vancomycin (Leung *et al.*, 2013; Berti *et al.*, 2015).

Broad-spectrum β-lactam antibiotics are frequently employed as initial empiric therapy, but their long-term use carries risks. Disruption of the normal gastrointestinal flora frequently results in nausea and diarrhea and can predispose a patient to CDI. It also selects for more resistant subpopulations of bacteria, even if these populations are not associated with the infection the clinician intends to treat. These subpopulations can persist as a reservoir of antibiotic resistance and may be a source of future infection. In females, disruption of the normal vaginal flora by broad-spectrum β-lactam use can result in overgrowth of yeast and vaginal candidiasis. Accordingly, antimicrobial stewardship programs should routinely evaluate the susceptibility trends of commonly used β-lactams in the institution concerned as well as track the superinfection rates that can occur with broad-spectrum agents.

De-escalation is a common objective of many antimicrobial stewardship programs. The broad-spectrum activity of some β-lactam antibiotics makes them

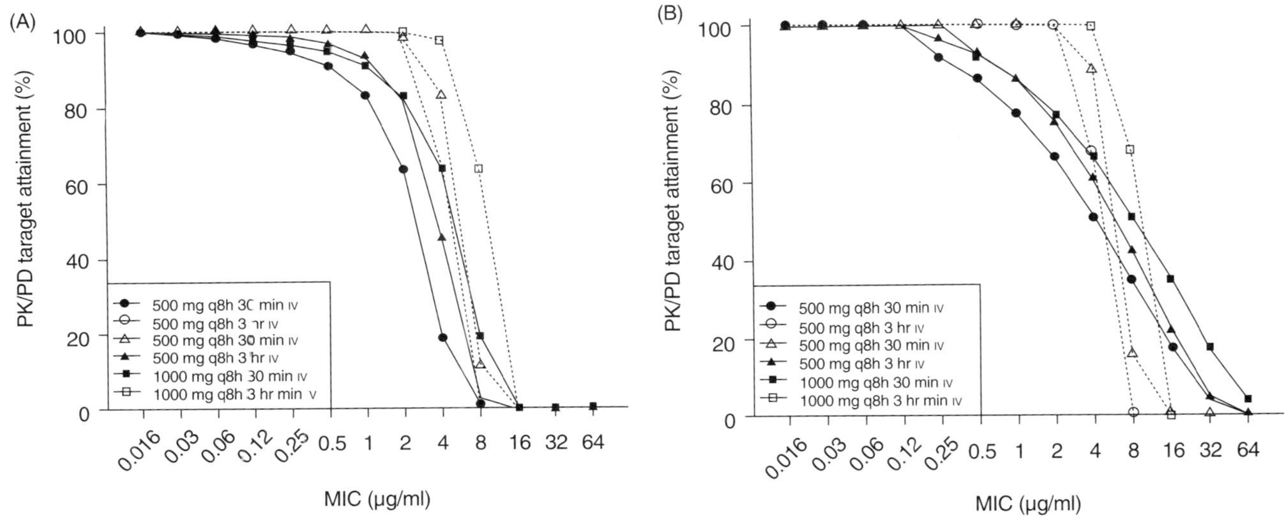

Fig. 20.1. Comparison of 30 min and 3 hr (intravenous (IV) infusion regimens of β-lactams in healthy adults evaluated by Monte Carlo simulation. Percentage probability of attaining a pharmacokinetic/pharmacodynamic (PK/PD) target of 40% of the time (T) above the minimum inhibitory concentration (MIC) of antibiotic, i.e., 40% T > MIC, for various dosing regimens of (A) imipenem–cilastatin and (B) merpenem. Adapted from Lee *et al.* (2010) with permission.

key targets for rapid de-escalations when antibiotic sensitivities dictate narrowing the scope of coverage or when antibiotics are no longer required. However, in practice, narrowing from broad-spectrum therapy may be difficult as the patient may show clinical improvement on broad-spectrum therapy, and there may have been provider handoff (i.e., handover or transfer of information during transitions in care) omissions or simple oversight of new results on susceptibility data. Antibiotic de-escalation requires a dedicated approach by an antibiotic stewardship team, and the reports of successful de-escalation efforts require multidisciplinary involvement. Multiple studies have targeted carbapenems for de-escalation as a means of reducing exposure. These studies have shown a benefit to reducing carbapenem use in a variety of settings, although the benefits for resistance patterns were minimal (Hibbard *et al.*, 2010; Ahmad *et al.*, 2014). This approach has also been successful in a resource-limited setting and for other broad-spectrum β-lactams (Apisarnthanarak *et al.*, 2013). De-escalation or consolidation of β-lactam therapies are prudent means to reducing unnecessary antibiotic exposure that may lead to reduction in resistance and adverse events in the patient. The potential for reducing antimicrobial resistance for the hospital or system is highly complex, and so this may require additional measures, including infection control and other restrictions, in order to see large improvements in susceptibility trends.

Conclusion

Many of the advances of modern medicine have been made possible because of the effectiveness of β-lactam antibiotics to prevent and treat infections. The development of resistance among different organisms to a variety of β-lactams is of significant concern due to the potential for rapid global spread and limited new agents available for treatment. β-Lactam antibiotic stewardship is an important principle for preserving the activity of this important antimicrobial class. Some of the initiatives described in this chapter, including de-escalation, antibiotic restriction, and prolonged infusion often require a multidisciplinary effort for full implementation. Other initiatives, such as renal dosing adjusted for β-lactams, can be maintained by pharmacists with physician support and oversight. These initiatives will help to preserve the utility of the available β-lactam antibiotics and ensure the judicious use of new β-lactams in development once they are presented in the clinical setting.

References

Ahmad, F., Pogue, J.M., Marchaim, D., Chopra, T., Bheemreddy, S., Lee, J., Mudegowdra, N.S., Chaudhry, A., and Kaye, K.S. (2014) Evaluation of the potential impact of a carbapenem de-escalation program in an academic healthcare system. *Journal of Infection and Public Health* 7, 50–53.

Apisarnthanarak, A., Bhooanusas, N., Yaprasert, A., and Mundy, L.M. (2013) Carbapenem de-escalation therapy in a resource-limited setting. *Infection Control and Hospital Epidemiology* 34, 1310–1313.

Asbel, L.E. and Levison, M.E. (2000) Cephalosporins, carbapenems, and monobactams. *Infectious Disease Clinics of North America* 14, 435–447, ix.

Berti, A.D., Hutson, P.R., Schulz, L.T., Webb, A.P., and Rose, W.E. (2015) Compatibility of cefepime and vancomycin during simulated Y-site administration of prolonged infusion. *American Journal of Health-System Pharmacy* 72, 390–395.

Bush, K. and Jacoby, G.A. (2010) Updated functional classification of β-lactamases. *Antimicrobial Agents and Chemotherapy* 54, 969–976.

CDC (2013) *Antibiotic Resistance Threats in the United States, 2013.* Centers for Disease Control and Prevention, Atlanta, Georgia. Available at: http://www.cdc.gov/drugresistance/threat-report-2013/pdf/ar-threats-2013-508.pdf (accessed 28 October 2014).

Chan, L.C., Basuino, L., Diep, B., Hamilton, S., Chatterjee, S.S., and Chambers, H.F. (2015) Ceftobiprole- and ceftaroline-resistant methicillin-resistant *Staphylococcus aureus*. *Antimicrobial Agents and Chemotherapy* 59, 2960–2963.

Cho, J.C., Fiorenza, M.A., and Estrada, S.J. (2015) Ceftolozane/tazobactam: a novel cephalosporin/β-lactamase inhibitor combination. *Pharmacotherapy* 35, 701–715.

Chow, K.M., Szeto, C.C., Hui, A.C., Wong, T.Y., and Li, P.K. (2003) Retrospective review of neurotoxicity induced by cefepime and ceftazidime. *Pharmacotherapy* 23, 369–373.

Craig, W.A. (1998) Pharmacokinetic/pharmacodynamic parameters: rationale for antibacterial dosing of mice and men. *Clinical Infectious Diseases* 26, 1–10; quiz 11–12.

Craig, W.A. and Andes, D.R. (2014) Cephalosporins. In: Bennett, J.E. (ed.) *Mandell, Douglas, and Bennett's Principles and Practice of Infectious Diseases, Volume* 1, 8th edn. Elsevier Saunders, Philadelphia, Pennsylvania, pp. 278–292.

Dauner, D.G., Nelson, R.E., and Taketa, D.C. (2010) Ceftobiprole: a novel, broad-spectrum cephalosporin with activity against methicillin-resistant *Staphylococcus aureus*. *American Journal of Health-System Pharmacy* 67, 983–993.

Dellit, T.H., Owens, R.C., Mcgowan, J.E., Jr., Gerding, D.N., Weinstein, R.A., Burke, J.P., Huskins, W.C.,

Paterson, D.L., Fishman, N.O., Carpenter, C.F. *et al.* (2007) Infectious Diseases Society of America and the Society for Healthcare Epidemiology of America guidelines for developing an institutional program to enhance antimicrobial stewardship. *Clinical Infectious Diseases* 44, 159–177.

Doi, Y. and Chambers, H.F. (2014a) Penicillins and β-lactamase inhibitors. In: Bennett, J.E. (ed.) *Mandell, Douglas, and Bennett's Principles and Practice of Infectious Diseases, Volume 1*, 8th edn. Saunders, pp. 263–277.

Doi, Y. and Chambers, H.F. (2014b) Other β-lactam antibiotics. In: Bennett, J.E. (ed.) *Mandell, Douglas, and Bennett's Principles and Practice of Infectious Diseases, Volume* 1, 8th edn. Elsevier Saunders, Philadelphia, Pennsylvania, pp. 293–297.

Dow, R.D., Rose, W.E., Fox, B.C., Thorpe, J.M., and Fish, J.T. (2011) Retrospective study of prolonged versus intermittent infusion piperacillin–tazobactam and meropenem in intensive care unit patients at an academic medical center. *Infectious Diseases in Clinical Practice* 19, 413–417.

Gonzalez-Estrada, A. and Radojicic, C. (2015) Penicillin allergy: a practical guide for clinicians. *Cleveland Clinic Journal of Medicine* 82, 295–300.

Hibbard, M.L., Kopelman, T.R., O'Neill, P.J., Maly, T.J., Matthews, M.R., Cox, J.C., Vail, S.J., Quan, A.N., and Drachman, D.A. (2010) Empiric, broad-spectrum antibiotic therapy with an aggressive de-escalation strategy does not induce Gram-negative pathogen resistance in ventilator-associated pneumonia. *Surgical Infections* 11, 427–432.

Lapuebla, A., Abdallah, M., Olafisoye, O., Cortes, C., Urban, C., Quale, J., and Landman, D. (2015) Activity of meropenem combined with RPX7009, a novel β-lactamase inhibitor, against Gram-negative clinical isolates in New York City. *Antimicrobial Agents and Chemotherapy* 59, 4856–4860.

Lee, L.S., Kinzig-Schippers, M., Nafziger, A.N., Ma, L., Sörgel, F., Jones, R.N., Drusano, G.L., and Bertino, J.S., Jr. (2010) Comparison of 30-min and 3-h infusion regimens for imipenem/cilastatin and for meropenem evaluated by Monte Carlo simulation. *Diagnostic Microbiology and Infectious Diseases* 68, 251–258.

Leung, E., Venkatesan, N., Ly, S.C., and Scheetz, M.H. (2013) Physical compatibility of vancomycin and piperacillin sodium–tazobactam at concentrations typically used during prolonged infusions. *American Journal of Health-System Pharmacy* 70, 1163–1166.

Liscio, J.L., Mahoney, M.V., and Hirsch, E.B. (2015) Ceftolozane/tazobactam and ceftazidime/avibactam: two novel β-lactam/β-lactamase inhibitor combination agents for the treatment of resistant Gram-negative bacterial infections. *International Journal of Antimicrobial Agents* 46, 266–271.

Lodise, T.P., Lomaestro, B.M., and Drusano, G.L. (2006) Application of antimicrobial pharmacodynamic concepts into clinical practice: focus on β-lactam antibiotics. Insights from the Society of Infectious Diseases Pharmacists. *Pharmacotherapy* 26, 1320–1332.

Macy, E. (2014) Penicillin and beta-lactam allergy: epidemiology and diagnosis. *Current Allergy and Asthma Reports* 14: 476.

Miller, A.D., Ball, A.M., Bookstaver, P.B., Dornblaser, E.K., and Bennett, C.L. (2011) Epileptogenic potential of carbapenem agents: mechanism of action, seizure rates, and clinical considerations. *Pharmacotherapy* 31, 408–423.

Munoz-Price, L.S., Poirel, L., Bonomo, R.A., Schwaber, M.J., Daikos, G.L., Cormican, M., Cornaglia, G., Garau, J., Gniadkowski, M., Hayden, M.K. *et al.* (2013) Clinical epidemiology of the global expansion of *Klebsiella pneumoniae* carbapenemases. *The Lancet Infectious Diseases* 13, 785–796.

Murray, K.M. and Keane, W.R. (1992) Review of drug-induced acute interstitial nephritis. *Pharmacotherapy* 12, 462–467.

Nicolau, D.P. (2015) Focus on ceftazidime–avibactam for optimizing outcomes in complicated intra-abdominal and urinary tract infections. *Expert Opinion on Investigational Drugs* 24, 1261–1273.

Patel, N., Scheetz, M.H., Drusano, G.L., and Lodise, T.P. (2010) Identification of optimal renal dosage adjustments for traditional and extended-infusion piperacillin–tazobactam dosing regimens in hospitalized patients. *Antimicrobial Agents and Chemotherapy* 54, 460–465.

Rodvold, K.A. and McConeghy, K.W. (2014) Methicillin-resistant *Staphylococcus aureus* therapy: past, present, and future. *Clinical Infectious Diseases* 58(Suppl 1), S20–S27.

Sader, H.S., Fritsche, T.R., and Jones, R.N. (2006) Accuracy of three automated systems (MicroScan WalkAway, Vitek, and Vitek 2) for susceptibility testing of *Pseudomonas aeruginosa* against five broad-spectrum beta-lactam agents. *Journal of Clinical Microbiology* 44, 1101–1104.

Sader, H.S., Farrell, D.J., Flamm, R.K., and Jones, R.N. (2014) Antimicrobial susceptibility of Gram-negative organisms isolated from patients hospitalized in intensive care units in United States and European hospitals (2009–2011). *Diagnostic Microbiology and Infectious Diseases* 78, 443–448.

Sauvage, E., Kerff, F., Terrak, M., Ayala, J.A., and Charlier, P. (2008) The penicillin-binding proteins: structure and role in peptidoglycan biosynthesis. *FEMS Microbiology Reviews* 32, 234–258.

Slimings, C. and Riley, T.V. (2014) Antibiotics and hospital-acquired *Clostridium difficile* infection: update of

systematic review and meta-analysis. *Journal of Antimicrobial Chemotherapy* 69, 881–891.
van Hal, S.J. and Paterson, D.L. (2011) New Gram-positive antibiotics: better than vancomycin? *Current Opinion in Infectious Diseases* 24, 515–520.
Yates, A.B. (2008) Management of patients with a history of allergy to beta-lactam antibiotics. *The American Journal of Medicine* 121, 572–576.
Zhanel, G.G., Ketter, N., Rubinstein, E., Friedland, I., and Redman, R. (2009) Overview of seizure-inducing potential of doripenem. *Drug Safety* 32, 709–716.

21 Current Approach to Optimal Use and Dosing of Vancomycin in Adult Patients

JOSEPH J. CARRENO[1], DMITRIY MARTIROSOV,[2] AND THOMAS P. LODISE[1*]

[1]*Albany College of Pharmacy and Health Sciences, Albany, New York, US;*
[2]*Henry Ford Hospital, Detroit, Michigan, US*

Introduction

Despite its approval nearly 60 years ago, vancomycin remains a mainstay in therapy for invasive infections due to Gram-positive bacteria (Rybak *et al.*, 2009; Liu *et al.*, 2011). In the US alone, it is estimated that clinicians administer over 100 million days of vancomycin therapy a year (Kirst *et al.*, 1998). The high rate of vancomycin use is related to the growing prevalence of infections due to methicillin-resistant *Staphylococcus aureus* (MRSA). In the past, MRSA was confined to critical care unit settings (Moran *et al.*, 2005, 2006, 2012), but now it is the most common *S. aureus* antibiotic susceptibility phenotype across all healthcare settings, including general hospital patient wards, skilled nursing homes and facilities, and dialysis centers (Capitano *et al.*, 2003; NNIS System, 2004; Reed *et al.*, 2005). Infections due to MRSA are also problematic in the community, where it is a predominant cause of invasive skin and soft tissue infections (Moran *et al.*, 2005; Davis *et al.*, 2007; Klevens *et al.*, 2007; Miller *et al.*, 2007; Terp *et al.*, 2014). Previously, MRSA was largely limited to patients with certain predisposing risk factors, such as prior antibiotic exposure, recent admission to a healthcare institution, presence of an indwelling central venous catheter, or dialysis (Lowy, 1998; Cosgrove *et al.*, 2003). However, more recent epidemiologic investigations suggest that the prevalence of infection due to MRSA is increasing among populations without these traditional risk factors (Culpepper *et al.*, 2001; Naimi *et al.*, 2003; Tobin-D'Angelo *et al.*, 2003; Baillargeon *et al.*, 2004; Fridkin *et al.*, 2005; Lee *et al.*, 2005; Moran *et al.*, 2005, 2006, 2012; Ochoa *et al.*, 2005; Zetola *et al.*, 2005; King *et al.*, 2006; Klevens *et al.*, 2007; Hidron *et al.*, 2009; David and Daum, 2010; Chua *et al.*, 2011; Skov *et al.*, 2012). In the community setting, MRSA prevalence has increased among young adults and children over the past 15–20 years and is directly related to the emergence of the new community-acquired MRSA strain, USA300 (Zetola *et al.*, 2005; David and Daum, 2010; Chua *et al.*, 2011; Skov *et al.*, 2012).

Given the high prevalence of MRSA infections across all patient settings, empirical anti-MRSA therapy is now commonplace for patients presenting with suspected or confirmed infections due to *S. aureus* (Lodise *et al.*, 2003; Rybak *et al.*, 2009; Bauer *et al.*, 2010; Wong *et al.*, 2012; Davis *et al.*, 2013; Terp *et al.*, 2014). This is particularly important for patients presenting with an invasive infection consistent with *S. aureus*. The benefits of administering early, appropriate therapy are well documented, and ensuring timely appropriate delivery is a fundamental pillar of all antimicrobial stewardship programs (ASPs) (Lodise *et al.*, 2003; Dellit *et al.*, 2007; Dellinger *et al.*, 2013). For most facilities and ASPs, "getting it right the first time" involves the administration of intravenous vancomycin, which is now used as first-line treatment for

*Corresponding author. E-mail: thomas.lodise@acphs.edu

both complicated and uncomplicated infections due to MRSA. These infections include: skin and soft tissue infections (Liu *et al.*, 2011; Stevens *et al.*, 2014), bacteremia and infective endocarditis (Liu *et al.*, 2011; Baddour *et al.*, 2005), pneumonia (ATS and IDSA, 2005), bone and joint infections (Liu *et al.*, 2011), and central nervous system infections (Tunkel *et al.*, 2004). Vancomycin is also the drug of choice for other complicated Gram-positive infections in patients with penicillin allergies (Rybak *et al.*, 2009; Liu *et al.*, 2011).

The primary purpose of this chapter is to provide an overview of contemporary vancomycin dosing approaches. The current understanding of the vancomycin pharmacokinetic/pharmacodynamic (PK/PD) profile is best summarized in the 2009 consensus review prepared by the Infectious Diseases Society of America (IDSA), the Society of Infectious Diseases Pharmacists (SIDP), and the American Society of Health-System Pharmacists (ASHP). Their statement represents the first set of national, evidence-based recommendations for vancomycin dosing and monitoring, and will be the focus of support for the information provided in this chapter (Rybak *et al.*, 2009). Although a consensus guideline for vancomycin dosing and monitoring exists, there are still many questions surrounding how to best optimize vancomycin dosing to maximize efficacy and minimize the risk of toxicity. As such, this chapter will also propose areas for future study and consideration. Lastly, this chapter is limited to reviewing the current approach to vancomycin dose optimization in adult patients.

Current Understanding of the Pharmacodynamics and Toxicodynamics of Vancomycin

Pharmacodynamics: efficacy

In order to optimize the dosing of any antimicrobial, a firm understanding of the drug exposure–effect link is required. The currently accepted PK/PD index for vancomycin is a value of ≥400 for the ratio of the area under the curve (AUC) to minimum inhibitory concentration (MIC) by broth microdilution (AUC/MIC_{BMD}) (Moise-Broder *et al.*, 2004; Rybak *et al.*, 2009; Kullar *et al.*, 2011a; Brown *et al.*, 2012; Holmes *et al.*, 2013; Lodise *et al.*, 2014). *In vitro* and *in vivo* PK/PD MRSA infection model studies have found that bactericidal activity (i.e., a 2 log reduction in bacterial inoculum) is achieved when the vancomycin AUC/MIC_{BMD} ratio exceeds 400. There are also mounting clinical data, albeit mostly retrospective in nature, in support of this PK/PD target for vancomycin (Moise-Broder *et al.*, 2004; Kullar *et al.*, 2011a; Brown *et al.*, 2012; Holmes *et al.*, 2013). In a retrospective cohort study of patients with lower respiratory tract infection due to *S. aureus*, Moise-Broder *et al.* (2004) found that an AUC/MIC_{BMD} ratio of ≥350 was independently associated with increased odds of clinical success (adjusted odds ratio (aOR) = 7.19, 95% confidence interval (CI) 1.91–27.3) after adjusting for methicillin-susceptible *S. aureus* (MSSA) as a pathogen, single lobe involvement, baseline serum albumin, and baseline creatinine clearance. In the time-to-event analyses, an AUC/MIC_{BMD} ratio of ≥400 was associated with improved time to pathogen eradication (P = 0.0046). Kullar *et al.* (2011a) have described similar findings in their retrospective cohort study of patients with *S. aureus* bacteremia (n = 320). Patients with a 24-hour AUC (i.e. AUC_{24}) to MIC_{BMD} ratio less than the classification and regression tree analysis derived breakpoint of 421 (where AUC_{24} is the AUC over 24 h) had higher failure rates (defined as 30 day mortality, persistent signs and symptoms of infection at the end of vancomycin therapy, or persistent bacteremia defined as >7 days of bacteremia) compared with those with an AUC_{24}/MIC_{BMD} ratio ≥ 421 (61.2 vs. 48.6%, P = 0.038).

Holmes *et al.* (2013) have also investigated the effect of vancomycin exposure on patient outcomes. In contrast to previous studies, these investigators utilized both BMD (MIC_{BMD}) and Etest® (MIC_{Etest}) methodologies to determine AUC/MIC ratios (the Etest® is a manual *in vitro* diagnostic device previously known as the Epsilometer test). In their retrospective cohort study, obtaining an AUC/MIC_{BMD} ratio ≥ 373 was associated with a significantly decreased incidence of 30 day all-cause mortality. The authors also found that the median AUC/MIC_{BMD} was statistically significantly different than the median AUC/MIC_{Etest} (436.1 vs. 271.5, P < 0.001). This finding emphasizes the importance of the MIC testing methodology used when determining the optimal AUC/MIC ratio for vancomycin, because the required AUC/MIC ratio is higher when measured by BMD than by the Etest® method. The 0.5–1 $\log_2$ variation between MIC values determined by BMD and Etest® (Swenson *et al.*, 2009) may, in part, explain some of the variability in AUC/MIC targets noted in the literature. Thus, the AUC/MIC_{Etest} should be viewed as a conservative estimate of the AUC/MIC ratio needed for optimal response by BMD. In a similarly designed retrospective cohort of patients with MRSA bloodstream infections, Jung *et al.* (2014) found that the 24 h vancomycin AUC/MIC ratio threshold associated with failure varied by MIC testing method (AUC_{24}/MIC_{Etest} < 430 vs. AUC_{24}/MIC_{BMD} < 398.5). Interestingly, the *S. aureus* vancomycin MIC distributions by BMD and Etest® were nearly identical in their study cohort. Lastly, a retrospective cohort study of 50 patients with complicated MRSA bloodstream infections and infective endocarditis by Brown *et al.* (2012) reported an increased incidence of attributable mortality in patients who did not achieve the critical AUC_{24}/MIC_{Etest} ratio of 211. Collectively, these data support the premise of targeting an AUC/MIC_{BMD} ratio of 400 for dosing vancomycin.

Toxicodynamics: acute kidney injury

With any drug, an understanding of its toxicodynamic profile is also required for optimal dosing. Although highly debated, the collective literature suggests that the risk of acute kidney injury (AKI) increases as a function of trough concentration (i.e., the lowest drug concentration before the next dose), especially when this is maintained above 15–20 mg/L. This is well described in a systematic literature review by van Hal *et al.* (2013). Overall, these authors found that the maintenance of trough concentrations in excess of 15 mg/L substantially increased the risk of a nephrotoxic event (odds ratio (OR) 2.74, 95% confidence interval (CI) 1.94–3.88; $P < 0.01$) relative to trough concentrations <15 mg/L. The probability of a nephrotoxic event was also found to increase as a function of treatment duration, with most episodes occurring after 7 days of therapy (van Hal *et al.*, 2013). There are also limited data to suggest that the risk of AKI increases along the vancomycin exposure continuum, especially when the daily AUC is in excess of 700–1300 mg-h/L (Lodise *et al.*, 2009; Suzuki *et al.*, 2012; Neely *et al.*, 2014). Combined, these data establish the upper safety trough threshold for vancomycin and strongly suggest that trough concentrations should not be maintained above 15–20 mg/L.

When evaluating the toxicodynamics of vancomycin, it is important to recognize other factors that may complicate or exacerbate the risk of AKI. In addition to large and prolonged vancomycin exposure, several additional risk factors for nephrotoxicity have been noted. Specifically, the host-related factors associated with nephrotoxicity include increased weight, (Lodise *et al.*, 2008, 2009; Hall *et al.*, 2013), preexisting renal dysfunction (Lodise *et al.*, 2008; Pritchard *et al.*, 2010; Minejima *et al.*, 2011), and critical illness. (Lodise *et al.*, 2008, 2009; Bosso *et al.*, 2011; Carreno *et al.*, 2014). Concurrent administration of nephrotoxic agents such as aminoglycosides (Farber and Moellering, 1983; Rybak *et al.*, 1990; Cano *et al.*, 2012), loop diuretics (Vance-Bryan *et al.*, 1994; Jeffres *et al.*, 2007), amphotericin B (Vance-Bryan *et al.*, 1994; Fullmer *et al.*, 2013), and vasopressors (Jeffres *et al.*, 2007; Minejima *et al.*, 2011) have been shown to increase the risk of kidney injury. Recently, piperacillin–tazobactam has also been reported to increase the risk of acute kidney injury in patients receiving vancomycin (Moenster *et al.*, 2013; Burgess and Drew, 2014; Gomes *et al.*, 2014; Meaney *et al.*, 2014). It is unclear whether the threshold for vancomycin-induced AKI varies according to these covariates, but clinicians should be mindful of the potential for additional risk when prescribing vancomycin to patients when these conditions are present.

Translation of the Current Understanding of Vancomycin PK/PD to the Bedside

Even though an AUC/MIC ratio of ≥400 is the clinically accepted vancomycin PK/PD target for serious infections due to MRSA, contemporary empiric vancomycin dosing schemes are designed to achieve a trough concentration of 15–20 mg/L. Currently, expert guidelines recommend monitoring trough concentrations as a surrogate marker for the AUC based on the historical difficulty of estimating AUC/MIC ratios in clinical practice (Rybak *et al.*, 2009; Liu *et al.*, 2011). In the past, the calculation of AUC in clinical practice involved the collection of multiple vancomycin serum concentrations during the same dosing interval, with subsequent calculation of the AUC using PK software that was not readily available at all institutions. As such, the guidelines accepted trough-directed dosing as a more practical alternative to AUC-guided dosing in clinical practice.

Of note, trough monitoring is not recommended for all patients. Patients are candidates for trough monitoring if they meet the following criteria: (i) they have unstable renal function; (ii) they will be on prolonged therapy; (iii) they are receiving aggressive dosing (i.e., targeting troughs of 15–20 mg/L); (iv) they are at high risk of nephrotoxicity (e.g., they are receiving concurrent nephrotoxins); (v) they are morbidly obese; or (vi) they have fluctuating volumes of distribution (Rybak *et al.*, 2009; Liu *et al.*, 2011). In patients with relatively uncomplicated infections (e.g., uncomplicated skin and soft tissue infections), trough monitoring is not routinely recommended (Rybak *et al.*, 2009; Liu *et al.*, 2011).

Despite its subsequent widespread integration in clinical practice (Davis *et al.*, 2013), the clinical benefits of maintaining higher vancomycin trough values have not been well described. Of the growing number of studies that evaluated the relationship between vancomycin exposure and outcomes, we are only aware of one retrospective study of patients with MRSA bloodstream infections which suggested that failure may be lower in patients with troughs of 15–20 mg/L relative to those with trough values of <15 mg/L (Kullar *et al.*, 2011a). From a PK/PD perspective, it is not surprising that there are limited clinical data to support the range of 15–20 mg/L. As highlighted in several vancomycin PD mathematical modeling studies, maintaining trough concentrations of 15–20 mg/L only ensures a high likelihood (>90%) of achieving an AUC/MIC_{BMD} ratio of >400 for *S. aureus* isolates with vancomycin MIC values of ≤1 mg/L (Mohr and Murray, 2007; Patel *et al.*, 2011). For MIC values in excess of 1 mg/L, the probability of achieving an AUC/MIC ratio ≥ 400 is less than 50%, even for the most aggressive intravenous (IV) dosing strategy (i.e., 2 g IV q12h) (see Fig. 21.1). Of note, the probability of target attainment (PTA) analyses are all predicated on the assumption that the vancomycin PK/PD target is truly an AUC/MIC ratio ≥ 400. However, the Monte Carlo simulation studies are highly suggestive that the current approach of maintaining vancomycin trough concentrations of 15–20 mg/L will have a suboptimal PTA profile when the vancomycin MIC value is in excess of 1 mg/L. Furthermore, the ability to enhance the PK/PD profile by dosing vancomycin to trough values in excess of 20 mg/L is negated by the growing number of investigations that have demonstrated higher rates of acute kidney injury with the maintenance of higher trough values (van Hal *et al.*, 2013).

Another potential concern with trough monitoring is that several recent studies have established that a trough value is a poor surrogate for the daily AUC value. As shown in Table 21.1, a wide range of AUCs can be generated by several different dosing regimens yielding isometric troughs and vice versa. (Neely *et al.*, 2012, 2014; Pai *et al.*, 2014). This should not be surprising—the AUC reflects the cumulative exposure over time. In contrast, the trough represents a single exposure point at the end of the dosing interval. The trough value will ensure a minimum AUC value. While achieving a minimum AUC is reassuring, there is considerable variability in the upper range of AUC values (see Fig. 21.2). This has clear implications for ASPs as a trough of 15–20 mg/L may lead to

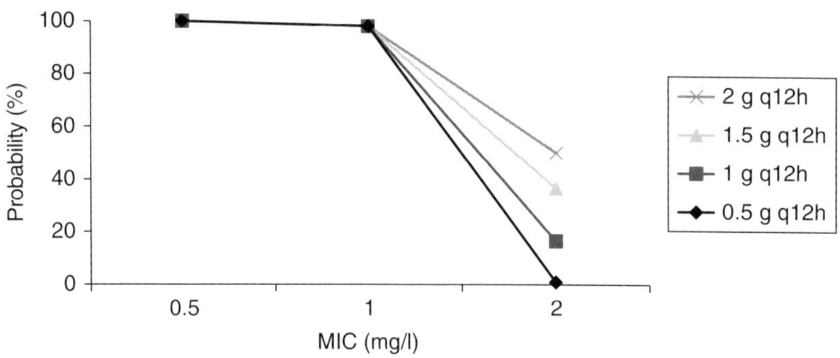

Fig. 21.1. Percentage probability of achieving ratio of area under the curve (AUC) to minimum inhibitory concentration (MIC) ≥400 for vancomycin regimens of varying intensity when minimum (trough) concentration (Cmin) values were 15–20 mg/L. Among the 9999 subjects simulated, the total number of subjects with Cmin values of 15–20 mg/L for the different regimens was: 0.5 g q12h, 406 subjects; 1 g q12h, 1100 subjects; 1.5 g q12h, 1190 subjects; and 2 g q12h,1096 subjects. Adapted from Patel *et al.*, 2011.

Table 21.1. Median (interquartile range) values for area under the curve over from 72 to 96 h (AUC_{72-96h}) and minimum (trough) concentration at 96 h ($Cmin_{96h}$) from the Monte Carlo simulation analysis for four vancomycin intravenous (IV) dosing regimens, stratified by creatinine clearance (CrCl). Adapted from Patel *et al.*, 2011.

CrCl	1.2 L/h	2.4 L/h	3.6 L/h	4.8 L/h	6.0 L/h	7.2 L/h
500 mg IV every 12 h						
AUC_{72-96h}	607 (414–846)	396 (265–563)	296 (199–429)	237 (158–346)	197 (132–292)	169 (112–251)
$Cmin_{96h}$	18.5 (11.6–27.3)	10.6 (5.9–16.7)	6.9 (3.5–11.7)	4.9 (2.1–8.7)	3.6 (1.3–6.8)	2.7 (0.9–5.5)
1000 mg IV every 12 h						
AUC_{72-96h}	1214 (827–1693)	792 (530–1127)	593 (398–857)	474 (315–692)	395 (262–581)	338 (224–503)
$Cmin_{96h}$	37.0 (23.1–54.6)	21.2 (11.9–33.3)	13.9 (6.9–23.4)	9.7 (4.2–17.5)	7.2 (2.7–13.7)	5.5 (1.7–11.1)
1500 mg IV every 12 h						
AUC_{72-96h}	1822 (1241–2539)	1188 (795–1690)	889 (596–1286)	710 (473–1039)	591 (396–875)	508 (337–754)
$Cmin_{96h}$	55.7 (34.7–82.0)	31.7 (17.8–50.5)	20.7 (10.3–34.9)	14.6 (6.4–26.5)	10.8 (4.1–20.5)	8.2 (2.6–16.7)
2000 mg IV every 12 h						
AUC_{72-96h}	2429 (1655–3385)	1588 (1076–2276)	1182 (793–1712)	947 (631–1385)	789 (528–1166)	677 (452–1010)
$Cmin_{96h}$	74.3 (46.3–109.3)	42.3 (23.8–67.3)	27.7 (13.7–46.5)	19.4 (8.5–35.3)	14.3 (5.4–27.4)	10.9 (3.4–22.2)

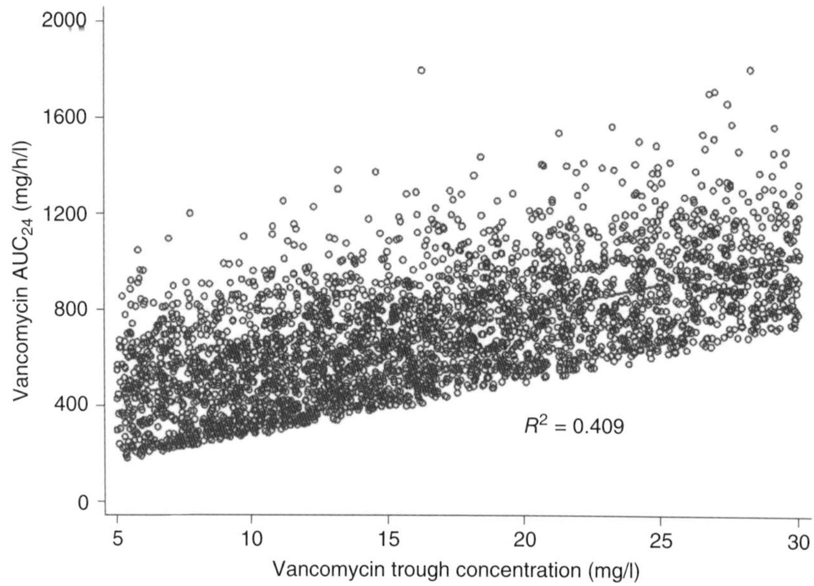

Fig. 21.2. Scatter and linear fit plot of vancomycin area under the curve over 24 h (AUC_{24}) versus trough vancomycin concentration from a 5000 subject Monte Carlo simulation. From Pai *et al.*, 2014.

suboptimal AUC/MIC ratios in a fair number of patients when the MIC value is in excess of 1 mg/L (Fig. 21.1). Conversely, troughs of 15–20 mg/L may elevate the risk of AKI in others as the intensity of the daily AUC has been shown to increase the risk of nephrotoxic events (Patel *et al.*, 2011). Although practical, the limitations surrounding trough-only monitoring suggest that it may not be sufficient to guide vancomycin dosing in all patients and that AUC-guided dosing may be warranted in certain populations (see "Considerations and Future Directions" near end of chapter).

Empiric Dosing of Vancomycin in Clinical Practice

Current expert guideline empiric dosing recommendations

Despite its noted limitations, trough monitoring has been widely adopted by most institutions (Davis *et al.*, 2013), and it is now the current standard for monitoring vancomycin as per expert guidelines. To help to facilitate the early attainment of troughs of 15–20 mg/L for patients with serious MRSA infections, several strategies can be employed. As global guidance, the expert guidelines recommend daily doses of 15–20 mg/kg (actual body weight) given every 8–12 h for patients with normal renal function, with subsequent dosing based on observed vancomycin trough concentrations. Whilst the data are limited, the guidelines recommend that initial dosages for obese patients should also be based on actual body weight, because traditional fixed-dosed regimens (e.g., 1 g every 12 h) may be inadequate for attaining therapeutic concentrations; thereafter, dosing should be similarly adjusted based on serum vancomycin concentrations (Rybak *et al.*, 2009; Liu *et al.*, 2011). However, the recent IDSA MRSA guidelines state that vancomycin maintenance doses should not exceed 2 g/dose.(Liu *et al.*, 2011). It is important to note that expert guidelines recommend weight-based dosing only for patients with complicated infections. The guidelines clearly indicate that traditional fixed dosing regimens with limited therapeutic monitoring can be utilized for uncomplicated skin and soft tissue infections in nonobese patients with normal renal function (Liu *et al.*, 2011).

Empiric dosing based on population estimates of a patient's PK parameters through use of PK equations

While the expert guidelines provide general guidelines for empiric vancomycin dosing, it is well established that vancomycin clearance (CL) is proportional to renal function, and that the volume of distribution at steady state (Vd) is proportional to weight. There are currently a number of published formulae describing the relationship been vancomycin population PK parameters and baseline patient covariates (see Tables 21.2 and 21.3) (Matzke *et al.*, 1984; Rodvold *et al.*, 1988; Burton *et al.*, 1989; Birt and Chandler, 1990; Yasuhara *et al.*, 1998; Winter, 2004; Buelga, 2005; Mulla and Pooboni, 2005; Llopis-Salvia and Jimenez-Torres, 2006; Murphy *et al.*, 2006; Staatz *et al.*, 2006; Bauer, 2008; Yamamoto *et al.*, 2009; Dolton *et al.*, 2010; Revilla *et al.*, 2010; Sanchez *et al.*, 2010; Tanaka *et al.*, 2010; Marsot *et al.*, 2012). In most instances, the authors have used the Cockcroft and Gault equation (Cockcroft and Gault, 1976) to estimate creatinine clearance (CrCl) when describing the relationship between renal function and vancomycin CL. Similarly, most authors have also used actual body weight when assessing the relationship between body size and Vd. These vancomycin population PK parameter estimation formulae, in conjunction with first or second-order pharmacokinetic equations, are frequently used in clinical practice for determining an empiric vancomycin regimen. Calculations are typically performed with calculators—either handheld or computerized physician order entry (CPOE) based—or by using one of the growing number of freely available software programs on a variety of websites (Neely *et al.*, 2012, 2014; Tatarinova *et al.*, 2013).

Recently, a paper that compared the predictive performance of the various published formulae used to empirically estimate vancomycin PK parameters (Murphy *et al.*, 2006) found the Matzke method (Matzke *et al.*, 1984) to have "the best combination of the least bias and best precision." However, the results clearly indicated that none of the methods were sufficiently reliable to replace

Table 21.2. Published prediction methods for pharmacokinetic parameters of vancomycin: vancomycin clearance (CLvancomycin) based on creatinine clearance (CrCl) and volume of distribution at steady state (Vd). Adapted from Murphy *et al.*, 2006.

Method	CLvancomycin	CL units	Vd	Vd units
Matzke *et al.* (1984)[a,b]	(CrCl × 0.689) + 3.66	mL/min	CrCl > 60 ml/min: 0.72 CrCl 10–60 ml/min: 0.89 CrCl < 10 ml/min: 0.9	L/kg
Rodvold *et al.* (1988)[a,c]	(CrCl × 0.79) + 15.7	mL/min	CrCl > 70 ml/min/70 kg: 0.5 CrCl 40–70 ml/min/70 kg: 0.59 CrCl 10–39 ml/min/70 kg: 0.64	L/kg
Birt and Chandler (1990)[a,c]	(CrCl × 0.674) + 13.45	mL/min	0.54	L/kg
Ambrose and Winter (2004)[a,d]	CrCl	mL/min	(0.17 × (age in years)) + (0.22 × (ABW in kg)) + 15	L
Burton *et al.* (1989)[e,f]	CrCl × 0.048	L/h	0.706	L/kg
Bauer (2008)[g]	(CrCl × 0.695) + 0.05	mL/min/kg	0.7	L/kg

[a]V estimated using actual body weight (ABW).
[b]CrCl estimated using the Cockcroft–Gault method (Cockcroft and Gault, 1976) and ABW.
[c]CrCl estimated using the Cockcroft–Gault (1976) method (Winter, 2004) but the weight to be used is not stated. As Cockcroft and Gault (1976) used ABW in their study, this approach is assumed and used.
[d]CrCl estimated using the Cockcroft–Gault method, (Winter, 2004) but the weight to be used is not stated. Elsewhere in the same textbook, it is recommended that an adjusted body weight (BWadj) be used, so this approach is assumed and used. The method for estimating BWadj is IBW[h] + 0.4(ABW – IBW), where IBW = ideal body weight.
[e]V estimated using BWadj if ABW > IBW; ABW used if ABW ≤ IBW. IBW = 0.73 × height (cm) – 59.42.
[f]Burton *et al.* (1989) developed this equation for CLvancomycin after feedback from measured vancomycin concentrations and the use of Bayesian iteration.
[g]The Bauer (2008) method uses the Salazar–Corcoran (Salazar and Corcoran, 1988) approach to estimate CrCl in obese patients (defined as ABW/IBW[h] ≥ 1.3). It uses the Cockcroft–Gault method and ABW to predict CrCl for nonobese patients. V estimated using ABW up to ABW/IBW = 1.3. After that, IBW is used.
[h]IBW is calculated using the formula of Devine (Pai and Paloucek, 2000) for all methods except Burton's: IBW (males) = 50 kg + 2.3(height (in) – 60) kg; IBW (females) = 45.5 kg + 2.3(height (in) – 60) kg

Table 21.3. Summary of published population pharmacokinetic models for vancomycin: total clearance (CL) based on creatinine clearance (CrCl) and volume of distribution at steady state (Vd). Adapted from Marsot *et al.*, 2012.

Study	CL (L/h)[a]	Units	Vd (L)[a]	Units
One compartment models				
Buelga (2005)	1.08 × CrCl[b]	L/h	0.98 × WT[c]	L/kg
Staatz *et al.* (2006)	2.97 × (1 + 0.0205 × (CrCl – CrCl median))[b]	L/h	1.24	L/kg
Tanaka *et al.* (2010)	0.875 × GFR[d]	L/h	0.864	L/kg
Revilla *et al.* (2010)	0.67 × CrCl + Age$^{-0.24}$	mL/min/kg	0.82 × 2.49^{A} (l/kg), where A = 0, if SCr[f] ≤ 1 mg/dl; A = 1, if SCr > 1 mg/dl	L
Two-compartment models				
Yasuhara *et al.* (1998)[g]	CrCl ≤ 85 ml/min: 0.0478 × CrCl CrCl > 85 ml/min: 3.51	L/h	60.7[h]	L
Mulla and Pooboni (2005)	Age < 1000 days: (2.4 + 0.0018 × age)/SCr Age >1000 days: 4.3/SCr	L/kg/h	Age < 4000 days: 0.45 Age > 4000 days: 0.37	L/kg
Llopis-Salvia and Jimenez-Torres (2006)	0.034 × CrCl + 0.015 × WT[b]	L	0.414 × WT	L/kg
Yamamoto *et al.* (2009)	CrCl ≥ 85 ml/min: 3.83[b] CrCl < 85 ml/min: 0.0322 × CrCl + 0.32[b]	L/h	0.206 × WT[i] 0.478 × WT[j]	L
Sanchez *et al.* (2010)	0.157 + 0.563 × CrCl	L/h	0.283 × WT	L

[a]Unless specified otherwise.
[b]CrCl calculated with the Cockcroft–Gault formula.
[c]WT= body weight (in kg).
[d]GFR= glomerular filtration rate (ml/min);
[e]CrCl calculated with the Levey formula.
[f]SCr = serum creatinine; mg/dL for Revilla *et al*; mmol/L for Mulla and Pooboni;
[g]In this study, the described pharmacokinetic parameters were vancomycin clearance, CL (details of the correlation with CrCl are given), distribution volume at steady state (Vss), k12 = first-order transfer rate constant from the central compartment to the peripheral compartment, k21 = first-order transfer rate constant from the peripheral compartment to the central compartment.
[h]Distribution volume at steady state Vss.
[i]In healthy subjects.
[j]In patients with Gram-positive infections.

therapeutic monitoring of vancomycin serum concentrations. Therefore, empiric estimation of PK parameters (i.e., Vd and CL) through the use of first- and second-order PK equations can be used to start an individualized vancomycin dosing regimen, although subsequent dosing should be guided by observed vancomycin trough concentrations.

A consideration when applying these equations is that it is assumed that there is a linear relationship between the vancomycin population PK parameter and the body size descriptor employed. While a wide variety of actual weight-based estimates of Vd (for example: 0.4–1 L/kg) have been reported in the literature (Rybak, 2006), mounting data suggest that it is not entirely accurate to describe vancomycin Vd as being proportional to body weight. As noted by a recent review of vancomycin PK in obesity(Grace, 2012), as weight increases, the coefficient used to calculate volume of distribution decreases. Blouin *et al.* (1982) reported a Vd proportionality coefficient of 0.29 L/kg (r^2 = 0.943, P = 0.005) in patients weighing greater than 190% of their assumed ideal body weight. Others have reported larger coefficients (e.g., 0.81 l/kg) in patients with weights greater than 120% of lean body weight (r^2 = 0.58) (Vance-Bryan *et al.*, 1993). Nevertheless, these findings suggest that vancomycin Vd may be better described by a logistic rather than a linear function. Alternatively, it may be that vancomycin does not substantially vary as a function of actual weight but by some other body size descriptor (Pai, 2012). At this point, dosing should be based on actual body weight, but more intensive therapeutic monitoring should be performed in obese patients (Rybak *et al.*, 2009).

Empiric dosing through the use of a nomogram

One way to operationalize empiric dosing of vancomycin in clinical practice based on PK equations and population estimates of a patient's PK parameters is through the use of dosing nomograms. In brief, a dosing nomogram is a two- to three-dimensional chart that provides empiric dosing recommendations based on patient-specific characteristics. As population estimates of vancomycin have largely been reported to be related to renal function and body weight, most vancomycin dosing nomograms provide an empiric dosing scheme for a given patient based on their estimated CrCl and body weight. An example of a widely used vancomycin dosing nomogram by Kullar *et al.* (2011b) is displayed in Table 21.4. Consistent with most vancomycin nomograms, the authors developed empiric dosing recommendations based on total body weight and CrCl estimated by the Cockcroft–Gault equation to achieve the target trough range of 15–20 mg/L.

Although nomograms offer a convenient mechanism for dosing vancomycin empirically in practice, one must recognize that subsequent therapeutic monitoring is still required as nomograms have limited predictive accuracy (44–58%) for definitive target trough attainment (Kullar *et al.*, 2011b; Devabhakthuni *et al.*, 2012; Thalakada *et al.*, 2012). In the nomogram developed by Kullar *et al.* (2011b), only 58% of patients had initial troughs within the range of 15–20 mg/L in near steady-state conditions. After initial dose adjustments, 154 patients (77%) had a trough within the range of 15–20 mg/L. However, the performance of the nomogram was improved to 80% accuracy when the dosing range was expanded to 13–22 mg/L. Devabhakthuni *et al.* (2012) have also evaluated a nomogram method of dosing. In their retrospective quasi-experimental study, the effect of a CPOE-based nomogram on the accuracy of the initial trough was examined. A total of 450 patients were included in the study: 225 preimplementation and 225 post-implementation. Despite a higher initial mean trough concentration post-implementation (17 vs. 14 mg/L, $P = 0.048$), the proportion of initial vancomycin troughs that were on target was unchanged post-implementation (44 vs. 45%, $P = 0.89$, pre- vs. post-implementation).

Thalakada *et al.* (2012) have developed a nomogram for achieving high-target vancomycin troughs. In their multicenter, retrospective cohort study, a training sample was used to develop a nomogram for vancomycin dosing to achieve target troughs of 14.5–20.5 mg/L. This study enrolled 169 patients from two separate hospitals into a training sample, and 105 patients into a validation sample whose troughs were used to validate the nomogram. After validation, the nomogram had a 56% predictive success rate. Lastly, Wesner *et al.* (2013) developed

Table 21.4. Vancomycin dosing nomogram. Doses ≥2 g should be infused over 2 h; doses of 1.5 g should be infused over 90 min. Weight refers to total weight. Creatinine clearance was calculated by using the Cockcroft–Gault equation. Adapted from Kullar *et al.*, 2011b.

Weight (kg)	Creatinine clearance (mL/min)						
	40–49	50–59	60–69	70–79	80–89	90–99	≥100
50–54	500 mg q12h	750 mg q12h	1000 mg q12h	750 mg q8h	1000 mg q8h	1000 mg q8h	1250 mg q8h
55–59	750 mg q12h	1000 mg q12h	1250 mg q12h	750 mg q8h	1000 mg q8h	1000 mg q8h	1250 mg q8h
60–64	750 mg q12h	1000 mg q12h	1250 mg q12h	750 mg q8h	1000 mg q8h	1250 mg q8h	1500 mg q8h
65–69	750 mg q12h	1000 mg q12h	1250 mg q12h	1000 mg q8h	1000 mg q8h	1250 mg q8h	1500 mg q8h
70–74	750 mg q12h	1250 mg q12h	750 mg q8h	1000 mg q8h	1250 mg q8h	1500 mg q8h	1500 mg q8h
75–79	1000 mg q12h	1250 mg q12h	750 mg q8h	1000 mg q8h	1250 mg q8h	1500 mg q8h	1750 mg q8h
80–84	1000 mg q12h	1250 mg q12h	1000 mg q8h	1250 mg q8h	1250 mg q8h	1500 mg q8h	1750 mg q8h
85–89	1000 mg q12h	1250 mg q12h	1000 mg q8h	1250 mg q8h	1500 mg q8h	1750 mg q8h	2000 mg q8h
90–94	1000 mg q12h	1500 mg q12h	1000 mg q8h	1250 mg q8h	1500 mg q8h	1750 mg q8h	2000 mg q8h
95–99	1250 mg q12h	1500 mg q12h	1000 mg q8h	1250 mg q8h	1500 mg q8h	1750 mg q8h	2000 mg q8h
100–104	1250 mg q12h	1500 mg q12h	1250 mg q8h	1500 mg q8h	1750 mg q8h	2000 mg q8h	2000 mg q8h
105–109	1250 mg q12h	1500 mg q12h	1250 mg q8h	1500 mg q8h	1750 mg q8h	2000 mg q8h	2250 mg q8h
≥110	1250 mg q12h	1500 mg q12h	1250 mg q8h	1500 mg q8h	1750 mg q8h	2000 mg q8h	2250 mg q8h

a novel nomogram to address the issue of higher vancomycin dosing. This prospective, randomized controlled trial compared a novel nomogram with traditional PK dosing in 473 patients. A statistically significant increase in initial target trough attainment (33 vs. 44%, $P = 0.014$) was noted in patients treated with the nomogram. There was also a statistically significant decrease in the incidence of troughs <10 mg/L (29.9 vs. 20.8%, $P = 0.037$) and no significant change in incidence of supratherapeutic troughs (8.1 vs. 9.6%, $P = 0.706$). Collectively, these data suggest that nomograms are useful for identifying empiric regimens for patients, but highlight the need for subsequent therapeutic monitoring to ensure that the patient is within the therapeutic trough range.

Continuous Infusion

Continuous infusion is another alternative empiric dosing strategy that may help specific patients to readily attain an AUC/MIC ratio of 400. By targeting a steady-state concentration of 15–20 mg/L, a daily AUC of 360–480 can be attained, resulting in a favorable PTA profile for MRSA infections with vancomycin MIC_{BMD} values ≤ 1 mg/L. While the ability to reliably achieve a target daily AUC is attractive, continuous infusion regimens are not currently recommended by expert guidelines for routine use as the activity of vancomycin is concentration dependent, and continuous infusions are typically reserved for time-dependent antibiotics like the β-lactams. (Rybak *et al.*, 2009; Liu *et al.*, 2011).

However, the pharmacokinetics of continuous infusions suggest that such regimens may be a reasonable alternative to conventional dosing and provide a convenient way to readily achieve the deserved vancomycin therapeutic range (i.e., 15–20 mg/L) throughout the entire dosing period. In short, continuous infusions allow clinicians to check a vancomycin concentration and quickly titrate the vancomycin dose to achieve a certain therapeutic concentration. This feature is highly desirable in critically ill patients with rapidly changing renal function because the exposure profiles associated with conventional dosing schemes are then often unpredictable. There are a number of examples in the literature that demonstrate the PK benefits of using continuous infusions of vancomycin (James *et al.*, 1996; Wysocki *et al.*, 2001; Boffi *et al.*, 2004; Vuagnat *et al.*, 2004; Hutschala *et al.*, 2009; Ingram *et al.*, 2009; Cataldo *et al.*, 2012; DiMondi and Rafferty, 2013; Saugel *et al.*, 2013; Schmelzer *et al.*, 2013; Saugel *et al.*, 2014). Overall, clinical outcomes were reported to be similar between continuous and intermittent dosing, and no differences in adverse outcomes were noted. Of note, continuous infusions of vancomycin require a dedicated IV line which can often be problematic in patients receiving multiple IV medications and with limited IV access.

Special Considerations in Empiric Dose Selection

Loading doses

As vancomycin is frequently used in patients with life-threatening infections due to MRSA, there is often a need for the rapid attainment of therapeutic trough concentrations. Cognizant of this, the vancomycin and MRSA guidelines recommend that loading doses (25–30 mg/kg of actual body weight) be considered in patients who are seriously ill. The use of prolonged infusion times (i.e., 2 h) and antihistamines may also be considered to mitigate the risk of red man syndrome (an infusion-related reaction peculiar to vancomycin) and anaphylaxis that may be associated with loading doses.

Although debatable, it appears that weight-based loading doses are preferred over fixed dose loads. In a prospective quasi-experimental study, patients treated with 15 mg/kg vancomycin had a significantly higher mean 1 h post-serum vancomycin concentration relative to those that received a 500 mg loading dose (19.1 mg/l vs. 10.4 mg/l) (Mohammedi *et al.*, 2006). Despite the more rapid attainment of the higher vancomycin serum concentrations, it is still unclear whether using a loading dose improves patient efficacy outcomes. More importantly, some literature suggests that loading doses may increase the risk of AKI (Francis *et al.*, 2013), even though these evaluations are at high risk of selection bias given the overrepresentation of critically ill patients in the loading dose group relative to the non-loading dose group. It is also important to note that the use of loading doses adds an additional layer of complexity to the medication use process and may propagate medication errors. If loading doses are implemented in ASPs, programs should cautiously monitor that only critically or seriously ill patients receive loading doses.

Dosing in patients with unstable renal function, including those on renal replacement therapies

As mentioned previously, vancomycin clearance is highly dependent on renal function. In order to select an appropriate empiric scheme, a clinician must be able to reasonably estimate a patient's renal function. However, patients with unstable kidney function must be handled differently because their renal function is not readily predictable. For patients with unstable renal function who are not receiving renal replacement therapy, it is advisable to administer a loading dose of vancomycin with subsequent dosing based on collected vancomycin concentrations. While there are no explicit recommendations in the expert guidelines, vancomycin concentrations should be collected 12–24 h after the initial vancomycin dose and redosing should occur when the vancomycin concentration falls below a designated threshold value. For most institutions, the vancomycin redosing threshold is 15 mg/L to ensure a daily AUC in excess of 400 mg-h/L. Upon initial signs of renal function recovery (i.e., urine output increase > 0.5 mL/kg/h or a decrease in serum creatinine), patients should be reevaluated for dosing appropriateness.

For patients receiving continuous or intermittent renal replacement therapies, vancomycin dosing is dependent on numerous factors, including: type and duration of renal replacement therapy, presence of residual renal function, patient weight, and indication for vancomycin therapy (Heintz *et al.*, 2009). Although a number of protocols have been published (Frazee *et al.*, 2012; Zelenitsky *et al.*, 2012; Paciullo *et al.*, 2013; Escobar *et al.*, 2014; Petejova *et al.*, 2014) (see Table 21.5) the approach to dosing vancomycin in patients receiving one of the continuous renal replacement modalities correlates closely with standard practice for patients with unstable renal function. These patients typically receive a loading dose (fixed or weight-based), with subsequent dosing based on collected vancomycin concentrations. In most instances, clinicians will redose vancomycin once the observed concentration falls below 15 mg/L. This dosing approach is also consistent with what is routinely done in patients receiving peritoneal dialysis (Gray *et al.*, 1985; Boyce *et al.*, 1988; Bailie *et al.*, 1995; Li *et al.*, 2010). For patients receiving intermittent thrice weekly high-flux hemodialysis (HD), most dosing protocols involve the administration of a vancomycin loading dose (fixed or weight-based), followed by administration of vancomycin post-HD. For almost all dosing protocols, the dose administered post-HD is based on the observed vancomycin concentration immediately prior to the administration of HD; the goal is to give a post-HD dose that maintains pre-HD vancomycin concentrations of 10–20 mg/L.

Table 21.5. Suggested dosing regimens for patients on renal replacement therapy.

Reference	Renal replacement type[a]	Weight group	Loading dose	Maintenance dose
Zelenitsky *et al.* (2012)	High-flux hemodialysis	<70 kg	1000 mg	500 mg
		70–100 kg	1250 mg	750 mg
		>100 kg	1500 mg	1000 mg
Heintz *et al.* (2009)	CAPD	N/A[b]	15–25 mg/kg	5–10 mg/kg after HD
	CVVH	N/A	15–25 mg/kg	10–15 mg/kg q12–48 h
	CVVHD	N/A	15–25 mg/kg	10–15 mg/kg q24h
	CVVHDF	N/A	15–25 mg/kg	7.5 mg–10 mg/kg q12h
Li *et al.* (2010)	Intermittent peritoneal dialysis (per exchange, once daily)	N/A	N/A	15–30 mg/kg q5–7 days
	Continuous (all exchanges)	N/A	1000 mg	25 mg/kg
	Automated peritoneal dialysis	N/A	30 mg/kg IP[c] in long dwell	15 mg/kg IP in long dwell every 3–5 days

[a]CAPD, continous ambulatory peritoneal dialysis; CVVH, continuous venovenous hemofiltration; CVVHD, continuous venovenous hemodialysis; CVVHDF, continuous venovenous hemodialfiltration; HD, hemodialysis.
[b]Not applicable.
[c]Intraperitoneal.

Individualizing Dosing Based on Observed Vancomycin Concentrations

Regardless of the empiric dosing strategy employed, it is important to emphasize that therapeutic monitoring is often required with vancomycin to ensure that the trough concentration is within the recommended therapeutic range. There are several things to consider when performing therapeutic monitoring for vancomycin. First, vancomycin exhibits linear pharmacokinetics (Winter, 2004; Rybak, 2006; Marsot *et al.*, 2012; Bauer, 2014). This is useful in clinical practice because if an observed vancomycin level is not on target, alterations of total daily dose will provide proportional changes in the observed concentrations. The caveat to this assumption is that the dosing interval is unchanged. The linear PK of vancomycin also enables clinicians to obtain individualized estimates of a patient's PK parameters based on observed vancomycin concentrations (see Box. 21.1). In the past, the estimation

Box 21.1. Simplified calculation of k_e, the elimination rate constant, and the area under the curve based on observed levels of vancomycin. Adapted from Pai *et al.*, 2014.

The following formula can be transformed to determine the elimination constant based on two observed vancomycin levels within a dosing interval:

$$C_{final} = C_{initial} \times e^{-k_e(T_2-T_1)}$$

After algebraic rearrangement, the k_e can be determined with the following method:

$$k_e = \frac{-\ln\left(\frac{C_{final}}{C_{initial}}\right)}{T_2 - T_1}$$

The area under the curve (AUC) between the start (same as C_t, the concentration at the start of infusion) and concentration at the theoretical end of infusion ($C_{eoi'}$) for a given infusion time (t') can be related as the area of a trapezoid:

$$AUC_{t0-t1} = (C_{eoi'} + C_t) \times 0.5 \times t'$$

The area under a mono-exponential curve from the time of the end of infusion ($t1$) to the time of the end of the dosing interval ($t2$) is:

$$AUC_{t1-t2} = \int_{t1}^{\infty} C_{eoi'}\, e^{-ke(t)}dt - \int_{t2}^{\infty} C_t e^{-k_e(t)}dt$$

$$AUC_{t1-t2} = \frac{C_{eoi'} - C_t}{k_e}$$

So the area under scenario 1 can be simplified to:

$$AUC_{t0-t2} = \frac{t' \times (C_{eoi'} + C_t)}{2} + \frac{C_{eoi'} - C_t}{k_e}$$

The above formula tends to underpredict the true AUC. An alternative to overcome this limitation would be to back-extrapolate the concentration to the theoretical start of infusion ($C_{soi'}$). Under this second scenario, the equation can be simplified to:

$$AUC_{t0-t2} = \int_{t1}^{\infty} C_{soi'}\, e^{-ke(t)}dt - \int_{t2}^{\infty} C_t e^{-ke(t)}dt$$

$$AUC_{t0-t2} = \frac{C_{soi'} - C_t}{k_e}$$

Under all these scenarios, the AUC_{24} will be a function of the number of identical doses administered during that interval of 24 h. To calculate the AUC_{24} the interval AUC is multiplied by the number of daily vancomycin doses.

of a patient's individualized PK parameters involved the use of first-order PK equations and handheld calculators, but computer software is now freely available to perform these calculations. Once estimated, the individualized PK parameters allow the clinician to adjust the dosing regimen to achieve the recommended therapeutic trough range of 15–20 mg/L (Neely *et al.*, 2012, 2014; Pai *et al.*, 2014).

Considerations and Future Directions

Several things should be considered when evaluating the current approach to dosing vancomycin. First, a clear understanding of the exposure targets associated with maximal effect and minimal toxicity are needed to optimally dose antibiotics in clinical practice. For vancomycin, dosing centers on achieving an AUC/MIC_{BMD} ratio ≥ 400 through the maintenance of trough values of 15–20 mg/L (Rybak *et al.*, 2009; Liu *et al.*, 2011). While an AUC/MIC_{BMD} ratio ≥ 400 is currently considered the optimal PK/PD "efficacy" target, it is important to recognize that this target is largely derived from retrospective, single-center, observational studies of patients with MRSA bloodstream infections (Moise-Broder *et al.*, 2004; Kullar *et al.*, 2011a; Brown *et al.*, 2012; Holmes *et al.*, 2013; Jung *et al.*, 2014). In the same vein, the vancomycin exposure target associated with AKI is unknown as there been have only been a few attempts, all retrospective in nature, to quantify the relationship between vancomycin exposure and probability of AKI (Lodise *et al.*, 2009; Suzuki *et al.*, 2012). Clearly, there is a need for hypothesis-driven, interventional, dose-optimized, multicentered studies across different infection types to truly delineate the vancomycin exposure associated with maximal response and minimal risk of AKI. In the absence of such trials, a definitive assertion on the optimal PK/PD targets for vancomycin efficacy and toxicity remains elusive. For now, all data should be interpreted cautiously and clinicians must recognize that our current thinking is subject to change as more robust data becomes available.

It is also important to recognize that most of the landmark clinical studies that established the contemporary efficacy PK/PD target relied on simple vancomycin CL formulas to estimate the vancomycin AUC. The data demonstrate that these CL formulas provide imprecise estimates of the AUC. This finding is not surprising as there is considerable inter-patient variability in vancomycin exposure profiles in clinical practice; it is not possible to generate valid estimates of exposure variables in a given individual based on CL formulas alone (Neely *et al.*, 2012, 2014; Pai *et al.*, 2014). Formerly, it was cumbersome to estimate AUCs in a clinical setting, but Neely *et al.* (2014) have recently demonstrated that Bayesian software programs can be used to generate accurate and reliable estimates of the daily AUC values with trough-only PK sampling. Using this validated Bayesian method to estimate the daily AUC for each patient in the study cohort, Lodise *et al.* (2014) found that outcomes were maximized when day 1 and 2 AUC/MIC_{BMD} ratios exceeded 521 and 650, respectively. Interestingly, these critical day 1 and 2 AUC/MIC_{BMD} ratios were higher than the current accepted AUC/MIC_{BMD} ratio of 400 (Moise-Broder *et al.*, 2004; Kullar *et al.*, 2011b; Holmes *et al.*, 2013). Although one retrospective study does not refute the prevailing thinking surrounding the pharmacodynamics of vancomycin, it highlights the importance of generating valid estimates of the AUC values through Bayesian modeling techniques when conducting vancomycin exposure–outcomes analyses in patients. These data further highlight the need for larger scale, multicentered clinical trials to determine the vancomycin PK/PD profile associated with maximal effect and minimal toxicity in order to optimize vancomycin dosing.

If future studies continue to demonstrate that AUC is indeed the driver of the efficacy and toxicity of vancomycin, clinicians will need to seriously reconsider the way that it is administered in clinical practice. Clearly, reliance on trough-only monitoring is insufficient. The most straightforward alternative to trough-only monitoring is AUC-guided dosing. In the past, AUC monitoring required the collection of multiple levels of vancomycin over the same dosing interval. With these data, a clinician would calculate the AUC using the linear-trapezoid rule. This approach requires precise timing and monitoring and, thus, has usually been impractical outside of a research setting. However, this is no longer the case—it is now possible to accurately estimate the AUC with trough-only data through the use of Bayesian software which can be used at the bedside to identify the optimal vancomycin dosage that readily achieves the predefined AUC target (Fuchs *et al.*, 2013). Bayesian software programs also provide the ability to devise innovative treatment schemas, such as front-loading doses with a lower maintenance dosing regimen, to rapidly achieve target concentrations.

Alternatively, the AUC can be accurately estimated based on the collection of two timed steady-state serum vancomycin concentrations and use of first-order PK equations (Pai *et al.*, 2014). The equations used to compute AUC from two samples are based in part on an original approach proposed by Begg *et al.* (1995) for aminoglycosides, and modified by Pai *et al.* (2014) (see Box 21.1). The major limitation of this approach is that, unlike the Bayesian approach, it is not adaptive in that it can only provide a snapshot of the AUC for the sampling period. This AUC calculation will not be correct if a physiologic change such as renal dysfunction occurs during or after the sampling period. It is also extremely difficult to estimate the vancomycin AUC_{24} with the equation-based method in patients who receive multiple dosing regimens within a 24 h period. Lastly, the equation-based method will provide the clinician with an AUC value and, ultimately, the clinician will need to use this information to devise a new vancomycin dosing scheme. However, this estimate of AUC is a clear step above trough-only or peak-only concentration interpretations.

Conclusions

Vancomycin remains a mainstay of therapy for serious infections due to MRSA. Despite continual evolution of dosing strategies over the last 60 years, the optimal dosing strategy to maximize efficacy and minimize toxicity remains unclear. Dosing of vancomycin currently centers on achieving an AUC/MIC_{BMD} ratio of ≥400 through the maintenance of trough values of 15–20 mg/L (Rybak *et al.*, 2009; Liu *et al.*, 2011). This is typically accomplished in clinical practice by determining an empiric dosing regimen for a given patient based on population estimates of their PK parameters, with subsequent dosing guided by observed trough concentrations. Alternative dosing strategies, such as loading doses, continuous infusions, and dosing by observed concentrations (i.e., dosing by levels) can be considered in specific populations. It is important to recognize that the current approach to dosing vancomycin is grounded in the evidence that the AUC/MIC_{BMD} ratio of ≥400 is the critical PK/PD index for vancomycin, though robust clinical data in support of this target are limited. If future studies continue to demonstrate that the AUC is indeed the driver of efficacy and toxicity, clinicians will need to seriously reconsider moving towards AUC-guided dosing. Although calculations of AUC values were historically difficult to accomplish in the clinical setting, it is now possible to accurately estimate the AUC with trough-only data through the use of Bayesian software. Alternatively, the AUC can be accurately estimated based on the collection of two timed steady-state serum vancomycin concentrations and use of first-order PK equations (Pai *et al.*, 2014), There will always be barriers to the implementation of these approaches, but they are "largely logistic and educational, but not technological"(Neely *et al.*, 2014). With any new methodology, the outcomes associated with these AUC-guided dosing approaches will need to be demonstrated in well-designed, multicentered prospective studies before widespread adoption into the clinical setting.

References

ATS (American Thoracic Society) and IDSA (Infectious Diseases Society of America) (2005) Guidelines for the management of adults with hospital-acquired, ventilator-associated, and healthcare-associated pneumonia. *American Journal of Respiratory and Critical Care Medicine* 171, 388–416.

Ambrose, P.J. and Winter, M.E. (2004) Vancomycin. In: Winter, M.E. (ed.) *Basic Clinical Pharmacokinetics*, 4th edn. Lippincott Williams & Wilkins, Philadelphia, Pennsylvania, pp. 451–476.

Baddour, L.M., Wilson, W.R., Bayer, A.S., Fowler, V.G., Jr., Bolger, A.F., Levison, M.E., Ferrieri, P., Gerber, M.A., Tani, L.Y., Gewitz, M.H. *et al.* (2005) Infective endocarditis: diagnosis, antimicrobial therapy, and management of complications: a statement for healthcare professionals from the Committee on Rheumatic Fever, Endocarditis, and Kawasaki Disease, Council on Cardiovascular Disease in the Young, and the Councils on Clinical Cardiology, Stroke, and Cardiovascular Surgery and Anesthesia, American Heart Association: endorsed by the Infectious Diseases Society of America. *Circulation* 111, e394–e434.

Bailie, G.R., Haqqie, S.S., Eisele, G., Gorman, T., and Low, C.L. (1995) Effectiveness of once-weekly vancomycin and once-daily gentamicin, intraperitoneally, for CAPD peritonitis. *Peritoneal Dialysis International* 15, 269–271.

Baillargeon, J., Kelley, M.F., Leach, C.T., Baillargeon, G., and Pollock, B.H. (2004) Methicillin-resistant *Staphylococcus aureus* infection in the Texas prison system. *Clinical Infectious Diseases* 38, e92–e95.

Bauer, K.A., West, J.E., Balada-Llasat, J.M., Pancholi, P., Stevenson, K.B., and Goff, D.A. (2010) An antimicrobial stewardship program's impact with rapid polymerase chain reaction methicillin-resistant *Staphylococcus aureus/S. aureus* blood culture test in patients with *S. aureus* bacteremia. *Clinical Infectious Diseases* 51, 1074–1080.

Bauer, L.A. (2008) *Applied Clinical Pharmacokinetics*, 2nd edn. McGraw-Hill Medical, New York.

Bauer, L.A. (2014) *Applied Clinical Pharmacokinetics*, 3nd edn. McGraw Hill Education, New York. Available at: http://accesspharmacy.mhmedical.com/content.aspx?bookid=1374&sectionid=74719392 (accessed 20 May 2016).

Begg, E.J., Barclay, M.L., and Duffull, S.B. (1995) A suggested approach to once-daily aminoglycoside dosing. *British Journal of Clinical Pharmacology* 39, 605–609.

Birt, J.K. and Chandler, M.H. (1990) Using clinical data to determine vancomycin dosing parameters. *Therapeutic Drug Monitoring* 12, 206–209.

Blouin, R.A., Bauer, L.A., Miller, D.D., Record, K.E., and Griffen, W.O., Jr. (1982) Vancomycin pharmacokinetics in normal and morbidly obese subjects. *Antimicrobial Agents and Chemotherapy* 21, 575–580.

Boffi El Amari, E., Vuagnat, A., Stern, R., Assal, M., Denormandie, P., Hoffmeyer, P., and Bernard, L. (2004) High versus standard dose vancomycin for osteomyelitis. *Scandinavian Journal of Infectious Diseases* 36, 712–717.

Bosso, J.A., Nappi, J., Rudisill, C., Wellein, M., Bookstaver, P.B., Swindler, J., and Mauldin, P.D. (2011) Relationship between vancomycin trough concentrations and nephrotoxicity: a prospective multicenter trial. *Antimicrobial Agents and Chemotherapy* 55, 5475–5479.

Boyce, N.W., Wood, C., Thomson, N.M., Kerr, P., and Atkins, R.C. (1988) Intraperitoneal (IP) vancomycin therapy for CAPD peritonitis—a prospective, randomized comparison of intermittent v continuous therapy. *American Journal of Kidney Diseases* 12, 304–306.

Brown, J., Brown, K., and Forrest, A. (2012) Vancomycin AUC24/MIC ratio in patients with complicated bacteremia and infective endocarditis due to methicillin-resistant *Staphylococcus aureus* and its association with attributable mortality during hospitalization. *Antimicrobial Agents and Chemotherapy* 56, 634–638.

Buelga, D.S., del Mar Fernandez de Gatta, M., Herrera, E.V., Dominguez-Gil, A., and Garcia, M.J. (2005) Population pharmacokinetic analysis of vancomycin in patients with hematological malignancies. *Antimicrobial Agents and Chemotherapy* 49, 4934–4941.

Burgess, L.D. and Drew, R.H. (2014) Comparison of the incidence of vancomycin-induced nephrotoxicity in hospitalized patients with and without concomitant piperacillin–tazobactam. *Pharmacotherapy* 34, 670–676.

Burton, M.E., Gentle, D.L., and Vasko, M.R. (1989) Evaluation of a Bayesian method for predicting vancomycin dosing. *DICP* 23, 294–300.

Cano, E.L., Haque, N.Z., Welch, V.L., Cely, C.M., Peyrani, P., Scerpella, E.G., Ford, K.D., Zervos, M.J., Ramirez, J.A., Kett, D.H., and the IMPACT-HAP Study Group (2012) Incidence of nephrotoxicity and association with vancomycin use in intensive care unit patients with pneumonia: retrospective analysis of the IMPACT-HAP Database. *Clinical Therapeutics* 34, 149–157.

Capitano, B., Leshem, O.A., Nightingale, C.H., and Nicolau, D.P. (2003) Cost effect of managing methicillin-resistant *Staphylococcus aureus* in a long-term care facility. *Journal of the American Geriatrics Society* 51, 10–16.

Carreno, J.J., Kenney, R.M., and Lomaestro, B. (2014) Vancomycin-associated renal dysfunction: where are we now? *Pharmacotherapy* 34, 1259–1268.

Cataldo, M.A., Tacconelli, E., Grilli, E., Pea, F., and Petrosillo, N. (2012) Continuous versus intermittent infusion of vancomycin for the treatment of Gram-positive infections: systematic review and meta-analysis. *Journal of Antimicrobial Chemotherapy* 67, 17–24.

Chua, K., Laurent, F., Coombs, G., Grayson, M.L., and Howden, B.P. (2011) Antimicrobial resistance: not community-associated methicillin-resistant *Staphylococcus aureus* (CA-MRSA)! A clinician's guide to community MRSA—its evolving antimicrobial resistance and implications for therapy. *Clinical Infectious Diseases* 52, 99–114.

Cockcroft, D.W. and Gault, M.H. (1976) Prediction of creatinine clearance from serum creatinine. *Nephron* 16, 31–41.

Cosgrove, S.E., Sakoulas, G., Perencevich, E.N., Schwaber, M.J., Karchmer, A.W., and Carmeli, Y. (2003) Comparison of mortality associated with methicillin-resistant and methicillin-susceptible *Staphylococcus aureus* bacteremia: a meta-analysis. *Clinical Infectious Diseases* 36, 53–59.

Culpepper, R., Nolan, R., Chapman, S., Kennedy, A., Currier, M., and CDC (Centers for Disease Control and Prevention) (2001) Methicillin resistant *Staphylococcus aureus* skin or soft tissue infections in a state prison – Mississippi, 2000 *Morbidity and Mortality Weekly Report (MMWR)* 50, 919–922.

David, M.Z. and Daum, R.S. (2010) Community-associated methicillin-resistant *Staphylococcus aureus*: epidemiology and clinical consequences of an emerging epidemic. *Clinical Microbiology Reviews* 23, 616–687.

Davis, S.L., Perri, M.B., Donabedian, S.M., Manierski, C., Singh, A., Vager, D., Haque, N.Z., Speirs, K., Muder, R.R., Robinson-Dunn, B. *et al.* (2007) Epidemiology and outcomes of community-associated methicillin-resistant *Staphylococcus aureus* infection. *Journal of Clinical Microbiology* 45, 1705–1711.

Davis, S.L., Scheetz, M.H., Bosso, J.A., Goff, D.A., and Rybak, M.J. (2013) Adherence to the 2009 consensus guidelines for vancomycin dosing and monitoring practices: a cross-sectional survey of U.S. Hospitals. *Pharmacotherapy* 33, 1256–1263.

Dellinger, R.P., Levy, M.M., Rhodes, A., Annane, D., Gerlach, H., Opal, S.M., Sevransky, J.E., Sprung, C.L., Douglas, I.S., Jaeschke, R. *et al.* (2013) Surviving Sepsis Campaign: international guidelines for management of severe sepsis and septic shock, 2012. *Intensive Care Medicine* 39, 165–228.

Dellit, T.H., Owens, R.C., McGowan, J.E., Jr., Gerding, D.N., Weinstein, R.A., Burke, J.P., Huskins, W.C., Paterson, D.L., Fishman, N.O., Carpenter, C.F. *et al.* (2007) Infectious Diseases Society of America and the Society for Healthcare Epidemiology of America guidelines for developing an institutional program to enhance antimicrobial stewardship. *Clinical Infectious Diseases* 44, 159–177.

Devabhakthuni, S., Gonzalez, J.P., Tata, A.L., Lee, S., Shah, P., Ibe Offurum, A., and Gulati, M. (2012) Evaluation of vancomycin dosing and monitoring in adult medicine patients. *Hospital Pharmacy* 47, 451–459.

DiMondi, V.P. and Rafferty, K. (2013) Review of continuous-infusion vancomycin. *Annals of Pharmacotherapy* 47, 219–227.

Dolton, M., Xu, H., Cheong, E., Maitz, P., Kennedy, P., Gottlieb, T., Buono, E., and McLachlan, A.J. (2010) Vancomycin pharmacokinetics in patients with severe burn injuries. *Burns* 36, 469–476.

Escobar, L., Andresen, M., Downey, P., Gai, M.N., Regueira, T., Bórquez, T., Lipman, J., and Roberts, J.A. (2014) Population pharmacokinetics and dose simulation of vancomycin in critically ill patients during high-volume haemofiltration. *International Journal of Antimicrobial Agents* 44, 163–167.

Farber, B.F. and Moellering, R.C., Jr. (1983) Retrospective study of the toxicity of preparations of vancomycin from 1974 to 1981. *Antimicrobial Agents and Chemotherapy* 23, 138–141.

Francis, J.K., Pogue, J.M., Mynatt, R.P., Casapao, A.M., Barber, K.E., and Rybak, M.J. (2013) Effect of vancomycin loading dose on clinical outcomes in patients with methicillin-resistant *Staphylococcus aureus* (MRSA) bacteremia. Presentation to: *16th Annual Making a Difference in Infectious Diseases Pharmacotherapy (MAD-ID)—16th Annual Meeting, Orlando, Florida, 9 May to 11* May, 2013. Society of Infectious Diseases Pharmacists, Austin, Texas.

Frazee, E.N., Kuper, P.J., Schramm, G.E., Larson, S.L., Kashani, K.B., Osmon, D.R., and Leung, N. (2012) Effect of continuous venovenous hemofiltration dose on achievement of adequate vancomycin trough concentrations. *Antimicrobial Agents and Chemotherapy* 56, 6181–6185.

Fridkin, S.K., Hageman, J.C., Morrison, M., Sanza, L.T., Como-Sabetti, K., Jernigan, J.A., Harriman, K., Harrison, L.H., Lynfield, R. and Farley, M.M. (2005) Methicillin-resistant *Staphylococcus aureus* disease in three communities. *The New England Journal of Medicine* 352, 1436–1444.

Fuchs, A., Csajka, C., Thoma, Y., Buclin, T., and Widmer, N. (2013) Benchmarking therapeutic drug monitoring software: a review of available computer tools. *Clinical Pharmacokinetics* 52, 9–22.

Fullmer, A., McCue, D., and Feng, C. (2013) Retrospective review of vancomycin-induced nephrotoxicity in patients with leukemia. *Journal of Oncology Pharmacy Practice* 20, 403–408.

Gomes, D.M., Smotherman, C., Birch, A., Dupree, L., Della Vecchia, B.J., Kraemer, D.F., and Jankowski, C.A. (2014) Comparison of acute kidney injury during treatment with vancomycin in combination with piperacillin–tazobactam or cefepime. *Pharmacotherapy* 34, 662–669.

Grace, E. (2012) Altered vancomycin pharmacokinetics in obese and morbidly obese patients: what we have learned over the past 30 years. *Journal of Antimicrobial Chemotherapy* 67, 1305–1310.

Gray, H.H., Goulding, S., and Eykyn, S.J. (1985) Intraperitoneal vancomycin and ceftazidime in the treatment of CAPD peritonitis. *Clinical Nephrology* 23, 81–84.

Hall, R.G., II, Hazlewood, K.A., Brouse, S.D., Giuliano, C.A., Haase, K.K., Frei, C.R., Forcade, N.A., Bell, T., Bedimo, R.J., and Alvarez, C.A. (2013) Empiric guideline-recommended weight-based vancomycin dosing and nephrotoxicity rates in patients with methicillin-resistant *Staphylococcus aureus* bacteremia: a retrospective cohort study. *BMC Pharmacology and Toxicology* 14: 12.

Heintz, B.H., Matzke, G.R., and Dager, W.E. (2009) Antimicrobial dosing concepts and recommendations for critically ill adult patients receiving continuous renal replacement therapy or intermittent hemodialysis. *Pharmacotherapy* 29, 562–577.

Hidron, A.I., Low, C.E., Honig, E.G., and Blumberg, H.M. (2009) Emergence of community-acquired meticillin-resistant *Staphylococcus aureus* strain USA300 as a cause of necrotising community-onset pneumonia. *The Lancet Infectious Diseases* 9, 384–392.

Holmes, N.E., Turnidge, J.D., Munckhof, W.J., Robinson, J.O., Korman, T.M., O'Sullivan, M.V., Anderson, T.L., Roberts, S.A., Warren, S.J., Gao, W. *et al.* (2013) Vancomycin AUC/MIC ratio and 30-day mortality in patients with *Staphylococcus aureus* bacteremia. *Antimicrobial Agents and Chemotherapy* 57, 1654–1663.

Hutschala, D., Kinstner, C., Skhirdladze, K., Thalhammer, F., Muller, M., and Tschernko, E. (2009) Influence of vancomycin on renal function in critically ill patients after cardiac surgery: continuous versus intermittent infusion. *Anesthesiology* 111, 356–365.

Ingram, P.R., Lye, D.C., Fisher, D.A., Goh, W.P., and Tam, V.H. (2009) Nephrotoxicity of continuous versus intermittent infusion of vancomycin in outpatient parenteral antimicrobial therapy. *International Journal of Antimicrobial Agents* 34, 570–574.

James, J.K., Palmer, S.M., Levine, D.P., and Rybak, M.J. (1996) Comparison of conventional dosing versus continuous-infusion vancomycin therapy for patients with suspected or documented Gram-positive infections. *Antimicrobial Agents and Chemotherapy* 40, 696–700.

Jeffres, M.N., Isakow, W., Doherty, J.A., Micek, S.T., and Kollef, M.H. (2007) A retrospective analysis of possible renal toxicity associated with vancomycin in patients with health care-associated methicillin-resistant *Staphylococcus aureus* pneumonia. *Clinical Therapeutics* 29, 1107–1115.

Jung, Y., Song, K.H., Cho, J., Kim, H.S., Kim, N.H., Kim, T.S., Choe, P.G., Chung, J.Y., Park, W.B., Bang, J.H. *et al.* (2014) Area under the concentration–time curve to minimum inhibitory concentration ratio as a predictor of vancomycin treatment outcome in methicillin-resistant *Staphylococcus aureus* bacteraemia. *International Journal of Antimicrobial Agents* 43, 179–183.

King, M.D., Humphrey, B.J., Wang, Y.F., Kourbatova, E.V., Ray, S.M., and Blumberg, H.M. (2006) Emergence of community-acquired methicillin-resistant *Staphylococcus aureus* USA 300 clone as the predominant cause of skin and soft-tissue infections. *Annals of Internal Medicine* 144, 309–317.

Kirst, H.A., Thompson, D.G., and Nicas, T.I. (1998) Historical yearly usage of vancomycin. *Antimicrobial Agents and Chemotherapy* 42, 1303–1304.

Klevens, R.M., Morrison, M.A., Nadle, J., Petit, S., Gershman, K., Ray, S., Harrison, L.H., Lynfield, R., Dumyati, G., Townes, J.M. *et al.* (2007) Invasive methicillin-resistant *Staphylococcus aureus* infections in the United States. *JAMA (The Journal of the American Medical Association)* 298, 1763–1771.

Kullar, R., Davis, S.L., Levine, D.P., and Rybak, M.J. (2011a) Impact of vancomycin exposure on outcomes in patients with methicillin-resistant *Staphylococcus aureus* bacteremia: support for consensus guidelines suggested targets. *Clinical Infectious Diseases* 52, 975–981.

Kullar, R., Leonard, S.N., Davis, S.L., Delgado, G., Jr., Pogue, J.M., Wahby, K.A., Falcione, B., and Rybak, M.J. (2011b) Validation of the effectiveness of a vancomycin nomogram in achieving target trough concentrations of 15–20 mg/L suggested by the vancomycin consensus guidelines. *Pharmacotherapy* 31, 441–448.

Lee, N.E., Taylor, M.M., Bancroft, E., Ruane, P.J., Morgan, M., McCoy, L., and Simon, P.A. (2005) Risk factors for community-associated methicillin-resistant *Staphylococcus aureus* skin infections among HIV-positive men who have sex with men. *Clinical Infectious Diseases* 40, 1529–1534.

Li, P.K., Szeto, C.C., Piraino, B., Bernardini, J., Figueiredo, A.E., Gupta, A., Johnson, D.W., Kuijper, E.J., Lye, W.C., Salzer, W. *et al.* (2010) Peritoneal dialysis-related infections recommendations: 2010 update. *Peritoneal Dialysis International* 30, 393–423.

Liu, C., Bayer, A., Cosgrove, S.E., Daum, R.S., Fridkin, S.K., Gorwitz, R.J., Kaplan, S.L., Karchmer, A.W., Levine, D.P., Murray, B.E. *et al.* (2011) Clinical practice guidelines by the Infectious Diseases Society of America for the treatment of methicillin-resistant *Staphylococcus aureus* infections in adults and children. *Clinical Infectious Diseases* 52, e18–e55.

Llopis-Salvia, P. and Jimenez-Torres, N.V. (2006) Population pharmacokinetic parameters of vancomycin in critically ill patients. *Journal of Clinical Pharmacy and Therapeutics* 31, 447–454.

Lodise, T.P., McKinnon, P.S., Swiderski, L., and Rybak, M.J. (2003) Outcomes analysis of delayed antibiotic treatment for hospital-acquired *Staphylococcus aureus* bacteremia. *Clinical Infectious Diseases*, 1418–1423.

Lodise, T.P., Lomaestro, B., Graves, J., and Drusano, G.L. (2008) Larger vancomycin doses (at least four grams per day) are associated with an increased incidence of nephrotoxicity. *Antimicrobial Agents and Chemotherapy* 52, 1330–1336.

Lodise, T.P., Patel, N., Lomaestro, B.M., Rodvold, K.A., and Drusano, G.L. (2009) Relationship between initial vancomycin concentration–time profile and nephrotoxicity among hospitalized patients. *Clinical Infectious Diseases* 49, 507–514.

Lodise, T.P., Drusano, G.L., Zasowski, E., Dihmess, A., Lazariu, V., Cosler, L., and McNutt, L.A. (2014) Vancomycin exposure in patients with methicillin-resistant *Staphylococcus aureus* bloodstream infections: how much is enough? *Clinical Infectious Diseases* 59, 666–675.

Lowy, F.D. (1998) *Staphylococcus aureus* infections. *The New England Journal of Medicine* 339, 520–532.

Marsot, A., Boulamery, A., Bruguerolle, B., and Simon, N. (2012) Vancomycin: a review of population pharmacokinetic analyses. *Clinical Pharmacokinetics* 51, 1–13.

Matzke, G.R., McGory, R.W., Halstenson, C.E., and Keane, W.F. (1984) Pharmacokinetics of vancomycin in patients with various degrees of renal function. *Antimicrobial Agents and Chemotherapy* 25, 433–437.

Meaney, C.J., Hynicka, L.M., and Tsoukleris, M.G. (2014) Vancomycin-associated nephrotoxicity in adult medicine patients: incidence, outcomes, and risk factors. *Pharmacotherapy* 34, 653–661.

Miller, L.G., Quan, C., Shay, A., Mostafaie, K., Bharadwa, K., Tan, N., Matayoshi, K., Cronin, J., Tan, J., Tagudar, G., and Bayer, A.S. (2007) A prospective investigation of outcomes after hospital discharge for endemic, community-acquired methicillin-resistant and -susceptible *Staphylococcus aureus* skin infection. *Clinical Infectious Diseases* 44, 483–492.

Minejima, E., Choi, J., Beringer, P., Lou, M., Tse, E., and Wong-Beringer, A. (2011) Applying new diagnostic criteria for acute kidney injury to facilitate early identification of nephrotoxicity in vancomycin-treated patients. *Antimicrobial Agents and Chemotherapy* 55, 3278–3283.

Moenster, R.P., Linneman, T.W., Finnegan, P.M., Hand, S., Thomas, Z., and McDonald, J.R. (2013) Acute renal failure associated with vancomycin and β-lactams for the treatment of osteomyelitis in diabetics: piperacillin–tazobactam compared to cefepime. *Clinical Microbiology and Infection* 20(6), O384–O389.

Mohammedi, I., Descloux, E., Argaud, L., Le Scanff, J., and Robert, D. (2006) Loading dose of vancomycin in critically ill patients: 15 mg/kg is a better choice than 500 mg. *International Journal of Antimicrobial Agents* 27, 259–262.

Mohr, J.F. and Murray, B.E. (2007) Point: vancomycin is not obsolete for the treatment of infection caused by methicillin-resistant *Staphylococcus aureus*. *Clinical Infectious Diseases* 44, 1536–1542.

Moise-Broder, P.A., Forrest, A., Birmingham, M.C., and Schentag, J.J. (2004) Pharmacodynamics of vancomycin and other antimicrobials in patients with *Staphylococcus aureus* lower respiratory tract infections. *Clinical Pharmacokinetics* 43, 925–942.

Moran, G.J., Amii, R.N., Abrahamian, F.M., and Talan, D.A. (2005) Methicillin-resistant *Staphylococcus aureus* in community-acquired skin infections. *Emerging Infectious Diseases* 11, 928–930.

Moran, G.J., Krishnadasan, A., Gorwitz, R.J., Fosheim, G.E., McDougal, L.K., Carey, R.B., and Talan, D.A. (2006) Methicillin-resistant *S. aureus* infections among patients in the emergency department. *The New England Journal of Medicine* 355, 666–674.

Moran, G.J., Krishnadasan, A., Gorwitz, R.J., Fosheim, G.E., Albrecht, V., Limbago, B., and Talan, D.A. (2012) Prevalence of methicillin-resistant *staphylococcus aureus* as an etiology of community-acquired pneumonia. *Clinical Infectious Diseases* 54, 1126–1133.

Mulla, H. and Pooboni, S. (2005) Population pharmacokinetics of vancomycin in patients receiving extracorporeal membrane oxygenation. *British Journal of Clinical Pharmacology* 60, 265–275.

Murphy, J.E., Gillespie, D.E., and Bateman, C.V. (2006) Predictability of vancomycin trough concentrations using seven approaches for estimating pharmacokinetic parameters. *American Journal of Health-System Pharmacy* 63, 2365–2370.

Naimi, T.S., LeDell, K.H., Como-Sabetti, K., Borchardt, S.M., Boxrud, D.J., Etienne, J., Johnson, S.K., Vandenesch, F., Fridkin, S., O'Boyle, C., *et al.* (2003) Comparison of community- and health care-associated methicillin-resistant *Staphylococcus aureus* infection. *JAMA (The Journal of the American Medical Association)* 290, 2976–2984.

Neely, M.N., van Guilder, M.G., Yamada, W.M., Schumitzky, A., and Jelliffe, R.W. (2012) Accurate detection of outliers and subpopulations with Pmetrics, a nonparametric and parametric pharmacometric modeling and simulation package for R. *Therapeutic Drug Monitoring* 34, 467–476.

Neely, M.N., Youn, G., Jones, B., Jelliffe, R.W., Drusano, G.L., Rodvold, K.A., and Lodise, T.P. (2014) Are vancomycin troughs adequate for optimal dosing? *Antimicrobial Agents and Chemotherapy* 58, 309–316.

NNIS System (2004) National Nosocomial Infections Surveillance (NNIS) System Report, data summary from January 1992 through June 2004, issued October 2004. *American Journal of Infection Control* 32, 470–485.

Ochoa, T.J., Mohr, J., Wanger, A., Murphy, J.R., and Heresi, G.P. (2005) Community-associated methicillin-resistant *Staphylococcus aureus* in pediatric patients. *Emerging Infectious Diseases* 11, 966–968.

Paciullo, C.A., Harned, K.C., Davis, G.A., Connor, M.J., Jr., and Winstead, P.S. (2013) Vancomycin clearance in high-volume venovenous hemofiltration. *Annals of Pharmacotherapy* 47(3): e14.

Pai, M.P. (2012) Drug dosing based on weight and body surface area: mathematical assumptions and limitations in obese adults. *Pharmacotherapy* 32, 856–868.

Pai, M.P. and Paloucek, F.P. (2000) The origin of the "ideal" body weight equations. *Annals of Pharmacotherapy* 34, 1066–1069.

Pai, M.P., Neely, M., Rodvold, K.A., and Lodise, T.P. (2014) Innovative approaches to optimizing the delivery of vancomycin in individual patients. *Advanced Drug Delivery Reviews* 77, 50–57.

Patel, N., Pai, M.P., Rodvold, K.A., Lomaestro, B., Drusano, G.L., and Lodise, T.P. (2011) Vancomycin: we can't get there from here. *Clinical Infectious Diseases* 52, 969–974.

Petejova, N., Martinek, A., Zahalkova, J., Duricova, J., Brozmannova, H., Urbanek, K., Grundmann, M., Plasek, J., and Kacirova, I. (2014) Vancomycin pharmacokinetics during high-volume continuous venovenous hemofiltration in critically ill septic patients. *Biomedical Papers, Palacký University, Faculty of Medicine and Dentistry, Olomouc, Czech Republic* 158, 65–72.

Pritchard, L., Baker, C., Leggett, J., Sehdev, P., Brown, A., and Bayley, K.B. (2010) Increasing vancomycin serum trough concentrations and incidence of nephrotoxicity. *The American Journal of Medicine* 123, 1143–1149.

Reed, S.D., Friedman, J.Y., Engemann, J.J., Griffiths, R.I., Anstrom, K.J., Kaye, K.S., Stryjewski, M.E., Szczech, L.A., Reller, L.B., Corey, G.R. *et al.* (2005) Costs and outcomes among hemodialysis-dependent patients with methicillin-resistant or methicillin-susceptible

Staphylococcus aureus bacteremia. *Infection Control and Hospital Epidemiology* 26, 175–183.

Revilla, N., Martin-Suarez, A., Perez, M.P., Gonzalez, F.M., and Fernandez de Gatta Mdel, M. (2010) Vancomycin dosing assessment in intensive care unit patients based on a population pharmacokinetic/pharmacodynamic simulation. *British Journal of Clinical Pharmacology* 70, 201–212.

Rodvold, K.A., Blum, R.A., Fischer, J.H., Zokufa, H.Z., Rotschafer, J.C., Crossley, K.B., and Riff, L.J. (1988) Vancomycin pharmacokinetics in patients with various degrees of renal function. *Antimicrobial Agents and Chemotherapy* 32, 848–852.

Rybak, M.J. (2006) The pharmacokinetic and pharmacodynamic properties of vancomycin. *Clinical Infectious Diseases* 42(Suppl 1), S35–S39.

Rybak, M.J., Albrecht, L.M., Boike, S.C., and Chandrasekar, P.H. (1990) Nephrotoxicity of vancomycin, alone and with an aminoglycoside. *Journal of Antimicrobial Chemotherapy* 25, 679–687.

Rybak, M.J., Lomaestro, B.M., Rotschafer, J.C., Moellering, R.C., Jr., Craig, W.A., Billeter, M., Dalovisio, J.R., and Levine, D.P. (2009) Vancomycin therapeutic guidelines: a summary of consensus recommendations from the Infectious Diseases Society of America, the American Society of Health-System Pharmacists, and the Society of Infectious Diseases Pharmacists. *Clinical Infectious Diseases* 49, 325–327.

Salazar, D.E. and Corcoran, G.B. (1988) Predicting creatinine clearance and renal drug clearance in obese patients from estimated fat-free body mass. *The American Journal of Medicine* 84, 1053–1060.

Sanchez, J.L., Dominguez, A.R., Lane, J.R., Anderson, P.O., Capparelli, E.V., and Cornejo-Bravo, J.M. (2010) Population pharmacokinetics of vancomycin in adult and geriatric patients: comparison of eleven approaches. *International Journal of Clinical Pharmacology and Therapeutics* 48, 525–533.

Saugel, B., Nowack, M.C., Hapfelmeier, A., Umgelter, A., Schultheiss, C., Thies, P., Phillip, V., Eyer, F., Schmid, R.M., and Huber, W. (2013) Continuous intravenous administration of vancomycin in medical intensive care unit patients. *Journal of Critical Care* 28, 9–13.

Saugel, B., Gramm, C., Wagner, J.Y., Messer, M., Lahmer, T., Meidert, A.S., Schmid, R.M., and Huber, W. (2014) Evaluation of a dosing regimen for continuous vancomycin infusion in critically ill patients: an observational study in intensive care unit patients. *Journal of Critical Care* 29, 351–355.

Schmelzer, T.M., Christmas, A.B., Norton, H.J., Heniford, B.T., and Sing, R.F. (2013) Vancomycin intermittent dosing versus continuous infusion for treatment of ventilator-associated pneumonia in trauma patients. *The American Surgeon* 79, 1185–1190.

Skov, R., Christiansen, K., Dancer, S.J., Daum, R.S., Dryden, M., Huang, Y.C., and Lowy, F.D. (2012) Update on the prevention and control of community-acquired meticillin-resistant *Staphylococcus aureus* (CA-MRSA). *International Journal of Antimicrobial Agents* 39, 193–200.

Staatz, C.E., Byrne, C., and Thomson, A.H. (2006) Population pharmacokinetic modelling of gentamicin and vancomycin in patients with unstable renal function following cardiothoracic surgery. *British Journal of Clinical Pharmacology* 61, 164–176.

Stevens, D.L., Bisno, A.L., Chambers, H.F., Dellinger, E.P., Goldstein, E.J., Gorbach, S.L., Hirschmann, J.V., Kaplan, S.L., Montoya, J.G., and Wade, J.C. (2014) Practice guidelines for the diagnosis and management of skin and soft tissue infections: 2014 update by the infectious diseases society of America. *Clinical Infectious Diseases* 59, 147–159.

Suzuki, Y., Kawasaki, K., Sato, Y., Tokimatsu, I., Itoh, H., Hiramatsu, K., Takeyama, M., and Kadota, J. (2012) Is peak concentration needed in therapeutic drug monitoring of vancomycin? A pharmacokinetic–pharmacodynamic analysis in patients with methicillin-resistant *Staphylococcus aureus* pneumonia. *Chemotherapy* 58, 308–312.

Swenson, J.M., Anderson, K.F., Lonsway, D.R., Thompson, A., McAllister, S.K., Limbago, B.M., Carey, R.B, Tenover, F.C., and Patel, J.B. (2009) Accuracy of commercial and reference susceptibility testing methods for detecting vancomycin-intermediate *Staphylococcus aureus*. *Journal of Clinical Microbiology* 47, 2013–2017.

Tanaka, A., Aiba, T., Otsuka, T., Suemaru, K., Nishimiya, T., Inoue, T., Murase, M., Kurosaki, Y., and Araki, H. (2010) Population pharmacokinetic analysis of vancomycin using serum cystatin C as a marker of renal function. *Antimicrobial Agents and Chemotherapy* 54, 778–782.

Tatarinova, T., Neely, M., Bartroff, J., van Guilder, M., Yamada, W., Bayard, D., Jelliffe, R., Leary, R., Chubatiuk, A., and Schumitzky, A. (2013) Two general methods for population pharmacokinetic modeling: non-parametric adaptive grid and non-parametric Bayesian. *Journal of Pharmacokinetics and Pharmacodynamics* 40, 189–199.

Terp, S., Krishnadasan, A., Bowen, W., Joo, J., Furoy, D., Chan, J., Moran, G., and Talan, D. (2014) Introduction of rapid methicillin resistant *Staphylococcus aureus* polymerase chain reaction testing and antibiotic selection among hospitalized patients with purulent skin infections. *Clinical Infectious Diseases* 58, e129–e132.

Thalakada, R., Legal, M., Lau, T.T., Luey, T., Batterink, J., and Ensom, M.H. (2012) Development and validation of a novel vancomycin dosing nomogram for achieving high-target trough levels at 2 Canadian teaching hospitals. *The Canadian Journal of Hospital Pharmacy* 65, 180–187.

Tobin-D'Angelo, M, Arnold, K., Lance-Parker, S., LaMarre, M., Bancroft, E., Jones, A., Itano, A., Chambers, M.,

Mascola, L., Clark, J. *et al.* (2003) Methicillin-resistant *Staphylococcus aureus* infections in correctional facilities—Georgia, California, and Texas, 2001–2003. *Morbidity and Mortality Weekly Report (MMWR)* 52, 992–996.

Tunkel, A.R., Hartman, B.J., Kaplan, S.L., Kaufman, B.A., Roos, K.L., Scheld, W.M., and Whitley, R.J. (2004) Practice guidelines for the management of bacterial meningitis. *Clinical Infectious Diseases* 39, 1267–1284.

van Hal, S.J., Paterson, D.L., and Lodise, T.P. (2013) Systematic review and meta-analysis of vancomycin-induced nephrotoxicity associated with dosing schedules that maintain troughs between 15 and 20 milligrams per liter. *Antimicrobial Agents and Chemotherapy* 57, 734–744.

Vance-Bryan, K., Guay, D.R., Gilliland, S.S., Rodvold, K.A., and Rotschafer, J.C. (1993) Effect of obesity on vancomycin pharmacokinetic parameters as determined by using a Bayesian forecasting technique. *Antimicrobial Agents and Chemotherapy* 37, 436–440.

Vance-Bryan, K., Rotschafer, J.C., Gilliland, S.S., Rodvold, K.A., Fitzgerald, C.M., and Guay, D.R. (1994) A comparative assessment of vancomycin-associated nephrotoxicity in the young versus the elderly hospitalized patient. *Journal of Antimicrobial Chemotherapy* 33, 811–821.

Vuagnat, A., Stern, R., Lotthe, A., Schuhmacher, H., Duong, M., Hoffmeyer, P., and Bernard, L. (2004) High dose vancomycin for osteomyelitis: continuous vs. intermittent infusion. *Journal of Clinical Pharmacy and Therapeutics* 29, 351–357.

Wesner, A.R., Brackbill, M.L., Coyle, L.L., and Kidd, R.S. (2013) Prospective trial of a novel nomogram to achieve updated vancomycin trough concentrations. *Interdisciplinary Perspectives on Infectious Diseases* 2013: 839456.

Winter, M.E. (ed.) (2004) *Basic Clinical Pharmacokinetics*, 4th edn. Lippincott Williams & Wilkins, Philadelphia, Pennsylvania.

Wong, J.R., Bauer, K.A., Mangino, J.E., and Goff, D.A. (2012) Antimicrobial stewardship pharmacist interventions for coagulase-negative staphylococci positive blood cultures using rapid polymerase chain reaction. *Annals of Pharmacotherapy* 46, 1484–1490.

Wysocki, M., Delatour, F., Faurisson, F., Rauss, A., Pean, Y., Misset, B., Thomas, F., Timsit, J.F., Similowski, T., Mentec, H. *et al.* (2001) Continuous versus intermittent infusion of vancomycin in severe staphylococcal infections: prospective multicenter randomized study. *Antimicrobial Agents and Chemotherapy* 45, 2460–2467.

Yamamoto, M., Kuzuya, T., Baba, H., Yamada, K., and Nabeshima, T. (2009) Population pharmacokinetic analysis of vancomycin in patients with gram-positive infections and the influence of infectious disease type. *Journal of Clinical Pharmacy and Therapeutics* 34, 473–483.

Yasuhara, M., Iga, T., Zenda, H., Okumura, K., Oguma, T., Yano, Y., and Hori, R. (1998) Population pharmacokinetics of vancomycin in Japanese adult patients. *Therapeutic Drug Monitoring* 20, 139–148.

Zelenitsky, S.A., Ariano, R.E., McCrae, M.L., and Vercaigne, L.M. (2012) Initial vancomycin dosing protocol to achieve therapeutic serum concentrations in patients undergoing hemodialysis. *Clinical Infectious Diseases* 55, 527–533.

Zetola, N., Francis, J.S., Nuermberger, E.L., and Bishai, W.R. (2005) Community-acquired meticillin-resistant *Staphylococcus aureus*: an emerging threat. *The Lancet Infectious Diseases* 5, 275–286.

22 Principles of Switching from Intravenous to Oral Administration

Jamie L. Wagner[1] and Susan L. Davis[2*]

[1]University of Mississippi, Jackson, Mississippi, US; [2]Wayne State University and Henry Ford Hospital, Detroit, Michigan, US

Introduction

Antimicrobials are the most widely used group of drugs in the hospital setting, with almost 33% of patients at any given time receiving this type of drug (Fraser *et al.*, 1997; McLaughlin *et al.*, 2005). Almost half of these patients are receiving antimicrobials that are incorrectly prescribed or inappropriately used (Dellit *et al.*, 2007). The overprescribing of antimicrobials has been shown to lead to increased microbial resistance, increased hospital length of stay, increased morbidity and mortality, and increased use of hospital resources. One common antimicrobial stewardship intervention for reducing unnecessary antimicrobial use and the associated cost is the implementation of an intravenous (IV) to oral (PO) conversion program or policy.

Creating an institutional IV to PO conversion policy is an easy way for antimicrobial stewardship programs to demonstrate value while improving patient outcomes, reducing resistance, reducing cost, and improving patient safety. This chapter will further discuss the reasons behind IV to PO conversion, and the evidence to support it, as well as address the controversies surrounding the implementation of an IV to PO conversion policy.

Rationale for IV to PO Programs

Systematic conversion of IV to PO antimicrobial regimens was suggested as early as the mid-1990s, based on the rationale that many antimicrobials with high oral availability could provide adequate serum and tissue concentrations (Przybylski *et al.*, 1997). The switch from IV to PO also came about because of the need to decrease the risk of infection and complications from IV catheter lines, increased patient satisfaction, and the option for earlier patient hospital discharge (MacGregor and Graziani, 1997). Switching from IV to PO antimicrobials also makes the administration of medication less labor intensive for the nursing staff and is less costly than continuing therapy with an IV regimen (Sevinç *et al.*, 1999). When implemented in patients meeting the appropriate criteria, conversion to oral therapy is associated with reduced length of stay, reduced costs, and reduced complications due to continued intravenous access (Ahkee *et al.*, 1997; Laing *et al.*, 1998; Mertz *et al.*, 2009). In addition, many studies have demonstrated the benefit of IV to PO conversion in multiple infection settings, including skin and soft tissue infections, community-acquired pneumonia, and urinary tract infections (Ahkee *et al.*, 1997; Laing *et al.*, 1998; Ramirez *et al.*, 1999; Rhew *et al.*, 2001; Mertz *et al.*, 2009). More recently, conversion to oral therapy after a short course of IV therapy for serious infections, including osteomyelitis, has demonstrated benefit, but this practice is considered by some to be controversial.

Cost Impact of Using PO over IV Antimicrobials

Most patients who receive IV antimicrobials are eligible for IV to PO switch based on clinical improvement, tolerance of oral medication, ability to absorb oral medication, and the absence of any clinical indication to continue IV therapy. By converting from IV to PO antimicrobials, both hospitals and patients can have a dramatic decrease in costs

*Corresponding author. E-mail: sldavis@wayne.edu

without compromising efficacy or patient safety. The cost implications of IV to PO conversion can most directly be applied to drug acquisition costs and the cost of ancillary supplies for the administration of intravenous therapy. Depending on the extent of the program and the costs of the drugs included, systematic IV to PO conversion programs can reduce a hospital's drug costs by hundreds of thousands of dollars annually (Conly and Shafran, 1994; Laing *et al.*, 1998; Wong-Beringer *et al.*, 2001; Mertz *et al.*, 2009; Jones *et al.*, 2012). When the subsequent impact on reduced length of stay is also considered, the estimated total annual cost reduction is much greater (Ehrenkranz *et al.*, 1992). Overall, the early conversion of IV antimicrobials to the oral formulation can result in significant nationwide cost savings to health systems and patients without impacting safety, efficacy, or patient satisfaction.

Patient Eligibility for IV to PO Conversion

While individual policies vary, it is generally recommended that IV to PO conversion interventions be targeted at patients meeting certain clinical and infection-related characteristics. Generally, these characteristics include clinical improvement, an appropriate infectious indication, and the absence of characteristics that would impair oral absorption. Examples of these characteristics are listed in Table 22.1. For patients hospitalized for an infection, a generalized consensus is that patients receive up to 3 days of IV antimicrobials before conversion to oral antimicrobials, although in certain populations a shorter duration of IV is possible (Siegel *et al.*, 1996). Clinical judgment and assessment is critical when evaluating patients for IV to PO conversion, as well as taking into account their gastrointestinal (GI) function and the individual pharmacokinetic and pharmacodynamic characteristics of the oral antimicrobial.

Many disease states can be treated effectively with oral antimicrobials, such as community-acquired pneumonia, skin and soft tissue infections, urinary tract infections, intra-abdominal infections, and even osteomyelitis. In community-acquired pneumonia patients, switching to oral antimicrobials, such as azithromycin and cefpodoxime, after up to 3 days of IV therapy, resulted in equal efficacy to that in patients treated with a full course of IV therapy (>95% cure rate), a shorter length of hospital stay while maintaining successful clinical outcomes, and reduced overall direct and indirect hospital costs. For patients with skin and soft tissue infections (SSTIs), oral antimicrobials are equally efficacious as IV formulations. In deeper SSTIs, PO antimicrobials are effective following initial IV therapy and when paired with surgical incision and drainage, and appropriate debridement. Oral antimicrobials used to treat pyelonephritis and cystitis have been shown to achieve 90% success in infection resolution when

Table 22.1. Suggested patient eligibility and exclusion criteria for intravenous (IV) to oral (PO) conversion.

Eligibility criteria

- Receiving an intravenous agent with high bioavailability (most policies list specific included agents)
- Clinical stability
  - Afebrile for at least 24 h
  - White blood cell count decreasing toward normal limits (if it was initially elevated)
  - Pretreatment signs and symptoms of infection improving
- Functioning gastrointestinal (GI) tract
  - Receiving other oral medication
  - Tolerating oral diet, enteral nutrition, or at least 1000 ml fluids/day

Exclusion criteria

- Patients with conditions precluding reliable absorption of oral medication
  - Severe nausea and vomiting
  - Motility disorder of the GI system, malabsorption, or short bowel syndrome
  - Continuous nasogastric suction
  - Severe mucositis
  - Strict NPO (nil per os) status for another reason
- Infectious indication in which oral therapy is not recommended (e.g., endocarditis, meningitis)
- Pediatric patients (should not be converted without consultation with primary provider)
- Neutropenic or transplant patients (should not be converted without consultation with primary provider)

compared with IV antimicrobials, while providing a 7.5-fold cost savings over IV regimens. In patients treated with PO ciprofloxacin for Gram-negative osteomyelitis, there were comparable rates of success with fewer days in the hospital, lower rates of treatment complications, and lower costs compared with patients treated with IV ciprofloxacin for the entire treatment duration.

Drug Considerations for IV to PO Conversion

The ideal drug characteristics for an oral conversion include an antimicrobial that has the appropriate spectrum of activity, achieves good absorption from the GI tract, and provides IV-equivalent blood levels, along with a low-frequency dosing schedule. Determination of the appropriate spectrum of activity is limited by the availability of culture data, and preferences for dosing frequency are somewhat subjective, although considerable data exist to support the pharmacokinetics of orally administered antimicrobials. Many factors can influence the systemic absorption of oral antimicrobials, such as bioavailability, gastrointestinal motility, protein binding, volume of distribution, and drug–drug interactions. Table 22.2 provides a summary of common intravenous antimicrobials and their oral equivalents, along with the factors that can play a role in the absorption of the orally administered antimicrobials.

Bioavailability

One of the first things to consider when switching an IV antimicrobial to a PO antimicrobial is the bioavailability of the oral agent. Bioavailability is the fraction of drug (F factor) absorbed into systemic circulation after administration. When drugs are administered through the IV route, the bioavailability is 100% (F = 1.00); however, when drugs are administered orally, transdermally, rectally, or subcutaneously, the bioavailability is oftentimes less than 100% (F < 1.00). There are several factors that alter the bioavailability of an oral agent. All oral agents have a vehicle surrounding the active drug that must be broken down to release the drug into the gastrointestinal (GI) tract, and this can influence the availability of the active drug. The released drug then must pass through the GI tract before entering the circulation. Some drugs are subjected to GI tract metabolism or are actively pumped back into the GI tract (e.g., P-glycoprotein), thereby limiting the concentration of drug available for metabolism by the liver. Once the drug passes through the GI tract, it then travels to the liver, where it is subjected to first-pass metabolism, before entering systemic circulation. Drugs that have high liver extraction ratios, or high first-pass effects, will have lower bioavailability than drugs that have lower liver extraction ratios. These factors combined influence the overall bioavailability of the oral agent.

Gastrointestinal absorption

There are other factors that can influence the systemic absorption of oral agents, including the location of GI absorption, which has specific implications in patients receiving antimicrobials via feeding tube. The pH of the duodenum, the site of absorption for most oral antimicrobials is, on average, around 6, while the pH of the jejunum, a less frequent absorption site, is typically 7–9. This can present a problem for patients who have had bowel surgery that affects the passage of materials through the duodenum. The pH of the site of absorption can also influence the ionization of the drug, and so limit the systemic absorption to only the unionized drug, and further limiting the amount of drug available for systemic absorption. Gastric motility can also influence the systemic absorption of oral antimicrobials. In patients who have impaired gastric motility (e.g., gastroparesis), systemic absorption can be delayed or nonexistent, and contribute to ineffective systemic absorption of oral antimicrobials.

Protein binding

Once the drug is in the systemic circulation, it has the opportunity to bind to proteins such as albumin within the plasma, thus limiting its ability to act upon the site of action, as only unbound drug is considered to be active. Antimicrobial dosing recommendations already account for systemic protein binding in healthy individuals; however, in patients with lower systemic protein levels, more unbound, free drug is available for use. Clinicians should keep this concept in mind when making dosing recommendations for patients with low serum albumin levels.

Volume of distribution

An oral antimicrobial agent's volume of distribution will also play a role in systemic absorption.

Table 22.2. Pharmacokinetic considerations in intravenous (IV) to oral (PO) conversion of common antimicrobials. Adapted from *Drug Information Handbook* (Lexi-Comp, 2015) and the Clinical Pharmacology database (2015).

IV agent	Approximate oral equivalent	F (bioavailability)	Protein binding	Volume of distribution (l/kg)[a]	Other considerations
Acyclovir	Acyclovir	10–20%	9–33%	0.8	Valacyclovir is another oral conversion option
Ampicillin	Amoxicillin	74–92%	20% amoxicillin	0.26–0.31	
Ampicillin–sulbactam	Amoxicillin–clavulanate		25% clavulanate		
Azithromycin	Azithromycin	37%	51% at 0.02 µg/ml, 7% at 2 µg/ml	31.1	F may be altered if coadministered with P-glycoprotein inhibitors
Cefazolin	Cephalexin	80–100%	6–20%	0.23–0.35	F decreased with food but this not clinically relevant
Ceftriaxone	Cefpodoxime	50%	22–33%	0.6–1.2	Prodrug; some conversion to active drug in acidic stomach; altered by acid-suppressive therapy
Ciprofloxacin	Ciprofloxacin	70%	20–40%	1.2–2.7	F decreased when coadministered with divalent cations
Clindamycin	Clindamycin	90%	92–94%	0.6–1.2	–
Doxycycline	Doxycycline	90–100%	23–93%	0.75	F decreased by antacids, divalent cations
Fluconazole	Fluconazole	90–100%	11–12%	0.722–1.069	–
Ganciclovir	Valganciclovir	60%[b]	1–2% of active ganciclovir	0.569–0.837	F increased with food and decreased on an empty stomach; rapid conversion to ganciclovir ≤6 h after administration
Linezolid	Linezolid	100%	31%[c]	40–50 l	
Metronidazole	Metronidazole	90–100%	10%	0.55	F increased with colonic inflammation
Moxifloxacin	Moxifloxacin	90%	30–50%	1.7–2.7	F decreased when coadministered with divalent cations
Penicillin G	Penicillin V	60%	75–89%	Unknown	–
Trimethoprim–sulfamethoxazole (TMP/SMX)	TMP/SMX	90–100%	44% TMP, 70% SMX	2 TMP, 0.36 SMX	–
Voriconazole	Voriconazole	96%	58%	4.6	–

[a]Unless otherwise indicated.
[b]Bioavailability with food.
[c]Stalker and Jungbluth, 2003.

The volume of distribution (Vd) is a theoretical number that relates the total amount of drug in the body to the amount of drug in the serum. The approximate volume of the plasma is 0.07 l/kg, that of the extracellular fluid is 0.13 l/kg, and that of the tissue is 0.6 l/kg. The combination of protein binding and volume of distribution can influence how and where an antimicrobial agent disperses. For example, penicillin V has a small volume of distribution, yet it can be found in many different tissues. This is because penicillin V has significant protein binding within the plasma but not within the tissues.

The combination of oral bioavailability, protein binding, and volume of distribution must be considered when evaluating the overall effectiveness of an oral antimicrobial agent. For example, azithromycin has an oral bioavailability of 38% (F = 0.38), protein binding of 7–51%, and a Vd of 31.1 l/kg. Studies have shown that the PO administration of azithromycin at the same dose as IV administration is effective in obtaining optimal bacterial killing of *Streptococcus pneumoniae* in patients with community-acquired pneumonia. Oral acyclovir has an absolute bioavailability of 20% (F = 0.20), protein binding of 9–33%, and a Vd of 0.8 l/kg. Because the drug has a low bioavailability, despite minimal protein binding and distribution into the tissues, optimal concentrations can be achieved by increasing the frequency of the dosing interval compared with the IV formulation (e.g., five times a day versus every 8 h).

Drug–drug interactions

Drug–drug interactions (DDIs) can occur with every step in the pharmacodynamic process, from absorption through to elimination. Alterations in the gastric pH, intestinal pH, or intestinal motility will affect the absorption of oral antimicrobials. For example, recommendations for the administration of doxycycline are to take with food, although divalent cations should be avoided. If a patient has mild gastroesophageal reflux disease (GERD) and takes calcium carbonate before each meal, the absorption of doxycycline would be drastically impaired as a result of the binding of the antimicrobial to the antacid. Once the antimicrobial is absorbed into systemic circulation, there is another opportunity for a DDI in protein binding. The displacement of an antimicrobial from systemic proteins can increase the systemic concentration of active drug, so putting the patient at risk of supratherapeutic concentrations. Conversely, antimicrobials can also displace other drugs from systemic proteins, thereby decreasing the overall systemic concentration of the antimicrobial. The major source of DDIs typically occurs during the metabolism phase. Most antimicrobials, and other drugs, are metabolized through the CYP450 (cytochrome P450) enzyme system, with some utilizing P-glycoprotein as an alternative method. Enzyme induction, enzyme inhibition, and competitive metabolism are mechanisms through which the systemic antibiotic concentration can be altered. Lastly, alterations in clearance through the kidney, mainly through secretion into the tubules, can either increase or decrease antimicrobial systemic concentrations. Clinicians should be aware of antimicrobials that interact with other medications so as to ensure the appropriate dose adjustment and subsequent monitoring.

Stepdown vs. Sequential Therapy

The simplest form of IV to PO switch is the “stepdown,” in which an IV therapy is replaced with a PO formulation of the same drug. However, patients receiving IV antimicrobial regimens that have no equivalent and reliable oral formulation should not necessarily be excluded from switching to PO antimicrobials. When the IV to PO switch is accomplished by providing a different drug, this is termed “sequential” therapy. One of the most common occasions on which sequential IV to PO therapy is seen is in community-acquired pneumonia. The initial IV therapy for community-acquired pneumonia is frequently the combination of a cephalosporin and a macrolide. Instead of converting to two separate oral agents, it is feasible to provide a similar spectrum of activity with a respiratory quinolone (Davis *et al.*, 2005). Switching between drug classes during IV to PO conversion does create a few challenges that need to be considered. If no organism is identified for susceptibility testing, the activity of the sequential agent must be inferred based on the local antibiogram. Also, drug- and class-specific patient intolerances or drug interactions need to be reevaluated for the new agent. For these reasons, some institutions do not include sequential therapy in automatic conversion without first consulting the primary provider.

Special Populations: IV to PO Conversion in the ICU

Critically ill patients present a unique challenge when considering IV to PO conversions of antimicrobials and close evaluation of many variables is required before the conversion is implemented. Some patient variables to consider include the functional capability of the GI tract, blood flow to non-vital organs, fluid shifts, protein binding, and the ability of the patient to tolerate other oral medications. Strong consideration of individual patient factors and the severity of the illness is recommended before converting from IV to PO antimicrobials.

Gastrointestinal tract

An important concept to consider when opting for oral antimicrobials in a critically ill patient is the ability of the GI tract to function appropriately. In the initial phase of illness, the GI tract may exhibit hypermotility, subsequently prohibiting the full absorption or delaying the absorption of oral antimicrobials. Patients who are more than 48 h past the acute phase of illness may experience fluctuations in organ function due to shunting of blood away from non-vital organs (e.g., the GI tract, kidneys, liver) to vital organs (e.g., the brain and heart). The GI tract would not receive an adequate blood supply or adequate oxygen to function and will begin to shut down, thus also limiting the amount of drug absorbed. Patients may have physical constraints to a functioning GI tract as well; these can include gastric obstruction, ileus (obstruction of the ileum), malabsorption syndrome, short-gut syndrome or other malady that requires the use of total parenteral nutrition (TPN), moderate-to-severe diarrhea, or persistent nausea and vomiting. Another factor to consider when changing patients to oral antimicrobials is the addition of an acid suppressive agent to prevent stress ulcers. The addition of these agents may reduce the pH-dependent absorption of medications from the stomach and intestine, thereby decreasing the overall systemic drug concentrations resulting from agents administered via the oral route. The use of opiates for pain control or sedation will also slow the absorption of oral antimicrobials and reduce systemic drug concentrations.

Hepatic and renal alterations

In the acute, hyperdynamic phase of illness, drug metabolism increases rapidly through the liver, and renal perfusion may likewise increase. The pharmacokinetic properties of antimicrobials will be altered and may require higher or more frequent dosing (Varghese *et al.*, 2011; Felton *et al.*, 2014). Liver dysfunction is present in up to 54% of critically ill patients (Krishnan and Murray, 2003). The more common liver dysfunction affecting antimicrobials in the critically ill, though, is impaired liver function. As blood flow and oxygen supply to the liver wanes, the rate of metabolism through the CYP enzymes, as well as other liver metabolic pathways, declines, so reducing the metabolism of antimicrobials. Drugs that require liver metabolism to inactivate the parent compound will remain in the system for longer periods of time, and so increase the likelihood of adverse drug events. Oral antimicrobials that undergo extensive first-pass metabolism may have significantly higher oral bioavailability, and when combined with reduced kidney function, may require dose reductions to prevent adverse effects. Additionally, as a patient's core body temperature declines into a state of hypothermia, the CYP enzymes have reduced function, thus slowing antimicrobial metabolism. Despite reduced liver metabolism, some metabolic processes are unaffected (Power *et al.*, 1998).

Similar processes are seen in the kidneys in patients with reduced renal perfusion caused by critical illness. As the kidney function declines, the body's ability to regulate systemic pH begins to fail. The altered systemic pH will impact the ionization of antimicrobials and hence alter the extent of absorption into other body compartments. Alteration of the systemic pH also influences the ability of acidic drugs, such as cephalosporins, penicillins, and sulfonamides, to bind to albumin, so further increasing active systemic drug concentration. If antimicrobials have more than 30% of unchanged drug excreted by the kidneys, such as aminoglycosides, they are likely to have longer half-lives and maintain therapeutic or supratherapeutic drug concentrations for a longer duration. In patients who develop acute renal failure that requires assistance, the renal replacement therapy modality must be considered when adjusting antimicrobial doses. Some antimicrobials, such as vancomycin and aminoglycosides, require supplemental doses after a hemodialysis session to replenish the drug levels to therapeutic concentrations. When continuous renal replacement therapy is utilized in critically ill patients, the type of filter, run time, flow rate, and ability of the drug to be removed by the system must be considered when adjusting doses. Antimicrobials that have 30% or more of their concentration removed through continuous renal replacement therapy (CRRT) must have a dose or frequency adjustment performed to maintain adequate drug concentrations (Roberts and Lipman, 2009). Therapeutic drug monitoring, when possible, is recommended to ensure adequate drug concentrations.

Volume of distribution

Patients who are experiencing severe acute illness are typically given extreme fluid volume resuscitation with normal saline to return their blood pressure to normal values. However, in the acute phase, patients may be experiencing fluid shifts caused by the systemic inflammatory process. Fluid shifts from the intracellular space to the extracellular space, otherwise known as third spacing, can drastically change drug behavior. Hydrophilic antimicrobials, such as β-lactams, aminoglycosides, and linezolid, will have a much larger Vd than in a healthy patient, whereas, hydrophobic antimicrobials, such as fluoroquinolones, macrolides, and clindamycin, will not see much change in their Vd. Because of this concept, the net increase in body weight secondary to fluid resuscitation should be added to the total Vd to provide adequate antimicrobial concentration within tissues and plasma. As the Vd increases, the half-life of the antimicrobial will also increase, providing the patient with active drug for longer periods of time. Clinicians must also consider the ability of antimicrobials to penetrate to the site of action. As a patient's severity of illness increases, tissue penetration becomes impaired; this is attributed to capillary leakage, tissue edema, and microvascular failure. In patients with septic shock, antimicrobial penetration into tissues is 5–10 times lower than in healthy volunteers (Varghese *et al.*, 2011). In particular, hydrophilic antimicrobials tend to have lower interstitial penetration, while hydrophobic antimicrobials are largely unaffected and retain their interstitial penetration. Additionally, most of the fluid given during resuscitation distributes to the tissues, subsequently diluting the concentration of the antimicrobial within the site of action, despite its higher distribution.

Protein binding

Hypoalbuminemia is a problem in critically ill patients as drug pharmacokinetic and pharmacodynamic properties can be severely altered, requiring dose and frequency adjustment. For example, as the level of albumin and other systemic proteins decrease, the Vd of highly bound drugs increases, making more unbound, active drug available. Aminoglycosides have a larger Vd, in part from fluid resuscitation and hypoalbuminemia, and would require a larger dose to compensate for the expanded Vd, despite more free drug being available. Decreased protein binding of acidic and basic drugs also allows for the increased clearance of antimicrobials.

Overall, patients who are acutely ill must be critically evaluated before appropriate conversion of antimicrobials from IV to PO. It is recommended that patients who are hemodynamically unstable should receive IV antimicrobials instead of PO antimicrobials until they are stable, along with appropriate therapeutic drug monitoring, when available. However, as these patients improve and reach clinical stability, they may no longer have contraindications to oral therapy and may be considered for conversion. Patients recovering from critical illness may continue to require nutrition provided through enteral feeding tubes. There are some challenges with the administration of antimicrobials through these tubes, such as the concurrent administration of enteral feeding and the antimicrobial, and binding to the enteral tube. A summary of these findings is displayed in Table 22.3. Several antimicrobials have different formulations that may influence the properties of feeding tube administration, and caution is advised when selecting appropriate antimicrobials for converting to PO for administration through an enteral feeding tube.

IV to PO Conversion for Antimicrobial Stewardship

Considerable data supports antimicrobial IV to PO conversion as a successful antimicrobial stewardship strategy that is associated with positive clinical, microbiological, and economic outcomes. Thus, the Infectious Disease Society of America (IDSA) recommends IV to PO conversion as an essential component for antimicrobial stewardship programs. Using the criteria described above, health systems can design policies and procedures to help guide personnel in appropriate antimicrobial IV to PO conversion.

Successful antimicrobial stewardship programs have incorporated many different techniques in IV to PO conversion policies to be successful. Electronic medical records with clinical decision support can be used to identify patients meeting criteria, and generate alerts and reminders to pharmacists, or

Table 22.3. Intravenous (IV) to oral (PO) considerations in patients with enteral feeding tubes. Adapted from Wohlt *et al*. (2009), Mitchell (2014), White and Bradnam (2015), and the Clinical Pharmacology database (2015).

IV agent	Approximate oral equivalent	Administration recommendations	Primary absorption site	Break in feeding required	Other considerations
Acyclovir	Acyclovir	Tablet: dilute with 10–50 ml water Capsule: open capsule and dilute with 10–50 ml of water Suspension: unknown	Duodenum	Hold tube feed to administer medication then restart	Suspension contains small amount of sorbitol (<7.5 g)
Ampicillin	Amoxicillin	Tablet: dilute with at least 10 ml water Capsule: open capsule and dilute with at least 10 ml water Suspension: mix with equal volume of water to reduce viscosity	Duodenum and jejunum	Hold tube feed to administer medication then restart	Cannot crush extended release formulations; suspension is hypermolar due to sucrose content and will cause gastrointestinal (GI) discomfort if not diluted with equal volume of water before administration
Ampicillin–sulbactam	Amoxicillin–clavulanate	Tablet: dilute with at least 1 ml water Suspension: dilute with 10–30 ml water		Hold tube feed to administer medication then restart	
Azithromycin	Azithromycin	Tablet: dilute with at least 10 ml water Suspension: unknown	Duodenum	Hold tube feed to administer medication then restart	Avoid giving with aluminum or magnesium containing products
Cefazolin	Cephalexin	Tablet: dilute with at least 10 ml water Capsule: open capsule and dilute powder with at least 10 ml water Suspension: mix with equal volume of water to reduce viscosity	Duodenum	Hold tube feed to administer medication then restart	Avoid crushing tablets or opening capsules; suspension is hypermolar due to sucrose content and will cause extreme GI discomfort if not diluted with equal volume of water before administration
Ceftriaxone	Cefpodoxime	Tablet: dilute with at least 10 ml water Suspension: unknown	Duodenum	Hold tube feed to administer medication then restart[a]	None
Ciprofloxacin	Ciprofloxacin	Tablet: dilute with 20 ml water Suspension: cannot be given via feeding tube	Duodenum primarily with some jejunum	Hold tube feeds for 1 h before and 2 h after administration	Cannot crush XR (extended-release) tablet formulation; must use higher doses when giving via feeding tube or intrajejunal administration; suspension adheres to tube; enteral dose must be 2× IV dose
Clindamycin	Clindamycin	Capsule: open capsule and dissolve powder with 30 ml water Solution: unknown	Unknown	Hold tube feed to administer medication then restart	None[b]

Doxycycline	Doxycycline	Tablet: dilute with at least 10 ml water Capsule: recommended to not open capsule as contents are an esophageal irritant Suspension: unknown	Duodenum	Hold tube feed for 1 h before and 2 h after administration	Cannot open delayed-release capsule; if no break in feeding is possible, must use higher doses
Fluconazole	Fluconazole	Tablet: recommended to use suspension for tube administration Suspension: no need for further dilution	Duodenum and jejunum	Hold tube feed to administer medication then restart	None[b]
Ganciclovir	Valganciclovir	Tablet: do not crush Solution: unknown	Duodenum, jejunum, ileum	Hold tube feed to administer medication then restart	Do not crush—medication is teratogenic and has irritant potential; avoid direct skin or mucous membrane contact with broken or crushed tablets, oral solution powder, and reconstituted oral solution
Linezolid	Linezolid	Tablet: recommended to use suspension for tube administration Suspension: no further dilution required	Unknown	Hold tube feed to administer medication then restart	Suspension and tablet have 100% bioavailability
Metronidazole	Metronidazole	Tablet: dilute with at least 10 ml water Capsule: open capsule and dilute powder with at least 10 ml water	Duodenum	Hold tube feed for 1 h before administration then restart	Cannot crush ER (extended-release) formulation
Moxifloxacin	Moxifloxacin	Tablet: dilute with at least 10 ml water	Unknown	Hold tube feed for 1 h before administration then restart	None
Penicillin G	Penicillin V	Tablet: recommended to use liquid formulation Solution: no further dilution required Suspension: no further dilution required	Duodenum	Hold tube feed(s) for 2 h before and 1–2 h after administration	Intrajejunal administration may result in higher concentrations; recommended to not crush tablet as powder may cross-contaminate or increase sensitization to drug
Trimethoprim–sulfamethoxazole (TMP/SMX)	TMP/SMX	Tablet: recommended to use liquid formulation Suspension: mix with equal volume of water to reduce viscosity	Unknown	Hold tube feed to administer medication then restart	After medication administration, rinse oral syringe with equal volume and administer to patient
Voriconazole	Voriconazole	Tablet: recommended to use liquid formulation Suspension: no further dilution required	Duodenum and jejunum	Hold tube feed for 1 h before and 2 h after administration	None

[a]According to the drug reference monograph in Micromedex.
[b]White and Bradnam, 2015.

directly to prescribers. In systems without this technology, the antimicrobial stewardship program may elect to use other screening methods to identify patients, and then intervene in person. Policies may also include language to delegate authority to clinical pharmacists to facilitate the ordering and monitoring of oral antimicrobials in patients meeting the prespecified criteria.

When developing an antimicrobial IV to PO conversion policy, it is recommended first to survey providers to determine their baseline understanding of and attitudes toward oral antimicrobial therapy, and to identify potential barriers to the conversion program. This will aid not only in designing computer-generated alerts for qualified patients, but also in designing the necessary educational programs to accompany policy implementation. Other barriers that can limit the success of an IV to PO conversion program are time constraints, limited staff, and insufficient training. It is a common misconception that insurance providers and other payers will not reimburse for hospital stay in patients on oral therapy. Indeed, if no other medical issues remain to be managed, antimicrobial conversion to oral therapy should be used to facilitate discharge or transfer. However, if other comorbid conditions necessitate longer hospital stay, receipt of an oral antimicrobial should not preclude a patient from remaining hospitalized. Educational efforts, including discussion with case management, should be planned if this dogma is a barrier to IV to PO conversion programs.

Just as many stewardship interventions use a multifaceted approach to implementation, a similar approach is useful in implementing IV to PO conversion programs. This might include prospective audit with intervention and feedback, patient-specific education materials, nursing and physician education materials, and physician reminders to assess further need for antimicrobial therapy. Key stakeholders should be included in the program planning and rollout, including opinion leaders from infectious diseases, nursing, case management, hospitalists (specialists in the practice of hospital medicine), and others. Employing a multifaceted, interdisciplinary approach supports the successful alteration of prescriber behavior and long-term success. It is also recommended that compliance with the policy be monitored routinely with the dissemination of results to all participating individuals. This will also provide evidence of clinical and economic benefit for continued justification of the program.

Summary

Intravenous to oral conversion of antimicrobial therapy is a cost-effective and widely utilized strategy to reduce antimicrobial cost, minimize complications related to IV therapy, and facilitate hospital discharge. When comparing IV and enteral antimicrobials in general, there is essentially no difference in antibiotic-associated adverse events, readmission, or in-hospital mortality, while oral therapy can improve patient satisfaction and reduce cost. Patients considered for IV to PO conversion should be clinically improving and tolerating an oral diet or medications. Medications appropriate for IV to PO conversion include a wide range of agents from a variety of classes. Stewardship program leaders should design IV to PO conversion programs around the needs of their patient population and the resources available in their health system. Regardless of the approach chosen, IV to PO conversion of antimicrobials can be an excellent opportunity for the stewardship team to demonstrate success at improving quality of antimicrobial use and reducing overall cost.

References

Ahkee, S., Smith, S., Newman, D., Ritter, W., Burke, J., and Ramirez, J.A. (1997) Early switch from intravenous to oral antibiotics in hospitalized patients with infections: a 6-month prospective study. *Pharmacotherapy* 17, 569–575.

Clinical Pharmacology (2015) Online database. Gold Standard (Elsevier), Tampa, Florida. Available at: http://www.clinicalpharmacology.com (accessed 23 May 2016).

Conly, J.M. and Shafran, S.D. (1994) Intravenous-to-oral conversion therapy for antimicrobials. *Canadian Journal of Infectious Diseases* 5, 15–16.

Davis, S.L., Delgado, G., Jr., and McKinnon, P.S. (2005) Pharmacoeconomic considerations associated with use of intravenous-to-oral moxifloxacin for community acquired pneumonia. *Clinical Infectious Diseases* 15(Suppl 2), S136–S143.

Dellit, T.H., Owens, R.C., McGowan, J.E., Jr., Gerding, D.N., Weinstein, R.A., Burke, J.P., Huskins, W.C., Paterson, D.L., Fishman, N.O., Carpenter, C.F. *et al.* (2007) Infectious Disease Society of America and the Society for Healthcare Epidemiology of America guidelines for developing an institutional program to enhance antimicrobial stewardship. *Clinical Infectious Diseases* 44, 159–177.

Ehrenkranz, N.J., Nerenberg, D.E., Shultz, J.M., and Slater, K.C. (1992) Intervention to discontinue parenteral antimicrobial therapy in patients hospitalized with pulmonary infections: effect on shortening patient stay. *Infection Control and Hospital Epidemiology* 13, 21–32.

Felton, T.W., Hope, W.W., and Roberts, J.A. (2014) How severe is antibiotic pharmacokinetic variability in the critically ill patients and what can be done about it? *Diagnostic Microbiology and Infectious Disease* 79, 441–447.

Fraser, G.L., Stogsdill, P., Dickens, J.D., Jr., Wennberg, D.E., Smith, R.P., Jr., and Prato, B.S. (1997) Antibiotic optimization: An evaluation of patient safety and economic outcomes. *Archives of Internal Medicine* 157, 1689–1694.

Jones, M., Huttner, B., Madaras-Kelly, K., Nechodom, K., Nielson, C., Goetz, M.B., Neuhauser, M.M., Samore, M.H., and Rubin, M.A. (2012) Parenteral to oral conversion of fluoroquinolones: low-hanging fruit for antimicrobial stewardship programs? *Infection Control and Hospital Epidemiology* 33, 362–367.

Krishnan, V. and Murray, P. (2003) Pharmacologic issues in the critically ill. *Clinics In Chest Medicine* 24, 671–688.

Laing, R.B.S., Mackenzie, A.R., Shaw, H., Gould, I.M., and Douglas, J.G. (1998) The effect of intravenous-to-oral switch guidelines on the use of parenteral antimicrobials in medical wards. *Journal of Antimicrobial Chemotherapy* 42, 107–111.

Lexi-Comp (2015) *Drug Information Handbook*, 24th edn. Lexi-Comp, Hudson, Ohio.

MacGregor, R.R. and Graziani, A.L. (1997) Oral administration of antibiotics: a rational alternative to the parenteral route. *Clinical Infectious Diseases* 24, 457–467.

McLaughlin, C.M., Bodasing, N., Boyter, A.C., Fenelon, C., Fox, J.G., and Seaton, R.A. (2005) Pharmacy-implemented guidelines on switching from intravenous to oral antibiotics: an intervention study. *QJM* 98, 745–752.

Mertz, D., Koller, M., Haller, P., Lampert, M.L., Plagge, H., Hug, B., Koch, G., Battegay, M., Flückiger, U., and Bassetti, S. (2009) Outcomes of early switching from intravenous to oral antibiotics on medical wards. *Journal of Antimicrobial Chemotherapy* 64, 188–199.

Mitchell, J. (2014) *Oral Dosage Forms That Should Not Be Crushed, Updated January 2014 [Do Not Crush List]*. Institute for Safe Medication Practices. Latest version available at: http://www.ismp.org/tools/donotcrush.pdf (accessed 23 May 2016).

Power, B.M., Forbes, A.M., van Heerden, P.V., and Ilett, K.F. (1998) Pharmacokinetics of drugs used in critically ill adults. *Clinical Pharmacokinetics* 34, 25–56.

Przybylski, K.G., Rybak, M.J., Martin, P.R., Weingarten, C.M., Zaran, F.K., Stevenson, J.G., and Levine, D.P. (1997) A pharmacist-initiated program of intravenous to oral antibiotic conversion. *Pharmacotherapy* 17, 271–276.

Ramirez, J.A., Vargas, S., Ritter, G.W., Brier, M.E., Wright, A., Smith, S., Newman, D., Burke, J., Mushtaq, M., and Huang, A. (1999) Early switch from intravenous to oral antibiotics and early hospital discharge: a prospective observational study of 200 consecutive patients with community-acquired pneumonia. *Archives of Internal Medicine* 159, 2449–2454.

Rhew, D.C., Tu, G.S., Ofman, J., Henning, J.M., Richards, M.S., and Weingarten, S.R. (2001) Early switch and early discharge strategies in patients with community-acquired pneumonia. *Archives of Internal Medicine* 161, 722–727.

Roberts, J.A. and Lipman, J. (2009) Pharmacokinetic issues for antibiotics in the critically ill patient. *Critical Care Medicine* 37, 840–851.

Sevinç, F., Prins, J.M., Koopmans, R.P., Langendijk, P.N.J., Bossuyt, P.M.M., Dankert, J., and Speelman, P. (1999) Early switch from intravenous to oral antibiotics: guidelines and implementation in a large teaching hospital. *Journal of Antimicrobial Chemotherapy* 43, 601–606.

Siegel, R.E., Halpern, N.A., Almenoff, P.L., Lee, A., Cashin, R., and Greene, J.G. (1996) A prospective randomized study of inpatient intravenous antibiotics for community-acquired pneumonia: the optimal duration of therapy. *Chest* 105, 1109–1115.

Stalker, D.J. and Jungbluth, G.L. (2003) Clinical pharmacokinetics of linezolid, a novel oxazolidinone antibacterial. *Clinical Pharmacokinetics* 42, 1129–1140.

Varghese, J.M., Roberts, J.A., and Lipman, J. (2011) Antimicrobial pharmacokinetic and pharmacodynamics issues in the critically ill with severe sepsis and septic shock. *Critical Care Clinics* 27, 19–34.

White, R. and Bradnam, V. (2015) *Handbook of Drug Administration via Enteral Feeding Tubes*, 3rd edn. Pharmaceutical Press, London.

Wohlt, P.D., Zheng, L., Gunderson, S., Balzar, S.A., Johnson, B.D., and Fish, T.J. (2009) Recommendations for the use of medications with continuous enteral nutrition. *American Journal of Health-System Pharmacy* 55, 1458–1467.

Wong-Beringer, A., Nguyen, K.H., and Razeghi, J. (2001) Implementing a program for switching from i.v. to oral antimicrobial therapy. *American Journal of Health-System Pharmacy* 58, 1146–1149.

23 The Role of Pharmacists in Antimicrobial Stewardship

HALEY J. MORRILL,[1,2*] MONICA DOROBISZ,[3] AND KERRY L. LAPLANTE[1,2,4]

[1]Veterans Affairs Medical Center, Providence, Rhode Island, US; [2]University of Rhode Island College of Pharmacy, Kingston, Rhode Island, US; [3]Department of Pharmacy, Kent Hospital, Warwick, Rhode Island, US; [4]Warren Alpert Medical School of Brown University, Providence, Rhode Island, US

Introduction

Antimicrobial stewardship cannot be described without the involvement of pharmacists. As antimicrobial stewardship inherently involves the appropriate selection, dosing, route of administration and duration of antimicrobial drug therapy, there is a direct link to the roles and responsibilities of the pharmacist. In fact, hospital pharmacists have been involved in practices to improve antimicrobial use, such as recommending antimicrobial dose adjustments for organ dysfunction and managing antimicrobial formularies, long before any formal recommendations for antimicrobial stewardship existed. These types of achievements have placed pharmacists in the forefront of antimicrobial stewardship. Pharmacists' knowledge of pharmacotherapy, pharmacokinetics, and pharmacodynamics, influence over antimicrobial use, and established membership on multidisciplinary committees puts them in a unique position of influence (ASHP, 2010a,b).

As such, the 2007 guidelines for developing an antimicrobial stewardship program (ASP) from the Infectious Diseases Society of America (IDSA) and Society of Healthcare Epidemiologists (SHEA) recommend that a clinical pharmacist with infectious disease (ID) training and an ID physician co-lead a multidisciplinary antimicrobial stewardship team (Dellit *et al.*, 2007). It is the responsibility of both the physician and pharmacist leaders to act as "antimicrobial stewardship champions," taking ownership over the mission for building the program and overseeing and advocating for antimicrobial stewardship throughout the hospital (Septimus and Owens, 2011). The US Centers for Disease Control and Prevention (CDC) too has recommended that pharmacists assume a prominent role in antimicrobial stewardship (CDC, 2014). As per CDC recommendations, a single pharmacist should be identified to co-lead the program. Additionally, the American Society of Health System Pharmacists (ASHP) has recognized the importance of pharmacists in promoting the optimal use of antimicrobials, and in a statement paper in 2010, the ASHP outlined the fundamental roles and responsibilities of the pharmacist in antimicrobial stewardship and promoting the optimal use of antimicrobials (ASHP, 2010b). The ASHP's white paper, *A Hospital Pharmacist's Guide to Antimicrobial Stewardship Programs* (2010a), covers similar ground, but in greater detail. The main points from these documents on the role of hospital pharmacists in antimicrobial stewardship are summarized in Table 23.1.

This chapter describes the roles (and responsibilities) of the whole pharmacy department in an effective ASP and summarizes the potential pharmacy-driven antimicrobial stewardship interventions that can be implemented. Additionally, the chapter describes formal and non-formal training opportunities for pharmacists practicing antimicrobial stewardship. Finally, it discusses the roles and responsibilities

*Corresponding author. E-mail: haley.morrill@va.gov

Table 23.1. American Society of Health System Pharmacists' (ASHP) statement and white paper on the role of pharmacists in antimicrobial stewardship. From ASHP, 2010a,b.

Role
Running a multidisciplinary ASP that analyzes patient outcomes to assess the effect of antimicrobial use policies throughout the institution.
Encouraging multidisciplinary collaboration to ensure that the prophylactic, empiric, and therapeutic use of antimicrobials result in optimal patient outcomes within the institution.
Making recommendations to improve the selection, dosing, timely initiation, proper monitoring and de-escalation of antimicrobials.
Assisting the Pharmacy and Therapeutics Committee to ensure that the antimicrobials available are appropriate for the patient population served and in developing restricted antimicrobial use procedures, therapeutic interchange, treatment guidelines, and care plans.
Collecting and analyzing antimicrobial use data and performing clinical and economic outcome analyses.
Collaborating with microbiology to ensure appropriate and timely antimicrobial susceptibility testing.
Collaborating with the microbiology, infectious disease (ID) specialists, and infection control to compile annual antibiograms.
Using information technology to streamline antimicrobial resistance surveillance, antimicrobial use and clinical outcome reporting, and the development of clinical decision support tools.
Using efficient and effective systems to reduce the risk of adverse drug events and medication errors to promote safe antimicrobial management practices.
Participating in efforts to prevent the transmission of infections among patients, healthcare workers, and others within the hospital.
Educating multidisciplinary health professionals, other pharmacists and pharmacists in training, patients, and the public in antimicrobial stewardship and infection.

of the pharmacist as an antimicrobial stewardship leader/champion in developing, implementing, and maintaining an effective ASP.

Role of the Pharmacy Department in Antimicrobial Stewardship

Pharmacy directors

For an ASP to be most effective, the hospital administration and leadership, such as the chief of staff, medical center director, director of operations, department chiefs, and hospital medical staff executive committees must support the program (Dellit *et al.*, 2007). Directors of pharmacy continuously interact with hospital administrators and other department heads. It is because of this essential involvement that pharmacy directors must commit to the ASP's mission and assume leadership roles in promoting the function of pharmacists as the antimicrobial use experts within their institutions (ASHP, 2010a). Pharmacy directors should team up with the ASP's physician and pharmacist leaders/champions, medical leaders from ID, and other medical staff champions, to lead the efforts of formally introducing antimicrobial stewardship to the organization concerned, and advocating for and obtaining support from the administration.

IDSA/SHEA guidelines for antimicrobial stewardship recommend that both the physician and pharmacist stewardship leaders be compensated appropriately for their time devoted to the duties and activities of antimicrobial stewardship (Dellit *et al.*, 2007). Therefore, pharmacy directors may also be involved in the decisions of hiring antimicrobial stewardship personnel, in particular the antimicrobial stewardship pharmacist leader. In some hospitals, especially smaller community hospitals with budgetary constraints, administrators may decide that the responsibility of co-leading an ASP should be incorporated into the pharmacy's existing services. However, it should be noted that in hospitals without a dedicated antimicrobial stewardship pharmacist, routine pharmacy services may take priority over antimicrobial stewardship activities, thus compromising the success of these efforts (ASHP, 2010a). Pharmacist time that is devoted to the program is essential to its success, as is the cooperation of the pharmacy staff that works with the antimicrobial stewardship pharmacist leader. While devoted pharmacist time is essential for all programs, the amount of full time equivalents (FTEs) that is necessary will vary based on each individual hospital, depending on factors such as the size of the hospital, its case mix, the patients served, and existing institutional resources.

Hospital pharmacists

All pharmacists are trained to be medication experts and are highly educated in a wide variety of pharmacotherapy issues, so they have an inherent responsibility to ensure that all patients at their institution are receiving the optimal antimicrobial therapy. Nonetheless, there are many different roles for pharmacists in hospitals, such as medication processing and dispensing, the optimization of medication therapy, and providing drug information and education. Additionally, there are many different specialized or clinical coordinator pharmacist positions available at many hospitals, and there are often student and resident pharmacists in training as well. The structure of pharmacy departments itself may differ from one institution to another, with some institutions having centralized and some decentralized pharmacies. Despite differences in job duties and pharmacy structure, all of the pharmacists involved can meaningfully contribute to antimicrobial stewardship, although the specific roles that each will play may differ. It is important to note that all recommendations made by hospital pharmacists should align with those of the antimicrobial stewardship team. To facilitate parallel recommendations, there should be regular communication between the two parties.

Due to their role in processing medication orders and familiarity with the hospital formulary, pharmacists whose primary responsibility is in processing and dispensing may meaningfully contribute to antimicrobial stewardship. These pharmacists may serve as "gatekeepers" for restricted antimicrobials by notifying prescribers that authorization is required for their use. Pharmacists can also assure that the right dose of an antimicrobial is ordered, and that the ordered antimicrobial can be safely used in the patient. In addition to monitoring for potential adverse drug events, pharmacists are crucial in detecting and preventing antimicrobial-related drug interactions. Processing and dispensing clinical pharmacists may also carry out most of the pharmacy-driven interventions that are outlined in following (main) section on "Pharmacy-driven Antimicrobial Stewardship Interventions."

In general, the broad responsibilities of pharmacists who process orders do not allow enough time for comprehensive reviews of antimicrobial regimens. Furthermore, these pharmacists may not have the adequate training in ID and antimicrobial stewardship to feel comfortable providing recommendations in complex cases, or approving the use of restricted antimicrobials. Even so, these pharmacists can still contribute by flagging complicated cases for review by either an ID physician or clinical pharmacist with ID training, or by discussing these cases with the ID and/or antimicrobial stewardship team.

Decentralized pharmacists and pharmacists rounding with medical teams should also do their part to ensure that their patients are receiving the most optimal antimicrobial regimens. These pharmacists can aid in central line removal by recommending intravenous (IV) to oral (PO) conversion of antimicrobials in patients that meet the criteria for conversion to oral treatment. They can also make recommendations for dose optimization, de-escalation, and the discontinuation of duplicate or unnecessary antimicrobials. As well as to making recommendations for stopping antimicrobials, pharmacists can recommend discontinuing other unnecessary drugs, such as acid suppressants, which increase a patient's risk of *Clostridium difficile* infection. They can also make recommendations to ensure that drug levels of antimicrobials with narrow therapeutic windows, such as vancomycin and aminoglycosides, are correctly obtained. De-centralized pharmacists and pharmacists participating in medical rounds are ideally suited to providing education on appropriate antimicrobial utilization to nurses and multidisciplinary providers because they interact closely with these practitioners on a daily basis and know what specific topics these healthcare providers may need training in. Lastly, they can help distribute the educational materials developed by the antimicrobial stewardship team, such as antimicrobial guidebooks, treatment algorithms, or posters.

Serving on multidisciplinary committees

Pharmacist involvement in multidisciplinary committees, namely the Pharmacy and Therapeutics (P&T) Committee and the Infection Control Committee, is instrumental to the success of the ASP (ASHP, 2010a). Pharmacists involved with these teams should make sure that antimicrobial stewardship efforts align with the P&T committee and infection control policies and practices. Antimicrobial stewardship efforts are unlikely to be successful without simultaneously effective infection prevention and control measures, and vice versa (ASHP, 2010b).

Pharmacy-driven Antimicrobial Stewardship Interventions

In 2014, the CDC provided hospitals with guidance on the core elements that should be present in an ideal ASP (CDC, 2014). In this publication, several pharmacy-driven interventions were highlighted as important contributions to antimicrobial stewardship (see Table 23.2).

Antimicrobial Stewardship Training Opportunities for Pharmacists

Formal training

According to the 2007 IDSA/SHEA guidelines, the pharmacist who serves as the leader of the ASP should be trained in ID (Dellit *et al.*, 2007). Formal, postgraduate residency and/or fellowship training in ID clinical practice and/or research are the most widely accepted methods to provide training in ID for pharmacists (Drew *et al.*, 2009; Ernst *et al.*, 2009; Crader, 2014). Specialized residencies and fellowships both provide the trainee with focused training in ID, although the educational objectives and clinical and research experiences provided by each training type generally differ significantly (Ernst *et al.*, 2009). ID residencies are clinically focused and prepare trainees to practice in the patient care setting. Fellowships are research-focused training programs that usually provide less clinical practice experience than specialized residencies (Ernst *et al.*, 2009).

Due to the limited number of ID pharmacy residency and fellowship training programs available, the number of ID pharmacists with formal training is limited (Drew *et al.*, 2009; Ernst *et al.*, 2009; Owens *et al.*, 2009; ASHP, 2010b; DiazGranados and Abd, 2011; Crader, 2014). While large academic medical centers with ample financial resources may be able to support formally trained ID pharmacists, smaller community hospitals with budgetary constraints and limited resources may not be able to recruit or even financially support an ID pharmacist with these credentials. Some institutions may be able to support a part-time position or use off-site expertise, while in others the burden of providing ASP time may fall on to a pharmacist without specialized infectious disease training. This should not be viewed as a barrier that should halt the implementation of an ASP (Allerberger *et al.*, 2009; ASHP, 2010b; DiazGranados and Abd, 2011).

Table 23.2. Pharmacy-driven interventions to improve antibiotic use recognized by the US Centers for Disease Control and Prevention (CDC). Adapted and expanded from CDC, 2014.

Intervention	Examples
Automatic changes from intravenous to oral antimicrobials	Appropriate for agents with high oral bioavailability such as: • Fluoroquinolones • Linezolid • Metronidazole • Clindamycin • Trimethoprim–sulfamethoxazole
Dose adjustments for organ dysfunction	• Renal dose adjustment • Hepatic dose adjustment
Dose optimization	• Dose adjustments based on patient characteristics (e.g., age, weight) • Optimizing therapy for the causative organism (e.g., higher dose or extended infusion for infections due to resistant organisms) • Dose adjustment to increase drug penetration at the site of infection (e.g., central nervous system, bone, and endocardium)
Automatic alerts for potential unnecessary therapeutic duplication	• Avoidance of dual anaerobic therapy (e.g., active meropenem and metronidazole orders in a patient)
Automatic stop orders	• Automatic stop orders for surgical prophylaxis
Detection and prevention of antibiotic-related drug interactions	• Interaction between vitamins and fluoroquinolones (pharmacists can intervene by counseling patients or nurses to administer the doses at least a few hours apart)

Informal training

There are various informal training methods that a non-ID trained pharmacist assuming the role of the antimicrobial stewardship pharmacist co-leader (with an ID physician) can utilize to enhance his/her knowledge of ID pharmacotherapy and antimicrobial stewardship (Crader, 2014). Certificate training programs in antimicrobial stewardship are available through the Society of Infectious Diseases Pharmacists (SIDP) and Making a Difference in Infectious Diseases (MAD-ID; see http://mad-id.org/). These programs typically provide education through traditional continuing education and skills training, and are structured into self-learning through webcasts and reading, live webinars and teleconferences, and a practicum component, in which the participant institutes some aspect of antimicrobial stewardship at their healthcare facility.

There are numerous other informal ID training opportunities that may be beneficial to a pharmacist without ID training, such as the identification of a physician or pharmacist mentor with expertise in ID for case discussions, attending infectious disease rounds, joining ID professional organizations, and/or attending ID professional meetings (ASHP, 2010a; Septimus and Owens, 2011). Other cost-effective ways of obtaining basic knowledge include obtaining and reviewing basic references (e.g., Clinical Laboratory Standards Institute (CLSI) antimicrobial susceptibility testing manuals) and reading published ID literature, such as review articles and practice guidelines.

The development of a collaborative workgroup or antimicrobial stewardship task force, or joining a previously established group, may be another means of informal antimicrobial stewardship training. The Veterans Affairs (VA) Antimicrobial Stewardship Taskforce, which was chartered in May 2011, promotes the development and expansion of antimicrobial stewardship activities throughout the VA system (Graber *et al.*, 2013). The VA Antimicrobial Stewardship Taskforce offers example policies and procedures, monthly educational webinars, and a SharePoint Site, which serves as a forum for communication (Graber *et al.*, 2013). In May 2011, the Rhode Island Antimicrobial Stewardship Taskforce was formed; this is a group that includes pharmacists from each of the state's 12 acute-care hospitals. Members meet for monthly meetings, where they provide educational presentations, develop educational tools and materials, and share ideas and experiences concerning antimicrobial stewardship.

Numerous medical applications, eBooks, and websites are also available that can provide an antimicrobial steward with a large amount of ID education. Through the use of smartphones and tablets, these resources can be used to obtain medical information right at a patient's bedside (Kullar and Goff, 2014).

Role of the Pharmacist Antimicrobial Stewardship Leader in Developing, Implementing, and Maintaining an ASP

The overall responsibility of the pharmacist antimicrobial stewardship leader is to encourage the appropriate use of antimicrobials by promoting multidisciplinary collaboration within the hospital, and guiding the selection, dosing, timing, de-escalation, and discontinuation of antimicrobial therapy (Dellit *et al.*, 2007; ASHP, 2010a,b; Septimus and Owens, 2011). Additionally, this pharmacist plays a role in the education of physicians and other healthcare workers, patients, and the public, and should be directly involved with infection control and prevention practices (Septimus and Owens, 2011). While the pharmacist and physician should ideally co-lead the multidisciplinary ASP together, the pharmacist is often the primary individual tasked with developing, implementing, and maintaining the program. Therefore, the pharmacist antimicrobial stewardship leader plays a very important role in the creation and continued success of the ASP.

Obtaining administrative support through proposal development

One of the most difficult barriers to overcome in the implementation of an ASP is gaining the necessary approval and financial support from administration (Johannsson *et al.*, 2011). A strong strategic plan and business case in the form of a compelling proposal with sound arguments and solid data is crucial in gaining the attention of hospital leadership (ASHP, 2010a). The proposal must seek adequate authority and financial support from administration for the ASP, and is generally developed by the antimicrobial stewardship physician and pharmacist leaders, with assistance from the director of pharmacy and/or pharmacy clinical coordinator, infection control leaders, and other key stakeholders. A commitment from hospital administration, and not just in the form of financial support, is critical to the success of the program (MacDougall and Polk, 2005).

The ASP proposal should contain several sections of important information, including a strong background, which demonstrates the need for the ASP, the goals and proposed activities of the program based on the scope of problems identified and available resources, and the justification for the program based on institutional costs and on the consequences of inappropriate antimicrobial use (ASHP, 2010a, 2013). Furthermore, the financial implications for the hospital in terms of reimbursement for hospital-acquired infections (HAIs), the impact on institutional image due to publicly reported infection-related national quality indicators, and quantification of the resource requirements and associated costs for the ASP are also sections that should be included (ASHP, 2010a).

To effectively demonstrate the need for the ASP, it is important to outline antimicrobial utilization and resistance problems within the hospital using institutional data. Data from previously published reports can also be used to help to strengthen the case for the ASP. Rates of infections caused by resistant organisms, bacterial susceptibilities, and antimicrobial use patterns and costs are all examples of data that could be included (ASHP, 2010a). Guidelines and recommendations from authoritative groups such as the IDSA, SHEA, CDC, The Joint Commission on Accreditation of Healthcare Organizations (https://www.jointcommission.org/), and, in some cases, the state's Department of Health can be cited to provide additional support for program implementation (Dellit *et al.*, 2007; SHEA *et al.*, 2012; CDC, 2014). For example, IDSA/SHEA and the CDC have called for ASPs to be implemented in all hospitals (SHEA *et al.*, 2012; CDC, 2014).

The proposal should also outline the goals of the ASP and the proposed activities of the program. Short- and long-term goals should be realistic, focusing first on activities that are inexpensive or easy to implement and readily achievable (known as "low-hanging fruit" interventions) in order to maximize results from what may initially be limited resources (ASHP, 2010a). This section of the proposal should also include details of the implications of not correcting these identified issues within the institution.

Discussion on HAI issues within the proposal can bridge the discussion to financial implications. Financial justification for the antimicrobial stewardship should be outlined using institutional costs and the consequences of inappropriate antimicrobial use, such as morbidity and mortality associated with resistant organisms and HAIs, the adverse effects of antimicrobials, and overall drug costs associated with the overuse of antimicrobials. Published estimates of the cost of HAI are available in literature (Anderson *et al.*, 2007; ASHP, 2010a).

Furthering the discussion around financial issues leads to the attempt to quantify the resource requirements and costs associated with the proposed ASP (ASHP, 2010a). The costs associated with personnel (e.g., compensation for the antimicrobial stewardship pharmacist and physician leaders) and information technology (e.g., data mining systems) may be substantial, and it is important to present these in the context of the potential for net cost savings or costs avoided (ASHP, 2010a). The latter can be estimated through a prediction of decreased antimicrobial usage, avoidance of the continuation of a recent trend toward increasing antimicrobial costs (if identified as a problem), and improved patient safety and clinical outcomes. The hospital or pharmacy finance personnel may be helpful in providing cost figures (especially for antimicrobial expenditures) for the proposal. Another option to which financial administrators may be more agreeable to is a pilot to test antimicrobial stewardship activities on a limited basis. Data on cost savings and/or avoidance from the pilot program may be used to help to justify funding for ASP staff and hospital-wide expansion of antimicrobial stewardship activities (ASHP, 2010a).

Assembling a multidisciplinary team

Once the ASP proposal is approved by hospital administration, a core team must be assembled to lead antimicrobial stewardship efforts and activities. As previously mentioned, a clinical pharmacist with ID training and an ID physician should co-lead the program (Dellit *et al.*, 2007). The antimicrobial stewardship pharmacist and physician leaders must collaborate with other multidisciplinary members of the core team, which should ideally include a clinical microbiologist, infection control professional, information system specialist, and hospital epidemiologist (Dellit *et al.*, 2007). Other important stakeholders should assist the core antimicrobial stewardship team as needed (Drew *et al.*, 2009). The core team should meet with representatives from various departments in the hospital (e.g., intensive care, internal medicine, emergency, surgery, etc.) that will be affected by the activities of the ASP (ASHP, 2010a). It may be important to involve influential physician leaders and other key staff from the various affected departments early on so as to prevent resistance to antimicrobial stewardship activities (Drew *et al.*, 2009).

Assessing current institutional practices and analyzing baseline data

Knowledge of current institutional practices and the analysis of baseline antimicrobial use and resistance data is essential to the success of an effective ASP (Dellit *et al.*, 2007). Consulting with ID physicians, microbiologists, infection control personnel, and information technology specialists can help to identify limitations or problems with current institutional practices (ASHP, 2010a). Robust antimicrobial and resistance data can also be used to uncover problems, namely, areas of antimicrobial overuse or high antimicrobial resistance.

Planning and implementing antimicrobial stewardship strategies

The findings from the assessment of current institutional practices and the analysis of baseline antimicrobial utilization and resistance data can be used to identify target areas for antimicrobial stewardship intervention and improvement in antimicrobial use. When initially starting a program, one or two target areas for improvement should be identified (GNYHA and UHF, 2011). To avoid becoming overwhelmed or ineffective, an ASP should establish specific goals and an implementation strategy to address target areas for improvement (GNYHA and UHF, 2011). Both process and outcomes goals should be developed (Dellit *et al.*, 2007). The antimicrobial stewardship pharmacist and physician leaders should work together to identify targets and goals, and consult with multidisciplinary core team members as needed.

After the target areas are identified and goals are developed, it should be determined which antimicrobial stewardship strategies will be most effective to address the problem issues and meet the agreed goals. To ensure that goals are met, a strategic plan should be devised to implement these strategies (ASHP, 2010a). A few items to consider here include whether the intervention should be implemented on a unit-by-unit basis or hospital wide, the impact of the intervention on available resources, and a realistic and manageable time frame (GNYHA and UHF, 2011).

Collecting and analyzing data

To determine the impact of an ASP, the collection and analysis of data are needed. The antimicrobial stewardship pharmacist tasked with data collection and analysis should consult with hospital epidemiologists, information technology, and other supportive personnel to increase the efficiency and accuracy of this process. Ideally, the data analysis should demonstrate a favorable impact of the ASP interventions on antimicrobial utilization, clinical outcomes, antimicrobial resistance, and/or healthcare costs (ASHP, 2010a). Data can be compared before and after ASP implementation, and also over time to document the effect of the program and to identify trends (ASHP, 2010a); data can be benchmarked with local hospitals and with published data as well (ASHP, 2010a).

An important component of any ASP is understanding how an intervention has affected antimicrobial use (Dodds Ashley *et al.*, 2014). The antimicrobial utilization metrics that have been discussed earlier in this book (days of therapy, DOT, and defined daily dose, DDD, per 1000 patient days) are often used to assess how effective an intervention has been at changing antimicrobial use. Other process metrics can also be used to demonstrate the impact and effectiveness of implemented antimicrobial stewardship activities (Dodds Ashley *et al.*, 2014; Septimus, 2014); these measures assess whether an ASP has accomplished a task, changed a behavior, or changed a process (Dodds Ashley *et al.*, 2014).

Outcomes metrics are important because they measure the degree to which the intended outcomes of an ASP are achieved (Dellit *et al.*, 2007). These metrics can generally be divided into improved patient outcomes, reduced collateral damage (e.g., adverse drug events or *C. difficile* infection), and reduced resistance (Septimus, 2014). While patient outcomes measures, such as length of stay, mortality rates and readmission rates are important, they may be too indirectly related to appropriate antimicrobial prescribing to be an accurate reflection of the performance of an ASP. Additionally, the assessment of clinical outcomes requires a tremendous amount of work to collect the data (Septimus, 2014). Nonetheless, clinicians consider patient-centered outcomes to be the most important determinants of success, and for this reason, it is crucial to analyze these metrics (Johannsson *et al.*, 2011). The rate of *C. difficile* infection is a key metric used to demonstrate the impact of an ASP on reduction of collateral damage.

Antibiograms may be useful for measuring the success of antimicrobial stewardship activities in reducing antimicrobial resistance. While measuring antimicrobial resistance is critical, the factors associated with this resistance are complex and involve many factors, such as infection control, antimicrobial use within and outside the hospital, and patient colonization and immune status; as such, it may be challenging for an ASP to demonstrate a favorable impact on antimicrobial resistance (Septimus, 2014). Moreover, it can take years before an implemented intervention has an effect on antimicrobial resistance, so the success of an ASP should not be based solely on resistance rates (ASHP, 2010a).

Reducing costs is a secondary goal to that of improving outcomes and reducing resistance. However, hospital administrators generally take great interest in the cost savings of an ASP (Dodds Ashley *et al.*, 2014; Septimus, 2014). ASPs usually require personnel and financial investments, and so administrators may require data demonstrating a return on their investments. Calculating cost savings from decreased antimicrobial exposures can demonstrate the value of an ASP and also justify requests for further resources, such as personnel, for the program. It is often helpful to measure cost savings for individual antimicrobials or for certain classes of antimicrobials. Reduction in the use of high-cost antimicrobials and overall antimicrobial use may also be useful to measure. Cost analyses should not just take into consideration the costs associated with antimicrobial use, but also the costs associated with antimicrobial resistance (e.g., more expensive agents and/or increased length of stay in hospital may be required) and HAIs (ASHP, 2010a; Septimus, 2014). In addition to reductions in antimicrobial use and resistance, ASPs may also be associated with reductions in length of stay (LOS) and improved patient outcomes (Ohl and Dodds Ashley, 2011). Therefore, broad cost-savings analyses, such as assessing the costs associated with reductions in LOS, decreased readmissions, prevention of adverse drug effects and medication errors, and/or infection avoidance, should be conducted when possible (Septimus, 2014).

Disseminating findings

It is crucially important to the maintenance and continued success of the antimicrobial stewardship team that the results of the data analysis be reported and disseminated throughout the hospital. Findings should be presented by the antimicrobial stewardship pharmacist and physician leaders to hospital administrators, the P&T committee, the infection control committee, and other key stakeholders, so that they are fully aware of the impact that the ASP is making (ASHP, 2010a). An annual report describing the program's activities and findings should be completed and programs should also consider publishing their results (ASHP, 2010a).

Overcoming pushback

The efforts of the ASP may potentially be met with resistance, often due to the "culture" of the institution, or from specific individuals who may harbor long-established inappropriate prescribing habits that are difficult to change. Prescribers may fear that the ASP, especially ASPs that utilize highly restrictive antimicrobial approval processes, jeopardizes their autonomy (Sunenshine *et al.*, 2004; Owens *et al.*, 2009). Misconceptions may arise that pin the ASP as a strictly pharmacy cost-driven initiative that seeks to eliminate clinical judgment. Thus, an approach to overcome these potential barriers must be developed before the ASP is formally launched (ASHP, 2010a).

The education of all staff impacted by the ASP, including administrators, supervisors, nursing staff, and, especially, physicians is a crucial step in avoiding these problems and ensuring a successful start to the program. The antimicrobial stewardship physician and pharmacist champions should lead the educational efforts, although the involvement of multidisciplinary key opinion leaders is crucially important (ASHP, 2013). Essential elements of the informational materials should include the rationale, goals, and components of the ASP in a way that is meaningful to the particular group being educated. The benefits of the program should be described. It is particularly important to stress how the program may improve patient safety and patient outcomes. Example formats for education include grand rounds, morbidity and mortality conferences, newsletters, posters, and one-on-one informational vignettes (Dellit *et al.*, 2007; Allerberger *et al.*, 2009).

Once the ASP is underway, uncooperative prescribers may still pose problems and resist the ASP's efforts. In these situations, it is helpful to have the support of hospital administrators, such as the Chief Medical Officer, who can help to enforce requirements of the ASP in a respectful and diplomatic way (ASHP, 2010a; Patel and MacDougall, 2010). Providers may also learn from and benefit from prescribing feedback relative to their peers (e.g., the number of carbapenem starts compared with other similar, but not identified, physicians) in order to see that they are outliers. Finally, every patient case and prescriber interaction should be taken as an opportunity to educate using a rational and evidence-based approach.

Summary

Pharmacists play an important role in antimicrobial stewardship. The whole pharmacy department, including the pharmacy director, pharmacists participating in processing and dispensing orders, and decentralized/rounding pharmacists, can all contribute to antimicrobial stewardship and the optimal use of antimicrobials through compliance with several different strategies, including the pharmacy-driven interventions recommended by the CDC. The antimicrobial stewardship pharmacist leader is responsible for advocating antimicrobial stewardship throughout the hospital, and plays an integral role in the development, implementation, and continued success of the multidisciplinary program.

References

Allerberger, F., Gareis, R., Jindrak, V., and Struelens, M.J. (2009) Antibiotic stewardship implementation in the EU: the way forward. *Expert Review of Anti-infective Therapy* 7, 1175–1183.

Anderson, D.J., Kirkland, K.B., Kaye, K.S., Thacker, P.A., II, Kanafani, Z.A., Auten, G., and Sexton, D.J. (2007) Under-resourced hospital infection control and prevention programs: penny wise, pound foolish? *Infection Control and Hospital Epidemiology* 28, 767–773.

ASHP (2010a) *A Hospital Pharmacist's Guide to Antimicrobial Stewardship Programs*. American Society of Health-System Pharmacists, Bethesda, Maryland. Available at: http://www.ashpadvantage.com/docs/stewardship-white-paper.pdf (accessed 24 May 2016).

ASHP (2010b) ASHP statement on the pharmacist's role in antimicrobial stewardship and infection prevention and control. *American Journal of Health-System Pharmacy* 67, 575–577. Full text available at: https://www.ashp.org/DocLibrary/BestPractices/SpecificStAntimicrob.aspx (accessed 24 May 2016).

ASHP (2013) *An Interprofessional Approach to Antimicrobial Stewardship: Implementing Team-Based Strategies that Impact Patient Outcomes*. American Society of Health-System Pharmacists, Bethesda, Maryland. Available at: http://www.ashpadvantagemedia.com/downloads/2013-asp-discussion-guide.pdf (accessed 24 May 2016).

CDC (2014) *Core Elements of Hospital Antibiotic Stewardship Programs*. Centers for Disease Control and Prevention, Atlanta, Georgia. Available at: http://www.cdc.gov/getsmart/healthcare/pdfs/core-elements.pdf (accessed 24 May 2016).

Crader, M.F. (2014) Development of antimicrobial competencies and training for staff hospital pharmacists. *Hospital Pharmacy* 49, 32–40.

Dellit, T.H., Owens, R.C., McGowan, J.E., Jr., Gerding, D.N., Weinstein, R.A., Burke, J.P., Huskins, W.C., Paterson, D.L., Fishman, N.O., Carpenter, C.F. *et al.* (2007) Infectious Diseases Society of America and the Society for Healthcare Epidemiology of America guidelines for developing an institutional program to enhance antimicrobial stewardship. *Clinical Infectious Diseases* 44, 159–177.

DiazGranados, C.A. and Abd, T.T. (2011) Participation of clinical pharmacists without specialized infectious diseases training in antimicrobial stewardship. *American Journal of Health-System Pharmacy* 68, 1691–1692.

Dodds Ashley, E.S., Kaye, K.S., DePestel, D.D., and Hermsen, E.D. (2014) Antimicrobial stewardship: philosophy versus practice. *Clinical Infectious Diseases* 59(Suppl 3), S112–S121.

Drew, R.H., White, R., MacDougall, C., Hermsen, E.D., and Owens, R.C., Jr. (2009) Insights from the Society of Infectious Diseases Pharmacists on antimicrobial stewardship guidelines from the Infectious Diseases Society of America and the Society for Healthcare Epidemiology of America. *Pharmacotherapy* 29, 593–607.

Ernst, E.J., Klepser, M.E., Bosso, J.A., Rybak, M.J., Hermsen, E.D., Segarra-Newnham, M., and Drew, R.H. (2009) Recommendations for training and certification for pharmacists practicing, mentoring, and educating in infectious diseases pharmacotherapy. *Pharmacotherapy* 29, 482–488.

Graber, C.J., Madaras-Kelly, K., Jones, M.M., Neuhauser, M.M., and Goetz, M.B. (2013) Unnecessary antimicrobial use in the context of *Clostridium difficile* infection: a call to arms for the Veterans Affairs Antimicrobial Stewardship Task Force. *Infection Control and Hospital Epidemiology* 34, 651–653.

GNYHA (Greater New York Hospital Administraion) and UHF (United Hospital Fund) (2011) *Antimicrobial Stewardship Toolkit—Best Practices from the GNYHA/UHF Antimicrobial Stewards Collaborative*. Available at: http://www.shea-online.org/Portals/0/GNYHA_Antimicrobial_Stewardship_Toolkit_FINALv2 Dec2011.pdf (accessed 24 May 2016).

Johannsson, B., Beekmann, S.E., Srinivasan, A., Hersh, A.L., Laxminarayan, R., and Polgreen, P.M. (2011) Improving antimicrobial stewardship: the evolution of programmatic strategies and barriers. *Infection Control and Hospital Epidemiology* 32, 367–374.

Kullar, R. and Goff, D.A. (2014) Transformation of antimicrobial stewardship programs through technology and informatics. *Infectious Disease Clinics of North America* 28, 291–300.

MacDougall, C. and Polk, R.E. (2005) Antimicrobial stewardship programs in health care systems. *Clinical Microbiology Reviews* 18, 638–656.

Ohl, C.A. and Dodds Ashley, E.S. (2011) Antimicrobial stewardship programs in community hospitals: the evidence base and case studies. *Clinical Infectious Diseases* 53(Suppl 1), S23–S28; quiz S29–S30.

Owens, R.C., Jr., Shorr, A.F., and Deschambeault, A.L. (2009) Antimicrobial stewardship: shepherding precious resources. *American Journal of Health-System Pharmacy* 66, S15–S22.

Patel, D. and MacDougall, C. (2010) How to make antimicrobial stewardship work: practical considerations for hospitals of all sizes. *Hospital Pharmacy* 45(11 Suppl 1), S10–S18.

Septimus, E. (2014) Antimicrobial stewardship—qualitative and quantitative outcomes: the role of measurement. *Current Infectious Disease Reports* 16: 433.

Septimus, E.J. and Owens, R.C., Jr. (2011) Need and potential of antimicrobial stewardship in community hospitals. *Clinical Infectious Diseases* 53(Suppl 1), S8–S14.

SHEA, IDSA, and PIDS (2012) Policy statement on antimicrobial stewardship by the Society for Healthcare Epidemiology of America (SHEA), the Infectious Diseases Society of America (IDSA), and the Pediatric Infectious Diseases Society (PIDS). *Infection Control and Hospital Epidemiology* 33, 322–327.

Sunenshine, R.H., Liedtke, L.A., Jernigan, D.B., Strausbaugh, L.J. and for Infectious Diseases Society of America Emerging Infections Network (2004) Role of infectious diseases consultants in management of antimicrobial use in hospitals. *Clinical Infectious Diseases* 38, 934–938.

24 Formulary Management and Economic Considerations: Bridging the Gap between Quality Care and Cost

DAYNA MCMANUS,[1] MICHAEL A. RUGGERO,[2] AND JEFFREY E. TOPAL[3*]

[1]*Inova Fairfax Hospital, Falls Church, Virginia, US;* [2]*Hospital of the University of Pennsylvania, Philadelphia, Pennsylvania, US;* [3]*Yale School of Medicine and Yale New Haven Hospital, New Haven, Connecticut, US*

Introduction

Traditionally, a driver for the formation of antimicrobial stewardship teams in the inpatient setting has been the aggregate cost reduction of antimicrobial therapy. However, one of the most important aims should be the optimization of patient care outcomes and prevention of the development of antimicrobial resistance (Fishman, 2006; Rybak, 2007). Unfortunately, these quality metrics are much more difficult to quantify than antibiotic expenditures, but they need to be studied and quantified in order to continue the growth and success of stewardship efforts. Often, there is much initial enthusiasm for antimicrobial stewardship efforts and a significant reduction of overall antibiotic expenditures can be demonstrated in the first year or two of such a program. Thereafter, and once the antibiotic prescribing process proper checks and balances have been implemented, reductions in antimicrobial expenditures are often more modest in nature (Fishman, 2006; Rybak, 2007; CDC, 2014). Nevertheless, maintaining the success of antimicrobial stewardship requires dedicated resources—both human and IT resources, the constant analysis of prescribing patterns of antibiotics, as well as monitoring changes in antimicrobial resistance (Owens *et al.*, 2004; Rybak, 2007; CDC, 2014). Studies have revealed that removing these dedicated resources results in increased costs and increased duration of antimicrobial therapy (Capelletty and Jacobs, 2013).

Antimicrobial stewardship is a dynamic process that must evolve to meet the changing economic realities in healthcare. Such challenges include: a reduction in healthcare reimbursement for inpatient stays; the introduction of novel, high-cost antibiotics whose costs exceed the prior daily drug costs by hundreds to thousands of dollars; the onslaught of multiple shortages of generic antimicrobials; and the continued pressure of increasing antimicrobial resistance (Owens *et al.*, 2004; Fishman, 2006; Rybak, 2007; CDC, 2014).

In this chapter, we will review the components of an antimicrobial stewardship program (ASP) that are required for the successful reduction of antimicrobial expenditures, especially through formulary management and the optimization of patient therapy and outcomes, with the final goal of reducing the burden of antimicrobial resistance.

Development/Creation of an Antimicrobial Formulary

The development of an antimicrobial formulary is an integral component for any antimicrobial stewardship program and should be one of the initial goals when starting any such program. The antimicrobial

*Corresponding author. E-mail: Jeffrey.Topal@ynhh.org

formulary should balance the need for specific agents for the appropriate treatment of specific patient populations and the clinical syndromes encountered within an institution. Therefore, the formulary decisions for antimicrobials and their use should be made with the local susceptibilities and antibiogram in mind (Fishman, 2006; Dellit *et al.*, 2007). The formulary should also strive to reach important patient safety goals, such as minimizing the development of antimicrobial resistance. Lastly, the formulary should achieve the fiduciary responsibility for cost-effective therapy in the current era of cost containment in healthcare (Woodward *et al.*, 1987; Fishman, 2006; Dellit *et al.*, 2007).

One of the first decisions that is faced when defining an antimicrobial formulary is whether that formulary should be open or closed. An open formulary allows for the prescribing of any antimicrobial that is approved by the US Food and Drug Administration (FDA) and is desired by a physician or licensed independent practitioner (Owens *et al.*, 2004; Rybak, 2007). Although such an approach preserves a practitioner's autonomy for what he or she believes may be best for a specific patient, it often results in multiple and similar agents being used within an institution. This approach then handicaps the antimicrobial stewardship efforts for standardization in terms of both the criteria for antibiotic use and reviewing for the appropriateness of use. An open formulary also results in yet more challenges for the stewardship program, not only to the program's efforts to control antimicrobial use, but also by the creation of difficulties in making consistent recommendations to stop or change antimicrobial therapy after it has already been started (Woodward *et al.*, 1987; Owens *et al.*, 2004).

Conversely, a closed formulary will allow only for the prescribing of selected antimicrobials within a specific institution. With a closed formulary, the therapeutic duplication of antimicrobial agents can be minimized, which will optimize the review process for stewardship efforts through standardization of the use of selected antimicrobials (Woodward *et al.*, 1987; Owens *et al.*, 2004; Patterson, 2006). More importantly, each antimicrobial on the formulary can be objectively assessed for its place in the antimicrobial armamentarium based on the patient populations treated at a specific institution, the clinical syndromes treated, current antimicrobial resistance patterns, and cost considerations (Owens *et al.*, 2004; Patterson, 2006; Khadem *et al.*, 2012). Also, the microbiology laboratory's susceptibility testing protocols can be optimized to test the specific agents used within an institution, which, in turn, can prevent mistakes in the interpretation of results by minimizing the extrapolation of results from antimicrobial to another (Berild *et al.*, 2001; Owens *et al.*, 2004).

Once the decision has been made to utilize a closed formulary, the objective review of the antimicrobial agents available for the formulary is an important first step for any stewardship program. Additionally, as new antimicrobials become available in the marketplace, such formulary reviews will help to define whether there is a role for such new agents based on an objective evaluation of peer-reviewed literature and other clinical data, as opposed to the prescriber's preference or pharmaceutical marketing campaigns. Creating and maintaining a closed antimicrobial formulary requires multidisciplinary input from the antimicrobial stewardship pharmacist, designated infectious disease physicians, representatives from the microbiology staff, infection preventionists, and pharmacy purchasing agents. When an antimicrobial formulary is delineated, it should be approved by the appropriate governing administrative body, such as an institution's Pharmacy & Therapeutics (P&T) Committee (Woodward *et al.*, 1987; Berild *et al.*, 2001; Patterson, 2006; Khadem *et al.*, 2012).

If a physician (MD) or licensed independent practitioner (LIP) desires to use an antimicrobial that is not available on the closed formulary, then a similar agent should be selected. If there is no similar agent on the formulary, then the practitioner would have to make an appeal to the pharmacy to obtain the agent, but such requests can then be reviewed prospectively for appropriateness by the stewardship team, which further aids in their efforts to implement an ASP (Woodward *et al.*, 1987; Berild *et al.*, 2001).

Minimizing therapeutic duplication

Many antimicrobials are similar in their therapeutic class in terms of antimicrobial coverage as well as in the clinical data supporting their use for the treatment of clinical infectious disease syndromes (e.g., as an empiric therapy for meningitis), and for specific types of infection based on the causative organism or site of infection. A closed formulary allows the selection of a single antimicrobial in a specific class for a specific coverage or indication (Wright, 1991; CDC, 2014). The many benefits of minimizing the therapeutic duplication of antimicrobials include:

1. A reduction in potential medication dosing errors by MDs/LIPs and pharmacy staff due to their familiarity with the agents that are on formulary.
2. A maximization of the efficiency of stewardship efforts as there are fewer antimicrobials to track on a daily basis.
3. A streamlining of pharmacy operations in tracking antimicrobial usage, maintaining an adequate inventory, and reordering.
4. A decrease in the potential for providers to utilize antimicrobials inappropriately as there are only certain indications for certain antimicrobials.

Although there may be FDA labeling differences for similar antimicrobial agents, review of the medical literature often provides the necessary support for therapeutic equivalence in the clinical use of these agents (Wright, 1991; CDC, 2014; Schultz *et al.*, 2014).

Using the formulary to determine therapeutic equivalence and its economic impact on the formulary

In addition to the clinical benefits of minimizing therapeutic duplication, the economic benefits also become apparent. If an institution deems specific antimicrobials to be therapeutically equivalent, then that institution can obtain competitive quotes for pricing of the agents among the respective pharmaceutical companies and then select the agent that is most financially advantageous. If such efforts are maximized, then a closed formulary may also be used to prevent inadvertent or inappropriate usage of the other agents that have been deemed to be therapeutically equivalent (CDC, 2014; Schultz *et al.*, 2014). Examples of possible therapeutically equivalent antimicrobial agents are shown in Table 24.1.

Development of Criteria for Appropriate Antibiotic Use

Once an institution's antimicrobial formulary has been delineated, the next step is the development of "criteria for use" for antimicrobial agents. Creating these "criteria for use" for antimicrobials allows an institution to define the acceptable use of a specific antimicrobial agent based on a patient's clinical syndrome (e.g., community-acquired pneumonia, CAP vs. healthcare-associated pneumonia, HCAP), local antimicrobial resistance patterns, and/or unique patient populations (e.g., patients with neutropenia and fever, with postoperative surgical prophylaxis, or patients status post (s/p) solid organ or stem cell transplant, etc.) (CDC, 2014). Although clinical guidelines can aid in the process of determining the appropriateness of a specific antimicrobial for a type of infection, there are often multiple therapeutic agents cited without a specific hierarchy for use due to the lack of comparative clinical trials between agents, or the use of non-inferior trials to compare new agents with a standard of care (Braxton *et al.*, 2010; Levin *et al.*, 2012; Malcolm *et al.*, 2013). Again, creating drug-specific "criteria for use," requires multidisciplinary input from the antimicrobial stewardship pharmacist, infectious disease physicians, and key physician stakeholders of the specific patient populations (Rybak, 2007; CDC, 2014).

One important distinction in the use of antimicrobials where criteria for use are helpful is to delineate the empiric, culture-documented, or surgical prophylaxis use of an agent. If criteria for the empiric use of broad-spectrum antimicrobial agents are defined, then a time-limited duration of use could be part of the initial order. For example, prescribing piperacillin–tazobactam for the empiric treatment of HCAP may be a reasonable use of the agent. However, to encourage de-escalation of such broad-spectrum empiric therapy, a specific predefined limit for the duration of the empiric order, such as 48 or 72 h, could be part of such an empiric order (Malcolm *et al.*, 2013; Murray *et al.*, 2014). This would help to prompt physicians to reevaluate whether such broad-spectrum coverage is still necessary 48 or 72 h after it started. This "automatic" timed-out review would help the efforts of the stewardship teams without requiring much extra work. Similarly, the criteria for the use of cefazolin for postoperative surgical prophylaxis could be automatically limited to a 24 h period, as opposed to the culture-documented treatment of a methicillin-sensitive *Staphylococcus aureus* (MSSA) infection. Examples of criteria for the use of selected antimicrobials (adapted from Yale New Haven Hospital) are shown in Table 24.2 (Guglielmo *et al.*, 2005; CDC, 2014; Murray *et al.*, 2014).

These "criteria for use" can be built into the electronic medical record (EMR) or replicated on a paper antibiotic order if the institution does not have an EMR. Specific criteria can be built for each individual antibiotic that are mandatory for providers to select when ordering these agents, and may help to guide the appropriate use of these

Table 24.1. Examples of possible therapeutically equivalent agents for various antimicrobial classes. From Schultz *et al.*, 2014.

Antimicrobial class	Agents with potential therapeutic equivalence	Significant differences between antimicrobials	Comments
Anti-staphylococcal penicillins	• Nafcillin IV • Oxacillin IV	No significant differences	
2nd Generation cephalosporins, cephamycins	• Cefoxitin IV • Cefotetan IV	Difference in dosing frequency (q6h vs. q 12h)	Cefoxitin may have lower rates of resistance to *Bacteroides fragilis* and non-*B. fragilis* spp.
3rd Generation cephalosporins	• Cefotaxime IV • Ceftriaxone IV	Difference in dosing frequency (q6 to 8h versus q 24h)	Ceftriaxone should not be used in neonates due to risk of hyperbilirunemia
Anti-pseudomonal cephalosporins	• Ceftazidime IV • Cefepime IV	Cefepime has greater Gram-positive coverage than ceftazidime, but need to balance this requirement with concomitant use of other antimicrobials such as vancomycin	
Anti-pseudomonal carbapenems	• Doripenem • Imipenem • Meropenem	Doripenem and meropenem are more active than imipenem in vitro against *Pseudomonas aeruginosa* Imipenem has the greatest risk of potential for seizures	
Fluoroquinolones	• Ciprofloxacin oral (PO) or intravenous (IV) and moxifloxacin PO/IV • Levofloxacin PO/IV	Ciprofloxacin and moxifloxacin provide the coverage of levofloxacin for clinical use	The use of two fluoroquinolones (i.e., ciprofloxacin and moxifloxacin) allows for greater delineation of appropriate use: Ciprofloxacin: treatment of urinary tract infection (UTI), healthcare-associated pneumonia (HCAP), and *Pseudomonas* Moxifloxacin: treatment of community-acquired pneumonia (CAP). May help to mitigate overuse of levofloxacin
Echinocandins	• Caspofungin IV • Micafungin IV • Anidulafungin IV	Caspofungin has drug–drug interactions with rifampin, cyclosporine and tacrolimus Caspofungin is the only echinocandin approved by the US Food and Drug Administration (FDA) for the treatment of *Aspergillus* infections Coverage against *Aspergillus* among the three is thought to be similar and has been studied clinically (Enoch *et al.*, 2014; Marr *et al.*, 2015).	

antimicrobial agents. This would also allow for the order to have a specific duration when "empiric" is chosen instead of "culture-documented." Reports can also be generated for the stewardship team to retrospectively track the appropriate selection of indication by providers to provide feedback and education if the agent(s) is not utilized appropriately. These reports can also be utilized by the stewardship team to prospectively prioritize what antimicrobial therapy needs review. For example, the stewardship team can focus on patients with indications for "empiric" therapy at 48 or 72 h

Table 24.2. Criteria for the use of selected antibiotics. Adapted from Yale New Haven Hospital practice and Grayson *et al.*, 2010.

Antibiotic/antibiotic combination	Type of use	Use
Piperacillin–tazobactam	Empiric use	• Empiric treatment of nosocomial pneumonia • Empiric treatment of *Pseudomonas aeruginosa* infections • Empiric treatment of an intra-abdominal infection • Empiric treatment of a mixed infection that includes Gram-negative bacteria resistant to ampicillin–sulbactam, *Staphylococcus aureus*, and anaerobes
	Culture-documented use	• Culture-documented treatment of *P. aeruginosa* infection • Culture-documented treatment of a polymicrobial infection with *P. aeruginosa*, *S. aureus*, and anaerobes • Therapy for patients with fever and neutropenia
Vancomycin	Empiric use	• Empiric treatment of a suspected central line infection • Empiric treatment of a suspected methicillin-resistant *S. aureus* (MRSA) infection in a patient known to be previously colonized/infected with MRSA • Empiric treatment of an enterococcal infection in a patient allergic to penicillin • Empiric treatment of a suspected Gram-positive infection in a patient allergic to penicillins and cephalosporins • Empiric treatment of a suspected Gram-positive infection of a prosthetic device (e.g., prosthetic heart valve, ventriculoperitoneal (VP) shunt, pacemaker) • Empiric treatment of a sternal wound infection following cardiothoracic surgery • Empiric treatment of a wound infection/meningitis following a neurosurgical procedure • Empiric treatment of meningitis when high-level penicillin-resistant *Streptococcus pneumoniae* is suspected
	Culture-documented use	• Treatment of an enterococcal infection: organism is resistant to ampicillin and susceptible to vancomycin OR the patient is allergic to penicillin • Treatment of an MRSA infection • Treatment of an *S. aureus* infection and patient is allergic to penicillins and cephalosporins • Treatment of a staphylococcal (non-*S. aureus*) infection (if treating positive blood cultures, the organism must be isolated from more than one blood culture) and organism is resistant to oxacillin OR the patient is allergic to penicillins and cephalosporins • Treatment of an *S. pneumoniae* meningitis and organism is highly resistant to penicillin OR the patient is allergic to penicillins and cephalosporins • Treatment of documented *Corynebacterium* infection

after the antibiotic has been started, because this is when culture results should be available; the team can then see if it is possible to optimize therapy (Guglielmo *et al.*, 2005; Levin *et al.*, 2012; CDC, 2014; Murray *et al.*, 2014).

Having criteria for the use of antimicrobials and "antibiotic timeouts," and reviewing these strategies for effectiveness are all recommended by the *Core Elements of Hospital Antibiotic Stewardship Programs* document from the US Center for Disease Control and Prevention (CDC, 2014). Because this is a strategy recommended by the CDC, the Centers for Medicaid and Medicare (CMS) may require criteria for use for antimicrobials and "antibiotic timeouts" as a Condition of Participation in the future (CDC, 2014; CMS, 2014).

Implementing the Antimicrobial Restriction Process

Following the creation of a defined antimicrobial formulary and the definition of criteria for the use of broad-spectrum and/or expensive agents, the antimicrobial stewardship program can play an integral role in controlling usage and then, by default, the expenditures on these drugs (Fishman, 2006).

In antimicrobial stewardship, there are three major methods to monitor and control the use of antibiotics. The first method is to require approval for the dispensing of specific antimicrobials at the point of prescribing. This requires real-time review by a pharmacist, ideally one who is infectious disease (ID) trained, an ID physician, and/or an ID fellow. This approach allows for direct feedback to the prescriber of the appropriateness of the antibiotic use based on the clinical information available at the time of the request. It also allows for real-time recommendations for alternative therapy if the request is not appropriate. Prior approval before dispensing prevents the routine overuse of selected antibiotics by eliminating the default prescribing position, which is the need to cover all organisms (Fishman, 2006; Rybak, 2007; Khadem *et al.*, 2012). For example, if a patient is known to be colonized with vancomycin-resistant enterococcus (VRE), this approach would prevent the use of an anti-VRE agent in the treatment of pneumonia, when such an organism would not be a likely pathogen (Berk and Verghese, 1989; Moellering, 1992). The use of predefined criteria for such restricted antimicrobials, as outlined previously, improves the likelihood that the approval process will be followed in a fair and consistent manner (Fishman, 2006; Khadem *et al.*, 2012; CDC, 2014).

In the upfront approval process, the use of EMRs can be useful by restricting who has access to enter restricted antimicrobials. Such features may potentially include restriction to pharmacist order-entry only vs. the designation of specific physicians to have such privileges, or a combination of the two. For example, some EMRs can be configured to allow a restricted antimicrobial to be entered by the primary physician caring for a patient, but the request requires cosigning by designated personnel, such as an ID attending or an ID pharmacist, before the order can be completed. The actual process of order entry for restricted antimicrobials needs to be delineated as part of the antimicrobial stewardship program, as do the human resources available for real-time review. The antimicrobials that require upfront review should be limited to avoid overwhelming the review process, which could delay care for patients (Rybak, 2007; Khadem *et al.*, 2012; CDC, 2014). For example, high-cost antimicrobials, antimicrobials with significant monitoring requirements, or antimicrobials with a high risk for adverse drug events or drug–drug interactions, may be considerations when deciding which antimicrobials to include in a real-time approval process. Finally, auditing the upfront restriction process is important to identify outliers in the approval process and/or changes in the practice of prescribing patterns. These audits may done as a drug use evaluation (DUE) process, with the results presented at an antimicrobial use or stewardship committee meeting (Fishman, 2006; Rybak, 2007; Khadem *et al.*, 2012).

Other modalities for antimicrobial restriction include prospective daily review of high-cost, high-use, and/or broad-spectrum antibiotics. This process is usually facilitated by targeting key antibiotics and having a team analyze the appropriateness of their use. The personnel needed for successful concurrent review require at a minimum a pharmacist dedicated to ID and/or the stewardship working in close concert with an attending ID specialist designated to the stewardship process. The most important aspect of the success of this review process is feedback to the prescriber on the appropriateness of the use of the targeted antibiotics, and offering the most cost-effective alternatives, narrower therapies, and/or more streamlined therapies, whenever possible. Delineation of the antibiotic review and the recommendations for therapy that are communicated to the prescriber may vary depending on the available resources and the numbers of patients reviewed. Direct verbal communication to the prescriber, and written communication via designated notes in the medical record, are often complementary in achieving the goal of having stewardship recommendations accepted. "Face to face" feedback may be one of the most effective ways to provide this information to prescribers, though the capacity to do this may be limited by time and resources (Owens *et al.*, 2004; Fishman, 2006; Rybak, 2007; CDC, 2014).

Finally, another effective method to reduce the duration of empiric use of high-cost and/or broad-spectrum antibiotics is to have a time-limited duration for empiric use of 48 or 72 h duration. Use after the initial empiric period then would require approval by the stewardship team (CDC, 2014). At 48 or 72 h from when the order is started, the stewardship team would have additional input from the patient's clinical course, laboratory data, radiology data, and microbiology data to determine if whether continued use is appropriate or not; if it is not, then alternative therapy could be recommended. Even though such a system is less burdensome than an upfront approval process or concurrent review, it requires close auditing of the antibiotic orders that are set to expire. The issue of an antibiotic order expiring and the prescriber recognizing the expiration must be balanced with the need that the patient may truly need the specific therapy to be continued. EMRs can flag such expiring orders, but the need for further efforts may be indicated by electronic "pop-up" alerts to both the patient's primary team and to pharmacy staff. So that antibiotic orders may be prevented from inadvertently "falling off" the patient medication profile at the time of expiration when they are now indicated to continue based on new culture results, a secondary review of orders set to expire should be part of the daily work flow for the pharmacist covering the patient. Again, ERMs can aid in the identification of such orders through the use of "flagging" and/or providing centralized reports of expiring orders to pharmacy staff for review (Braxton *et al.*, 2010; Levin *et al.*, 2012; Malcolm *et al.*, 2013).

Conclusion

Although the reduction of antimicrobial expenditures is often one of the driving impetuses for the creation of an antimicrobial stewardship team in the inpatient setting, one must recognize that the reduction in antibiotic expenditures occurs as a result of the stewardship process, which includes the formulary management. By optimizing a patient's therapy, cost reductions may be a result of several factors: shorter durations of empiric high-cost and/or broad-spectrum therapy; the use of narrower, less expensive therapy; the avoidance of prolonged duration of therapy; minimizing therapeutic duplication; and minimizing adverse events, such as *Clostridium difficile* infection (Owens *et al.*, 2004; Fishman, 2006; Rybak, 2007; CDC, 2014).

The many components needed for optimizing the formulary for stewardship that have been delineated in this chapter provide a framework for the stewardship process from the administrative level to the level of the patient. While robust antimicrobial stewardship requires intensive and multilayered processes, the return on investment for healthcare institutions is broad based, with reduction in antibiotic expenditures, improvement in patient care, and aid to the fight against the rising rates of antimicrobial resistance (Berild *et al.*, 2001; Fishman, 2006; Capelletty and Jacobs, 2013; CDC, 2014).

References

Berild, D., Ringertz, S.H., Lelek, M., and Fosse, B. (2001) Antibiotic guidelines lead to reductions in the use and cost of antibiotics in a university hospital. *Scandinavian Journal of Infectious Diseases* 33, 63–67.

Berk, S.L. and Verghese, A. (1989) Emerging pathogens in nosocomial pneumonia. *European Journal of Clinical Microbiology* 8, 11–14.

Braxton, C.C., Gerstenberger, P.A., and Cox, G.G. (2010) Improving antibiotic stewardship: order set implementation to improve prophylactic antimicrobial prescribing in the outpatient surgical setting. *The Journal of Ambulatory Care Management* 33, 131–140.

Capelletty, D. and Jacobs, D. (2013) Evaluating the impact of a pharmacist's absence from an antimicrobial stewardship team. *American Journal of Health-System Pharmacy* 70, 1065–1069.

CDC (2014) *Core Elements of Hospital Antibiotic Stewardship Programs*. Centers for Disease Control and Prevention, Atlanta, Georgia. Available at: http://www.cdc.gov/getsmart/healthcare/pdfs/core-elements.pdf (accessed 8 January 2015).

CMS (2014) *Hospital Infection Control Worksheet*. Centers for Medicare & Medicaid Services, Baltimore, Maryland. Available at: http://go.cms.gov/1B6NCSV (accessed 25 May 2016).

Dellit, T.H., Owens, R.C., McGowan, J.E., Jr., Gerding, D.N., Weinstein, R.A., Burke, J.P., Huskins, W.C., Paterson, D.L., Fishman, N.O., Carpenter, C.F. *et al.* (2007) Infectious Diseases Society of America and the Society for Healthcare Epidemiology of America guidelines for developing an institutional program to enhance antimicrobial stewardship. *Clinical Infectious Diseases* 44, 159–177.

Enoch, D.A., Idris, S.F., Aliyu, S.H., Micallef, C., Sule, O., and Karas, J.A. (2014) Micafungin for the treatment of invasive aspergillosis. *Journal of Infection* 68, 507–526.

Fishman, N. (2006) Antimicrobial stewardship. *The American Journal of Medicine* 119(6 Suppl 1), S53–S61.

Grayson, L.I., Crowe, S.U., McCarthy, J.A., Mills, J., Mouton, J.W., Norrby, S.R., Paterson, D.L., and Pfaller, M.A. (eds) (2010) *Kucers' The Use of Antibiotics*, 6th edn. CRC Press, Taylor & Francis Group, Boca Raton, Florida.

Guglielmo, B.J., Dudas, V., Maewal, I., Young, R., Hills, A., Villmann, M., Gibbs, L., Gropper, M., and Jacobs, R. (2005) Impact of a series of interventions in vancomycin prescribing on use and prevalence of vancomycin-resistant enterococci. *The Joint Commission Journal on Quality and Patient Safety* 31, 469–475.

Khadem, T.M., Dodds Ashley, E., Wrobel, M.J., and Brown, J. (2012) Antimicrobial stewardship: a matter of process or outcome? *Pharmacotherapy* 32, 688–706.

Levin, P.D., Idrees, S., Sprung, C.L., Weissman, C., Weiss, Y., Moses, A.E., and Benenson, S. (2012) Antimicrobial use in the ICU: indications and accuracy – an observational trial. *Journal of Hospital Medicine* 7, 672–678.

Malcolm, W., Nathwani, D., Davey, P., Cromwell, T., Patton, A., Reilly, J., Cairns, S., and Bennie, M. (2013) From intermittent antibiotic point prevalence surveys to quality improvement: experience in Scottish hospitals. *Antimicrobial Resistance and Infection Control* 2(1): 3.

Marr, K.A., Schlamm, H.T., Herbrecht, R., Rottinghaus, S.T., Bow, E.J., Cornely, O.A., Heinz, W.J., Jagannatha, S., Koh, L.P., Kontoyiannis, D.P. *et al.* (2015) Combination antifungal therapy for invasive aspergillosis: a randomized trial. *Annals of Internal Medicine* 162, 81–89.

Moellering, R.C., Jr. (1992) Emergence of *Enterococcus* as a significant pathogen. *Clinical Infectious Diseases* 14, 1173–1178.

Murray, C., Shaw, A., Lloyd, M., Smith, R.P., Fardon, T.C., Schembri, S., and Chalmers, J.D. (2014) A multidisciplinary intervention to reduce antibiotic duration in lower respiratory tract infections. *Journal of Antimicrobial Chemotherapy* 69, 515–518.

Owens, R.C., Jr., Fraser, G.L., and Stogsdill, P. (2004) Antimicrobial stewardship programs as a means to optimize antimicrobial use: insights from the Society of Infectious Diseases Pharmacists. *Pharmacotherapy* 24, 896–898.

Patterson, D.L. (2006) The role of antimicrobial management programs in optimizing antibiotic prescribing within hospitals. *Clinical Infectious Diseases* 42, S90–S95.

Rybak, M. (2007) Antimicrobial stewardship. *Pharmacotherapy* 27(10 Pt 2), 131S–135S.

Schultz, L., Lowe, T.J., Srinivasan, A., Neilson, D., and Pugliese, G. (2014) Economic impact of redundant antimicrobial therapy in US hospitals. *Infection Control and Hospital Epidemiology* 35, 1229–1235.

Woodward, R.S., Medoff, G., Smith, M.D., and Gray, J.L. III (1987) Antibiotic cost savings from formulary restrictions and physician monitoring in a medical-school-affiliated hospital. *The American Journal of Medicine* 83, 817–823.

Wright, D.B. (1991) Antimicrobial formulary management: a case study in a teaching hospital. *Pharmacotherapy* 11(1P2), 27S–31S.

25 Approaches to Benchmarking Antibiotic Use

Emily L. Heil* and Harold C. Standiford

University of Maryland School of Pharmacy, Baltimore, Maryland, US, University of Maryland Medical Center and University of Maryland School of Medicine, Baltimore, Maryland, US

Introduction

Measurement is the first step that leads to control and eventually to improvement. If you can't measure something, you can't understand it. If you can't understand it, you can't control it. If you can't control it, you can't improve it.

H. James Harrington, Harrington Institute Inc., California

Benchmarking is a process of comparing performance. The American Society for Quality defines it as "A technique in which a company measures its performance against that of best in class companies, determines how those companies achieved their performance levels and uses the information to improve its own performance. Subjects that can be benchmarked include strategies, operations, and processes." (ASQ, 2007). There is no standardized methodology or process in benchmarking that has been universally adopted, although there are multiple texts that provide approaches to benchmarking, including Boxwell's *Benchmarking for Competitive Advantage* and Camp's 12-stage approach to benchmarking (Boxwell, 1994; Camp, 2006). Benchmarking can be done internally or externally as a means of comparing an organization's performance with other performance standards or organizations.

In healthcare organizations, the concept of benchmarking is a cornerstone of performance improvement initiatives and the identification of best practices. The goals of benchmarking in healthcare include enhancing patient outcomes, streamlining patient care delivery systems, reducing costs, and complying with local, state or national initiatives (Aviles, 2012). Benchmarking activity supports practitioners in a continuous cycle of comparison and sharing (Ellis, 2000). As it relates to infectious diseases, benchmarking has been utilized to identify outliers so that best practice strategies can be identified and implemented for a variety of healthcare-associated infections (HAIs) as well as sepsis and community-acquired pneumonia (Platt, 2005; Harris and McGregor, 2008; Uçkay *et al.*, 2008; Edwards *et al.*, 2009; Waterer *et al.*, 2011; Dudeck *et al.*, 2013). The 2007 guidelines for antimicrobial stewardship from the Society for Healthcare Epidemiology of America (SHEA) and the Infectious Diseases Society of America (IDSA) recommend that hospitals measure antimicrobial drug use so that they can compare their use of antimicrobials with that of other hospitals (Dellit *et al.*, 2007). Benchmarking in antimicrobial stewardship is different in that there is no external "gold standard" for antimicrobial usage.

Metrics of Antibiotic Use for Benchmarking

There are a number of ways that antimicrobial use can be quantified at an institution-specific level. Acquisition costs are often utilized, as they are easy to obtain; however, due to differences in contractual agreements, purchasing volume, and formularies, acquisition costs are not an ideal metric for benchmarking against other institutions. Two more commonly used metrics for quantifying antibiotic use are defined daily dose (DDD) and antibiotic days of

*Corresponding author. E-mail: eheil@rx.umaryland.edu

  Antimicrobial Stewardship: Principles and Practice (eds K. LaPlante *et al.*)

therapy (DOT). The key advantages and disadvantages of these metrics are summarized in Table 25.1, along with those for a third metric, length of therapy, or LOT.

The best metric to quantify antibiotic use for the purposes of benchmarking has yet to be determined. The World Health Organization (WHO) promotes the use of DDD, which is based on the "assumed average maintenance dose per day for a drug used for its main indication in adults" (WHOCC, 2009). To estimate antibiotic use, the total number of grams of antibiotic used over the period of interest are divided by the WHO-assigned DDD. Comparisons with antibiotic use by other hospitals can then be done, usually by expressing DDDs per 1000 patient days (PD). For example, the WHO DDD for levofloxacin is 500 mg, so 7 days of a dose of 500 mg of levofloxacin would be 7 DDD (500 mg dose/500 mg DDD × 7 days = 7 DDD); and a dose of levofloxacin of 750 mg for 7 days would be 10.5 DDD (750 mg dose/500 mg DDD × 7 days = 1.5 DDD × 7 days = 10.5 DDD). An alternate to the DDD method is DOT, where a DOT represents the administration of a single agent on a given day regardless of the number of doses administered or the dose strength. There are advantages and disadvantages for each methodology.

Expressing antibiotic use as DDDs per 1000 patient days allows hospitals with differences in location, formularies, and patient census to benchmark drug use to one another. The DDD methodology was a major advance in identifying a standardized metric because it can be used by hospitals without sophisticated IT capabilities. However, DDD methods underestimate antibiotic exposure when doses are reduced for patients with impaired renal function (Zagorski *et al.*, 2002). DDDs may also not be available for a new antibacterial agent, or may change over time as new dosages are approved for existing drugs—which can complicate comparisons over time. Additionally, DDDs cannot be used to measure pediatric antibiotic use. In contrast, antibiotic DOT

Table 25.1. Antibiotic consumption metrics.

Consumption metric	Advantages	Disadvantages
Expenditures	• Purchase data is easy to obtain • Easily understood by administrators • Favorably viewed in offsetting the costs of a stewardship program	• Least accurate measure—only a surrogate marker for what the patient actually receives • Affected by changes in costs, contracts and formulary
Defined daily dose (DDD)	• Can be used to benchmark within and between hospitals, regions, or even countries • Relatively easy to obtain and calculate in the absence of computerized records by using purchasing data	• DDD values may change over time • DDD values may not reflect typical doses • Not valid for pediatric antibiotic use • Inaccurate for patients with renal insufficiency • Does not account for formulary mix when used as a measure of overall antibiotic use rather than as a measure of the use of a specific antibiotic
Days of therapy (DOT)	• Not affected by change in dosing • Allows multiple patient populations to be compared accurately • Endorsed by the Centers for Disease Control and Prevention (CDC) and the CDC's National Healthcare Safety Network (NHSN)	• Difficult to obtain without sophisticated computerized records • Favors those that use broad-spectrum monotherapy vs. narrow-spectrum combination therapy. For example, piperacillin–tazobactam × 7 days = 7 DOTs whereas ceftriaxone + metronidazole × 7 days = 14 DOTs
Length of therapy (LOT)	• Provides a more accurate assessment of treatment duration than DOT • Ratio of DOT/LOT can be used to assess frequency of combination antibiotic therapy vs. monotherapy	• Cannot be used to compare the use of different drugs

can be used to measure antimicrobial use in children and is not influenced by changes in WHO DDDs. The disadvantages of DOT are that it can overestimate use for drugs given in multiple dosages per day and can be difficult to measure in institutions without robust computerized pharmacy records. DOT also does not provide an accurate measure of length of therapy because combination therapy will increase the total DOTs (Polk *et al.*, 2007). For example, a patient being treated with levofloxacin and metronidazole for 10 days will generate 20 DOTs compared with a patient receiving monotherapy with piperacillin–tazobactam, who will generate 10 DOTs (Ibrahim and Polk, 2012).

DDD and DOT were compared in a study of over a million antibacterial drug doses across 130 US hospitals from August 2002 to July 2003 (Polk *et al.*, 2007). Aggregate hospital antibiotic use by DDDs per 1000 PD and DOTs per 1000 PD was discordant for many frequently used antibiotic drugs. For some antimicrobials, such as levofloxacin and linezolid, there was no significant difference between DDDs per 1000 PD and DOTs per 1000 PD because the administered daily dosage was equivalent to the DDD. However, for many agents, where the administered dosage exceeded the DDD, such as for ampicillin–sulbactam or cefepime, the estimates of use based on DDDs per 1000 PD significantly exceeded those based on DOTs per 1000 PD. Furthermore, agents such as piperacillin–tazobactam and ceftriaxone had lower dosages than the DDD, and estimates of use expressed as DDDs per 1000 PD were significantly lower than those based on DOTs per 1000 PD. This study demonstrates one of the main criticisms of DDD methodology. During the first half of the investigation, the approved dose of levofloxacin for most hospital infections was 500 mg, which is equivalent to the DDD, but in 2002, the FDA approved a 750 mg daily dose of levofloxacin for the treatment of hospital-acquired pneumonia. If hospitals began to treat all patients with this higher dosage, the DOTs would remain unchanged but the DDD measure would increase by 50%.

A newer metric is length of therapy (LOT) which refers to the number of days that a patient receives systemic antimicrobial agents irrespective of the number of different antibiotic agents. LOT will always be less than or equal to DOT and the ratio of DOT/LOT can be used as a proxy measurement of the average number of antibacterial agents used (Berrington, 2010; Ibrahim and Polk, 2012).

To allow comparisons between other hospitals or in the same hospital over time, regardless of the metric (DDD, DOT, LOT) used, the numerator should be divided by a metric that reflects the hospital size or census. This is usually patient days or admissions multiplied by 100 or 1000. Whether hospital days or number of admissions is used can be significant, particularly as length of stay has decreased over time for many hospitals. Thus, a decreasing number of days, even though the number of admissions stays the same or increases, may increase DDDs, DOTs, or LOT/1000 PD.

The choice of metric will depend on an individual hospital's goals and available resources. The DDD methodology is the one that is most commonly used, and it is promoted by the WHO and the IDSA/SHEA stewardship guidelines (Dellit *et al.*, 2007), but many US hospitals with the necessary IT infrastructure are moving toward DOT. The CDC's National Healthcare Safety Network (NHSN) antimicrobial use and resistant (AUR) option utilizes DOTs per 1000 PD (CDC, 2016). Utilization of only one metric, such as DDD, may be insufficient because DDD, DOT, and LOT provide different and complementary information, and it is unknown which one of these indices is best (Berrington, 2010).

The source of antibiotic use data used for benchmarking metrics is also important. Antibiotic use data is most accurate when derived from the electronic medication administration record (eMAR) or ideally from the bar coding medication administration (BCMA) record. The NHSN AUR requires antimicrobial data as documented in the eMAR and/or BCMA, and does not allow a manual entry option (see CDC, 2016). Data from prescribing (ordering) is less precise but can be helpful for internal monitoring purposes if the eMAR or BCMA is not available. Purchasing data is the least accurate form of information. The major emphasis that there is for hospitals to obtain a computerized database and electronic medical records should make it easier to obtain the necessary eMAR or BCMA data for benchmarking.

Risk Adjustment for Antibiotic Usage

The need to risk adjust antimicrobial usage numbers to account for differences in patient populations and hospital characteristics remains a challenge in the benchmarking of antibiotic use. Patient mix and severity of illness (SOI) have a major impact on antibiotic use. Patients with comorbid conditions and greater SOI are more likely to receive antibiotic therapy whether it is prophylactic, empiric or definitive treatment (Rogues *et al.*, 2004). This was well illustrated in a study involving 70 US academic medical centers that participate in the Clinical Database/Resource Manager of the University HealthSystem Consortium (UHC, now part of Vizient) in 2009. The range in antibacterial drug use ranged broadly from a low of around 600 DOTs/1000 PD to a high of around 1100 DOTs/1000 PD. The data demonstrate that academic medical centers are not similar enough to allow the comparison of unadjusted data (Polk *et al.*, 2011).

Multiple models have been proposed for risk adjusting for infection control and antimicrobial use, but the indirect standardization model using the standardized infection ratio (SIR) has shown some advantages and has gained popularity (Gustafson, 2006). The method of using SIR parallels that of the standardized mortality ratio (SMR), which has been widely used in public health to analyze mortality data (Malpiedi *et al.*, 2013). In HAI data analysis, the SIR compares the actual number of HAIs in a locality or state with the baseline US experience (i.e., the standard population), adjusting for several risk factors that have been found to be most associated with differences in infection rates. With the SIR, an observed/expected (O/E) ratio of greater than 1.0 indicates that more HAIs were observed than expected, thus accounting for differences in the types of patients followed. Conversely, an SIR of less than 1.0 indicates that fewer HAIs were observed than predicted (Malpiedi *et al.*, 2013). Although there is some controversy, the SIR appears to be the best tool available for the public reporting of infection rates as it is the most practical statistic for benchmark comparisons, intrahospital comparisons over time and interhospital comparisons (Gustafson, 2006).

In the specific case of antimicrobials, the indirect standardization method using SIR to risk adjust antimicrobial use was applied by Polk *et al.* (2011) for 70 hospitals participating in the UHC. The authors determined O/E ratios as measured by DOT and LOT per 1000 patient days. Additionally, antimicrobial drug use was determined for 35 clinical service lines (CSLs) in each hospital based on clustering patients with similar medical severity diagnosis-related groups (DRGs). This allowed them to determine the O/E ratio for specific patients in each hospital based on similar DRGs. By this method, the observed O/E ratio was 1.44 for one hospital, 44% greater than expected. When benchmarked against bone marrow transplant (BMT) patients, this hospital's antimicrobial use was nearly three times that expected when it was benchmarked similar populations across all the hospitals. Also, the O/E ratio was high for a number of additional CSLs for the hospital. Thus, this hospital's antibacterial drug use exceeded the use in similar patients when benchmarked across a large number of CSLs, including BMT. Using this type of information, an antimicrobial stewardship program can identify whether a hospital's drug use is high because of a high number of patients in a specific CSL, the number of antimicrobials administered, excessive LOTs, or an excessive number of antimicrobials used (Polk *et al.*, 2011). While the methodology for the new AUR module for the NHSN has not been publically defined by the CDC, it is likely the CDC will use the indirect standardization and SIR for data analysis in order to keep a level playing field for risk adjustment and for ease in reporting (Standiford, personal communication).

The ultimate goal of using local and national benchmark data is to elicit practice changes. Project Intensive Care Antimicrobial Resistance Epidemiology (ICARE) was a collaborative study between the Division of Healthcare Quality Promotion at the CDC and the Rollins School of Public Health of Emory University that monitored antibiotic use over a 4 year study period in the National Nosocomial Infectious Surveillance (NNIS) system. One analysis of data from Project ICARE identified that changes in prescriber practice in intensive care units (ICUs) were associated with significant decreases in vancomycin use and prevalence of vancomycin-resistant enterococci (VRE) compared with vancomycin use and VRE prevalence in those ICUs not using unit-specific changes in practice. When local monitoring data was presented in the context of a risk-adjusted benchmark, hospitals were able to direct their efforts to reduce excessive use of antimicrobial drugs and reduce antimicrobial resistance (Fridkin *et al.*, 2002). Local data compared with local, national, and international benchmarks can be shared with a variety of hospital committees, such as the Pharmacy and Therapeutics (P&T) committee, along with hospital personnel such as ICU physicians, infectious diseases (IC) practitioners and house staff to spur changes in antimicrobial use.

Quality-based Benchmarking

None of the techniques described for benchmarking, including indirect standardization, address "appropriate" antimicrobial use directly. Antimicrobial stewardship programs (ASPs) have utilized quality markers already in place in hospitals such as 30 day mortality, length of stay (LOS), and readmissions, though the sensitivity of these measures in being directly related to appropriate antimicrobial therapy is poorly defined (Standiford *et al.*, 2012).

An expert panel from Canada and the US evaluated metrics that could be used to aid quality improvement efforts for evaluating antimicrobial stewardship programs. Five metrics were selected by the panel: (i) days of therapy per 1000 PD; (ii) number of patients with specific organisms that are drug resistant; (iii) mortality related to antimicrobial-resistant organisms; (iv) conservable days of therapy among patients with community-acquired pneumonia (CAP), skin and soft tissue infections (SSTI), sepsis or bloodstream infections (BSI); (v) and unplanned hospital readmission within 30 days after discharge from the hospital in which the most responsible diagnosis was one of CAP, SSTI, sepsis or BSI (Morris *et al.*, 2012).

Additionally, stewardship programs can utilize quality-based metrics already in place, such as guideline adherence for the Surgical Care Improvement Project (SCIP) and pneumonia-based core measures, and for HAI surveillance for the Centers for Medicaid and Medicare (CMS). The frequency of resistant organisms, including *Clostridium difficile*, as quality markers has been used by individual antimicrobial stewardship programs but not in benchmarking projects across programs (Fishman, 2006). However, resistant organisms within a hospital patient population are very complex, as they relate to infection control procedures and initiatives over time as well as to antimicrobial stewardship efforts. Measuring the quality of antibiotic use in addition to antibiotic consumption will be an important future direction for benchmarking in stewardship programs (Jacob and Gaynes, 2010). Finally, while benchmarking in an inpatient environment has been the emphasis of this chapter, as a new healthcare system is implemented, there will be even more need to include the outpatient population, as this is where the majority of antimicrobials are consumed.

Summary

Benchmarking their antibiotic use against that of other hospitals can help hospitals to identify areas where antibiotic use can be improved. To be valid, antibiotic use must be risk adjusted for patient mix and severity of illness. There is a need to have a national standard for methods of measurement so that benchmarking can be used to promote rational antibiotic use. Areas of future need include the benchmarking of antifungal drug use, antiviral use, and antimicrobial use in pediatric and outpatient populations, and in the developing world.

References

ASQ (2007) Quality Glossary. American Society for Quality, Milwaukee, Wisconsin Available at: http://asq.org/glossary/ (accessed 26 May 2016).

Aviles, M.R. (ed.) (2012) *Benchmarking in Health Care*, 2nd edn. Joint Commission International, Oak Brook, Illinois.

Berrington, A. (2010) Antimicrobial prescribing in hospitals: be careful what you measure. *Journal of Antimicrobial Chemotherapy* 65, 163–168.

Boxwell, R.J., Jr. (1994) *Benchmarking for Competitive Advantage*. McGraw-Hill, New York.

Camp, R.C. (2006) *Benchmarking: The Search for Industry Best Practices That Lead to Superior Performance*. Productivity Press, New York.

CDC (2016) *Antimicrobial Use and Resistance (AUR) Module*. Centers for Disease Control and Prevention, Atlanta, Georgia. Available at: http://www.cdc.gov/nhsn/PDFs/pscManual/11pscAURcurrent.pdf (accessed 27 April 2016).

Dellit, T.H., Owens, R.C., McGowan, J.E., Jr., Gerding, D.N., Weinstein, R.A., Burke, J.P., Huskins, W.C., Paterson, D.L., Fishman, N.O., Carpenter, C.F. *et al.* (2007) Infectious Diseases Society of America and the Society for Healthcare Epidemiology of America guidelines for developing an institutional program to enhance antimicrobial stewardship. *Clinical Infectious Diseases* 44, 159–177.

Dudeck, M.A., Horan, T.C., Peterson, K.D., Allen-Bridson, K., Morrell, G., Antilla, A., Pollock, D.A., and Edwards, J.R. (2013) *National Healthcare Safety Network (NSHN) Report, Data Summary for 2011, Device-associated Module*. Centers for Disease Control and Prevention, Atlanta, Georgia.. Available at: http://www.cdc.gov/nhsn/PDFs/dataStat/NHSN-Report-2011-Data-Summary.pdf (accessed 26 May 2016).

Edwards, J.R., Peterson, K.D., Mu, Y., Banerjee, S., Allen-Bridson, K., Morrell, G., Dudeck, M.A., Pollock, D.A., and Horan, T.C. (2009) National Healthcare Safety Network (NHSN) report: data summary for

2006 through 2008, issued December 2009. *American Journal of Infection Control* 37, 783–805.

Ellis, J. (2000) Sharing the evidence: clinical practice benchmarking to improve continuously the quality of care. *Journal of Advanced Nursing* 32, 215–225.

Fishman, N. (2006) Antimicrobial Stewardship. *The American Journal of Medicine* 119(6 Suppl 1), S53–S61.

Fridkin, S.K., Lawton, R., Edwards, J.R., Tenover, F.C., McGowan, J.E., Jr., Gaynes, R.P., ICARE, and NNIS (2002) Monitoring antimicrobial use and resistance: comparison with a national benchmark on reducing vancomycin use and vancomycin-resistant enterococci. *Emerging Infectious Diseases Journal* 8, 702–707.

Gustafson, T.L. (2006) Three uses of the standardized infection ratio (SIR) in infection control. *Infection Control and Hospital Epidemiology* 27, 427–430.

Harris, A.D. and McGregor, J.C. (2008) The importance of case-mix adjustment for infection rates and the need for more research. *Infection Control and Hospital Epidemiology* 29, 693–694.

Ibrahim, O.M. and Polk, R.E. (2012) Benchmarking antimicrobial drug use in hospitals. *Expert Review of Anti-infective Therapy* 10, 445–457.

Jacob, J.T. and Gaynes, R.P. (2010) Emerging trends in antibiotic use in US hospitals: quality, quantification and stewardship. *Expert Review of Anti-infective Therapy* 8, 893–902.

Malpiedi, P.J., Peterson, K.D., Soe, M., Edwards, J.R., Scott, R.D., II, Wise, M.E., Baggs, J., Yi, S.H., Dudeck, M.A., Arnold, K.E. *et al.* (2013) *2011 National and State Healthcare-Associated Infections Standardized Infection Ratio Report, Using Data Reported to the National Healthcare Safety Network as of September 4, 2012*. Centers for Disease Control and Prevention, Atlanta, Georgia. Available at: http://www.cdc.gov/hai/pdfs/SIR/SIR-Report_02_07_2013.pdf (accessed 26 May 2016).

Morris, A.M., Brener, S., Dresser, L., Daneman, N., Dellit, T.H., Avdic, E., and Bell, C.M. (2012) Use of a structured panel process to define quality metrics for antimicrobial stewardship programs. *Infection Control and Hospital Epidemiology* 33, 500–506.

Platt, R. (2005) Toward better benchmarking. *Infection Control and Hospital Epidemiology* 26, 433–434.

Polk, R.E., Fox, C., Mahoney, A., Letcavage, J., and MacDougall, C. (2007) Measurement of adult antibacterial drug use in 130 US hospitals: comparison of defined daily dose and days of therapy. *Clinical Infectious Diseases* 44, 664–670.

Polk, R.E., Hohmann, S.F., Medvedev, S., and Ibrahim, O. (2011) Benchmarking risk-adjusted adult antibacterial drug use in 70 US academic medical center hospitals. *Clinical Infectious Diseases* 53, 1100–1110.

Rogues, A.M., Placet-Thomazeau, B., Parneix, P., Vincent, I., Ploy, M.C., Marty, N., Merillou, B., Labadie, J.C., and Gachie, J.P. (2004) Use of antibiotics in hospitals in south-western France. *Journal of Hospital Infection* 58, 187–192.

Standiford, H.C., Chan, S., Tripoli, M., Weekes, E., and Forrest, G.N. (2012) Antimicrobial stewardship at a large tertiary care academic medical center: cost analysis before, during, and after a 7-year program. *Infection Control and Hospital Epidemiology* 33, 338–345.

Uçkay, I., Ahmed, Q.A., Sax, H., and Pittet, D. (2008) Ventilator-associated pneumonia as a quality indicator for patient safety? *Clinical Infectious Diseases* 46, 557–563.

Waterer, G.W., Rello, J., and Wunderink, R.G. (2011) Management of community-acquired pneumonia in adults. *American Journal of Respiratory and Critical Care Medicine* 183, 157–164.

WHOCC (2009) DDD: Definition and General Considerations. WHO Collaborating Centre for Drug Statistics Methodology, Norwegian Institute of Public Health, Oslo. Available at: http://www.whocc.no/ddd/definition_and_general_considera/ (accessed 26 May 2016).

Zagorski, B.M., Trick, W.E., Schwartz, D.N., Wisniewski, M.F., Hershow, R.C., Fridkin, S.K., and Weinstein, R.A. (2002) The effect of renal dysfunction on antimicrobial use measurements. *Clinical Infectious Diseases* 35, 1491–1497.

26 Development and Execution of Stewardship Interventions

AMY HANSON AND CHRISTOPHER W. CRANK*

Rush University Medical Center, Chicago, Illinois, US

Introduction

If the etymology of the word "stewardship" is examined, it is found that it is often used within a religious context and represents "doing in good faith" for others by giving back. It is also a concept that can be broadly applied to the environment, economics, health, property, information, theology, etc. Stewardship is generally recognized as the assignment of responsibility to shepherd and safeguard the valuables of others. This chapter discusses key stewardship concepts and principles that clinicians can use to improve antimicrobial utilization. We also discuss antimicrobial stewardship strategies that influence each step of antimicrobial prescribing, and have an impact at multiple levels (Fig. 26.1).

The Infectious Diseases Society of America (IDSA) and the Society for Healthcare Epidemiology of America (SHEA) guidelines for developing an effective antimicrobial stewardship program (ASP) in hospitals were published 2007 (Dellit *et al.*, 2007). These guidelines outlined several principles that can be used to guide the development of an effective ASP, including: (i) obtaining administrative support upfront; (ii) utilizing a team-based approach; (iii) appyling evidence-based core mechanisms to control antimicrobial use; and (iv) exploring additional stewardship interventions. The chapter starts by discussing these four principles, and then goes on to briefly consider the roles of technology advancements, collaboration with the microbiology laboratory, and benchmarking in the development of an ASP. Lastly, it summarizes the barriers to ASPs and ways to overcome them.

Obtain Administrative Support Upfront

Hospital administration should support the ASP's mission, and allow flexibility in implementing the necessary initiatives to meet program goals in optimizing patient care and, secondarily, in reducing expenditures (Goldmann *et al.*, 1996). Hospital administration "buy-in" is often the decisive factor that allows ASPs to grow and transform practices. The support should not only be for development and maintenance of the program, but also for the necessary infrastructure to track antimicrobial consumption and measure outcomes data. A strong business plan is crucial in gaining the attention of hospital leadership. The process of completing an ASP business plan can be broken down into the following nine steps (Perencevich *et al.*, 2007).

Step 1: Define the problem and develop potential solutions

Hospital administrators are unlikely to comprehend the extent of antibiotic use and resistance problems within an institution unless they are clearly demonstrated. Therefore, the business plan should identify important issues and how an ASP can address these issues using evidence-based strategies. The plan should highlight problem areas and frame the benefits of the ASP around patient safety, quality, and cost-efficient care.

Step 2: Meet with key administrators

ASP leaders should schedule initial in-person meetings with key stakeholders, such as the chief of staff, medical center director, director of operations, department

*Corresponding author. E-mail: christopher_w_crank@rush.edu

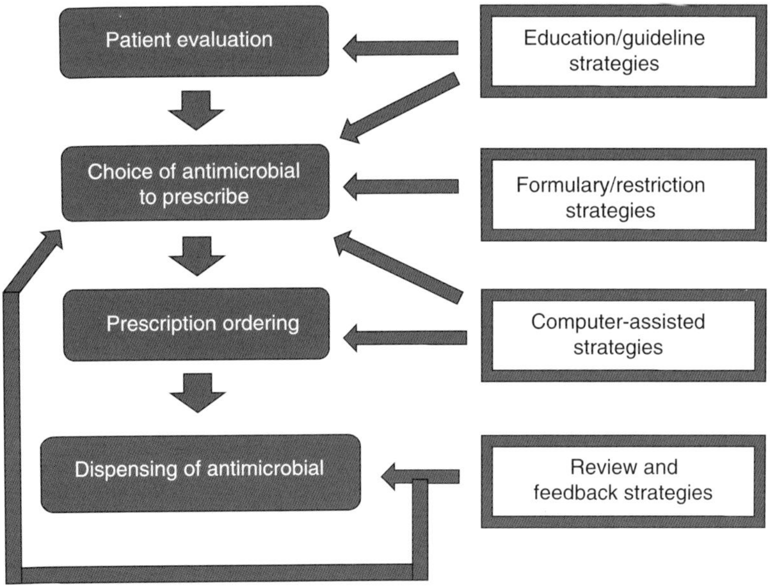

Fig. 26.1. The antimicrobial prescribing process and antimicrobial stewardship strategies. From MacDougall and Polk, 2005.

chiefs, and hospital medical staff executive committees in order to gain initial leadership support for the program.

Step 3: Determine the cost of the ASP

Establishing an ASP often requires a financial investment. ASP leaders should estimate the initial start-up costs and expected annual costs for the program. Examples of costs that a program may incur include a dedicated physician/pharmacist full time equivalent (FTE), data analyst time, special software or computer decision support systems, rapid diagnostic testing equipment, education/training, and printing fees.

Step 4: Determine which costs the ASP can lower

The business plan should layout the expected "return on investment" from implementing an ASP. Financial justification for the ASP could include institutional and published costs related to inappropriate antimicrobial use, such as increased length of stay and readmissions, unintended consequences of antimicrobial use, such as adverse drug effects and *Clostridium difficile* infection (CDI), and overall drug costs associated with antimicrobial overuse. Literature on these costs exists for a variety of facilities from small community hospitals to large academic medical centers.

Step 5: Determine the costs associated with healthcare-associated infections

Resistant healthcare-associated infections (HAIs) are a significant burden to hospital administration. Therefore, most leaders will be interested in the exact costs related to HAIs that are impacting their institution. Many prior studies have described the estimated costs of HAIs, and these can be used to project savings if no local data exist. However, institution-specific costs are often best received because they reflect a higher likelihood of accurate projections.

Step 6: Calculate the financial benefits of the ASP

To complete the analysis, the estimated cost savings and other financial benefits that result from the ASP must be calculated after the total costs of the program have been deducted.

Step 7: Include additional financial or health benefits of the ASP

The interventions implemented by an ASP can have benefits for a number of outcomes. For example, a new prospective audit and feedback program, in addition to decreasing antibiotic use and costs, could also improve patient outcomes, such as decreasing mortality or improving cure rates, reduce resistance rates, and improve patient safety indicators such as adverse drug events or rates of CDI. It is important to consider all of these when determining cost-effectiveness.

Step 8: Make the business case for the ASP

Additional meetings with administration after the initial "buy-in" meeting may be necessary to further make the case for an ASP through presentation of all the financial and health benefit estimates.

Step 9: Prospectively collect cost and outcome data

After the program is up and running, it is important to report back to administration the tracked progress. To foster long-term administrative support and program expansion, the ASP should consider meeting with administration at least annually to share the successes of the program. Additionally, representatives from the ASP should consider serving on the hospital's Pharmacy and Therapeutics (P&T) and Infection Control committees, so that other hospital committees are aware of ongoing stewardship initiatives.

Use of a Team-based Approach

The backbone of an ASP includes both an infectious diseases (ID) physician and a clinical pharmacist with ID training (MacDougall and Polk, 2005; Dellit *et al.*, 2007; Tamma and Cosgrove, 2011). An ID physician supporting the ASP can aid in both garnering acceptance for the program and compliance with ASP interventions by other physicians. The ID physician should demonstrate that the ASP is multidisciplinary and patient centered and is not a pharmacy-driven cost-savings initiative. ID physician support can also be useful in coordinating recommendations from the ID department on new clinical guidelines or pathways, and for getting these tools accepted at medical staff meetings such as those of the P&T committee to facilitate hospital-wide dissemination.

While guidelines recommend that a clinical pharmacist with ID training co-leads the program with an ID physician, pharmacists without formal ID training can also play an integral role in stewardship interventions. There are several online ASP certification programs for interested pharmacists, such as the Society of Infectious Disease Pharmacists (SIDP) certificate program (http://www.proce.com/activities/activity_detail?id=45) or the Making A Difference in Infectious Diseases (MAD-ID) basic program (http://mad-id.org/antimicrobial-stewardship-programs/antimicrobial-stewardship-programs-basic-program/). Pharmacy generalists can aid in triaging cases that may require ID physician intervention to the ASP physician, perform intravenous (IV) to oral (PO) conversions, monitor for appropriate and optimized dosing, and make recommendations to providers on overlapping antibiotic spectra or based on microbiology results. However, having an ID-trained pharmacist devoted to stewardship efforts is ideal because the majority of pharmacists whose primary role is processing and dispensing medication orders do not have the time to perform full chart reviews to assess the appropriateness of antimicrobial therapy, and also may not feel comfortable making these recommendations in some of the more complex cases, owing to a lack of specialized ID training.

Collaboration of the ASP with clinical microbiologists is extremely important (see section on Collaboration with the Microbiology Laboratory). Other important members of the ASP team include information technology (IT) specialists, and infection control and hospital epidemiology professionals. An IT specialist is vital for implementing interventions that revolve around the electronic medical record (EMR). Infection control efforts must be synergistic with those of the ASP. Drug and non-drug strategies that are directed towards decreasing the emergence and transmission of multidrug resistant organisms (MDROs) should be in place (Talbot, 2012). Two of the biggest concerns are the transmission of MDROs and development of CDI, and a successful ASP can have a tremendous impact on both of these. The partnership between ASPs and infection control (IC) teams is described in Fig. 26.2. Hospital epidemiologists can help measure the impact of the ASP on several outcomes, including antibiotic use, resistance, and patient outcomes and safety.

Apply Core, Evidence-based Mechanisms to Control Antimicrobial Consumption

The two core antimicrobial stewardship strategies are prospective audit and feedback, and formulary restriction. These strategies are not mutually exclusive and instead can be used synergistically to improve antimicrobial use.

Prospective audit with intervention and feedback

Reduction of inappropriate antimicrobial use can be achieved through prospective auditing of antimicrobial orders, with intervention and feedback to prescribing physicians. Prospective audit with intervention and feedback requires a regular review (usually daily) of antimicrobial orders to be performed by either an ID physician or a clinical pharmacist with ID training. The review can range from being as broad as all antimicrobials ordered to a narrowed list of targeted agents that may be overused or have problems with resistance. Follow-up communication to providers to recommend alternative agents should be performed, and can also serve as a means to reinforce the stewardship education that has been conducted. The follow-up communication can be done by verbal communication (i.e., face-to-face communication or a telephone call) or written communication (i.e., a note in the patient's chart).

A benefit of prospective audit and feedback is that it allows physicians to maintain autonomy. Prospective audit and feedback programs have been shown to decrease days of antimicrobial therapy, including reductions in IV and broad-spectrum use, and to demonstrate cost savings, while having no adverse impact on clinical response compared with control groups (Fraser *et al.*, 1997; Solomon *et al.*, 2001; LaRocco, 2003). A multidisciplinary antibiotic management program was associated with sustained reductions in antibiotic use and costs, along with decreases in rates of CDI and nosocomial infection caused by MDROs (Carling *et al.*, 2003). Although recommending de-escalation from broad- to narrow-spectrum agents is an important facet of many audit and feedback programs, recommending discontinuation of therapy when therapy is no longer indicated is also important. In one study, drug discontinuation made up 30% of stewardship interventions (Hecker *et al.*, 2003). In a small 120 bed community

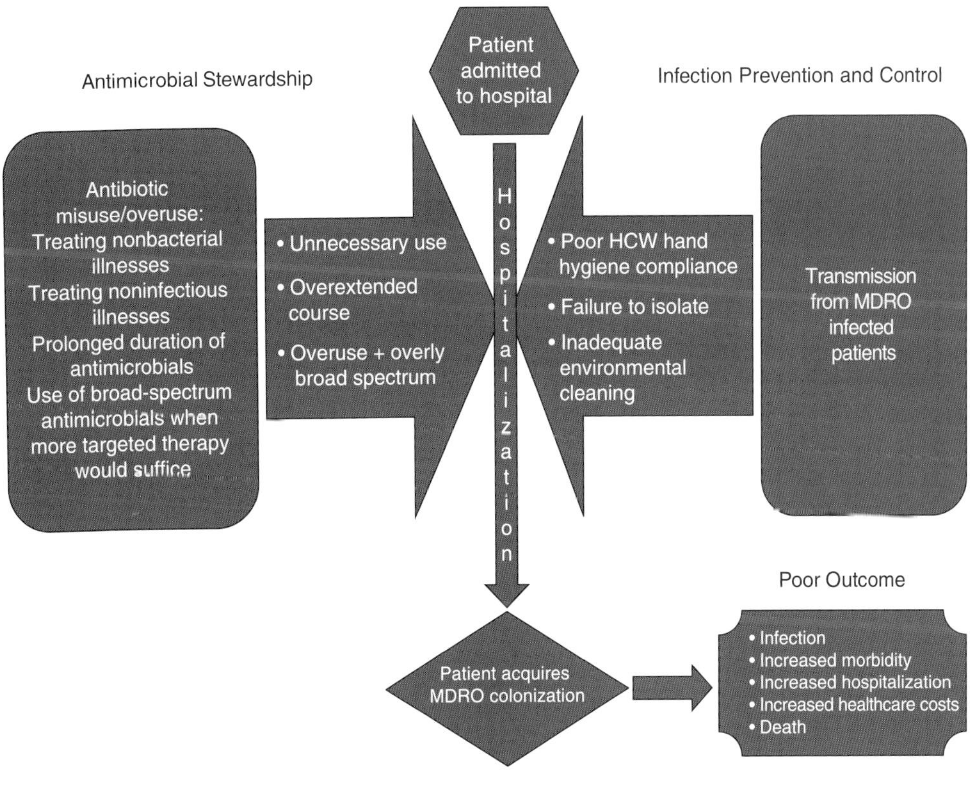

Fig. 26.2. Partnership between the antimicrobial stewardship program (ASP) and infection prevention and control to reduce the transmission of multidrug-resistant organisms (MDROs). HCW, healthcare worker.

hospital with limited resources, a scaled down model of prospective audit with feedback only 3 days a week was associated with a 19% reduction in antimicrobial expenditures compared with the preintervention period (LaRocco, 2003).

Providers may be hesitant to accept antibiotic discontinuation recommendations. It is important that physicians understand that the recommendations made by the ASP are voluntary and are typically made based solely on a chart review. It is the responsibility of the patient's primary treating team to provide bedside evaluations and to evaluate a patient's clinical status. However, markers such as procalcitonin (PCT) can help in difficult situations to give providers more comfort in following recommendations to discontinue antimicrobials (Fazili *et al.*, 2012). As systemic bacterial infections cause rapid release of PCT, it has been of growing interest to trend procalcitonin levels and use PCT algorithms to delineate when it is appropriate to "strongly encourage discontinuation" or "encourage reduction or discontinuation" of antimicrobials. In a recent meta-analysis of data from 14 trials with >4000 patients with upper or lower acute respiratory tract infections, PCT-guided treatment was associated with a statistically significant shorter duration of antimicrobials for patients admitted to the intensive care unit (ICU) (3.2 days), and for patients with community-acquired pneumonia (CAP) (3.3 days) or ventilator-associated pneumonia (VAP) (2.2 days) (Schuetz *et al.*, 2012). While data thus far has suggested that biomarkers such as PCT can serve as reliable markers and tools for the future of ASPs, additional studies are warranted.

Formulary restriction and preauthorization

Formulary restriction and preauthorization involves limiting the use of certain antimicrobials so that they are only available based on a specific set of criteria. When prescribers order restricted agents that do not meet their criteria for use, redirection feedback to the prescribers recommending a more appropriate, often non-restricted agent, can lend an educational opportunity so that the next time prescribers order that antimicrobial, the order is more appropriate. Clinical pharmacists with ID training are often in charge of approving orders for restricted antimicrobials. Prescribers may view formulary restriction as a loss of autonomy, but they should be aware that they always have the option of making a formal ID consultation when they feel that the use of restricted agents that do not meet criteria for use is still necessary. This provides a build-in back-up mechanism to the system in times of a "tough call" where there is need for a second ID opinion. In addition to approving a restricted agent request, the ID pharmacist can also make recommendations on appropriate dosing, route of administration, duration, drug interactions, adverse drug reactions to monitor for, and the interpretation of microbiology results.

Several studies have demonstrated that implementing formulary restriction and preauthorization programs is associated with significant and sustained decreases in antimicrobial use and costs without negatively impacting length of therapy or mortality (Britton *et al.*, 1981; Seligman, 1981; Hayman and Sbravati, 1985; Woodward *et al.*, 1987; Coleman *et al.*, 1991; Maswoswe and Okpara, 1995; Frank *et al.*, 1997; White *et al.*, 1997). These studies have demonstrated annual antimicrobial cost savings after implementing restrictive policies ranging upwards of $800,000. Limited evidence suggests that these programs can also reduce resistance rates. One study showed a statistically significant decrease in rates of *Stenotrophomonas* and methicillin-resistant *Staphylococcus aureus* (MRSA) colonization or infection after the implementation of a restriction program (Frank *et al.*, 1997). Another study demonstrated that β-lactam and fluoroquinolone susceptibility rates improved after a restricted drug approval program was implemented, without an effect on clinical outcomes (White *et al.*, 1997). However, most other studies have either found no impact on resistance rates or have not evaluated this outcome.

There are several different ways to set up a restriction program, depending on the resources available to the institution. If clinical pharmacy services in ID are only available during the day, daytime restricted agent review hours can be set. During "after hours" or weekends, prescribers can obtain restricted antimicrobials and then retrospective review can occur by the ID pharmacists on the next business day. The most proactive approach would be coverage of the restricted pager and/or on-call program 24/7. To accomplish this, an institution would likely need a clinical pharmacist on-call program. If there are limited personnel resources at a facility, this program could be staffed by rotating pharmacy residents. Weekend coverage can either default to the pharmacist on call, or can be a shared responsibility with the ID fellow physician on service. Clear criteria are needed to guide the appropriate approval of restricted agents. Another strategy that has been implemented is to break down drugs into the categories of unrestricted agents, controlled agents, and restricted agents (Woodward *et al*., 1987). Controlled agents are approved for a limited time, but require the stewardship team review to approve continuation if being prescribed past 72 h. For restricted agents, approval prior to verification is required. As with prospective audit and feedback, different methods have been used to provide prescribers with feedback, including one-on-one conversations in person or via telephone calls, notes in medical charts, or electronic notification.

Explore Additional Stewardship Interventions

There are several supplemental interventions that should be explored in addition to the core antimicrobial stewardship strategies (see Table 26.1).

Education

There are a variety of approaches to educating clinicians on optimizing the use of antimicrobials. Education can come in the form of staff conferences, medical grand rounds, clinical pharmacy consultations, drug utilization evaluations, newsletters, ID department meetings, use of the hospital intranet, and/or informational sessions by physician offices. Pocket guides containing recommendation for antimicrobials based on disease state can be effective, and should be rolled out with order sets that parallel these new clinical pathways and guidelines when possible. Education alone is not a sustainable intervention, and must be supplemented with feedback strategies to have a sustained impact on prescribing patterns (Bantar *et al*., 2003). With the turnover in medical staff as a natural component of academic medical center training, reeducation is often key to keeping the messages of stewardship alive. Education is further discussed in Section I, Chapter 4 "The Importance of Education in Antimicrobial Stewardship."

Development of guidelines and clinical pathways

One key to successful adherence to published national guidelines is developing and implementing strategies such as clinical pathways. ASPs can facilitate a multidisciplinary development of evidence-based practice guidelines that incorporate local resistance patterns (from an antibiogram) to make the guidelines institution specific. Implementing clinical guidelines and pathways can reduce length of stay and hospital admissions, and can decrease antimicrobial utilization and therapy duration, while having positive effects on clinical outcomes and mortality reduction. One example of the effects that clinical pathways can have on improved patient care is from a multicenter controlled clinical trial in which 19 hospitals were assigned either to conventional management (n = 10) or to the implementation of a CAP clinical pathway (n = 9) (Marrie *et al*., 2000). The pathway use was associated with a 1.7 days reduction in number of bed days per managed patient (4.4 vs. 6.1 days, $P = 0.04$) and an 18% decrease in the admission rate of low-risk patients (31 vs. 49%, $P = 0.01$). Pathway-driven patients received an average of 1.7 days fewer of IV antimicrobials (4.6 vs. 6.3 days, $P = 0.01$) and were more likely to receive treatment with a single class of antimicrobials (64 vs. 27%, $P < 0.001$). The implementation of clinical pathways in a surgical ICU was associated with improvements in clinical improvement or cure, persistent infection and death (Price *et al*., 1999). Additionally, after the implementation of the clinical

Table 26.1. Core and supplemental strategies for antimicrobial stewardship programs (ASPs) recommended by the Infectious Diseases Society of America (IDSA) and the Society for Healthcare Epidemiology of America (SHEA).

Core strategies	Supplemental strategies
• Formulary restriction and preauthorization requirements • Prospective audit with direct intervention and feedback	• Education • Evidence-based guidelines and clinical pathways • Antimicrobial cycling • Antimicrobial order forms • Combination therapy • Streamlining or de-escalation of therapy • Dose optimization • Intravenous (IV) to oral (PO) conversion

pathways, antimicrobial costs were reduced from \$676.54 per patient to \$157.88 per patient (P = .001). Guidelines and clinical pathways are most effective if they are implemented simultaneously and with education and feedback on antimicrobial use.

Antimicrobial order forms

One intervention that can go hand in hand with the development of clinical guidelines and pathways is the use of antimicrobial order forms. The use of these order forms can have a significant impact on prescribing patterns by developing a more uniform approach to antimicrobial prescribing. Clinical guidelines and intuition-specific resistance patterns should be used to develop antimicrobial selection for various infectious disease states. The use of order sets (a group of related orders) in early goal-directed sepsis to aid in appropriate antimicrobial selection based on the suspected source of infection is of dire importance and can have tremendous impact on patient outcomes. Another popular trend is to make surgical prophylaxis antimicrobial selection order-set driven. In one study, the use of perioperative prophylaxis order forms with auto-discontinuation at day 2 resulted in a decrease in the duration of therapy (from 4.9 to 2.4 days) and a lower percentage of patients that received prophylaxis for >2 days (a reduction from 85 to 44%) (Durbin *et al.*, 1981). The rate of inappropriate initial antimicrobials was also decreased from 30 to 11% with the use of order forms. In addition to ensuring that the appropriate drug and duration are selected, antimicrobial order forms can also assist prescribers in selecting an appropriate dose and route for their patient.

There are several barriers to overcome in the implementation of order sets, such as physician buy in and administrative support, ensuring adherence to the guidelines, costs, and IT-related issues. Overall, studies have demonstrated the success of order sets in decreasing antimicrobial utilization over time, increasing the appropriateness of the antimicrobials prescribed, decreasing the use of restricted antimicrobials, and decreasing pharmacy costs (Kuti *et al.*, 2002; Deuster *et al.*, 2010).

De-escalation of empiric antimicrobial therapy and limiting duration of therapy

Streamlining or de-escalation of antibiotic therapy from broad-spectrum empiric coverage to narrower spectrum targeted therapy is an important strategy for reducing the overuse of broad-spectrum antibiotics and conserving their effectiveness. As part of streamlining or de-escalation, inappropriate or redundant empiric antimicrobials should be discontinued based on culture and susceptibility results, which are usually available from 72–96 h after starting therapy. Streamlining or de-escalation can lead to a reduction in antibiotic use and cost-savings. In the first 7 months after the implementation of a de-escalation program led by a pharmacist and an ID physician, streamlining recommendations were made in 340 of 625 patients (54%) reviewed (Briceland *et al.*, 1988). The projected annual savings were \$107,637. ASP personnel should collaborate with the microbiology lab to facilitate timely reporting and follow-up on culture and susceptibility results.

Intravenous (IV) to oral (PO) conversion

Most patients admitted to the hospital with serious infections are started on IV antimicrobial therapy initially. For agents with good oral bioavailability, once the patient's clinical status permits, the patient can be converted from an IV to a PO agent. Examples of agents with good bioavailability include fluoroquinolones, azithromycin, oxazolidinones, metronidazole, clindamycin, trimethoprim–sulfamethoxazole, fluconazole, and voriconazole. To facilitate IV to PO conversions, the ASP should develop an IV to PO switch policy outlining defined clinical criteria for conversion, and consider obtaining P&T approval for the policy to allow the pharmacy to automatically change the route for patients meeting these criteria. These simple interventions can result in reduced length of hospital stay, lower healthcare costs and fewer potential complications from IV therapy access. A pharmacist-initiated IV to PO program was associated with a decreased length of hospital stay of 1.53 days and 12 month cost savings of \$15,149 and \$161,072 for drug acquisition and reduced length of hospital stay, respectively (Przybylski *et al.*, 1997). Another study demonstrated that a median duration of IV therapy was reduced from 3 to 2 days after implementing an IV to PO program (McLaughlin *et al.*, 2005).

Dose optimization

Dose optimization involves making antimicrobial dose adjustments based on individual patient characteristics (such as age, weight, or renal function), the causative organism, site of infection, and the pharmacokinetic and pharmacodynamic (PK/PD) characteristics of the antimicrobial. Examples of optimizing PK/PD principles include the use of extended-interval dosing for aminoglycosides, because aminoglycosides exhibit concentration-dependent killing, and the use of prolonged infusion of β-lactams, as β-lactams exhibit time-dependent killing. As prolonged infusions maximize the amount of time that the free drug concentration exceeds the minimum inhibitory concentration (MIC) of the pathogen more than do short infusions, prolonged infusion may offer a clinical benefit for the treatment of infections due to resistant pathogens, although clinical data on this is lacking. There are data available on using extended (over 3 to 4 h) or continuous (over the entire dosing interval) infusion for several β-lactams, including piperacillin–tazobactam, cefepime, meropenem, and doripenem. In a single-center retrospective study, extended interval dosing of piperacillin–tazobactam was associated with a lower mortality rate compared with intermittent infusion dosing in patients that were critically ill (APACHE II score > 17) with *Pseudomonas aeruginosa* infections (Lodise *et al.*, 2007). This study also demonstrated that extended interval dosing may translate into cost savings through a reduction in the tonnage of piperacillin–tazobactam used. However, other data have demonstrated that prolonged infusion offers no advantage over intermittent infusion in treatment success, mortality, or length of stay (Arnold *et al.*, 2013).

Consider Technology Advancements

The Leapfrog Group, an organization focused on improving patient safety in hospitals nationwide, has identified computerized physician order entry (CPOE) as one of the most important "leaps" that organizations can take to substantially improve patient safety (Doolan and Bates, 2002). Recommendations issued by this organization appear to have stimulated the market to increase the number of commercially available systems, as well as to increase the number of hospitals wanting to implement CPOE in the next few years. CPOE allows for standardization of the order entry process, numerous real-time patient safety features, portability through a variety of technology devices, streamlined data collection with reporting capabilities, and increased communications and access to records.

There has also been a surge in computer-decision support systems (CDSS), which allows for an electronic review of patient's allergies, potential drug–drug interactions, and dose recommendations based on an individual patient's renal and hepatic function. It also allows more efficient workflow for ASP interventions that may have already been in place, for example IV to PO conversions. The real-time identification of positive blood cultures with the help of a CDSS can allow the ASP team to make more timely interventions, which in turn decreases the time to appropriate antimicrobial therapy. Studies that have used CDSS as part of their ASP in ICUs have demonstrated significant reductions in orders for drugs to which the patients had reported allergies, excess drug dosages based on renal function, adverse drug events, antimicrobial-susceptibility mismatches, antimicrobial costs, and length of hospital stay (Evans *et al.*, 1986, 1998). The other advantage of a CDSS is from the infection prevention and control perspective. In one study, computer surveillance identified 90% of confirmed nosocomial infections, compared with 76% of such infections identified by manual surveillance, and allowed infection control practitioners to reduce the time required for such activities by 65% (Rubinstein *et al.*, 1988). Computer-based surveillance can facilitate good stewardship by more efficient targeting of antimicrobial interventions, tracking of antimicrobial resistance patterns, and the identification of nosocomial infections and adverse drug events.

Collaboration with the Microbiology Laboratory

Collaboration of the ASP with a microbiologist is a crucial relationship to help in initiating core stewardship strategies. Microbiologists should aid the stewardship team in developing the annual institution antibiogram. Annual antibiograms should be disseminated throughout the healthcare system to aid prescribers in selection of the appropriate empiric antimicrobial therapy. Antibiograms can also be used to track institution-specific resistance trends from year to year. Cascade reporting from the microbiology lab is also a way to encourage the use of narrow-spectrum antimicrobials first as a result of stepwise reported susceptibilities. Most microbiology labs have the capability to suppress the reporting of certain antimicrobials from the initial susceptibility reports. These "hidden" susceptibility results can be made available upon request in cases where toxicity, allergy, coinfections or other considerations make first line therapy suboptimal. In an era of growing technology, which is allowing for more rapid identification of organisms and their susceptibilities to antibiotics, immediate action on this information is often not taken by clinicians without concomitant recommendations for appropriate therapy changes by the stewardship team. As an example, the use of a rapid test to identify methicillin-sensitive *Staphylococcus aureus* (MSSA) isolates was associated with a shorter time to optimal therapy, but only with active pharmacy/ID intervention (Nevrekar *et al.*, 2013). Section III of this book further details the role of the microbiology lab in antimicrobial stewardship.

Table 26.2. Selected potential barriers to antimicrobial stewardship programs (ASPs) with proposed possible solutions (Ohl *et al.*, 2012).

Potential barriers	Possible solutions
Lack of funding for personnel	• Create a business plan to present to hospital leadership to request additional funds. An ASP saves money and pays for itself • Implement an unfunded pilot for 6 months using available clinical pharmacist and physician champion volunteers to show potential cost savings to justify the budget • Use guidelines/clinical pathways or change antimicrobial bundles to affect physician prescribing. Monitor compliance with clinical pharmacists or nurses and communicate this information to medical staff leadership
Inability to recruit an infectious diseases (ID) physician champion team leader	• Contract with an ID physician at another hospital to provide prospective audit and feedback using telemedicine and/or electronic medical records • Contract with a hospitalist or other specialty clinician to fill this role
Inability to recruit an ID clinical pharmacist	• Use or hire a non-ID clinical pharmacist and provide supplemental training (various online stewardship training programs are available) • Use clinical pharmacists in other specialty areas as extensions of the physician champion or sole ID clinical pharmacist
Lack of medical staff/clinician support	• Work with medical staff leadership early in the process • Enlist the help of the hospital quality/patient safety officer • Enlist and develop leaders from "problem" specialties with help from the antimicrobial stewardship committee • Employ large-group education that highlights problems with antimicrobial resistance or *Clostridium difficile* infections • Rely more on prospective audit and feedback, which may be accepted better than formulary restriction
Outlier physicians	• Measure and audit usage by these physicians • Benchmark antimicrobial utilization by these physicians against other physicians in the practice or group • Work with these physicians to understand their antimicrobial needs and patient usage • Have the physician champion work with the outlier's medical staff department head or chief medical officer • Work with the outlier's medical staff credentialing committee
Inadequate IT resources to obtain needed microbiology and drug utilization data	• Decrease the number of antimicrobials or the time period required for measurement • Obtain an full-time equivalent (FTE) portion for an IT staff member • Include representatives on the hospital's clinical IT committee
Small hospital that is part of a larger hospital system	• Use and adapt the resources and system-wide guidelines and policies of the larger system, and use a local physician champion and clinical pharmacist • Concentrate on prospective audit and feedback rather than preauthorization • Pick "low-hanging fruit," such as following up with patients admitted with pneumonia or urinary tract infections who are found to have alternative diagnoses without subsequent stopping of antimicrobials

Benchmarking Work and Trending Progress

Both process measures (i.e., did the intervention result in the desired change in antimicrobial use?) and outcome measures (i.e., did the process implemented reduce or prevent resistance or other unintended consequences of antimicrobial use?) are useful in determining the impact of an ASP on antimicrobial use, resistance patterns, and other clinical outcomes. The hospital administration will often want to see that the ASP is decreasing antimicrobial use and antimicrobial costs. However, it is clinicians that are generally most interested in seeing improvements in patient outcomes and reductions in patient safety indicators. Benchmarking involves the comparison of process and outcome metrics with groups of similar hospitals. Approaches for benchmarking are discussed in Chapter 25.

Identification of and Overcoming Barriers to ASPs

There are a number of barriers that might be encountered along the way that can hinder the implementation or performance of an ASP, which are outlined in Table 26.2. Despite these barriers, it should be noted that ASPs have consistently demonstrated a decrease in antimicrobial use (of 22–36%) and annual savings of $200,000–900,000 (Dellit *et al.*, 2007).

Conclusions

This chapter has reviewed important principles in the development and implementation of antimicrobial stewardship interventions. In the beginning stages, constructing a multidisciplinary team committed to supporting the ASP and obtaining administrative buy in are crucial. Once this is secured, a thoughtful analysis of institution-specific stewardship interventions should be developed with the multidisciplinary team. The two core mechanisms to improve antimicrobial use include prospective audit with intervention and feedback, and formulation restriction with preauthorization. Additional stewardship strategies, such as prescriber education and pathways, should also be explored. Further stewardship strategies in hospitals have led to reduced antimicrobial use and costs, improved quality of care and patient safety, reduced rates of *C. difficile* infection, and slowed emergence of resistance.

References

Arnold, H.M., Hollands, J.M., Skrupky, L.P. Smith, J.R., Juang, P.H., Hampton, N.B., McCormick, S., Reichley, R.M., Hoban, A., Hoffmann, J. *et al.* (2013) Prolonged infusion antibiotics for suspected Gram-negative infections in the ICU: a before-after study. *Annals of Pharmacotherapy* 47, 170–180.

Bantar, C., Sartori, B., Vesco, E. Heft, C., Saúl, M., Salamone, F., and Oliva, M.E. (2003) A hospitalwide intervention program to optimize the quality of antibiotic use: impact on prescribing practice, antibiotic consumption, cost savings, and bacterial resistance. *Clinical Infectious Diseases* 37, 180–186.

Briceland, L.L., Nightingale, C.H., Quintiliani, R., Cooper, B.W., and Smith, K.S. (1988) Antibiotic streamlining from combination therapy to monotherapy utilizing an interdisciplinary approach. *Archives of Internal Medicine* 148, 2019–2022.

Britton, H.L., Schwinghammer, T.L., and Romano, M.J. (1981) Cost containment through restriction of cephalosporins. *American Journal of Hospital Pharmacy* 38, 1897–1900.

Carling, P., Fung, T., Killion, A., Terrin, N., and Barza, M. (2003) Favorable impact of a multidisciplinary antibiotic management program conducted during 7 years. *Infection Control and Hospital Epidemiology* 24, 699–706.

Coleman, R.W., Rodondi, L.C., Kaubisch, S., Granzella, N.B., and O'Hanley, P.D. (1991) Cost-effectiveness of prospective and continuous parenteral antibiotic control: experience at the Palo Alto Veterans Affairs Medical Center from 1987 to 1989. *The American Journal of Medicine* 90, 439–444.

Dellit, T.H., Owens, R.C., McGowan, J.E., Jr., Gerding, D.N., Weinstein, R.A., Burke, J.P., Huskins, W.C., Paterson, D.L., Fishman, N.O., Carpenter, C.F. *et al.* (2007) Infectious Diseases Society of America and the Society for Healthcare Epidemiology of America: guidelines for developing an institutional program to enhance antimicrobial stewardship. *Clinical Infectious Diseases* 44, 159–177.

Deuster, S., Roten, I., and Muehlebach, S. (2010) Implementation of treatment guidelines to support the judicious use of antimicrobial therapy. *Journal of Clinical Pharmacy and Therapeutics* 35, 71–78.

Doolan, D.F. and Bates, D.W. (2002) Computerized physician order entry systems in hospitals: mandates and incentives. *Health Affairs* 21, 180–188.

Durbin, W.A., Jr., Lapidas, B., and Goldmann, D.A. (1981) Improved antibiotic usage following introduction of a novel prescription system. *JAMA (The Journal of the American Medical Association)* 246, 1796–1800.

Evans, R.S., Larsen, R.A., Burke, J.P., Gardner, R.M., Meier, F.A., Jacobson, J.A., Conti, M.T., Jacobson, J.T., and Hulse, R.K. (1986) Computer surveillance of hospital-acquired infections and antibiotic use. *JAMA*

(The Journal of the American Medical Association) 256, 1007–1011.

Evans, R.S., Pestotnik, S.L., Classen, D.C., Clemmer, T.P., Weaver, L.K., Orme, J.F., Jr., Lloyd, J.F., and Burke, J.P. (1998) A computer-assisted management program from antibiotics and other antiinfective agents. *The New England Journal of Medicine* 338, 232–238.

Fazili, T., Endy, T., Javaid, W., and Maskey, M. (2012) Role of procalcitonin in guiding antibiotic therapy. *American Journal of Health-System Pharmacy* 69, 2057–2061.

Frank, M.O., Batteiger, B.E., Sorensen, S.J., Hartstein, A.I., Carr, J.A., McComb, J.S., Clark, C.D., Abel, S.R., Mikuta, J.M., and Jones, R.B. (1997) Decrease in expenditures and selected nosocomial infections following implementation of an antimicrobial-prescribing improvement program. *Clinical Performance and Quality Healthcare* 5, 180–188.

Fraser, G.L., Stogsdill, P., Dickens, J.D., Jr., Wennberg, D.E., Smith, R.P., and Prato, S. (1997) Antibiotic optimization: an evaluation for patient safety and economic outcomes. *Archives of Internal Medicine* 157, 1689–1694.

Goldmann, D.A., Weinstein, R.A., Wenzel, R.P. Tablan, O.C., Duma, R.J., Gaynes, R.P., Schlosser, J., and Martone, W.J. (1996) Strategies to prevent and control the emergence and spread of antimicrobial-resistant microorganisms in hospitals: a challenge to hospital leadership. *JAMA (The Journal of the American Medical Association)* 275, 234–240.

Hayman, J.N. and Sbravati, E.C. (1985) Controlling cephalosporins and aminoglycoside costs through pharmacy and therapeutics committee restrictions. *American Journal of Hospital Pharmacy* 42, 1343–1347.

Hecker, M.T., Aron, D.C., Patel, N.P. Lehmann, M.K., and Donskey, C.J. (2003) Unnecessary use of antimicrobials in hospitalized patients. *Archives of Internal Medicine* 163, 972–978.

Kuti, J.L., Capitano, B., and Nicolau, D.P. (2002) Cost-effective approaches to the treatment of community-acquired pneumonia in the era of drug resistance. *Pharmacoeconomics* 20, 513–528.

LaRocco, A., Jr. (2003) Concurrent antibiotic review programs—a role for infectious diseases specialists at small community hospitals. *Clinical Infectious Diseases* 37, 742–743.

Lodise, T.P., Jr., Lomaestro, B., and Drusano, G.L. (2007) Piperacillin–tazobactam for *Pseudomonas aeruginosa* infection: clinical implications of an extended-infusion dosing strategy. *Clinical Infectious Diseases* 44, 357–363.

MacDougall, C. and Polk, R.E. (2005) Antimicrobial stewardship programs in health care systems. *Clinical Microbiology Reviews* 18, 638–656.

Marrie, T.J., Lau, C.Y., Wheeler, S.L., Wong, C.J., Vandervoort, M.K., and Feagan, B.G. (2000) A controlled trial of a critical pathway for treatment of community-acquired pneumonia. *JAMA (The Journal of the American Medical Association)* 283, 749–755.

Maswoswe, J.J. and Okpara, A.U. (1995) Enforcing a policy for restricting antimicrobial use. *American Journal of Health-System Pharmacy* 52, 1433–1435.

McLaughlin, C.M., Bodasing, N., Boyter, A.C., Fenelon, C., Fox, J.G., and Seaton, R.A. (2005) Pharmacy-implemented guidelines on switching from intravenous to oral antibiotics: an intervention study. *The Quarterly Journal of Medicine* 98, 745–752.

Nevrekar, S.N., Wang, S.K., Zamora Gonzales, J., Crank, C.W., and Singh, K. (2013) Use of a penicillin-binding protein 2a (PBP2a) latex agglutination assay to reduce vancomycin usage in patients with methicillin-sensitive *Staphylococcus aureus* (MSSA) infections. ICAAC. Poster presentation D587 to: *53rd Interscience Conference on Antimicrobial Agents and Chemotherapy (ICAAC), September 10–13, 2013, Denver,* Colorado. USA. American Society for Microbiology, Washington, DC

Ohl, C.A., Beardsley, J.R., and Whitehurst, S. (2012) *Module 4 Creating Change: Using an Antimicrobial Stewardship Program to Improve Antimicrobial Usage*. Joint Commission Resources, Oak Brook, Illinois.

Perencevich, E.N. Stone, P.W., Wright, S.B., Carmeli, Y., Fisman, D.N., and Cosgrove, S.E. (2007) SHEA Guideline, Raising standards while watching the bottom line: making a business case for infection control. *Infection Control and Hospital Epidemiology* 28, 1121–1133.

Price, J., Ekleberry, A., Grover, A. Melendy, S., Baddam, K., McMahon, J., Villalba, M., Johnson, M., and Zervos, M.J. (1999) Evaluation of clinical practice guidelines on outcome of infection in patients in the surgical intensive care unit. *Critical Care Medicine* 27, 2118–2124.

Przybylski, K.G., Rybak, M.J., Martin, P.R. Weingarten, C.M., Zaran, F.K., Stevenson, J.G., and Levine, D.P. (1997) A pharmacist-initiated program of intravenous to oral antibiotic conversion. *Pharmacotherapy* 17, 271–276.

Rubinstein, E., Barzilai, A., Segev, S., Samra, Y., Modan, M., Dickerman, O., and Haklai, C. (1988) Antibiotic cost reduction by providing cost information. *European Journal of Clinical Pharmacology* 35, 269–272.

Schuetz, P., Briel, M., Christ-Crain, M., Stolz, D., Bouadma, L., Wolff, M., Luyt, C.E., Chastre, J., Tubach, F., Kristoffersen, K.B. *et al.* (2012) Procalcitonin to guide initiation and duration of antibiotic treatment in acute respiratory infections: an individual patient data meta-analysis. *Clinical Infectious Diseases* 55, 651–662.

Seligman, S.J. (1981) Reduction in antibiotic costs by restricting use of an oral cephalosporin. *The American Journal of Medicine* 71, 941–944.

Solomon, D.H., Van Houten, L., Glynn, R.J., Baden, L., Curtis, K., Schrager, H., and Avorn, J. (2001) Academic detailing to improve use of broad-spectrum antibiotics at an academic medical center. *Archives of Internal Medicine* 161, 1897–1902.

Talbot, T.R. (2012) *Antimicrobial Stewardship Initiative Toolkit*. Joint Commission Resources, Oak Brook, Illinois.

Tamma, P.D. and Cosgrove, S.E. (2011) Antimicrobial stewardship. *Infectious Disease Clinics of North America* 25, 245–260.

White, A.C., Jr., Atmar, R.L., Wilson, J., Cate, T.R., Stager, C.E., and Greenberg, S.B. (1997) Effects of requiring prior authorization for selected antimicrobials: expenditures, susceptibilities, and clinical outcomes. *Clinical Infectious Diseases* 25, 230–239.

Woodward, R.S., Medoff, G., Smith, M.D., and Gray, J.L., III (1987) Antibiotic cost savings from formulary restrictions and physician monitoring in a medical-school-affiliated hospital. *The American Journal of Medicine* 83, 817–823.

27 Technologic Support for Antimicrobial Stewardship

Renée-Claude Mercier[1]* and Carla Walraven[2]

[1]University of New Mexico College of Pharmacy, Albuquerque, New Mexico, US; [2]University of New Mexico Hospitals Department of Pharmacy Services, Albuquerque, New Mexico, US

Introduction

Over the last 25 years, advances in technology have significantly changed practice within the healthcare setting. From literature searching, e-mail communication, electronic medical records, clinical decision support systems to mobile devices, every component has helped to enhance access to data, and communication with the patient and amongst providers. Technological advances have been of great benefit for most specialties within the hospital, including the antimicrobial stewardship team. Important aspects of stewardship involve medical chart reviewing and documentation, literature searching, education and communication with healthcare professionals. Technologies that have contributed to facilitating the work of the stewardship team are the focus of this chapter.

The healthcare technology field is dynamic and constantly in search of new avenues for improving patient care and outcomes using electronic tools. Most new technologies have been embraced by healthcare professionals and are utilized in daily activities. One could not function in today's healthcare without the presence of these technological advances, and it is easy to become completely dependent upon them. The acceptance of all these changes has been mostly limited by healthcare resources and the risk surrounding the loss of patient privacy and lack of protection of information. However, many improvements have been made to ensure privacy in the utilization of these technologies, and their economic benefits have been documented, with the aim of justifying to the healthcare system that investment in them not only improves patient outcomes but can also yield cost savings.

Health System Surveillance Tools

Electronic medical records

Recent advances in healthcare technology include the advent of the electronic medical record (EMR), which allows fragmented healthcare data from paper-based medical charts to be combined into a single, organized electronic record from which providers can retrieve more comprehensive patient data and medical history in real time. While EMRs have greatly empowered providers to make more informed clinical decisions that would otherwise be difficult or even impossible, they also provide significant tracking and trending capabilities, particularly for antimicrobial stewardship.

Currently, all EMR systems provide standard pharmaceutical safety reviews for patients, such as cross-checking patient drug allergies with new medication orders, reviewing dose ranges for weight-based drug therapy, and identifying any drug–drug interactions (Mekhjian *et al.*, 2002). Now, for the first time, EMRs include the capability for providers to view patient-specific trends over time. This means that providers can engage in active surveillance of current and readmitted patients who have positive microbiological cultures for multidrug-resistant organisms (MDROs), methicillin-resistant *Staphylococcus aureus* (MRSA), carbapenem-resistant *Enterobacteriaceae*

*Corresponding author. E-mail: rmercier@salud.unm.edu

  Antimicrobial Stewardship: Principles and Practice (eds K. LaPlante *et al.*)

(CRE), or *Clostridium difficile*-associated diarrhea, all of which require infection control precautions to be implemented in order to prevent the spread of antimicrobial resistance in hospitals. As a result of EMRs now providing patient-specific trends, providers are better able to guide empiric antimicrobial therapy, not only for patients, but also for individual institutions.

Perhaps more significant than the EMR's capability to provide patient-specific trends over time is its capability to report data and evaluate medical trends in real time at an institutional level. The use of EMRs to create institutional antibiograms provides the foundation for and is the most recognized culmination of antimicrobial resistance trends. In addition, given the customizable functions of EMRs, reports can now be further refined to a specific patient population (e.g., critically ill patients and pediatric patients) or hospital unit (e.g., inpatient rehabilitation, hematology/oncology).

The EMR is often integrated with other databases, such as the computerized physician order entry (CPOE) interface that providers use to order medications or laboratory tests for a particular patient. Medication orders are received and processed by the pharmacy using another integrated system, such as a closed loop medication ordering and administration system). Orders that have been verified by the pharmacist are then displayed in the electronic medication administration record (eMAR) for nursing staff to review and use as needed; for instance, the eMAR can indicate when to administer a particular drug and allows nurses to document the exact time that medications were administered.

Most EMR systems can be customized to meet the needs of individual institutions, including programming some of the basic antimicrobial stewardship activities into the system. For example, formulary restrictions can be integrated into the CPOE, thereby preventing physicians from ordering non-formulary or restricted antimicrobial agents. Should a physician feel compelled to order a restricted or non-formulary antimicrobial, he or she would be prompted to contact the pharmacy directly, at which time a pharmacist could discuss alternative drug options that may be applicable. In addition, automatic therapeutic substitutions, antimicrobial stop dates, and therapeutic dosing interchanges (e.g., extended infusions of β-lactam antibiotics) can also be programmed into the EMR system to help enforce appropriate drug dosing and duration.

Formulary management within institutions may vary according to the operational hours of the antimicrobial stewardship program (ASP). In hospitals where the ASP is not operational 24 h a day, 7 days a week, formulary restrictions can, in part, be managed with electronic stop dates to ensure that restricted antimicrobial agents are not continued indefinitely. In the absence of an ASP, hospital pharmacies can concede by sending out a few doses of the restricted antimicrobial agent with a hard stop that forces providers to seek out the appropriate approvals prior to continuation.

Similarly, institutional clinical pathways can be prebuilt into the CPOE for commonly seen infectious disease states (Pestotnik *et al.*, 1996; Evans *et al.*, 1998). These clinical pathways can be bundled as a group of laboratory tests, medications, and diagnostic tests that, for the majority of patients presenting with certain infectious disease states, would help to standardize patient care and enable hospitals to document compliance with The Joint Commission's Core Measure Sets (https://www.jointcommission.org/core_measure_sets.aspx) performance measurements. Care bundles may be implemented in conjunction with clinical decision programs and include other medical considerations, such as the removal of intravenous catheters, the repetition of blood cultures to document the clearance of candidemia, appropriate durations of therapy, ophthalmology examinations, and de-escalation of therapy once identification is known (Antworth *et al.*, 2013; Guarascio *et al.*, 2013).

Clinical surveillance software programs

When combined with the recent enhancements to the EMR system, the inclusion of clinical surveillance programs can offer significant advantages to ASPs. Software-based clinical surveillance programs use prebuilt, customizable rules that alert users to clinically relevant information. While most commercially available programs come with a core set of antimicrobial stewardship functions (see Table 27.1), including preprogrammed rules for "bug–drug" mismatches, overlapping antibiotic coverage, and opportunities for intravenous (IV) to oral (PO) conversions, de-escalation, or discontinuation of therapy, user-defined rules can also be built into the system to capture a larger range of patient interventions. Some rules, for example, can target patients on linezolid who are also on selective serotonin receptor inhibitors for serotonin syndrome, or patients who have a positive *Clostridium difficile* test who are also on a proton pump inhibitor (PPI).

Sample screen shots of the various capabilities of a computer decision-support program are shown in Figures 27.1–27.3. Additional software modules for infection control or non-infectious disease pharmacy can be integrated with existing clinical surveillance programs based on institutional need and budgetary allowances.

Specialized surveillance programs include clinical decision software programs that can flag broad-spectrum antibiotics for de-escalation when culture results become available. While rapid diagnostic tests, including those for differentiating between coagulase-negative staphylococci, methicillin-susceptible *S. aureus* (MSSA), and MRSA bacteremia, are used by antimicrobial stewardship to de-escalate empiric therapy, the value of these tests depends on the ability to promptly communicate the results to the provider. As clinical decision support programs can make alerts in real time, they are ideally situated to notifying the provider of new microbiology results. Using a clinical decision software program, Patel *et al.* (2012) identified piperacillin–tazobactam as a potential drug for de-escalation based on an internal analysis of its use. Over the course of a year, de-escalation rates of piperacillin–tazobactam increased from 40 to 90%, with de-escalation commonly being performed by clinical prompting of the provider before a pharmacist could intervene. Bauer *et al.* (2010) also demonstrated quicker de-escalation to the appropriate antibiotic, decreased length of hospital stay, and decreased mortality with a clinical decision-support system, although these results did not reach statistical significance.

Table 27.1. Clinical surveillance program functionality.

Core functions	Additional functions
• Produce real-time antibiograms	• Automated local and national surveillance reporting
• Provide prebuilt clinical rules and alerting ○ "Bug–drug" mismatches ○ Intravenous (IV) to oral (PO) conversions ○ Targeted medication alerts ○ Duplicate antibiotic therapy ○ Lack of antimicrobial therapy with a positive microbiology culture ○ Renal dosing adjustments ○ Critical laboratory value alerts	• Customizable clinical rules and alerting
• Enable antimicrobial benchmarking	• Clinical intervention documentation and tracking

Clinical outcomes

The goal of any clinical surveillance program is to positively affect patient outcomes. Pestotnik *et al.* (1996) were among the first to implement an in-house clinical decision-support system aimed at improving antimicrobial prescribing and patient care at a 520 bed teaching hospital. Over the course of their 7 year study period, the physician acceptance rate of computer-prompted alerts improved from 30 to 99.9%. This translated to improved Gram-negative drug susceptibilities, including those of *Pseudomonas aeruginosa* isolates, to tobramycin, imipenem, and ceftazidime, a reduction of antibiotic adverse events (from 26.9 to 18.4%), and a statistically significant reduction in mortality rates among patients receiving an antibiotic (from 3.65 to 2.65%). Financially, these improvements also corresponded to significant cost savings, despite over half of the patient population receiving antibiotics.

Yong *et al.* (2010) also demonstrated the return of susceptibilities for common Gram-negative bacteria in a 24 bed intensive care unit (ICU) over a 7 year study period. Before implementation of a clinical decision-support system, the antimicrobial susceptibility of *P. aeruginosa* was declining on average by 9.1% each year. After the implementation of a clinical decision-support program to curtail the consumption of broad-spectrum antibiotics, *P. aeruginosa* susceptibility rates started trending in the opposite direction, resulting in a net return to susceptibility of 18.3% a year by the end of the study period.

After education and the implementation of a computer decision-support program for ICU

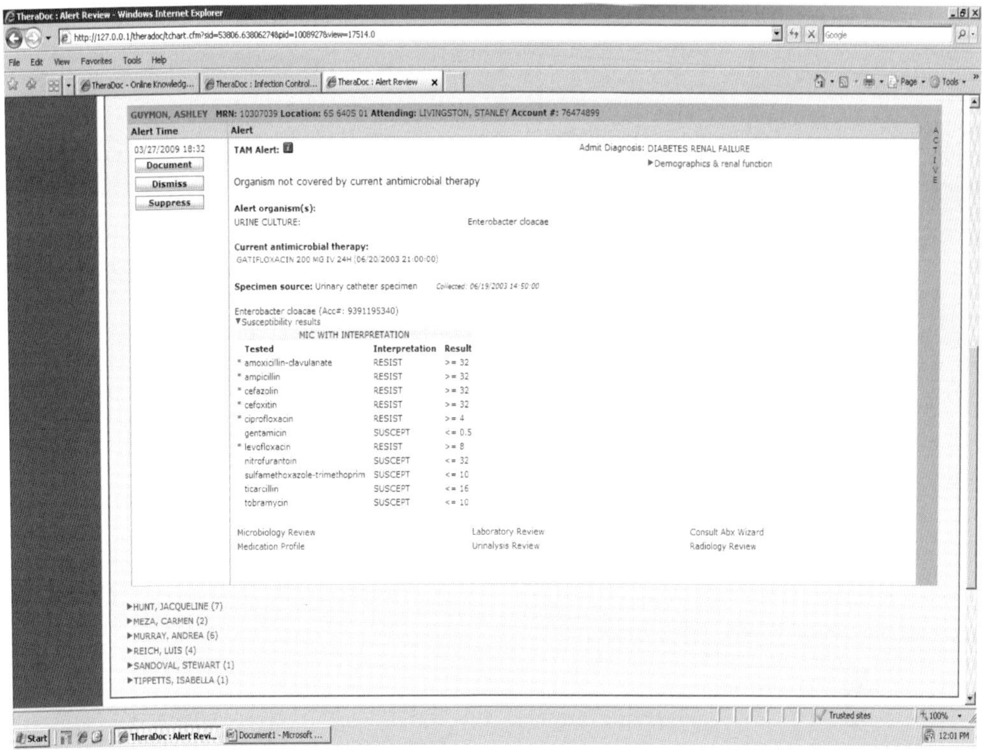

Fig. 27.1. Example of a therapeutic antibiotic alert. Here, the computer decision-support program has identified that a patient is receiving an antibiotic to which the pathogen is not susceptible. Image courtesy of TheraDoc Clinical Decision Support Software.

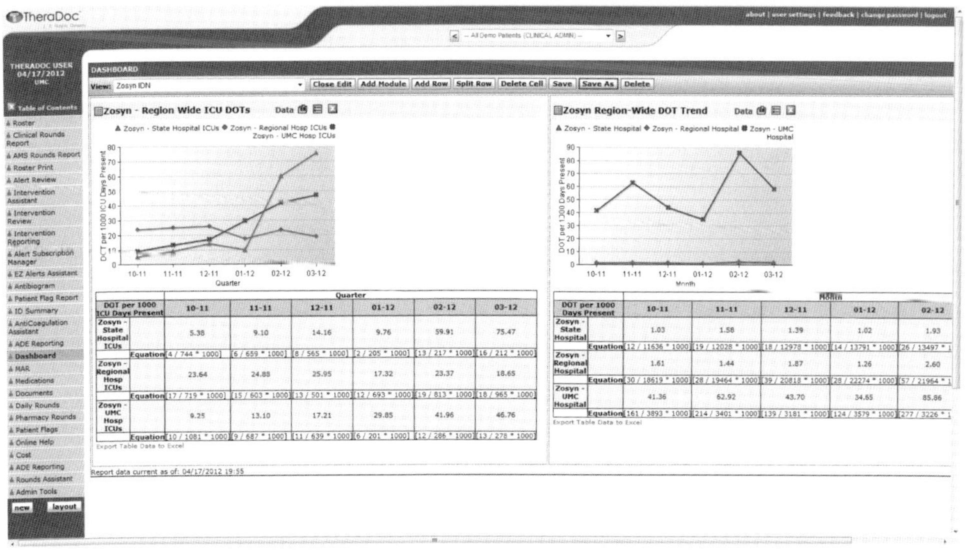

Fig. 27.2. Screen shot of antimicrobial use in terms of duration of therapy (DOT). Image courtesy of TheraDoc Clinical Decision Support Software.

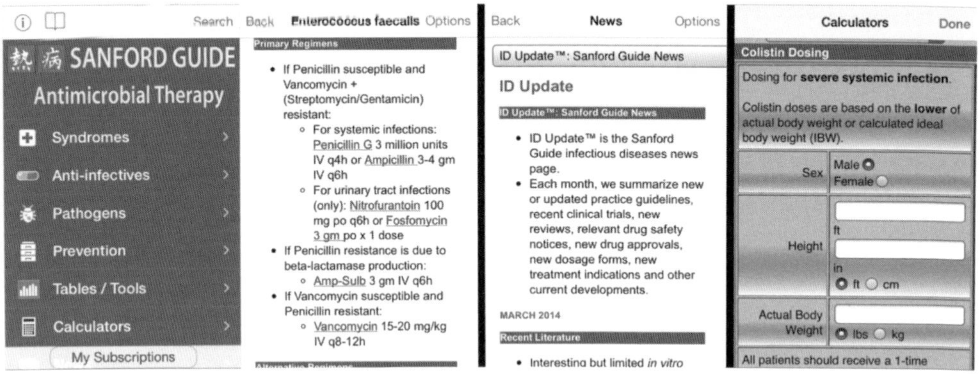

Fig. 27.3. Screenshots of the Sanford Guide App from iPhone.

patients, Thursky *et al.* (2006) found that providers accessed the computer decision-support program on average three times a day per patient, and that this became the preferred method for reviewing a patient's microbiology results, based on the effective presentation. As a result, providers significantly reduced their use of broad-spectrum antibiotics, particularly third-generation cephalosporins and carbapenems, even after adjustments were made for severity of illness and the presence of positive microbiology results.

Although education is a large component of antimicrobial stewardship efforts, its impact is oftentimes short lived, as Buising *et al.* (2008) found when comparing traditional educational intervention with the impact of a clinical decision-support system. To improve antibiotic prescribing for patients diagnosed with community-acquired pneumonia (CAP), these authors discovered that clinical decision-support system computer alerts triggered at the time of order entry enhanced appropriate empiric antibiotic prescribing and even decreased the time to antibiotic administration. Checking that antibiotic prescribing is appropriate, i.e., that it takes account of any patient allergies, drug–drug interactions, dosing adjustments for comorbid conditions, and compliance with quality measures, can delay antibiotic administration. However, as Buising *et al.* reported, the consistency of clinical alerts timed in conjunction with order entry provides real-time clinical knowledge at the point of care when providers are most receptive to interventions.

Based on the studies outlined above, the benefits associated with clinical surveillance programs are highlighted in Table 27.2.

Table 27.2. Clinical benefits directly associated with clinical surveillance programs.

- Increased appropriateness of antimicrobial therapy (drug, dose, and duration) (Thursky *et al.*, 2006; Buising *et al.*, 2008; Heil *et al.*, 2012; Patel *et al.*, 2012)
- Decreased antimicrobial adverse events (Pestotnik *et al.*, 1996; Buising *et al.*, 2008)
- Decreased antimicrobial resistance (Pestotnik *et al.*, 1996; Yong *et al.*, 2010)
- Decreased hospital length of stay (Heil *et al.*, 2012)
- Decreased patient mortality (Pestotnik *et al.*, 1996)
- Decreased healthcare-associated costs (Pestotnik *et al.*, 1996; Heil *et al.*, 2012)

Reporting

Aside from their clinical benefits, clinical surveillance programs also provide significant reporting capabilities. Typically, such programs contain large amounts of individual patient and institutional data, which can be used for data mining, trending, or reporting purposes. At large institutions, where patient review is often limited by the sheer volume of patients, clinical surveillance programs have proven to enhance the effectiveness of ASPs and their activities. In a landmark study by Evans *et al.* (1998), their clinical surveillance software program was able to pull together clinical information in 3.5 s compared with the average 14 min that it would take an infectious disease (ID) specialist to retrieve the same clinical information. Likewise, clinical surveillance software programs can also systematically and rapidly process large amounts of data from existing information systems. At the University of Maryland, Standiford *et al.* (2012) estimated that their ASP, combined with a clinical surveillance program, saved them an additional $600,000 annually compared with their ASP alone.

Clinical decision-support programs offer a variety reporting features, depending on the specific program concerned, which have been effective in benchmarking antimicrobial usage for stewardship programs. Both internal and external benchmarking have been employed to track antimicrobial stewardship interventions and highlight where future interventions may be necessary. Normalized antimicrobial utilization data is often reported in terms of antimicrobial days of therapy (DOT) or the defined daily dose (DDD) per patient days or hospital bed occupancy used by the World Health Organization (WHO).

Antimicrobial barcoding administration and eMAR data can also be integrated into clinical decision-support programs in order to more accurately report actual antimicrobial usage. Furthermore, most programs have the capability to help institutions meet regulatory compliance measures by offering automated public reporting of antimicrobial resistance data to national agencies, such as the Centers for Disease Control and Prevention (CDC) National Healthcare Safety Network (NHSN).

Drawbacks

Within most ASPs, interventions and acceptance rates are tracked using a separate program or database. Currently, clinical decision-support programs lack the means to track antimicrobial stewardship-initiated interventions that are made on behalf of a patient. This feature may be integrated with clinical decision-support programs in the future; however, at present, separate programs for intervention tracking are still required.

The biggest drawback to using clinical decision-support programs is that they require both financial and information technology (IT) commitments upon installation and for the duration of the subscription. These commitments can be substantial barriers to an institution with limited resources. In addition, some tasks, such as benchmarking antimicrobial use, can be accomplished using other data information systems (e.g., purchasing data is often used as a surrogate marker of antimicrobial consumption). Nonetheless, the interventions identified by clinical decision-support programs have demonstrated significant cost savings above those that an antimicrobial stewardship program can accomplish alone. In some cases, the cost savings may cover or offset the cost of purchasing and maintaining a clinical decision-support program.

Information Technologies

Mobile devices

Mobile devices such as smartphones and tablets are increasingly being used by healthcare providers. The *Wolters Kluwer Health* 2013 *Physician Outlook Survey* conducted with over 300 physicians in April 2013 showed that 80% used smartphones in daily practice and that 60% used a tablet. More than half used both a smartphone and a tablet. The most common use for the smartphones was to access drug information, and for the tablet, medical research was the most common reason cited. Interestingly, search engines such as Yahoo and Google were the most frequently used to access medical information that could directly impact the diagnosis, treatment, and care of patients (Wolters Kluwer Health, 2013).

According to a different survey conducted with 1063 physicians and mid-level providers in 2013, 86% of the healthcare professionals used a smartphone in their professional activities. This was an 8% increase from the previous year's data. Similarly, there was an increase in tablet use in daily practice among the clinicians from 34% in 2012 to 53% in 2013 (Epocrates, 2013; Terry, 2013; Wicklund, 2013). In 2013, the annual Manhattan Research report "Taking the Pulse" revealed that in the US 72% of the 2950 physicians surveyed in 2013 used a tablet in their professional activities, an increase compared to 62% the previous year (Comstock, 2013).

Overall, the tablet has been responsible for the significant increase in mobile devices in the healthcare field in the last few years. Tablets are being used by clinicians to interact with electronic health records, write notes and e-prescribe (Epocrates, 2013; Wicklund, 2013). The Apple iPad has the greatest share of the tablet market owing to the quality of the graphics and the choice of apps available. According to the online publication "iMedicalApps" (http://www.imedicalapps.com/) in 2014, no tablet comes close to the iPad's medical section for healthcare professionals.

Pharmacists have also adopted mobile technologies as a way of accessing clinical references and as point-of-care tools (Fox *et al.*, 2007; Dasgupta *et al.*, 2009, 2010; Aungst, 2013). Clinical interventions made by pharmacists are being documented in real time using these technologies, and the clinical decision-support systems, as mentioned above, are more commonly now used as part of ASPs. The availability of these support systems on digital assistants has led to an increase in clinical interventions being documented and significant healthcare cost savings (Calloway *et al.*, 2013). In addition, ASPs also benefit from digital assistants as they provide access to current infectious diseases clinical reference sources such as the Sanford Guides (http://www.sanfordguide.com/), the Infectious Diseases Society of America (IDSA) Practice Guidelines (http://www.idsociety.org/Guidelines_mobile/), drug databases such as those from Lexicomp (http://www.wolterskluwercdi.com/lexicomp-online/databases/) and Micromedex (http://micromedex.com/mobile1), local antibiograms, ASP institutional guidelines, and the Johns Hopkins guide to antibiotics (http://www.hopkinsguides.com/hopkins/ub).

One of the important functions of ASPs is to provide education to healthcare professionals. Means to accomplish such tasks are not easily disseminated across a hospital or healthcare system. The utilization of medical apps has allowed broader access to medical information related to antimicrobial agents, microbiology, diagnosis and the treatment of infectious diseases, thereby alleviating some of the educational barriers observed. Hospitals that use the evidence-based clinical decision-support resource UpToDate® (http://www.uptodate.com/home), either as an app or on desktop, partook in direct healthcare benefits. For example, patients had shorter inpatient length of stay and lower risk-adjusted 30 day mortality rates, with improved quality performance measures compared with hospitals without access to UpToDate (Isaac *et al.*, 2012).

Advantages

The utilization of mobile technologies by healthcare professionals as time-saving tools to access clinical references at the point of care has been shown to optimize medical research ability, expand access, and increase productivity (Isaac *et al.*, 2012; Phua *et al.*, 2012). These technologies also have the potential to improve the quality of care, as it has been estimated that their use is linked to the avoidance of more than 27 million potentially dangerous drug interactions each year (Wicklund, 2013). Furthermore, mobile technologies have the potential to inform, educate, and empower patients. Access to information on concerns about the overuse of antibiotics may lead to decreased antimicrobial resistance through a decrease in misuse or overuse via a limiting of consumers' requests to receive antimicrobials during medical visits (Harmon, 2010).

Another benefit of tablets is the ability to provide bedside education to patients. A computer on wheels is already utilized to provide anticoagulation education to patients who are new to oral or injectable anticoagulants. It would be feasible to also provide this type of education to patients receiving long-term antimicrobial therapy.

In an issue of the *The Washington Post* dated January 12, 2014, Ritu Agarwal, founder and director of the Center for Health Information at the University of Maryland's Robert H. Smith School of Business, stated that mobile technologies should help to address one of the persistent problems in healthcare to date—patient activation and engagement. With greater access to healthcare through electronic medical records, medical literature, mobile apps and social networking, healthcare consumers are more likely to be informed and actively participate in their own healthcare decisions (Beyers, 2014).

Other important advantages of tablets include features that help to organize all patient monitoring forms in one place. This helps to cut down on paper that healthcare professionals, including house staff, print off on a daily basis. Another possible use of tablets may be the sharing of handoff reports between providers. As an example, healthcare institutions are already trying such endeavors by transitioning to the cache system in Cerner's PowerChart® (http://www.cerner.com/solutions/Hospitals_and_Health_Systems/Acute_Care_EMR/PowerChart/?LangType=2057). This allows for real-time, up to date patient information to be communicated between all medical teams without having to make phone calls or tracking personnel down to pass along the information verbally.

Drawbacks

Perhaps some of the mobile technologies may save time for providers, but there is also a fear that access to digital communications may create an increase in workload and become a time burden for providers (Harmon, 2010). Of course, another concern related to having patient information on a personal digital assistant is concern about possible violations of HIPPA (the US Health Insurance Portability and Accountability Act of 1996). Much of the work done currently related to the use of mobile devices for patient care has to do with ensuring securely protected health information. Compliance with secured measures must be closely monitored across all institutions and healthcare professionals (FDA, 2015).

Another concern about the use of mobile devices is the potential for spread of bacterial infections due to colonization of these devices. Studies have demonstrated that 1–40% of mobile phones used by healthcare providers are colonized with bacteria such as MRSA, vancomycin-resistant enterococci (VRE), coliform bacilli and *Pseudomonas* and *Acinetobacter* species (Brady *et al.*, 2009). Tekerekoğlu *et al.* (2011) compared the bacterial colonization rate of mobile phones between phones belonging to patients, visitors, and healthcare workers. The results showed that phones belonging to patients and visitors were more likely to be colonized with pathogenic bacteria than phones belonging to healthcare workers (39.2 vs. 20.6%, respectively; $P = 0.02$). These results are encouraging and support the use of mobile devices by healthcare workers within healthcare settings. However, close monitoring and infection control measures must be in place to avoid mobile devices serving as reservoir of bacteria and promoting the further spread of bacterial infections.

Successful decontamination of mobile devices is not easily accomplished as most are affected by exposure to liquids and high temperature. Here, an ASP may serve a role in increasing education on the promotion of hand washing and the use of disinfectant wipes on mobile devices to limit bacterial colonization. Ultraviolet irradiation has also been proposed as a mode of disinfection of these devices (Tekerekoğlu *et al.*, 2011).

Infectious diseases-related apps

Most clinical apps have been developed and reviewed with physicians in mind, e.g., orthopedic surgeons and cardiologists (see http://www.imedicalapps.com/). Several apps are available as drug references ranging from the more basic to those with enhanced features that have been of great value for most medical providers, including pharmacists. Care must be taken when searching for medical apps, as so many are poorly categorized, and others have not been reviewed and their data content may be questionable. It is important to look at the developer of the app as well as at the last review date to make sure that the app provides updates. Several apps also have free content which is limited, with payment required for broad access and more valuable information.

Several reviews of ID-related apps for smartphones and tablets are available and of value to healthcare professionals who wish to compare and contrast the different programs. Figure 27.3 illustrates four screen shots from the Sanford Guide as an example of such an app (Burdette *et al.*, 2012; Goff, 2012; Husain, 2012; Mooley *et al.*, 2013), while Table 27.3 summarizes a few of the commonly used and beneficial apps for the antimicrobial stewardship team. Other apps also have an infectious disease focus such as “Bugs and Drugs” from Epocrates (which has recently been removed from the Epocrates app store), “Antibiograms” from Portable Databases, and the “Antibiotic Guide” produced by Dr Farookh Jishi, although the information provided by these apps has not been formally reviewed (for instance, data accessed on “Bugs and Drugs” for specific areas such as New Mexico could not be validated).

Social media

The utilization of social media to enhance healthcare outcomes has not been researched thoroughly despite the increasing popularity and accessibility of these means of communication. As most providers are now employing mobile devices in their professional activities, it is reasonable to postulate that communications could be increased within health systems and between providers and their patients if social media avenues were exploited by the healthcare field. Hospital portals are not always easy to navigate, and information often gets lost during dissemination. In the future, for example, health systems or single institutions could use Twitter to transmit small messages to the house staff reminding them to wash their hands, or alerting them of flu cases being seen within the institution/community. Alerts through social media could also help to remind staff about upcoming lectures, newly approved antibiograms, and drug shortages, just to name a few items. Podcasts are also utilized to provide education materials. The University of South

Table 27.3. Medical apps for antimicrobial stewardship programs.

App category	App	Description
Drug reference sources	Micromedex Drug Reference (http://micromedex.com/mobile1)	Simple user interface, logins not required, pharma influence not a concern, off-label use
	Epocrates Rx (http://www.epocrates.com/products/features)	Easy to use, limited information available, dosing calculation helpful, news and reviews done by John Bartlett, MD
	Lexicomp (http://www.wolterskluwercdi.com/drug-reference/apps/)	Expanded drug information, international names listed, FDA (US Food and Drug Administration) warnings
	Sanford Guide (http://www.sanfordguide.com)	Comprehensive clinical reference (expanded coverage compared with print edition), includes clinical conditions, antimicrobial drug information, news and calculators
	Johns Hopkins Antibiotic Guide (http://www.hopkinsguides.com/hopkins/mobile)	Comprehensive antimicrobial guide, easy to use
Clinical reference sources	UpToDate® (http://www.uptodate.com/home/uptodate-mobile-access)	Evidence-based, comprehensive clinical reference, peer-reviewed information
	Medscape (http://www.medscape.com/public/applanding)	Contains disease states information, interaction checker, news and CME (continuing medical education) in addition to drug information, medical literature with reference links
	ID Compendium (http://edgydoc.com/apps-mobile-and-more/)	Easy to use, provides drug and disease information, no link to guidelines, designed for house staff and students
Calculators	QxMd (http://www.qxmd.com/apps/calculate-by-qxmd)	Comprehensive clinical formula/calculations, including over 150 unique calculations and decision-support tools, simple and easy
	MedCalc Pro (http://medcalx.ch/)	Easy access to over 300 formulas, scores and scales, storage and sharing capacity
	Antibiotic Kinetics© (http://www.rxkinetics.com/abpk.html)	Pharmacokinetic dosing of antimicrobial drugs, serum levels monitoring for vancomycin and gentamicin, including once-daily dosing, graphing properties
Infection prevention/control	iScrub Lite (https://compepi.cs.uiowa.edu/iscrub/)	Records hand hygiene compliance efficiently, intuitive interface
	HealthMap: Outbreaks Near Me (http://www.healthmap.org/outbreaksnearme/)	Real-time disease outbreak information including H1N1 influenza, alerts available, reporting capacity
	Flu Tracker (http://www.cdc.gov/flu/apps/fluview-mobile-app.html)	Flu related news from CDC (US Centers for Disease Control and Prevention), including education, map feature with weekly update on influenza activity

Florida releases an Infectious Diseases Podcast on a regular basis which contains relevant and up-to-date information for its clinicians. Several other podcasts related to infectious diseases and bacterial resistance are also readily available for mobile devices. Such easy and open access allows for busy healthcare providers to stay connected at all times without too much effort.

A healthcare system constitutes an interprofessional milieu with diverse professionals coming together to care for patients. Reaching out to this diverse community has great challenges that used to be ignored. Interprofessional Education (IPE) is now considered to be at the core of healthcare education, in which each discipline must understand and respect what other professions contribute to patient care. More so than other services within the healthcare system, an ASP requires many healthcare providers to come together to improve and optimize antimicrobial use within an institution or healthcare system. Social networking has been used to effectively educate students from different professional curricula. The same models of social networking education could be utilized within an ASP to educate providers. Of course, this model of education should not be the sole mode used, but it could complement already existing methods. Furthermore, a recently published study demonstrated that in order to be effective, education through social networks needs to be performed within a facilitated environment and curriculum (Pittenger, 2013).

The unsecure nature of current social media does not allow its use as a professional means for communication in the healthcare field, but it has the potential to quickly transform healthcare communications. However, this would require a commitment from healthcare systems to develop secure and private networks, and from researchers to investigate the utility and merit of such a method of communication.

Future Directions

National surveillance systems

Large, multihospital healthcare systems, such as the Veterans Health Administration (VHA) and Kaiser Permanente, have long been utilizing a single electronic health record that can be accessed by any healthcare professional within the network. There has been a similar move towards the adoption of electronic personal health records (PHRs), whereby a patient's entire healthcare record, from infant vaccinations to laboratories for chronic disease states, is accessible to providers, regardless of their affiliation with a particular healthcare system (Tang *et al.*, 2006).

Because antimicrobial drug resistance is considered a public health threat, there needs to be better surveillance of antimicrobial prescribing and resistance trends to address outbreaks as they occur, rather than implementing damage control after the fact. Both England and Scotland have implemented national ASPs to track antimicrobial prescribing habits and resistance patterns. Through their national stewardship programs, they have been able to use existing regulatory frameworks to influence prescribing practices aimed at improving patient outcomes (Nathwani *et al.*, 2011; Ashiru-Oredope *et al.*, 2013). Scotland's national ASP specifically targeted healthcare-associated infections (HAIs), and since the implementation of the program in 2008, *Clostridium difficile* infections (CDIs) have been significantly reduced to in excess of the original goal of 30% (Nathwani *et al.*, 2012).

In the US, the CDC's NHSN is the largest national surveillance database that tracks HAIs. More than 12,000 healthcare facilities contribute data to the NHSN, and this also serves to fulfill compliance with the Centers for Medicare and Medicaid Services (CMS) infectious reporting requirements (CDC, 2014). Data on national antimicrobial resistance trends are analyzed and reported on a regular basis and can be compared by local, regional, and state antibiotic use. Frequently, national efforts to curtail inappropriate antibiotic prescribing habits backed by financial incentives to compliant healthcare facilities may be implemented to effect a positive change.

Telehealth

To date, antimicrobial stewardship has been primarily accomplished at the local hospital or health system level and little has been done on a larger scale. The impact of inappropriate antimicrobial use and the burden of resistance should be addressed on a more global level. The challenges associated with such a project would include reaching out to providers within the community and in rural areas, as well as removing boundaries between healthcare delivery systems and countries, for example. National health surveillance systems, the electronic data capture of antimicrobial consumption, and digital medical records may help in the establishment of such global programs. In addition, telehealth has become a novel way to practice medicine and provide care on a global level. Telehealth programs such as the Project ECHO® developed at the University of New Mexico have been very successful in providing treatment for complex diseases to an expanded number of patients (Arora *et al.*, 2011, 2012). Conventional medical care approaches have been incapable of delivering care to all those patients who need it, and telehealth has allowed an increased access to healthcare. This platform has also been used as a way to disseminate education on the appropriate use of antimicrobial agents. Telehealth has the potential of bringing community providers, rural providers, and even international providers together to combine efforts in order to decrease the misuse of antibiotics, promote safer and more effective ways to treat infectious diseases, and globalize antimicrobial stewardship.

Conclusions

While ASPs have been successful at enhancing patient care by optimizing antibiotic therapy, significant technological advances such as EMRs, clinical surveillance programs, and the use of mobile devices and their applications can be utilized to further improve many stewardship functions so that the entire system—patient, physician, pharmacist, and institution—benefits as a whole.

Not only do clinical decision-support programs allow limited personnel to affect more antimicrobial clinical interventions using real-time computer alerts, but most of these programs can be further customized to meet the needs of individual institutions, whether those needs include flagging broad-spectrum antibiotics for de-escalation or automated reporting to the NHSN. As many of the aforementioned studies have shown, these applications provide greater efficiencies as well as cost savings, both in patient care and in meeting regulatory compliance.

Mobile technologies have drastically improved access to drug information and clinical reference sources. They have facilitated communications and interactions amongst healthcare providers and between the providers and their patients. Mobile devices and apps are facilitating the work of ASPs by allowing easier access to patient information at the point of care, real-time documentation of interventions and recommendations, and the dissemination of educational information to healthcare professionals and patients.

Future challenges from technology support for ASPs are light in comparison with the unlimited possibilities afforded by new and upcoming advances. The possibilities are endless and only limited by our creativity and resources.

References

Antworth, A., Collins, C.D., Kunapuli, A., Klein, K., Carver, P., Gandhi, T., Washer, L., and Nagel, J.L. (2013) Impact of an antimicrobial stewardship program comprehensive care bundle on the management of candidemia. *Pharmacotherapy* 33, 137–143.

Arora, S., Thornton, K., Murata, G., Deming, P., Kalishman, S., Dion, D., Parish, B., Burke, T., Pak, W., Dunkelberg, J. *et al.* (2011) Outcomes of treatment for hepatitis C virus infection by primary care providers. *The New England Journal of Medicine* 364, 2199–2207.

Arora, S., Thornton, K., and Bradford, A. (2012) Access to care: management of hepatitis C viral infection in remote locations. *Clinical Liver Disease* 1, 87–90.

Ashiru-Oredope, D. and Hopkins, S. (2013) Antimicrobial stewardship: English Surveillance Programme for Antimicrobial Utilization and Resistance (ESPAUR). *Journal of Antimicrobial Chemotherapy* 68, 2421–2423.

Aungst, T.D. (2013) Medical applications for pharmacists using mobile devices. *Annals of Pharmacotherapy* 47, 1088–1095.

Bauer, K.A., West, J.E., Balada-Llasat, J.-M., Pancholi, P., Stevenson, K.B., and Goff, D.A. (2010) An antimicrobial stewardship program's impact. *Clinical Infectious Diseases* 51, 1074–1080.

Beyers, D. (2014) Information technology is changing [the] health-care system. *The Washington Post*, January 12, 2014. Available at: http://www.washingtonpost.com/business/capitalbusiness/information-technology-is-changing-health-care-system/2014/01/10/0b66e8c6-7250-11e3-9389-09ef9944065e_story.html (accessed 3 March 2014).

Brady, R.R.W., Verran, J., Damani, N.N., and Gibb, A.P. (2009) Review of the mobile communications devices reservoirs of nosocomial pathogens. *Journal of Hospital Infection* 71, 295–300.

Buising, K.L., Thursky, K.A., Black, J.F., MacGregor, L., Street, A.C., Kennedy, M.P., and Brown, G.V. (2008) Improving antibiotic prescribing for adults with community acquired pneumonia: does a computerized decision support system achieve more than academic detailing along? A time series analysis. *BMC Medical Informatics and Decision Making* 8:35.

Burdette, S.D., Trotman, R., and Cmar, J. (2012) Mobile infectious disease references: from the bedside to the beach. *Clinical Infectious Diseases* 55, 114–125.

Calloway, S., Akilo, H.A., and Bierman, K. (2013) Impact of a clinical decision support system on pharmacy clinical interventions, documentation efforts, and costs. *Hospital Pharmacy* 48, 744–752.

CDC (2014) National Healthcare Safety Network (NHSN). Centers for Disease Control and Prevention, Atlanta, Georgia. Available at: http://www.cdc.gov/nhsn (accessed 5 March 2014).

Comstock, J. (2013) Manhattan: 72% of physicians have tablets. Available at: http://mobihealthnews.com/21733/manhattan-72-percent-of-physicians-have-tablets/ (accessed 25 March 2014).

Dasgupta, A., Sansgiry, S.S., Sherer, J.T., Wallace, D., and Sikri, S. (2009) Application of the extended technology acceptance model in predicting pharmacists' intention to use personal digital assistants. *Journal of the American Pharmacists Association* 49, 792–796.

Dasgupta, A., Sansgiry, S.S., Sherer, J.T., Wallace, D., and Sikri, S. (2010) Pharmacists' utilization and interest in usage of personal digital assistants in their professional responsibilities. *Health Information Library Journal* 27, 37–45.

Epocrates (2013) *Epocrates 2013 Mobile Trends Report: Maximizing Multi-Screen Engagement among Clinicians*. Available at: http://www.epocrates.com/oldsite/statistics/2013%20Epocrates%20Mobile%20Trends%20Report_FINAL.pdf (accessed 13 March 2014).

Evans, R.S., Pestotnik, S.L., Classen, D.C., Clemmer, T.P., Weaver, L.K., Orme, J.F., Jr., Lloyd, J.F., and Burke, J.P. (1998) A computer-assisted management program for antibiotics and other antiinfective agents. *The New England Journal of Medicine* 338, 232–238.

FDA (2015) Mobile Medical Applications. US Food and Drug Administration, Silver Spring, Maryland. Available at: http://www.fda.gov/medicaldevices/digitalhealth/mobilemedicalapplications/default.htm (accessed 30 May 2016).

Fox, B.I., Felkey, B.G., Berger, B.A., Krueger, K.P., and Rainer, R.K., Jr. (2007) Use of personal digital assistants for documentation of pharmacists' interventions: a literature review. *American Journal of Health-System Pharmacy* 64, 1516–1525.

Goff, D.A. (2012) iPhones, iPads, and medical applications for antimicrobial stewardship. *Pharmacotherapy* 32, 657–661.

Guarascio, A.J., Slain, D., McKnight, R., Petros, K., Parker, J., Wilson, A., Defazio, C.M., and Sarwari, A.R. (2013) A matched-control evaluation of an antifungal bundle in the intensive care unit at a university teaching hospital. *International Journal of Clinical Pharmacy* 35, 145–148.

Harmon, K. (2010) The future of your medical data – How health data is becoming better connected, a la *Facebook*, changing the way patients, doctors and institutions interact. *Scientific American*, April 12, 2010. Available at: http://www.scientificamerican.com/article/future-of-medical-data/ (accessed 10 January 2014).

Heil, E.L., Daniels, L.M., Long, D.M., Rodino, K.G., Weber, D.J., and Miller, M.B. (2012) Impact of a rapid peptide nucleic acid fluorescence in situ hybridization assay on treatment of *Candida* infections. *American Journal of Health-System Pharmacy* 69, 1910–1914.

Husain, I. (2012) Top 10 free iPad medical apps for healthcare professionals. Available at: http://www.imedicalapps.com/2012/06/top-free-ipad-medical-apps/ (accessed 13 March 2014).

Isaac, T., Zheng, J., and Jha, A. (2012) Use of *UpToDate* and outcomes in US hospitals. *Journal of Hospital Medicine* 7, 85–90.

Mekhjian, H.S., Kumar, R.R., Kuehn, L., Bentley, T.D., Teater, P., Thomas, A., Payne, B., and Ahmad, A. (2002) Immediate benefits realized following implementation of physician order entry at an academic medical center. *Journal of the American Medical Informatics Association* 9, 529–539.

Mooley, A., Mangino, J.E., and Goff, D.A. (2013) Review of infectious diseases applications for iPhone/iPad and android: from pocket to patient. *Clinical Infectious Diseases* 57, 1145–1154.

Nathwani, D., Sneddon, J., Malcolm, W., Wiuff, C., Patton, A., Hirding, S., Eastaway, A., Seaton, R.A., Watson, E., Gillies, E. *et al*. (2011) Scottish Antimicrobial Prescribing Group (SAPG): development and impact of the Scottish National Antimicrobial Stewardship Programme. *International Journal of Antimicrobial Agents* 38, 16–26.

Nathwani, D., Sneddon, J., Patton, A., and Malcolm, W. (2012) Antimicrobial stewardship in Scotland: impact of a national programme. *Antimicrobial Resistance and Infection Control* 1:7.

Patel, J., Esterly, J.S., Scheetz, M.H., Bolon, M.K., and Postelnick, M.J. (2012) Effective use of a clinical decision-support system to advance antimicrobial stewardship. *American Journal of Health-System Pharmacy* 69, 1543–1544.

Pestotnik, S.L., Classen, D.C., Evans, R.S., and Burke, J.P. (1996) Implementing antibiotic practice guidelines through computer–assisted decision support: clinical and financial outcomes. *Annals of Internal Medicine* 124, 884–890.

Phua, J., See, K.C., Khalizah, H.J., Low, S.P., and Lim, T.K. (2012) Utility of the electronic information resource UpToDate for clinical decision-making at bedside rounds. *Singapore Medical Journal* 53, 116–120.

Pittenger, A.L. (2013) The use of social networking to improve the quality of interprofessional education. *American Journal of Pharmaceutical Education* 77, 1–9.

Standiford, H.C., Chan, S., Tripoli, M., Weekes, E., and Forrest, G.N. (2012) Antimicrobial stewardship at a large tertiary care academic medical center: cost analysis before, during, and after a 7-year program. *Infection Control and Hospital Epidemiology* 33, 338–345.

Tang, P., Ash, J., Bates, D., Overhage, J., and Sands, D. (2006) Personal health records: definitions, benefits, and strategies for overcoming barriers to adoption. *Journal of the American Medical Informatics Association* 13, 121–126.

Tekerekoğlu, M.S., Duman, Y., Serindağ, A., Cuğlan, S.S., Kaysadu, H., Tunc, E., and Yakupogullari, Y. (2011) Do mobile phones of patients, companions and visitors carry multidrug-resistant hospital pathogens? *American Journal of Infection Control* 39, 379–381.

Terry, K. (2013) 47% of doctors use smartphone, tablet and PC. *Information Week*, August 14, 2013. Available at: http://www.informationweek.com/mobile/47--of-doctors-use-smartphone-tablet-and-pc/d/d-id/1111170 (accessed 5 March 2014).

Thursky, K.A., Buising, K.L., Bak, N., Macgregor, L., Street, A.C., Macintyre, C.R., Presneill, J.J., Cade, J.F., and Brown, G.V. (2006) Reduction of broad-spectrum antibiotic use with computerized decision support in an intensive care unit. *International Journal for Quality in Health Care* 18, 224–231.

Wicklund, E. (2013) Epocrates study cites advantages to providers and patients in using drug reference apps 2013. mHealthNews archive, February 01, 2013. Available at: http://mobihealthnews.com/news/epocrates-study-cites-advantages-providers-and-patients-using-drug-reference-apps (accessed 30 May 2016).

Wolters Kluwer Health (2013) *Wolters Kluwer Health 2013 Physician Outlook Survey*. Wolters Kluwer Health, Alphen aan den Rijn, The Netherlands. Available at: http://www.wolterskluwerhealth.com/News/Documents/White%20Papers/Wolters%20Kluwer%20Health%20Physician%20Study%20Executive%20Summary.pdf (accessed 20 March 2014).

Yong, M.K., Buising, K.L., Cheng, A.C., and Thursky, K.A. (2010) Improved susceptibility of Gram-negative bacteria in an intensive care unit following implementation of a computerized antibiotic decision support system. *Journal of Antimicrobial Chemotherapy* 65, 1062–1069.

28 Role of Guidelines and Statistical Milestones for Antimicrobial Stewardship

Damary C. Torres*

Winthrop-University Hospital, State University of New York at Stony Brook, Mineola, New York, US and College of Pharmacy and Health Sciences, St. John's University

Introduction

Guidelines play an important role in directing practitioners toward appropriate antimicrobial stewardship programs (ASPs). By using guidelines, institutions can broadly implement the initiatives of an ASP with strong corroboration and backing. Guidelines can be disease specific, such as those from the American Thoracic Society (ATS) and the Infectious Diseases Society of America (IDSA) for the management of adults with hospital-acquired, ventilator-associated, and healthcare-associated pneumonia (ATS and IDS, 2005), and for the management of community-acquired pneumonia (CAP) (Mandell *et al.*, 2007). They can also be specifically written for ASPs, such as the IDSA/SHEA (Society for Healthcare Epidemiology of America) guidelines for developing an institutional program to enhance antimicrobial stewardship (Dellit *et al.*, 2007), and the guidelines on the core elements of hospital antibiotic stewardship programs from the US Centers for Disease Control and Prevention (CDC). Using both types of guidelines as literature support for an ASP will enhance its acceptability and the reception of the recommendations made by members of the program. The guidelines can also be adapted to individual institutions.

The IDSA and SHEA Guidelines

The guidelines for developing an institutional program to enhance antimicrobial stewardship developed by IDSA and SHEA were originally published in 2007, and although they are currently being updated, they still provide a very helpful reference to develop and improve ASPs. The guidelines begin by discussing the members of the ASP team. The essential members are an infectious diseases (ID) physician and a clinical pharmacist with infectious diseases training. Other members, including an infection control (IC) specialist and a hospital epidemiologist, can enhance the vital relationship between the ASP, infection control and the institution's pharmacy and therapeutics (P&T) committee or its equivalents. The guidelines also emphasize the need for financial and administrative support, and for compensation for the team members from the institution's leadership (Dellit *et al.*, 2007).

The remainder of the guidelines document focuses on the elements of the ASP (see Table 28.1). Some measures have not been proven to be effective and are not recommended, such as antimicrobial cycling, scheduled antimicrobial switches, and using redundant combination therapy to prevent resistance. The guidelines do recommend the use of established guidelines, or the use of evidence from the literature to develop institution-specific pathways or guidelines for antimicrobial use. The use of guidelines is associated with a decreased length of stay in hospital, a decreased rate of admission of low-risk pneumonia patients, fewer days of intravenous (IV) therapy, a decrease in antimicrobial use and cost, and decreased emergence of resistant

*E-mail: castanhd@stjohns.edu

pathogens. These advantages are seen without an increased risk of mortality, complications, or readmission (Dellit *et al.*, 2007).

The CDC Guidelines

The CDC has been advocating ASPs for many years, and recently published a resource entitled *Core Elements of Hospital Antibiotic Stewardship Programs* (CDC, 2015a). This is part of the CDC's "Get Smart for Healthcare Campaign" launched in 2009, which included improving antibiotic use as a key strategy to addressing antibiotic resistance. The CDC lists seven core elements of a hospital ASP: (i) leadership commitment to allocate sufficient human, financial and technological resources; (ii) accountability through a single leader for the ASP; (iii) drug expertise of a single pharmacist to improve antibiotic use; (iv) action by implementing at least one recommended policy or intervention; (v) the tracking or monitoring antibiotic prescribing and resistance patterns; (vi) regular reporting of antibiotic use and resistance to doctors and other staff; and (vii) educating clinicians on resistance and optimal prescribing. Interventions can be broad, pharmacy driven, and/or infection and syndrome specific. Infection and syndrome-specific interventions will frequently rely on published guidelines to target problematic areas such as the diagnosis and treatment duration of CAP, treating urinary tract infections (UTIs) but not asymptomatic bacturia, appropriate antibiotic choice and duration for skin and soft tissue infections (SSTIs), stopping empiric coverage of methicillin-resistant *Staphylococcus aureus* (MRSA) infections if the infection is not by MRSA or the organism is not methicillin resistant, stopping unnecessary antibiotics if a patient has *Clostridium difficile* infection (CDI), and the treatment of culture proven infections. The CDC has also published a checklist of these core elements to help institutions establish and improve their ASPs (CDC, 2015b). The checklist allows hospitals and healthcare systems to "assess key elements and actions to ensure optimal antibiotic prescribing and limit overuse and misuse of antibiotics in hospitals." The CDC recommends that all hospitals implement an ASP, while at the same time recognizing that all elements may not be practical or necessary in every institution.

Table 28.1. Elements of an antimicrobial stewardship program (ASP). From Dellit *et al.* (2007).

Active antimicrobial stewardship practices
Prospective audit with intervention and feedback
Formulary restriction and preauthorization requirements for specific agents
Supplemental antimicrobial stewardship strategies
Education
Guidelines and clinical pathways
Streamlining or de-escalation of therapy
Dose optimization
Conversion from parenteral (IV) to oral (PO) therapy
Computer surveillance and decision support
Microbiology laboratory testing
Monitoring of process and outcome measures
Creating a multidisciplinary ASP

The IDSA/ATS Guidelines on the Management of CAP

In a retrospective cohort study of 209 adults treated for CAP at Denver Health Medical Center (Jenkins *et al.*, 2013), clinical practice deviated from the IDSA/ATS guidelines on CAP (Mandell *et al.*, 2007) in the decision to hospitalize, in microbiologic testing, in oral antibiotic selection, and in the duration of therapy. In this study, 79% of patients (166) were treated in a medical ward and the remaining 21% (43) were treated in the intensive care unit (ICU), while according to the IDSA/ATS guidelines, 29% would have been recommended for outpatient therapy. Unnecessary blood cultures were performed in up to 81% of non-ICU cases, of which 57% of positive cultures were false-positives, which can lead to unnecessary therapy, and increase length of stay and expenditure. When switching to oral antibiotics, 66% of patients were switched to a new drug class, which is inconsistent with the IDSA/ATS guidelines, contributes to fluoroquinolone overuse, and exposes the patient to a third drug class which can cause resistance and other side effects. The guidelines also recommend 5–7 days of antibiotic therapy, but 56% of patients in the study received treatment for 10 days or longer. This study highlights the discordance between clinical practice and published guidelines, and highlights for the institution what areas to target to improve patient care. By evaluating the problems that an institution or healthcare system is having before the implementation of an ASP, one can ensure that the ASP is designed to target the problems that exist and will also help to develop the outcomes and milestones that should be reached by the ASP.

Guidelines for ASP Implementation: Targets, Outcomes, Measures

Doron and Davidson (2011) reviewed the necessary elements of an ASP, and included clinical guideline development and implementation as a vital part of a successful ASP. Guideline development can begin with nationally published guidelines and incorporate local expertise, local trends of antimicrobial resistance, and hospital targets for decreased and appropriate antibiotic use to create hospital-specific or healthcare network-specific guidelines. By evaluating areas where the institution or network is not in accordance with published guidelines, the ASP can be individualized to target these areas for improvement. Studies evaluating the implementation of ventilator-associated pneumonia treatment guidelines have shown improved rates of appropriate initial therapy. The development and implementation of guidelines can also include non-infectious disease specialists, such as emergency department clinicians, who are often on the front line of antibiotic prescribing.

ASPs have shown annual savings of \$200,000–\$900,000. Developing a business plan that includes baseline expenditures, including personnel costs, and proposed interventions with their presumed cost savings. Targeting outcome data to follow, including defined daily doses (DDDs) or days of therapy (DOT), can help to track the benefits of an ASP, including cost savings. The defined daily dose, or DDD, is defined by the World Health Organization (WHO) as "the assumed average maintenance dose per day for a drug used for its main indication in adults" (WHOCC, 2009), and the DDDs used are calculated from the total number of grams of antimicrobial agent used divided by the number of grams in an average daily dose (DDD) as defined by WHO (WHOCC, 2015). The DOT is defined as the administration of a single agent on a given day regardless of the number of doses or strength. DDD does not account for changes in dosing based on organ function, age, or pharmacokinetic dosing, but facilitates in benchmarking institutions. DOT does not reflect actual doses and can underrepresent antimicrobials that are dosed multiple times a day, but can be more helpful in comparing the use of different antimicrobial classes within an institution (Doron and Davidson, 2011).

Statistical milestones can help determine whether a program is meeting its projected outcomes and demonstrate benefits to the institution. If standardized, they can also be used to compare different institutions in the same or different healthcare systems.

The IDSA guidelines discuss the different types of measures to be established. These include process and outcome goals and measures. A process goal can be to change the use of specific antimicrobials or classes, and the process measure would be to compare preimplementation levels with levels after the ASP is implemented to determine the level of success and identify areas for improvement. These are associated with an outcome goal of decreasing or preventing resistance or other adverse outcomes. The outcome measure would evaluate the level of antimicrobial resistance, adverse events and cost, and would include the measurement of unintended outcomes, such as increased use of other, non-targeted antimicrobials and rates of CDI (Dellit *et al.*, 2007).

The CDC's resources, including its *Checklist for Core Elements of Hospital Antibiotic Stewardship* (CDC, 2015b), can help to establish targets and measures for an ASP. The CDC campaign resource also discusses antibiotic use measures, including evaluating whether prescribers accurately apply diagnostic criteria, obtain cultures, and apply other relevant tests prior to starting therapy, and whether they use recommended agents, document the indication and planned duration of therapy, and modify the antimicrobials used based on microbiologic findings. Standardized audit tools, such as those found on the CDC website, can help to facilitate data collection. It is also important to document the acceptance or refusal of interventions. Antibiotic use measures using DDD or DOT have been described above. Through its National Healthcare Safety Network (NHSN), the CDC has also developed an "Antibiotic Use (AU) Option," through which monthly DOT data and reports can be automatically collected for institutions that have information systems that can submit electronic medication administration records (eMARs) and/or barcoding medication records using an HL7 clinical document architecture. To participate in the Option, institutions can work with the information technology department and software providers to enable their systems to generate standard formatted files for the NHSN. As more institutions participate, benchmarks for antibiotic use can be created. The AU Option is part of the CDC's Antibiotic Use and Resistance (AUR) module (CDC, 2016).

The CDC also recommends tracking clinical outcome measures such as rates of CDI, reductions in antibiotic resistance measured by tracking pathogens cultured after admission and tracking the incidence of super-infections, and annual drug and overall cost savings. Cost savings can help to increase support for the ASP, but after an initial period of significant cost savings, these will stabilize. Monitoring the rate of rise of the costs of antibiotic use prior to implementation of the ASP can help to advocate for continued support. Education is also an important part of an ASP and can help to motivate prescribers to continue to improve prescribing patterns and to continue to support the ASP (CDC, 2015a).

A recent Cochrane review evaluated interventions to improve antibiotic prescribing practices for hospital inpatients from the perspective of antibiotic stewardship, and included studies up to 2006 (Davey *et al.*, 2013). The Cochrane review categorized hospital-based ASP interventions in several ways, and various outcome measures were used to assess their effectiveness. However, few of the 89 studies included reported patient outcomes.

Later, and using a different approach, Wagner *et al.* (2014) performed a systematic review of 37 ASP studies in inpatient hospital settings in the US published up to 2013 to evaluate their effects on patient, prescribing and microbial outcomes. This later review summarizes evidence on ASPs either not reviewed in, or published since, the last Cochrane review. The ASP strategies identified included audit and feedback programs (38%), formulary restriction or preauthorization (14%), guideline implementation with feedback (11%), guideline implementation with no feedback (11%), computerized decision support (11%), and protocol or policy implementation (11%). Of the 29 studies in which patient outcomes were reported, 60% found no difference in outcomes of mortality, length of stay, readmission, or incidence of CDI. Of the 31 studies that included antimicrobial outcomes, improvement in either antimicrobial use, selection, timing, or duration was seen in 74% of studies. Only nine studies measured microbial outcomes, and of these, 78% saw improvements in either institutional resistance or study population resistance. No studies acknowledged any barriers to implementing an ASP. Most study periods were short, and offered little advice for program sustainability. Additionally, no programs compared different interventions or elements, so it is not possible to draw conclusions about which is the best approach. Most programs were in large, university-affiliated institutions, which may limit the generalizability of the findings to other settings. The studies showed ASPs can decrease antimicrobial use without apparent harm, but that further benefits will require large, multicenter trials with specific outcomes to be measured.

A descriptive cost analysis of an ASP at the University of Maryland Medical Center over 7 years showed significant cost savings (Standiford *et al.*, 2012). The authors compared utilization costs before fiscal year (FY) 2001, during the 7 year program (FYs 2002–2008), and after the program was terminated (FYs 2009–2010) to allow the resources to increase infectious disease consults as an alternative ASP. During the 7 year ASP, the team provided active, real-time review of antimicrobial orders for preselected restricted antimicrobials and provided active intervention if needed. A computer decision-support system was a vital part of the program. The annual cost of antimicrobials before the ASP was $7,774,588. In the first year of the ASP, the cost decreased by $35,974. By the end of the program, the annual cost was $4,776,663, a reduction of $2,997,925 over 8 years. After the program was terminated, annual cost increased once again to $5,869,764 in the first year and to $6,742,948 in the second year. Much of the cost savings over the period reported came from a reduction of antifungal use in the cancer center. Antifungal costs remained low after program discontinuation, but antibacterial use increased by $1,568,509 in the 2 years post ASP. There were no significant differences in the quality indicators measured, including length of stay, readmissions and mortality, although there was a trend toward shorter length of stay and lower mortality, which continued after the ASP was terminated. The ASP therefore showed significant cost savings without adverse outcomes, and based on this information, it was restarted with a small modification. Now, instead of a preauthorization requirement for restricted antimicrobials, an automatic infectious disease consult is generated. By proving the benefit of the program, the authors were able to obtain the necessary funding for the program.

Bevilacqua *et al.* (2011) reported outcomes and milestones for a 15 year program at the Nancy Teaching Hospital in France. The ASP was first established in the 1990s, and then reinforced starting in 2006. The ASP consisted of prescription guidelines developed by itself, a specific prescription form for antibiotics, and an infectious disease team to target orders that did not conform to the guidelines or were for broad-spectrum, costly or IV fluoroquinolone antibiotics. This observational study showed that 80% of the infectious disease team's recommendations were accepted. A drop of 34% in overall annual antibiotic cost was observed over the 3 years from 2005 to 2008, which lowered antimicrobial therapy costs by €1,308,902 (approximately $1,500,000). Antimicrobial consumption in DDD/1000 patient days (PDs) decreased by 10% and included declines in all classes. IV fluoroquinolone use decreased by 15% over the 3 years. This study showed the sustainability and opportunities for improvement in long-standing programs. The goal of the reinforcing adjustments was to improve antibiotic stewardship and decrease the use of extended broad-spectrum antimicrobials, expensive antimicrobials, and IV antibiotics, while decreasing costs. These are all significant and valid markers and milestones for an ASP.

Two studies have evaluated the impact of an ASP on specific milestones, one by Cisneros *et al.* (2014), and the other by Talpaert *et al.* (2011). Cisneros *et al.* (2014) evaluated the impact of an ASP in a tertiary teaching hospital center in Seville, Spain. The ASP used locally developed guidelines and clinical reviews (conducted by counselling interviews) of random orders on the appropriateness of antimicrobial use and consumption. Prior to the ASP, which started in January 2011, 53% of antimicrobial prescriptions were inappropriate. After the first year, only 26.4% of prescriptions were inappropriate based on clinical review—a statistically significant decrease (relative risk (RR) 0.38; 95% confidence interval (CI), 0.23–0.43; $P < 0.001$). Antimicrobial consumption decreased by 42%, from 1150 DDDs/1000 PDs to 852 DDDs/1000 PDs.

Talpaert *et al.* (2011) evaluated the impact of an ASP in adult medical and surgical wards in an acute general hospital in London. The ASP consisted of locally developed guidelines and the development of an ASP team to review high-risk and perform ward rounds on the incidence of CDI. Education through teaching sessions, personal discussions, pocket cards, and posters for the wards was also a vital component. A comparison of antibiotic use and CDI incidence for 12 months before and after implementation of the ASP (in April 2006) revealed a 58.5% drop in fluoroquinolone use and a 45.8% drop in cephalosporin use. A corresponding significant increase in the use of low-risk antibiotics was noted during the intervention period. A significant decrease in CDI rates associated with the ASP was seen (incidence rate ratio 0.34; 95% CI, 0.20–0.58; $P < 0.0001$). The study did not show a decrease in DDD. Team recommendations were accepted 98.7% of the time. A subanalysis of elderly patients with UTIs or lower respiratory tract infections showed a significant decrease in mortality in the 80 days after ASP implementation and compared with the 80 days before (3.6 vs. 5%, respectively; odds ratio (OR), 0.49; $P = 0.004$).

Upon examining the results of the above two studies, one can see that establishing goals, outcomes and milestones for an ASP program can influence the results obtained. As the second study did not include a decrease in DDD after ASP implementation, it was neither a focus for the ASP team nor a significant outcome. Accordingly, it is important for institutions to determine which outcomes to focus on and design the ASP to achieve those milestones.

Several studies have evaluated the role and intensity of ASP measures on specific outcomes. Dumartin *et al.* (2011) performed a 5 year longitudinal survey of hospitals in France to study the relationship between ASPs and antibiotic use. Surveillance methods have been stable since 2005 in the study region. The survey results showed that hospitals with more ASP measures had stable or decreasing antibiotic use, including fluoroquinolone use. Hospitals with the two particular ASP measures had a significant and independently-associated decrease in antibiotic use on logistic regression analysis. These measures were carrying out practice audits (OR, 6.1; 95% CI, 1.2–32.4; $P = 0.03$), and having an antibiotic advisor who spent a minimum of 0.5 days a week for at least 2 years on this task (OR, 5.7; 95% CI, 1.4–22.9; $P = 0.01$).

Pope *et al.* (2009) also performed a survey, this time evaluating how many institutions have established the IDSA/SHEA guidelines on ASPs. A total of 3500 surveys were sent out to hospital practitioners in various hospital settings and 357 were returned (rate of 10%). The most frequent ASP strategy used was "prospective monitoring of prescribing and appropriateness after the first dose of a targeted antibiotic" (66%). Local antibiograms and tracking of resistance patterns were commonly used to guide the ASP in 95% and 76% of responses, respectively. Common outcome measures monitored were direct drug costs (58%), pharmacy savings (43%), length of stay (26%), and mortality (26%). This study was hampered by the low response rate, but it is concerning that only a quarter of respondents measured the vital clinical outcomes of mortality and length of stay for their ASPs. It is sometimes difficult to correlate changes in these outcomes with implementation of an ASP, but they are vital milestones. However, as this survey was sent out in 2008, only one year after the guidelines were published, it may not reflect the most current practice in 2015.

Mehta *et al.* (2014) compared the impact of two types of ASP measures, one before and one after a change in ASP strategy, in a tertiary care academic medical center in Philadelphia, Pennsylvania. The first measure was formulary restriction with prior authorization, which required approval from an ASP member for selected agents over a period of 2 years; this was considered the "preintervention period." The second measure, which was considered the "postintervention period," and was in place for the 2 subsequent years, was prospective audit with feedback to prescribers; this involved review after the order was written to recommend changes in the agent, in dosing, or in the duration of therapy. The outcomes assessed were antimicrobial use measured in DOT and length of stay. In the postintervention period, antimicrobial consumption increased for all broad-spectrum Gram-negative antibiotics, with an increase of 0.8 DOT/1000 PDs a month over the preintervention rate, in which it had been decreasing by 4 DOT/1000 PDs per month, giving an increase of 4.8 DOT/1000 PDs a month ($P = 0.001$). The largest increase in use was seen for piperacillin–tazobactam and cefepime—an increase to 1.35 DOT/1000 PDs a month postintervention vs. a decrease of 1.86 DOT/1000 PDs a month preintervention (an increase of 3.21 DOT/1000 PDs a month; $P = 0.003$). No significant increase was seen in the use of the other targeted broad-spectrum Gram-negative agents (aztreonam, ceftazidime, levofloxacin, and meropenem) in the postintervention period. However, the overall use of all systemic antimicrobial agents increased significantly from −9.75 to −0.10 DOT/1000 PDs a month (an increase of 9.65 DOT/1000 PDs a month; $P < 0.001$). Likewise, the use of non-audited antimicrobials increased significantly from −7.43 preintervention to −1.87 DOT/1000 PDs a month postintervention (an increase of 5.56 DOT/1000 PDs a month; $P < 0.001$). Length of stay for patients receiving [any] antimicrobials had been decreasing at a rate of −1.57 days/1000 PDs a month preintervention, but increased in the postintervention period by 1.94 days/1000 PDs a month ($P = 0.16$). The results of this study demonstrate that prospective audit was superior to retrospective review for the measured milestones of antimicrobial consumption and length of stay, although further studies are needed to confirm these results.

Conclusion

In conclusion, the use of guidelines can help to direct not only the operations of an ASP, but also its goals. Adapting national standards to local problems and establishing the milestones for a program to reach can enhance its success rate. Further research is needed to evaluate the role of ASPs in community, nonteaching hospitals, in long-term care and rehabilitation facilities, and in the ambulatory setting.

References

ATS (American Thoracic Society) and IDSA (Infectious Diseases Society of America) (2005) Guidelines for the management of adults with hospital-acquired, ventilator-associated, and healthcare-associated pneumonia. *American Journal of Respiratory and Critical Care Medicine* 171, 388–416.

Bevilacqua, S., Demore, B., Boschetti, E., Doco-Lecompte, T., May, I., May, T., Rabaud, C., and Thilly, N. (2011) 15 years of antibiotic stewardship policy in the Nancy Teaching Hospital. *Médecine et Maladies Infectieuses* 41, 532–539.

CDC (2015a) *Core Elements of Hospital Antibiotic Stewardship Programs*. Centers for Disease Control and Prevention, Atlanta, Georgia. Available at: http://www.cdc.gov/getsmart/healthcare/implementation/core-elements.html (accessed 3 January 2015).

CDC (2015b) *Checklist for Core Elements of Hospital Antibiotic Stewardship Programs*. Centers for Disease Control and Prevention, Atlanta, Georgia. Available at: http://www.cdc.gov/getsmart/healthcare/implementation/checklist.html (accessed 3 January 2015).

CDC (2016) *Antimicrobial Use and Resistance (AUR) Module*. Centers for Disease Control and Prevention, Atlanta, Georgia. Available at: http://www.cdc.gov/nhsn/PDFs/pscManual/11pscAURcurrent.pdf (accessed 27 April 2016).

Cisneros, J.M., Neth, O., Gil-Navarro, M.V., Lepe, J.A., Jiménez-Parrilla, F., Cordero, E., Rodríguez-Hernández, M.J., Amaya-Villar, R., Cano, J. Gutiérrez-Pizarraya, A. *et al.* (2014) Global impact of an educational antimicrobial stewardship programme on prescribing practice in a tertiary hospital centre. *Clinical Microbiology and Infection* 20, 82–88.

Davey, P., Brown, E., Charani, E., Fenelon, L., Gould, I.M., Holmes, A., Ramsay, C.R., Wiffen, P.J., and Wilcox, M. (2013) Interventions to improve antibiotic prescribing practices for hospital inpatients. *Cochrane Database of Systematic Reviews* 4:CD003543.

Dellit, T.H., Owens, R.C., Mcowan, J.E., Jr., Gerding, D.N., Weinstein, R.A., Burke, J.P., Huskins, W.C., Paterson, D.L., Fishman, N.O., Carpenter, C.F. *et al.* (2007) The Infectious Diseases Society of America and the Society for Healthcare Epidemiology of America guidelines for developing and institutional program to enhance antimicrobial stewardship. *Clinical Infectious Diseases* 44, 159–177.

Doron, S. and Davidson, L.E. (2011) Antimicrobial stewardship. *Mayo Clinic Proceedings* 86, 1113–1123.

Dumartin, C., Rogues, A., Amadéo, B., Péfau, M., Venier, A.G., Parneix, P., and Maurain, C. (2011) Antibiotic usage in south-western French hospitals: trends and association with antibiotic stewardship practices. *Journal of Antimicrobial Chemotherapy* 66, 1631–1637.

Jenkins, T.C., Stella, S.A., Cervantes, L., Knepper, B.C., Sabel, A.L., Price, C.S., Shockley, L., Hanley, M.E., Mehler, P.S., and Burman, W.J. (2013) Targets for antibiotic and health care resource stewardship in inpatient community-acquired pneumonia: a comparison of management practices with National Guideline Recommendations. *Infection* 41, 135–144.

Mandell, L.A., Wunderink, R.G., Anzueto, A., Bartlett, J.G., Campbell, G.D., Dean, N.C., Dowell, S.F., File, T.M., Jr., Musher, D.M., Niederman, M.S. *et al.* (2007) Infectious Diseases Society of America/American Thoracic Society consensus guidelines on the management of community-acquired pneumonia in adults. *Clinical Infectious Diseases* 44(Suppl 2), S27–S72.

Mehta, J.M., Haynes, K., Wileyto, E.P., Gerber, J.S., Timko, D.R., Morgan, S.C., Binkley, S., Fishman, N.O., Lautenbach, E., and Zaoutis, T. (2014) Comparison of prior authorization and prospective audit with feedback for antimicrobial stewardship. *Infection Control and Hospital Epidemiology* 35, 1092–1099.

Pope, S.D., Dellit, T.H., Owens, R.C. and Hooton, T.M. (2009) Results of survey on implementation of Infectious Diseases Society of America and Society for Healthcare Epidemiology of America guidelines for developing an institutional program to enhance antimicrobial stewardship. *Infection Control and Hospital Epidemiology* 30, 97–98.

Standiford, H.C., Chan, S., Tripole, M., Weekes, E., and Forrest, G.N. (2012) Antimicrobial stewardship in a large tertiary care academic medical center: cost analysis before, during, and after a 7-year program. *Infection Control and Hospital Epidemiology* 33, 338–345.

Talpaert, M.J., Rao, G.G., Cooper, B.S., and Wade, P. (2011) Impact of guidelines and enhanced antimicrobial stewardship on reducing broad-spectrum antibiotic usage and its effect on incidence of *Clostridium difficile* infection. *Journal of Antimicrobial Chemotherapy* 66, 2168–2174.

Wagner, B., Filice, G.A., Drekonja, D., Greer, N., MacDonald, R., Rutks, I., Butler, M., and Wilt, T.J. (2014) Antimicrobial stewardship programs in inpatient hospital settings: a systematic review. *Infection Control and Hospital Epidemiology* 35, 1209–1228.

WHOCC (2009) DDD: Definition and General Considerations. WHO Collaborating Centre for Drug Statistics Methodology, Norwegian Institute of Public Health, Oslo. Available at: http://www.whocc.no/ddd/definition_and_general_considera/ (accessed 26 May 2016).

WHOCC (2015) ATC/DDD Index 2015. WHO Collaborating Centre for Drug Statistics Methodology, Norwegian Institute of Public Health, Oslo. Available at: http://www.whocc.no/atc_ddd_index/ (accessed 20 January 2015).

29 Economic Considerations of Antimicrobial Stewardship Programs

PANAYIOTIS D. ZIAKAS*

The Warren Alpert School of Medicine of Brown University and Rhode Island Hospital, Providence, Rhode Island, US

Introduction

The cost of pathogen resistance and the role of antimicrobial stewardship programs (ASPs)

The need for effective control of antimicrobial agents was recognized over 50 years ago by H. Reimann, who summarized the importance of antimicrobial stewardship programs (ASPs) in one single sentence: "Much needless expense, untoward effect, harm, and disappointment can be prevented by better judgment in the use of antimicrobials for therapy and prophylaxis." (Reimann, 1961). Antimicrobial stewardship refers to the responsible use of antimicrobials in order to improve clinical outcomes, limit adverse events and prevent unintentional consequences such as *Clostridium difficile* infections and the emergence of antibiotic resistance. ASPs are essential parts of quality care, patient safety, and public health, and antimicrobial stewardship is regarded by members of the Society for Healthcare Epidemiology of America (SHEA) as being among the most important scientific questions in healthcare epidemiology, second only to the problem of multidrug-resistant Gram-negative bacteria (Research Committee of SHEA and Sinaii, 2010).

Antimicrobial agents are unique agents inasmuch as their use or misuse in a patient can compromise their efficacy for another patient, and they are prescribed by physicians with different clinical disciplines—medical microbiology and infectious diseases represent just two of more than 50 medical specialties. In practice, antibiotics are universally prescribed by all doctors as opposed to other drugs (chemotherapy agents, antipsychotics, etc.), for which prescribing is limited to a single specialty (Charani *et al.*, 2010). The inappropriate selection of broad-spectrum antibiotics and longer administration than necessary are common mistakes; they aim to ensure effective and rapid therapy but, in fact, they cover gaps of knowledge and uncertainty of outcome, hence their use as "drugs of fear" (Kunin *et al.*, 1973). ASPs help healthcare professionals to select and administer the appropriate agent for a given patient with a suspected or proven infection (Goff, 2011).

ASP interventions have an impact on both healthcare and the economy. Antimicrobials can account for up to 30% of nosocomial budgets. Only in 1985, $4 billion was spent on medications in US hospitals, $1.4 billion of which was paid out for antimicrobial medications (Avorn *et al.*, 1988). The exponential rise in the use of antibiotics a decade later has led to annual costs exceeding $7 billion for antimicrobials, and more than $4 billion for drug-resistant bacteria (Office of Technology Assessment US Congress, 1995). The gravity and magnitude of antimicrobial resistance and the necessity for ASPs led to published recommendations on the optimal use of antimicrobial agents by the Infectious Disease Society of America (IDSA) and SHEA more than two decades ago (McGowan, 1983; Marr *et al.*, 1988). The current IDSA/SHEA guidelines have examined the economic impact of ASPs (Dellit *et al.*, 2007) using drug cost reduction as the primary quantitative measure of economic success, based on the premise

*E-mail: Panagiotis_Ziakas@brown.edu

that up to 50% of antimicrobial consumption may be unjustified (John and Fishman, 1997). The guidelines state that ASPs result in annual cost savings ranging between $200,000 and $900,000 as revealed by six pertinent studies on the topic (Schentag *et al.*, 1993; Ansari *et al.*, 2003; Carling *et al.*, 2003; LaRocco, 2003; Lutters *et al.*, 2004; Ruttimann *et al.*, 2004).

Given that a proportion of expenditures for antimicrobials results from unnecessary use, it is not surprising that ASPs may be self-sustaining (Dellit *et al.*, 2007). Shifting from intravenous (IV) to oral (PO) treatment, adjusting dosing for renal insufficiency, de-escalating from broad-spectrum to pathogen-specific therapy, discontinuing redundant therapies and eliminating their use when the patient is not infected are measures with direct and measurable reduction in costs (Owens, 2008). Despite the recognized benefits of ASPs, a survey across 357 practitioners completed in 2008 found that only 48% of responders had an active ASP in their institutions, and that 26% were considering or developing ASPs. Furthermore, only one quarter of responders from institutions with ASPs in place reported that health-related outcomes, including length of stay and mortality, were measured. As obstacles to establishing an ASP in the US, a survey by Pope *et al.* (2009) showed that economic considerations ranked second (36%) to personnel shortages (55%). A similar survey in German hospitals showed that >80% of 355 intensive care units (ICUs) that responded had some kind of restriction policy, including control of antibiotic prescription, access to resistance data, and reports from the pharmacy on antibiotic use and costs. Notably, policies differed in their surveillance of pathogen resistance, antimicrobial use, and the employment of infectious disease specialists; non-profit and public hospitals, as well as hospitals with their own laboratory facilities, were twice as likely to participate in cross-institutional networks for bacterial resistance than were private hospitals and hospitals with external laboratories (Maechler *et al.*, 2013).

Economic Measures of ASP Performance

The applicability of economics to antimicrobial resistance and ASPs extends far beyond the obvious association of resistance with healthcare and societal costs; it is also influenced by factors such as the behavior of healthcare providers, distributors, patients, and consumers (Eggleston *et al.*, 2010). ASPs that focus solely on pharmaceutical cost reductions impartially address their economic impact; cost-efficiency strategies accompanied by appropriate outcome data are warranted to support the viability of such programs (Stevenson *et al.*, 2012). Unfortunately, as concerns most ASPs, the measuring of antimicrobial usage and cost before and after the initiation of the program remains their sole justification, as these are the easiest outcomes to calculate (Drew *et al.*, 2009). Other costs associated with program implementation—suboptimal treatment, the emergence of resistance, and drug-related adverse events—cannot be dismissed and should also be calculated in terms of costs and savings. For example, a restrictive policy in an antibiotic agent may decrease its consumption and pharmacy cost at the expense of emerging resistance to another class of agent (the "squeezing the balloon" effect) (Burke, 1998), and ASPs should also monitor for these outcomes (Goff, 2011). A notable case was the reported 44% reduction in ceftazidime-resistant *Klebsiella* infections due to 80% restriction of cephalosporin use, which was accompanied by an increase of 69% in imipenem-resistant *Pseudomonas aeruginosa* infections (Rahal *et al.*, 1998). Moreover, savings from the restriction of use of an agent account for less than 10% of patient hospitalization; they may be high during the first years of an ASP but eventually they reach a nadir as they are necessary drugs, and the goal is not to eliminate them, but to justify their use (Paladino, 2004).

The initial goal of ASPs was to rationalize the allocation of resources for the reason that antimicrobials constitute a significant proportion of healthcare budgets, and this has led many hospitals to adopt such strategies owing to their potential for cost savings. However, it should be noted that this holds true only when focusing solely on pharmacy drug costs; when attention is shifted to the costs of treating an infection, the drug costs are comparatively small relative to the cost of care, as seen with infections caused by drug-resistant pathogens vs. susceptible organisms (Cosgrove *et al.*, 2002; Cosgrove, 2006). With the emergence of multidrug-resistant bacteria and the shortage of new antimicrobials, the primary target of ASPs has moved to preserve their utility (averting the "post-antibiotic era") and cost dominated as a goal. Although it is plausible to think that ASPs will reduce resistance, it may not be all that evident because the development of resistance can progress rapidly but declines slowly. Besides, improving hospital resistance and preserving the rational use of antimicrobials is only one aspect of the problem, given that inappropriate use in the community is another determinant of resistance. A 10 year follow up on the Swedish effort for outpatient control of antibiotics, which started in 1994, led to a significant decline in their use and rates of occurrence of penicillin-resistant *Streptococcus pneumoniae*, highlighting the need of such successful efforts for a multidisciplinary approach and time (Mölstad *et al.*, 2008). That is why ASPs can pragmatically focus only on improving patient outcomes while restricting inappropriate use and costs (Dellit *et al.*, 2007). A similar effort in Canada to restrain outpatient use of antibiotics took 15 years to display a measurable benefit in antibiotic consumption, with the most striking effect being recorded for broad-spectrum penicillins. Cost was constrained by $13 million a year for the decade 2000–2010 (Finley *et al.*, 2013).

It becomes apparent that morbidity and cost are considered to be the most sensitive estimates to quantifying the impact of antimicrobial resistance and efficacy of ASPs. Economic considerations include hospital costs (related to inpatient morbidity and mortality), third-party payer costs (for inpatient and outpatient healthcare), costs to the patient (related to loss of work, functioning and performance status decline, lack of antimicrobial compounds) and costs to society itself (the total healthcare costs of resistance, including the lack of antimicrobial classes) (Cosgrove, 2006). The societal cost may outweigh all other parameters—it was estimated as billions of dollars almost two decades ago. The decisions that are made may depend on which of these views is considered or whose view is the most influential. Hospitals measure both pharmaceutical costs and labor costs when providing healthcare; third-party payers focus on the cost of reimbursing hospital bills. Patients and society count mortality, quality of life, and loss of productivity as significant outcomes (Roberts *et al.*, 2009).

Objective economic analysis relies on both cost and healthcare outcomes, and includes cost–benefit analysis, cost-effectiveness analysis, and cost–utility analysis. Cost–benefit analysis deals only with the monetary impact of competing strategies, and neglects healthcare outcomes (Zilberberg and Shorr, 2010). This means that a strategy involving less cost prevails over a strategy with higher cost. Cost-effectiveness analysis evaluates intervention strategies by examining monetary cost relative to a valid measure of effect (improved survival, risk reduction, hospital days saved) (Perencevich *et al.*, 2007). A similar approach, defined as cost–utility analysis, defines quality of life as a measure of effectiveness and uses quality-adjusted life-years (QALY) (Sassi, 2006). Cost-effectiveness and cost–utility analysis should be considered the appropriate models to address the impact of ASPs (Zilberberg and Shorr, 2010; Stevenson *et al.*, 2012).

ASPs Usually Lack a State-of-the-art Economic Analysis

The ideal ASP would lead to changes in antibiotic consumption (both quantitative and qualitative), declines in infection rates, less resistance and improved survival. In simple terms, "DRIM" (Drugs, Resistance, Infections, Mortality) is the key element of performance, along with economic impact. I recently reviewed 40 articles published from 2011 to September 2013 that clearly refer to ASPs, and in these both a lack of comprehensive economic analysis and a relative lack of healthcare outcome data was evident. A summary of the studies is given in Table 29.1.

Almost all (88%) of the studies reviewed had changes in antimicrobial use as their major goal, and reported variable success in reducing consumption, shifting to other classes, or reducing the length of administration. The impact on infection is reported in less than 50% of studies, even though there are data to prove a decline, particularly in targeted infections (such as MRSA and *C. difficile* infections), or shortening the length of nosocomial stay. Mortality data are also reported in <50% of studies, which uniformly suggest that ASPs lack a measurable impact on mortality. Microbial resistance is the major initiative for deploying ASPs, but only 15% of studies reviewed reported pertinent data, and only one claimed an improvement in microbial susceptibility. Economic analysis is limited to cost–benefit data (estimated in 45% of studies), while cost-effectiveness data are absent. Across the studies that present data on an economic impact, cost savings from antibiotic constraints are the preferred outcome (25% of all studies), followed by patient and hospital savings; societal costs are not evaluated.

Table 29.1. A summary of recent (2011–September 2013) studies of antimicrobial stewardship programs, with brief details of the settings in which they were done, and their aims, health-related outcomes and economic impacts.

	Health-related aims and outcomes[a]		
Study	Setting; time period; and aim	Outcome	Economic impact
Aldeyab *et al.*, 2013	Hospital and community; 2008–2010; to reduce MRSA	Reduction of MRSA incidence	Not reported
Avdic *et al.*, 2012	Hospital; 4 months; treatment of community acquired pneumonia (CAP)	Fewer antibiotic days	Not reported
Bevilacqua *et al.*, 2011	Hospital; 2005–2008; to control antimicrobial use	Decline from 701 to 634 DDDs/1000 PDs	€1,308,902 savings in drug costs
Boyce *et al.*, 2013	Hospital; 2009–2011; to discontinue vancomycin for suspected MRSA pneumonia	Vancomycin discontinued for MRSA-negative nasal and throat cultures	Not reported
Cairns *et al.*, 2013	Hospital; 18 months; to control antimicrobial use	Reduction from 1022 DDDs/1000 OBDs to 937 in ICU	Not reported
Chalfine *et al.*, 2012	Hospital; 2000–2009; to reduce hospital-acquired MRSA	Reduction in antibiotic use from 812.4 to 561.8 DDDs/1000 PDs; reduction in MRSA colonization and infection	Not reported
Chan *et al.*, 2011	Hospital; 2003–2009; to control antimicrobial use	Cephalosporin, fluoroquinolone, glycopeptide use decreased, carbapenem use increased; mortality unaffected; effect on infections variable, but MRSA decreased	Not reported
Cisneros *et al.*, 2013	Hospital;12 months; to limit antimicrobial use	Reduction from 1150 to 852 DDDs/1000 OBDs	€1,012,560 savings annually in drug costs
Dancer *et al.*, 2013	Hospital; 16 months; ban on use of ceftriaxone and ciprofloxacin to reduce MRSA; hospital-acquired *Clostridium difficile*, ESBL-producing coliforms	Reduction of 95% in ceftrizaxone use, 72.5% ciprofloxacin use, incidence rates reduced	Not reported
Dubrovskaya *et al.*, 2012	Hospital; 4 months; to reduce intra-abdominal infections	Decline in use of cefoxitin, cefazolin, ciprofloxacin; no difference in LOS	Not reported
Dumartin *et al.*, 2011	Hospitals; 2005–2009; survey of ASP implementation and antibiotic use in 74 centers	Stability in antibiotic use	Not reported
Guarascio *et al.*, 2013	ICU; 6 months; to limit echinocandin use	Fewer days of caspofungin therapy	$1013 savings per patient
Heil *et al.*, 2012	Hospital; 10 months; use of rapid test in treatment of *Candida* infections	Frequent switching from fluconazole to echinocandin; no difference in LOS or mortality	$415 savings per patient
Jump *et al.*, 2012	Long-term care; 18 months; to reduce antimicrobial use, *C. difficile* infections	Reduction from 175 to 122 DOT/1000 PDs; *C. difficile* decreased	Not reported
Katsios *et al.*, 2012	ICU; 2 months; to control antimicrobial use	Reduction from 3592 to 3100 DDDs/1000 PDs	$16,744 total savings in drug costs
Kim *et al.*, 2013	Hospital; 2006–2011; to reduce MRSA bacteremia	Decline from 750 to 652 DDD/1000 PDs; reduced bacteremia	Not reported

Continued

Table 29.1. Continued.

Study	Health-related aims and outcomes[a]		Economic impact
	Setting; time period; and aim	Outcome	
Kullar *et al.*, 2013	Level I trauma; 2008–2011; to reduce MRSA bacteremia by daptomycin switch	Lower length of stay and inpatient care; no difference in mortality	No gain; $19,755 vs. $18,385 for vancomycin cost
Liew *et al.*, 2012a	Hospital; 2008–2010; to reduce antimicrobial use	Reduction in antibiotic use; shorter LOS; fewer readmissions; same mortality	Not reported
Liew *et al.*, 2012b	Hospital; 12 months; to test utility of procalcitonin use in malignancy	Reduction in carbapenem use; no difference in mortality	Not reported
Lima *et al.*, 2011	Hospital; 2006; control of carbapenem use	Decline in carbapenem use; improved *Pseudomonas aeruginosa* and *Klebsiella pneumoniae* susceptibilities to other antibiotics; no differences in mortality	Not reported
Lopez-Medrano *et al.*, 2013	Hospital; 12 months; to limit antifungal use	Less voriconazole and caspofungin use, more amphotericin B use; no difference in infection and mortality	$370,681 hospital savings
Magedanz *et al.*, 2012	Hospital; 2005–2008; to promote rational use of antimicrobials	Reduction of 49 to 37 DDDs/100 PDs monthly; reduction of carbapenem-resistant *P. aeruginosa*; increase in ceftazidime resistance	~$21,000 monthly savings from antibiotics
Malani *et al.*, 2013	Hospital; 12 months; to limit antimicrobial use	Decrease in antimicrobial use from 215.7 to 160.8 DDDs/1000 PDs; fewer *C. difficile* infections	$164,553 savings on antibiotics
Mondain *et al.*, 2013	Hospital; 2002–2011; to control antimicrobial use	Stable use at ~400 DDDs/1000 PDs	Not reported
Niwa *et al.*, 2012	Hospital; 2009–2011; to limit intravenous (IV) antibiotics	Reduction in IV antibiotic use from 210 to 193 DDDs/1000 PDs; MRSA reduction	$3,915,913 annual hospital savings; $247,000 savings from antibiotics
Pate *et al.*, 2012	Hospital; 5 months; to control antimicrobial use	Decline from 29 to 20 DDDs/1000 PDs; shorter LOS	$159,580 savings from drug costs
Perez *et al.*, 2013	Hospital; 4 months; rapid identification of Gram-negative bloodstream infections	Shorter length of stay	$19,547 hospital savings per patient
Ramsamy *et al.*, 2013	Trauma unit; 12 months; to reduce bacterial resistance	93.5% appropriate prescriptions; differences in resistance not reported	Not reported
Rimawi *et al.*, 2013	Hospital; 5 months; penicillin allergy testing	Differences in drug use not reported	$82,000 savings annually
Schabas *et al.*, 2012	Hospital; 2 years; to control *C. difficile*	Fewer *C. difficile* infections	Not reported
Slain *et al.*, 2011	ICU; 2004-2010; to control *P. aeruginosa*	Decline in IV ciprofloxacin and ceftazidime use in 2001–2007; increased resistance in 2007–2010	Not reported

Continued

Table 29.1. Continued.

Study	Health-related aims and outcomes[a]		Economic impact
	Setting; time period; and aim	Outcome	
Slekovec *et al.*, 2012	Outpatient; 20 months; to institute guidelines on urinary tract infections in women	Less norfloxacin and more nitrofurantoin and fosfomycin trometamol prescribed	Not reported
Smith *et al.*, 2012	Hospital; 33 months; to institute antimicrobial guidelines for community acquired pneumonia in children	Increase in ampicillin use; reduction in ceftriaxone; no difference in mortality and LOS	Not reported
Standiford *et al.*, 2012	Hospital; 7 years; to control antimicrobial use	Reduction of 439 DDDs/1000 PDs	$2,949,705 savings in 3 year hospital costs
Storey *et al.*, 2012	Hospital; 16 months; to control antimicrobial use	All agents reduced from 1109 to 929 DDDs/1000 PDs	$6 saving per PD
Suryadevara *et al.*, 2012	Hospital; 15 months; to assess appropriateness of vancomycin monitoring in pediatric patients and evaluate effectiveness of pharmacy drug monitoring and healthcare provider education	Reduced inappropriate monitoring	$7848 total hospital savings
Teo *et al.*, 2012	Hospital; 12 months; to reduce inappropriate antimicrobial use	10% decrease in antibiotic use; no difference in mortality; 70.5% of interventions on use accepted	$198,575 total drug cost savings; plus $91,194 total savings in patient costs
Wong *et al.*, 2012	Hospital; 4 months; to control coagulase negative *Staphylococcus aureus*	Shorter antistaphylococcal courses; shorter LOS	$17,884 total savings in hospital cost, $8338 in infection cost
Yam *et al.*, 2012	Hospital; 13 months; to target and control use of 6 antimicrobials with high potential for misuse	Reduction in *C. difficile* infections	~$7000 savings/1000 PDs on antibiotics
Yeo *et al.*, 2012	Hospital; 2009–2010; to control and intervene in antimicrobial use in a hematology–oncology unit	Decrease in antibiotic use; resistance unaffected	$41,342 savings from patients and third party payers (insurance)

[a]DDD, daily defined dose; DOT, days of therapy; ESBL, extended-spectrum β-lactamase; ICU, intensive care unit; LOS, length of stay; MRSA, methicillin resistant *Staphylococcus aureus*; OBD, occupied bed days, PDs, patient days.

A previous analysis of 70 interventions for infection control yielded similar results (Stone *et al.*, 2005a). When studies were evaluated using the 23 items from the Harvard Cost-Effectiveness analysis, it was clear that the studies did not meet adequate standards, and the majority were deemed as simple economic analyses (Neumann *et al.*, 2005; Stone *et al.*, 2005b; Stevenson *et al.*, 2012). Studies that integrate standard economic analysis are a rarity. Among notable exceptions was a cost–utility analysis published in 2009 that evaluated the impact of ASPs on patients with bacteremia. The authors estimated a gain of 0.08 QALYs and $2367 per QALY, suggesting that ASP improved bacteremia care and was a cost-effective strategy (Scheetz *et al.*, 2009). IDSA and SHEA, in their recent ASP policy statement (Fishman *et al.*, 2012), acknowledge the lack of cost-effectiveness data and state that patient-centered outcomes are needed to evaluate the effectiveness and cost-effectiveness of ASP interventions. So far, these efforts have been partially hampered by the absence of standardized definitions and inadequate study design. Such drawbacks include selection bias, lack of confounder adjustments, variability in duration, and lack of statistical power.

ASPs in the Era of Limited Resources

In a hectic era for healthcare budgets, reporting only a reduction in antimicrobial use will not suffice as justification for an ASP; specific documentation must also be presented that demonstrates that such ASPs are both effective in their healthcare aims and cost-effective. Reducing cost is only one of the four measurable outcomes of ASPs, which include improved patient outcomes, greater patient safety, and reduced drug resistance (McGowan, 2012), and it may be secondary to patient safety and quality of care; however, for healthcare providers and administrators, cost reduction is a measure of great interest. For example, resistance rates may be a core outcome of ASP policies, but it is likely to be several years before these decrease substantially. Conversely, ASPs that focus on specific issues that are known to contribute to increased cost (*C. difficile*-associated diarrhea or MRSA infections, for example) may stand a better chance of being adopted, but both improve patient care and add significant institutional savings at the same time. ASPs also require funding to compensate the personnel involved in the design and monitoring of such programs.

Although the percentage of institutions reported to have ASPs in place has increased over the last decade, small community hospitals are less likely to implement such programs given that lack of funding for such programs presents an obstacle; healthcare administrators need more cost reduction data to support such programs (Johannsson *et al.*, 2011). Jacob and Gaynes (2010) evaluated the data from a Cochrane review of such programs (Davey *et al.*, 2005). Based on this evaluation, they suggested that the cost-effectiveness of ASPs is still unclear, and questioned whether the cost of such interventions is balanced by the marginal cost savings (Jacob and Gaynes, 2010). All ASPs will require some allocation of economic resources, and funding of the core ASP members is critical for their viability (Owens *et al.*, 2009).

There are difficulties in applying economic research to the standard practice of ASPs. First, the applicability of economic research is an issue of concern when ASPs apply to a certain setting or a homogenous population, and physicians may question its use in their patient populations and their standard of practice. Secondly, healthcare outcome data may not be easily interpreted as managers and physicians are not always acquainted with this type of data, and important findings may be omitted or misinterpreted. Some specialized healthcare personnel may not be able to access data collection and modeling because of certain constraints (Stevenson *et al.*, 2012), and large academic centers may have access whereas community-based programs do not. Ready access to economic and patient data is an important parameter of analysis, and manual collection of data in the absence of electronic means may preclude such analysis (Stevenson *et al.*, 2012).

Institutional budget constraints are not always flexible in adopting new ASP policies, particularly if the perceived benefit extends far beyond the "early discharge" of a patient. Robust and applicable programs may be discarded if their perceived benefit is evident within a short time frame (Neumann, 2004; Stone *et al.*, 2005b). An ASP cannot be long lived if it is not endorsed by the hospital administration. Funding for initiating and sustaining ASPs implies commitment of the hospital leadership and willingness to invest resources in the program (Goldmann *et al.*, 1996; Tamma and Cosgrove, 2011). ASPs may compete with other programs, thus causing their funding to be postponed. For example, an ASP initiative may be ranked second to a program of anticoagulation and thrombosis prevention (Lenchus, 2011); however, has been previously noted (Owens *et al.*, 2009), we are losing antimicrobials, not anticoagulants, and gross misuse of antimicrobials is well documented. Deploying business case models may overcome such limitations. SHEA has published a seminal report outlining the key elements of a successful business case model to control infections, which also applies to ASP. There are nine discrete steps required. These include: (i) framing the problem and developing a hypothesis for solutions; (ii) meeting the key administrators; (iii) extrapolating the annual cost; (iv) evaluating avoided cost through lower infection rates; (v) estimating the local costs of the targeted infection; (vi) measuring the financial impact; (vii) calculating the additional economic or health benefits; (viii) delineating the case; and (ix) prospectively gathering both cost and outcome data (Perencevich *et al.*, 2007).

The general idea behind business case models is to perform the analysis from a business perspective, namely that of the hospital. The aim is to look at the costs and benefits of an ASP intervention to justify its implementation to hospital administrators. Healthcare outcomes such as morbidity and mortality are not considered unless they have a perceivable economic impact (gain or loss) for the hospital. A business case exists when the investor (hospital) expects a financial return from the intervention within a reasonable time frame. That can be realized as direct profit, decreasing losses, or avoidance of cost. Of note, a business model may differ from an economic analysis. The two models converge when there is financial gain over a reasonable time, but may diverge when there is a significant benefit in morbidity and mortality, and the perceived benefit is expected several years later. The investor is unlikely to realize that benefit (Leatherman *et al.*, 2003). The ASP at Ohio State University Medical Center is a paradigm of a successful business model based on the implementation of the previous steps that promoted financial return to the investors, demonstrated continued cost reductions related to ASP interventions, and provided credible data to ensure confidence and support from the medical staff and hospital administrators (Stevenson *et al.*, 2012).

Notably, these strategies may be applicable in high-income countries but are absent in developing countries, where hospitals lack the appropriate infrastructure, and ASPs are usually defined by individual initiatives to establish a program as part of physician activities. As ASPs become more complex, they require the involvement of more than one discipline (infectious disease specialists, microbiologists, information system trained staff, clinical pharmacists), including the commitment of the hospital administration (Curcio, 2010). Supplemental strategies to the core components of an ASP (namely formulary restriction or prospective audit) include education and compliance with guidelines, the use of order forms, alteration of strategies (on the selection, combination, and route of administration of antimicrobials), and computer facility support. In that context, ASPs may be constrained by limited resources, but some simple measures can be implemented, even in such a setting. These include a switch from intravenous to oral administration (Lau *et al.*, 2011), antimicrobial batching to avoid wasting the excess drug, as in the example of caspofungin and daptomycin from the Ohio State University Wexner Medical Center, and automatic therapeutic substitution of (expensive) vancomycin capsules by the (inexpensive) vancomycin solution prepared from the IV powder for the treatment of *C. difficile* infections (Goff *et al.*, 2012).

The Future of ASPs

An ASP should focus on both clinical and economic considerations, and financial justification relies on the diligent collection of pertinent data. ASPs become inevitably more complex as they shift from a technical or clinical expert approach to a more sophisticated strategic approach at hospital or organization level. Infection control was initially a problem for few expert clinicians, but now ASPs are integrated into the infrastructure of the organization, and affect the relationship between prescribers, infection experts, healthcare personnel, and managers. Utilizing economic models is imperative to assessing existing or proposed ASPs, particularly when different hospital strategies compete for limited healthcare resources. Cost and health benefit data have to be integrated to influence decisions on new ASPs or to sustain the viability of older ones. ASP leaders need to understand organizational culture if they are to be influential at that level; integration with structures and processes will provide access to financial and physical resources, along with support for changing practices (Murray and Holmes, 2012). As the crisis is still fueled by both antimicrobial resistance and the paucity of new and effective drugs, ASPs must remain a top priority for healthcare systems and public health.

References

Aldeyab, M.A., Scott, M.G., Kearney, M.P., Alahmadi, Y.M., Magee, F.A., Conlon, G., and McElnay, J.C. (2013) Impact of an enhanced antibiotic stewardship on reducing methicillin-resistant *Staphylococcus aureus* in primary and secondary healthcare settings. *Epidemiology and Infection* 142, 494–500.

Ansari, F., Gray, K., Nathwani, D., Phillips, G., Ogston, S., Ramsay, C., and Davey, P. (2003) Outcomes of an intervention to improve hospital antibiotic prescribing: interrupted time series with segmented regression analysis. *Journal of Antimicrobial Chemotherapy* 52, 842–848.

Avdic, E., Cushinotto, L.A., Hughes, A.H., Hansen, A.R., Efird, L.E., Bartlett, J.G., and Cosgrove, S.E. (2012) Impact of an antimicrobial stewardship intervention on shortening the duration of therapy for community-acquired pneumonia. *Clinical Infectious Diseases* 54, 1581–1587.

Avorn, J., Soumerai, S.B., Taylor, W., Wessels, M.R., Janousek, J., and Weiner, M. (1988) Reduction of incorrect antibiotic dosing through a structured educational order form. *Archives of Internal Medicine* 148, 1720–1724.

Bevilacqua, S., Demore, B., Boschetti, E., Doco-Lecompte, T., May, I., May, T., Rabaud, C., and Thilly, N. (2011) 15 years of antibiotic stewardship policy in the Nancy Teaching Hospital. *Médecine et Maladies Infectieuses* 41, 532–539.

Boyce, J.M., Pop, O.F., Abreu-Lanfranco, O., Hung, W.Y., Fisher, A., Karjoo, A., Thompson, B., and Protopapas, Z. (2013) A trial of discontinuation of empiric vancomycin therapy in patients with suspected methicillin-resistant *Staphylococcus aureus* health care-associated pneumonia. *Antimicrobial Agents and Chemotherapy* 57, 1163–1168.

Burke, J.P. (1998) Antibiotic resistance – squeezing the balloon? *JAMA (The Journal of the American Medical Association)* 280, 1270–1271.

Cairns, K.A., Jenney, A.W., Abbott, I.J., Skinner, M.J., Doyle, J.S., Dooley, M., and Cheng, A.C. (2013) Prescribing trends before and after implementation of an antimicrobial stewardship program. *The Medical Journal of Australia* 198, 262–266.

Carling, P., Fung, T., Killion, A., Terrin, N., and Barza, M. (2003) Favorable impact of a multidisciplinary antibiotic management program conducted during 7 years. *Infection Control and Hospital Epidemiology* 24, 699–706.

Chalfine, A., Kitzis, M.D., Bezie, Y., Benali, A., Perniceni, L., Nguyen, J.C., Dumay, M.F., Gonot, J., Rejasse, G., Goldstein, F. *et al.* (2012) Ten-year decrease of acquired methicillin-resistant *Staphylococcus aureus* (MRSA) bacteremia at a single institution: the result of a multifaceted program combining cross-transmission prevention and antimicrobial stewardship. *Antimicrobial Resistance and Infection Control* 1(1):18.

Chan, Y.Y., Lin, T.Y., Huang, C.T., Deng, S.T., Wu, T.L., Leu, H.S., and Chiu, C.H. (2011) Implementation and outcomes of a hospital-wide computerised antimicrobial stewardship programme in a large medical centre in Taiwan. *International Journal of Antimicrobial Agents* 38, 486–492.

Charani, E., Cooke, J., and Holmes, A. (2010) Antibiotic stewardship programmes – what's missing? *Journal of Antimicrobial Chemotherapy* 65, 2275–2277.

Cisneros, J.M., Neth, O., Gil-Navarro, M.V., Lepe, J.A, Jiménez-Parrilla, F., Cordero, E., Rodríguez-Hernández, M.J., Amaya-Villar, R., Cano, J., Gutiérrez-Pizarraya, A. *et al.* (2013) Global impact of an educational antimicrobial stewardship programme on prescribing practice in a tertiary hospital centre. *Clinical Microbiology and Infection* 20, 82–88.

Cosgrove, S.E. (2006) The relationship between antimicrobial resistance and patient outcomes: mortality, length of hospital stay, and health care costs. *Clinical Infectious Diseases* 42(Suppl 2), S82–S89.

Cosgrove, S.E., Kaye, K.S., Eliopoulous, G.M., and Carmeli, Y. (2002) Health and economic outcomes of the emergence of third-generation cephalosporin resistance in *Enterobacter* species. *Archives of Internal Medicine* 162, 185–190.

Curcio, D. (2010) Antibiotic stewardship: the "real world" when resources are limited. *Infection Control and Hospital Epidemiology* 31, 666–668.

Dancer, S.J., Kirkpatrick, P., Corcoran, D.S., Christison, F., Farmer, D., and Robertson, C. (2013) Approaching zero: Temporal effects of a restrictive antibiotic policy on hospital-acquired *Clostridium difficile*, extended-spectrum β-lactamase-producing coliforms and meticillin-resistant *Staphylococcus aureus*. *International Journal of Antimicrobial Agents* 41, 137–142.

Davey, P., Brown, E., Fenelon, L., Finch, R., Gould, I., Hartman, G., Holmes, A., Ramsay, C., Taylor, E., Wilcox, M., and Wiffen, P. (2005) Interventions to improve antibiotic prescribing practices for hospital inpatients. *Cochrane Database of Systematic Reviews* 4:CD003543.

Dellit, T.H., Owens, R.C., McGowan, J.E., Jr., Gerding, D.N., Weinstein, R.A., Burke, J.P., Huskins, W.C., Paterson, D.L., Fishman, N.O., Carpenter, C.F. *et al.* (2007) Infectious Diseases Society of America and the Society for Healthcare Epidemiology of America guidelines for developing an institutional program to enhance antimicrobial stewardship. *Clinical Infectious Diseases* 44, 159–177.

Drew, R.H., White, R., MacDougall, C., Hermsen, E.D., and Owens, R.C., Jr. (2009) Insights from the Society of Infectious Diseases Pharmacists on antimicrobial stewardship guidelines from the Infectious Diseases Society of America and the Society for Healthcare Epidemiology of America. *Pharmacotherapy* 29, 593–607.

Dubrovskaya, Y., Papadopoulos, J., Scipione, M.R., Altshuler, J., Phillips, M., and Mehta, S.A. (2012) Antibiotic stewardship for intra-abdominal infections: early impact on antimicrobial use and patient outcomes. *Infection Control and Hospital Epidemiology* 33, 427–429.

Dumartin, C., Rogues, A.M., Amadeo, B., Pefau, M., Venier, A.G., Parneix, P., and Maurain, C. (2011) Antibiotic usage in south-western French hospitals: trends and association with antibiotic stewardship measures. *Journal of Antimicrobial Chemotherapy* 66, 1631–1637.

Eggleston, K., Zhang, R., and Zeckhauser, R.J. (2010) The global challenge of antimicrobial resistance: insights from economic analysis. *International Journal of Environmental Research and Public Health* 7, 3141–3149.

Finley, R., Glass-Kaastra, S.K., Hutchinson, J., Patrick, D.M., Weiss, K., and Conly, J. (2013) Declines in outpatient antimicrobial use in Canada (1995–2010). *PloS ONE* 8(10): e76398.

Fishman, N., Patterson, J., Saiman, L., Srinivasan, A., Trivedi, K.K., van Schooneveld, T., Lynfield, R., Gerding, D., Septimus, E., Schwartz, D. *et al.* (2012) Policy statement on antimicrobial stewardship by the Society for Healthcare Epidemiology of America (SHEA), the Infectious Diseases Society of America (IDSA), and the Pediatric Infectious Diseases Society (PIDS). *Infection Control and Hospital Epidemiology* 33, 322–327.

Goff, D.A. (2011) Antimicrobial stewardship: bridging the gap between quality care and cost. *Current Opinion in Infectious Diseases* 24(Suppl 1), S11–S20.

Goff, D.A., Bauer, K.A., Reed, E.E., Stevenson, K.B., Taylor, J.J., and West, J.E. (2012) Is the "low-hanging fruit" worth picking for antimicrobial stewardship programs? *Clinical Infectious Diseases* 55, 587–592.

Goldmann, D.A., Weinstein, R.A., Wenzel, R.P., Tablan, O.C., Duma, R.J., Gaynes, R.P., Schlosser, J., and Martone, W.J. (1996) Strategies to prevent and control the emergence and spread of antimicrobial-resistant microorganisms in hospitals. A challenge to hospital leadership. *JAMA (The Journal of the American Medical Association)* 275, 234–240.

Guarascio, A.J., Slain, D., McKnight, R., Petros, K., Parker, J., Wilson, A., Defazio, C.M., and Sarwari, A.R. (2013) A matched-control evaluation of an antifungal bundle in the intensive care unit at a university teaching hospital. *International Journal of Clinical Pharmacy* 35, 145–148.

Heil, E.L., Daniels, L.M., Long, D.M., Rodino, K.G., Weber, D.J., and Miller, M.B. (2012) Impact of a rapid peptide nucleic acid fluorescence in situ hybridization assay on treatment of *Candida* infections. *American Journal of Health-System Pharmacy* 69, 1910–1914.

Jacob, J.T. and Gaynes, R.P. (2010) Emerging trends in antibiotic use in US hospitals: quality, quantification and stewardship. *Expert Review of Anti-infective Therapy* 8, 893–902.

Johannsson, B., Beekmann, S.E., Srinivasan, A., Hersh, A.L., Laxminarayan, R., and Polgreen, P.M. (2011) Improving antimicrobial stewardship: the evolution of programmatic strategies and barriers. *Infection Control and Hospital Epidemiology* 32, 367–374.

John, J.F., Jr. and Fishman, N.O. (1997) Programmatic role of the infectious diseases physician in controlling antimicrobial costs in the hospital. *Clinical Infectious Diseases* 24, 471–485.

Jump, R.L., Olds, D.M., Seifi, N., Kypriotakis, G., Jury, L.A., Peron, E.P., Hirsch, A.A., Drawz, P.E., Watts, B., Bonomo, R.A., and Donskey, C.J. (2012) Effective antimicrobial stewardship in a long-term care facility through an infectious disease consultation service: keeping a LID on antibiotic use. *Infection Control and Hospital Epidemiology* 33, 1185–1192.

Katsios, C.M., Burry, L., Nelson, S., Jivraj, T., Lapinsky, S.E., Wax, R.S., Christian, M., Mehta, S., Bell, C.M., and Morris, A.M. (2012) An antimicrobial stewardship program improves antimicrobial treatment by culture site and the quality of antimicrobial prescribing in critically ill patients. *Critical Care* 16(6): R216.

Kim, Y.C., Kim, M.H., Song, J.E., Ahn, J.Y., Oh, D.H., Kweon, O.M., Lee, D., Kim, S.B., Kim, H.W., Jeong, S.J. *et al.* (2013) Trend of methicillin-resistant *Staphylococcus aureus* (MRSA) bacteremia in an institution with a high rate of MRSA after the reinforcement of antibiotic stewardship and hand hygiene. *American Journal of Infection Control* 41, e39–e43.

Kullar, R., Davis, S.L., Kaye, K.S., Levine, D.P., Pogue, J.M., and Rybak, M.J. (2013) Implementation of an antimicrobial stewardship pathway with daptomycin for optimal treatment of methicillin-resistant *Staphylococcus aureus* bacteremia. *Pharmacotherapy* 33, 3–10.

Kunin, C.M., Tupasi, T., and Craig, W.A. (1973) Use of antibiotics. A brief exposition of the problem and some tentative solutions. *Annals of Internal Medicine* 79, 555–560.

LaRocco, A., Jr. (2003) Concurrent antibiotic review programs – a role for infectious diseases specialists at small community hospitals. *Clinical Infectious Diseases* 37, 742–743.

Lau, B.D., Pinto, B.L., Thiemann, D.R., and Lehmann, C.U. (2011) Budget impact analysis of conversion from intravenous to oral medication when clinically eligible for oral intake. *Clinical Therapeutics* 33, 1792–1796.

Leatherman, S., Berwick, D., Iles, D., Lewin, L.S., Davidoff, F., Nolan, T., and Bisognano, M. (2003) The business case for quality: case studies and an analysis. *Health Affairs* 22, 17–30.

Lenchus, J.D. (2011) Strategies for venous thromboembolism prophylaxis programs. *Postgraduate Medicine* 123, 91–101.

Liew, Y.X., Lee, W., Loh, J.C., Cai, Y., Tang, S.S., Lim, C.L., Teo, J., Ong, R.W., Kwa, A.L., and Chlebicki, M.P. (2012a) Impact of an antimicrobial stewardship programme on patient safety in Singapore General Hospital. *International Journal of Antimicrobial Agents* 40, 55–60.

Liew, Y.X., Lee, W., Cai, Y.Y., Teo, J., Tang, S.S., Ong, R.W., Lim, C.L., Lingegowda, P.B., Kwa, A.L., and Chlebicki, M.P. (2012b) Utility and safety of procalcitonin in an antimicrobial stewardship program (ASP) in patients with malignancies. *European Journal of Clinical Microbiology and Infectious Diseases* 31, 3041–3046.

Lima, A.L.L.M., Oliveira, P.R.D. de, Paula, A.P. de, Dal-Paz, K., Almeida, J.N. de, Jr., Félix Cda, S., and Rossi, F. (2011) Carbapenem stewardship: positive impact on hospital ecology. *The Brazilian Journal of Infectious Diseases* 15, 1–5.

Lopez-Medrano, F., San Juan, R., Lizasoain, M., Catalan, M., Ferrari, J.M., Chaves, F., Lumbreras, C., Montejo, J.C., Herroros de Tejada, A., and Aguado, J.M. (2013) A non-compulsory stewardship programme for the management of antifungals in a university-affiliated hospital. *Clinical Microbiology and Infection* 19, 56–61.

Lutters, M., Harbarth, S., Janssens, J.P., Freudiger, H., Herrmann, F., Michel, J.P., and Vogt, N. (2004) Effect of a comprehensive, multidisciplinary, educational program on the use of antibiotics in a geriatric university hospital. *Journal of the American Geriatrics Society* 52, 112–116.

Maechler, F., Schwab, F., Geffers, C., Meyer, E., Leistner, R., and Gastmeier, P. (2013) Antibiotic stewardship in Germany: a cross-sectional questionnaire survey of 355 intensive care units. *Infection* 42, 119–125.

Magedanz, L., Silliprandi, E.M., and dos Santos, R.P. (2012) Impact of the pharmacist on a multidisciplinary team in an antimicrobial stewardship program: a quasi-experimental study. *International Journal of Clinical Pharmacy* 34, 290–294.

Malani, A.N., Richards, P.G., Kapila, S., Otto, M.H., Czerwinski, J., and Singal, B. (2013) Clinical and economic outcomes from a community hospital's antimicrobial stewardship program. *American Journal of Infection Control* 41, 145–148.

Marr, J.J., Moffet, H.L., and Kunin, C.M. (1988) Guidelines for improving the use of antimicrobial agents in hospitals: A statement by the Infectious Diseases Society of America. *The Journal of Infectious Diseases* 157, 869–876.

McGowan, J.E., Jr. (1983) Antimicrobial resistance in hospital organisms and its relation to antibiotic use. *Reviews of Infectious Diseases* 5, 1033–1048.

McGowan, J.E. (2012) Antimicrobial stewardship – the state of the art in 2011: focus on outcome and methods. *Infection Control and Hospital Epidemiology* 33, 331–337.

Mölstad, S., Erntell, M., Hanberger, H., Melander, E., Norman, C., Skoog, G., Lundborg, C.S., Söderström, A., Torell, E., and Cars, O. (2008) Sustained reduction of antibiotic use and low bacterial resistance: 10-year follow-up of the Swedish Strama programme. *The Lancet Infectious Diseases* 8, 125–132.

Mondain, V., Lieutier, F., Dumas, S., Gaudart, A., Fosse, T., Roger, P.M., Bernard, E., Farhad, R., and Pulcini, C. (2013) An antibiotic stewardship program in a French teaching hospital. *Médecine et Maladies Infectieuses* 43, 17–21.

Murray, E. and Holmes, A. (2012) Addressing healthcare-associated infections and antimicrobial resistance from an organizational perspective: progress and challenges. *Journal of Antimicrobial Chemotherapy* 67(Suppl 1), i29–i36.

Neumann, P.J. (2004) Why don't Americans use cost-effectiveness analysis? *The American Journal of Managed Care* 10, 308–312.

Neumann, P.J., Greenberg, D., Olchanski, N.V., Stone, P.W., and Rosen, A.B. (2005) Growth and quality of the cost-utility literature, 1976–2001. *Value in Health* 8, 3–9.

Niwa, T., Shinoda, Y., Suzuki, A., Ohmori, T., Yasuda, M., Ohta, H., Fukao, A., Kitaichi, K., Matsuura, K., Sugiyama, T. *et al*. (2012) Outcome measurement of extensive implementation of antimicrobial stewardship in patients receiving intravenous antibiotics in a Japanese university hospital. *International Journal of Clinical Practice* 66, 999–1008.

Office of Technology Assessment, US Congress (1995) *Impacts of Antibiotic-resistant Bacteria*. Government Printing Office, Washington, DC.

Owens, R.C., Jr. (2008) Antimicrobial stewardship: concepts and strategies in the 21st century. *Diagnostic Microbiology and Infectious Disease* 61, 110–128.

Owens, R.C., Jr., Shorr, A.F., and Deschambeault, A.L. (2009) Antimicrobial stewardship: shepherding precious resources. *American Journal of Health-System Pharmacy* 66, S15–S22.

Paladino, J.A. (2004) Economics of antibiotic use policies. *Pharmacotherapy* 24(12P2), 232S–238S.

Pate, P.G., Storey, D.F., and Baum, D.L. (2012) Implementation of an antimicrobial stewardship program at a 60-bed long-term acute care hospital. *Infection Control and Hospital Epidemiology* 33, 405–408.

Perencevich, E.N., Stone, P.W., Wright, S.B., Carmeli, Y., Fisman, D.N., and Cosgrove, S.E. (2007) Raising standards while watching the bottom line: making a business case for infection control. *Infection Control and Hospital Epidemiology* 28, 1121–1133.

Perez, K.K., Olsen, R.J., Musick, W.L., Cernoch, P.L., Davis, J.R., Land, G.A., Peterson, L.E., and Musser, J.M. (2013) Integrating rapid pathogen identification and antimicrobial stewardship significantly decreases hospital costs. *Archives of Pathology and Laboratory Medicine* 137, 1247–1254.

Pope, S.D., Dellit, T.H., Owens, R.C., and Hooton, T.M. (2009) Results of survey on implementation of Infectious Diseases Society of America and Society for Healthcare Epidemiology of America guidelines for developing an institutional program to enhance antimicrobial stewardship. *Infection Control and Hospital Epidemiology* 30, 97–98.

Rahal, J.J., Urban, C., Horn, D., Freeman, K., Segal-Maurer, S., Maurer, J., Mariano, N., Marks, S., Burns, J.M., Dominick, D., and Lim, M. (1998) Class restriction of cephalosporin use to control total cephalosporin resistance in nosocomial *Klebsiella*. *JAMA (The Journal of the American Medical Association)* 280, 1233–1237.

Ramsamy, Y., Muckart, D.J., and Han, K.S. (2013) Microbiological surveillance and antimicrobial stewardship minimise the need for ultrabroad-spectrum combination therapy for treatment of nosocomial infections in a trauma intensive care unit: an audit of an evidence-based empiric antimicrobial policy. *South African Medical Journal* 103, 371–376.

Reimann, H.A. (1961) The misuse of antimicrobics. *Medical Clinics of North America* 45, 849–856.

Research Committee of SHEA and Sinaii, N. (2010) Charting the course for the future of science in healthcare epidemiology: results of a survey of the membership of the Society of Healthcare Epidemiology of America. *Infection Control and Hospital Epidemiology* 31, 669–675.

Rimawi, R.H., Cook, P.P., Gooch, M., Kabchi, B., Ashraf, M.S., Rimawi, B.H., Gebregziabher, M., and Siraj, D.S. (2013) The impact of penicillin skin testing on clinical practice and antimicrobial stewardship. *Journal of Hospital Medicine* 8, 341–345.

Roberts, R.R., Hota, B., Ahmad, I., Scott, R.D., II, Foster, S.D., Abbasi, F., Schabowski, S., Kampe, L.M.,

Ciavarella, G.G., Supino, M. *et al.* (2009) Hospital and societal costs of antimicrobial-resistant infections in a Chicago teaching hospital: implications for antibiotic stewardship. *Clinical Infectious Diseases* 49, 1175–1184.

Ruttimann, S., Keck, B., Hartmeier, C., Maetzel, A., and Bucher, H.C. (2004) Long-term antibiotic cost savings from a comprehensive intervention program in a medical department of a university-affiliated teaching hospital. *Clinical Infectious Diseases* 38, 348–356.

Sassi, F. (2006) Calculating QALYs, comparing QALY and DALY calculations. *Health Policy and Planning* 21, 402–408.

Schabas, A., Fisman, D.N., and Schabas, R. (2012) Control of *Clostridium difficile*-associated diarrhea by antibiotic stewardship in a small community hospital. *Canadian Journal of Infectious Diseases and Medical Microbiology* 23, 82–83.

Scheetz, M.H., Bolon, M.K., Postelnick, M., Noskin, G.A., and Lee, T.A. (2009) Cost-effectiveness analysis of an antimicrobial stewardship team on bloodstream infections: a probabilistic analysis. *Journal of Antimicrobial Chemotherapy* 63, 816–825.

Schentag, J.J., Ballow, C.H., Fritz, A.L., Paladino, J.A., Williams, J.D., Cumbo, T.J., Ali, R.V., Galletta, V.A., Gutfeld, M.B., and Adelman, M.H. (1993) Changes in antimicrobial agent usage resulting from interactions among clinical pharmacy, the infectious disease division, and the microbiology laboratory. *Diagnostic Microbiology and Infectious Disease* 16, 255–264.

Slain, D., Sarwari, A.R., Petros, K.O., McKnight, R.L., Sager, R.B., Mullett, C.J., Wilson, A., Thomas, J.G., Moffett, K., Palmer, H.C., and Dedhia, H.V. (2011) Impact of a multimodal antimicrobial stewardship program on *Pseudomonas aeruginosa* susceptibility and antimicrobial use in the intensive care unit setting. *Critical Care Research and Practice* 2011, 416426.

Slekovec, C., Leroy, J., Vernaz-Hegi, N., Faller, J.P., Sekri, D., Hoen, B., Talon, D., and Bertrand, X. (2012) Impact of a region wide antimicrobial stewardship guideline on urinary tract infection prescription patterns. *International Journal of Clinical Pharmacy* 34, 325–329.

Smith, M.J., Kong, M., Cambon, A., and Woods, C.R. (2012) Effectiveness of antimicrobial guidelines for community-acquired pneumonia in children. *Pediatrics* 129, e1326–e1333.

Standiford, H.C., Chan, S., Tripoli, M., Weekes, E., and Forrest, G.N. (2012) Antimicrobial stewardship at a large tertiary care academic medical center: cost analysis before, during, and after a 7-year program. *Infection Control and Hospital Epidemiology* 33, 338–345.

Stevenson, K.B, Balada-Llasat, J.M., Bauer, K., Deutscher, M., Goff, D., Lustberg, M., Pancholi, P., Reed, E., Smeenk, D., Taylor, J., and West, J. (2012) The economics of antimicrobial stewardship: the current state of the art and applying the business case model. *Infection Control and Hospital Epidemiology* 33, 389–397.

Stone, P.W., Braccia, D., and Larson, E. (2005a) Systematic review of economic analyses of health care-associated infections. *American Journal of Infection Control* 33, 501–509.

Stone, P.W., Hedblom, E.C., Murphy, D.M., and Miller, S.B. (2005b) The economic impact of infection control: making the business case for increased infection control resources. *American Journal of Infection Control* 33, 542–547.

Storey, D.F., Pate, P.G., Nguyen, A.T., and Chang, F. (2012) Implementation of an antimicrobial stewardship program on the medical-surgical service of a 100-bed community hospital. *Antimicrobial Resistance and Infection Control* 1: 32.

Suryadevara, M., Steidl, K.E., Probst, L.A., and Shaw, J. (2012) Inappropriate vancomycin therapeutic drug monitoring in hospitalized pediatric patients increases pediatric trauma and hospital costs. *The Journal of Pediatric Pharmacology and Therapeutics* 17, 159–165.

Tamma, P.D. and Cosgrove, S.E. (2011) Antimicrobial stewardship. *Infectious Disease Clinics of North America* 25, 245–260.

Teo, J., Kwa, A.L., Loh, J., Chlebicki, M.P., and Lee, W. (2012) The effect of a whole-system approach in an antimicrobial stewardship programme at the Singapore General Hospital. *European Journal of Clinical Microbiology and Infectious Diseases* 31, 947–955.

Wong, J.R., Bauer, K.A., Mangino, J.E., and Goff, D.A. (2012) Antimicrobial stewardship pharmacist interventions for coagulase-negative staphylococci positive blood cultures using rapid polymerase chain reaction. *Annals of Pharmacotherapy* 46, 1484–1490.

Yam, P., Fales, D., Jemison, J., Gillum, M., and Bernstein, M. (2012) Implementation of an antimicrobial stewardship program in a rural hospital. *American Journal of Health-System Pharmacy* 69, 1142–1148.

Yeo, C.L., Chan, D.S., Earnest, A., Wu, T.S., Yeoh, S.F., Lim, R., Jureen, R., Fisher, D., and Hsu, L.Y. (2012) Prospective audit and feedback on antibiotic prescription in an adult hematology–oncology unit in Singapore. *European Journal of Clinical Microbiology and Infectious Diseases* 31, 583–590.

Zilberberg, M.D. and Shorr, A.F. (2010) Understanding cost-effectiveness. *Clinical Microbiology and Infection* 16, 1707–1712.

30 Pharmacoeconomic Implications of Antimicrobial Adverse Events

CHESTON B. CUNHA*

Rhode Island Hospital, Providence, Rhode Island, US

Introduction

Antibiotic stewardship programs (ASPs) have several objectives that result in the optimal use of antimicrobials. Implied in optimal antimicrobial therapy are the notions of using the best available drug for any given infectious disease diagnosis, optimizing the dosing of the selected drug, and using that drug for the shortest duration of therapy that is clinically effective (Cunha *et al.*, 2013). Optimal drug therapy also implies decreased use of inappropriate antimicrobials. It is hoped that optimal antimicrobial therapy may minimize the potential for antibiotic resistance, particularly among aerobic Gram-negative bacilli (GNBs), which can be multidrug-resistant organisms (MDROs) (Cunha, 2001; Mach *et al.*, 2007). ASPs require infectious disease leadership and a coordinated effort with microbiology, with pharmacy, and with infection control to be effective (Goff *et al.*, 2012). They have been shown to be cost-effective and are an added benefit to hospitals that have them (Nicolau, 2009; Signorovitch *et al.*, 2010; Sick *et al.*, 2013; Zimlichman *et al.*, 2013). Because changes in established habits are required to implement ASPs, an extensive educational program by infectious disease clinicians is necessary for their success (Leung *et al.*, 2011; Teo *et al.*, 2012).

Pharmacoeconomic Perspectives

The justification for ASPs is straightforward, i.e., that antimicrobial therapy should be optimized, and that this should result in improved outcomes with the efficient and effective utilization of antimicrobial therapies. Clearly, the optimization of antimicrobial therapy has important pharmacoeconomic implications (Roberts *et al.*, 2009; Mansouri *et al.*, 2011; Goff *et al.*, 2012; Niwa *et al.*, 2012; Teo *et al.*, 2012; Malani *et al.*, 2013; Sick *et al.*, 2013; Zimlichman *et al.*, 2013). For the same indication, the use of a less expensive antimicrobial over another that is equally efficacious may result in considerable cost savings (Paladino *et al.*, 2007; Mokabberi *et al.*, 2010). The optimization of regimens utilizing less frequent dosing, when this is pharmacokinetically feasible, is an important cost-containing measure (Nicolau, 2009; Mertz *et al.*, 2010; Kaier, 2012; Cunha *et al.*, 2013). Using lower but effective doses for less severe infectious diseases is another cost-saving strategy (Cunha *et al.*, 2013). A shortened course of therapy also has pharmacoeconomic implications, not only in terms of decreased drug costs, but also as decreased hospital length(s) of stay (LOS) (Raut *et al.*, 2009; File, 2012; Gray *et al.*, 2012; Rubinstein and Keynan, 2013).

Pharmacoeconomically, one of the key components of ASPs is a parenteral (IV) to oral (PO) switch program. Because antibiotics, at any given dosage, are less expensive administered PO than IV, the cost savings potential of an IV to PO switch program is considerable (Cunha, 1997, 2006; Colli *et al.*, 2007). Aside from the acquisition cost of the necessary antibiotics, there is an additional cost to administering antibiotics intravenously. For antibiotic IV administration, the typical cost of each antibiotic dose administered intravenously is $10.00/dose. In an IV to PO switch program, not only is the cost of the oral antimicrobial less than its IV counterpart, but the IV administration costs are saved

*E-mail: ccunha@lifespan.org

(Yen *et al.*, 2012). So not only are acquisition costs less with PO antibiotics, there is also no IV administration cost to the hospital. Table 30.1 presents some example cost savings.

In addition to the cost advantages using PO administration, IV-related side effects, e.g., phlebitis and central venous catheter (CVC)-associated bacteremias are virtually eliminated when PO therapy is used for most of the hospital stay (Vrijens *et al.*, 2010). Early IV to PO switch therapy also permits earlier discharge and better bed utilization (Cunha, 1997, 2006; Colli *et al.*, 2007; Frei *et al.*, 2009; Restrepo and Frei, 2010; Niwa *et al.*, 2012). With less time spent in the hospital, the risk of acquiring nosocomial MDROs or other hospital-associated complications, e.g., *Clostridium difficile* diarrhea, is decreased (Nicolau, 2009; Monge *et al.*, 2013) (see Table 30.2).

Indirect Costs of Antimicrobial Therapy

Obligatory other drug therapy cost estimates are difficult to come by, and may not be obvious. If one attempts to determine the cost of using metronidazole to treat intra-abdominal infections (IAIs), it immediately becomes apparent that metronidazole should not be used alone for IAIs and must be used in combination with an anti-aerobic GNB antibiotic (Cunha *et al.*, 2013). The true cost of using metronidazole for IAIs, which at first appears relatively inexpensive when used by itself, is far greater. The actual cost is the acquisition cost of metronidazole

Table 30.1. Estimated treatment costs for intravenous (IV) and oral (PO) antibiotics based on average US pricing.

IV antibiotics	Usual adult dose (70 kg)	Cost per dose	IV administrative cost ($10/dose)	Drug cost per day	Lowest wholesale price per dose (average wholesale price, AWP, where known per dose, as per REDBOOK™ 2016)
Ciprofloxacin	400 mg (IV) q12h	$3.44	$20	$6.88 + $20 = $26.88	$3.44 ($3.48)
Clindamycin	600 mg (IV) q8h	$5.09	$30	$15.27 + $30 = $45.27	$3.48
Doxycycline	100 mg (IV) q12h	$16.80	$20	$33.60 + $20 = $53.60	$18.20
Levofloxacin	500mg (IV) q24h	$4.50	$10	$14.50	$8.45 ($24.2)
	750 mg (IV) q24h	$7.49	$10	$17.49	$25.2 ($29.2)
Linezolid	600 mg (IV) q12h	$112.74	$20	$225.48 + $20 = $245.48	$96.00 ($599.49)
Metronidazole	500 mg (IV) q8h	$1.08	$30	$3.24 + $30 = $33.24	$2.34
Minocycline	100 mg (IV) q12h	$71.18	$20	$142.36 + $20 = $162.36	$65.30 ($120.89)
Trimethoprim–sulfamethoxazole	160/800 mg (IV) q6h	$8.70	$40	$34.80 + $40 = $74.80	$5.35 ($12.73)

PO antibiotics	Usual adult dose (70 kg)	Cost per dose	IV administrative cost ($10/dose)	Drug cost per day	Average wholesale price (AWP) per dose as per REDBOOK™ 2016
Ciprofloxacin	500 mg (PO) q12h	$0.11	$0	$0.22	$3.44
	750 mg (PO) q12h	$0.35	$0	$0.70	$5.17
Clindamycin	300 mg (PO) q6h	$1.28	$0	$3.85	$1.45
Doxycycline	100 mg (PO) q12h	$2.37	$0	$4.74	$0.08 ($3)
Levofloxacin	500 mg (PO) q24h	$0.50	$0	$0.50	$15.60
	750 mg (PO) q24h	$1.05	$0	$1.05	$24.60
Linezolid	600 mg (PO) q12h	$111.95	$0	$223.90	$183.00
Metronidazole	500 mg (PO) q8h	$0.65	$0	$1.95	$0.69
Minocycline	100 mg (PO) q12h	$0.20	$0	$0.40	$1.67
Trimethoprim–sulfamethoxazole	1 DS tablet (PO) q6h	$1.15	$0	$4.60	$0.49

Table 30.2. Beyond antibiotic acquisition costs: the indirect costs of antibiotics.

1. Cost of IV administration ($10/dose):
   (i) If clinically acceptable, other things being equal, preferentially use antibiotics with *less frequent dosing intervals*;
   (a) For uncomplicated streptococcal cellulitis, both cefazolin and ceftriaxone are equally effective;
   (b) Acquisition drug costs aside, as cefazolin is given q8h (3 doses × $10/dose = $30/day), cefazolin *costs more to administer* than ceftriaxone given q24h (1 dose × $10/dose = $10/day).
2. Cost of obligatory other drug with "double drug" therapy:
   (i) For intra-abdominal infections (IAIs): metronidazole *plus another anti-aerobic Gram negative bacillus (GNB) antibiotic* = the actual cost of using metronidazole for IAIs;
   (ii) For community-acquired pneumonia (CAP): ceftriaxone *plus azithromycin* = the actual cost of using ceftriaxone for CAP.
3. Cost of monitoring:
   (i) Serial serum creatinine determinations to monitor renal dysfunction and to calculate creatinine clearances;
   (ii) Liver function tests (LFTs) to monitor hepatic drug reactions;
   (iii) Drug level measurements to monitor drug efficacy or toxicity.
4. Cost of antibiotic resistance from multidrug-resistant organisms (MDROs):
   (i) Length of stay (LOS) up;
   (ii) Economic implications of suboptimal bed utilization and infection control contaminant measures;
   (iii) Needless treatment of bacterial colonization (vs. infection), which often predisposes to MDROs;
   (iv) Cost of retreatment with another, usually more expensive, effective antibiotic.
5. Cost of therapeutic failure:
   (i) Length of stay (LOS) up;
   (ii) Cost of consultants/tests to determine the cause of persistent fever on treatment;
   (iii) Cost of retreatment with another, usually more expensive, effective antibiotic.
6. Cost of adverse effects:
   (i) Seizures;
   (ii) Non-*Clostridium difficile* diarrhea;
   (iii) *C. difficile* diarrhea/colitis;
   (iv) Phlebitis;
   (v) Drug fever;
   (vi) Drug rashes;
   (vii) Hypersensitivity reactions (anaphylaxis).

(plus the IV administration charge) combined with the cost for the other antibiotic that is necessary when metronidazole is used as part of a double-drug regimen for IAIs (Cunha, 2003, 2013). In these cases, monotherapy with an agent that has comparable antimicrobial coverage may ultimately prove to be a less expensive option.

In addition, some antimicrobials are potentially hepatotoxic or nephrotoxic, and their use may require monitoring. In the case of potential hepatotoxicity, serial liver function tests (LFTs) may be required. So because of the hepatotoxic potential of any drug, the cost of serial LFTs should be added to the actual cost of the drug itself. It is the institution and healthcare system that bears these indirect costs. Similarly, with potentially nephrotoxic drugs, serial serum creatinine determinations are necessary to properly dose the drug according to the creatinine clearance, as well as to detect early signs of renal damage/insufficiency (Cunha, 2003, 2013). The cost implications for patients requiring a prolonged course of antimicrobials should not be underestimated.

Still other antibiotics utilize measured serum levels for drug monitoring, e.g., vancomycin. The cost of determining serial antibiotic levels is considerable and is another "hidden charge" to add to that of a drug that may appear to be relatively inexpensive to administer IV, thus adding to the real world cost of that drug. Furthermore, serum levels are relatively expensive to measure, and multiple testing is the rule in hospitalized patients; these hidden costs must also be borne directly by the hospital and indirectly by the patient/healthcare system (Cunha, 2003, 2013).

There are still other costs which are difficult to quantitate but certainly have "real world" pharmacoeconomic effects. There is a real but often overlooked cost of antibiotic resistance (Mach *et al.*, 2007; Mauldin *et al.*, 2010). Antibiotic resistance, particularly with GNB MDROs, has pharmacoeconomic consequences in terms of increased LOS, impaired bed utilization from cohorting patients (Cartaxo Salgado *et al.*, 2001; Roberts *et al.*, 2009; Wilke, 2010), and the need for MDRO testing and monitoring by infection control. The therapy of these organisms may require use of an antibiotic with a higher likelihood of toxicity, thereby further increasing monitoring costs, LOS, etc.

There is also a cost of therapeutic failure which has important medical, as well as economic implications (Cunha and Ortega, 1995). Therapeutic failure results in consultants being called in to determine the cause of the failure. If therapeutic failure is due to an antimicrobial, e.g., wrong drug, dose, duration, etc., the patient will need to be retreated with other antimicrobials for effective therapy (Sánchez García, 2009). Therefore, the cost of the drug involved in the therapeutic failure (i.e., increased LOS, additional consults and laboratory tests needed by the consultant, as well as the costs of the effective drug needed to eradicate the infection) must be added to the cost of the effective retreatment (Cunha and Ortega, 1995; Ott *et al.*, 2012).

All of these "hidden costs" impact on the cost of antimicrobial therapy to the hospital and to healthcare systems. Exact numbers are difficult to come by but it is entirely clear that the cost of antimicrobial therapy involves many factors, and that both apparent and often hidden costs must be added to the acquisition costs. The acquisition cost of the antibiotic, by itself, is a "misleading indicator" of the true cost of the antibiotic (Cunha and Ortega, 1995) (see Table 30.2).

Costs of Adverse Effects

The most difficult hidden pharmacoeconomic factor to quantitate in antimicrobial therapy is the cost of adverse effects (AEs) (Cunha *et al.*, 2013). Antimicrobial AEs all have important pharmacoeconomic implications for the hospital costs of antimicrobial therapy (Sick *et al.*, 2013). Antimicrobial AEs may be viewed as common and non-life threatening, or uncommon and potentially life threatening. In both cases, there are important economic consequences of antimicrobial AEs. In hospitalized patients, perhaps the commonest adverse effects of antimicrobial therapy include phlebitis (for IV antibiotics only), drug fever and drug rash (Hamblin *et al.*, 2013). These are examples of common, but usually non-life threatening, AEs. In contrast, seizures are relatively uncommon, but are obviously important complications with multiple direct/indirect costs. Another very important cause of increased indirect costs as a result of AEs is *C. difficile* diarrhea/colitis, which has multiple and significant economic implications in terms of the cost of testing, cost of precautions, cost of treatment for the infection, and the economic consequences of increased LOS (Dubberke and Wertheimer, 2009; Bouza, 2012; McGlone *et al.*, 2012; Campbell *et al.*, 2013; Nistico *et al.*, 2013).

Antimicrobial AEs may be viewed according to their pharmacoeconomic implications. First, the initial cost to be considered is the cost of the clinical consultants and laboratory tests needed to make the diagnosis of an AE. Secondly, increased AEs are associated with and of vital importance in increasing potential LOS (Cunha *et al.*, 2013). Even the mildest AEs will increase LOS to some degree, and increased LOS is associated with an attendant decrease in reimbursement and suboptimal bed utilization (Fig. 30.1 and Table 30.3). Lastly, there are medicolegal implications with AEs, particularly the infrequent but serious AEs, such as *C. difficile* colitis (Wiegand *et al.*, 2012). If severe, aside from medicolegal liability-related costs, AEs can harm an institution's reputation and result in adverse publicity (Goldenberg *et al.*, 2012).

Conclusions

In summary, antibiotic adverse events are not usually considered as a "hidden cost" to hospitals of antimicrobial therapy (Eagye *et al.*, 2009; Cunha *et al.*, 2013). However, the impact of these adverse events should not be underestimated. Careful selection of drugs with equal efficacy, good safety profiles, and minimal adverse effect potential should be preferentially selected by an antimicrobial stewardship

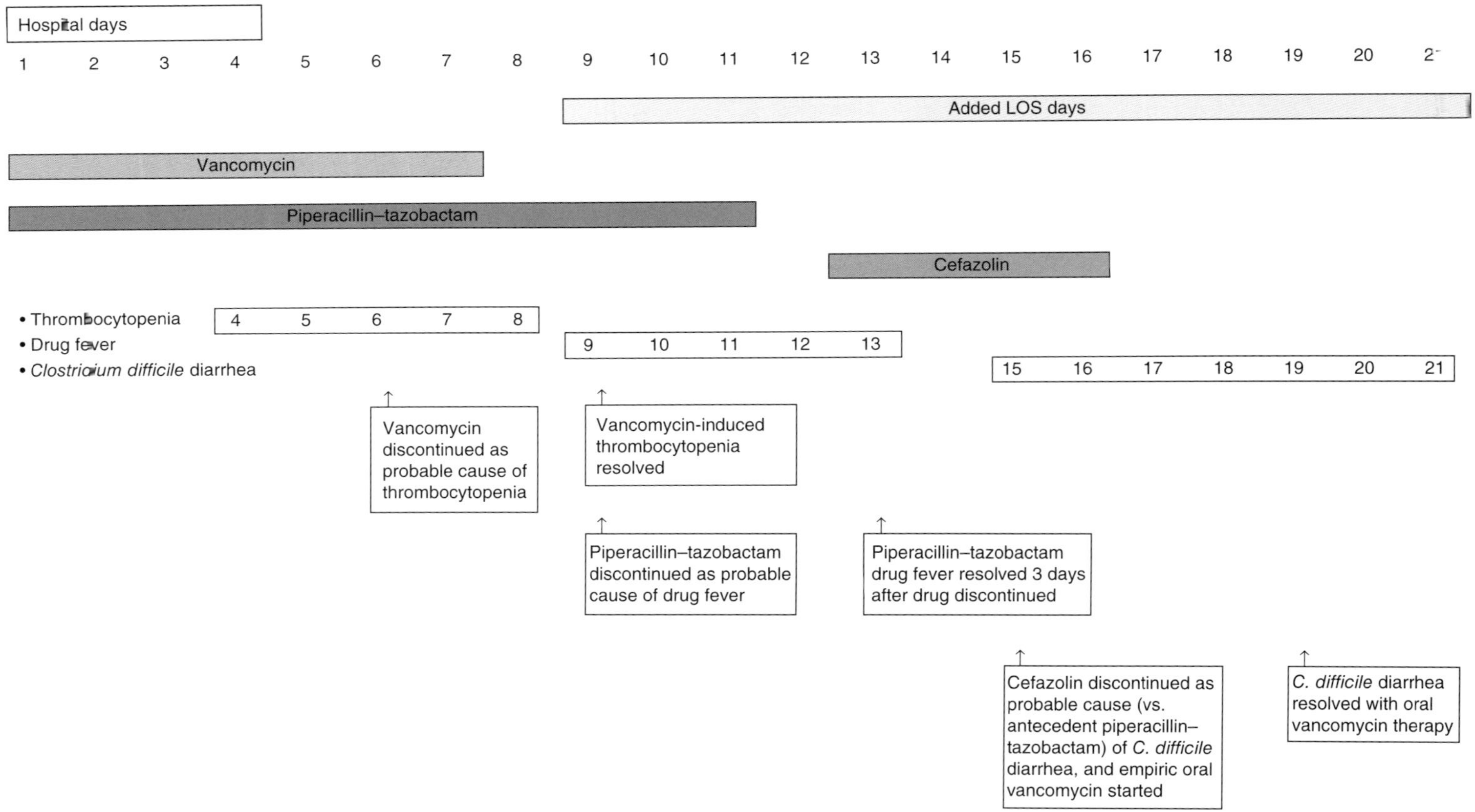

Fig. 30.1. Pharmacoeconomic implications of antimicrobial adverse effects (AEs) on hospital costs and length of stay (LOS): a schematic representation of the therapy of lower extremity cellulitis and potential AEs.

Table 30.3. Antimicrobial adverse events: pharmacoeconomic implications.

Antimicrobial adverse effects	Pharmacoeconomic parameters[a]		
	Economic potential (added costs of laboratory tests, subspecialty consultants)	Potential for increased length of stay (LOS) (prolonged LOS with reduced reimbursement and suboptimal bed utilization)	Medicolegal potential (cost of lawsuits and attendant adverse institutional publicity and damaged reputation)
Drug fevers	+++	+++	–
Drug rashes			
Maculopapular rashes	+	++	+
Stevens–Johnson syndrome	+	+++	+++
Anaphylaxis	+	–	+++
Thrombocytopenia	+	+++	++
Leukopenia	++	+	–
Anemia	+	–	+
Pancytopenia	++	+	+
Elevated transaminase	++	+	+
Elevated creatinine	0	+	+
Interstitial nephritis	+++	++	–
Phlebitis	++	+	–
Seizures	+++	+	–
Encephalopathy	+++	++	–
Diarrhea			
Non-*Clostridium difficile* diarrhea	+	–	–
C. difficile diarrhea	++	++	+
C. difficile colitis	+++	+++	+++
Oropharyngeal candidiasis (thrush)	–	++	–

[a]–, little/no impact; +, minimal impact; ++, moderate impact; +++, high impact.

program to minimize the possibility of adverse events and their attendant pharmacoeconomic consequences (Milkovich, 2009; Malani *et al.*, 2013; Sick *et al.*, 2013).

References

Bouza, E. (2012) Consequences of *Clostridium difficile* infection: understanding the healthcare burden. *Clinical Microbiology and Infection* 18, 5–12.

Campbell, R., Dean, B., Nathanson, B., Haidar, T., Strauss, M., and Thomas, S. (2013) Length of stay and hospital costs among high-risk patients with hospital-origin *Clostridium difficile*-associated diarrhea. *Journal of Medical Economics* 16, 440–448.

Cartaxo Salgado, F.X., Carneiro Gonçalves, J., Monteiro De Souza, C., Barbosa Da Silva, N., Gavilanes Sánchez, T.E., and Gomes de Oliveira Karnikowski, M. (2011) Cost of antimicrobial treatment of patients infected with multidrug-resistant organisms in the intensive care unit. *Medicina (B Aires)* 71, 531–535.

Colli, A., Campodonico, R., and Gherli, T. (2007) Early switch from vancomycin to oral linezolid for treatment of Gram-positive heart valve endocarditis. *Annual Thoracic Surgery* 84, 87–91.

Cunha, B.A. (1997) Intravenous-to-oral antibiotic switch therapy. A cost-effective approach. *Postgraduate Medicine* 101, 111–112.

Cunha, B.A. (2001) Effective antibiotic-resistance control strategies. *The Lancet* 357, 1307–1308 and 358, 1101.

Cunha, B.A. (2003) Factors in antibiotic formulary selection: antibiotic costs. *Pharmacy and Therapeutics* 28, 662–665.

Cunha, B.A. (2006) Oral antibiotic therapy of serious systemic infections. *Medical Clinics of North America* 90, 1197–1222.

Cunha, B.A. and Ortega, A.M. (1995) Antibiotic failure. *Medical Clinics of North America* 79, 663–672.

Cunha, C.B. (2013) Antibiotic overiew. In: Cunha, B.A. (ed.) *Antibiotic Essentials*, 12th edn. Jones & Bartlett, Sudbury, Massachusetts, pp. 1–15.

Cunha, C.B., Varughese, C.A., and Mylonakis, E. (2013) Antimicrobial stewardship programs (ASPs): the devil is in the details. *Virulence* 4, 147–149.

Dubberke, E.R. and Wertheimer, A.I. (2009) Review of current literature on the economic burden of *Clostridium difficile* infection. *Infection Control and Hospital Epidemiology* 30, 57–66.

Eagye, K.J., Nicolau, D.P., and Kuti, J.L. (2009) Impact of superinfection on hospital length of stay and costs in patients with ventilator-associated pneumonia. *Seminars in Respiratory Critical Care Medicine* 30, 116–123.

File, T.M., Jr. (2012) Duration and cessation of antimicrobial treatment. *Journal of Hospital Medicine* 7, S22–S33.

Frei, C.R., Jaso, T.C., Mortensen, E.M., Restrepo, M.I., Raut, M.K., Oramasionwu, C.U., Ruiz, A.D., Makos, B.R., Ruiz, J.L., Attridge, R.T. *et al.* (2009) Medical resource utilization among community-acquired pneumonia patients initially treated with levofloxacin 750 mg daily versus ceftriaxone 1000 mg plus azithromycin 500 mg daily: a US-based study. *Current Medical Resident Opinion* 25, 859–868.

Goff, D.A., Bauer, K.A., Reed, E.E., Stevenson, K.B., Taylor, J.J., and West, J.E. (2012) Is the "low-hanging fruit" worth picking for antimicrobial stewardship programs? *Clinical Infectious Diseases* 55, 587–592.

Goldenberg, S.D., Volpé, H., and French, G.L. (2012) Clinical negligence, litigation and healthcare-associated infections. *The Journal of Hospital Infection* 81, 156–162.

Gray, A., Dryden, M., and Charos, A. (2012) Antibiotic management and early discharge from hospital: an economic analysis. *Journal of Antimicrobial Chemotherapy* 67, 2297–2302.

Hamblin, S., Rumbaugh, K., and Miller, R. (2013) Prevention of adverse drug events and cost savings associated with PharmD interventions in an academic Level I trauma center: an evidence-based approach. *Journal of Trauma and Acute Care Surgery* 73, 1484–1490.

Kaier, K. (2012) Economic implications of the dynamic relationship between antibiotic use and hospital-acquired infections. *Value Health* 15, 87–93.

Leung, V., Gill, S., Sauve, J., Walker, K., Stumpo, C., and Powis, J. (2011) Growing a "positive culture" of antimicrobial stewardship in a community hospital. *The Canadian Journal of Hospital Pharmacy* 64, 314–320.

Mach, R., Vlcek, J., Prusova, M., Batka, P., Rysavy, V., and Kubena, A. (2007) Impact of a multidisciplinary approach on antibiotic consumption, cost and microbial resistance in a Czech hospital. *Pharmacy World and Science* [now *International Journal of Clinical Pharmacy*] 29, 565–572.

Malani, A.N., Richards, P.G., Kapila, S., Otto, M.H., Czerwinski, J., and Singal, B. (2013) Clinical and economic outcomes from a community hospital's antimicrobial stewardship program. *American Journal of Infection Control.* 41, 145–148.

Mansouri, M.D., Cadle, R.M., Agbahiwe, S.O., and Musher, D.M. (2011) Impact of an antibiotic restriction program on antibiotic utilization in the treatment of community-acquired pneumonia in a Veterans Affairs Medical Center. *Infection* 39, 53–58.

Mauldin, P.D., Salgado, C.D., Hansen, I.S., Durup, D.T., and Bosso, J.A. (2010) Attributable hospital cost and length of stay associated with health care-associated infections caused by antibiotic-resistant Gram-negative bacteria. *Antimicrobial Agents and Chemotherapy* 54, 109–115.

McGlone, S.M., Bailey, R.R., Zimmer, S.M., Popovich, M.J., Tian, Y., Ufberg, P., Muder, R.R., and Lee, B.Y. (2012) The economic burden of *Clostridium difficile*. *Clinical Microbiology and Infection* 18, 282–289.

Mertz, D., Plagge, H., Bassetti, S., Battegay, M., and Widmer, A.F. (2010) How much money can be saved by applying intravenous antibiotics once instead of several times a day? *Infection* 38, 479–482.

Milkovich, G. (2009) Future pharmacoeconomic criteria for the treatment of infections. *International Journal of Antimicrobial Agents* 34, S12–S14.

Mokabberi, R., Haftbaradaran, A., and Ravakhah, K. (2010) Doxycycline vs. levofloxacin in the treatment of community-acquired pneumonia. *Journal of Clinical Pharmacy and Therapeutics* 35, 195–200.

Monge, D., Millán, I., González-Escalada, A., and Asensio, A. (2013) [The effect of *Clostridium difficile* infection on length of hospital stay. A cohort study]. *Enfermedades Infecciosas Microbiología Clínica* 31, 660–664.

Nicolau, D.P. (2009) Containing costs and containing bugs: are they mutually exclusive? *Journal of Managed Care Pharmacy* 15(2, Suppl), S12–S17.

Nistico, J.A., Hage, J.E., Schoch, P.E., and Cunha, B.A. (2013) Unnecessary repeat *Clostridium difficile* PCR testing in hospitalized adults with *C. difficile*-negative diarrhea. *European Journal of Clinical Microbiology and Infectious Diseases* 32, 97–99.

Niwa, T., Shinoda, Y., Suzuki, A., Ohmori, T., Yasuda, M., Ohta, H., Fukao, A., Kitaichi, K., Matsuura, K., Sugiyama, T. *et al.* (2012) Outcome measurement of extensive implementation of antimicrobial stewardship in patients receiving intravenous antibiotics in a Japanese university hospital. *International Journal of Clinical Practice* 66, 999–1008.

Ott, S.R., Hauptmeier, B.M., Ernen, C., Lepper, P.M., Nüesch, E., Pletz, M.W., Hecht, J., Welte, T., and Bauer, T.T. (2012) Treatment failure in pneumonia: impact of antibiotic treatment and cost analysis. *European Respiratory Journal* 39, 611–618.

Paladino, J.A., Adelman, M.H., Schentag, J.J., and Iannini, P.B. (2007) Direct costs in patients hospitalised with community-acquired pneumonia after non-response to outpatient treatment with macrolide antibacterials in the US. *Pharmacoeconomics* 25, 677–683.

Raut, M., Schein, J., Mody, S., Grant, R., Benson, C., and Olson, W. (2009) Estimating the economic impact of a half-day reduction in length of hospital stay among patients with community-acquired pneumonia in the US. *Current Medical Research and Opinion* 25, 2151–2157.

Restrepo, M.I. and Frei, C.R. (2010) Health economics of use [of] fluoroquinolones to treat patients with community-acquired pneumonia. *The American Journal of Medicine* 123(4, Suppl), S39–S46.

Roberts, R.R., Hota, B., Ahmad, I., Scott, R.D., II, Foster, S.D., Abbasi, F., Schabowski, S., Kampe, L.M., Ciavarella, G.G., Supino, M. *et al.* (2009) Hospital and societal costs of antimicrobial-resistant infections in a Chicago teaching hospital: implications for antibiotic stewardship. *Clinical Infectious Diseases* 49, 1175–1184.

Rubinstein, E. and Keynan, Y. (2013) Short-course therapy for severe infections. *International Journal of Antimicrobial Agents* 42, S22–S24.

Sánchez García, M. (2009) Early antibiotic treatment failure. *International Journal of Antimicrobial Agents* 34, S14–S19.

Sick, A.C., Lehmann, C.U., Tamma, P.D., Lee, C.K., and Agwu, A.L. (2013) Sustained savings from a longitudinal cost analysis of an Internet-based preapproval antimicrobial stewardship program. *Infection Control and Hospital Epidemiology* 34, 573–580.

Signorovitch, J.E., Sheng Duh, M., Sengupta, A., Gu, A., Grant, R., Raut, M., Mody, S.H., Schein, J., Fisher, A.C., and Ng, D. (2010) Hospital visits and costs following outpatient treatment of CAP with levofloxacin or moxifloxacin. *Current Medical Research and Opinion* 26, 355–363.

Teo, J., Kwa, A.L., Loh, J., Chlebicki, M.P., and Lee, W. (2012) The effect of a whole-system approach in an antimicrobial stewardship programme at the Singapore General Hospital. *European Journal of Clinical Microbiology and Infectious Diseases* 31, 947–955.

Vrijens, F., Hulstaert, F., Van de Sande, S., Devriese, S., Morales, I., and Parmentier, Y. (2010) Hospital acquired, laboratory confirmed bloodstream infections: linking national surveillance data to clinical and financial hospital data to estimate increased length of stay and healthcare costs. *The Journal of Hospital Infection* 75, 158–162.

Wiegand, P.N., Nathwani, D., Wilcox, M.H., Stephens, J., Shelbaya, A., and Haider, S. (2012) Clinical and economic burden of *Clostridium difficile* infection in Europe: a systematic review of healthcare-facility-acquired infection. *The Journal of Hospital Infection* 81, 1–14.

Wilke, M.H. (2010) Multiresistant bacteria and current therapy – the economical side of the story. *European Journal Medical Research* 15, 571–576.

Yen, Y.H., Chen, H.Y., Wuan-Jin, L., Lin, Y.M., Shen, W.C., and Cheng, K.J. (2012) Clinical and economic impact of a pharmacist-managed i.v.-to-p.o. conversion service for levofloxacin in Taiwan. *International Journal of Clinical Pharmacology and Therapeutics* 50, 136–141.

Zimlichman, E., Henderson, D., Tamir, O., Franz, C., Song, P., Yamin, C.K., Keohane, C., Denham, C.R., and Bates, D.W. (2013) Health care-associated infections: a meta-analysis of costs and financial impact on the US health care system. *JAMA Internal Medicine* 173, 2039–2046

31 Antimicrobial Stewardship Programs in Areas of Increased Pathogen Resistance

MICHAEL SAMARKOS AND NIKOLAOS V. SIPSAS*

Medical School, National and Kapodistrian University of Athens and University Hospital of Athens "Laikon," Athens, Greece

Introduction

Antimicrobial resistance is considered among the major threats to human health (CDC, 2013) while very few new antibiotics are entering clinical practice or are in the production pipeline (Boucher *et al.*, 2009). As a result, the containment of antimicrobial resistance has become a global priority (WHO, 2012).

Although microorganisms can naturally produce antimicrobials, the use of antimicrobial agents in humans and animals is the most important factor driving resistance (Shlaes *et al.*, 1997; Costelloe *et al.*, 2010). Therefore, restriction in the use of antibiotics should be the principal target in the fight against resistance (Bartlett, 2011). This is not an impossible task. For instance, a study on the identification of targets for quality improvement in antimicrobial prescribing that covered 172 hospitals in 25 European countries, Zarb *et al.* (2011), using the ESAC (European Surveillance of Antimicrobial Consumption) PPS (point prevalence surveys) methodology, estimated that over a third of prescriptions for antimicrobials in hospitals do not comply with the local guidelines.

Goals of Antimicrobial Stewardship in Areas of Increased Pathogen Resistance

Antimicrobial stewardship programs (ASPs) must be *a fiduciary responsibility* for all healthcare institutions, and they should be implemented in all healthcare facilities (SHEA *et al.*, 2012). However, they face special problems and challenges in areas with increased pathogen resistance, when they need to implement strategies to:

- *Reduce resistance rates*: This is an overarching target for all ASPs, but when resistance is endemic, the need to reduce resistance rates becomes more pressing.
- *Control outbreaks*: In epidemics by resistant pathogens, ASPs should be part of a multidisciplinary program for outbreak containment.
- *Improve selection of the initial antimicrobial therapy*: Infections caused by resistant pathogens have worse outcomes than those caused by susceptible pathogens (Cosgrove, 2006). In areas of increased pathogen resistance, there is a substantial risk of either overtreatment or suboptimal treatment with antimicrobials (Morata *et al.*, 2012). Empiric overtreatment may occur in settings where physicians, being aware of the presence of increased resistance, are treating patients who are not infected by resistant pathogens (Livermore and Wain, 2013). Suboptimal or inappropriate treatment may result when physicians prescribe the usual empiric therapy for a patient who actually has a resistant pathogen. This is illustrated by the suggestion that a novel way to express resistance data is the "likelihood of inadequate treatment" (Burgmann *et al.*, 2009). To assess the probability of infection by resistant pathogens, ASPs can use individual patient data, unit-based and combination antibiograms, and data from surveillance studies (Paterson, 2008).

*Corresponding author. E-mail: nsipsas@med.uoa.gr

- *Reduce the duration of initial empiric antimicrobial therapy*: Relevant strategies include structural changes in workflow of the microbiology laboratory, the use of rapid non-culture techniques (e.g., PCR of clinical specimens), and the use of biomarkers to identify patients with bacterial infections (Davey *et al.*, 2013). These strategies should be integrated into ASPs for appropriate antimicrobial selection and de-escalation.
- *Optimize the efficacy of antimicrobials*: When resistant pathogens are prevalent, physicians have a narrower selection of antimicrobials. The challenge here is to optimize the efficacy of the given antimicrobial treatment by employing pharmacokinetic/pharmacodynamic (PK/PD) and susceptibility data to select the most appropriate dose and schedule or antimicrobial combination. Traditional antimicrobial stewardship activities such as prospective audit and feedback, as well as the use of a clinical decision support system (CDSS) can contribute to achieving this goal.

Interventions to Achieve ASP Targets

Interventions to improve antibiotic prescribing have been classified as persuasive, restrictive, and structural. Persuasive interventions include the dissemination of educational material, local consensus processes, educational meetings, reminders, audits, and feedback. Restrictive interventions include the use of compulsory antibiotic order forms, the requirement for expert approval, formulary restrictions, therapeutic substitutions (e.g., a change from intravenous [IV] to oral [PO] administration), automatic stop orders, and antibiotic cycling. Structural interventions include changing from paper to computerized records, changes to laboratory turnaround times, the introduction of new laboratory technology and the use of CDSSs (Davey *et al.*, 2013).

There are several methodological problems in studies on the effect of ASPs on antimicrobial resistance (Nijssen *et al.*, 2006; Davey *et al.*, 2013; Tacconelli *et al.*, 2014). These difficulties probably arise from the fact that the prevalence of resistance in hospitals is affected by multiple factors, such as antibiotic use, the introduction or transmission of resistance, patient turnover, "colonization pressure", cross-transmission between patients, and infection control practices. Under these conditions, it is not always easy to attribute results (such as a decrease in the resistance rates) to a specific intervention.

Core activities of ASPs

The two core strategies of ASP as described by the IDSA/SHEA guidelines are formulary restriction and preauthorization, and prospective audit with intervention and feedback (Dellit *et al.*, 2007). Formulary restriction and preauthorization are used in restrictive or "front-end" programs, which do not allow the prescription of a range of antimicrobials unless this has been approved by the stewardship team. Prospective audit with intervention and feedback are included in persuasive or "back-end" programs, which work by reviewing patients on antimicrobial treatment and offering recommendations to physicians. In the broader context, persuasive interventions include also guidelines and clinical pathways, reminders, the dissemination of educational materials, and educational sessions.

Restrictive strategies

The effect of restrictive interventions on controlling resistance is not big, and the IDSA/SHEA Guidelines have graded them as "B-II" (Dellit *et al.*, 2007). Davey *et al.* (2013) identified nine studies reporting microbiological outcomes for antibiotic-resistant Gram-negative bacteria (colonization or infection). Of these studies, eight included restrictive interventions, seven of which reported positive effects ranging from 25 to 96% (median 47%). Four out of seven studies with microbial outcomes for antibiotic-resistant Gram-positive bacteria included restrictive strategies, and these had uniformly positive results with statistically significant intervention effects and a mean effect size of 24%. Bradley *et al.* (1999) reported that a reduction in ceftazidime use by 61% was associated with 25% reduction in vancomycin-resistant enterococci (VRE).

Despite the fact that most studies have shown that restrictive strategies have a positive effect on resistance, there are concerns that antimicrobial stewardship efforts reduce the diversity of prescribing and may thus be harmful, because diversity of prescribing is actually a defense against resistance (Sandiumenge *et al.*, 2006). Restriction of one antimicrobial class usually leads to increased use of another and the emergence of resistance. This phenomenon has been termed "squeezing the resistance balloon" (Burke, 1998).

Antibiotic cycling is the practice of restricting antimicrobials sequentially, each for a given period of time, and it was introduced as a solution to the problem of "squeezing the balloon." The rationale is that alternating antimicrobial selection pressure would prevent the emergence of resistance. Although antibiotic cycling seemed initially to be effective in controlling resistance, it might have limited efficacy (Brown and Nathwani, 2005) and the IDSA/SHEA Guidelines do not recommend it (grading it as "C-II") (Dellit *et al.*, 2007). Interestingly, ecological modelling has suggested that cycling is unlikely to be effective (Bergstrom *et al.*, 2004). In another study, Toltzis *et al.* (2002) concluded that antibiotic cycling in a neonatal intensive care unit (ICU) had no detectable effect in decreasing the reservoir of resistant Gram-negative bacilli. Still, some authors have suggested that even the antibiotic heterogeneity alone that is provided by this practice, might justify its use (Sarraf-Yazdi *et al.*, 2012). There are several open issues with antibiotic cycling, such as the duration of rotation periods, and the selection and ordering of the antibiotics given (Tacconelli *et al.*, 2014). Recently, Goulart *et al.* (2013) suggested that alternating structurally similar antibiotics ("classic" cycling alternates between different antibiotic classes) can restore susceptibility to antibiotics after resistance has evolved.

Persuasive strategies

There are few studies on the effect of persuasive interventions on Gram-negative resistance. The Cochrane Review by Davey *et al.* (2013) included only one study of persuasive interventions which showed a statistically significant reduction in infections by ceftazidime-resistant *Enterobacteriaceae* (Carling *et al.*, 2003). Persuasive interventions seem to be effective in reducing the rates of Gram-positive resistance. There are several studies on methicillin-resistant *Staphylococcus aureus* (MRSA) and VRE which have shown positive effect sizes ranging from 2% to 87% (Carling *et al.*, 2003; Madaras-Kelly *et al.*, 2006). A notable study is that by Madaras-Kelly *et al.* (2006), which showed that a 50% reduction in levofloxacin use was associated with 21% reduction in MRSA infections.

Antibiotic optimization

The progress in our understanding of the PK/PD of antimicrobial drugs has given us the opportunity to optimize their use in patients with resistant organisms (Ambrose *et al.*, 2007). The most studied case of antibiotic optimization is the extended or continuous infusions of piperacillin–tazobactam and carbapenems. Falagas *et al.* (2013) studied extended dosing of piperacillin–tazobactam and carbapenems together, and concluded that there was a mortality benefit to this type of administration. The question of whether extended infusions might select resistant mutant strains has not been answered (Soman *et al.*, 2013). A recent modelling study suggested that extended infusion schedules might not be better than bolus schedules at suppressing resistance (Felton *et al.*, 2013).

Antibiotic optimization strategies are also especially relevant in infections by multidrug-resistant organisms (MDROs), when the selection of antibiotics is very limited. An illustrative example is infections by carbapenem-resistant *Enterobacteriaceae*. Population PK studies of colistin dosing in critically ill patients support the administration of a loading dose and subsequent use of high doses at extended intervals (Plachouras *et al.*, 2009). The effectiveness of this strategy has been confirmed by clinical studies (Dalfino *et al.*, 2012). Similarly, Tzouvelekis *et al.* (2012), in a review of clinical studies of infections by carbapenemase-producing *Enterobacteriaceae*, suggested that a combination of at least two active drugs should be used, as monotherapy was associated with worse clinical outcomes.

Parenteral to oral conversion

The conversion from parenteral to oral therapy, when the conditions (i.e., type of infection, microbiological confirmation of the pathogen, patient profile) allow (Hamilton-Miller, 1996), is a recommended element of ASPs, and is graded as an "A-I" intervention (Dellit *et al.*, 2007). This strategy might be especially relevant in areas where infections by MRSA are prevalent, because for moderate-to-severe MRSA infections (e.g., complicated skin and soft tissue infections), an IV to PO switch using linezolid can reduce the length of stay and overall cost (Schürmann *et al.*, 2009), and may also improve hospital bed use (Owens, 2008). There are no data on any effect of the IV to PO switch strategy on resistance rates, though the consequent reduction in the length of hospitalization might reduce new infections by MDROs and *Clostridium difficile* (Owens, 2008).

Risk assessment for infections by resistant pathogens

On an individual patient level, numerous variables have been reported as risk factors for infection by a resistant pathogen, including recent antibiotic use, prolonged hospitalization, hospitalization in the ICU, the prolonged presence of invasive medical devices (e.g., catheters), the presence of previously isolated pathogens, colonization by resistant pathogens, immunosuppression, underlying diseases such as diabetes mellitus, and age (Bisson *et al.*, 2002; Sakka *et al.*, 2008; Swaminathan *et al.*, 2013). These factors can vary by pathogen and by study, depending on the patient population, the setting (community vs. healthcare institution) and the study design.

However, the quantitative impact of each one of these factors has not been assessed and there are no formal clinical prediction rules for infections by resistant pathogens. Standardization of the risk-assessment for resistant pathogens using clinical prediction rules or other similar tools should probably be one of the future goals of ASPs in endemic settings. In fact, when the prevalence of resistance is increased, it is expected that the positive predictive value of these tools will increase as well, thus allowing the clinicians to tailor the initial empiric therapy to individual patients.

Cumulative antibiograms and surveillance data

Unit-based antibiograms are cumulative susceptibility reports which are based on data from a unit (e.g., one ward) rather than the whole hospital (Paterson, 2008). A study by Binkley *et al.* (2006) showed that unit-based antibiograms were more informative than hospital-wide antibiograms for decisions on empiric antimicrobial therapy. Another study suggested that unit-based antibiograms better represent the susceptibility pattern of hospital-acquired strains, while hospital-wide antibiograms reflect the patterns of community-acquired strains (Lamoth *et al.*, 2010).

Combination antibiograms are susceptibility reports that provide information for antimicrobial cross-resistance, and they are used when an empiric combination treatment is planned (Paterson, 2008). Their influence on the selection of empiric treatment seems to be modest, although they can provide information to avoid antimicrobial combinations that are not superior to monotherapy (Pogue *et al.*, 2011). Clinicians do not usually use cumulative antibiograms when making decisions about antimicrobial treatment, as shown in a study by Mermel *et al.* (2008), in which 64% of house officers never used this type of antibiogram.

Data from antimicrobial resistance surveillance programs at a national or international level have been used traditionally in making decisions on initial empiric antimicrobial treatment. These data can inform clinicians about increased rates of resistance to antimicrobials so as to avoid their empiric use (Paterson, 2008), but it is difficult to establish an association between guidance using surveillance data and the appropriateness of treatment at an individual patient level (Masterton, 2000).

Structural interventions

Structural interventions include changes to laboratory turnaround time that require substantive changes to work patterns of the microbiology laboratory, the introduction of new technology, or the introduction of computerized decision support (Davey *et al.*, 2013). Specific interventions include rapid microbiology testing using automated systems, and rapid reporting by changing from paper to computerized records with appropriate alerts or by phoning through the initial results. Data on the effectiveness of these interventions are conflicting (Davey *et al.*, 2013). Resistance rates have been reported in only one study, in which no differences were found, except for a higher frequency of carbapemen-resistant *Pseudomonas aeruginosa* in the intervention arm in comparison with historical controls (Galar *et al.*, 2012). It should be noted that in some reports, laboratory usage was lower and there was cost reduction in the intervention arm (Doern *et al.*, 1994; Galar *et al.*, 2012).

Biomarkers

The role of biomarkers is to distinguish patients with bacterial infection from those with viral infection or non-infectious diseases, thereby obviating the need for antimicrobial therapy. In addition, biomarkers can be used as an objective measure of clinical improvement. From the antimicrobial stewardship perspective, the determination of biomarkers can reduce antimicrobial exposure by guiding clinicians not to prescribe antimicrobials to patients with viral infections and by allowing them to shorten the duration of treatment in patients with bacterial infections (Livermore and Wain, 2013). The traditionally used biomarkers are white blood cell count and differential and C-reactive protein; however, these lack specificity for bacterial infections (Kruger and Welte, 2012). The determination of procalcitonin has a better specificity and can reduce antimicrobial exposure in critically ill patients with sepsis and in patients with pulmonary infections, without compromising clinical outcomes (Schuetz *et al.*, 2012).

Non-cultural methods for rapid pathogen identification

As the turnaround time of cultures cannot be reduced beyond a certain limit, more rapid microbiological diagnosis can be achieved using non-cultural methods. Livermore and Wain (2013) suggested the use of these methods and distinguished three levels on which they can be applied: (i) single or multiplex PCR seeking specific pathogens; (ii) PCR able to detect a range of pathogens; and (iii) genomic sequencing for pathogen and resistance gene identification. Single PCR tests for specific pathogens or specific resistance genes (e.g. *mecA* for MRSA) have been established in clinical practice, but they are not much relevant to ASPs, as they are used mostly for difficult-to-culture pathogens, to detect colonization, and for infection control purposes (Paule *et al.*, 2003). In contrast, multiplex PCR assays can simultaneously detect a number of pathogens and can support clinical decisions (Livermore and Wain, 2013).

The major limitation of the above methods is that they can identify only a narrow range of pathogens. Therefore, cultures should run in parallel. According to Livermore and Wain (2013), these systems are suitable for identifying infections by MDROs and are more useful on individual patient level rather than as tools for ASPs. We need studies to establish the impact of these methods on the outcomes of ASPs, such as the use of appropriate antimicrobial therapy, mortality, length of stay and, possibly, rates of resistance.

Matrix-assisted laser desorption-ionization time-of-flight (MALDI-TOF) mass spectrometry is a unique method of identifying species. The integration of MALDI-TOF into the clinical microbiology laboratory can have significant effects in time to appropriate or to optimal therapy, on the selection of therapy, on length of stay in the hospital or the ICU, and on total costs (Vlek *et al.*, 2012; Clerc *et al.*, 2013; Perez *et al.*, 2013; Tamma *et al.*, 2013). There is also evidence that the integration of antimicrobial stewardship with MALDI-TOF can have additive effects on outcomes (Tamma *et al.*, 2013).

Clinical decision support systems

CDSSs are "active knowledge systems which use two or more items of patient data to generate case-specific advice in contrast to passive knowledge systems in which the user conducts the search through the system's knowledge" (Wyatt and Spiegelhalter, 1991). The increasing implementation of electronic medical records (EMRs) has allowed CDSS to be used to improve decisions on antimicrobial therapy. A CDSS can support antimicrobial prescribing decisions either empirically or when they are pathogen guided, and provide assessments of infection risk and possible antibiotic resistance profiles, and advice on the choice of therapy and alerts for its initiation, and the modification or monitoring of therapy (Sintchenko *et al.*, 2008). In terms of antimicrobial stewardship, when pathogen resistance is frequent, a CDSS can improve the initial empiric therapy by, for example, detecting patients with risk factors for resistance. A CDSS can also shorten the time to initiation of antimicrobial therapy by detecting discrepancies between prescribed antibiotics and susceptibility results, and help to optimize dosing by suggesting adjustments or by detecting drug interactions.

Integration into the EMR allows the CDSS to support antimicrobial stewardship activities such as prospective audit and review (by using specially designed antibiotic order forms in the EMR), automatic stop orders, and IV to PO switching (by using the nutrition order report from the EMR). An integrated CDSS can also prioritize patients needing stewardship review by employing a variety of data from the EMR (e.g., laboratory results, drug levels); this is a high-value feature for ASPs with limited resources, or with a high workload, as it is the case of the areas with pathogen resistance (Kullar *et al.*, 2013).

The benefits of a CDSS for antibiotic stewardship are supported by studies showing more appropriate antimicrobial use, shortened length of stay, decreased costs, and decreased adverse events

(Paul *et al.*, 2006; Hermsen *et al.*, 2012; Patel *et al.*, 2012; Filice *et al.*, 2013). However, although evidence is accumulating, it has been noted that the methodology used to evaluate CDSSs is not always appropriate (Kaplan, 2001). In the recent Cochrane review by Davey *et al.* (2013), there was only one randomized trial which showed that the use of a CDSS improved the rate of appropriate empiric antibiotic treatment while reducing the use of broad-spectrum antibiotic treatment and antibiotic costs (Paul *et al.*, 2006). The two major barriers to the adoption of a CDSS are the cost of the software, which ranges from $100,000 to $500,000 a year (Kullar *et al.*, 2013), and the time required by ASP and Information Technology (IT) personnel to create, validate, and implement different rules and alerts, as well as the time that ASP personnel spend reviewing alerts which are not always actionable (Hermsen *et al.*, 2012; Kullar *et al.*, 2013).

National and international initiatives

National and international initiatives are indispensable for the sustained success of ASP efforts. Their functions include the design of the framework and the coordination of antimicrobial stewardship activities, the engagement of all key stakeholders, the ensuring of an integrated approach to antimicrobial stewardship within the public health agenda, the securing of the allocation of the necessary funds to an ASP, and the promotion of the goals of ASP to the public (Allerberger *et al.*, 2008; Nathwani *et al.*, 2011; O'Brien and Gould, 2013). Examples of national initiatives are those of the Scottish Antimicrobial Prescribing Group and the Antibiotic Policy Coordination Committee of Belgium, which have been successful in reducing the rates of resistance at least for some pathogens (Goossens *et al.*, 2008; Nathwani *et al.*, 2011).

International organizations such as the World Health Organization (WHO) and the European Centre for Disease Prevention and Control (ECDC) have launched initiatives to tackle the problem of resistance. These initiatives can mobilize the necessary expertise and resources in order to support a concerted effort to prevent and control antimicrobial resistance (WHO, 2012). The need for concerted efforts is underlined by the "Chennai Declaration," which even though it is technically a national initiative from India, stresses that resistance should be tackled on a global scale (Ghafur *et al.*, 2013).

Performance Measures

ASPs are quality improvement projects; therefore, establishing metrics for the evaluation of their impact is absolutely essential (Varkey *et al.*, 2007). ASPs have many and diverse elements and measuring their impact could help to determine which elements are more appropriate in a particular setting. Conversely, each ASP should set specific targets, depending on the existing problems in a particular setting, such as the overprescription of antimicrobials for respiratory infections, overuse of broad-spectrum antimicrobials, or increased resistance rates of specific pathogens (e.g. MRSA).

Performance measures include "process measures" and "outcome measures." In the case of antimicrobial stewardship, process measures include all indicators relating to the degree of compliance with the ASP policies (e.g., percentage of patients receiving the appropriate surgical prophylaxis antibiotic regimen) or to the consumption of antimicrobials (Berrington, 2010). Outcome measures include patient outcome measures (e.g., mortality rates, readmission rates), safety outcomes (e.g., adverse drug events), resistance indicators and cost indicators (McGowan, 2012).

In areas of increased pathogen resistance, the obvious choice of measures would include antimicrobial resistance indicators. However, this is a matter of debate. Some authors have suggested that resistance rates should not be used as an outcome measure for ASPs because they are influenced, apart from antimicrobial consumption, by many other variables (Griffith *et al.*, 2012), and because it might take years before an intervention results in reduced antimicrobial resistance rates (Ohl and Luther, 2011). Other authors, in contrast, argue that because ASPs have now been functioning for several years, we should be able to see their effect on resistance rates (McGowan, 2012). Another issue in using resistance rates as an indicator is whether we should be counting rates of colonization or rates of infection by resistant pathogens. Infection rates are used by a large number of ASPs because they are easier and less expensive to obtain, although they represent only "the tip of the iceberg," so colonization rates should be measured whenever this is feasible (Nijssen *et al.*, 2006). When an ASP is targeting a particular resistant pathogen or group of pathogens, the measurable outcome is resistance to this pathogen expressed as either colonization or infection rate. We have previously mentioned the "squeezing the balloon" of resistance effect; to avoid this, apart from pathogen-specific resistance rates, a measure of "global" resistance trends would also be useful. One such measure could be the "Drug resistance index", developed by Laxminarayan and Klugman (2011). This is a composite index that expresses aggregate resistance to various antibiotics and its trends across time. Despite these issues, resistance rates should be used, if not as an outcome measure for ASPs, then as a guide to the whole effort against resistance.

Summary and Conclusions

Antibiotic stewardship programs in areas of increased pathogen resistance face a difficult task. They should aim at the choice of the optimal empiric therapy, which, in the face of increased resistance, might require the use of broad-spectrum or combination antimicrobials, while at the same time they should attempt to reduce resistance rates. To achieve these targets, ASPs in areas of resistance should develop validated rules for the choice of empiric antimicrobial therapy and attempt to reduce its duration by changing the workflow of the microbiology laboratory and employing newer technologies for the faster identification of the causative pathogens. Additionally ASPs should aim at the optimal use of antimicrobials by using PK/PD data. In this effort, the microbiology laboratory and IT support (in the form of a CDSS) are indispensable.

In short, as I.M. Gould has written: "What is really required is better diagnosis (of course including better diagnostics), better severity assessment and, where antibiotics are indicated, better knowledge of pharmacokinetic/pharmacodynamic dosing schedules to optimise outcome and delay development of resistance" (Gould, 2009).

References

Allerberger, F., Frank, A., and Gareis, R. (2008) Antibiotic stewardship through the EU project "ABS International". *Wiener klinische Wochenschrift* 120, 256–263.

Ambrose, P.G., Bhavnani, S.M., Rubino, C.M., Louie, A., Gumbo, T., Forrest, A., and Drusano, G.L. (2007) Pharmacokinetics–pharmacodynamics of antimicrobial therapy: it's not just for mice anymore. *Clinical Infectious Diseases* 44, 79–86.

Bartlett, J.G. (2011) A call to arms: the imperative for antimicrobial stewardship. *Clinical Infectious Diseases* 53(Suppl 1), S4–S7.

Bergstrom, C.T., Lo, M., and Lipsitch, M. (2004) Ecological theory suggests that antimicrobial cycling will not reduce antimicrobial resistance in hospitals. *Proceedings of the National Academy of Sciences of the United States of America* 101, 13285–13290.

Berrington, A. (2010) Antimicrobial prescribing in hospitals: be careful what you measure. *Journal of Antimicrobial Chemotherapy* 65, 163–168.

Binkley, S., Fishman, N.O., LaRosa, L.A., Marr, A.M., Nachamkin, I., Wordell, D., Bilker, W.B., and Lautenbach, E. (2006) Comparison of unit-specific and hospital-wide antibiograms: potential implications for selection of empirical antimicrobial therapy. *Infection Control and Hospital Epidemiology* 27, 682–687.

Bisson, G., Fishman, N.O., Patel, J.B., Edelstein, P.H., and Lautenbach, E. (2002) Extended-spectrum beta-lactamase-producing *Escherichia coli* and *Klebsiella* species: risk factors for colonization and impact of antimicrobial formulary interventions on colonization prevalence. *Infection Control and Hospital Epidemiology* 23, 254–260.

Boucher, H.W., Talbot, G.H., Bradley, J.S., Edwards, J.E., Gilbert, D., Rice, L.B., Scheld, M., Spellberg, B., and Bartlett, J. (2009) Bad bugs, no drugs: no ESKAPE! An update from the Infectious Diseases Society of America. *Clinical Infectious Diseases* 48, 1–12.

Bradley, S.J., Wilson, A.L., Allen, M.C., Sher, H.A., Goldstone, A.H., and Scott, G.M. (1999) The control of hyperendemic glycopeptide-resistant *Enterococcus* spp. on a haematology unit by changing antibiotic usage. *Journal of Antimicrobial Chemotherapy* 43, 261–266.

Brown, E.M. and Nathwani, D. (2005) Antibiotic cycling or rotation: a systematic review of the evidence of efficacy. *Journal of Antimicrobial Chemotherapy* 55, 6–9.

Burgmann, H., Stoiser, B., Heinz, G., Schenk, P., Apfalter, P., Zedtwitz-Liebenstein, K., Frass, M., and Carmeli, Y. (2009) Likelihood of inadequate treatment: a novel approach to evaluating drug-resistance patterns. *Infection Control and Hospital Epidemiology*, 30, 672–677.

Burke, J.P. (1998) Antibiotic resistance – squeezing the balloon? *JAMA (The Journal of the American Medical Association)* 280, 1270–1271.

Carling, P., Fung, T., Killion, A., Terrin, N., and Barza, M. (2003) Favorable impact of a multidisciplinary antibiotic management program conducted during 7 years. *Infection Control and Hospital Epidemiology* 24, 699–706.

CDC (2013) *Antimicrobial Resistance Threats in the United States, 2013*. Centers for Disease Control and Prevention, Atlanta, Georgia. Available at: http://www.cdc.gov/drugresistance/pdf/ar-threats-2013-508.pdf (accessed 3 June 2016).

Clerc, O., Prod'hom, G., Vogne, C., Bizzini, A., Calandra, T., and Greub, G. (2013) Impact of matrix-assisted laser desorption ionization time-of-flight mass spectrometry on the clinical management of patients with Gram-negative bacteremia: a prospective observational study. *Clinical Infectious Diseases* 56, 1101–1107.

Cosgrove, S.E. (2006) The relationship between antimicrobial resistance and patient outcomes: mortality, length of hospital stay, and health care costs. *Clinical Infectious Diseases* 42(Suppl 2), S82–S89.

Costelloe, C., Metcalfe, C., Lovering, A., Mant, D., and Hay, A.D. (2010) Effect of antibiotic prescribing in primary care on antimicrobial resistance in individual patients: Systematic review and meta-analysis. *BMJ* 340:c2096.

Dalfino, L., Puntillo, F., Mosca, A., Monno, R., Spada, M.L., Coppolecchia, S., Miragliotta, G., Bruno, F., and Brienza, N. (2012) High-dose, extended-interval colistin administration in critically ill patients: is this the right dosing strategy? A preliminary study. *Clinical Infectious Diseases* 54, 1720–1726.

Davey, P., Brown, E., Charani, E., Fenelon, L., Gould, I.M., Holmes, A., Ramsay, C.R., Wiffen, P.J., and Wilcox, M. (2013) Interventions to improve antibiotic prescribing practices for hospital inpatients. *Cochrane Database of Systematic Reviews* 4:CD003543.

Dellit, T.H., Owens, R.C., McGowan, J.E., Jr., Gerding, D.N., Weinstein, R.A., Burke, J.P., Huskins, W.C., Paterson, D.L., Fishman, N.O., Carpenter, C.F. *et al.* (2007) Infectious Diseases Society of America and the Society for Healthcare Epidemiology of America guidelines for developing an institutional program to enhance antimicrobial stewardship. *Clinical Infectious Diseases* 44, 159–177.

Doern, G.V., Vautour, R., Gaudet, M., and Levy, B. (1994) Clinical impact of rapid in vitro susceptibility testing and bacterial identification. *Journal of Clinical Microbiology* 32, 1757–1762.

Falagas, M.E., Tansarli, G.S., Ikawa, K., and Vardakas, K.Z. (2013) Clinical outcomes with extended or continuous versus short-term intravenous infusion of carbapenems and piperacillin/tazobactam: a systematic review and meta-analysis. *Clinical Infectious Diseases* 56, 272–282.

Felton, T.W., Goodwin, J., O'Connor, L., Sharp, A., Gregson, L., Livermore, J., Howard, S.J., Neely, M.N., and Hope, W.W. (2013) Impact of bolus dosing versus continuous infusion of piperacillin and tazobactam on the development of antimicrobial resistance in *Pseudomonas aeruginosa*. *Antimicrobial Agents and Chemotherapy* 57, 5811–5819.

Filice, G.A., Drekonja, D.M., Thurn, J.R., Rector, T.S., Hamann, G.M., Masoud, B.T., Leuck, A.M., Nordgaard, C.L., Eilertson, M.K., and Johnson, J.R. (2013) Use of a computer decision support system and antimicrobial therapy appropriateness. *Infection Control and Hospital Epidemiology* 34, 558–565.

Galar, A., Leiva, J., Espinosa, M., Guillén-Grima, F., Hernáez, S., and Yuste, J.R. (2012) Clinical and economic evaluation of the impact of rapid microbiological diagnostic testing. *Journal of Infection* 65, 302–309.

Ghafur, A., Mathai, D., Muruganathan, A., Jayalal, J.A., Kant, R., Chaudhary, D., Prabhash, K., Abraham, O.C., Gopalakrishnan, R., Ramasubramanian, V. *et al.* (2013) The Chennai Declaration: a roadmap to tackle the challenge of antimicrobial resistance. *Indian Journal of Cancer* 50, 71–73.

Goossens, H., Coenen, S., Costers, M., De, C.S., De, S.A., Gordts, B., Laurier, L., and Struelens, M. (2008) Achievements of the Belgian Antibiotic Policy Coordination Committee (BAPCOC). *Eurosurveillance* 13(46):19036.

Goulart, C.P., Mahmudi, M., Crona, K.A., Jacobs, S.D., Kallmann, M., Hall, B.G., Greene, D.C. and Barlow, M. (2013) Designing antibiotic cycling strategies by determining and understanding local adaptive landscapes. *PLoS One* 8(2), e56040.

Gould, I.M. (2009) Controversies in infection: infection control or antibiotic stewardship to control healthcare-acquired infection? *Journal of Hospital Infection* 73, 386–391.

Griffith, M., Postelnick, M., and Scheetz, M. (2012) Antimicrobial stewardship programs: methods of operation and suggested outcomes. *Expert Review of Anti-infective Therapy* 10, 63–73.

Hamilton-Miller, J.M. (1996) Switch therapy: the theory and practice of early change from parenteral to non-parenteral antibiotic administration. *Clinical Microbiology and Infection* 2, 12–19.

Hermsen, E.D., VanSchooneveld, T.C., Sayles, H., and Rupp, M.E. (2012) Implementation of a clinical decision support system for antimicrobial stewardship. *Infection Control and Hospital Epidemiology* 33, 412–415.

Kaplan, B. (2001) Evaluating informatics applications – clinical decision support systems literature review. *International Journal of Medical Informatics* 64, 15–37.

Kruger, S. and Welte, T. (2012) Biomarkers in community-acquired pneumonia. *Expert Review of Respiratory Medicine* 6, 203–214.

Kullar, R., Goff, D.A., Schulz, L.T., Fox, B.C., and Rose, W.E. (2013) The "epic" challenge of optimizing antimicrobial stewardship: the role of electronic medical records and technology. *Clinical Infectious Diseases* 57, 1005–1013.

Lamoth, F., Wenger, A., Prod'hom, G., Vallot, Y., Pluss-Suard, C., Bille, J. and Zanetti, G. (2010) Comparison of hospital-wide and unit-specific cumulative antibiograms in hospital- and community-acquired infection. *Infection* 38(4), 249–253.

Laxminarayan, R. and Klugman, K.P. (2011) Communicating trends in resistance using a drug resistance index. *BMJ Open* 1(2):e000135.

Livermore, D.M. and Wain, J. (2013) Revolutionising bacteriology to improve treatment outcomes and antibiotic stewardship. *Journal of Infection and Chemotherapy* 45, 1–10.

Madaras-Kelly, K.J., Remington, R.E., Lewis, P.G., and Stevens, D.L. (2006) Evaluation of an intervention designed to decrease the rate of nosocomial methicillin-resistant *Staphylococcus aureus* infection by encouraging decreased fluoroquinolone use. *Infection Control and Hospital Epidemiology* 27, 155–169.

Masterton, R.G. (2000) Surveillance studies: how can they help the management of infection? *Journal of Antimicrobial Chemotherapy* 46(Suppl B), 53–58.

McGowan, J.E. (2012) Antimicrobial stewardship – the state of the art in 2011: focus on outcome and methods. *Infection Control and Hospital Epidemiology* 33, 331–337.

Mermel, L.A., Jefferson, J. and Devolve, J. (2008) Knowledge and use of cumulative antimicrobial susceptibility data at a university teaching hospital. *Clinical Infectious Diseases* 46(11): 1789.

Morata, L., Cobos-Trigueros, N., Martinez, J.A., Soriano, A., Almela, M., Marco, F., Sterzik, H., Nunez, R., Hernandez, C., and Mensa, J. (2012) Influence of multidrug resistance and appropriate empirical therapy on the 30-day mortality rate of *Pseudomonas aeruginosa* bacteremia. *Antimicrobial Agents and Chemotherapy* 56, 4833–4837.

Nathwani, D., Sneddon, J., Malcolm, W., Wiuff, C., Patton, A., Hurding, S., Eastaway, A., Seaton, R.A., Watson, E., Gillies, E. *et al.* (2011) Scottish Antimicrobial Prescribing Group (SAPG): development and impact of the Scottish National Antimicrobial Stewardship Programme. *International Journal of Antimicrobial Agents* 38, 16–26.

Nijssen, S., Bootsma, M., and Bonten, M. (2006) Potential confounding in evaluating infection-control interventions in hospital settings: Changing antibiotic prescription. *Clinical Infectious Diseases* 43, 616–623.

O'Brien, D.J. and Gould, I.M. (2013) Maximizing the impact of antimicrobial stewardship: the role of diagnostics, national and international efforts. *Current Opinion in Infectious Diseases* 26, 352–358.

Ohl, C.A. and Luther, V.P. (2011) Antimicrobial stewardship for inpatient facilities. *Journal of Hospital Medicine* 6(Suppl 1), S4–S15.

Owens, R.C., Jr. (2008) Antimicrobial stewardship: concepts and strategies in the 21st century. *Diagnostic Microbiology and Infectious Disease* 61, 110–128.

Patel, J., Esterly, J.S., Scheetz, M.H., Bolon, M.K., and Postelnick, M.J. (2012) Effective use of a clinical decision-support system to advance antimicrobial stewardship. *American Journal of Health-System Pharmacy* 69, 1543–1544.

Paterson, D.L. (2008) Impact of antibiotic resistance in Gram-negative bacilli on empirical and definitive antibiotic therapy. *Clinical Infectious Diseases* 47(Suppl 1), S14–S20.

Paul, M., Andreassen, S., Tacconelli, E., Nielsen, A.D., Almanasreh, N., Frank, U., Cauda, R., and Leibovici, L. (2006) Improving empirical antibiotic treatment using TREAT, a computerized decision support system: cluster randomized trial. *Journal of Antimicrobial Chemotherapy* 58, 1238–1245.

Paule, S.M., Trick, W.E., Tenover, F.C., Lankford, M., Cunningham, S., Stosor, V., Cordell, R.L., and Peterson, L.R. (2003) Comparison of PCR assay to culture for surveillance detection of vancomycin-resistant enterococci. *Journal of Clinical Microbiology* 41, 4805–4807.

Perez, K.K., Olsen, R.J., Musick, W.L., Cernoch, P.L., Davis, J.R., Land, G.A., Peterson, L.E., and Musser, J.M. (2013) Integrating rapid pathogen identification and antimicrobial stewardship significantly decreases hospital costs. *Archives of Pathology and Laboratory Medicine* 137, 1247–1254.

Plachouras, D., Karvanen, M., Friberg, L.E., Papadomichelakis, E., Antoniadou, A., Tsangaris, I., Karaiskos, I., Poulakou, G., Kontopidou, F., Armaganidis, A. *et al.* (2009) Population pharmacokinetic analysis of colistin methanesulfonate and colistin after intravenous administration in critically ill patients with infections caused by Gram-negative bacteria. *Antimicrobial Agents and Chemotherapy* 53, 3430–3436.

Pogue, J.M., Alaniz, C., Carver, P.L., Pleva, M., Newton, D., and DePestel, D.D. (2011) Role of unit-specific combination antibiograms for improving the selection of appropriate empiric therapy for Gram-negative pneumonia. *Infection Control and Hospital Epidemiology* 32, 289–292.

Sakka, V., Tsiodras, S., Galani, L., Antoniadou, A., Souli, M., Galani, I., Pantelaki, M., Siafakas, N., Zerva, L., and Giamarellou, H. (2008) Risk-factors and predictors of mortality in patients colonised with vancomycin-resistant enterococci. *Clinical Microbiology and Infection* 14, 14–21.

Sandiumenge, A., Diaz, E., Rodriguez, A., Vidaur, L., Canadell, L., Olona, M., Rue, M., and Rello, J. (2006) Impact of diversity of antibiotic use on the development of antimicrobial resistance. *Journal of Antimicrobial Chemotherapy* 57, 1197–1204.

Sarraf-Yazdi, S., Sharpe, M., Bennett, K.M., Dotson, T.L., Anderson, D.J., and Vaslef, S.N. (2012) A 9-year retrospective review of antibiotic cycling in a surgical intensive care unit. *Journal of Surgical Research* 176(2), e73–e78.

Schuetz, P., Briel, M., Christ-Crain, M., Stolz, D., Bouadma, L., Wolff, M., Luyt, C.E., Chastre, J., Tubach, F., Kristoffersen, K.B. *et al.* (2012) Procalcitonin to guide initiation and duration of antibiotic treatment in acute respiratory infections: an individual patient data meta-analysis. *Clinical Infectious Diseases* 55, 651–662.

Schürmann, D., Sorensen, S.V., De Cock, E., Duttagupta, S., and Resch, A. (2009) Cost-effectiveness of linezolid versus vancomycin for hospitalised patients with complicated skin and soft-tissue infections in Germany. *The European Journal of Health Economics* 10, 65–79.

Shlaes, D.M., Gerding, D.N., John, J.F., Jr., Craig, W.A., Bornstein, D.L., Duncan, R.A., Eckman, M.R., Farrer, W.E., Greene, W.H., Lorian, V. *et al.* (1997) Society for Healthcare Epidemiology of America and Infectious Diseases Society of America Joint Committee on the Prevention of Antimicrobial Resistance: guidelines for the prevention of antimicrobial resistance in hospitals. *Clinical Infectious Diseases* 25, 584–599.

Sintchenko, V., Coiera, E., and Gilbert, G.L. (2008) Decision support systems for antibiotic prescribing. *Current Opinion in Infectious Diseases* 21, 573–579.

SHEA, IDSA and PIDS (2012) Policy statement on antimicrobial stewardship by the Society for Healthcare Epidemiology of America (SHEA), the Infectious Diseases Society of America (IDSA), and the Pediatric Infectious Diseases Society (PIDS). *Infection Control and Hospital Epidemiology* 33, 322–327.

Soman, R., Gupta, N., Shetty, A., and Rodrigues, C. (2013) Are prolonged/continuous infusions of β-lactams for all? *Clinical Infectious Diseases* 57, 323.

Swaminathan, M., Sharma, S., Poliansky, B.S., Patel, G., Banach, D.B., Phillips, M., LaBombardi, V., Anderson, K.F., Kitchel, B., Srinivasan, A., and Calfee, D.P. (2013) Prevalence and risk factors for acquisition of carbapenem-resistant *Enterobacteriaceae* in the setting of endemicity. *Infection Control and Hospital Epidemiology* 34, 809–817.

Tacconelli, E., Cataldo, M.A., Dancer, S.J., De Angelis, G., Falcone, M., Frank, U., Kahlmeter, G., Pan, A., Petrosillo, N., Rodriguez-Bano, J. *et al.* (2014) ESCMID guidelines for the management of the infection control measures to reduce transmission of multidrug-resistant Gram-negative bacteria in hospitalized patients. *Clinical Microbiology and Infection* 20(Suppl 1), 1–55.

Tamma, P.D., Tan, K., Nussenblatt, V.R., Turnbull, A.E., Carroll, K.C., and Cosgrove, S.E. (2013) Can matrix-assisted laser desorption ionization time-of-flight mass spectrometry (MALDI-TOF) enhance antimicrobial stewardship efforts in the acute care setting? *Infection Control and Hospital Epidemiology* 34, 990–995.

Toltzis, P., Dul, M.J., Hoyen, C., Salvator, A., Walsh, M., Zetts, L., and Toltzis, H. (2002) The effect of antibiotic rotation on colonization with antibiotic-resistant bacilli in a neonatal intensive care unit. *Pediatrics* 110, 707–711.

Tzouvelekis, L.S., Markogiannakis, A., Psichogiou, M., Tassios, P.T., and Daikos, G.L. (2012) Carbapenemases in *Klebsiella pneumoniae* and other *Enterobacteriaceae*: an evolving crisis of global dimensions. *Clinical Microbiology Reviews* 25, 682–707.

Varkey, P., Reller, M.K., and Resar, R.K. (2007) Basics of quality improvement in health care. *Mayo Clinic Proceedings* 82, 735–739.

Vlek, A.L., Bonten, M.J., and Boel, C.H. (2012) Direct matrix-assisted laser desorption ionization time-of-flight mass spectrometry improves appropriateness of antibiotic treatment of bacteremia. *PLoS ONE* 7(3): e32589.

WHO (2012) *The Evolving Threat of Antimicrobial Resistance: Options for Action*. World Health Organization, Geneva, Switzerland. Available at: http://apps.who.int/iris/bitstream/10665/44812/1/9789241503181_eng.pdf (accessed 3 June 2016).

Wyatt, J. and Spiegelhalter, D., (1991) Field trials of medical decision-aids: potential problems and solutions. *Proceedings of the 15th Annual Symposium on Computer Application in Medical Care (SCAMC), Washington, DC, November 1991*. American Medical Informatics Association (AMIA), Bethesda, Maryland, pp. 3–7. Available at: http://www.ncbi.nlm.nih.gov/pmc/articles/PMC2247484/pdf/procascamc00004-0028.pdf (accessed 3 June 2016).

Zarb, P., Amadeo, B., Muller, A., Drapier, N., Vankerckhoven, V., Davey, P., and Goossens, H. (2011) Identification of targets for quality improvement in antimicrobial prescribing: the web-based ESAC Point Prevalence Survey 2009. *Journal of Antimicrobial Chemotherapy* 66, 443–449.

32 Role of Antimicrobial Stewardship in Pediatrics

JENNIFER L. GOLDMAN[1*] AND JASON G. NEWLAND[2]

[1]University of Missouri-Kansas City and Children's Mercy Hospital, Missouri, Kansas City, US; [2]Washington University School of Medicine in St. Louis and St. Louis Children's Hospital, Missouri, US

Introduction

Antimicrobial stewardship has been an effective strategy to improve antibiotic prescribing in pediatrics. A significant reduction in antibiotic use has been observed in freestanding children's hospitals with active antimicrobial stewardship programs (ASPs) and in outpatient settings where stewardship principles have been implemented (Andrews *et al.*, 2012; Gerber *et al.*, 2013a; Hersh *et al.*, 2015). This chapter will describe current knowledge regarding antimicrobial use in children, the negative consequences of excess antimicrobial prescribing, the factors unique to pediatric antimicrobial stewardship, and the successful stewardship strategies implemented in the inpatient and outpatient settings.

Antimicrobial Use in Children

Inpatient use

Antimicrobials are frequently prescribed drugs in hospitalized children. The use of broad-spectrum antimicrobials (e.g., carbapenems and linezolid) has significantly increased, while that of narrow-spectrum agents (e.g., cefazolin and gentamicin) has decreased (Pakyz *et al.*, 2009). At 40 freestanding children's hospitals in the US, approximately 60% of hospitalized children are prescribed at least one antimicrobial. Significant variability in antimicrobial prescribing exists among these pediatric institutions, with the proportion of children receiving antibiotics ranging from 38 to 72%, despite controlling for factors that would account for different levels of patient severity (Gerber *et al.*, 2010). A further breakdown of the antimicrobial use in these hospitals found that 43% of the days of therapy were used in surgical patients. The four most common clinical conditions for antibiotic use were pneumonia, appendicitis, cystic fibrosis, and skin and soft tissue infections, which together accounted for approximately 13% of all antimicrobial use (Gerber *et al.*, 2013b).

Point prevalence surveys have also been used to quantify the extent of inpatient antimicrobial use. Grohskopf *et al.* (2005) noted that 71% of children admitted to pediatric intensive care units (PICU) and 43% of those admitted to neonatal intensive care units (NICU) received antimicrobial therapy. In the PICU, the antimicrobials were mainly used empirically and were broad spectrum, but in the NICU the neonates were more likely to receive more than one antimicrobial. A study performing point prevalence surveys in the fall and winter in the cardiac and medical–surgical critical care units (CCCU-PICU) of a pediatric hospital in Toronto, Canada identified that 70 and 79% of the children, respectively, were receiving an antimicrobial. In the winter, 28% of the patients were receiving four or more antimicrobials, and the most common indication at this time was for empiric use (Blinova *et al.*, 2013).

An international point prevalence survey of hospitalized children saw the percentage of children receiving an antimicrobial in non-European hospitals being greater than European hospitals (44 vs. 39%). Additionally, the northern European region

*Corresponding author. E-mail: jgnewland@wustl.edu

had a significantly lower antibiotic prevalence rate in the pediatric wards than the southern European region (30 vs. 38%). The hospital units that experienced the greatest use were hematology/oncology (64%) and the PICU (57%) (Versporten *et al.*, 2013).

Although data are available on overall antimicrobial use, information on the frequency of inappropriate use in pediatrics is limited. Bolon *et al.* (2005a) reported that in a freestanding children's hospital, 35% of initial vancomycin courses and 72% of prolonged courses were inappropriate. The most common reasons for inappropriate prescribing were continued use despite the identification of a β-lactam susceptible bacterium and perioperative prophylaxis in children not allergic to a β-lactam. In a PICU and cardiac intensive care setting, inappropriate use was as high as 62%. Unnecessary broad-spectrum therapy, absence of infection, and prolonged length of therapy were the most commonly encountered reasons for inappropriate use (Blinova *et al.*, 2013).

Outpatient use

The majority of antibiotic prescribing in pediatrics occurs in the outpatient setting. In the US, an estimated 49 million antimicrobial prescriptions a year are provided for children in the ambulatory setting. Half of the prescriptions are for broad-spectrum antimicrobials, with azithromycin being the most common. While overall the data have suggested that antimicrobial use is decreasing in the outpatient setting, the use of azithromycin continues to increase (Hicks *et al.*, 2011). Almost three-quarters of the use is for the treatment of acute respiratory tract infections, of which approximately 25% do not need to be treated with an antibiotic as the etiology is presumed to be viral (Hersh *et al.*, 2011). Studies have suggested that 11 to 12 million excess antibiotic prescriptions are provided to children with acute respiratory tract infections yearly (Nyquist *et al.*, 1998; Kronman *et al.*, 2014).

Outpatient parenteral antimicrobial therapy (OPAT) is an important and often overlooked aspect of outpatient pediatric antimicrobial use. OPAT occurs in approximately 3% of hospital discharges (Madigan and Banerjee, 2013; Knackstedt *et al.*, 2015). Complications with OPAT occur frequently, affecting up to 50% of patients (Ruebner *et al.*, 2006; Barrier *et al.*, 2012; Madigan and Banerjee, 2013; Akar *et al.*, 2014). The most common complications are due to a problem with the central line (e.g., dislodgement) or an adverse event from an antibiotic. The involvement of infectious diseases in OPAT ranges from 45 to 90% of cases, and inappropriate use of OPAT is as high as 43% (Akar *et al.*, 2014; Knackstedt *et al.*, 2015).

Antibiotic Resistance in Children

The overall impact of antibiotic use is not trivial as antibiotic resistance is a known consequence of antimicrobial use. In 2013, a threat report on antimicrobial resistance from the US Centers for Disease Control and Prevention (CDC, 2013) estimated that 23,000 Americans die annually from antibiotic-resistant infections and that an additional 2 million are infected with an antibiotic-resistant infection. It is unclear what proportions of these infections were in children. However, the community-associated methicillin-resistant *Staphylococcus aureus* (MRSA) era was first recognized in children. Herold *et al.* (1998) demonstrated a significant increase in children with MRSA infections who had no previously recognized risk factors for the acquisition of MRSA. Interestingly, this resistance resulted in a significant increase in clindamycin prescribing for *S. aureus* infections among children admitted to freestanding children's hospitals in the US (Herigon *et al.*, 2010).

Another important pathogen in children is *Streptococcus pneumoniae*. *S. pneumoniae* is the most common cause of community-acquired pneumonia (CAP), sinusitis, and acute otitis media. In regions of the US with the highest use of antimicrobials, non-susceptible invasive *S. pneumoniae* isolates occurred at a higher frequency than in regions that used fewer antimicrobials. Additionally, cephalosporin and macrolide antibiotic prescribing have been associated with multidrug-resistant *S. pneumoniae* (Hicks *et al.*, 2011). Europe has observed a similar phenomenon, with high antibiotic-consuming countries in the south and east having higher antibiotic resistance than the northern European countries that have less overall antibiotic consumption (Albrich *et al.*, 2004).

Antimicrobial-related Adverse Drug Reactions in Children

Adverse drug reactions (ADRs) are yet another negative consequence of antimicrobial use. An ADR describes a response to a medicine that is noxious and unintended, and occurs with the normal drug dose (Edwards and Biriell, 1994; Nebeker *et al.*, 2004; WHO, 2008). The majority of ADRs are dose dependent (type A), although ~20% are completely unpredictable and an immune-mediated process is usually assumed (type B) (Riedl and Casillas, 2003). An example of a type A ADR is the infusion-related presentation of red man syndrome from vancomycin exposure. Type B ADRs include Stevens–Johnson syndrome following the administration of trimethoprim–sulfamethoxazole or anaphylaxis associated with the use of penicillin.

It is estimated that the incidence of ADRs in hospitalized children is approximately 9.5%, with 2% of pediatric hospital admissions related to ADRs, and an average incidence of 1.5% ADRs in the outpatient pediatric setting (Impicciatore *et al.*, 2001; Jonville-Béra *et al.*, 2002). Antimicrobials are one of the most common drug classes associated with ADRs in children in the hospital, the emergency department and the outpatient setting (Le *et al.*, 2006; Smyth *et al.*, 2012). These undesired drug effects result in significant use of healthcare. An estimate of over 100,000 emergency department visits are due to antibiotic-associated ADRs annually. Penicillins and cephalosporins are most frequently implicated; but trimethoprim–sulfamethoxazole and clindamycin have a higher actual rate of associated ADRs (Shehab *et al.*, 2008). The extent of undesired drug effects related to antibiotics in children is likely under-recognized and under-reported, because documenting suspected reactions is voluntary and practitioners may view the process as time-consuming or lack knowledge about the reporting system (Eland *et al.*, 1999; Hazell and Shakir, 2006). ADRs can result in devastating consequences to affected individuals and their families, in addition to causing a significant economic burden to the medical care system (Goettler *et al.*, 1997). Hence, minimizing unnecessary antimicrobial usage is the most critical step in avoiding adverse reactions as they are both unpredictable and undesired.

Other unintended consequences of antimicrobial exposure also require evaluation and serve as key motivators for active stewardship programs. *Clostridium difficile*-associated disease is a well-recognized sequela of antibiotic exposure and the evidence suggests that there is an increase in the burden of *C. difficile* among children (Kim *et al.*, 2008). Pediatric *C. difficile* disease is associated with increased mortality, higher medical care costs and increased length of stay in hospitalized children (Sammons *et al.*, 2013). Thus, *C. difficile* infections serve as a potential outcome measure of effective stewardship as well as an effective driver for implementing an ASP in efforts to limit antimicrobial-associated disease.

Pediatric Antimicrobial Stewardship Programs

The combination of the excess use of antibiotics, the increases in antibiotic resistance observed, and the high number of ADRs in children make performing antimicrobial stewardship in pediatrics an essential part of providing safe care. While many of the stewardship strategies described in the Infectious Diseases Society of America and the Society for Healthcare Epidemiology of America (IDSA/SHEA) ASP guidelines (Dellit *et al.*, 2007) can be used for both adults and children, unique factors specific to children must be considered when implementing stewardship in pediatric care settings. Many of the disease processes, infectious pathogens, antimicrobial pharmacokinetic/pharmacodynamic (PK/PD) parameters, and antimicrobial tolerability profiles differ between children and adults. A paucity of data often surrounds these topics when considering children, and therefore adult data are often extrapolated to pediatrics. However, when considering the implementation of a pediatric ASP, the differences between children and adults must be acknowledged to assure appropriate targeted interventions, especially in institutions that care for both children and adults.

Hospital-based ASPs have been increasing among children's hospitals. A recent survey found that among 38 freestanding children's hospitals in the US, 16 had a formal ASP for which the hospital provided financial support for a physician, pharmacist and/or a data analyst. Furthermore, 13 of these programs were started after 2007, the year that the IDSA/SHEA guidelines were published. Finally, in all of the hospitals surveyed, antimicrobial stewardship strategies were being implemented even in the absence of dedicated financial support for an ASP (Newland *et al.*, 2014), although the hospitals that provided financial support for their ASPs were more likely to monitor process and outcome data, and experienced a significantly greater reduction in antimicrobial use (Newland *et al.*, 2014; Hersh *et al.*, 2015). Table 32.1 provides a list of suggested process and outcome data for pediatric ASPs to monitor (Agwu *et al.*, 2008; Metjian *et al.*, 2008;

Table 32.1. Suggested process and outcome data for pediatric antimicrobial stewardship programs (ASPs).

Process measures	References
Days of therapy (DOT)/1000 patient days[a]	Newland *et al.*, 2012; Hersh *et al.*, 2015
Length of therapy (LOT)/1000 patient days[b]	Newland *et al.*, 2012
Doses/1000 patient days	Di Pentima *et al.*, 2011
Compliance with ASP recommendations	Metjian *et al.*, 2008; Newland *et al.*, 2012
Outcome measures	**References**
Costs	Agwu *et al.*, 2008; Metjian *et al.*, 2008
Length of hospital stay	–
Antibiotic resistance	Di Pentima *et al.*, 2011
Clostridium difficile rates of infection	–
Readmissions	–
Avoidable adverse drug reactions	

[a]All antibiotics are included in this metric. For example, a patient receiving two antibiotics for 5 days would have a DOT of 10.
[b]Focuses only on length of antibiotic therapy. For example, a patient receiving two antibiotics for 5 days would have a LOT of 5.

Di Pentima *et al.*, 2011; Newland *et al.*, 2012; Hersh *et al.*, 2015).

Pediatric ASPs, like adult ASPs, require a core group of dedicated personnel to design and successfully implement the program. Ideally, key personnel would include a clinical pharmacist, a pediatric infectious diseases physician and a data analyst. Recent data from pediatric ASPs demonstrated the median full-time equivalent (FTE) for a pharmacist was 0.5 and that for a physician was 0.25. Only one hospital in this study provided support for a data analyst (Newland *et al.*, 2014). Other important members of any ASP team include IT (information technology) personnel, clinical microbiologists, and hospital infection preventionists (Dellit *et al.*, 2007).

Antimicrobial Stewardship Strategies for Hospitalized Children

Two core antimicrobial stewardship strategies are recommended by the IDSA/SHEA antimicrobial stewardship guidelines: prospective audit with intervention and feedback, and formulary restriction with preauthorization for specific agents (Dellit *et al.*, 2007). No studies in pediatrics have compared these two strategies to determine which one is the most effective. However, at an adult institution that had a preauthorization program, a significant increase in the use of antibiotics was observed after switching to a prospective audit strategy (Mehta *et al.*, 2014). Importantly, these two strategies are not mutually exclusive when starting an ASP and, therefore, a combination of these core strategies is often used (Newland *et al.*, 2014). A preference for one approach or the other, or for a combination of both is dependent upon which strategy(ies) will be successful for a specific hospital or healthcare facility and what resources are readily available.

Prospective audit

The general structure of a prospective audit program incorporates the evaluation of selected antimicrobials by an ASP following a prescription by an ordering clinician. The ASP reviews the clinical indication of the prescribed antimicrobial on a case-by-case basis and provides the necessary recommendations to optimize treatment therapy based on the clinical information provided, microbiologic data, resistance patterns, and the clinical literature. The most common recommendations reported from pediatric prospective audit programs are: to discontinue antimicrobial therapy (12–44%), adjust the dose (7–10%), obtain an infectious disease (ID) consult (6–12%), and convert from intravenous (IV) to peroral (PO) formulation (6–10%) (Di Pentima *et al.*, 2009, 2011; Newland *et al.*, 2012). Initial agreement with ASP recommendations was 80% in one study. (Newland *et al.*, 2012), and compliance with the recommendations made has been reported to be 83–92% (Di Pentima *et al.*, 2011; Newland *et al.*, 2012; Goldman *et al.*, 2015).

These prospective-audit programs have had a significant impact on antimicrobial use, outcomes and physician satisfaction. Di Pentima *et al.* (2011) noted an overall antimicrobial dose administration decline of 38%, and a 21% decline in their ASP-targeted antimicrobial doses, despite a 7% increase in patient acuity. Furthermore, this study demonstrated no significant increase in the development of resistance among common pathogenic bacteria. In another prospective-audit ASP at a freestanding children's hospital, a 17%/month average decline in days of therapy (DOT)/1000 patient days was observed in the antibiotics monitored by the program, and a 7% average monthly decline in all antibiotics when compared with other like children's hospitals (Newland *et al.*, 2012). This program also surveyed the clinicians impacted by the program: 82% agreed that the ASP improved the quality of care for the patients, and 84% agreed that it reduced inappropriate antibiotic prescribing (Stach *et al.*, 2012).

Prospective audit provides the advantage of allowing the prescribing physician to determine the initial plan of care without any impact on their sense of autonomy. Additionally, more information (i.e., culture results, disease progression, etc.) is available at the time of ASP review to enhance stewardship recommendations. A disadvantage of the prospective audit strategy is the significant amount of time and labor required to actively audit antimicrobial use and provide daily interventions. Furthermore, the window of opportunity to optimize antimicrobial therapy from the moment of antimicrobial initiation is lost. For example, the decision to initiate the treatment of tracheitis or asymptomatic bacteriuria cannot be impacted early in the treatment course.

Preauthorization

Preauthorization is another core strategy for ASPs. This strategy typically integrates antimicrobial restriction and preauthorization with a hospital's pharmacy and therapeutics committee. Specific antimicrobials are selected to be restricted and require approval (preauthorization) prior to administration. The antimicrobial preauthorization process should directly involve members of the ASP team, because this approach has demonstrated effective antimicrobial management (Gross *et al.*, 2001). Metjian *et al.* (2008), in a prospective observational study in a large pediatric hospital, noted that half of the requested antimicrobial preauthorizations required an ASP recommendation. The most common recommendations by the ASP were to: obtain an ID consult or provide advice on dose, antimicrobial selection, or duration of therapy (43%); optimize antimicrobial treatment (33%); target the known or suspected pathogen (20%); and stop antimicrobial treatment (4%). The overall compliance with the ASP recommendations was observed to be good at 89%, and no negative patient consequences were observed. In addition to improved antimicrobial use, the cost saving in drug acquisition costs over the 4 month study period was $50,090. In an alternative approach at another institution, an innovative web-based preauthorization program integrated into the electronic ordering system improved frontline clinician satisfaction while providing a sustained reduction of antibiotic costs of $103,787 a year from 2005–2011 (Agwu *et al.*, 2008; Sick *et al.*, 2013).

The advantage of preauthorization is that the ASP can intervene when empiric treatment strategies are being considered. Therefore, when a patient has presumed tracheitis without worsening respiratory symptoms or asymptomatic bacteriuria, the ASP can disallow the initiation of the antibiotics. Preauthorization can also prove advantageous in decreasing broad-spectrum antimicrobial exposure in outbreak settings with highly resistant organisms, and also in times of drug shortages (Quale *et al.*, 1996). Studies have shown that restricting the use of third- and fourth-generation cephalosporins in NICUs can significantly decrease the number of extended-spectrum β-lactamase (ESBL)-producing resistant Gram-negative infections and decrease the total number of multidrug-resistant Gram-negative bacteria isolated (de Araujo *et al.*, 2007). Finally, the preauthorization of ASPs is felt to be less resource intensive than prospective audit programs as less information is available to the ASP. Furthermore, the prescribers are engaging the ASP, thereby making it more efficient to communicate with the prescriber about the ASP recommendation. A potential disadvantage of the preauthorization strategy is the greater perceived loss of autonomy felt by the prescribing physician, which is a perceived barrier for ASP implementation among pediatric ID physicians (Hersh *et al.*, 2009). In some instances, prescribing clinicians can circumvent the preauthorization process by selectively ordering antimicrobials during non-preauthorization hours (LaRosa *et al.*, 2007). At the time of preauthorization and antimicrobial initiation, clinical data are often limited, so that culture results and data on the progression of the illness cannot be incorporated into the ASP's decision process.

Supplemental strategies

The IDSA/SHEA antimicrobial stewardship guidelines provide a number of supplemental strategies that have been effective in pediatric ASPs (Table 32.2) (Metjian *et al.*, 2008; Di Pentima *et al.*, 2009; Fraser *et al.*, 2010; Di Pentima *et al.*, 2011; Byington *et al.*, 2012; Newman *et al.*, 2012; Sick *et al.*, 2014; Keren *et al.*, 2015). The strategy in pediatrics that has seen frequent success is the use of clinical practice guidelines or pathways. The use of CAP guidelines in children's hospitals has demonstrated significant increases in the use of narrow-spectrum penicillin/ampicillin in hospitalized children with uncomplicated CAP (Newman *et al.*, 2012; Smith *et al.*, 2012; Ambroggio *et al.*, 2013). When evaluating the presence of a CAP guideline in 41 children's hospitals in the US, those with a guideline were more likely to use the recommended narrow-spectrum ampicillin or penicillin therapy (46% of children) than those without a guideline (24% of children) (Neuman *et al.*, 2012). Byington *et al.* (2012) implemented a Care Process Model for febrile infants that resulted in a 7 and 9% reduction in antibiotic use in the inpatient and outpatient setting, respectively, among the low-risk infants. Furthermore, a 15% increase in use of the recommended antibiotics was observed after the implementation of the care process model. Education, dose optimization, de-escalation, combination therapy, and IV to PO conversion are successful supplemental stewardship strategies implemented in pediatric ASPs. Conversely, antibiotic order forms and antibiotic cycling have been shown to be ineffective strategies in pediatrics for reducing antibiotic use and decreasing antibiotic resistance (Toltzis *et al.*, 2002; Bolon *et al.*, 2005b).

ASPs in Special Populations

Certain patient populations within hospitals require additional thought when developing stewardship strategies. Special populations that likely have different stewardship needs include the PICU, NICU, hematology/oncology, patients who have had a solid organ or bone marrow transplant, and those patients with cystic fibrosis. Each of these special populations provides unique challenges for ASPs.

The most data that are available among special populations that describe how ASPs might improve antimicrobial prescribing are for the NICU (Cantey and Patel, 2014). Inappropriate use of antibiotics in the NICU is associated with increases in cases of necrotizing enterocolitis, late-onset sepsis, and death (Cotton, 2010; Kuppala *et al.*, 2011). Additionally, third-generation cephalosporin use in the NICU has been associated with an increase in invasive candidiasis and an increase in the number of infants colonized with bacteria that are resistant to empiric therapy (de Man *et al.*, 2000; Benjamin *et al.*, 2010).

Patel *et al.* (2009) have identified areas within the NICU where inappropriate antibiotic prescribing occurred. The most common reason was failing to target the identified pathogen with a narrower spectrum antibiotic, followed by prolonged duration of therapy for surgical prophylaxis, and the treatment of colonization. Additionally, inappropriate

Table 32.2. Effective supplemental strategies for pediatric antimicrobial stewardship programs (ASPs).

Education
Passive activities, including lectures, conferences, written guidelines. Most effective when implemented in conjunction with other strategies (Patel *et al.*, 2007; Davey *et al.*, 2013)

Guidelines and clinical pathways
Incorporate evidence-based guidelines with local antibiogram data to recommend selection of appropriate antimicrobial agent and dose (Byington *et al.*, 2012; Newman *et al.*, 2012)

Combination therapy
Provide broad spectrum of activity for empiric treatment of critically ill patients (Sick *et al.*, 2014)

Streamlining or de-escalation of therapy
Tailor antimicrobial selection to match the targeted pathogen, and discontinue unnecessary therapy (Metjian *et al.*, 2008; Di Pentima *et al.*, 2011)

Dose optimization
Consider pharmacokinetic and pharmacodynamic principles to select the most safe and effective dose (Metjian *et al.*, 2008; Di Pentima *et al.*, 2009)

Conversion from parenteral to oral therapy
Administer agents with high oral bioavailability orally rather than intravenously (Fraser *et al.*, 2010; Keren *et al.*, 2015)

antimicrobial therapy was most likely to occur after the first 72 h, when decisions to stop antimicrobial therapy are being made.

Antimicrobial stewardship strategies that have been effective include restriction, with preauthorization, of cephalosporins (mentioned above), and the use of clinical guidelines. Chiu *et al.* (2011) implemented a vancomycin use guideline in a NICU that entailed empirically treating infants with suspected late-onset sepsis with nafcillin. They observed a significant reduction (35%) of vancomycin starts and a significant increase in justified vancomycin use.

Outpatient Stewardship

As previously discussed, most of the antibiotic prescribing for children occurs in the outpatient or ambulatory setting. Therefore, it is imperative that antimicrobial stewardship extends into the outpatient setting. Different types of stewardship strategies have been utilized here, including education, guidelines, audit and feedback, and public campaigns (Finkelstein *et al.*, 2008; Gerber *et al.*, 2013a; Vodicka *et al.*, 2013).

Gerber *et al.* (2013a) utilized education and audit and feedback to improve antibiotic prescribing in a network of outpatient clinics affiliated with a major children's hospital. This cluster-randomized trial showed that stewardship interventions decreased overall broad-spectrum prescribing by 12%, with a reduction in broad-spectrum prescribing for pneumonia and sinusitis of 11 and 20%, respectively. Unfortunately, a later study by the same authors showed that when the audit and feedback was removed from the clinic setting, broad-spectrum antibiotic prescribing returned to preintervention levels (Gerber *et al.*, 2014).

A systematic review of antimicrobial stewardship interventions that targeted the prescribing of antibiotics in the treatment of respiratory tract infection showed that parental education and physician behavior change led to a reduction of 6–21% in antibiotic prescribing. An interactive booklet used by clinicians to assess the parental expectation of receiving an antibiotic, in addition to providing parent education, proved the most effective approach. Parental education alone has not resulted in a significant reduction in antibiotic prescribing (Vodicka *et al.*, 2013).

A community intervention that involved physician behavior change and a community-wide educational campaign observed a modest reduction in antibiotic prescribing. Finkelstein *et al.* (2008) performed a cluster-randomized trial in 16 communities in Massachusetts. The physician intervention involved providing parental educational material to the physician office, educational sessions and ongoing materials throughout the intervention for the physicians, and prescribing data. The parental intervention involved mailing educational materials to families with children less than 6 years of age as well as providing ongoing newsletters and educational materials in pharmacies and day care settings. This complex community intervention resulted in a significant decrease in antibiotic prescribing for children of 24–47 months and 48–72 months of age of 4.2 and 6.7%, respectively.

Conclusions and Future Needs

Antimicrobial stewardship programs developed for hospitalized children have been successful. The core strategies of preauthorization and prospective audit have been effective in reducing antibiotic use in freestanding children's hospitals. Furthermore, supplemental strategies such as the implementation of clinical guidelines, education, and/or intravenous to peroral conversion have performed best in conjunction with a core strategy. Future research should focus on better quantitating the amount of inpatient prescribing that is truly inappropriate. Additionally, data on the impact of stewardship programs on antimicrobial resistance and patient outcomes such as duration of hospital stay, readmission, and adverse drug reactions are needed. Finally, the most effective and efficient methods for implementing antimicrobial stewardship programs for children hospitalized in community hospitals and large academic centers need to be described.

Most antimicrobial prescribing for children occurs in the outpatient setting and here stewardship strategies such as education and/or audit and feedback have been successful. However, the sustainability of these techniques has not been demonstrated. Research in the outpatient setting should focus on developing stewardship interventions that will create sustained improvement in antimicrobial prescribing.

References

Agwu, A.L., Lee, C.K., Jain, S.K., Murray, K.L., Topolski, J., Miller, R.E., Townsend, T., and Lehmann, C.U. (2008) A World Wide Web-based antimicrobial stewardship program improves efficiency, communication, and user satisfaction and reduces cost in a tertiary care pediatric medical center. *Clinical Infectious Diseases* 47, 747–753.

Akar, A., Singh, N., and Hyun, D.Y. (2014) Appropriateness and safety of outpatient parenteral antimicrobial therapy in children: opportunities for pediatric antimicrobial stewardship. *Clinical Pediatrics* 53, 1000–1003.

Albrich, W.C., Monnet, D.L., and Harbarth, S. (2004) Antibiotic selection pressure and resistance in *Streptococcus pneumoniae* and *Streptococcus pyogenes. Emerging Infectious Diseases* 10, 514–517.

Ambroggio, L., Thomson, J., Murtagh Kurowski, E., Courter, J., Statile, A., Graham, C., Sheehan, B., Iyer, S., Shah, S.S., and White, C.M. (2013) Quality improvement methods increase appropriate antibiotic prescribing for childhood pneumonia. *Pediatrics* 131, e1623–e1631.

Andrews, T., Thompson, M., Buckley, D.I., Heneghan, C., Deyo, R., Redmond, N, Lucas, P.J., Blair, P.S., and Hay, A.D. (2012) Interventions to influence consulting and antibiotic use for acute respiratory tract infections in children: a systematic review and meta-analysis. *PLoS ONE* 7(1): e30334.

Barrier, A., Williams, D.J., Connelly, M., and Creech, C.D. (2012) Frequency of peripherally inserted central catheter complications in children. *The Pediatric Infectious Disease Journal* 31, 519–521.

Benjamin, D.K., Jr., Stoll, B.J., Gantz, M.G., Walsh, M.C., Sánchez, P.J., Das, A., Shankaran, S., Higgins, R.D., Auten, K.J., Miller, N.A. *et al.* (2010) Neonatal candidiasis: epidemiology, risk factors, and clinical judgment. *Pediatrics* 126, e865–e873.

Blinova, E., Lau, E., Bitnun, A., Cox, P., Schwartz, S., Atenafu, E., Yau, Y., Streitenberger, L., Parshuram, C.S., Marshall, J., and Seto, W. (2013) Point prevalence survey of antimicrobial utilization in the cardiac and pediatric critical care unit. *Pediatric Critical Care Medicine* 14, e280–e288.

Bolon, M.K., Arnold, A.D., Feldman, H.A., Rehkopf, D.H., Strong, E.F., Goldmann, D.A., and Wright, S.B. (2005a) Evaluating vancomycin use at a pediatric hospital: new approaches and insights. *Infection Control and Hospital Epidemiology* 26, 47–55.

Bolon, M.K., Arnold, A.D., Feldman, H.A., Goldmann, D.A., and Wright, S.B. (2005b) An antibiotic order form intervention does not improve or reduce vancomycin use. *The Pediatric Infectious Disease Journal* 24, 1053–1058.

Byington, C.L., Reynolds, C.C., Korgenski, K., Sheng, X., Valentine, K.J., Nelson, R.E., Daly, J.A., Osguthorpe, R.J., James, B., Savitz, L. *et al.* (2012) Costs and infant outcomes after implementation of a care process model for febrile infants. *Pediatrics* 130, e16–e24.

Cantey, J.B. and Patel, S.J. (2014) Antimicrobial stewardship in the NICU. *Infectious Disease Clinics of North America* 28, 247–261.

CDC (2013) *Antimicrobial Resistance Threats in the United States, 2013*. Centers for Disease Control and Prevention, Atlanta, Georgia. Available at: http://www.cdc.gov/drugresistance/pdf/ar-threats-2013-508.pdf (accessed 3 June 2016).

Chiu, C.H., Michelow, I.C., Cronin, J., Ringer, S.A., Ferris, T.G., and Puopolo, K.M. (2011) Effectiveness of a guideline to reduce vancomycin use in the neonatal intensive care unit. *The Pediatric Infectious Disease Journal* 30, 273–278.

Cotton, C.M. (2010) Early, prolonged use of postnatal antibiotics increased the risk of necrotising enterocolitis. *Archives of Disease in Childhood: Education and Practice* 95(3), 94.

Davey, P., Brown, E., Charani, E., Fenelon, L., Gould, I.M., Holmes, A., Ramsay, C.R., Wiffen, P.J., and Wilcox, M. (2013) Interventions to improve antibiotic prescribing practices for hospital inpatients. *Cochrane Database of Systematic Reviews* 4:CD003543.

de Araujo, O.R., da Silva, D.C., Diegues, A.R., Arkader, R., Cabral, E.A., Afonso, M.R., Louzada, M.E., and Albertoni Ade, C. (2007) Cefepime restriction improves Gram-negative overall resistance patterns in neonatal intensive care unit. *Brazilian Journal of Infectious Diseases* 11, 277–280.

de Man, P., Verhoeven, B.A., Verbrugh, H.A., Vos, M.C., and van den Anker, J.N. (2000) An antibiotic policy to prevent emergence of resistant bacilli. *The Lancet* 355, 973–978.

Dellit, T.H., Owens, R.C., McGowan, J.E., Jr., Gerding, D.N., Weinstein, R.A., Burke, J.P., Huskins, W.C., Paterson, D.L., Fishman, N.O., Carpenter, C.F. *et al.* (2007) Infectious Diseases Society of America and the Society for Healthcare Epidemiology of America guidelines for developing an institutional program to enhance antimicrobial stewardship. *Clinical Infectious Diseases* 44, 159–177.

Di Pentima, M.C., Chan, S., Eppes, S.C., and Klein, J.D. (2009) Antimicrobial prescription errors in hospitalized children: role of antimicrobial stewardship program in detection and intervention. *Clinical Pediatrics* 48, 505–512.

Di Pentima, M.C., Chan, S., and Hossain, J. (2011) Benefits of a pediatric antimicrobial stewardship program at a children's hospital. *Pediatrics* 128, 1062–1070.

Edwards, I.R. and Biriell, C. (1994) Harmonisation in pharmacovigilance. *Drug Safety* 10, 93–102.

Eland, I.A., Belton, K.J., van Grootheest, A.C., Meiners, A.P., Rawlins, M.D., and Stricker, B.H.(1999) Attitudinal survey of voluntary reporting of adverse drug reactions. *British Journal of Clinical Pharmacology* 48, 623–627.

Finkelstein, J.A., Huang, S.S., Kleinman, K., Rifas-Shiman, S.L., Stille, C.J., Daniel, J., Schiff, N., Steingard, R., Soumerai, S.B., Ross-Degnan, D. *et al.* (2008) Impact of a 16-community trial to promote judicious antibiotic use in Massachusetts. *Pediatrics* 121, e15–e23.

Fraser, J.D., Aguayo, P., Leys, C.M., Keckler, S.J., Newland, J.G., Sharp, S.W., Murphy, J.P., Snyder, C.L., Sharp, R.J., Andrews, W.S. *et al.* (2010) A complete course of intravenous antibiotics vs a combination of intravenous and oral antibiotics for perforated appendicitis in children: a prospective, randomized trial. *Journal of Pediatric Surgery* 45, 1198–1202.

Gerber, J.S., Newland, J.G., Coffin, S.E., Hall, M., Thurm, C., Prasad, P.A., Feudtner, C., and Zaoutis, T.E. (2010) Variability in antibiotic use at children's hospitals. *Pediatrics* 126, 1067–1073.

Gerber, J.S., Prasad, P.A., Fiks, A.G., Localio, A.R., Grundmeier, R.W., Bell, L.M., Wasserman, R.C., Keren, R., and Zaoutis, T.E. (2013a) Effect of an outpatient antimicrobial stewardship intervention on broad-spectrum antibiotic prescribing by primary care pediatricians: a randomized trial. *JAMA (The Journal of the American Medical Association)* 309, 2345–2352.

Gerber, J.S., Kronman, M.P., Ross, R.K., Hersh, A.L., Newland, J.G., Metjian, T.A., and Zaoutis, T.E. (2013b) Identifying targets for antimicrobial stewardship in children's hospitals. *Infection Control and Hospital Epidemiology* 34, 1252–1258.

Gerber, J.S., Prasad, P.A., Fiks, A.G., Localio, A.R., Bell, L.M., Keren, R., and Zaoutis, T.E. (2014) Durability of benefits of an outpatient antimicrobial stewardship intervention after discontinuation of audit and feedback. *JAMA (The Journal of the American Medical Association)* 312, 2569–2570.

Goettler, M., Schneeweiss, S., and Hasford, J. (1997) Adverse drug reaction monitoring – cost and benefit considerations. Part II: cost and preventability of adverse drug reactions leading to hospital admission. *Pharmacoepidemiology and Drug Safety* 6(Suppl 3), S79–S90.

Goldman, J.L., Lee, B.R., Hersh, A.L., Yu, D., Stach, L.M., Myers, A.L., Jackson, M.A., Day, J.C., McCulloh, R.J., and Newland, J.G. (2015) Clinical diagnoses and antimicrobials predictive of pediatric antimicrobial stewardship recommendations: a program evaluation. *Infection Control and Hospital Epidemiology* 36, 673–680.

Grohskopf, L.A., Huskins, W.C., Sinkowitz-Cochran, R.L., Levine, G.L., Goldmann, D.A., Jarvis, W.R., and the Pediatric Prevention Network (2005) Use of antimicrobial agents in United States neonatal and pediatric intensive care patients. *The Pediatric Infectious Disease Journal* 24, 766–773.

Gross, R., Morgan, A.S., Kinky, D.E., Weiner, M., Gibson, G.A., and Fishman, N.O. (2001) Impact of a hospital-based antimicrobial management program on clinical and economic outcomes. *Clinical Infectious Diseases* 33, 289–295.

Hazell, L. and Shakir, S.A. (2006) Under-reporting of adverse drug reactions: a systematic review. *Drug Safety* 29, 385–396.

Herigon, J.C., Hersh, A.L., Gerber, J.S., Zaoutis, T.E. and Newland, J.G. (2010) Antibiotic management of *Staphylococcus aureus* infections in US children's hospitals, 1999–2008. *Pediatrics* 125, e1294–e1300.

Herold, B.C., Immergluck, L.C., Maranan, M.C., Lauderdale, D.S., Gaskin, R.E., Boyle-Vavra, S., Leitch, C.D., and Daum, R.S. (1998) Community-acquired methicillin-resistant *Staphylococcus aureus* in children with no identified predisposing risk. *JAMA (The Journal of the American Medical Association)* 279, 593–598.

Hersh, A.L., Beekmann, S.E., Polgreen, P.M., Zaoutis, T.E., and Newland, J.G. (2009) Antimicrobial stewardship programs in pediatrics. *Infection Control and Hospital Epidemiology* 30, 1211–1217.

Hersh, A.L., Shapiro, D.J., Pavia, A.T., and Shah, S.S. (2011) Antibiotic prescribing in ambulatory pediatrics in the United States. *Pediatrics* 128, 1053–1061.

Hersh, A.L., De Lurgio, S.A., Thurm, C., Lee, B.R., Weissman, S.J., Courter, J.D., Brogan, T.V., Shah, S.S., Kronman, M.P., Gerber, J.S., and Newland, J.G. (2015) Antimicrobial stewardship programs in free-standing children's hospitals. *Pediatrics* 135, 33–39.

Hicks, L.A., Chien, Y.W., Taylor, T.H., Jr., Haber, M., and Klugman, K.P. (2011) Outpatient antibiotic prescribing and nonsusceptible *Streptococcus pneumoniae* in the United States, 1996–2003. *Clinical Infectious Diseases* 53, 631–639.

Impicciatore, P., Choonara, I., Clarkson, A., Provasi, D., Pandolfini, C., and Bonati, M. (2001) Incidence of adverse drug reactions in paediatric in/out-patients: a systematic review and meta-analysis of prospective studies. *British Journal of Clinical Pharmacology* 52, 77–83.

Jonville-Béra, A.P., Giraudeau, B., Blanc, P., Beau-Salinas, F., and Autret-Leca, E. (2002) Frequency of adverse drug reactions in children: a prospective study. *British Journal of Clinical Pharmacology* 53, 207–210.

Keren, R., Shah, S.S., Srivastava, R., Rangel, S., Bendel-Stenzel, M., Harik, N., Hartley, J., Lopez, M., Seguias, L., Tieder, J. *et al.* (2015) Comparative effectiveness of intravenous vs oral antibiotics for postdischarge treatment of acute osteomyelitis in children. *JAMA Pediatrics* 169, 120–128.

Kim, J., Smathers, S.A., Prasad, P., Leckerman, K.H., Coffin, S., and Zaoutis, T. (2008) Epidemiological features of *Clostridium difficile*-associated disease among inpatients at children's hospitals in the United States, 2001–2006. *Pediatrics* 122, 1266–1270.

Knackstedt, E.D., Stockmann, C., Davis, C.R., Thorell, E.A,, Pavia, A.T., and Hersh, A.L. (2015) Outpatient parenteral antimicrobial therapy in pediatrics: an opportunity to expand antimicrobial stewardship. *Infection Control and Hospital Epidemiology* 36, 222–224.

Kronman, M.P., Zhou, C., and Mangione-Smith, R. (2014) Bacterial prevalence and antimicrobial prescribing trends for acute respiratory tract infections. *Pediatrics* 134, e956–e965.

Kuppala, V.S., Meinzen-Derr, J., Morrow, A.L., and Schibler, K.R. (2011) Prolonged initial empirical antibiotic treatment is associated with adverse outcomes in premature infants. *Journal of Pediatrics* 159, 720–725.

LaRosa, L.A., Fishman, N.O., Lautenbach, E., Koppel, R.J., Morales, K.H., and Linkin, D.R. (2007) Evaluation of antimicrobial therapy orders circumventing an antimicrobial stewardship program: investigating the strategy of "stealth dosing". *Infection Control and Hospital Epidemiology* 28, 551–556.

Le, J., Nguyen, T., Law, A.V., and Hodding, J. (2006) Adverse drug reactions among children over a 10-year period. *Pediatrics* 118, 555–562.

Madigan, T. and Banerjee, R. (2013) Characteristics and outcomes of outpatient parenteral antimicrobial therapy at an academic children's hospital. *The Pediatric Infectious Disease Journal* 32, 346–349.

Mehta, J.M., Haynes, K., Wileyto, E.P., Gerber, J.S., Timko, D.R., Morgan, S.C., Binkley, S., Fishman, N.O., Lautenbach, E., and Zaoutis, T. (2014) Comparison of prior authorization and prospective audit with feedback for antimicrobial stewardship. *Infection Control and Hospital Epidemiology* 35, 1092–1099.

Metjian, T.A., Prasad, P.A., Kogon, A., Coffin, S.E., and Zaoutis, T.E. (2008) Evaluation of an antimicrobial stewardship program at a pediatric teaching hospital. *The Pediatric Infectious Disease Journal* 27, 106–111.

Nebeker, J.R., Barach, P., and Samore, M.H. (2004) Clarifying adverse drug events: a clinician's guide to terminology, documentation, and reporting. *Annals of Internal Medicine* 140, 795–801.

Neuman, M.I., Hall, M., Hersh, A.L., Brogan, T.V., Parikh, K., Newland, J.G., Blaschke, A.J., Williams, D.J., Grijalva, C.G., Tyler, A., and Shah, S.S. (2012) Influence of hospital guidelines on management of children hospitalized with pneumonia. *Pediatrics* 130, e823–e830.

Newland, J.G., Stach, L.M., De Lurgio, S.A., Hedican, E., Yu, D., Herigon, J.C., Prasad, P.A., Jackson, M.A., Myers, A.L., and Zaoutis, T.E. (2012) Impact of a prospective-audit-with-feedback antimicrobial stewardship program at a children's hospital. *Journal of the Pediatric Infectious Diseases Society* 1, 179–186.

Newland, J.G., Gerber, J.S., Weissman, S.J., Shah, S.S., Turgeon, C., Hedican, E.B., Thurm, C., Hall, M., Courter, J., Brogan, T.V. *et al.* (2014) Prevalence and characteristics of antimicrobial stewardship programs at freestanding children's hospitals in the United States. *Infection Control and Hospital Epidemiology* 35, 265–271.

Newman, R.E., Hedican, E.B., Herigon, J.C., Williams, D.D., Williams, A.R., and Newland, J.G. (2012) Impact of a guideline on management of children hospitalized with community-acquired pneumonia. *Pediatrics* 129, e597–e604.

Nyquist, A.C., Gonzales, R., Steiner, J.F., and Sande, M.A. (1998) Antibiotic prescribing for children with colds, upper respiratory tract infections, and bronchitis. *JAMA (The Journal of the American Medical Association)* 279, 875–877.

Pakyz, A.L., Gurgle, H.E., Ibrahim, O.M., Oinonen, M.J., and Polk, R.E. (2009) Trends in antibacterial use in hospitalized pediatric patients in United States academic health centers. *Infection Control and Hospital Epidemiology* 30, 600–603.

Patel, S.J., Larson, E.L., Kubin, C.J., and Saiman, L. (2007) A review of antimicrobial control strategies in hospitalized and ambulatory pediatric populations. *The Pediatric Infectious Disease Journal* 26, 531–537.

Patel, S.J., Oshodi, A., Prasad, P., Delamora, P., Larson, E., Zaoutis, T., Paul, D.A., and Saiman, L. (2009) Antibiotic use in neonatal intensive care units and adherence with Centers for Disease Control and Prevention 12 Step Campaign to Prevent Antimicrobial Resistance. *The Pediatric Infectious Disease Journal* 28, 1047–1051.

Quale, J., Landman, D., Saurina, G., Atwood, E., DiTore, V., and Patel, K. (1996) Manipulation of a hospital antimicrobial formulary to control an outbreak of vancomycin-resistant enterococci. *Clinical Infectious Diseases* 23, 1020–1025.

Riedl, M.A. and Casillas, A.M. (2003) Adverse drug reactions: types and treatment options. *American Family Physician* 68, 1781–1790.

Ruebner, R., Keren, R., Coffin, S., Chu, J., Horn, D., and Zaoutis, T.E. (2006) Complications of central venous catheters used for the treatment of acute hematogenous osteomyelitis. *Pediatrics* 117, 1210–1215.

Sammons, J.S., Localio, R., Xiao, R., Coffin, S.E., and Zaoutis, T. (2013) *Clostridium difficile* infection is associated with increased risk of death and prolonged hospitalization in children. *Clinical Infectious Diseases* 57, 1–8.

Shehab, N., Patel, P.R., Srinivasan, A., and Budnitz, D.S. (2008) Emergency department visits for antibiotic-associated adverse events. *Clinical Infectious Diseases* 47, 735–743.

Sick, A.C., Lehmann, C.U., Tamma, P.D., Lee, C.K.K., and Agwu, A.L. (2013) Sustained savings from a longitudinal cost analysis of an internet-based preapproval antimicrobial stewardship program. *Infection Control and Hospital Epidemiology* 34, 573–580.

Sick, A.C., Tschudin-Sutter, S., Turnbull, A.E., Weissman, S.J., and Tamma, P.D. (2014) Empiric combination therapy for Gram-negative bacteremia. *Pediatrics* 133, e1148–e1155.

Smith, M.J., Kong, M., Cambon, A., and Woods, C.R. (2012) Effectiveness of antimicrobial guidelines for community-acquired pneumonia in children. *Pediatrics* 29, e1326–e1333.

Smyth, R.M., Gargon, E., Kirkham, J., Cresswell, L., Golder, S., Smyth, R., and Williamson, P. (2012) Adverse drug reactions in children – a systematic review. *PLoS ONE* 7(3): e24061.

Stach, L.M., Hedican, E.B., Herigon, J.C., Jackson, M.A., and Newland, J.G. (2012) Clinicians' attitudes towards an antimicrobial stewardship program at a children's hospital. *Journal of the Pediatric Infectious Diseases Society* 1, 190–197.

Toltzis, P., Dul, M.J., Hoyen, C., Salvator, A., Walsh, M., Zetts, L., and Toltzis, H. (2002) The effect of antibiotic rotation on colonization with antibiotic-resistant bacilli in a neonatal intensive care unit. *Pediatrics* 110, 707–711.

Versporten, A., Sharland, M., Bielicki, J., Drapier, N., Vankerckhoven, V., and Goossens, H. (2013) The antibiotic resistance and prescribing in European Children project: a neonatal and pediatric antimicrobial web-based point prevalence survey in 73 hospitals worldwide. *The Pediatric Infectious Disease Journal* 32, e242–e253.

Vodicka, T.A., Thompson, M., Lucas, P., Heneghan, C., Blair, P.S., Buckley, D.I., Redmond, N., and Hay, A.D. (2013) Reducing antibiotic prescribing for children with respiratory tract infections in primary care: a systematic review. *British Journal of General Practice* 63, e445–54.

WHO (2008) *Medicines: Safety of Medicines – Adverse Drug Reactions*. Fact sheet No. 293, World Health Organization, Geneva, Switzerland. Updated October 2008. Available at: http://www.wiredhealthresources.net/resources/NA/WHO-FS_MedicinesSafetyAdverseReactions.pdf (accessed 6 June 2016).

33 Antimicrobial Stewardship in the Intensive Care Unit

Mahlet Tadele[1,2], Hugh G. Boothe,[2] Melvin E. Stone Jr.,[1,2] and John McNelis[1,2*]

[1]*Albert Einstein College of Medicine, Bronx New York, US;* [2]*Jacobi Medical Center, Bronx, New York, US*

Introduction

In the past two decades there has been an alarming rise in both the rate of hospital-acquired resistant pathogens to commonly prescribed antibiotics and the emergence of multidrug-resistant organisms (MDROs) in the community (Roberts *et al.*, 2009). This rise has been attributed to a significant increase in the use of broad-spectrum antibiotics, often inappropriately. Examples of inappropriate use include use of the wrong antibiotic, prolonged use of certain antibiotics beyond the recommended duration of therapy, treating viral syndromes and colonization with antibiotics, and using the wrong dosage and treating very susceptible bacteria with broad-spectrum antibiotics. The above examples, coupled with the decline in development of newer antibiotic agents, have made the judicial use of antibiotics critical. Indeed, the Infectious Diseases Society of America (IDSA) and the Society for Healthcare Epidemiology of America (SHEA), in their guidelines for developing institutional programs for antimicrobial stewardship, estimated that 30–50% of antibiotic use in hospitals is inappropriate or unnecessary (Dellit *et al.*, 2007). In a sample of 1391 patients, antimicrobial-resistant infections were associated with increased medical costs per patient of $18,588–29,069, excess hospital stay of 6.4–12.7 days, and increased mortality (attributable mortality 6.5%) for infected patients; the societal costs of this excess mortality were $10.7–15 million (Roberts *et al.*, 2009).

Antibiotics differ from other medications in that their effects often extend beyond the intended individual, so that prescribing for one individual possibly impacts the broader population in general. This is particularly significant in intensive care unit (ICU) patients who have high acuity of illness and complexities due to the nature of their illness, comorbidities, and debilitated status. They are vulnerable to various infections and to colonization by resistant bacteria due to their decreased defense mechanisms, high frequency of ventilator use, their use of multiple invasive devices, and the procedures that they undergo, such as the insertion of central venous and urinary catheters. Further compounding the problem is the design of most modern ICUs, wherein patients are cohorted within a confined space, often sharing resources and personnel. As a result, ICU patients have a high risk of acquiring nosocomial infections, and 20–30% of all hospital admissions result in infections with resistant organisms that are associated with increased morbidity, mortality, and cost (Kollef and Fraser, 2001).

Thus, infection remains among the most important concerns in critically ill patients. Early and reliable diagnosis of infection is often difficult in this setting. The high-risk nature of these patients promotes the use of antibiotic therapy and increases the risk of excessive antibiotic exposure. Indeed, the majority of patients admitted to the ICU will receive one or more antibiotics sometime during their hospitalization. An international study of 1200 ICUs demonstrated that about 71% of the patients in the ICU receive some form of antibiotics (Vincent *et al.*, 2009). The overuse of antimicrobial agents is also an ongoing problem in most ICUs.

*Corresponding author. E-mail: john.mcnelis@nbhn.net

Factors leading to inappropriate prescribing include diagnostic uncertainty, lack of knowledge, lack of availability of microbiologic support or infectious disease (ID) specialists, pharmaceutical marketing pressure, and the fear of missing or undertreating a life-threatening infection. Despite well-established guidelines to treat certain clinical syndromes and infections, ICU patients are often continued on antibiotics beyond the recommended duration of time. Inappropriate antimicrobial use is associated with poor patient outcomes, including increased morbidity, mortality from adverse reactions and toxicity, superinfection, development of and colonization with MDROs through alteration of the gut flora, and the development of *Clostridium difficile*. The economic impact is also reflected in an increased length of stay and the cost of the associated complications. This reinforces the need for a tighter rein on antibiotic dispensing.

The combination of rising rates of antimicrobial resistance, a rapidly dwindling effective antimicrobial armamentarium, and increasing financial pressures has sparked new interest in one intervention that addresses all of these concerns. As a result, many institutions have developed an antimicrobial stewardship program (ASP). A recent systematic review of the results of implementing ASPs in the ICU showed a reduction of antibiotic utilization (by 11–38% defined daily doses/1000 patient days), decreased costs, shorter duration of therapy, less inappropriate use, and fewer adverse events related to antibiotics (Kaki *et al.*, 2011). A number of organizations have recognized the importance of implementing stewardship interventions and programs in hospitals, with major efforts being spearheaded by the US Centers for Disease Control and Prevention (CDC) and SHEA. The creation of SHEA's Antimicrobial Stewardship Taskforce and the launch of CDC's "Get Smart for Healthcare" campaign in 2009 marked the start of a new era of a nationally coordinated effort to promote inpatient antibiotic stewardship in the US (CDC, 2016). The growing importance of ASPs is reflected in the inclusion of several antibiotic quality measures in the new "inpatient infection control worksheet" currently being piloted by the Center for Medicare and Medicaid Services (CMS, 2011). While an ASP may not independently improve antibiotic use in hospitals, or solve problems associated with unregulated usage, it is hoped that such programs will both raise awareness and help to lay a foundation to accomplish the goal of optimal antibiotic usage.

The success of ASPs centers on a multidisciplinary approach. The ASP group develops a series of coordinated interventions designed to improve and measure the appropriate use of antimicrobial agents. It promotes the selection of an optimal antimicrobial drug regimen, including dosing, duration of therapy, and route of administration. Core members of an antibiotic stewardship group in the ICU should include an intensivist, an ID physician, an ID-trained pharmacist, and representatives from microbiology, infection prevention, and information technology (Dellit *et al.*, 2007; Bartlett, 2011) (see Fig. 33.1). The goal of any ASP is to optimize the clinical outcomes while avoiding adverse events associated with antibiotics. The ASP programs also work to reduce the emergence of resistant organisms, including *C. difficile*, as these often occur as secondary outcomes to the unnecessary and excessive use of antibiotics. Overall, the impact of ASPs in improving patient outcomes and quality of care, reducing the length of hospitalization, and saving cost has been well documented. Recent published studies have demonstrated the positive impact of ASPs in cost saving and the consequences of turning away from such interventions (Standiford *et al.*, 2012).

One of the most vexing challenges of ASPs in ICUs is implementation, buy-in and culture change. There is a paucity of literature on ASPs and ICUs that specifically deals with the challenges of implementing successful and sustainable programs in that situation. Success hinges on the ability to convince all stakeholders that the implementation of ASPs in ICUs is of utmost importance in improving outcomes and decreasing costs. Therapeutic intervention with appropriate antibiotics with proper dosing and duration maximizes patient clinical outcomes and reduces the costs associated with the improper management of antibiotics. A robust ASP program also serves to limit the unintended and adverse consequences of antibiotics, such as toxicity from overdosing, subtherapeutic treatment from underdosing, the emergence of MDROs and drug-resistant *C. difficile*, the complications of which can impact beyond the patent ward. The success of ASPs in ICUs depends upon the support and cooperation of frontline prescribers, and requires a champion ICU physician in a leadership role who recognizes the benefit of antibiotic stewardship. The clinical education of prescribers, antibiotic restriction, prospective audit with intervention, de-escalation, and appropriate

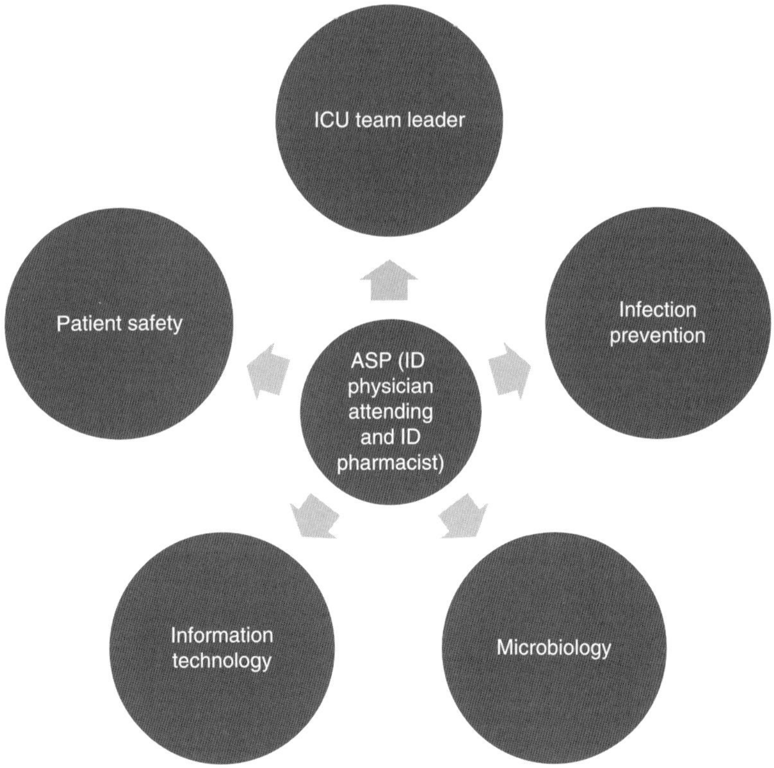

Fig. 33.1. Intensive care unit (ICU) model of the antimicrobial stewardship program (ASP) as a hub for the optimization of patient care. ID, infectious diseases.

duration of treatment based on established guidelines are the common principles of ASPs in the ICU (Standiford *et al.*, 2012).

Resistant Organisms in the Intensive Care Units (ICUs)

Antibiotic resistance is a major public health threat. According to the CDC, approximately 2 million people are infected annually with resistant organisms, with an estimated 23,000 resultant deaths (CDC, 2013). Further exacerbating the problem is the migration of resistant organisms from healthcare facilities into the community. This has become increasingly challenging as carbapenem-resistant *Enterobacteriaceae* (CRE) are on the rise, often transmittable from asymptomatic carriers. Approximately 140,000 healthcare-associated infections by *Enterobacteriaceae* occur in the US annually, and 9300 of them are caused by CRE. Nearly half of all hospitalized patient with CRE bloodstream infections die (Jacob *et al.*, 2013). The association between resistant organisms such as *Staphylococcus aureus*, enterococci, and Gram-negative bacilli, and a dramatic increase in mortality, morbidity, hospital length of stay and cost is well documented. It has been estimated that approximately $6,000–30,000 is spent per patient on infections due to resistant organisms compared with patients infected with susceptible organisms (Cosgrove, 2006).

In addition to the increase in antibiotic resistance and the emergence of difficult-to-treat Gram-negative organisms, excessive antibiotic usage has also led to an increase in invasive candidiasis in the ICU. These problems have been attributed to the random use of very broad-spectrum antibiotics in the ICU. Unfortunately, this usage, while effective for most susceptible bacteria, has led to conditions that select for resistant bacteria. Despite the many advances in the medical and surgical field in treating complex diseases and critically ill patients, we are exhausting antibiotic choices to treat life-threatening multidrug-resistant Gram-negative infections. The acronym ESKAPE (for *Enterococcus faecium*, *Staphylococcus aureus*, *Klebseilla pneumoniae*, *Acinetobacter baumannii*, *Pseudomonas aeruginosa*, and *Enterobacter* species) has been developed by IDSA to describe the alarmingly resistant organisms arising in hospitalized patients and those coming from long-term care facilities (Snitkin *et al.*, 2012). The lack of development of new drugs to combat these resistant organisms is worrisome; there will be no new drug to combat ESKAPE until 2018 at the earliest (Bartlett, 2011; Bow, 2013).

The spread of resistant organisms beyond the ICU has been documented in multiple studies using molecular methods of genetic tracing (Kollef and Fraser, 2001). The global spread of carbapenemase-producing *Enterobacteriaceae* is especially alarming (Nordmann *et al.*, 2011). Nationally, the use of antibiotics in the treatment of infections due to MDROs has risen sharply in the past 10 years. An important consequence of antibiotic resistance is the growing awareness that serious infections in hospitalized patients, particularly those in ICUs, may not be treated adequately. Several studies have demonstrated that resistance frequently leads to a delay in administering microbiologically effective therapy (Ibrahim *et al.*, 2000; Lautenbach *et al.*, 2001).

The cornerstone of countering conditions leading to these types of outbreaks lies in the work of infection control, and sustainable proper antibiotic utilization through antibiotic stewardship (Sandora and Goldmann, 2012). Leadership by ASP team members in proper antibiotic use is critical in combating the conditions that promote the development of emergent resistant organisms. Antibiotic choice should be based on the resistance patterns of specific ICUs, and on pharmacokinetics, pharmacodynamics, potential adverse events, and clinical guidelines. A close working relationship with the microbiology laboratory to immediately alert ICU teams of the emergence of resistant organisms is also critical to implementing proper infection control measures in a timely manner.

Clostridium difficile in the ICU

The single most important identified risk factor in the development of *C. difficile* infection (CDI) is exposure to antibiotics. This is compounded by lapses in simple infection precautions such as hand hygiene. Approximately 85% of patients with *C. difficile* infection have a documented antibiotic exposure within 28 days of infection (Owens *et al.*, 2008). Current reports from the CDC (2013) attribute 250,000 infections and 14,000 deaths a year to CDI, with over $1 billion dollars in medical costs a year. Death attributed to *C. difficile* increased by 400% from 2000 to 2007 (CDC, 2013). This increase may result from increased incidence and recognition, but it is also believed to be a consequence of a new NAP1 strain of *C. difficile* (Bartlett, 2011). The development of this new strain has been attributed to the widespread and often indiscriminate use of fluoroquinolones in healthcare facilities in Canada, and a similar conclusion has been reached to account for the increase in CDI in the US (Hall *et al.*, 2012).

Patients with life-threatening severe CDI are not uncommon in the ICU. Approximately 20% of patients admitted to the ICU with CDI were being administered antibiotics in the absence of any evidence of infection (Marra *et al.*, 2007). One approach proposed for reducing the rate of CDI is to identify and restrict the use of the high-risk antibiotics that are implicated in the development of CDI. Restriction and review for the appropriateness of use of high-risk antibiotics such as fluoroquinolones, clindamycin, and third-generation cephalosporins by an ASP team has been shown to reduce the incidence of CDI (Aldeyab *et al.*, 2012). In addition, the ASP team can also implement guidelines for the treatment of severe CDI. This has been shown to have a positive impact on the management of CDI. A recent paper by Kim *et al.* (2013) demonstrated that a multidisciplinary approach led by an ASP team in the treatment of severe *C. difficile* colitis resulted in complete resolution in 70% of the patients without surgical intervention.

Strategies of Antimicrobial Stewardship Programs in ICUs

The basic principles and strategies of ASPs are particularly challenging in ICU patients given the unique hemodynamic, pharmacokinetic, and pharmacodynamics issues of the critically ill. An ICU physician champion leading the ASP is critical to the success of ASP management in the ICU. The implementation of an ASP in the ICU involves the use of several techniques and strategies, which together make up an antibiotic care bundle.

Antibiotic care bundles (ACBs)

The use of an antibiotic care bundle (ACB) in the ICU can provide a helpful framework for tailoring appropriate antibiotic utilization (Cooke and Holmes, 2007). Its basic components include:

1. Proper documentation for indication of the use of antibiotic. This involves an understanding of underlying disease pathology as well as the institutional antibiogram.
2. Obtaining appropriate sterile site culture. This is often overlooked as clinicians have a tendency to perform random cultures, and often culture pus.
3. Appropriate selection of empiric antibiotic based on the ICU susceptibility and institutional guidelines. This again this hinges on an understanding of the institution's antibiogram.
4. Documentation of start and anticipated stop dates of antibiotic administration. Therapy should be initiated with a documented end date. This will limit prolonged usage of unnecessary antibiotics.
5. De-escalation of therapy based on sensitivity results. This is critical to changing from a broad-based empiric therapy to a more focused organism-specific therapy.
6. Antibiotic "time out" to reassess the need for continuation of an antibiotic or determine the length of treatment. This will force clinicians to reevaluate on a daily basis the need for ongoing antibiotic therapy, in much the same manner that invasive devices are reevaluated on a daily basis.
7. Remove devices and catheters that are not needed and serve as a nidus for infection.

Incorporating ACBs into daily ICU rounds will provide a consistent strategy to ensure that each patient receives the optimal and appropriate antibiotic treatment. The bundles will ensure appropriate antibiotic selection, dosage, and route of administration. This helps to promote compliance with ASP recommendations and creates a framework for the optimal use of antibiotics (see Fig. 33.2).

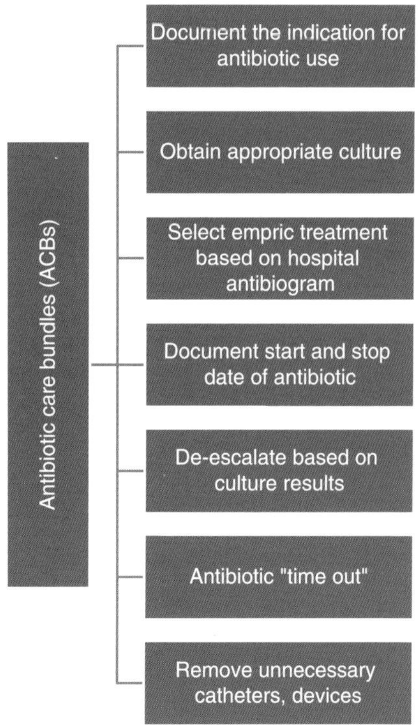

Fig. 33.2. Antibiotic care bundles (ACBs).

Antimicrobial "time out"

Many ICUs often start to treat infections with empiric broad-spectrum antibiotics. One component of effective antimicrobial stewardship is the antimicrobial "time out" that prompts and alerts team to reassess antibiotic use on a daily basis. Facilities that use electronic medical records (EMRs) can incorporate antibiotic "time out" into the EMR to include the name, date of initiation, and indication of the specific antibiotic used, as well as the planned discontinuation date. This will facilitate the daily review and assessment of the need for continuation, based on clinical response of the patient.

ASP-driven drug level monitoring of vancomycin and aminoglycosides

In general, antibiotic dosing regimens are extrapolated and derived from non-critically ill volunteers with normal renal function. This can be misleading, as ICU patients have complex pathophysiology with altered pharmacokinetics. In the critically ill patient, the volume of distribution of drugs into the extracellular water changes because of increased capillary permeability, and changes in oncotic pressures (Lipman, 2000). For example, vancomycin, a poorly protein-bound antibiotic, leaks into the extravascular space in critically ill patients, with a resultant high volume of distribution. Hence, the recommended maximum daily dose calculated from healthy volunteers may not reach an adequate serum level and trough concentration to treat the target infection. Hall *et al.* (2008) reported that a substantial proportion of obese patients receiving vancomycin intravenously did not receive adequate delivery into the tissue. In a similar manner, aminoglycosides are also poorly protein bound. Aminoglycoside distribution into the extravascular space will depend on the nature of the disease process, "third space losses," and vascular permeability. Aminoglycosides require increased clearance and require higher peak levels to be effective. Large once-daily dosing of aminoglycosides, as opposed to traditional three-times daily dosing, is currently advocated as more effective and less toxic method. ASP-driven monitoring of drug levels and dosing will ensure the achievement of therapeutic levels and appropriate dosing. It will also eliminate inappropriate and wasteful ordering of by providers. Ultimately, it will be a safety net for patients through the avoidance of the unnecessary iatrogenic complications associated with toxic levels of antibiotics (Cano *et al.*, 2012).

Vancomycin stop orders

The emergence of *S. aureus* with vancomycin resistance, coupled with the inappropriate use of the antibiotic, makes it an important target for ASP provider intervention. The increasing number of ICU patients receiving vancomycin make the continual evaluation of use critical. Current recommendations include reevaluation of vancomycin use every 48 h to curb unnecessary use of the drug, and prevent toxic levels occurring.

Antibiotic dose optimization: ASP pharmacokinetic consultations in the ICU

Dose optimization interventions and calculations of pharmacokinetics are major ASP functions. Although formerly viewed as a means to efficiently trim excess drug exposure secondary to renal dysfunction, the modern application of pharmacodynamics principles is important in order to maximize drug penetration against organisms with elevated minimum inhibitory concentrations (MICs), patients with excess body mass indices, and for closed-space or otherwise difficult-to-penetrate sites of infection (e.g., meningitis, endocarditis, pneumonia, bone and joint infections). Ambrose *et al.* (2007) have provided an in-depth review of the subject, which serves as a primer for all ASPs incorporating optimal dosing strategies. Even though they approached dosing strategies from the belief that patients would receive downward dose adjustments for renal impairment, they and others found that there was a significant number of patients who actually required increased drug exposure (Owens *et al.*, 2004; Fraser *et al.*, 2005). Other examples of pharmacodynamic dose optimization include regimens intended to treat higher MIC pathogens more effectively, such as continuous or prolonged infusion of short half-life lactams (e.g., piperacillin–tazobactam, cefepime, meropenem), and extended-interval aminoglycoside dosing. Lodise *et al.* (2007) found decreased mortality in critically ill patients (with APACHE II scores >17) when prolonged infusions of piperacillin–tazobactam were used vs. standard infusions of piperacillin–tazobactam. Most recently, a double-blind, randomized, controlled trial demonstrated that doripenem infused over 4 h maintained an adequate time above the MIC (T > MIC) for higher MIC pathogens (such as *Pseudomonas aeruginosa*), compared with standard infusion imipenem in patients with ventilator-associated pneumonia (VAP) (Chastre *et al.*, 2008).

Prospective audit, feedback, and de-escalation

One of the core strategies in any successful stewardship program is the prospective auditing of antibiotic use with ongoing direct feedback to prescribers. ICUs have the highest use of broad-spectrum antibiotics, often for prolonged periods of time. One barrier to implementing this particular strategy is often the practice patterns in the ICU for antibiotic ordering. To promote collaboration, an exemption from preauthorization and restriction over the first 72 h of treatment is recommended. This maintains the autonomy of ICU prescribers and avoids delays in instituting appropriate therapy in at-risk patients. A mandatory reevaluation is then undertaken at 72 h. An ongoing ICU/ID educational round promotes continual reevaluation of ICU antibiotic usage. A collaboration and good working relationship between the individual ICU attendant nursing staff and the ID physicians in charge of ASPs is important for the success of this initiative.

Intravenous to oral conversion of antibiotics

Patients who are receiving intravenous antibiotics that have the same bioavailability as oral formulations should be switched to the oral formulation as long as they meet certain clinical criteria. This can be implemented in ICU patients as well.

Order sets

Computerized or standardized physician order sets have been shown to improve the management of septic shock and severe bacteremic sepsis. A recent study by Thiel *et al.* (2009) showed improvement in appropriate antibiotic choice, decreased incidence of organ failure, and improved survival as a result of adopting standardized order sets into the EMR.

Role of procalcitonin

One of the challenges faced by ICU providers concerns the duration of therapy in the treatment of infections. Recommendations for antibiotic duration are generally unclear and vary with specialty and even with physician experience. Procalcitonin (PCT) has been reported to help in the evaluating the direct duration of antibiotic therapy. It is a 116-amino acid precursor of calcitonin that is elevated in several inflammatory conditions, bacterial infections, sepsis, malaria, and pancreatitis. It starts to elevate within 2–4 h of the inflammatory condition, and peaks on the second day. The level falls off rapidly during the recovery period (Agarwal and Schwartz, 2011). There are circumstances in which the PCT level may be elevated secondary to non-bacterial infections, such as massive stress from trauma, surgery, burns, and cardiac shock. However, in the absence of active infection, PCT levels go down after the inciting event resolves (Smith *et al.*, 2013). Review of the available randomized control studies on PCT has shown the utility of PCT, specifically in ICU patients, in shortening the duration of treatment or in the complete cessation of antibiotics if PCT levels are within normal limits, without compromising patient safety. While initial reports are promising, the cost-effectiveness of PCT-directed antibiotic therapy remains unclear (Heyland *et al.*, 2011; CDC, 2013).

Daily review of positive blood cultures

A close working relationship between the microbiology laboratory and ASP teams can facilitate the optimal care of ICU patients by providing real-time information about positive blood cultures and preliminary culture sensitivity results. ASP teams will assure the proper match of microorganism with antibiotic(s) according to preliminary sensitivities.

Guidelines and clinical pathways

The creation of clinical guidelines and pathways for specific syndromes in the ICU (such as VAP) based on local microbiology data can be very effective. Studies have shown significant reductions in antimicrobial use, duration, cost and antimicrobial resistance without affecting mortality or length of stay when clinical algorithms and guidelines are used (Singh *et al.*, 2000; Chastre *et al.*, 2003).

Antibiotic rotation in the ICU

The rotation of antibiotics is a modality that is in use in many ICUs. Barie *et al.* (2005) demonstrated the efficacy of rotating antibiotics in several studies, showing that such rotation is associated with a high rate of appropriate antibiotic therapy as well as improved resistance patterns. The routine rotation of antibiotics, however, has not received routine acceptance in ICUs, and was discouraged in one literature review (Brown and Nathwani, 2005). The use of antibiotic rotation remains controversial.

Optimization of Antibiotics by ASPs in Specific Situations for ICU Patients

One of the challenges of managing critically ill patients is the question of determining the proper dose of antibiotics to use. In the face of organ dysfunction and significantly altered pharmacokinetics this can become a vexing task. As stated above, dosing regimens of antibiotics in the ICU are derived from healthy volunteers. This can often give a misleading picture of the dosage required for adequate tissue penetration. In order to optimize treatment and patient outcomes in ICU patients, it is imperative to understand the pathophysiological changes that occur in critically ill patients. This also necessitates closer monitoring and understanding of the pharmacokinetic/pharmacodynamics of specific antibiotics. The majority of patients admitted to ICUs have altered antibiotic pharmacokinetics and pharmacodynamics secondary to the inflammatory responses that both follow and result in the release of cytokines and other mediators. Endothelial damage and increased capillary permeability result in fluid shifts with a resultant altered volume of distribution and clearance.

Alterations in volume of distribution primarily affect hydrophilic antibiotics such as glycopeptides, aminoglycosides, β-lactams, carbapenems, linezolid, daptomycin, and colistin. This may necessitate increasing the initial loading dose of the drug as well as the interval dosing. Hypoalbuminemia and hypoproteinemia are very common in critically ill patients, with albumin being disproportionately distributed into the extravascular space compared with the intravascular space. This results in lower drug-albumin binding, which leads to low plasma drug concentration, and higher concentrations of unbound drug, which is more easily eliminated. Services provided by the ASP team, particularly the ID pharmacist, assist in proper dosing in patients with hypoalbuminemia, hypoproteinemia, and other derangements in volume of distribution. Including ID pharmacists in daily rounds with the ICU team provides an excellent method of communication of the proper dosing recommendation as well as the monitoring for potential drug interactions.

Antibiotic dosing in burn ICU patients

There are complex changes and fluctuations in the pharmacokinetics of antibiotic metabolism, clearance, and distribution in burn patients (Jamal *et al.*, 2012). During the early phase of post-burn injury, hypovolemia, hypoalbuminemia, and derangements in glomerular filtration predominate. In the later post-burn phase (>48–72 h), a hyperdynamic state predominates with increased renal clearance and liver dysfunction. This results in burn patients having a significantly faster clearance of medications than patients with normal renal function. Conventional dosing of antimicrobials such as aminoglycosides, fluoroquinolones, glycopeptides and β-lactams can result in suboptimal drug levels, therapeutic failure, and the development of resistance if the dosing regimen is not adjusted based on the individual patient's needs. Increased clearance and a decreased half-life of aminoglycosides may necessitate more frequent dosing or using higher doses to achieve the appropriate target concentrations for the indicated infections. Vancomycin may need higher loading doses of 15–20 mg/kg, as the conventional dosing of 1 g may not be adequate to achieve adequate tissue penetration. β-Lactams will also not reach the effective concentration and will result in subtherapeutic treatment. Hence, β-lactams require an increase in the dose, frequency, and duration of infusion.

Unfortunately, for most antimicrobials, optimal dosing still remains more of an art than a science. This is especially true for burn patients, and is a challenge that can be addressed by the ASP team. With the help of the pharmacokinetic, clinical support software tool "Kinetidex," the ASP can better dose antibiotics in critical care patients, including burn patients. Because renal function does not account for the extremely high volumes of distribution in this patient population, plasma levels of routinely measureable drugs (aminoglycosides and vancomycin) are used to approximate the excretion rates of those antibiotics that are not routinely measured and so provide a more accurate dosing recommendation. Services provided by ASPs in the ICU can include daily pharmacokinetic consults to ensure that patients are getting adequate and therapeutic levels of antibiotics.

Antibiotic dosing in renal replacement therapy

Critically ill patients admitted to the ICU often develop some degree of renal insufficiency up to and including acute kidney injury (AKI). Approximately 5% of ICU patients who sustain AKI will require renal replacement therapy (RRT) (Aitken *et al.*, 2013). In AKI patients with infection and or sepsis, timely recognition and appropriate antibiotic administration is critical to reduced morbidity and mortality. Unfortunately, there is great variability in terms of the degree of illness and organ dysfunction among ICU patients, and multiple problems can often be simultaneously at play. Not infrequently, multiple medications are required to address multiple issues, and all require adequate dosing.

One of the challenges for treating infections specific to patients receiving RRT is their rapidly changing clinical status, hemodynamics, and disrupted pharmacokinetics. Therefore, the dosing of antibiotics must be corrected and adjusted for those changes to sustain adequate penetration and decreased elimination. During the process of removing toxic wastes efficiently, hydrophilic antibiotics (particularly those with low plasma protein binding) are also removed (Pea *et al.*, 2007). There are certain concerns associated with the use of 'semi-continuous' high-efficiency treatments, such as sustained low-efficiency dialysis (Fliser and Kielstein, 2006), or pulse high-volume hemofiltration (Ratanarat *et al.*, 2006), which alternate between periods of high clearance and of no or low clearance that make antibiotic dosing very difficult. This will result in subtherapeutic levels of antibiotics, especially with β-lactam antibiotics, whose level should remain above the MIC for as long as possible (Schetz, 2007). One possible solution is to provide continuous infusion of the antibiotic adapted according to the flow of RRT. Hence, antibiotic dosing should be individualized because of variability from patient to patient, and should be in consultation with an ID pharmacist who understands the basic pharmacokinetic principles and can optimize appropriate dosing in RRT patients. The ASP team that monitors ICU antibiotic use will often identify inappropriate dosing, and this will have a positive impact on patient outcomes and safety.

Antibiotic dosing in extracorporeal membrane oxygenation (ECMO)

Patients who are on ECMO for severe lung or heart dysfunction are also affected by the alteration that occurs in their pharmacokinetics, specifically an increase in the volume of distribution and binding of certain antibiotics to the ECMO circuit (Shekar *et al.*, 2012). Lipophilic drugs such as fluoroquinolones, macrolides, lincosamides, tigecycline, and clindamycin are absorbed into the membranes and tubing of the ECMO, which, in turn, reduces the amount of drug concentration in the plasma (Mehta *et al.*, 2007). Individualizing these antibiotic regimens is suggested by recent systematic review as opposed to guidelines due to the variability of pharmacokinetics (Mousavi *et al.*, 2011). Clinicians often order standard doses of antibiotics, which will put patients at risk of treatment failure because of their subtherapeutic levels or their toxicity.

Catheter-associated urinary tract infections (CAUTIs)

CAUTIs are one of the most common healthcare-associated infections in ICUs. Nearly all patients admitted to ICUs will have urinary catheters to monitor their urinary output. Infection prevention strategies help to decrease the rates of occurrence of CAUTIs. This is also an area where ASPs can intervene in educating providers about evidence-based treatment duration for CAUTIs, which can be as short as 3 days in women <65 years of age if the catheter is removed (Hooton *et al.*, 2010; CDC, 2013).

Transplant Patients and the Role of Antibiotic Stewardship

Significant advances in the treatment modalities of solid organ transplantation and hematopoietic cell transplantation have resulted in ever-increasing numbers of post-transplant patients in the ICU. These patients are at greatest risk of infection due to their severe immune-compromised state. They are often on antibiotic prophylaxis or on broad-spectrum empiric antibiotics because of suspected infections. High rates of antibiotic use in oncology and bone marrow transplant (BMT) patients have been well documented, as well as empiric therapy in neutropenic fever. Consequently, as a result of the prolonged use of antibiotics in transplant patients, there is increased rate of occurrence of MDROs and *C. difficile* in transplant patients (Herati and Blumberg, 2012; Dubberke *et al.*, 2013; van Duin *et al.*, 2013). As a result, ASPs are playing an ever-increasing and crucial role in the transplant field (Aitken *et al.*, 2013).

The basic principles of ASP intervention in the transplant field, as elsewhere, include formulary restriction, antibiotic choice, dosage, duration, prospective audit/feedback, and de-escalation. However, another crucial role played by ASPs is in the area of drug optimization, focusing on drug interactions between antimicrobials and immunosuppressive agents, which could potentially be harmful by exposing patients to subtherapeutic or supratherapeutic doses of each medication. An example of one such drug interaction is the inhibition by the antifungal agent voriconazole of the hepatic metabolism of the immunosuppressant tacrolimus. This interaction necessitates a preemptive dose adjustment before starting the antifungal medication. Lupei *et al.* (2010) demonstrated the link between inadequate or improper therapy with antibiotics and higher mortality in transplant patients. The lack of effective antibiotics against particular Gram-negative antibiotics will have dire consequences for patient outcome (Brown and Nathwani, 2005). In addition, the education of transplant physicians, physician assistants, nurses, and nurse practitioners is essential, and the ASP can facilitate the optimization of antibiotics during the transition of patients from ICUs to medical floors and vice versa. ASPs can also play a role in drafting guidelines based on local susceptibility patterns, and assisting in alternative prophylaxis regimens at times when there is a specific drug shortage.

Surgical Site Infection and Antibiotic Stewardship

Out of approximately 30 million surgical procedures performed in the US, approximately 2–5% result in surgical site infection (SSI) each year. It is estimated that up to 60% of these infections could be prevented (File *et al.*, 2011). According to a recent prevalence study, SSIs account for 31% of all healthcare-associated infections and are a major public health burden (Magill *et al.*, 2012). Despite the advances in infection prevention, antibiotic prophylaxis, and surgical techniques, the morbidity and mortality associated with SSIs still exceed 3% (Awad, 2012).

The Surgical Care Improvement Project (SCIP) initiated by the Institute for Healthcare Improvement (IHI) to prevent SSIs has four measures that focus on antibiotic use (Bratzler *et al.*, 2013). These include: (i) the initiation of prophylactic antibiotics; (ii) appropriate antibiotic selection and dosing; (iii) discontinuation of antibiotics within 24 h of surgery (48 h for cardiac surgery patients); and (iv) appropriate redosing for patients undergoing prolonged procedures (CDC, 2013). The ASP plays an important role in ensuring that antibiotic selection conforms to best practice guidelines and limits unnecessary use of broad-spectrum coverage. Stewardship optimizes appropriate dosing, giving special consideration to obese patients and to prolonged procedures that may result in significant blood loss, so requiring redosing of antibiotics. ASPs will work closely with patient safety officers in the implementation of and compliance with these measures (Stullberg *et al.*, 2010).

Education

Active education strategies on proper antibiotic prescribing based on local susceptibility patterns as well as on case-based educational sessions with ICU teams is yet another beneficial role for an ASP. Creating guidelines to treat certain syndromes based on the local ICU susceptibility pattern is important, but changing prescriber behavior can be challenging. It has been shown that face-to-face encounters with and teaching of prescribers has been more effective in changing prescriber behavior than didactic teaching and the use of printed materials (Avorn and Soumeriai, 1983). Implementing guidelines requires active participation from the prescribers, and regular educational conferences between ID specialists and ICU team members based on actual cases and management could be beneficial.

Conclusion

Antimicrobial stewardship in the ICU is an organized approach among the various groups of ID specialists, pharmacists, infection prevention leaders, microbiologists, and ICU team leaders to provide adequate and appropriate antibiotic therapy, minimize adverse effects from antibiotics, and reduce the rate of treatment of non-sterile site cultures that may signify colonization. Although antibiotics are lifesaving medications, their misuse can be detrimental to patients in breeding multidrug-resistant Gram-negative infections, invasive candidiasis, *C. difficile*, iatrogenic complication-like nephrotoxicity, and drug reactions and interactions, which can all result in poor patient outcomes. Infections with these types of organisms have been associated with higher mortality as well as an increased length of hospital stay and cost of hospitalization. As the most complex and fragile patients are admitted to the ICU, it is of utmost importance that patients receive timely and appropriate antibiotics. As a result their illnesses, the volume of distribution of antibiotics as well as the clearance of antibiotics is altered in these patients. In order to achieve an optimal dosing of antibiotics in ICUs, these specific pharmacokinetic dynamics must be taken into consideration when ordering a specific dose of antibiotic. The role of the ASP in this setting is critical in optimizing antibiotic dose by looking at the altered pharmacokinetic and kill characteristics of antibiotics, and individualizing the dose based upon the patient's prior antibiotic use history to reduce morbidity and mortality.

Special attention and daily intervention from ASP teams and pharmacokinetic consults is likely to benefit ICU patients who have fluctuating hemodynamics and difficult pharmacokinetics due to their type of injury, and specifically for patients on continuous RRT (CRRT), ECMO, and with significant burn injuries. Reassessing the need for broad-spectrum antibiotics on day 3 and closely working with ID specialists to optimize patient care is of great importance. In circumstances where ASP recommendations are not followed through, it is important to identify non-compliant physicians and to reach out directly to gain an understanding of the major obstacles that have hindered the implementation of the recommendations. Continuing one-on-one education and creating guidelines should be an integral part of antibiotic stewardship.

Advances in medical and surgical therapy, such as complex cardiothoracic surgery, joint replacement, organ transplantation, cancer treatment, and immunosuppressant medications leave patients at higher risk of becoming infected and dependent on the availability of potent antibiotics. Without the proper armamentarium to fight infections, these lifesaving and improving modern devices, procedures, and medications may become useless. All healthcare providers need to understand the importance of appropriate antibiotic use and the importance of working together with their ASPs.

References

Agarwal, R. and Schwartz, D.N. (2011) Procalcitonin to guide duration of antimicrobial therapy in intensive care units. *Clinical Infectious Diseases* 53, 379–387.

Aitken, S.L., Palmer, H.R., Topal, J.E., Gabardi, S., and Tichy, E. (2013) Call for antimicrobial stewardship in solid organ transplantation. *American Journal of Transplantation* 13, 2499.

Aldeyab, M.A., Kearney, M.P., Scott, M.G., Aldiab, M.A., Alahmadi, Y.M., Darwish Elhajji, F.W., Magee, F.A., and McElnay, J.C. (2012) An evaluation of the impact of antibiotic stewardship on reducing the use of high-risk antibiotics and its effect on the incidence of *Clostridium difficile* infection in hospital settings. *Journal of Antimicrobial Chemotherapy* 67, 2988–2996.

Ambrose, P.G., Bhavnani, S.M., Rubino, C.M., Louie, A., Gumbo, T., Forrest, A., and Drusano, G.L. (2007) Pharmacokinetics–pharmacodynamics of antimicrobial therapy: it's not just for mice anymore. *Clinical Infectious Disease* 44, 79–86.

Avorn, J. and Soumeriai, S. (1983) Improving drug-therapy decisions through educational outreach: a randomized controlled trial of academically based "detailing". *The New England Journal of Medicine* 308, 1457–1461.

Awad, S.S. (2012) Adherence to surgical care improvement project measures and post-operative surgical site infections. *Surgical Infection* 13, 234–237.

Barie, P.S., Hydo, L.J., Shou, J., Larone, D.H., and Eachempati, S.R. (2005) Influence of antibiotic therapy on mortality of critical surgical illness caused or complicated by infection. *Surgical Infection* 6, 41–54.

Bartlett, J.G. (2011) A call to arms: the imperative for antimicrobial stewardship. *Clinical Infectious Diseases* 53(Suppl 1), S4–S7.

Bow, E.J. (2013) There should be an ESKAPE for febrile neutropenic cancer patients: the dearth of effective antibacterial drugs threatens anticancer efficacy. *Journal of Antimicrobial Chemotherapy* 68, 492–495.

Bratzler, D.W., Dellinger, E.P., Olsen, K.M., Perl, T.M., Auwaerter, P.G., Bolon, M.K., Fish, D.N., Napolitano, L.M., Sawyer, R.G., Slain, D. *et al.* (2013) Clinical practice guidelines for antimicrobial prophylaxis in surgery. *American Journal of Health-System Pharmacy* 70, 195–283.

Brown, E.M. and Nathwani, D. (2005) Antibiotic cycling or rotation: a systematic review of the evidence of efficacy. *Journal of Antimicrobial Chemotherapy* 55, 10–21.

Cano, E.L., Haque, N.Z., Welch, V.L., Cely, C.M., Peyrani, P., Scerpella, E.G., Ford, K.D., Zervos, M.J., Ramirez, J.A., Kett, D.H., and the IMPACT-HAP Study Group (2012) Incidence of nephrotoxicity and association with vancomycin use in intensive care units with pneumonia: retrospective analysis of the IMPACT-HAP Database. *Clinical Therapeutics* 34, 149–157.

CDC (2013) *Antimicrobial Resistance Threats in the United States, 2013*. Centers for Disease Control and Prevention, Atlanta, Georgia. Available at: http://www.cdc.gov/drugresistance/pdf/ar-threats-2013-508.pdf (accessed 3 June 2016).

CDC (2016) Get Smart for Healthcare. Centers for Disease Control and Prevention, Atlanta, Georgia. Available at: http://www.cdc.gov/getsmart/healthcare (accessed 7 June 2016).

CMS (2011) *Survey & Certification Focus on Patient Safety and Quality – Draft Surveyor Worksheets*. Centers for Medicare and Medicaid Services, Baltimore, Maryland. Available at: https://www.cms.gov/Surveycertificationgeninfo/downloads/SCLetter12_01.pdf (accessed 7 June 2016).

Chastre, J., Wolff, M., Fagon, J.-Y., Chevret, S., Thomas, F., Wermert, D., Clementi, E., Gonzalez, J., Jusserand, D., Asfar, P. *et al.* (2003) PneumA trial group: comparison of 8 vs. 15 days of antibiotic therapy for ventilator-associated pneumonia in adults: a randomized trial. *JAMA (The Journal of the American Medical Association)* 290, 2588–2598.

Chastre, J., Wunderink, R., Prokocimer, P., Lee, M., Kaniga, K., and Friedland, I. (2008) Efficacy and safety of intravenous infusion of doripenem versus imipenem in ventilator-associated pneumonia: a multicenter, randomized study. *Critical Care Medicine* 36, 1089–1096.

Cooke, F.J. and Holmes, A.H. (2007) The missing care bundle: antibiotic prescribing in hospitals. *International Journal of Antimicrobial Agents* 30, 25–29.

Cosgrove, S.E. (2006) The relationship between antimicrobial resistance and patient outcomes: mortality, length of hospital stay, and health care costs. *Clinical Infectious Diseases* 42(Suppl 2), S82–S89.

Dellit, T.H., Owens, R.C., McGowan, J.E., Jr., Gerding, D.N., Weinstein, R.A., Burke, J.P., Huskins, W.C., Paterson, D.L., Fishman, N.O., Carpenter, C.F. *et al.* (2007) Infectious Diseases Society of America and the Society for Healthcare Epidemiology of America guidelines for developing an institutional program to enhance antimicrobial stewardship. *Clinical Infectious Diseases* 44, 159–177.

Dubberke, E.R., Burdette, S.D., and the AST Infectious Disease Community of Practice (2013) *Clostridium difficile* infections in solid organ transplantation. *American Journal of Transplantation* 13(Suppl 4), 42–49.

File, T.M., Solomkin, J.S., and Cosgrove, S.E. (2011) Strategies for improving antimicrobial use and the role of antimicrobial stewardship programs. *Clinical Infectious Disease* 53(Suppl 1), S15–S22.

Fliser, D. and Kielstein, J.T. (2006) Technology insight: treatment of renal failure in the intensive care unit with extended dialysis. *Nature Clinical Practice Nephrology* 2, 32–39.

Fraser, G.L., Stogsdill, P. and Owens, R.C., Jr. (2005) Antimicrobial stewardship initiatives: a programmatic approach to optimizing antimicrobial use. In: Owens, R.C., Jr., Ambrose, P.G., and Nightingale, C.H. (eds) *Antibiotic Optimization Concepts and Strategies in Clinical Practice)*, 1st edn. Marcel Dekker, New York, pp. 261–326.

Hall, A.J., Curns, A.T., McDonald, L.C., Parashar, U.D., and Lopman, B.A. (2012) The roles of *Clostridium difficile* and norovirus among gastroenteritis-associated deaths in the United States, 1999–2007. *Clinical Infectious Disease* 55, 216–223.

Hall, R.G., II, Payne, K.O., Bain, A.M., Rahman, A.P., Nguyen, S.T., Eaton, S.A., Busti, A.J., Vu, S.L., and Bedimo, R. (2008) Multi center evaluation of vancomycin dosing: emphasis on obesity. *American Journal of Medicine* 121, 515–518.

Herati, R.S. and Blumberg, E.A. (2012) Losing ground: multidrug-resistant bacteria in solid-organ transplantation. *Current Opinion in Infectious Diseases* 25, 445–449.

Heyland, D.K., Johnson, A.P., Reynolds, S.C., and Muscedere, J. (2011) Procalcitonin for reduced antibiotic exposure in the critical care settings a systematic review and an economic evaluation. *Critical Care Medicine* 39, 1792–1799.

Hooton, T.M., Bradley, S.F., Cardenas, D.D., Colgan, R., Geerlings, S.E., Rice, J.C., Saint, S., Schaeffer, A.J., Tambayh, P.A., Tenke, P. *et al.* (2010) Diagnosis, prevention, and treatment of catheter-associated urinary tract infection in adults: 2009 International Clinical Practice Guidelines from the Infectious Diseases Society of America. *Clinical Infectious Diseases* 50, 625–663.

Ibrahim, E.H., Sherman, G., Ward, S., Fraser, V. and Kollef, M.H. (2000) The influence of inadequate antimicrobial treatment of bloodstream infections on patient outcomes in the ICU selling. *Chest* 118, 146–155.

Jacob, J.T., Klein, E., Laxminarayan, R., Beldavs, Z., Lynfield, R., Kallen, A.J., Ricks, P., Edwards, J., Srinivasan, A., Fridkin, S. *et al.* (2013) Vital signs: carbapenem-resistant Enterobacteriaceae. *Morbidity and Mortality Weekly Report (MMWR)* 62, 165–170.

Jamal, J.A., Economou, C.J., Lipman, J., and Roberts, J.A. (2012) Improving antibiotic dosing in special situations in the ICU: burns, renal replacement therapy and extracorporeal membrane oxygenation. *Current Opinion in Critical Care* 18, 460–471.

Kaki, R., Elligsen, M., Walker, S., Simor, A., Palmay, L., and Daneman, N. (2011) Impact of antimicrobial stewardship in critical care: a systematic review. *Journal of Antimicrobial Chemotherapy* 66, 1223–1230.

Kim, P.K., Huh, H.C., Cohen, H.W., Feinberg, E.J., Ahmad, S., Coyle, C., Teperman, S., and Boothe, H. (2013) Intracolonic vancomycin for severe *Clostridium difficile* colitis. *Surgical Infections* 14, 532–539.

Kollef, M.H. and Fraser, V.J. (2001) Antibiotic resistance in the intensive care unit. *Annals of Internal Medicine* 134, 298–314.

Lautenbach, E., Patel, J.B., Bilker, W.A., Edelstein, P.H., and Fishman, N.O. (2001) Extended-spectrum beta-lactamase-producing *Escherichia coli* and *Klebsiella pneumoniae*: risk factors for infection and impact of resistance on outcomes. *Clinical Infectious Diseases* 32, 1162–1171.

Lipman, J. (2000) Clinical practice review towards better ICU antibiotic dosing. *Critical Care and Resuscitation* 2, 282–289.

Lodise, T.P., Jr., Lomaestro, B., and Drusano, G.L. (2007) Piperacillin-tazobactam for *Pseudomonas aeruginosa* infection: clinical implications of an extended-infusion dosing *strategy. Clinical Infectious Disease* 44, 357–363.

Lupei, M.I., Mann, H.J., Beilman, G.J., Oancea, C., and Chipman, J.G. (2010) Inadequate antibiotic therapy in solid organ transplant recipients is associated with a higher mortality rate. *Surgical Infections* 11, 33–39.

Magill, S.S., Hellinger, W., Cohen, J., Kay, R., Bailey, C., Boland, B., Carey, D., de Guzman, J., Dominguez, K., Edwards, J. *et al.* (2012) Prevalence of healthcare-associated infections in acute care facilities. *Infection Control and Hospital Epidemiology* 33(3), 283–291.

Marra, A.R., Edmond, M.B., Wenzel, R.P. and Bearman, G.M.L. (2007) Hospital-acquired *Clostridium difficile*-associated disease in the intensive care unit setting: epidemiology, clinical course and outcome. *BMC Infectious Diseases* 7:42.

Mehta, N.M., Halwick, D.R., Dodson, B.L., Thompson, J.E., and Arnold, J.H. (2007) Potential drug sequestration during extracorporeal membrane oxygenation; results from an ex vivo experiment. *Intensive Care Medicine* 33, 1018–1024.

Mousavi, S., Levcovich, B., and Mojtahedzadeh, M. (2011) A systematic review of pharmacokinetic changes in critically ill patients: role of extracorporeal membrane oxygenation *DARU Journal of Pharmaceutical Sciences* 19, 312–321.

Nordmann, P., Naas, T. and Poirel, L. (2011) Global spread of carbapenemase-producing *Enterobacteriaceae*. *Emerging Infectious Diseases* 17, 1791–1798.

Owens, R.C., Jr., Fraser, G.L. and Stogsdill, P. (2004) Antimicrobial stewardship programs as a means to optimize antimicrobial use. Insights from the Society of Infectious Diseases pharmacists. *Pharmacotherapy* 24, 896–908.

Owens, R.C., Jr., Donskey, C.J., Gaynes, R.P., Loo, V.G., and Muto, C.A. (2008) Antimicrobial-associated risk factors for *Clostridium difficile* infection. *Clinical Infectious Diseases* 46(Suppl 1), S19–S31.

Pea, F., Viale, P., Pavan, F., and Furlanut, M. (2007) Pharmacokinetic considerations for antimicrobial therapy in patients receiving renal replacement therapy. *Clinical Pharmacokinetics* 46, 997–1038.

Ratanarat, R., Brendolan, A., Ricci, Z., Salvatori, G., Nalesso, F., de Cal, M., Cazzavillan, S., Petras, D., Bonello, M., Bordoni, V. *et al.* (2006) Pulse high-volume hemofiltration in critically ill patients: a new approach for patients with septic shock. *Seminars in Dialysis* 19, 69–74.

Roberts, R.R., Hota, B., Ahmad, I., Scott, R.D., II, Foster, S.D., Abbasi, F., Schabowski, S., Kampe, L.M., Ciavarella, G.G., Supino, M. *et al.* (2009) Hospital and societal costs of antimicrobial-resistant infections in a Chicago teaching hospital: implications for antibiotic stewardship. *Clinical Infectious Diseases* 49, 1175–1184.

Sandora, T.J. and Goldmann, D. (2012) Preventing lethal hospital outbreaks of antibiotic-resistant bacteria. *The New England Journal of Medicine* 367, 2168–2170.

Schetz, M. (2007) Drug dosing in continuous renal replacement therapy: general rules. *Current Opinion in Critical Care* 13, 645–651.

Shekar, K., Fraser, J.F., Smith, M.T., and Roberts, J.A. (2012) Pharmacokinetic changes in patients receiving extracorporeal membrane oxygenation. *Journal of Critical Care* 27, 741.

Singh, N., Rogers, P., Atwood, C.W., Wagener, M.M., and Yu, V.L. (2000) Short-course empiric antibiotic therapy for patients with pulmonary infiltrates in the intensive care unit. A proposed solution for indiscriminate antibiotic prescription. *American Journal of Respiratory and Critical Care Medicine* 162, 505–511.

Smith, K.J., Wateska, A., Nowalk, M.P., Raymund, M., Lee, B.Y., Zimmerman, R.K., and Fine, M.J. (2013) Cost-effectiveness of procalcitonin-guided antibiotic use in community acquired pneumonia. *Journal of General Internal Medicine* 28, 1157–1164.

Snitkin, E.S., Zelanzny, A.M., Thomas, P.J., Stock, F., NISC Comparative Sequencing Program Group, Henderson, D.K., Palmore, T.N., Segre, J.A. (2012) Tracking a hospital outbreak of carbapenmase resistant *Klebsiella pneumonia* with whole-genome sequencing. *Science Translational Medicine* 4(148): 148ra116–148ra116.

Standiford, H.C., Chan, S., Tripoli, M., Weekes, E., and Forrest, G.N. (2012) Antimicrobial stewardship at a large tertiary care academic medical center: cost analysis before, during, and after a 7 year program. *Infection Control and Hospital Epidemiology* 33, 338–345.

Stullberg, J.J., Delaney, C.P., Neuhauser, D.V., Aron, D.C., Fu, P., and Koroukian, S.M. (2010) Adherence to surgical care improvement project measures and the association with postoperative infections. *JAMA (The Journal of the American Medical Association)* 303, 2479–2485.

Thiel, S.W., Asghr, E.A., Micek, S.T., Reichley, R.M., Doherty, J.A., and Kollef, M.H. (2009) Hospital wide impact of a standardized order set for the management of bacteremic severe sepsis. *Critical Care Medicine* 37, 819–824.

van Duin, D., van Delden, C., and the AST Infectious Diseases Community of Practice (2013) Multidrug-resistant Gram-negative bacteria infections in solid organ transplantation. *American Journal of Transplantation* 13(Suppl 4), 31–41.

Vincent, J.-L., Rello, J., Marshall, J., Silva, E., Anzueto, A., Martin, C.C., Moreno, R., Lipman, J., Gomersall, C., Sakr, Y., and Reinhart, K. (2009) International study of the prevalence and outcomes of infection in the intensive care units. *JAMA (The Journal of the American Medical Association)* 302, 2323–2329.

34 Role of Antimicrobial Stewardship in a Community Hospital

Sandy J. Estrada* and Aileen Martinez

Lee Memorial Health System, Fort Myers, Florida, US; Martin Health System, Stuart, Florida, US

Introduction

Antimicrobial stewardship programs (ASPs) are often associated with larger academic medical centers, although the majority of hospitals in the US are actually community hospitals. According to 2012 data from the National Healthcare Safety Network (NHSN) of the Centers for Disease Control and Prevention (CDC), out of 3742 acute care hospitals submitting data, 70% defined themselves as non-teaching institutions (Dudeck *et al.*, 2013). Thus, in order to make a significant impact on antimicrobial usage and resistance in community hospitals, the implementation of ASP strategies is vital.

Antimicrobial stewardship guidelines were first published in 2007, viz., the guidelines for developing an institutional program to enhance microbial stewardship from the Infectious Diseases Society of America (IDSA) and the Society for Healthcare Epidemiology of America (SHEA) (Dellit *et al.*, 2007). These were followed by additional publications that highlighted the advantages of ASPs, and the literature supporting the advantages of ASPs in all hospital settings, including community hospitals, has prompted many institutions to develop ASPs (Ohl and Dodds Ashley, 2011; Septimus and Owens, 2011). Despite this, formalized ASPs have developed more rapidly in large teaching institutions than in community hospitals. A survey conducted in 2009 by the IDSA Emerging Infections Network (EIN) of 1044 infectious diseases physicians (522 respondents) noted that ASPs were in place in 79% of university teaching hospitals, in 65% of non-university teaching hospitals, but in only 40% of community hospitals. It also noted that small community hospitals were least likely to have paid stewardship physicians on their staff (Johannsson *et al.*, 2011).

There remains then a need for ASPs to be implemented at all hospitals regardless of teaching status or size. Hospitals have been increasingly adopting these programs, but the greatest potential for additional growth resides in community hospitals. This chapter will discuss evidence for and outcome studies of ASPs in community hospitals, both large and small, as well as strategies for the implementation of an ASP in a community hospital.

Necessary Elements

The IDSA/SHEA stewardship guidelines review the ideal elements of an ASP. Not all of the recommended elements will be feasible to implement in the community setting. An infectious diseases (ID) physician is recommended as an essential part of an antimicrobial stewardship team and, ideally, this specialist should be compensated for his/her time. However, this may not be a reality for many community hospitals where an ID physician may not routinely see patients in the hospital. If an ID physician serves as a consultant in the hospital, he/she could be offered compensation on a per diem basis as a "medical director" of antimicrobial stewardship. In such a position, the ID physician might fill the role(s) of reviewing policies and procedures, providing expert opinion on formulary and restriction concerns, and providing consultations to the pharmacist or other stewardship staff in complicated cases. In lieu of an ID physician, an alternative physician champion should be sought. An intensivist (critical care physician) or pulmonologist with

*Corresponding author. E-mail: Sandy.Estrada@LeeMemorial.org

an interest in infectious diseases could be considered. Many non-teaching hospitals employ hospitalists (specialists in the practice of hospital medicine) to oversee the care of their patients. Hospitalists often have strict standards for patient quality initiatives and patient safety, and these goals may often align with the goals of an institution's ASP. Rohde *et al.* (2013) published a review of ASPs conducted by hospitalists, which included a review of the abstracts submitted to the Society of Hospital Medicine (SHM) annual meetings. The aspects discussed in this review included antibiotic appropriateness for specific disease states, the development of treatment guidelines, and the usage of order sets to guide therapy of antimicrobials. Similar strategies to these could be implemented in the community hospital setting when resources are limited.

An ID-trained pharmacist is not always available at the community hospital level. In this case, a generalist clinical pharmacist, critical care pharmacist, or other pharmacist with an interest in infectious diseases may fill this role. Several training programs are currently available to help non-ID trained pharmacists to expand their knowledge in the area of antimicrobial stewardship (see Table 34.1). Participation in one of the three currently available antimicrobial stewardship certificate programs accredited by the Accreditation Council for Pharmacy Education (ACPE) allows the pharmacist didactic training in ID principles, live interaction with faculty, and the opportunity to participate in a project to apply newly developed skills into practice. Pharmacists also have frequent opportunities for education through professional societies such as the Society of Infectious Diseases Pharmacists (SIDP), American College of Clinical Pharmacy (ACCP) and American Society of Health Systems Pharmacists (ASHP), which provide education in antimicrobial stewardship at their conferences and via other methods, such as webinars. Information is available at the respective websites of these organizations.

One of the advantages of ASPs is the increased collaboration between various disciplines on the healthcare team. Newly developing programs should consider evaluating which other professionals are available to participate in the development, implementation, or evaluation of the program. Consideration should be given to clinicians, infection preventionists and healthcare epidemiologists, and to quality improvement, microbiology, information technology, and nursing personnel. Some of these professionals can be invited to participate on an as-needed basis. For example, microbiologists are instrumental in the determination of protocols and policies for antimicrobial susceptibility testing and reporting.

When preparing to start or expand an ASP program, the program coordinators may refer to the CDC *Core Elements for Stewardship Programs* (CDC, 2014a). This compendium provides the basis for a needs assessment or gap analysis at the institution level in order to assess where the greatest needs may be. It reviews seven core elements that are present in successful stewardship programs. These seven elements are: leadership commitment, accountability, drug expertise, action, tracking, reporting, and education. Detail is provided on each element for specific considerations that the institution should evaluate in the companion checklist from the CDC (2014b) (see Table 34.2).

Strategy

In a community hospital setting with limited resources, the first step to setting up an ASP should be to develop a multidisciplinary team involving as many departments as feasible within the institution

Table 34.1. Antimicrobial stewardship certificate programs.

Program name[a]	Live programming	Online modules	Webinar/online lessons/ teleconference	Project	CE hours[b]	Time	Cost
MAD-ID Basic	No	Yes	Yes	Yes	19	12 months	$500[c]
MAD-ID Advanced	Yes	No	No	Yes	15.5	12 months	$300
SIDP	No	Yes	Yes	Yes	36–39	12 months	$750[d]

[a]MAD-ID, Making a Difference in Infectious Diseases Pharmacotherapy; SIDP, Society of Infectious Diseases Pharmacists.
[b]Continuing education hours.
[c]$350 for residents or fellows in training and $300 for participants from developing nations.
[d]$550 for trainees.

Table 34.2. Core elements of hospital antibiotic stewardship programs. Adapted from the CDC *Checklist for Core Elements of Hospital Antibiotic Stewardship Programs* (CDC, 2014b).

Leadership support
- Does your facility have a formal, written statement of support from leadership that supports efforts to improve antibiotic use (antibiotic stewardship)?
- Does your facility receive any budgeted financial support for antibiotic stewardship activities (e.g., support for salary, training, or IT)?

Accountability
- Is there a physician leader responsible for outcomes of stewardship activities at your facility?

Drug expertise
- Is there a pharmacist leader responsible for working to improve antibiotic use at your facility?
- Do any of the following staff work with stewardship leaders to improve antibiotic use?
  - Clinicians?
  - Infection prevention and healthcare epidemiology?
  - Quality improvement?
  - Microbiology laboratory?
  - Information technology (IT)?
  - Nursing?

Actions to support optimal antibiotic use
- Policies
  - Does your facility have a policy that requires prescribers to document in the medical record or during order entry a dose, duration, and indication for all antibiotic prescriptions?
  - Does your facility have facility-specific treatment recommendations, based on national guidelines and local susceptibilities, to assist with antibiotic selection for common clinical conditions?
- Broad interventions
  - Is there a formal procedure for all clinicians to review the appropriateness of antibiotics 48 h after the initial order (antibiotic time out)?
  - Do specified antibiotic agents need to be approved by a physician or pharmacist prior to dispensing (preauthorization) at your facility?
  - Does a physician or pharmacist review courses of therapy for specified antibiotic agents (prospective audit with feedback) at your facility?
- Pharmacy-driven interventions
  - Are there automatic changes from intravenous to oral antibiotic therapy in appropriate situations?
  - Are there dose adjustments in cases of organ dysfunction?
  - Is there dose optimization (pharmacokinetics/pharmacodynamics) to optimize the treatment of organisms with reduced susceptibility?
  - Are there automatic alerts in situations where therapy might be unnecessarily duplicative?
  - Are there time-sensitive automatic stop orders for specified antibiotic prescriptions?

Diagnosis and infection-specific interventions
- Does your facility have specific interventions in place to ensure optimal use of antibiotics to treat the following common infections?
  - Community-acquired pneumonia?
  - Urinary tract infection?
  - Skin and soft tissue infections?
  - Surgical prophylaxis?
  - Empiric treatment of methicillin-resistant *Staphylococcus aureus* (MRSA)?
  - Non-*Clostridium difficile* infection (CDI) antibiotics in new cases of CDI?
  - Culture-proven invasive (e.g., bloodstream) infections?

Tracking: monitoring antibiotic prescribing, use and resistance
- Process measures
  - Does your stewardship program monitor adherence to a documentation policy (dose, duration, and indication)?
  - Does your stewardship program monitor adherence to facility-specific treatment recommendations?
  - Does your stewardship program monitor compliance with one or more of the specific interventions in place?
- Antibiotic use and outcome measures
  - Does your facility track rates of CDI?
  - Does your facility produce an antibiogram (cumulative antibiotic susceptibility report)?

Continued

Table 34.2. Continued.

- Does your facility monitor antibiotic use (consumption) at the unit- and/or facility-wide level by one of the following metrics?
  - By counts of antibiotic(s) administered to patients per day (days of therapy, DOT)?
  - By number of grams of antibiotics used (defined daily dose, DDD)?
  - By direct expenditure for antibiotics (purchasing costs)?

- Reporting information to staff on improving antibiotic use and resistance
  - Does your stewardship program share facility-specific reports on antibiotic use with prescribers?
  - Has a current antibiogram been distributed to prescribers at your facility?
  - Do prescribers ever receive direct, personalized communication about how they can improve their antibiotic prescribing?
- Education
  - Does your stewardship program provide education to clinicians and other relevant staff on improving antibiotic prescribing?

(pharmacy, infection prevention, microbiology, information technology). If the staff is not present to structure a separate formal team, the team could be a subset of the Pharmacy and Therapeutics (P&T) Committee. In addition to the CDC checklist previously discussed, the ASHP has also published a guidance statement on the role of the pharmacist in antimicrobial stewardship that can serve as a reference for a new program (ASHP, 2010).

Two core strategies of ASPs are discussed in the guidelines, and in most community hospital settings, a combination of both of these will need to be utilized. The two strategies are prior authorization and prospective audit with feedback. Prior authorization is often known as antimicrobial restriction, and is generally the least time-consuming of the two. Antimicrobials can be restricted to a service or to an indication, or to a combination of both. Limitations commonly cited include provider acceptance and inaccuracies in the justification that is used to access an antimicrobial.

In addition to the two core strategies, there are many supplemental elements of an ASP that can be considered. These include education, guidelines and pathway development, order form development, streamlining/de-escalation, dose optimization and intravenous (IV) to oral (PO) conversion. Many institutions may hesitate to develop an ASP due to limited time and staff. Goff *et al.* (2012) reviewed the importance of giving consideration to the "low hanging fruit" strategy, and when limited resources are the largest challenge of a community hospital, this should be considered. Several strategies are judged to be low hanging fruit, including IV to PO conversions, the batching of IV antimicrobials, therapeutic substitution, and formulary restriction. These can often be done at the system level, and therefore require less time on a day-to-day basis than other more time-intensive strategies. Many institutions may already have instituted one or more of these strategies, although not under the scope of a formal ASP. Another approach is to target high-cost or high-risk medications that are frequently misused. For example, decreased misuse of high-cost antimicrobials with the implementation of a criteria-based restriction is a valuable outcome to justify the cost of an ASP to administration in the early development stages. Simple interventions that have high impact should be the goal in community hospitals with minimal resources (Wilde and Gross, 2013).

Community hospitals often serve as a site for the training of PharmD candidates, and pharmacy residency programs have also expanded significantly into the community setting in recent years. Under the guidance of a pharmacist and/or physician preceptor (teacher/instructor), these trainees can make significant contributions to the daily work of an ASP, thus allowing more time for the preceptor to handle other responsibilities. Students and residents can also play a role in the evaluation of the success of a stewardship program through retrospective data collection projects. Chahine *et al.* (2014) propose this strategy and discuss opportunities for pharmacy trainees to be actively involved in an ASP.

As the focus in healthcare continues to shift from inpatient to outpatient care and decreasing readmissions, many community hospitals have opened outpatient infusion centers for the delivery of IV antibiotics. Antimicrobial stewardship is important in this setting as well, and allows the stewardship team to have a role in the transition of care of patients with infectious diseases. A model practice for incorporating antimicrobial stewardship principles in an outpatient infusion center is discussed in a paper by Gordon *et al.* (2011).

Building a Case

When speaking with hospital administration to justify an ASP, sharing data from a similar institution can be helpful. It is also important to focus on one or two outcomes for the first year of the program, especially if resources are limited. While limiting or decreasing antimicrobial resistance may be an outcome of an ASP, it may take years before this outcome is noted. Additionally, it must be understood that the development or reversal of antimicrobial resistance is dependent on many different factors that often change simultaneously with the development of an ASP. Significant cost savings may be possible in the first year or two of an ASP, especially if no antimicrobial restrictions were in place prior to the implementation of the program. This is primarily attributed to decreased usage of antimicrobials. Caution must be used when sharing this information with hospital administration, as this initial cost savings will level out over time. If it is possible for existing staff to start an ASP at this point, then the cost savings generated can be utilized to expand the program in the future. Adherence to national guidelines and protocols are important outcomes that can be easily evaluated for an ASP, even when resources are limited.

Hospital-acquired infections, *Clostridium difficile* infection (CDI), and immunization for pneumococcal disease are all currently the focus of one or more regulatory agencies. The NHSN currently requires reporting of the rates of *C. difficile*, methicillin-resistant *Staphylococcus aureus* (MRSA), carbapenem-resistant Enterobacteriaceae (CRE), and multidrug-resistant (MDR) *Acinetobacter*. Therefore, these are areas in which an ASP can be beneficial through working in collaboration with the infection prevention team to track these infections. When developing an ASP, focusing on one or more of these areas will be beneficial to the individual patient as well as to the institution as a whole. The NHSN now requires the electronic submission of data on days of therapy as a means of benchmarking between hospitals. This is a first step towards a more formal evaluation of outcomes of patients with infectious diseases. Accountable Care Organizations (ACOs, groups of doctors, hospitals, and other healthcare providers, who come together to give coordinated high-quality care to their Medicare patients) will likely tie similar outcomes to reimbursement strategies. This further emphasizes the importance that ASPs focus on patient outcomes as the primary objective and move away from the strategy of cost of antimicrobials as the primary measure of success.

Stewardship Outcomes Studies in Community Hospitals

Small community hospital

Combined, the accounts of the two studies that follow provide evidence to support that even with limited time and resources, small community hospitals are able to make a large difference in the appropriate usage of antimicrobials.

A 100 bed community hospital in Dallas, Texas implemented an ASP on the medical–surgical service (43 beds, including eight intensive care beds) (Storey *et al.*, 2012). Stewardship interventions consisted of prospective audit and feedback on all patients receiving 2 days or longer of antimicrobial therapy. Written recommendations were placed in the patients' charts. Recommendations were made by a two-person ID physician/clinical pharmacist team. One hour twice weekly was devoted to the program, and it resulted in a 22% reduction in the monthly use of antimicrobial agents and a 32% reduction in cost per admission.

Glenwood Regional Medical Center, a 120 bed community hospital in West Monroe, Los Angeles, implemented an Antibiotic Support Team consisting of an ID physician, a dedicated clinical pharmacist, and an infection control representative (LaRacco, 2003). The pharmacist performed concurrent chart reviews 3 days a week that targeted patients receiving multiple, prolonged, or high-cost antibiotic therapy. Urgent recommendations were communicated to the physicians verbally, while non-urgent matters were communicated utilizing a form that was placed in the chart temporarily, but did not become part of the official medical record. The types of interventions made included: recommending discontinuation of therapy due to duplicate coverage, inappropriate use, or excessive duration; IV to PO conversion; the substitution or addition of antibiotic therapy, or a change in dosage. In a 1 year period, 488 recommendations were made with an acceptance rate of 69%. Antibiotic costs were decreased from $18.21 to $14.77 per patient day (a cost reduction of 19%), with a total estimated savings of $177,000. The antibiotic support team (AST) required ~8–12 h a week of the ID specialist's time. The study concluded that in addition to the financial and clinical benefits, concurrent antibiotic utilization programs could help smaller hospitals to attract ID specialists to settings where lower patient volumes may not support traditional consultative practices.

Medium-sized community hospital

The longest period of observation in antimicrobial stewardship data was conducted at Carney Hospital, a 200 bed community teaching hospital in Boston, Massachusetts over a 7 year period (Carling *et al.*, 2003). The study evaluated the impact of an interventional multidisciplinary antibiotic management program on expenditures for antibiotics and on the incidence of nosocomial infections caused by *C. difficile* and other antibiotic-resistant pathogens. The hospital established an antimicrobial management team, which included an ID physician and a dedicated clinical pharmacist with special training in infectious diseases. A clinical pharmacist shares the responsibility for evaluating each patient prospectively and for making recommendations with an ID physician who serves as team leader.

The team targeted all patients receiving parenteral third-generation cephalosporins, aztreonam, parental fluoroquinolones, and imipenem. Following joint assessment by the pharmacist and an ID physician on the staff of all evaluated patients, an informal consultation form was placed in the front of the patient's chart for review by the physician caregivers.

Other aspects of overall antibiotic management included: formulary management by a pharmacy committee; automatic 7 day stop orders for all antibiotics; a limitation on the routine reporting of the susceptibility of *Enterobacteriaceae* to first-generation cephalosporins, ampicillin–sulbactam, aminoglycosides, and ciprofloxacin; and ongoing educational programs for both resident and staff physicians. Institutional support for the program included the funding of a full-time pharmacist and a one-quarter-time ID physician.

As a result of the ongoing involvement of the pharmacy committee and the educational activities preceding the implementation of the program in late 1991, compliance with recommendations exceeded 85% of interventions within 6 months, and reached 98% by 1993. Interventions included discontinuing antibiotics after 2–3 days in patients who were not immunocompromised and in whom diagnostic evaluations failed to confirm infection, streamlining therapy based on cultures and sensitivities, and IV to PO conversion. The acquisition costs of antibiotics decreased $200,000 to $250,000 a year. The program also realized a statistically significant reduction in the number nosocomial infections caused by *C. difficile* (36%) and antibiotic-resistant *Enterobacteriaceae*.

Another study examined an ASP in a 500 bed community hospital that restricted access to all cephalosporins in an effort to reduce the incidence of colonization and infection by cephalosporin-resistant *Klebsiella* spp. Over a 1 year period, there was an 80.1% hospital-wide reduction in cephalosporin use, which correlated with a hospital-wide 44% reduction in the incidence of clinical cultures positive for *Klebsiella* spp. ($P < 0.01$), including a 70.9% reduction in use of the intensive care unit (ICU; $P < 0.001$). It should be noted, however, that because of a concomitant increase in the use of imipenem in the same hospital, there was an inadvertent 68.7% hospital-wide increase in the incidence of infection by imipenem-resistant *Pseudomonas aeruginosa* ($P < 0.01$). Accordingly, such unintended consequences should be evaluated when intensive restriction interventions are implemented (Rahal *et al.*, 1990).

Large community hospital

St. Joseph Mercy Hospital (SJMH), located in Ann Arbor, Michigan, is a 535 bed non university-affiliated community teaching hospital that is a member of a seven-hospital network, the St. Joseph Mercy Health System. The objective of this study was to describe the clinical and economic outcomes from the first year of implementation of a new ASP and to compare them with the outcomes from a similar period before the implementation of the ASP (Malani *et al.*, 2013). The ASP team at SJMH includes two ID physicians (0.3 full-time equivalent (FTE) total) and three critical care clinical pharmacists (with no dedicated stewardship time). On Monday through Friday, the team prospectively audits the initial orders, the use within 24 h, and the weekly use (if the duration of therapy is 7 days) of eight antimicrobials: aztreonam, caspofungin, daptomycin, ertapenem, linezolid, meropenem, tigecycline, and voriconazole. These agents were targeted based on the susceptibility patterns of clinical isolates, the spectrum of activity, risk of misuse, and cost. Together, these eight antimicrobials accounted for almost one third of SJMH's 2008 antimicrobial budget.

An informational real-time database developed by the internal decision support/health outcomes team, called the Quality Institute, is populated by daily use data (Monday through Friday) from the pharmacy utilization information system. The ASP pharmacist evaluates each order by reviewing the patient's electronic medical record (EMR). The pharmacist then makes a recommendation and, if necessary, provides feedback to the provider, using the following categories: approve, stop, deescalate to an agent with a narrower spectrum, or obtain an ID consult. The ID physician provides clinical support and decision-making on complex cases, aiding with feedback and communication to providers.

During the first year of the ASP, a total of 510 orders were reviewed. Of these, 323 (63%) were considered appropriate, 94 (18%) prompted de-escalation to a more narrow-spectrum antibiotic, 61 (12%) were denied, and 27 (5%) led to a formal ID consultation in 23 patients. Of the 23 consults (addressing 27 orders), eight orders were considered appropriate, 11 led to de-escalation, and another eight led to discontinuation. Five orders (1%) were continued against the advice of the ASP. The budget for the target antimicrobials decreased by 35.6%, from $462,404 to $297,851, for a saving of $164,553. There was a 25.4% decrease in utilization of the target antimicrobials, from 215.7 to 160.8 defined daily doses (DDDs)/1000 patient days. With limited resources (no dedicated pharmacist time) and targeting only eight agents, the ASP has, therefore, produced significant reductions in antimicrobial utilization and costs. Implementation of the ASP was associated with a 13.3% decrease in antimicrobial cost per patient day and a 15.2% decrease in the total antimicrobial budget, resulting in savings of $228,911 despite increases in purchase price from any of the target antimicrobials.

The study was unable to demonstrate an association between the implementation of the ASP and decreases in mortality, readmission rate, or length of stay. The experience at SJMH also demonstrates that a community ASP can be successful with limited resources, but that dedicated resources, such as pharmacy and ID physician FTEs, are critical to long-term success (Storey *et al.*, 2012).

Conclusion

When implementing an ASP, one size does not fit all as community hospitals vary widely from place to place. Each institution has its own challenges and must develop specific strategies to overcome those challenges. Also, ASP implementation and program development may be specifically challenging for community hospitals due to their size, staffing, personnel, financial support, information technology, and infrastructure. In a recent nationwide survey (Doron *et al.*, 2013), among all respondents, regardless of the presence or absence of a formal ASP, 96.4% (351 of 364) were using some form of antimicrobial stewardship technique. The most common techniques being used by hospitals without formal ASPs were guidelines/clinical pathways (65.6%, 118 of 180) and IV to PO conversion programs (63.9%, 115 of 180). This finding suggests that even without a formal ASP program, stewardship strategies can be implemented at the community hospital level. Of the hospitals without a formal ASP program, 63% had considered implementing one. For these institutions, consideration can be given to seeking additional training for stewardship personnel and reviewing additional stewardship strategies that are not time-consuming.

References

ASHP (2010) ASHP statement on the pharmacist's role in antimicrobial stewardship and infection prevention and control. *American Journal of Health-System Pharmacy* 67, 575–577.

Carling, P., Fung, T., Killion, A., Terrin, N., and Barza, M. (2003) Favorable impact of a multidisciplinary antibiotic management program conducted during 7 years. *Infection Control and Hospital Epidemiology* 24, 699–706.

CDC (2014a) *Core Elements of Hospital Antibiotic Stewardship Programs*. Centers for Disease Control and Prevention, Atlanta, Georgia Available at: http://www.cdc.gov/getsmart/healthcare/implementation/core-elements.html (accessed 24 March 2014).

CDC (2014b) *Checklist for Core Elements of Hospital Antibiotic Stewardship Programs*. Centers for Disease Control and Prevention, Atlanta, Georgia Available at: http://www.cdc.gov/getsmart/healthcare/

implementation/checklist.html (accessed 24 March 2014).

Chahine, E.B., El-Lababidi, R.M., and Sourial, M. (2014) Engaging pharmacy students, residents, and fellows in antimicrobial stewardship. *Journal of Pharmacy Practice* 28, 585–591.

Dellit, T.H., Owens, R.C., McGowan, J.E., Jr., Gerding, D.N., Weinstein, R.A., Burke, J.P., Huskins, W.C., Paterson, D.L., Fishman, N.O., Carpenter, C.F. *et al.* (2007) Infectious Diseases Society of America and the Society for Healthcare Epidemiology of America guidelines for developing an institutional program to enhance antimicrobial stewardship. *Clinical Infectious Diseases* 44, 159–177.

Doron, S., Nadkami, L., Price, L.L., Lawrence, K., Davidson, L.E., Evans, J., Garber, C., and Snydman, D.R. (2013) A nationwide survey of antimicrobial stewardship practices. *Clinical Therapeutics* 35, 758–785.

Dudeck, M.A., Weiner, L.M., Allen-Bridson, K., Malpiedi, P.J., Peterson, K.D., Pollock, D.A., Sievert, D.M., and Edwards, J.R. (2013) National Healthcare Safety Network (NHSN) report, data summary for 2012, Device-associated module. *American Journal of Infection Control* 41, 1148–1166.

Goff, D.A., Bauer, K.A., Reed, E.E., Stevenson, K.B., Taylor, J.J., and West, J.E. (2012) Is the "low-hanging fruit" worth picking for antimicrobial stewardship programs? *Clinical Infectious Diseases* 55, 587–592.

Gordon, S.M., Shrestha, N.K., and Rehm, S.J. (2011) Transitioning antimicrobial stewardship beyond the hospital: the Cleveland clinic's community-based parenteral anti-infective therapy (CoPAT program). *Journal of Hospital Medicine* 6(Suppl 1), S24–S30.

Johannsson, B., Beekmann, S.E., Srinivasan, A., Hersh, A.L., Laxminarayan, R., and Polgreen, P.M. (2011) Improving antimicrobial stewardship: the evolution of programmatic strategies and barriers. *Infection Control and Hospital Epidemiology* 32, 367–374.

LaRacco, A., Jr. (2003) Concurrent antibiotic review programs—a role for infectious diseases specialists at small community hospitals. *Clinical Infectious Diseases* 37, 742–743.

Malani, A.N., Richards, P.G., Kapila, S., Otto, M.H., Czerwinski, J., and Singal, B. (2013) Clinical and economic outcomes from a community hospital's antimicrobial stewardship program. *American Journal of Infection Control* 41, 145–148.

Ohl, C.A. and Dodds Ashley, E.S. (2011) Antimicrobial stewardship programs in community hospitals: the evidence base and case studies. *Clinical Infectious Diseases* 53(Suppl 1), S23–S28.

Rahal, J.J., Urban, C., Horn, D., Freeman, K., Segal-Maurer, S., Maurer, J., Mariano, N., Marks, S., Burns, J.M., Dominick, D., and Lim, M. (1990) Class restriction of cephalosporin use to control total cephalosporin resistance in nosocomial *Klebsiella*. *JAMA (The Journal of the American Medical Association)* 280, 1233–1237.

Rohde, J.M., Jacobson, D., and Rosenberg, D.J. (2013) Role of the hospitalist in antimicrobial stewardship: a review of work completed and description of a multisite collaborative. *Clinical Therapeutics* 6, 751–757.

Septimus, E.J. and Owens, R.J., Jr. (2011) Need and potential of antimicrobial stewardship in community hospitals. *Clinical Infectious Diseases* 53(Suppl 1), S8–S14.

Storey, D.F., Pate, P.G., Nguyen, A.T., and Chung, F. (2012) Implementation of an antimicrobial stewardship program on the medical–surgical service of a 100-bed community hospital. *Antimicrobial Resistance and Infection Control* 1, 1–8.

Wilde, A.M. and Gross, A.E. (2013) A new practitioner's guide to antimicrobial stewardship. *American Journal of Health-System Pharmacy* 70, 2180–2183.

35 Outpatient Parenteral Antimicrobial Therapy

Christopher J. Graber[1,2*] and Armen Arshakyan[2,3]

[1]*David Geffen School of Medicine at University of California, Los Angeles (UCLA), Los Angeles, California, US; [2]VA (Veterans Affairs) Greater Los Angeles Healthcare System, Los Angeles, California, US; [3]Saban Community Clinic, Los Angeles, California, US*

Introduction

Outpatient parenteral antimicrobial therapy (OPAT) allows patients to be given intravenous antibiotics in the community rather than as an inpatient. OPAT was first developed in the 1970s, initially in the management of children with cystic fibrosis (Rucker and Harrison, 1974), but has since been utilized in a wide variety of practice settings, including physicians' offices, hospital clinics, specialized infusion centers and, most frequently, patients' homes (Tice *et al.*, 2004). In the early 1990s, it was estimated that 1 in 1000 US residents received OPAT on a yearly basis (Winters *et al.*, 1992); this figure is likely higher in the current day. OPAT offers the potential for substantial cost savings, while increasing patient convenience and promoting patient-centered care. We estimate that our OPAT program at the VA Greater Los Angeles Healthcare System (VAGLA), which was established in 2003, saves approximately $4 million/year based on acute care bed days saved (Lai *et al.*, 2013); this calculation does not account for reductions in nosocomial infections, patient deconditioning, and other functional outcomes that are reduced by virtue of patients not residing in the hospital for extended time frames. In this chapter, we provide an overview of how an OPAT can best be employed to promote safe, effective treatment of infection that requires extended courses of intravenous therapy, based on guidelines from the Infectious Diseases Society of America (IDSA) (Tice *et al.*, 2004), our experience at VAGLA, and other OPAT programs that are described in the literature.

The OPAT Team

Having a dedicated OPAT team that communicates well with patients, their families and/or caregivers, home health service providers, contract pharmacists, and one another is crucial in providing safe, effective care. A clinical pharmacist, preferably trained in infectious diseases (ID), is typically "the eyes and ears" of the OPAT program and is often responsible for its day-to-day operation, including determining that the therapy planned is the right antibiotic at the right dose for the right indication, and ensuring that proper monitoring for treatment response and safety is in place. Clinical pharmacists involved in OPAT, particularly if they do not have formal training in ID, should also consider participating in certificate programs in antimicrobial stewardship, the principles of which are critical in establishing and maintaining an OPAT program. Antimicrobial stewardship certificate programs are available through the Society of Infectious Disease Pharmacists (SIDP) and the Making a Difference in Infectious Diseases Pharmacotherapy (MAD-ID) organization.

A physician should be involved in the OPAT program to serve as the final authority on whether or not OPAT is indicated in a particular place, and to provide overall clinical guidance and support to the clinical pharmacist. Preferably, this physician should be trained in ID or have ready access to an ID physician, and should have protected time to dedicate to the program. This physician should be familiar with the operating procedures of the hospital

*Corresponding author. E-mail: christopher.graber@va.gov

  Antimicrobial Stewardship: Principles and Practice (eds K. LaPlante *et al.*)

and work with hospital stakeholders in establishing protocols for OPAT referral, the insertion of peripherally inserted central venous catheters (PICCs) for the purposes of OPAT, and proper follow-up.

Home care coordinators and/or social workers are critical in determining whether or not OPAT is feasible for any particular patient, and in arranging for the proper services to be provided to the patient. They should be particularly involved in determining whether the environment in which OPAT is to be given is adequate to promote healing and safety, and whether the patient and his/her family and/or caregivers can participate effectively in OPAT.

Patient Eligibility for OPAT

In the majority of cases, the patient's diagnosis is made by the primary physician or the team taking care of that patient while hospitalized. In these cases, it is important to involve the OPAT team as early as possible for an evaluation of the patient's eligibility and appropriateness for OPAT, and for recommendations for the choice, route, and duration of antibiotic therapy. It is highly recommended that the OPAT team (or an ID consultant) evaluate and approve all OPAT requests prior to hospital discharge.

Suggested inclusion and exclusion criteria for OPAT are listed in Table 35.1. The three main tenets of suitability for OPAT are: (i) medical stability for discharge from the acute care environment with appropriate intravenous access in place; (ii) having a proper medical indication for OPAT; and (iii) having a safe environment in which OPAT can be administered and monitored. These factors will be discussed below.

Medical stability for OPAT

The OPAT-eligible patient should be hemodynamically stable, and the signs and symptoms of

Table 35.1. Inclusion and exclusion criteria for outpatient parenteral antimicrobial therapy (OPAT).

Inclusion criteria

- Patient is medically stable for discharge from acute care hospitalization
  - Patient is hemodynamically stable
  - Signs and symptoms of primary infectious process are improved from admission
  - Patient's mental status is at or near baseline
  - All other comorbid medical conditions (especially those involving renal or hepatic dysfunction) are stable and do not require further diagnostic work-up or treatment that would necessitate ongoing hospitalization
  - Appropriate intravenous access is in place
  - Patient has received at least one dose of proposed therapy to be given via OPAT without immediate complication
- Patient has a condition for which ongoing parenteral antimicrobial therapy is needed, including (but not limited to)
  - Endocarditis
  - *Staphylococcus aureus* bacteremia
  - Prosthetic device-associated infection
  - Acute osteomyelitis or osteoarticular infection
  - Moderate-to-severe diabetic foot infection (with or without associated osteomyelitis)
  - Severe or slow-to-resolve skin–soft tissue infection for which surgical intervention is not required or has already been done
  - Unresolved but clinically stable intra-abdominal infection/abscess
  - Complicated infection caused by multidrug-resistant pathogen for which oral options are not feasible/available
- Patient will be able to receive OPAT in a stable environment to which medication and ancillary services/supplies can be adequately delivered and administered
- Appropriate clinical follow-up (including laboratory monitoring and follow-up with an infectious diseases physician or primary care provider) is in place, and rapid and reliable communication mechanisms are in place if/when problems arise

Exclusion criteria

- Patient's medical care needs exceed resources available at the proposed site of care
- Proposed site of care is unstable or unsafe or the patient and his caregivers are unwilling or unable to participate in safe OPAT administration
- Patient actively injects illicit drugs (patients with a remote history of actively injecting illicit drugs who currently do not engage in such activities may potentially be eligible for OPAT on a case-by-case basis)

the primary infectious process should be improved from the time of hospital admission. The patient's mental status should be at or near baseline, particularly if the patient is to be involved in administering OPAT. All other comorbid medical conditions should be stable and should not require further diagnostic work-up or treatment that would necessitate ongoing hospitalization. In particular, renal and hepatic function should be stable, especially if the patient will be receiving medication through OPAT that is cleared by either of these routes.

Appropriate intravenous access should also be in place by the time of hospital discharge. Routes of intravenous access are discussed in a later section of this chapter. It is also highly recommended that the patient receive the first dose(s) of the proposed OPAT therapy while still an inpatient to ensure that there are no immediate adverse effects.

Proper medical indications for OPAT

Conditions for which OPAT is typically indicated include endocarditis, uncomplicated and complicated *Staphylococcus aureus* bacteremia, prosthetic device-associated infection, acute osteomyelitis or osteoarticular infection, moderate-to-severe diabetic foot infection (with or without associated osteomyelitis), severe or slow-to-resolve skin–soft tissue infection for which surgical intervention is not required or has already been done, unresolved but clinically stable intra-abdominal infection or abscess, and complicated infections caused by multidrug-resistant pathogens for which oral options are not readily feasible or available. The duration of parenteral therapy for all of the above conditions can be variable but should be concordant with available guidelines where appropriate. The IDSA maintains practice guidelines for several common infectious conditions on its website (http://www.idsociety.org/IDSA_Practice_Guidelines/).

The management of osteomyelitis with parenteral therapy deserves special mention. It has long been held that osteomyelitis requires an extended course of parenteral antimicrobial therapy (i.e., at least 6 weeks), but more recent data, as summarized in a 2012 review in *Clinical Infectious Diseases*, challenges this notion and suggests that, at least for chronic osteomyelitis, oral therapy may suffice (Spellberg and Lipsky, 2012). It has been the policy at our institution for osteomyelitis associated with diabetic foot infection to continue parenteral therapy until the signs and symptoms of acute infection resolve, then to transition to oral therapy that is typically continued for 6 weeks or more of total therapy. Vertebral osteomyelitis typically requires a more aggressive approach (often 8 weeks or more of parenteral therapy) owing to high rates of complications and clinical failures (Dombrowski and Winston, 2008; Livorsi *et al.*, 2008).

If an antibiotic that has equivalent or near-equivalent bioavailability when given orally vs. parenterally is being prescribed, it is generally recommended that it be given orally for patients who are compliant, are otherwise stable for discharge from the acute care setting, can tolerate oral medication, and for whom there is no concern for malabsorption. Medications in this category include fluoroquinolones, linezolid, azithromycin, metronidazole, and fluconazole.

Safety for OPAT

Finally, the patient should be able to receive OPAT in a stable environment to which medication and ancillary services/supplies can be adequately delivered and administered. A definitive post-discharge plan should be in place that includes where exactly the patient will be staying and for how long, who is assuming day-to-day care for the patient (and the degree to which that person will be able to participate in administering OPAT), what ancillary services (i.e., wound care, home nursing) will need to be arranged, as well as which healthcare providers are responsible for the patient's follow-up, when appointments with them are scheduled, and how the patients will get to said appointments. It is generally recommended that OPAT patients follow up with both their primary care provider and an ID physician, though the OPAT team may determine that ID follow-up may not necessarily be required for select cases in which a defined course of parenteral therapy for a straightforward indication is planned. A clear plan for how laboratory monitoring will be achieved and how the results of this laboratory monitoring will be made available to the responsible provider should also be in place, and rapid and reliable communication mechanisms should be established between all stakeholders (the patient, caregivers, the OPAT team, pharmacy, nursing team, case managers, etc.) to ensure that any problems that arise during OPAT are appropriately triaged.

Risk Considerations for Patients Receiving OPAT

If a patient's medical care needs exceed the resources available at the proposed site of care, or if the proposed site of care is unstable or unsafe, OPAT should be reconsidered. OPAT should not be given if the patient and his/her caregivers are unwilling or unable to participate in safe OPAT administration. Typically, patients who actively inject illicit drugs are not candidates for OPAT, although those patients with a remote history of actively injecting illicit drugs who currently do not engage in such activities may potentially be eligible for OPAT on a case-by-case careful review by the OPAT team. Active injected drug use not only represents a safety risk in that provision of intravenous access in the OPAT setting allows for the easy introduction of nonsterile and dangerous substances systemically, which can lead to overdose or infective complications, but it also serves as a major impediment to compliance with the care plan. The latter statement holds true for other types of substance abuse as well, and careful assessment is required in at-risk patients.

Special considerations for OPAT referrals from another healthcare facility

The OPAT program at a "closed" healthcare system will frequently receive OPAT requests when a patient that typically belongs to that system is admitted to an outside facility. It is highly recommended that the OPAT team has a clear, standardized protocol for dealing with these requests. The OPAT team must clearly communicate with the requestor at the outside facility that several requirements need to be in order before the OPAT request can be considered: the information on the indication for OPAT needs to be complete; the patient must have appropriate intravenous access in place and should have received at least one dose of the proposed OPAT therapy without complication; the patient's social situation needs to have been assessed to ensure OPAT safety; and the patient's follow-up plan must be clearly delineated, particularly who will be primarily responsible for follow-up of laboratory monitoring and which physicians will be seen in outpatient clinic follow-up. The initial prescribers of the therapy that will ostensibly be given via OPAT must be readily available to discuss the appropriateness of that therapy and whether or not alternate strategies to the proposed therapy may be indicated. If there is any question as to the indications and safety of the OPAT plan that is proposed which cannot be resolved via communication with the outside facility, the OPAT team should recommend intra-hospital transfer so that a more careful evaluation for OPAT eligibility can be done.

At our facility, all outside requestors of OPAT must submit a completed form that we have developed that delineates the infection for which OPAT is being requested, all available microbiology and susceptibility data, and other relevant clinical information (echocardiography, imaging, etc.), the specific medications and dosages that are being requested, and medical team and patient contact information. We also include a statement that all OPAT requests are subject to approval by the medical director or associate medical director of our OPAT program, and may be declined if incomplete information is received. We also request that the patient and a physician member of the care team sign a statement that: (i) the requesting physician has assessed both the patient and the discharge environment and attests that discharge is safe and that the treatment plan with OPAT is safe; (ii) the patient is agreeable to the treatment plan (which, except in unusual circumstances, involves follow-up with our ID outpatient clinic); (iii) all changes in antibiotic treatment must be coordinated through our OPAT service; and (iv) failure to comply with any of the above may result in termination of our service to provide OPAT.

OPAT Intravenous Access Options

As stated above, the patient should have adequate intravenous access in place before OPAT can be considered. If OPAT is only to be given for a short time period (i.e., less than 2 weeks), a peripheral intravenous line may be an option, especially for patients with good veins in whom it is anticipated that replacing this intravenous line will not be difficult if a problem arises. It is typically recommended that peripheral intravenous lines be changed every 72 to 96 h to reduce the risk of infection and phlebitis (O'Grady *et al.*, 2011), although evidence for this approach (particularly in the outpatient setting) is lacking. Midline catheters, which range from 7.5 to 20 cm in length and are typically inserted in an upper arm vein, may also be an option for relatively shorter time frames of administration (Tice *et al.*, 2004). Peripheral and midline catheters should be avoided in OPAT that involves potentially venotoxic medications (e.g., amphotericin B, nafcillin, quinupristin–dalfopristin).

The most common intravenous access option for OPAT is the PICC, as it can typically be used for 6 weeks or more without complication. PICCs can also be used with programmable pumps, which can be useful for antibiotics with frequent dosing schedules (i.e., every 4–6 h). When a PICC is placed, a chest radiograph should be performed to ensure that the catheter tip resides in the superior vena cava or right atrium. In order to avoid unnecessary placement of PICCs in patients for whom OPAT may not be indicated, we have a policy at our institution that the placement of PICCs for the purpose of giving intravenous antibiotics requires approval by the ID service or the OPAT team.

In patients with renal dysfunction who are either on hemodialysis or will require hemodialysis in the near future, consultation with a nephrologist and/or vascular surgeon may be necessary when PICC placement is being considered, as the PICC placement may interfere with plans for the placement of fistulas or other hemodialysis access options. PICC placement can often be avoided in these cases by selecting medications with relatively long half-lives that can be administered via existing hemodialysis access following a hemodialysis session. Example medications that can be administered in this fashion include vancomycin, daptomycin, cefazolin, and aminoglycosides.

Tunneled central catheters and ports are also used for longer term access, for the infusion of irritative agents, and for patients from whom it is necessary to obtain blood samples frequently. If the tunneled catheter or port was initially placed for a different indication than OPAT, care must be taken to ensure that OPAT does not interfere with the initial indication.

Intravenous Access Care and Complications

Care of intravenous access varies according to type and should be based on available guidelines (O'Grady *et al.*, 2011). Patients or their caregivers should inspect the access site on a daily basis, and healthcare professionals should inspect the site on a weekly basis. The OPAT team should be informed of any evidence of local phlebitis, induration, erythema, tenderness, or leakage (Tice *et al.*, 2004). The development of neck or arm edema in association with intravenous access should prompt evaluation for deep venous thrombosis, which, if present, typically requires device removal.

Occasionally, PICC lumens can be occluded, either by fibrin, by crystallized or precipitated medication, or by mechanical factors. Flushing the catheter with saline or a heparinized solution will frequently relieve the occlusion, but occasionally a small amount of a fibrinolytic agent can be administered for more difficult cases. Consultation with persons highly experienced in catheter management is recommended.

If there is evidence of infection at the catheter insertion site (induration, erythema, pain, tenderness, etc.), along the subcutaneous tract of a tunneled catheter, or in the subcutaneous pocket of a totally implanted intravascular device, catheter removal is typically required. Blood cultures should be performed in these situations and also when systemic signs of infection (fever, chills, new-onset hypotension, etc.) are present. The management of intravascular catheter-related infections is very well encapsulated in its specific IDSA guidelines document (Mermel *et al.*, 2009). In particular, the IDSA guidelines delineate when catheter salvage with the combination of systemic and antibiotic lock therapy may be considered. Antibiotic lock therapy, in which a solution containing a small amount of antibiotic (typically with heparin) is allowed to dwell in the catheter lumen, is often considered for select catheter-related bloodstream infections that are otherwise uncomplicated (i.e., there is no associated suppurative thrombophlebitis, endocarditis, or osteomyelitis), and are not caused by *S. aureus* or *Candida* spp. (Mermel *et al.*, 2009).

Setting for OPAT Administration

OPAT can be administered in a number of different settings, but is most commonly delivered at the patient's home, provided that this is a safe environment, as detailed above. At least initially, OPAT should be administered by a healthcare professional, but depending on their ability and willingness to do so, the patient and/or caregiver may be able to assume this task. Antibiotics that require frequent administration (i.e., every 4–6 h) may be given via a programmable pump, depending on the intravenous access used. If the patient and/or caregiver is administering OPAT at home, it is the policy at our institution that home health services maintain contact with the patient and visit at least weekly to inspect the catheter insertion site and ensure that no problems have arisen.

OPAT receipt at an infusion center or urgent care facility can be considered in select situations, particularly when a patient's home is not amenable to OPAT administration. Antibiotics that have dosing intervals compatible with times that the facility is open should be chosen, and a backup plan should be in place should unanticipated logistics issues arise.

Choice of Antibiotic Therapy, Laboratory Monitoring, and Clinical Follow-up

The choice of antibiotic therapy used in OPAT depends on a multitude of different factors, including the infecting organism and its susceptibility, drug pharmacokinetics and pharmacodynamics, drug stability, and the need for monitoring for therapeutic effectiveness and safety. The question of drug stability at room temperature (both after initial reconstitution and when diluted in solution) is often not well appreciated, but is particularly important, as it may be difficult and labor intensive to administer multiple daily doses of an antibiotic that has a short duration of stability in solution via OPAT (Williams and Raymond, 1998).

Commonly prescribed OPAT agents are detailed in Table 35.2, with typical dosing intervals, stability of the diluted drug at room temperature, and suggested laboratory monitoring. Close monitoring of the OPAT patient is required, particularly in the first 1–2 weeks post-hospitalization, to ensure that renal and/or hepatic function are stable, no side effects are occurring, dosing remains appropriate, and the patient is responding clinically. It must be clearly delineated who is responsible for any particular aspect of laboratory monitoring, including the drawing of blood, delivery to the clinical laboratory, following up on laboratory results, ensuring that the appropriate clinician is notified of the

Table 35.2. Commonly prescribed antimicrobials for outpatient parenteral antimicrobial therapy (OPAT).

Drug	Usual dosing interval[a]	Stability at 25°C[b]	Suggested laboratory monitoring[c,d]
Amphotericin B (liposomal)	24 h	6 h[e]	CBC, LFT weekly; Chem 7, magnesium at least 2–3× weekly
Ampicillin	4–6 h	8 h	CBC, Chem 7, LFT weekly
Ampicillin–sulbactam	6 h	8 h	CBC, Chem 7, LFT weekly
Caspofungin	24 h	1 day	CBC, Chem 7, LFT weekly
Cefazolin	8 h	1 day	CBC, Chem 7, LFT weekly
Cefepime	8–12 h	1 day	CBC, Chem 7 weekly
Ceftriaxone	12–24 h	2 days	CBC, Chem 7, LFT weekly
Daptomycin	24 h	12 h	CBC, Chem 7, CK weekly
Ertapenem	24 h	6 h	CBC, Chem 7 weekly
Gentamicin	8–24 h	30 days	CBC, Chem 7, peak, trough at least weekly, audiology monitoring
Imipenem–cilastatin	6–8 h	4 h	CBC, Chem 7, LFT weekly[f]
Meropenem	8 h	1 h	CBC, Chem 7, LFT weekly[f]
Nafcillin	4–6 h	3 days	CBC, Chem 7, LFT weekly
Oxacillin	4–6 h	2 days	CBC, Chem 7, LFT weekly
Penicillin G	4 h	1 day	CBC, Chem 7 weekly
Piperacillin–tazobactam	6–8 h	1 day	CBC, Chem 7, LFT weekly
Vancomycin	12 h	3–84 days[g]	CBC, Chem 7, trough at least weekly

[a]Assumes normal renal and/or hepatic function.
[b]Stability when diluted in normal saline, derived from drug monographs on Micromedex.
[c]CBC, complete blood count with differential; Chem 7, sodium, potassium, chloride, bicarbonate, blood urea nitrogen, creatinine, glucose; CK, creatine kinase; LFT, liver function tests—aspartate aminotransferase, alanine aminotransferase, alkaline phosphatase, total bilirubin.
[d]Prothrombin time/international normalized ratio (INR) should also be monitored in any patient on concurrent warfarin.
[e]Liposomal amphotericin should only be reconstituted in sterile water and diluted in 5% dextrose in water (D5W).
[f]Also consider monitoring antiepileptic drug levels, as can lower seizure threshold.
[g]Depends on formulation.

laboratory results, and altering therapy as a result of the laboratory monitoring. Clear lines of communication must exist between all parties involved in the process. For patients who are on warfarin, communication with the parties responsible for the monitoring and dosing of warfarin should be notified when any patient receives OPAT, as any antibiotic may have an effect on prothrombin time. Oversight of the laboratory monitoring process by an ID-trained physician is highly recommended.

The frequency of outpatient clinic follow-up of the OPAT patient can be variable and depends on patient stability and the disease process, but, in general, should occur within 2 weeks of OPAT initiation. Outpatient follow-up with an ID specialist is typically recommended, although it may not be necessary in select indications (particularly for uncomplicated infections with a defined duration of therapy). When possible, any tests, imaging, and/or procedures to monitor treatment response should be done shortly before scheduled outpatient follow-up, so that the results can be acted upon promptly. Decisions must also be made at the end of OPAT as to whether or not the patient needs to be transitioned to an oral antibiotic for maintenance/suppressive therapy, depending on the underlying condition and treatment response. For those instances in which when OPAT becomes unfeasible for social or medical reasons, consultation with an ID specialist as to whether the next course of action should be potential conversion to oral therapy, readmission, etc., may be in order.

Conclusions

Many barriers to OPAT still exist in the US, most notably limited reimbursement for OPAT-related activities in several payer models and an increasingly fragmented healthcare system that discourages effective communication. However, the establishment of an effective OPAT program can often serve as a springboard to other antimicrobial stewardship activities within an institution when its benefits are highlighted to different stakeholders. In particular, the administrative leadership will recognize the potential for immense cost savings (some of which can be allocated to other antimicrobial stewardship ventures). Clinical staff will see the opportunity for improving hospital throughput and lowering daily patient census levels by involving the OPAT team early. Most importantly, though, in an era of patient-centered care, a well-run OPAT program allows patients to receive treatment in a convenient environment that fits their needs and encourages healing. In particular, older patients may benefit from reductions in delirium by being in a more familiar environment. Thus, when possible, OPAT should be a critical part of any antimicrobial stewardship program.

References

Dombrowski, J.C. and Winston, L.G. (2008) Clinical failures of appropriately-treated methicillin-resistant *Staphylococcus aureus* infections. *The Journal of Infection* 57, 110–115.

Lai, A., Tran, T., Nguyen, H.M., Fleischmann, J., Beenhouwer, D.O., and Graber, C.J. (2013) Outpatient parenteral antimicrobial therapy at large Veterans Administration medical center. *The American Journal of Managed Care* 19, e317–e324.

Livorsi, D.J., Daver, N.G., Atmar, R.L., Shelburne, S.A., White, A.C., Jr., and Musher, D.M. (2008) Outcomes of treatment for hematogenous *Staphylococcus aureus* vertebral osteomyelitis in the MRSA ERA. *The Journal of Infection* 57, 128–131.

Mermel, L.A., Allon, M., Bouza, E., Craven, D.E., Flynn, P., O'Grady, N.P., Raad, I.I., Rijnders, B.J., Sherertz, R.J., and Warren, D.K. (2009) Clinical practice guidelines for the diagnosis and management of intravascular catheter-related infection: 2009 update by the Infectious Diseases Society of America. *Clinical Infectious Diseases* 49, 1–45.

O'Grady, N.P., Alexander, M., Burns, L.A., Dellinger, E.P., Garland, J., Heard, S.O., Lipsett, P.A., Masur, H., Mermel, L.A., Pearson, M.L. *et al.* (2011) Guidelines for the prevention of intravascular catheter-related infections. *Clinical Infectious Diseases* 52, e162–e193.

Rucker, R.W. and Harrison, G.M. (1974) Outpatient intravenous medications in the management of cystic fibrosis. *Pediatrics* 54, 358–360.

Spellberg, B. and Lipsky, B.A. (2012) Systemic antibiotic therapy for chronic osteomyelitis in adults. *Clinical Infectious Diseases* 54, 393–407.

Tice, A.D., Rehm, S.J., Dalovisio, J.R., Bradley, J.S., Martinelli, L.P., Graham, D.R., Gainer, R.B., Kunkel, M.J., Yancey, R.W., and Williams, D.N. (2004) Practice guidelines for outpatient parenteral antimicrobial therapy. *Clinical Infectious Diseases* 38, 1651–1672.

Williams, D.N. and Raymond, J.L. (1998) Community-based parenteral anti-infective therapy (CoPAT). Pharmacokinetic and monitoring issues. *Clinical Pharmacokinetics* 35, 65–77.

Winters, R.W., Parver, A.K., and Sansbury, J.D. (1992) *Home Infusion Therapy: A Service and Demographic Profile. A Report for the National Alliance for Infusion Therapy*. National Alliance for Infusion Therapy, Washington, DC.

36 Importance of Interdisciplinary Collaboration in Antimicrobial Stewardship: Immersion of Future Healthcare Professionals

Jacob Morton[1*] and Kerry L. LaPlante[1,2,3]

[1]Providence VA (Veterans Affairs) Medical Center, Rhode Island, US; [2]Brown University, Providence, Rhode Island, US; [3]University of Rhode Island College of Pharmacy, Kingston, Rhode Island, US

Introduction

To effectively facilitate patient care and the optimization of antimicrobial regimens, stewardship efforts are essential. According to the Infectious Diseases Society of America (IDSA) and the Society for Healthcare Epidemiology of America (SHEA), the entire concept of antimicrobial stewardship is founded upon interdisciplinary collaboration (Dellit *et al.*, 2007). Interdisciplinary collaboration, also referred to as interprofessional collaboration, is defined as "Occasions when two or more professions learn with, from and about each other to improve collaboration and the quality of care" (Barr *et al.*, 2006). When considering antimicrobial stewardship, this collaboration primarily occurs between those cited as core members of an antimicrobial stewardship team. These include an infectious diseases physician, a clinical pharmacist with infectious diseases training, a clinical microbiologist, an information system (IT) specialist, an infection control professional, and a hospital epidemiologist (Dellit *et al.*, 2007). Together, these individuals work to achieve the goals of antimicrobial stewardship, which include minimizing the development and spread of multidrug resistant organisms (MDROs) and maximizing cost-effective patient care (Dellit *et al.*, 2007).

Despite the documented benefits from various research studies and guideline recommendations, interdisciplinary collaboration in the realm of antimicrobial stewardship has proven to be a challenge (Greiner and Knebel, 2003). The reasons for this difficulty, however, are not exclusive to stewardship activities. Lack of communication and trust, differing views about professionalism and values, interdisciplinary competition, and misconceptions about the roles of other professions in patient care are all factors that reduce the effectiveness of healthcare teams, and thus, patient care (Greiner and Knebel, 2003).

One may argue that unless the barriers to operating as an efficient interdisciplinary team are overcome, antimicrobial stewardship efforts are not likely to be optimal. Overcoming and minimizing these barriers as early as possible will help to ensure that patients are not negatively affected. Therefore, the best time to implement strategies aimed at fostering interdisciplinary collaboration is prior to patient interaction, during the classroom training of healthcare professions students.

The History of Healthcare Interdisciplinary Collaboration

Perhaps one of the first major, organized interdisciplinary efforts in antimicrobial stewardship began

*Corresponding author. E-mail: mjacob.unc@gmail.com

in the 1980s at a large New England hospital (Briceland *et al.*, 1988), although the origin of interdisciplinary teams extends well beyond the scope of the past three decades. It should first be noted that the origins of interdisciplinary collaboration are not entirely attributed to one individual or location. In fact, the use of medical teams to provide optimal patient care was a concept that was being used in multiple areas of the world after 1900 (Baldwin, 2007). In 1915, Massachusetts General Hospital cared for patients in the outpatient setting with the philosophy that optimal care was achieved through the "teamwork of the doctor, the educator, and the social worker" (Baldwin, 2007). Across the Atlantic Ocean, the Peckham Experiment at London's Pioneer Health Centre, along with the Dawson report, were pivotal in the advocacy for a team-based model of care in 1920s Great Britain (Baldwin, 2007). In the 1940s, a New York Hospital utilized a team-based model composed of physicians, nurses, and social workers for the provision of home care services. This hospital outreach program is generally considered one of the foundational pieces for the development of interdisciplinary collaboration, particularly in the US (Baldwin, 2007).

In the 1950s, one of the most influential collaborative efforts was undertaken. The University of Washington Child Health Center began an interdisciplinary collaborative practice in family healthcare by enlisting practitioners from multiple disciplines (Baldwin, 2007). These disciplines included medicine, nursing, social work, psychiatry, nutrition, and medical technology, among others (Baldwin, 2007). The impact of this effort was dependent upon its focus as a health model rather than a medical model. Furthermore, this was one of the first collaborative efforts that incorporated interdisciplinary practice into student education (Baldwin, 2007). Not only did the work at the University of Washington establish a foundation for interdisciplinary student education, it also set a precedence for the importance of preventative care in medicine (Baldwin, 2007).

Following the interdisciplinary efforts achieved in New York, government involvement facilitated the establishment of interdisciplinary collaboration. The Office of Economic Opportunity provided funding for community clinics that utilized interdisciplinary teams, as well as funding for seminars on implementing team-based clinic activities (Baldwin, 2007). The Education for All Handicapped Children Act of 1975 required that an interdisciplinary assessment be performed on handicapped children to determine special education needs (Baldwin, 2007).

It was clear that by the 1970s, interdisciplinary collaboration was a goal that many organizations were hoping to achieve. In order to effectively equip future healthcare professionals with the necessary skills to work as a healthcare team, the Health Manpower Education Initiative Awards (HMEIA) funded one of the largest interdisciplinary education efforts to date (Baldwin, 2007). These federally funded awards were initially given to the universities of Nevada, Michigan State, North Carolina, Washington, Utah, and California at San Francisco (Baldwin, 2007). Based on the priorities of the funded site, research efforts included clinical and academic innovations for developing a healthcare team. In addition to the HMEIA funds, the ES Department of Veterans Affairs (VA) has also played a major role in the development of interdisciplinary collaboration. In 1979, the Interdisciplinary Team Training in Geriatrics program was established to provide better continuity of care and provide more comprehensive services (Baldwin, 2007).

Since the emergence of legislation in favor of interprofessional teams, there has been a large focus on interdisciplinary collaboration in the healthcare setting (Schmitt, 1994; Greiner and Knebel, 2003). The limiting factor in the development of interdisciplinary innovation has often been funding, as collaborative programs have rarely outlasted their funding (Schmitt, 1994). Regardless, there are several trends that have emerged since the initiation of interdisciplinary collaborative efforts that remain true today. First, interdisciplinary collaboration is not easy, and secondly, the current model of separate classroom educational experiences for healthcare professionals is not effectively preparing students for the interdisciplinary environment that is becoming more of a reality (Greiner and Knebel, 2003; Baldwin, 2007).

Challenges Associated with Interdisciplinary Education

In 2001, the Institute of Medicine reported that "The hierarchy in which physicians dominate and the emphasis on assuming individual responsibility for decision making result in a reliance on personal accountability and a failure to solicit the contributions of others who could bring added insight and relevant information" (Greiner and Knebel, 2003). Unfortunately, any negative impact associated with this hierarchy is most often felt by the patient and his or her caretakers. Because of inefficient interdisciplinary collaboration, continuity of care is diminished; time is wasted in clarifying contradictory treatment plans, and the resulting unnecessary adverse outcomes and costs are then passed on to the patient (Greiner and Knebel, 2003). When placing this situation in the context of antimicrobial stewardship, these adverse outcomes include increased morbidity and mortality, increased length of stay, and adverse drug events, particularly *Clostridium difficile* infections (CDIs), and increased development of MDROs (Bartlett, 2011).

As mentioned, the discovery of challenges in implementing interdisciplinary collaboration is not recent. The question then becomes, why have schools offering healthcare degree programs not incorporated interdisciplinary collaboration into the curriculum? To this point, efforts to provide interdisciplinary education have been limited primarily to experiential activities, but have not been introduced into the classroom setting on a large scale (Schmitt, 1994; Greiner and Knebel, 2003). Baldwin (2007) refers to ten potential limitations to the incorporation of interdisciplinary education into current healthcare curricula (Schmitt, 1994):

1. Insufficient continuity of learning based on the time required for the necessary didactic, curricular additions.
2. Lack of collaborative role models in the didactic and clinical practice settings.
3. Inelasticity of the current educational model's linear progression.
4. Competition and territorial tendencies regarding clinical roles.
5. Differing educational accreditation standards among healthcare professions.
6. Resistance to change of the current and long-established hierarchical medical model.
7. Administrative resistance to changes in education models.
8. Challenges in matching multiple academic schedules and student skill levels.
9. Expense of implementation and sustainability.
10. Resistance of currently established programs.

With the implementation of any new major procedural change, the natural instinct by administrators is to determine the cost and the source of funding. The primary challenge would be the determination of the cost to effectiveness ratio, as computing a monetary value for the establishment of these changes in curriculum would be very difficult, and likely an inaccurate representation of the actual expenses. Furthermore, determining the effectiveness of this collaboration would also be challenging, and the ability to reproduce and sustain these results may not be guaranteed (Baldwin, 2007). In addition to the aforementioned challenges, there is also another major situational challenge, in that there are differing numbers of medical, pharmacy, nursing, social work, and other healthcare programs in the US, many of which are located on different campuses. The larger universities may be able to develop an entire collaborative curriculum internally, without relying on healthcare programs at other universities. However, smaller universities or private programs may have to rely on the larger universities for collaboration, which further complicates successful implementation into the current model.

Lastly, a certain extent of confusion exists surrounding the competence of other healthcare professionals in their roles. This becomes pertinent in antimicrobial stewardship, as physicians and pharmacists are key members of the stewardship team (Dellit *et al.*, 2007). When considering the pharmacy profession, this is largely due to the shift from a Bachelor's of Science in Pharmacy (BSPharm) to the Doctor of Pharmacy (PharmD) degree in the 1990s and early 2000s (ACPE, 2006). This shift resulted in a broad expansion of the clinical knowledge and capabilities of pharmacists (Ahmed and Hassali, 2008). More recently, this change is also being seen with the establishment of the Doctor of Nursing Practice, which not only expands a nurse's ability to care for patients, but also prepares nurses to participate in and apply outcomes research that leads to improved patient care (AACN, 2014). As their education changes, so too will the roles of these healthcare practitioners. With the current education model separating healthcare professionals in the classroom, each profession struggles to stay up-to-date on the others' scope of expertise.

Perceptions of Interdisciplinary Collaboration and Antimicrobial Stewardship

As previously mentioned, analysis of historical perceptions concerning interdisciplinary collaboration reveals an inadequate understanding of the responsibilities of other health professionals (MacDonnell *et al.*, 2012). For example, one study involving non-pharmacy healthcare students showed that providing education about pharmacist capabilities led to a significant increase in the awareness of pharmacist capabilities (Vrontos *et al.*, 2011). What is more, there has long been the belief that the physician serves as the primary decision-maker in the clinical setting (Greiner and Knebel, 2003). From the standpoint of interdisciplinary collaboration, this may be considered detrimental, as "Students educated in a culture where the dominant belief is that physicians are all-knowing will likely not value shared decision-making with patients regardless of whether they are taught to do so .... The institutional norms are such that authoritarianism, boundaries of practice, and silos among professional disciplines are strictly enforced .... In such settings, the value of interdisciplinary teams will not likely be grasped by students" (Greiner and Knebel, 2003).

When considering this statement, "value" is a very broad term, and must therefore be defined in terms of the ultimate goal of interdisciplinary collaboration. Increasing the number of individuals examining and treating a patient does not necessarily improve the level of care. Rather, the true value of interdisciplinary collaboration is found in the ability of multiple specialized teams creating a culture of comprehensive care (Schmitt, 1994). As an antimicrobial stewardship pharmacist, one of the primary responsibilities is to encourage interdisciplinary collaboration, leading to optimal antimicrobial regimens (Septimus and Owens, 2011).

Successful antimicrobial stewardship requires a strong foundation of interdisciplinary collaboration in order to be successful (Dellit *et al.*, 2007). However, students that are not exposed to the concept of stewardship prior to entering professional practice may not be as likely to fully comprehend the benefits. A recent study of 317 fourth-year medical students from three schools of medicine was performed to assess their perceptions, attitudes, and knowledge about antimicrobial use and resistance (Abbo *et al.*, 2013). The survey found that only 40% of students were familiar with the term antimicrobial stewardship. Medical students from all three institutions demonstrated sufficient competencies in the risks associated with antimicrobial use and the treatment of community-acquired pneumonia, but performed poorly on questions of urinary tract infections, antimicrobial resistance and spectrum of activity, and the recognition of *C. difficile* (Abbo *et al.*, 2013). Also documented in the survey results was the fact that medical students that referred to either a physician, pharmacist, or the IDSA guidelines for reference materials performed significantly better on case questions than those students utilizing other resources (Abbo *et al.*, 2013).

The results of this study, while limited in their external validity, showed that current education surrounding antimicrobial stewardship is inadequate. Not only was education in antimicrobial stewardship perceived to be inadequate, but also education in infectious diseases in general, as 90% of medical students surveyed wanted more education about antimicrobials, and 79% wanted more education about antimicrobial resistance (Abbo *et al.*, 2013).

Similar research, utilizing a survey modified from the study assessing medical student knowledge by Abbo *et al.* (2013) was recently conducted among pharmacy students (Justo *et al.*, 2014). Based on the results from the 579 pharmacy students who responded to the survey, the findings were in agreement with those of medical students' knowledge of antimicrobial stewardship. An overwhelming majority of students agreed that antimicrobial resistance and overuse were significant problems (Justo *et al.*, 2014). Despite this recognition, only 73% felt that their didactic education in pharmacy prepared them to effectively monitor patients for safety and efficacy of antimicrobial therapy, and only 54% of the 579 students felt comfortable streamlining or de-escalating therapy (Justo *et al.*, 2014). This correlated with 60% of students correctly identifying antimicrobial spectrum of activity, and 50% of students correctly identifying the potential for unnecessary or inappropriate antimicrobial use in case vignette scenarios (Justo *et al.*, 2014).

Further studies are needed to assess the comprehensive knowledge base about antimicrobial stewardship on a national level, but regardless, the two studies described above bring to light aspects of healthcare for future practitioners that are concerning. While there is a great recognition of antimicrobial misuse and resistance, the didactic training of healthcare practitioners appears to be inadequate to appropriately manage this epidemic (Bartlett, 2011; Justo *et al.*, 2014).

When analyzing the attitudes of pharmacist and physician students towards interdisciplinary collaboration, the majority of studies have revealed that student pharmacists have a more positive attitude toward collaboration than their medical counterparts (Van Winkle *et al.*, 2011), although a study performed in Turkey provided evidence to the contrary, in that medical students held a more positive view towards collaboration with nurses (Van Winkle *et al.*, 2011). Despite the differing conclusions of the aforementioned studies, there is evidence to suggest that upon giving nursing, pharmacy, and medical students insight into each other's professions, there was an increase in the development of positive attitudes and self-confidence in the healthcare team (MacDonnell *et al.*, 2012). Furthermore, research showed that students enjoy learning with each other, and that interventions designed to encourage interdisciplinary learning led to significant increases in the number of students who feel that interdisciplinary learning should be required (MacDonnell *et al.*, 2012).

Collaborative Working Relationships

Interdisciplinary or not, collaboration and trust amongst individuals requires time and effort. It would be unrealistic to expect a working relationship to be established overnight with any new acquaintance. The same is true of interdisciplinary collaboration for healthcare providers. In fact, one might argue that trust is built more slowly in the healthcare setting because the health and welfare of a patient is now at stake.

It might be beneficial, then, to divide collaboration into several stages. The collaborative working relationship (CWR) model proposes five stages that show the development and progression of collaboration as it grows (Brock and Doucette, 2004). For the purposes of description, the model will consider the CWR between physicians and pharmacists.

- **Stage Zero** is known as "Professional Awareness" (Brock and Doucette, 2004). In this stage, exchange between individuals is considered to be minimal. All efforts at collaboration are those that are considered to be safe, and do not endanger one's own professional activities. Clear borders are present, and expectations are made known. Stage Zero is the baseline and lowest stage of interdisciplinary collaboration (Brock and Doucette, 2004).
- **Stage One** is "Professional Recognition" (Brock and Doucette, 2004). By now, a pharmacist is proactively encouraging and communicating with physicians. The pharmacist is slowly working to build a relationship of trust with the physician through small changes and effective communication (Brock and Doucette, 2004). An example of professional recognition would be a pharmacist contacting a physician in a hospital setting to change the dose of an antibiotic for the treatment of a patient's infection.
- **Stage Two** brings about "Exploration and Trial" (Brock and Doucette, 2004). The physician, having had several interactions with the pharmacist now, begins to test their expectations and trust. The physician might recall the pharmacist's clinical abilities in previous encounters. If the physician has a medication-related question, they might consider contacting the pharmacist. If the pharmacist has previously provided trustworthy and appropriate responses, the physician might engage the pharmacist for a more complicated request. Stage Two is primarily about considering risks and benefits (Brock and Doucette, 2004).
- **Stage Three** involves "Expansion of the Professional Relationship" (Brock and Doucette, 2004). At this point, both the pharmacist and the physician have begun to realize the benefits of interdisciplinary collaboration. The pharmacist now feels more comfortable contacting the physician about patient care or diagnostic issues, and the physician now feels confident in the pharmacist's abilities as the medication expert (Brock and Doucette, 2004).
- **Stage Four** is known as "Commitment to the Collaborative Working Relationship" (Brock and Doucette, 2004). The key to this final stage of the CWR is that the physician feels that the benefits of utilizing a pharmacist's abilities on a regular basis outweighs any potential risks. Three components are necessary for this stage: moderate input levels from both parties, interactions that are of a significant length, and collaboration on a consistent basis (Brock and Doucette, 2004). When applied to antimicrobial stewardship, this is the full utilization of both a physician's and pharmacist's capabilities to optimize patient care, prevent unnecessary antibiotic administration, and provide education to other members of the healthcare team.

Advocacy for Interdisciplinary Collaboration

Consider the following statement: "When one considers that nurses, physicians, and other health care professionals must work together in the same environment, it would seem feasible that learning together through allocated learning activities as an interprofessional group would enhance educational outcomes" (Dufrene, 2012). Regardless of profession, interdisciplinary collaboration has been shown to be an essential part of the current healthcare motif. The increasing complexity of drug–drug interactions, the greater possibility of medical errors, the rapid achievements in healthcare, the increased costs of drug-related morbidity, and other rapidly rising healthcare costs speak to the ever-growing number of concerns that require attention (Van Winkle *et al.*, 2011). It makes sense, then, that educating students as a team would effectively prepare them to serve in their respective roles upon entrance into clinical practice.

It should be noted that the concerns listed encompass a wide range of challenges within the healthcare field. As such, it makes sense that interdisciplinary collaboration should be utilized to specifically target and correct these problems. The purpose of interdisciplinary collaboration is to "ensure that all health care practitioners effectively use their specialized training and expertise to optimize patient care and improve therapeutic outcomes" (Van Winkle *et al.*, 2011). In achieving this purpose, the hope is that the following goals will consistently be achieved: (i) meeting the patient's needs; (ii) joint communication; and (iii) respecting the unique qualifications of all members of the healthcare team (Van Winkle *et al.*, 2011).

Developing Interdisciplinary Collaboration in the Setting of Antimicrobial Stewardship

In order for collaborative working relationships to ever go beyond Stage Zero, changes in interdisciplinary collaboration and interactions among healthcare professionals are needed. Likely the first needed change is the agreement upon the definition of a healthcare professional when designing an interdisciplinary team, as this is often inconsistent. For the purposes of antimicrobial stewardship, the IDSA explicitly names the recommended members of an interdisciplinary team (Dellit *et al.*, 2007). In comparison, other commissions charged with evaluating current inadequacies in healthcare are not all inclusive. For example, one organization's report from 2010 discusses the implementation of changes to improve interdisciplinary collaboration for physicians, nurses, and public health workers, but does not address the role of pharmacists in this same healthcare framework (Frenk *et al.*, 2010).

When examining the IDSA antimicrobial stewardship guidelines, advocacy for interdisciplinary collaboration is seen within the authorship, which includes both physicians and pharmacists (Dellit *et al.*, 2007). The guidelines serve as a model for healthcare professionals collaborating to achieve better, comprehensive, patient care. This comprehensive care must be in line with the National Patient Safety Goals (NPSG) established by The Joint Commission as part of its accreditation process for health systems in the US (Rosenberg, 2012). As part of reducing the risk of healthcare-associated infections (Goal 7 of the NPSGs), hospitals are charged with the prevention of MDRO, central line, and catheter-associated infections through the implementation of evidence-based practice (Rosenberg, 2012).

Antimicrobial stewardship serves as the vehicle for which evidence-based practice is implemented to decrease the risk of healthcare-associated infections. Physicians and pharmacists trained in infectious diseases work together to provide education and local guidelines for the health system, recommend formulary restrictions, and utilize computerized technology to review patient information on the appropriateness of antibiotic prescribing (Rosenberg, 2012). This collaboration requires a collaborative working relationship that would be considered Stage Three or Stage Four of the CWR model, as both parties must agree that collaboration is beneficial for themselves as well as for the patient. While the relationship between the infectious diseases physician and pharmacist is important, it is not the only collaborative relationship that must be fostered. The relationship between the infectious diseases pharmacist and the other physicians within the health system must also be developing as a collaborative working relationship.

Consider the relationship between an infectious diseases pharmacist and an internal medicine physician. At Stage Zero, the interaction between the two professionals is considered to be minimal and safe; clear boundaries are established (Brock and Doucette, 2004). In this setting, an attempt by the pharmacist to provide education on antibiotic prescribing might be considered passive—the pharmacist might leave informational material for the physician to review. In a more developed CWR, both parties have comfortable interactions and appreciate the benefits of each other's services

(Brock and Doucette, 2004). As such, the pharmacist would likely contact the physician directly via telephone or a face-to-face interaction to discuss antibiotic selection. When considering the impact on patient care, the passive and indirect educational efforts are probably ineffective and would not lead to changes in practice (Rosenberg, 2012).

Developing Interdisciplinary Collaboration in the Classroom

Returning briefly to the challenges associated with implementing interdisciplinary collaboration in the classroom setting, one of the most difficult challenges to overcome is most likely the comfortable design of current professional training. Classroom learning for healthcare professionals often tends to occur in silos, preventing exposure to the responsibilities and thought processes of other professions (Graybeal *et al.*, 2010). Overcoming these "silos" is an essential step in the implementation of interdisciplinary collaboration.

As with the implementation of any new educational method, sufficient evidence must be present to show that this new method is at least no worse than the present method of instruction. Despite the progressive movement toward collaborative work in the clinical practice setting, the classroom setting has lagged behind in the healthcare field (Graybeal *et al.*, 2010). Graybeal *et al.* (2010) have proposed that one reason for this lag is the challenge of establishing the true benefit of interdisciplinary education. Likely one of the most significant publications on the benefit of interdisciplinary education in recent years was the report by a group of international authors that examined over 100 studies involving interdisciplinary education (Barr *et al.*, 2006). The analysis included studies of both health professions students and post-licensure individuals; those studies focusing on students included end points primarily related to attitudes and perceptions toward other health professions, while studies involving licensed healthcare professionals examined the benefits in patient care (Barr *et al.*, 2006).

The evidence in both pre- and post-licensure settings points toward utilization of interdisciplinary collaboration in a stepwise fashion. Utilization begins with the personal preparation of students in the pre-licensure setting, during which the focus of interdisciplinary education is changing attitudes, perceptions and behaviors (Barr *et al.*, 2006). This training prepares students for effective teamwork to develop a culture of change-oriented practice; upon entering clinical practice, the individual is now able to incorporate these attitudes and perceptions into their working environment (Barr *et al.*, 2006). Over time, as the practitioner gains a foothold in their practice setting, patient care is changed (Barr *et al.*, 2006).

According to the World Health Organization (WHO), students learning to participate in a collaborative team "is a key step in moving health systems from fragmentation to a position of strength" (Gilbert *et al.*, 2010). To accomplish this key step, it is necessary to have faculty and administrative buy-in (Graybeal *et al.*, 2010). From a faculty standpoint, the collaboration of multiple professions is needed for curriculum development, while administrative support is necessary to justify resource allocation (Graybeal *et al.*, 2010).

While faculty and administrative support are necessary as part of building an interdisciplinary culture, healthcare students may also have a significant impact on the process. Student organizations may serve as influential voices in advocating for the furtherance of interdisciplinary educational opportunities (Graybeal *et al.*, 2010). This requires that students fully realize the implications and potential benefits of interprofessional education in the process of becoming a licensed healthcare practitioner.

Teaching Antimicrobial Stewardship in the Classroom

Several positive concepts have been identified in the development of interdisciplinary collaboration in the classroom. These include a curriculum that engages students prior to licensure, case studies incorporating applicable skills for multiple disciplines, and the involvement of practicing clinicians that implement the skills taught on a daily basis (Graybeal *et al.*, 2010). For example, a case study that implements daily stewardship activities by each profession would provide exposure to the actual demands and expectations of the antimicrobial stewardship team. These activities include prospective patient audit with interventions, the utilization of example formulary restrictions and preauthorizations, staff education, dose optimization, antibiotic de-escalation, and parenteral to oral conversions (Dellit *et al.*, 2007).

One integral piece of this education would be information technology, as these cases would provide the opportunity for exposure to electronic medical records (EMRs). Each profession would have the opportunity to learn how an EMR functions, and would also learn their respective roles in patient care documentation. A nursing student might document patient care, the medications administered, or any complications that arise, including increased oxygen requirements, low blood pressure, or fever. A medical student may document the daily examination of the patient, results and interpretations from labs or imaging, as well as a plan of care. From an infectious diseases standpoint, the medical student may document the duration of antibiotics, the plan for de-escalation, or the need for surgical interventions. The pharmacy student would have the opportunity to complete a chart review, noting the duration of therapy, signs of clinical improvement or deterioration, and any evidence-based recommendations on the therapeutic optimization of antimicrobials.

As medical professionals, it is important for each discipline to examine all patient encounters and documentation. For example, the pharmacy student should be aware of what the medical student and nursing student documents, the medical student should be aware of pharmacy and nursing documentation, and the nursing student should note what recommendations and plans of care have been offered by the medical and pharmacy students.

Other methods of interaction among healthcare professionals will also need to be developed in the classroom. In addition to reviewing electronic documentation, students must become comfortable making recommendations, asking questions of other healthcare professionals, and discussing a patient plan of care. Developing collaborative working relationships is only effective when active communication occurs. This is true both as students and as practicing healthcare providers. As such, it makes sense that a CWR must be effectively modeled to students in the classroom, not just in experiential education settings. Therefore, physician, pharmacist, and nurse educators must work together in designing these classroom cases, as well as in providing real examples of collaboration in action.

The Future of Interdisciplinary Education

To this point, data on the benefit of interdisciplinary education for multiple outcomes have proven to be insufficient (Reeves *et al.*, 2013). A meta-analysis was conducted in 2013 on studies published between 2006 and 2011, which analyzed patient outcomes, adherence rates to guidelines, patient satisfaction, clinical processes, collaborative behavior, practitioner competencies, and error rates (Reeves *et al.*, 2013). For all of the outcomes reviewed, the benefits of interdisciplinary education could not be adequately assessed owing to several limitations. Of the studies reviewed, the majority focused on patient outcomes, and concluded that interdisciplinary education may lead to improved patient outcomes (Reeves *et al.*, 2013). However, this conclusion was based on only six studies with a low quality of evidence (Reeves *et al.*, 2013). Furthermore, these six studies, along with the other nine studies included in the meta-analysis, were limited in their external validity because of selection and reporting bias, incomplete data, and other unreported biases (Reeves *et al.*, 2013).

While this meta-analysis was limited in its inclusiveness, it did provide a relatively accurate depiction of the current healthcare landscape from the aspect of the benefit of interdisciplinary education on outcomes. There have been no definitive studies providing a high level of evidence that any certain intervention or interdisciplinary educational strategy has proven beneficial (Reeves *et al.*, 2013). Rather, interventions such as antimicrobial stewardship, which is a collection of post-licensure interdisciplinary interventions, have shown to improve patient and health systems outcomes (Dellit *et al.*, 2007; Bartlett, 2011; Rosenberg, 2012).

The meta-analysis was clear in its review of the available data, in that more studies with minimized bias are needed to establish the true benefit of interdisciplinary education (Reeves *et al.*, 2013). Additionally, it will be necessary to delineate the differences between the implementation of pre-licensure education and post-licensure interventions on outcomes, as the currently available results are inconsistent (Dellit *et al.*, 2007; Bartlett, 2011; Rosenberg, 2012; Reeves *et al.*, 2013). When considering antimicrobial stewardship, although there has only been one definitive study on the pre-licensure knowledge of antibiotic optimization, it is clear that the awareness of antimicrobial resistance and the prevention of adverse drug events such as CDIs must be improved (Bartlett, 2011; Abbo *et al.*, 2013; Justo *et al.*, 2014).

Conclusion

Antimicrobial stewardship requires a high level of interdisciplinary collaboration to truly be effective.

Active communication, along with mutual respect, trust, and awareness of other healthcare professionals' scopes of practice are meant to facilitate improved patient care. The influence of interdisciplinary teams that began in the early 20th century has led to the involvement of organizational funding, government legislation, and the development of accreditation standards that support and focus on the incorporation of interdisciplinary collaboration into practice as well as educational settings. Despite this support, there are still major challenges associated with implementing interdisciplinary collaboration into health professions curricula. In order for patient care to be optimized, these challenges must be overcome, and education must adapt to the rapidly changing healthcare landscape. Interdisciplinary collaboration must be further incorporated into the didactic training of medical, pharmacy, nursing, and other allied health students. Along with interdisciplinary collation, antimicrobial stewardship provides a valid opportunity and area of teaching to implement new educational strategies. Providing an increased focus on stewardship and antimicrobial resistance appeals to students' desires (based on survey data) in healthcare education, and also prepares students to be practitioners in a team-based setting with the training necessary to make appropriate adjustments to antimicrobial regimens.

Government support is also essential to the development of an interdisciplinary education with a focus on antimicrobial stewardship. On September 18, 2014, President Barack Obama issued an executive order designed to improve the detection of antibiotic-resistant pathogens, and to also develop strategies aimed at the prevention of antimicrobial resistance through increased awareness and preservation of current and future antimicrobial spectrums of activity (Obama, 2014). Academicians are called to assist in the facilitation of this order (Obama, 2014). As such, early awareness of stewardship through didactic efforts are warranted to bring about the realization of this executive order.

References

AACN (2014) AACN *Key Differences between DNP and PhD/DNS programs[: Contrast Grid]*. American Association of Colleges of Nursing, Washington, DC. Available at: http://www.aacn.nche.edu/dnp/ContrastGrid.pdf (accessed 9 June 2016).

Abbo, L.M., Cosgrove, S.E., Pottinger, P.S., Pereyra, M., Sinkowitz-Cochran, R., Srinivasan, A., Webb, D.J., and Hooton, T.M. (2013) Medical students' perceptions and knowledge about antimicrobial stewardship: how are we educating our future prescribers? *Clinical Infectious Diseases* 57, 631–638.

ACPE (2006) Accreditation Standards and Guidelines: Professional Program in Pharmacy Leading to the Doctor of Pharmacy Degree. Accreditation Council for Pharmacy Education, Chicago, Illinois. Available at: https://www.acpe-accredit.org/standards/ (accessed 9 June 2016).

Ahmed, S.I. and Hassali, M.A.A. (2008) The controversy of PharmD degree. *American Journal of Pharmceutical Education* 72(3):71.

Baldwin, D.C., Jr. (2007) Some historical notes on interdisciplinary and interprofessional education and practice in health care in the USA. *Journal of Interprofessional Care* 21(Suppl. 1), 23–27.

Barr, H., Freeth, D., Hammick, M., Koppel, I., and Reeves, S. (2006) The evidence base and recommendations for interprofessional education in health and social care. *Journal of Interprofessional Care* 20, 75–78.

Bartlett, J.G. (2011) A call to arms: the imperative for antimicrobial stewardship. *Clinical Infectious Diseases* 53(Suppl 1), S4–S7.

Briceland, L.L., Nightingale, C.H., Quintiliani, R., Cooper, B.W., and Smith, K.S. (1988) Antibiotic streamlining from combination therapy to monotherapy utilizing an interdisciplinary approach. *Archives of Internal Medicine* 148, 2019–2022.

Brock, K.A. and Doucette, W.R. (2004) Collaborative working relationships between pharmacists and physicians: An exploratory study. *Journal of the American Pharmacists Association* 44, 358–365.

Dellit, T.H., Owens, R.C., McGowan, J.E., Jr., Gerding, D.N., Weinstein, R.A., Burke, J.P., Huskins, W.C., Paterson, D.L., Fishman, N.O., Carpenter, C.F. *et al.* (2007) Infectious Diseases Society of America and the Society for Healthcare Epidemiology of America guidelines for developing an institutional program to enhance antimicrobial stewardship. *Clinical Infectious Diseases* 44, 159–177.

Dufreno, C. (2012) Health care partnerships: a literature review of interdisciplinary education. *The Journal of Nursing Education* 51, 212–216.

Frenk, J., Chen, L., Bhutta, Z.A., Cohen, J., Crisp, N., Evans, T., Fineberg, H., Garcia, P., Ke, Y., Kelley, P. *et al.* (2010) Health professionals for a new century: transforming education to strengthen health systems in an interdependent world. *The Lancet* 376, 1923–1958.

Gilbert, J.H.V., Yan, J., and Hoffman, S.J. (2010) A WHO report: framework for action on interprofessional education and collaborative practice. *Journal of Allied Health* 39(Suppl 1), 196–197.

Graybeal, C., Long, R., Scalise-Smith, D. and Zeibig, E. (2010) The art and science of interprofessional education. *Journal of Allied Health* 39(Suppl. 1), 232–237.

Greiner, A.C. and Knebel, E. (eds) (2003) *Health Professions Education: A Bridge to Quality*. The National Academies Press, Washington, DC.

Justo, J.A., Gauthier, T.P., Scheetz, M.H., Chahine, E.B., Bookstaver, R.B., Gallagher, J.C., Hermsen, E.D., DePestel, D.D., Ernst, E.J., Jacobs, D.M. *et al*. (2014) Knowledge and attitudes of doctor of pharmacy students regarding the appropriate use of antimicrobials. *Clinical Infectious Diseases* 59(Suppl 3), S162–S169.

MacDonnell, C.P., Rege, S.V., Misto, K., Dollase, R. and George, P. (2012) An introductory interprofessional exercise for healthcare students. *American Journal of Pharmaceutical Education* 76(8): 154.

Obama, B. (2014) *Executive Order—Combating Antibiotic-Resistant Bacteria*. Office of the Press Secretary, The White House, Washington, DC. Available at: https://www.whitehouse.gov/the-press-office/2014/09/18/executive-order-combating-antibiotic-resistant-bacteria (accessed 9 June 2016).

Reeves, S., Perrier, L., Goldman, J., Freeth, D., and Zwarenstein, M. (2013) Interprofessional education: effects on professional practice and healthcare outcomes (update) [Version 3]. *Cochrane Database of Systematic Reviews* 3:CD002213.

Rosenberg, D.J. (2012) Infections, bacterial resistance, and antimicrobial stewardship: the emerging role of hospitalists. *Journal of Hospital Medicine* 7(Suppl 1), S34–S43.

Schmitt, M. (1994) USA: focus on interprofessional practice, education, and research. *Journal of Interprofessional Care* 8, 9–17.

Septimus, E.J. and Owens, R.C., Jr. (2011) Need and potential of antimicrobial stewardship in community hospitals. *Clinical Infectious Diseases* 53(Suppl 1), S8–S14.

Van Winkle, L.J., Fjortoft, N., and Hojat, M. (2011) Validation of an instrument to measure pharmacy and medical students' attitudes toward physician–pharmacist collaboration. *American Journal of Pharmaceutical Education* 75(9): 178.

Vrontos, E.B., Kuhn, C.H., and Brittain, K.L. (2011) Impact of interprofessional activities on health professions students' knowledge of community pharmacists' role and services. *American Journal of Pharmaceutical Education* 75(8): 152.

37 Antimicrobial Stewardship and the Importance of Working with the Government and Pharmaceutical Industry

ELIZABETH D. HERMSEN[1]* AND NICOLE M. MAHONEY[2]

[1]*Global Population Health, Merck & Co., Inc., Kenilworth, New Jersey, US;*
[2]*Global Regulatory Affairs, Merck & Co., Inc., Rockville, Maryland, US*

Introduction

Antimicrobial stewardship programs (ASPs) have been implemented in many healthcare institutions because of increased awareness of the need for the judicious use of antimicrobials (Dellit *et al.*, 2007; SHEA, IDSA and PIDS, 2012). Further ASP implementation will likely be accelerated by recent calls to action by federal and local governing bodies (PCAST, 2014; The White House, 2014a,b,c). Although the impetus for ASPs should be the patient and public health (Owens, 2008; Drew *et al.*, 2009), pharmacy cost savings are often the focus due to the relative ease of measurement, lack of consensus on quality indicators for ASPs, and prevalence of under-resourced programs in terms of personnel, information technology, and research/outcomes measurement infrastructure (Johannsson *et al.*, 2011; Ohl and Dodds Ashley, 2011; Dodds Ashley *et al.*, 2014). These factors, in combination with the impending requirement for ASP implementation, create a tenuous scenario for the practice of antimicrobial stewardship, with danger of promulgating cost-minimization strategies as opposed to initiatives focused on patient safety and quality of care.

Collaboration between the government, policymakers, pharmaceutical industry, and clinicians focused on antimicrobial stewardship is needed in order to achieve the shared goals of minimizing and preventing inappropriate antimicrobial use, improving patient outcomes, and combating the development of resistance. Meeting these goals requires careful forethought in legislation and policy, advocacy for patient-centered antimicrobial stewardship, the education of clinicians, the public, and policymakers, and outcomes-focused, real-world research. This chapter is intended to provide an overview of legislative and policy efforts to bolster antimicrobial stewardship across the globe, with examples of how the government, policymakers, the pharmaceutical industry, and clinicians have worked together for antimicrobial stewardship-related advocacy, education, and research. However, the information provided is not comprehensive and represents only a fraction of the many ongoing stewardship initiatives around the world. Additionally, although equally important with regard to combating antimicrobial resistance, stimulating innovation in antimicrobial development and ensuring access to effective antimicrobials are beyond the scope of this chapter.

Government Efforts in Antimicrobial Stewardship

Despite the rise in antimicrobial resistance in recent years, the delivery of new treatments has not kept pace, leaving some patients to suffer infections that cannot be treated adequately—if at all. Responding

*Corresponding author. E-mail: elizabeth.hermsen@merck.com

to increasing global concern over the lack of treatments, governments and health organizations are developing plans and mobilizing resources to tackle this looming public health threat. A multipronged strategy will be required, built around the four main actions outlined by the US Centers for Disease Control and Prevention (CDC, 2013) and other organizations: infection control, surveillance, ensuring the development of new treatments and diagnostics, and improving the prescribing and use of antibiotics to maximize the lifespan of these critical drugs (i.e., antimicrobial stewardship). Although the focus of this chapter is antimicrobial stewardship, we realize that none of these actions are more important than the others—they must be implemented in concert in order to have the best chance of meeting the "superbug" challenge of today and the future.

There is much activity at the global, national, and local levels, but we will emphasize government activity in the US. The specific policy instruments discussed include new laws or legislative proposals, regulations, recommendations, and tools to either mandate or support stewardship in various healthcare and market settings. As this field is dynamic and the sense of urgency for implementing solutions so great at the present time, the specifics of the information below may change. However, the use of government policy to provoke sustainable stewardship practices is powerful and expected to increase over time.

Coordinating a global response to antibiotic resistance

Recognizing the need for global coordination across the health and agricultural sectors, the World Health Organization (WHO) adopted Resolution WHA67.25 on antimicrobial resistance at the 67th World Health Assembly (WHA) in May of 2014 (WHO, 2014a). The goal of the resolution is to slow resistance through focused efforts to improve knowledge and understanding of antimicrobial resistance, conserve the effectiveness of treatments and foster stewardship, reduce infections, and stimulate the development of new treatments and diagnostics. The resolution points to practical steps that WHO member states may use to develop national policies, drive global stewardship standards, improve data collection and surveillance, and implement infection prevention and control practices. Importantly, the WHA resolution mandated that the WHO develops a global action plan to combat antimicrobial resistance and metrics to measure progress (Shallcross and Davies, 2014).

Stewardship is a critical pillar of the action plan, with the objective of optimizing the use of antimicrobial medicines in human and animal health. Draft recommendations to this end include: ensuring that the distribution and dispensing of antibiotics is carried out by accredited health or veterinary professionals; establishing the laboratory capacity to identify pathogens and their antibiotic susceptibility to guide optimal use; and reducing the nontherapeutic use of antimicrobial medicines in animal health. As member states have different priorities, capacities and needs, the plan is organized as a series of building blocks so that countries can select the most relevant elements for incorporation into their national plans to address resistance.

US efforts to foster antimicrobial stewardship

In September of 2014, the Obama Administration released its *National Strategy on Combating Antibiotic-Resistant Bacteria* (The White House, 2014c), a 5 year plan to address resistance, and the President signed an Executive Order directing government agencies to take specific actions (The White House, 2014a). At the same time, the President's Council of Advisors on Science and Technology (PCAST) also released a report on *Combating Antibiotic Resistance* (PCAST, 2014), outlining recommendations to improve antibiotic stewardship and spur the development of new treatments, among other things.

The Executive Order established a new government task force for combating antibiotic-resistant bacteria co-chaired by the Secretaries of Defense, Agriculture, and Health and Human Services (HHS), which was charged with outlining specific government actions to implement the National Strategy and address the recommendations made by PCAST. It also called for improved national surveillance to detect and track resistance, and the identification of ways to accelerate the development of "next generation" therapies and diagnostics. Further, the Executive Order directed the Secretary of HHS, in consultation with the Secretary of Agriculture, to establish a Presidential Advisory Council on Combating Antibiotic-Resistant

Bacteria (PACCARB), composed of nongovernmental experts to provide advice, information, and recommendations on the programs and policies needed to address antibiotic resistance. The PACCARB was established in March 2015, appointments to it were announced in mid-September 2015, and the inaugural meeting held soon afterwards (PACCARB, 2015).

With respect to antimicrobial stewardship, the Administration's Executive Order and National Strategy direct government agencies to propose new regulations and other actions to implement and improve ASPs across healthcare settings and in the community according to best practices, including those defined by the CDC (CDC, 2014a). Federal departments and agencies are required to lead by example by implementing ASPs across government facilities, such as healthcare facilities operated by the Veterans Administration (VA). The Administration also calls for the elimination of the use of medically important antibiotics for growth promotion in animals and to bring other in-feed uses of antibiotics for the treatment or disease prevention under veterinary oversight. According to the National Strategy, by 2020: all states will have implemented ASPs across the continuum of healthcare according to best practices, and at least 95% of hospitals will report antibiotic use data to the National Healthcare Safety Network (NHSN) of the CDC; inappropriate inpatient antibiotic use for monitored conditions and agents will be reduced by 20% from 2014 levels; inappropriate outpatient antibiotic use for monitored conditions and agents will be reduced by 50% from 2010 levels; and the use of medically important antibiotics for growth promotion in animals will be eliminated (The White House, 2014c).

To improve stewardship in healthcare settings, PCAST recommended that the Centers for Medicare & Medicaid Services (CMS) use reimbursement incentives to drive stewardship policy (PCAST, 2014). The CMS has now released proposed rules requiring that US hospitals and long-term care facilities implement ASPs as conditions of participation (COPS) for reimbursement (CMS, 2016). PCAST also recommended that for the outpatient setting, CMS would expand the Physician Quality Reporting System to include quality measures that discourage inappropriate use of antibiotics, such as antibiotic prescriptions for non-bacterial respiratory infections. Finally, the PCAST report recommended that the CMS should require antibiotic usage reporting and require quality measures based on data reported by healthcare facilities to the NHSN Antimicrobial Use and Resistance (AUR) module as part of the Inpatient Quality Reporting program (CDC, 2016), and as of 2016 the CMS has issued a proposed standard for this. These quality measures should be validated and endorsed and ready for implementation by 2018, according to PCAST. To this end, The Joint Commission on Accreditation of Healthcare Organizations (JCAHO) now has a proposed standard for antimicrobial stewardship that is slated to go into effect January 2017 (JCAHO, 2015).

The government's actions to combat antibiotic resistance will roll out according to a time line stipulated in the Executive Order (The White House, 2014a). The task force must submit a 5 year action plan detailing how the National Strategy will be implemented, including goals, milestones, time lines, and metrics to measure progress. The task force is also required to update the President within 6 months of releasing the action plan, and every subsequent year, on progress made in implementing the National Strategy and plans to address barriers to progress or new recommendations.

Antimicrobial stewardship mandates at the state level

California was the first state with a legislative mandate for healthcare facilities to implement ASPs. Signed into law in 2008, California Senate Bill 739 would establish the Hospital Infectious Disease Control Program, which would require the California Department of Public Health (CDPH) to conduct surveillance, prevention, and public reporting of hospital-acquired infections in general acute care hospitals. In addition, the law mandated that, by January 2008, all general acute care hospitals would develop processes for evaluating the judicious use of antibiotics and monitor results using quality improvement committees. In September 2014, California Senate Bill 1311 was signed into law, further requiring hospitals to adopt and implement stewardship policies in accordance with the guidelines established by the federal government and professional organizations. It also requires hospitals to establish a physician-supervised, multidisciplinary antimicrobial stewardship committee including at least one physician or pharmacist with specific stewardship training.

Between May 2010 and September 2011, California conducted a statewide survey of ASPs to assess existing ASPs and strategies in general acute care hospitals and to gauge the impact of the legislation (Trivedi and Rosenberg, 2013). Of the facilities that responded to the survey, approximately 50% had some form of ASP in place, with cited barriers to implementation including lack of staffing and/or funding. Where these efforts existed, facilities reported improved antibiotic usage and susceptibility patterns as important outcomes. In addition, the CDPH Healthcare-Associated Infections Program has launched the Antimicrobial Stewardship Program Initiative (CDPH, 2014) to provide support and resources for facilities to implement ASPs, and the Antimicrobial Stewardship Program Collaborative, funded through a grant from the CDC (Epson and Nennig, 2014).

Government resources to support antimicrobial stewardship

STEWARDSHIP CORE ELEMENTS In 2014, the CDC published recommendations for the seven core elements for ASPs in its *Core Elements of Hospital Antibiotic Stewardship Programs* publication (CDC, 2014a). These are: commitment of leadership including adequate human, financial and information technology resources; appointment of a single leader responsible for program outcomes to ensure accountability; appointment of staff with specific drug expertise such as a pharmacist leader; implementation of at least one stewardship action such as systemic evaluation of ongoing treatment need after a set period of initial treatment (i.e., an "antibiotic time out" after 48 hours); regular reporting on antibiotic use and resistance to hospital staff, including doctors and nurses; and educating clinicians about resistance and optimal prescribing. The CDC also issued a checklist for facilities to "systematically assess key elements and actions to ensure optimal antibiotic prescribing and limit overuse and misuse of antibiotics in hospitals" (CDC, 2014b). The checklist is intended to be a companion to the agency's *Core Elements of Hospital Antibiotic Stewardship Programs* (CDC, 2014a).

CDC NATIONAL HEALTHCARE SAFETY NETWORK ANTIMICROBIAL USE AND RESISTANCE MODULE The CDC's NHSN AUR Module provides a mechanism for facilities to report and analyze antibiotic use and/or resistance as part of stewardship efforts, as well as the capability for benchmarking usage patterns across facilities (CDC, 2016). The module includes two components: an antibiotic usage module and an antibiotic resistance reporting module. The module is fully automated, capturing antibiotic prescription and drug susceptibility test results electronically, but currently requires the use of third-party software. The US National Strategy aims for 95% of hospitals to report on antibiotic usage through this system by 2020 (The White House, 2014c).

The primary objective of the AUR module is to provide risk-adjusted inter- and intra-facility benchmarking of antibiotic usage with the secondary goal of evaluating trends in usage over time at the facility and national level. The primary metric reported to this module is antimicrobial days per 1000 patient days present. An antimicrobial day (also known as day of therapy) is defined by any amount of a specific antimicrobial agent administered in a calendar day to a particular patient as documented in the facility's electronic medication administration record (eMAR) and/or bar coding medication record (BCMA).

The resistance module allows for reporting of resistance via a standardized method in order to provide: local practitioners with data on antibiotic resistance to aid in clinical decision-making and prioritize transmission prevention efforts; facility-specific measures in the context of regional and national statistics (i.e., benchmarking); and a regional and national assessment of resistant pathogens in order to determine the infection burden. Antibiotic resistance data are reported as a proportion and a rate. The proportion of resistant organisms is defined as the number of resistant isolates divided by the number of isolates tested for the specific antibiotic being evaluated. In comparison, the antimicrobial resistance rate is defined as the number of resistant isolates per 1000 patient days.

The latest development is that the Standardized Antimicrobial Administration Ratio (SAAR) has now been endorsed by the National Quality Forum (NQF) as the new standardized metric of antibiotic use (AU) available to facilities reporting to the AU option of the NHSN (van Santen, 2016). The SAAR is a ratio of the observed antibiotic use (days of therapy reported for a particular category of antimicrobial agents in a particular location(s) to the predicted antibiotic use (days of therapy predicted by negative binomial regression modeling for the same agents in the same location(s)).

Policies to promote the development of new antibiotics and diagnostics

Because resistance is inevitable and erodes the utility of antibiotics over time, a steady supply of new treatments is needed to ensure that patients can be treated adequately. Over the past few decades, both the number of new drugs reaching patients and the number of large pharmaceutical companies developing antibiotics has declined. Drug development takes over 10 years and costs between $800 million and $1.7 billion, on average (DiMasi *et al.*, 2003; Paul *et al.*, 2010). At best, only one of five drug candidates tested in people will ultimately be approved for patients (Hay *et al.*, 2014). Furthermore, antibiotic developers face several unique scientific and regulatory challenges compared with those in other therapeutic areas. For example, the clinical testing of antibiotics to treat highly resistant bacterial infections is particularly difficult given that, initially, only small numbers of patients contract such infections and meet the regulatory requirements to participate in registration trials. The economic considerations are also unfavorable because of the low potential return on investment for antibiotics, as they are generally taken as short-course treatments, and the pricing and reimbursement of new antibiotics does not reflect the true value of these lifesaving drugs. In addition, uptake of new antibiotics may be slow owing to stewardship restrictions that limit the use of new drugs (Mossialos *et al.*, 2009).

ADDRESSING THE ECONOMIC CHALLENGES OF ANTIBIOTIC DEVELOPMENT Policymakers around the world have recognized that the market failure of antibiotics must be addressed in order to ensure a steady supply of treatments over time. This is reflected in the WHO Draft Global Action Plan to combat resistance (WHO, 2014b,c), and the UK government commission's inquiry on how to tackle antibiotic resistance (O'Neil, 2014), the final report from which has now been published (O'Neil, 2016), with several supporting documents.

US lawmakers took an important step to address the economic challenges of antibiotic development in 2012 when the Generating Antibiotic Incentives Now (GAIN Act) was signed into law as part of the US Food and Drug Administration (FDA) Safety and Innovation Act (FDASIA) (US Congress, 2012). The GAIN Act provides incentives for the development of antibiotics for serious or life-threatening infections, including those caused by drug-resistant pathogens. Antibiotics designated to receive benefits under the law (Qualified Infectious Disease Products, or QIDPs) get an additional 5 years of data exclusivity beyond existing periods, are eligible for fast-track designation from the FDA, and receive a priority review from the agency. In speeding the development and review process, and providing extra time where such antibiotics are free from generic competition, GAIN is meant to serve as an economic incentive to developers. As of 2014, four antibacterial drugs were approved under this important legislation.

US policymakers are seeking additional economic incentives to spur antibiotic development, including enhanced reimbursement for qualifying drugs. The Developing an Innovative Strategy for Antimicrobial Resistant Microorganisms (DISARM) Act (H.R., 4187) was introduced in 2014 and seeks to recognize the costs of new antimicrobial drugs under the Medicare payment system for inpatient services and provide for additional payment for their use (US Congress, 2014). The PCAST report also cited a number of economic incentives for antibiotic developers, which will be evaluated by the US government for feasibility and possible implementation.

STREAMLINING THE REGULATORY PATHWAY FOR ANTIBIOTICS Recognizing the inherent challenges of the clinical development of antibiotics—particularly those for multidrug resistant pathogens—the FDA in the US and the European Medicines Agency (EMA) in the European Union (EU) have increased efforts to streamline antibiotic clinical trials and provide timely guidance to developers. The FDA has issued several new guidance documents (US FDA, 2015), and in 2012 and 2013, the EMA updated its guidance to companies on how to conduct studies to demonstrate the benefits and risks for new antibiotics (EMA, 2012, 2013b). In 2012, the FDA created the Antibacterial Drug Development Task Force to support the development of new antibiotics. This multidisciplinary group, which comprises scientists and clinicians, is working with external experts from academia, regulated industry, professional societies, patient advocacy groups, and government agencies to facilitate development (US FDA, 2012b). The EMA, along with the European Commission, also hosted a workshop to discuss regulatory options to support antibiotic development and appropriate use (EMA, 2013c). The FDA and EMA are both represented on the Transatlantic Taskforce on Antimicrobial Resistance (TATFAR), which was created in 2009 with the goal of improving cooperation between the US and the EU on: (i) the appropriate therapeutic use of antimicrobial drugs in medical and veterinary communities; (ii) the prevention of healthcare and community-associated antibiotic-resistant infections; and (iii) strategies for spurring the development of new antibiotics (CDC, 2009).

LEGISLATIVE PROPOSALS TO STIMULATE ANTIMICROBIAL DEVELOPMENT Two related bills designed to streamline the regulatory pathway for antibiotics that address unmet needs in limited patient populations have been introduced in the US, modeled after the recommendations of the 2012 PCAST *Report to the President on Propelling Innovation in Drug Discovery, Development and Evaluation* (PCAST, 2012). The Antibiotic Development to Advance Patient Treatment (ADAPT) Act (H.R., 3742) was introduced in 2013 (US Congress, 2013c). ADAPT gives the FDA the explicit authority to approve antibiotics for limited patient populations with few or no other treatment options. Drugs approved via this pathway would presumably be evaluated using a different benefit–risk assessment than is used for products indicated for broader patient groups. They would also require special labeling to clearly identify the limitations of data used to support approval and the intended population for the drug. It may be challenging to ensure that drugs approved based on limited safety and efficacy data for a limited population are not used widely, and advocates of this pathway believe that special labeling would support judicious use. The PATH Act proposes a similar pathway, but is not limited to antibiotics (US Senate, 2015).

Besides creating a new regulatory pathway for limited population antibiotics, the 2013 ADAPT Act includes language to streamline the process for setting and updating antibiotic susceptibility interpretive criteria, or antibiotic "breakpoints." Setting accurate breakpoints is critical for patient care and good stewardship. If breakpoints are set too high, bacteria may be categorized as susceptible to an antibiotic when they are actually resistant, and there is a risk that patients will receive an ineffective therapy. If a breakpoint is set too low, bacteria may be labeled as resistant to an antibiotic when that drug could, in fact, be effective. At a time when antibiotic resistance is limiting the options that patients have, it is in the best interest of public health to optimize the use of the currently available antibiotic armamentarium.

STIMULATING THE DEVELOPMENT OF RAPID DIAGNOSTICS The US government has also announced plans to award a $20 million prize, co-sponsored by the National Institutes of Health (NIH) and the Biomedical Advanced Research and Development Authority (BARDA), to facilitate the development of rapid, point-of-care diagnostic tests to identify bacterial infections (The White House, 2014b). PCAST recommended this action, noting that antibiotic stewardship would be advanced by the development and widespread use of tests providing information on the infecting pathogen and its resistance profile. This announcement followed on the announcement of the UK Longitude Prize, a £10 million prize fund for a diagnostic tool that can rule out antibiotic use or help identify an effective antibiotic to treat a patient, which came in June 2014 (Longitude Prize, 2014).

POLICIES TO ADDRESS INAPPROPRIATE USE OF ANTIBIOTICS IN AGRICULTURE There have been many efforts to reduce antibiotic use in animal agriculture—particularly for those drugs that are critical for human medicine—in order to slow antibiotic resistance in humans. Antibiotics are used extensively in agriculture to treat or prevent infections, and in some regions are also used in feed for growth promotion. In 2006, the EU banned the authorization of antibiotics as growth promoters under Regulation (EC) No 1831/2003 (OJ, 2003). The US National Strategy aims to eliminate the use of antibiotics for growth promotion in food animals by 2020. This may be accomplished via voluntary actions of regulated industry in accordance with recent FDA guidance. Two FDA guidance documents, including Guidance #209 and Guidance #213, direct manufacturers to voluntarily phase out the use of medically important antibiotics in food animals for growth promotion purposes, and to ensure that licensed veterinarians oversee other uses of such drugs for the treatment, control, or prevention of disease (US FDA, 2012a, 2013). PCAST and other groups assert that it will be important to monitor how these voluntary changes affect antibiotic use and stewardship in agriculture. In 2014, the FDA announced that all 26 antibiotic manufacturers affected by FDA Guidance #213 have agreed to phase out the use of medically important antibiotics in food-producing animals for growth promotion, and to rely on veterinary oversight for the therapeutic uses of such drugs by 2016.

The introduction of the Preservation of Antibiotics for Medical Treatment Act of 2013, H.R.1150, and the Preventing Antibiotic Resistance Act of 2013, S.1256, represent two recent attempts to regulate the use of antibiotics in agriculture (US Congress, 2013a,b). These bills, which have not yet been signed into law, aim to phase out the nontherapeutic use of medically important antibiotics in food animals, among other things. The drugs covered include those widely used in people, including penicillins and tetracyclines, or those deemed critically important antibiotics by the WHO.

Advocacy

Individuals can influence government stewardship policy in a number of ways, and the following offers just a few suggestions.

Several healthcare professional societies have active stewardship efforts, participating in everything from developing and validating stewardship measures, to lobbying for change by state or federal policymakers. Members generally drive the policy platforms of professional societies, and the participation on working groups and committees can be influential. Organizations such as the Infectious Diseases Society of America (IDSA), which represents physicians, scientists, and other infectious disease professionals, are actively promoting the establishment of ASPs in every healthcare facility across the US. Members are encouraged to take action to support the organization's policies through letters or in-person visits to legislators and policymakers. The IDSA and other groups solicit the input of members to draft reports and policy statements, and these can have a significant influence on policy. The Society for Healthcare Epidemiology of America (SHEA), the IDSA, and the Pediatric Infectious Diseases Society (PIDS) have together issued a policy statement on antibiotic stewardship that is referenced as a guideline on how to conduct such programs in various settings (SHEA, IDSA and PIDS, 2012).

In the US, draft regulations and guidance documents are typically released for public comment and consideration before they are finalized and implemented. Professional societies and trade associations often comment on relevant policies on behalf of their members, but individuals can also submit comments, which are published and reviewed via a transparent process. It is likely that any regulations to require stewardship programs in healthcare facilities, as directed by the US National Strategy, will be vetted publicly. Notices of government action can be found on the website of the US Federal Register (https://www.federalregister.gov/), and comments are generally collected at Regulations.gov (https://www.regulations.gov/#!home). These two websites also provide notice of public meetings and workshops to discuss the development of scientific standards and gather feedback on selected topics. Often, time is reserved for public comment during open meetings. Organizations outside the US similarly encourage stakeholder input. For example, the EMA has held meetings on antibiotic resistance (EMA, 2013a), and the WHO conducted a public online consultation to gather feedback on its Global Action Plan to address antimicrobial resistance (WHO, 2014b).

Pharmaceutical Industry Efforts in Antimicrobial Stewardship

The scope of work being undertaken by global, federal, and local policymakers on antimicrobial stewardship and antibiotic development is impressive and highlights the importance and impact that clinician awareness and activism have. Likewise, the pharmaceutical industry has made significant contributions to the advancement of antimicrobial stewardship through education and research.

Education

Education is one of the stewardship strategies listed in the guidelines from the IDSA and SHEA and should be considered foundational to any ASP (Dellit *et al.*, 2007). Among other sources, the ability of the pharmaceutical industry to serve as a resource for education on infectious diseases and antimicrobial stewardship should not be overlooked. Education through pharmaceutical industry can happen in a variety of ways, including sponsorship of professional meetings and conferences, educational grants for independent medical education, such as The Joint Commission Antimicrobial Stewardship–Pfizer IGLC (Independent Grants for Learning & Change) Funded Project (The Joint Commission and Pfizer IGLC, 2014), industry-sponsored symposia at professional conferences, or the direct provision of educational resources or opportunities.

In addition to registration fees, professional meetings and conferences are made possible by financial support from pharmaceutical companies. Two of the major infectious diseases conferences are the Interscience Conference on Antimicrobial Agents and Chemotherapy (ICAAC) and ID Week. Exhibitors that were listed for ICAAC 2015 included Astellas Pharma, Cubist Pharmaceuticals (now Merck), Forest Pharmaceuticals, Merck, and The Medicines Company (ICAAC, 2015). The exhibitors that were listed for ID Week 2015 included Astellas Pharma and Cubist Pharmaceuticals (ID Week, 2015). Without such exhibits, these conferences would not be possible unless there were exorbitant registration fees.

A recent example of an independent medical education grant is exemplified by the grant from the Forest Research Institute (a subsidiary of Forest Laboratories) to ASHP Advantage, a division of the American Society of Health System Pharmacists (ASHP) to develop an educational initiative. This initiative, entitled "Implementing Antimicrobial Stewardship Programs in Health Systems: An Interprofessional Team Approach", was developed by a steering committee of antimicrobial stewardship experts for multidisciplinary healthcare professionals—including physicians, pharmacists, physician assistants, and nurse practitioners—interested in implementing an ASP, or those with an already-established ASP (ASHP Advantage, 2014). This tiered learning initiative includes live and on-demand webinars, discussion guides, and virtual posters, with continuing education credit available for all of the above-mentioned healthcare providers. ASHP Advantage has a history of offering antimicrobial stewardship-related education as a result of educational grants from the pharmaceutical industry, with a pharmacist's guide to ASPs supported by Ortho-McNeil, and an antimicrobial stewardship mentorship program whereby five institutions were selected for pharmacist mentors to visit the selected institutions to help to evaluate strategies in an effort to assist the programs with goal achievement (ASHP, 2010; ASHP Intersections, 2011).

Several pharmaceutical companies offer direct provision of education. For example, Cubist Pharmaceuticals, now Merck, provided the CASES Collaborative Antimicrobial Stewardship Education Symposia. These symposia were 2–4 h programs, hosted at local health systems, with topics tailored toward the host institution's specific educational needs for antimicrobial stewardship. As part of the program, an external expert was invited to speak on a topic of the host institution's choice.

Research

The SHEA, IDSA, and PIDS have stated that ASP research is vital to the continued progression of ASP practice (SHEA, IDSA, and PIDS, 2012). Further, the CDC recommends that ASP outcomes should be tracked and reported, and declares that research is required to evaluate the benefits of ASP interventions (CDC, 2014a). The pharmaceutical industry can be a resource for support of this much needed research.

In contrast with the support of investigator-initiated studies as described above, a recent publication summarized the experiences and challenges associated with a pharmaceutical industry-directed antimicrobial stewardship data collection and analysis tool (AS-DCAT) (Hermsen *et al.*, 2014). The tool was developed to assist ASPs with monitoring and reporting clinical and economic outcomes of therapy for a range of Gram-positive and Gram-negative infections, including methicillin-resistant *Staphylococcus aureus* (MRSA), vancomycin-resistant enterococci (VRE), and *Pseudomonas aeruginosa*. The authors describe a number of benefits of this research collaboration between the pharmaceutical industry and clinicians, including, but not limited to, robust data resulting from multiple participating sites, access to real-world data on prescribing practices, self-evaluation of ASP impact on a local level, ability for benchmarking and best practice sharing amongst the research network, and access to resources that may otherwise be unavailable or in short supply.

Conclusion

The rising threat of antimicrobial resistance, coupled with the decline in commercialization of novel antimicrobials, has led to increased interest in and awareness of the need for antimicrobial stewardship. Furthermore, in the near term, ASPs are likely to be required in a variety of healthcare settings, as recommended by the President's Council of Advisors on Science and Technology and US National Strategy. Given the impending explosion in ASP implementation, clinicians must be aware of relevant legislation and policy as well as the resources available to assist with the advancement of antimicrobial stewardship.

In order to meet the shared goals of minimizing and preventing inappropriate antimicrobial use, improving patient outcomes, and combating the development of resistance, the government, policy makers, the pharmaceutical industry, and clinicians must engage in partnerships and cooperate with each other. Such relationships will help to ensure alignment, consideration of various stakeholder perspectives, and ultimately, an accelerated path to optimal antimicrobial stewardship practice and research.

References

ASHP (2010) *A Hospital Pharmacist's Guide to Antimicrobial Stewardship Programs*. American Society of Health-System Pharmacists, Bethesda, Maryland. Available at: http://www.ashpadvantage.com/docs/stewardship-white-paper.pdf (accessed 18 December 2014).

ASHP Advantage (2014) Implementing Antimicrobial Stewardship Programs in Health Systems: An Interprofessional Team Approach. Available at: http://www.leadstewardship.org/resources.php (accessed 18 December 2014).

ASHP Intersections (2011) Antimicrobial Stewardship Initiative Welcomes Applicants. Available at: http://www.ashpintersections.org/2011/06/antimicrobial-stewardship-initiative-welcomes-applicants/ (accessed 18 December 2014).

CDC (2009) Transatlantic Task Force on Antimicrobial Resistance (TATFAR). Available at: http://www.cdc.gov/drugresistance/tatfar/index.html (accessed 10 February 2015).

CDC (2013) Antibiotic resistance threats in the United States, 2013. Centers for Disease Control and Prevention, Atlanta, Georgia. Available at: http://www.cdc.gov/drugresistance/threat-report-2013/pdf/ar-threats-2013-508.pdf (accessed 16 December 2014).

CDC (2014a) *Core Elements of Hospital Antibiotic Stewardship Programs*. Centers for Disease Control and Prevention, Atlanta, Georgia. Available at: http://www.cdc.gov/getsmart/healthcare/implementation/core-elements.html (accessed 16 December 2014).

CDC (2014b) *Checklist for Core Elements of Hospital Antibiotic Stewardship Programs*. Centers for Disease Control and Prevention, Atlanta, Georgia. Available at: http://www.cdc.gov/getsmart/healthcare/pdfs/checklist.pdf (accessed 16 December 2014).

CDC (2016) *Antimicrobial Use and Resistance (AUR) Module*. Centers for Disease Control and Prevention, Atlanta, Georgia. Available at: http://www.cdc.gov/nhsn/PDFs/pscManual/11pscAURcurrent.pdf (accessed 9 June 2016).

CDPH (2014) The California Antimicrobial Stewardship Program Initiative. California Department of Public Health, Sacramento, California Available at: http://www.cdph.ca.gov/programs/hai/Pages/AntimicrobialStewardshipProgramInitiative.aspx (accessed 16 December 2014).

CMS (2016) CMS Issues Proposed Rule that Prohibits Discrimination, Reduces Hospital-Acquired Conditions, and Promotes Antibiotic Stewardship in Hospitals. Centers for Medicare & Medicaid Services, Baltimore, Maryland. Available at: https://www.cms.gov/Newsroom/MediaReleaseDatabase/Fact-sheets/2016-Fact-sheets-items/2016-06-13.html (accessed 16 June 2016).

Dellit, T.H., Owens, R.C., McGowan, J.E., Jr., Gerding, D.N., Weinstein, R.A., Burke, J.P., Huskins, W.C., Paterson, D.L., Fishman, N.O., Carpenter, C.F. *et al.* (2007) Infectious Diseases Society of America and the Society for Healthcare Epidemiology of America guidelines for developing an institutional program to enhance antimicrobial stewardship. *Clinical Infectious Diseases* 44, 159–177.

DiMasi, J., Hansen, R.W. and Grabowski, H.G. (2003) The price of innovation: new estimates of drug development costs. *Journal of Health Economics* 22, 151–185.

Dodds Ashley, E.S., Kaye, K.S., DePestel, D.D., and Hermsen, E.D. (2014) Antimicrobial stewardship: philosophy versus practice. *Clinical Infectious Diseases* 59(Suppl 3), S154–S161.

Drew, R.H., White, R., MacDougall, C., Hermsen, E.D. and Owens, R.C., Jr. (2009) Insights from the Society of Infectious Diseases Pharmacists on antimicrobial stewardship guidelines from the Infectious Diseases Society of America and the Society for Healthcare Epidemiology of America. *Pharmacotherapy* 29, 593–607.

EMA (2012) Guideline on the evaluation of medicinal products indicated for treatment of bacterial infections. European Medicines Agency, London. Available at: http://www.ema.europa.eu/docs/en_GB/document_library/Scientific_guideline/2009/09/WC500003417.pdf (accessed 9 June 2016).

EMA (2013a) European Medicines Agency to host workshop on regulatory options in the fight against antimicrobial resistance. European Medicines Agency, London. Available at: http://www.ema.europa.eu/ema/index.jsp?curl=pages/news_and_events/news/2013/09/news_detail_001899.jsp&mid=WC0b01ac058004d5c1 (accessed 10 February 2015).

EMA (2013b) Addendum on the note for guidance on the evaluation of medicinal products indicated for treatment of bacterial infections European Medicines Agency, London. Available at: http://www.ema.europa.eu/docs/en_GB/document_library/Scientific_guideline/2013/11/WC500153953.pdf (accessed 9 June 2016).

EMA (2013c) Best use of medicines legislation to bring new antibiotics to patients and combat the resistance problem. European Medicines Agency, London. Available at: http://www.ema.europa.eu/ema/index.jsp?curl=pages/news_and_events/events/2013/09/event_detail_000781.jsp&mid=WC0b01ac058004d5c3 (accessed 10 February 2015).

Epson, E. and Nennig, I.P. (2014) Invitation to the antimicrobial stewardship program collaborative. California Department of Public Health and Healthcare-Associated Infections Program, California Department of Health, Sacramento, California. Available at: http://www.cdph.ca.gov/programs/hai/Documents/ASPCollaborativeInviteToHospitalsFinal111714.pdf (accessed 16 December 2014).

Hay, M., Thomas, D.W., Craighead, J.L., Economides, C., and Rosenthal, J. (2014) Clinical development success rates for investigational drugs. *Nature Biotechnology* 32, 40–51.

Hermsen, E.D., McDaneld, P.M., Eiland, E.H., Destache, C.J., Lusardi, K., Estrada, S.J., Mercier, R.C., DePestel, D.D., Lamp, K.C., Anderson, E. *et al.* (2014) Breaking down the barriers: challenges with development and implementation of an industry-sponsored antimicrobial stewardship data collection and analysis tool. *Clinical Infectious Diseases* 59(Suppl 3), S179–S184.

ICAAC (2015) 2015 Participating Exhibitors. Interscience Conference on Antimicrobial Agents and Chemotherapy, American Society for Microbiology, Washington, DC. Available at: http://www.icaac.org/index.php/component/content/article/94-icaac-2015/338-2015-participating-exhibitors (accessed 5 January 2015).

ID Week (2015) ID Week 2015 Exhibitor List. Was available at: http://www.idweek.org/exhibitor-list/ (accessed 5 January 2015).

JCAHO (2015) *Proposed Standard for Antimicrobial Stewardship in AHC* [ambulatory healthcare]*, CAH* [critical access hospital]*, HAP* [hospital accreditation program]*, NCC* [nursing care center]*, and OBS* [office-based surgery]. The Joint Commission, Oak Brook, Illinois. Available at: https://jointcommission.az1.qualtrics.com/CP/File.php?F=F_5tDHGzIVDMHenDn (accessed 16 June 2016).

Johannsson, B., Beekmann, S.E., Srinivasan, A., Hersh, A.L., Laxminarayan, R., and Polgreen, P.M. (2011) Improving antimicrobial stewardship: the evolution of programmatic strategies and barriers. *Infection Control and Hospital Epidemiology* 32, 367–374.

Longitude Prize (2014) The Longitude Prize. Run by Nesta, London. Details available at: https://longitude-prize.org/ (accessed 10 February 2015).

Mossialos, E., Morel, C.M., Edwards, S., Berenson, J., Gemmill-Toyama, M., and Brogan, D. (2009) *Policies and Incentives for Promoting Innovation in Antibiotic Research*. World Health Organization on behalf of European Observatory on Health Systems and Policies, WHO Regional Office for Europe, Copenhagen. Available at: http://www.euro.who.int/__data/assets/pdf_file/0011/120143/E94241.pdf (accessed 2 June 2014).

Ohl, C.A. and Dodds Ashley, E.S. (2011) Antimicrobial stewardship programs in community hospitals: the evidence base and case studies. *Clinical Infectious Diseases* 53(Suppl 1), S23–S28.

OJ (2003) Regulation (EC) No 1831/2003 of the European Parliament and of the Council of 22 September 2003 on additives for use in animal nutrition. *Official Journal of the European Union* L268, 46, 29–43 Available at: http://eur-lex.europa.eu/legal-content/EN/TXT/?qid=1416418929102&uri=CELEX:32003R1831 (accessed 10 February 2015).

O'Neil, J. (Chair) (2014) Review on Antimicrobial Resistance. Tackling drug-resistant infections globally. Commissioned by the UK Government and the Wellcome Trust. Available at: http://amr-review.org/ (accessed 16 December 2014).

O'Neil, J. (Chair) (2016) *Tackling Drug-resistant Infections Globally: Final Report and Recommendations. Review on Antimicrobial Resistance*. Commissioned by the UK Government and the Wellcome Trust. Available at: http://amr-review.org/sites/default/files/160525_Final%20paper_with%20cover.pdf (accessed 16 June 2016).

Owens, R.C., Jr. (2008) Antimicrobial stewardship: concepts and strategies in the 21st century. *Diagnostic Microbiology and Infectious Diseases* 61, 110–128.

PACCARB (2015) Presidential Advisory Council on Combating Antibiotic-Resistant Bacteria (PACCARB). Details available at: http://www.hhs.gov/ash/advisory-committees/paccarb (accessed 20 June 2016).

Paul, S.M., Mytelka, D.S., Dunwiddie, C.T., Persinger, C.C., Munos, B.H., Lindborg, S.R., and Schacht, A.L. (2010) How to improve R&D productivity: the pharmaceutical industry's grand challenge. *Nature Reviews Drug Discovery* 9, 203–214.

PCAST (2012) *Report to the President on Propelling Innovation in Drug Discovery, Development, and Innovation*. Executive Office of the President, President's Council of Advisors on Science and Technology, The White House, Washington, DC. Available at: http://www.whitehouse.gov/sites/default/files/microsites/ostp/pcast-fda-final.pdf (accessed 10 February 2015).

PCAST (2014) *Report to the President on Combating Antibiotic Resistance*. Executive Office of the President, President's Council of Advisors on Science and Technology, The White House, Washington, DC. Available at: http://www.whitehouse.gov/sites/default/files/microsites/ostp/PCAST/pcast_carb_report_sept2014.pdf (accessed 10 February 2015).

Shallcross, L.J. and Davies, S.C. (2014) The World Health Assembly resolution on antimicrobial resistance. *Journal of Antimicrobial Chemotherapy* 69, 2883–2885.

SHEA, IDSA and PIDS (2012) Policy statement on antimicrobial stewardship by the Society for Healthcare Epidemiology of America (SHEA), the Infectious Diseases Society of America (IDSA), and the Pediatric Infectious Diseases Society (PIDS). *Infection Control and Hospital Epidemiology* 33, 322–327.

The Joint Commission and Pfizer (2014) Antimicrobial Stewardship – Pfizer IGLC Funded Project. Available at: http://www.jointcommission.org/antimicrobial_stewardship_-_pfizer_iglc_funded_project_/ (accessed 10 February 2015).

The White House (2014a) Executive Order 13676—Combating Antibiotic-Resistant Bacteria. Office of the Press Secretary, The White House, Washington, DC. Available at: https://www.whitehouse.gov/the-press-office/2014/09/18/executive-order-combating-antibiotic-resistant-bacteria (accessed 9 June 2016).

The White House (2014b) Factsheet: Obama Administration Takes Actions to Combat Antibiotic-Resistant Bacteria. Office of the Press Secretary, The White House, Washington, DC. Available at: http://www.whitehouse.gov/the-press-office/2014/09/18/fact-sheet-obama-administration-takes-actions-combat-antibiotic-resistan (accessed 10 February 2015).

The White House (2014c) *National Strategy for Combating Antibiotic-resistant Bacteria*. The White House, Washington, DC. Available at: http://www.whitehouse.gov/sites/default/files/docs/carb_national_strategy.pdf (accessed 10 February 2015).

Trivedi, K.K. and Rosenberg, J. (2013) The state of antimicrobial stewardship programs in California. *Infection Control and Hospital Epidemiology* 34, 379–384.

US Congress (2012) Public law 112-144. Food and Drug Administration Safety and Innovation Act, July 9, 2012. Available at: http://www.gpo.gov/fdsys/pkg/PLAW-112publ144/pdf/PLAW-112publ144.pdf (accessed 16 December 2014).

US Congress (2013a) H.R.1150 – Preservation of Antibiotics for Medical Treatment Act of 2013. Available at: https://www.congress.gov/bill/113th-congress/house-bill/1150 (accessed 10 February 2015).

US Congress (2013b) S.1256 – Preventing Antibiotic Resistance Act of 2013. Available at: https://www.congress.gov/bill/113th-congress/senate-bill/1256?q=%7B%22search%22%3A%5B%22Preventing+Antibiotic+Resistance+Act+of+2013%2C+S.+1256%22%5D%7D (accessed 10 February 2015).

US Congress (2013c) Antibiotic Development to Advance Patient Treatment (ADAPT) Act of 2013. Available at: https://beta.congress.gov/113/bills/hr3742/BILLS-113hr3742ih.pdf (accessed 10 February 2015).

US Congress (2014) Developing an Innovative Strategy for Antimicrobial Resistant Microorganisms (DISARM) Act (H.R. 4187) (2014) Available at: https://beta.congress.gov/113/bills/hr4187/BILLS-113hr4187ih.pdf (accessed 10 February 2015).

US FDA (2012a) *[Guidance] #209. Guidance for Industry. The Judicious Use of Medically Important Antimicrobial Drugs in Food-Producing Animals*. US Food and Drug Administration, Silver Springs, Maryland. Available at: http://www.fda.gov/downloads/AnimalVeterinary/GuidanceComplianceEnforcement/GuidanceforIndustry/UCM216936.pdf (accessed 10 February 2015).

US FDA (2012b) Antibacterial Drug Development Task Force. US Food and Drug Administration, Silver Springs, Maryland. Available at: http://www.fda.gov/Drugs/DevelopmentApprovalProcess/DevelopmentResources/ucm317207.htm (accessed 10 February 2015).

US FDA (2013) *[Guidance] #213. Guidance for Industry. New Animal Drugs and New Animal Drug Combination Products Administered in or on Medicated Feed or Drinking Water of Food-Producing Animals: Recommendations for Drug Sponsors for Voluntarily Aligned Product Use Conditions with GFI #209*. US Food and Drug Administration, Silver Springs, Maryland. Available at: http://www.fda.gov/downloads/AnimalVeterinary/GuidanceComplianceEnforcement/GuidanceforIndustry/UCM299624.pdf (accessed 10 February 2015).

US FDA (2015) Clinical/Antimicrobial Guidances. US Food and Drug Administration, Silver Springs, Maryland. Available at: http://www.fda.gov/Drugs/GuidanceComplianceRegulatoryInformation/Guidances/ucm064980.htm (accessed 10 February 2015).

US Senate (2015) Promise for Antibiotics and Therapeutics for Health (PATH) Act. Available at: http://www.hatch.senate.gov/public/_cache/files/8032151a-ab4d-4cfc-ae46-d554343d34c1/TAM15060.pdf (accessed 10 February 2015).

van Santen (2016) *The Standardized Antimicrobial Administration Ratio (SAAR), NHSN Training, March 4, 2016*. Centers for Disease Control and Prevention, Atlanta, Georgia. Available at: https://www.cdc.gov/nhsn/pdfs/training/2016/au-saar-vansanten.pdf (accessed 16 June 2016).

WHO (2014a) *Sixty-Seventh World Health Assembly, Agenda item 16.5. Antimicrobial Resistance*. Resolution WHA 67.25, World Health Assembly. World Health Organization, Geneva, Switzerland. Available at: http://apps.who.int/gb/ebwha/pdf_files/WHA67/A67_R25-en.pdf (accessed 16 December 2014).

WHO (2014b) Comments submitted during an online consultation on the draft Global action plan on antimicrobial resistance, 4 July–1 September 2014. World Health Organization, Geneva, Switzerland. Available at: http://www.who.int/drugresistance/global_action_plan/amr-action-plan-comments/en/ (accessed 10 February 2015).

WHO (2014c) *Antimicrobial Resistance. Draft Global Action Plan on Antimicrobial Resistance. Report by the Secretariat*. Executive Board 136th Session, Provisional Agenda item 8.1. [Document] EB136/20, 12 December 2014, World Health Organization, Geneva, Switzerland. Available at: http://apps.who.int/gb/ebwha/pdf_files/EB136/B136_20-en.pdf (accessed 16 December 2014).

38 A Hospitalist Perspective on the Role of Antimicrobial Stewardship

SAJEEV HANDA*

Rhode Island Hospital and The Warren Alpert Medical School of Brown University, Providence, Rhode Island, US

Introduction

Each year, nearly 2 million patients in the US acquire an infection in hospital and approximately 90,000 succumb to these infections. More than 70% of the bacteria that cause hospital-acquired infections are resistant to at least one commonly used agent. Persons infected with drug-resistant organisms have longer hospital stays, higher mortality rates, and increased healthcare costs (Kisuule *et al.*, 2008). It has been estimated that >50% of medical inpatients in the US are being cared for by physicians known as 'hospitalists' (Pile, 2011), and that this number continues to grow annually.

The Role of the Hospitalist

> An eighty-one year old male presents to the Emergency Department having been referred from a rehabilitation center with a pyrexia of 38.5°C, myalgias, arthralgias, a dry-non-productive cough of twelve hours duration. He was recently discharged following a left hip arthroplasty for a fracture sustained when he slipped on ice. His past medical history is remarkable for hypertension for which he takes enalapril. He is unsure if he received the seasonal Influenza vaccination. On examination he is normotensive, pyrexic and mildly tachypnoeic. Finger pulse oximetry is 93% on room air. He appears ill but not toxic. His chest examination demonstrates normal chest expansion, diminished air-entry at the left base with trace crepitations that clear with coughing. Breath sounds are broncho-vesicular in nature. The surgical incision site is noted to be well opposed, mildly tender but without significant erythema and discharge. He has moderate range of motion of the left hip with some discomfort. White blood cell count is 14×10^9/l with a slight left shift and his chemistry demonstrates a serum creatinine of 2.7 mg/dl. Chest radiograph obtained demonstrates mild attenuation at the left base suggestive of early air-space disease and/or atelectasis. The Emergency Room physician obtains two sets of blood cultures and orders 4.5 g of intravenous (IV) piperacillin–tazobactam, 400 mg of IV ciprofloxacin, 1 g of IV vancomycin hydrochloride and then contacts the on-call hospitalist to arrange for admission to the general medical service.
>
> A hypothetical case study

Since the landmark article entitled "The emerging role of "hospitalists" in the American health care system" was published in *The New England Journal of Medicine* in 1996 (Wachter and Goldman, 1996), the identification and growth of this cohort of specialist inpatient physicians has grown exponentially. This has been exemplified by their membership within the Society of Hospital Medicine (SHM), whose growth in just over a decade has gone from several hundred to 12,000 in 2014 (J. Miller, SHM, Pennsylvania, 2014, personal communication), and there may be well over 40,000 hospitalists practising in this country alone (AHA, 2013).

As defined by the SHM, a hospitalist is a physician who specializes in the practice of hospital medicine. Following medical school, hospitalists typically undergo residency training in general internal medicine, general pediatrics or family practice, but may also receive training in other disciplines.

*E-mail: shanda@lifespan.org

Not foreign by any means to the European healthcare environment, SHM further defines hospital medicine as a medical specialty dedicated to the delivery of comprehensive medical care to hospitalized patients. Practitioners of hospital medicine include physicians ("hospitalists") and non-physician providers who engage in clinical care, teaching, research or leadership in the field of general hospital medicine. In addition to their core expertise in managing the clinical problems of acutely ill hospitalized patients, hospital medicine practitioners work to enhance the performance of hospitals and healthcare systems by:

- Prompt and complete attention to all patient care needs, including diagnosis, treatment and the performance of medical procedures (within their scope of practice).
- Employing quality and process improvement techniques.
- Collaboration, communication and coordination with all physicians and healthcare personnel caring for hospitalized patients.
- Safe transitioning of patient care within the hospital and from the hospital to the community, which includes oversight of care in post-acute care facilities.
- Efficient use of hospital and healthcare resources.

(SHM, 2009)

The Hospitalist and Antimicrobial Stewardship

The hospitalist sees the patient in the Emergency Room conducts a full history and physical, reviews all available laboratory findings and radiographic examination and is now faced with several decisions including a diagnostic dilemma: does the patient, in fact, have a health care associated pneumonia? Is there an infection brewing at the surgical site or could this be a case of acute Influenza? If so, what presumptive antimicrobial treatment should be instituted? Therapeutic decisions may also have unintended consequences if not thought out correctly. Continuing broad spectrum antimicrobial treatment instituted by the Emergency Room for presumptive health care associated pneumonia may result in the later development of *Clostridium difficile* colitis. Incorrect dosing of these agents may result in toxicity. A lack of clinical acumen may lead to a missed case of Influenza. The hospitalist in the Emergency Room now has several additional admissions to be seen and so decisions need to be made. On this particular patient a general medicine order set is used; the patient is placed in respiratory isolation and a respiratory viral panel ordered to test for Influenza A & B, Parainfluenza, Metapneumovirus, Rhinovirus and Adenovirus. The antimicrobials started in the Emergency Room are continued, dosing piperacillin–tazobactam at 4.5 gm IV every six hours, vancomycin hydrochloride at 1 g IV once a day, ciprofloxacin at 400 mg IV every 12 hours. The hospitalist adds oseltamivir phosphate 75 mg orally twice daily as there is concern for acute influenza. A legionella urinary antigen is also ordered and the patient is placed on Intravenous normal saline. The enalapril is held.

A hypothetical case study

From the moment a decision to institute antimicrobial treatment is made, the concept of stewardship takes effect. Antimicrobial stewardship has been defined as the coordinated effort to optimize antimicrobial usage (the right agent, at the right time, at the correct dose, for an appropriate duration), with goals of improving patient outcomes, reducing antimicrobial resistance and decreasing healthcare costs (Rhode *et al.*, 2013).

Practitioner factors shaping prescribing include:

- knowledge of the current best care practices;
- self-efficacy, which determines whether a provider is confident in his or her knowledge to adequately treat a specific infection; this may be augmented or facilitated by large group education;
- habit, which results in providers selecting a narrow repertoire of antimicrobials when treating an infection; and
- fear of liability.

Environmental factors include:

- published guidelines (including those involving surgical prophylaxis and institutional antibiograms);
- patient-driven factors such as affordability, compliance with dosing regimens, side effects and drug–drug interactions;
- peer influence such that providers are reluctant to change a prescription commenced by another provider (e.g., the Emergency Room Physician); and
- formulary restrictions/preauthorization (Kisuule *et al.*, 2008).

Other institutional factors that may exist include:

- computerized physician order entry, including reviewed and updated order sets;
- best practice alerts;

- suppressed microbiology susceptibility testing;
- rapid diagnostics;
- prospective audit and feedback; and
- antimicrobial cycling.

Although the Infectious Diseases Society of America (IDSA) promulgates, in its 2007 guidelines (Dellit *et al.*, 2007), that core members of a multidisciplinary antimicrobial stewardship team should include an infectious diseases physician, a clinical pharmacist (with infectious diseases training), a clinical microbiologist, an information system specialist, an infection control professional and a hospital epidemiologist, most institutions cannot do this. The logic from IDSA is that effective antimicrobial stewardship programs can be financially self-supporting and improve patient care. Comprehensive programs have consistently demonstrated a decrease in antimicrobial use with a potential annual savings of $200,000 to $900,000 in larger institutions (Dellit *et al.*, 2007). Problems with this scenario include the following: many institutions do not necessarily have enough infectious diseases physicians, or an attending person interested or available to do this work, let alone the ability to recruit a clinical pharmacist with the correct skill set, all of whom should be compensated for their time. Also, budgets in institutions are typically sequestered such that savings in one particular area do not directly translate into funding in another.

Consequently, institutions have "*fragments*" of an ideal stewardship program, which leads to a less than ideal situation, as exemplified by the example that follows.

In surveying a group of 67 hospitalists (response rate of 60%, *n* = 40) across two local institutions (719 and 247 beds) about antimicrobial stewardship, 97% felt it helpful, 18% perceived that it may delay care, but more remarkable, 49% were unaware as to whom, if anyone, was responsible for stewardship in their respective institutions; 41% felt that it fell under the realm of the infectious diseases physician, and 11%, the pharmacist. Furthermore, the concept of stewardship varied from one individual to the next, ranging from pharmacokinetic assistance, to judicious usage, to reduction of the risk of resistance, to cost reduction, to the assistance of usage with certain clinical scenarios. In reality, the definition of stewardship, as noted above, is all encompassing. That 3% of the respondents felt it to be unhelpful and 20% were concerned about possible delays in care (for example, awaiting antimicrobial release for formulary restricted antimicrobials) is cause for concern on the local front. This speaks to the necessity of further education and explanation of the role and necessity of antimicrobial stewardship, as well as an evaluation of the efficiency of the current system. Of note, four of the respondents indicated that they simply did not know or were unfamiliar with the concept of antimicrobial stewardship. Both of the institutions' stewardship programs involved infection control, published guidelines, restricted antibiotic release by the infectious diseases physicians, and sporadic input from pharmacy.

> While in the process of admitting another patient, the hospitalist receives a call from the Pharmacy Department advising that the dosing of the antimicrobials is incorrect based on the patient's calculated creatinine clearance of 23 ml/min. The hospitalist revises the orders accordingly: piperacillin–tazobactam 2.25 g IV every six hours, ciprofloxacin 400 mg IV once daily, oseltamivir phosphate 75 mg orally once daily, and the vancomycin is held pending a random level ordered with the morning labs. The pharmacist suggests to the hospitalist that an Infectious Diseases consultation may be warranted.
>
> A hypothetical case study

For the hospitalized patient, the inappropriate usage of antimicrobials may be classified into five categories. First, administration may occur when not indicated, for example, antibiotics for viral infections. Secondly, broad-spectrum antibiotics such as piperacillin–tazobactam and ciprofloxacin may be overused in the treatment of common infections. Such over-usage increases selective pressure for antimicrobial resistance, exposes patients to their attendant side effects, and increases the risk of developing *Clostridium difficile* colitis. Thirdly, providers may prescribe intravenous antimicrobials when the bioavailability and efficacy of oral agents may be similar. Fourthly, even though the correct antimicrobial is chosen, the incorrect dosage, schedule and/or duration of treatment can be detrimental to the patient. Lastly, 'bug–drug' mismatch occurs when susceptibility studies indicate that the antimicrobial used is ineffective or marginally effective. Beyond safety issues, these practices also increase costs to both the patient and the institution, and remain a public health concern necessitating quality improvement initiatives in the delivery of health care (Kisuule *et al.*, 2008).

The increase in the number of multi-drug-resistant (MDR) Gram-negative bacteria (including carbapenem-resistant *Enterobacteriaceae*) within the healthcare setting is of concern and cannot be overemphasized to the inpatient teams. Too frequently, clinicians in the US encounter Gram-negative bacterial species that are resistant to many, and occasionally all, antimicrobials. For many of these MDR Gram-negative pathogens, the antimicrobials that potentially remain active (e.g., aminoglycosides and polymyxins) are often more toxic and less efficacious for certain infections. Compounding this is the fact that the pharmaceutical industry's developmental pipeline for new antimicrobials with novel mechanisms of action has virtually come to a standstill. Even if an investigational drug was in Phase 2 or 3 trials, or has entered into the US Food and Drug Administration (FDA) fast track developmental program, it would be at least 10 or 15 years before that drug would be available in the US (Ohi and Luther, 2011).

Difficulties faced by the admitting hospitalist in the Emergency Room include the fact that the data needed to make a rational, informed decision about specific antimicrobial therapy are usually unavailable. For many infections, therapy cannot be delayed while waiting for microbiology or other findings, and broad-spectrum empiric therapy is begun on the basis of educated guesses made from the patient's presentation and characteristics, as well as local hospital antibiograms. Additionally, any delay in antimicrobial administration to treat an infection will increase patient morbidity and mortality. Furthermore, many hospitalists work in a shift-based model resulting in handovers that may occur once or sometimes two to three times during a patient's stay, particularly if there are dedicated admitting shifts. A comprehensive effective antimicrobial stewardship strategy would indubitably optimize the care of patients for whom antimicrobials are being considered.

The 2007 IDSA guidelines for institutional development of an antimicrobial stewardship program make it clear that the primary goal is to focus on patient care; that is, to optimize clinical outcomes while minimizing the unintended consequences of antimicrobial use in conjunction with a comprehensive infection control program (Dellit *et al.*, 2007).

Antimicrobial stewardship should be considered a medical staff function rather than a primary hospital function, as the former group, with their leadership, is most adept in employing the three 'Cs' (conceptualization, communication and coercion) when constructing, implementing and operating such a program:

- Conceptualization deals with understanding what needs to be done and how to do it.
- Communication ensures that all providers receive and understand the information.
- Coercion, though potentially construed a strong term, refers to the pressure exerted by leaders and others involved in the process to get things done within the institution involving all units or departments collaboratively (Ohi and Luther, 2011).

If an infectious diseases physician or clinical pharmacist with formal infectious diseases training is not available at a given institution, the team will need to include others ready to assume a greater leadership role. Although not mentioned in the guidelines, hospital medicine specialists and hospitalists may be well suited to adopt this role, can be integral to the leadership of a multidisciplinary team, and are clear stakeholders. Hospitalists have knowledge of the institution in which they support a wide range of services. In addition, hospital leadership and administration more often reach out to hospitalists to oversee patient quality care and safety improvement projects, the realm to which antimicrobial stewardship belongs. In the US, hospitalists are the dominant prescribers of antimicrobials and as such provide potentially a pivotal role where stewardship is concerned, both as prescribers and leaders (Pile, 2011). Additionally, there may be an evolving role for multisite hospitalist group leaders that oversee the care that their hospitalists provide in multiple institutions, such that best practices may be shared and recommendations made, particularly when there is a variation in formularies between those institutions.

On hospital day number one the patient's care is handed over to another hospitalist. He is clinically improved though remains pyrexic to 38°C. Oximetry is 95% on 2 liters of oxygen via nasal cannula. He is no longer tachypnoeic though respiratory examination reveals crepitations at the left base that do not clear on coughing. His white blood cell count is 12×10^9/l, serum creatinine is 1.7 mg/dl and a repeat chest radiograph demonstrates consolidation of the left lower lobe. Legionella urinary antigen is negative and blood cultures do not demonstrate growth. The respiratory viral panel obtained is negative. He is seen in consultation with Infectious Diseases and the intravenous ciprofloxacin is discontinued as is the oral oseltamivir. On hospital day number two the patient defervesces; a repeat white cell count is 9×10^9/l and the patient's serum creatinine is 1.0 mg/dl. His enalapril is resumed. The hospitalist consults case management in anticipation of a potential discharge if medically stable in twenty-four to forty eight hours.

A hypothetical case study

Another critical point to consider when discussing antimicrobial stewardship is that hospitalists serve as gatekeepers for patients leaving the hospital on either oral or intravenous antimicrobials. In interacting with a large number of community-based primary care physicians, hospitalists can play a key role in improving antimicrobial stewardship, not only within but also beyond the hospital setting. Patient care does not and should not end once the patient is discharged from the hospital.

Although uncharted territory in many institutions, a systems-based approach for antimicrobial stewardship for Community-Based Parenteral Anti-Infective Therapy (CoPAT; also known as OPAT, see Chapter 35) has been in operation at the Cleveland Clinic, a 1200 bed hospital in Cleveland, Ohio. CoPAT refers to the practice of administering antimicrobial therapy in the home or other outpatient settings. Infectious disease consultation is required for every patient being considered for discharge home, or to a sub-acute nursing facility, or to a long-term acute care center, who requires parenteral antimicrobials. Such consultation is necessary to ensure that the antimicrobial treatment is actually warranted and if so, the optimal type, route and duration; vascular access needs to be evaluated and optimized, and arrangements for the monitoring of clinical and laboratory aspects need to be made. The program is a bundled process, and once a patient is considered appropriate for CoPAT, a form is completed and a follow-up appointment made before the patient is discharged. Information exchange is transmitted via the patient's electronic health record, a system that is visible to anyone who has access to it and provides real-time information with a structured form that includes information such as the microorganisms being treated, the antimicrobials prescribed (start, stop time, etc.), and any concurrent oral antimicrobials and laboratory testing. The repository of information available from the CoPAT program allows the Clinic to identify areas for improvement. This includes information about patient demographics, antibiotic administration and indication, as well as the pathogenic microorganisms involved, if available. In addition, data collected allows the Clinic to identify areas for improvement for example, outcomes of CoPAT in patients with bone and joint infections, the incidence of *C. difficile* infection in patients undergoing CoPAT and so forth. Reasons for rehospitalization may be examined in detail too (Gordon *et al.*, 2011).

The Centers for Disease Control and Prevention Perspective

The Division of Healthcare Quality Promotion at the Centers for Disease Control and Prevention (CDC) in Atlanta, Georgia recognizes the importance of hospitalists in stewardship, particularly in those institutions that do not have dedicated infectious diseases physicians and clinical pharmacists. This has led to the concept of a practical "stewardship implementation framework," resulting with the CDC partnering with the Institute of Healthcare Improvement (IHI) in Cambridge, Massachusetts. The IHI uses the "Driver Diagram" to conceptualize an issue and determine its system components, which will then create a pathway to attain a goal that is the Primary Outcome. This is fed by Primary Drivers (system components that contribute to moving the Primary Outcome), which are in turn fed by Secondary Drivers (elements associated with the Primary Driver that can be used to create projects). The external experts convened, including a hospitalist, determined the Primary Drivers as:

1. Timely and appropriate initiation of antibiotics.
2. Appropriate administration and de-escalation of therapy.
3. Data monitoring and transparency (measuring and feeding back to clinicians, data on antimicrobial use and resistance).
4. Improving stewardship infrastructure.

Secondary Drivers have been identified, and the CDC and IHI have worked on a number of pilot projects, initially with the intent of being led by hospitalists (Srinivasan, 2011). Eight hospitals were chosen to support testing: two small and one large community hospital, four community teaching hospitals, and one large academic center, in varying geographic locations. Two of the hospitals (both community hospitals) did not have infectious diseases physicians on their staff and even though all had hospitalists, the majority did not engage them in this pilot testing, which was a pharmacy-driven process. Each hospital was committed to improving antibiotic stewardship and willing to test strategies in areas where knowledge and experience were limited.

The four key drivers described above were tested over a 12 month period as a modified collaborative model, with an initial face-to-face meeting, twice weekly conference calls to discuss progress, and challenges and barriers related to the key drivers, and a final face-to-face meeting to summarize the results and findings. The key barriers identified included: the ability to engage frontline providers (hospitalists, intensivists—critical care physicians, surgeons and nurses); embedding antibiotic review into the process of care; competing priorities and improvement initiatives; the availability of information technology support; and the time required, as well as the ability, to collect data. The conclusion of the pilot emphasized the importance of moving from a pharmacy-driven focus to embedding stewardship strategies into the process of daily care provided by the frontline clinicians. The current phase of the initiative is focused on testing an antimicrobial stewardship program led by hospitalists to explore interventions that can be readily incorporated into their daily work flow (Rhode *et al.*, 2013).

Nonetheless, while the work continues on this front, four immediate antibiotic quality improvement projects for hospitalists have been identified:

1. All antibiotic orders should include a dose, duration, and indication (too frequently, particularly during transitions of care, as well as during handovers from one hospitalist to another, the reason why the patient is on antimicrobials is unclear).
2. Patients with a positive blood culture should be on appropriate therapy (this, of course, pertains to those individuals where a positive isolate is deemed not to be a contaminant in the correct clinical setting).
3. An intervention should be developed to encourage the reassessment of patients who are started on antibiotics for community-acquired pneumonia (CAP) (pressure to institute treatment in cases of CAP or presumed CAP may result in overtreatment, necessitating de-escalation or inappropriate treatment).
4. Urinary tract infections in hospitalized patients must be properly diagnosed and treated (Srinivasan, 2011).

Hospitalists are and have been engaged in identifying institutional problems that effective stewardship programs may mitigate or solve in their entirety, as is illustrated by abstracts presented at the SHM annual meetings. Examples include a retrospective analysis of 501 patients with community-acquired methicillin-resistant *Staphylococcus aureus* (CA-MRSA). Performed in a community hospital, this demonstrated that only 52% of patients received the most appropriate antibiotics once cost was accounted for (Albanto *et al.*, 2009). A second example presented looked at MRSA bloodstream infections, and revealed that 77.6% of patients received effective antimicrobial therapy within the first 3 days following culture and sensitivity results (Herzke *et al.*, 2011). Lastly, another review performed by a group of hospitalists showed that out of 210 patients who had urine cultures, 54% lacked any guideline-accepted reason for obtaining them (Hartley *et al.*, 2011).

Summary

Antimicrobial stewardship is critical to every institution. In an era where hospitals and healthcare systems are not looking to expend *additional* resources, evaluation of *current* resources should be made. Hospitalists are situated ideally in any institution to assist in the construction of an antimicrobial stewardship framework, as it is they who will then also be responsible for executing the program; however, for the most part, they continue to remain an untapped resource. This is particularly important in those hospitals where Infectious Diseases/Clinical Pharmacy stewardship support is unavailable.

Hospitalists are and will continue to be the main prescribers of antimicrobials in the US, and they need to be engaged in improvement, as well as educational efforts. They are often proactive, knowledgeable, and are not uncommonly called upon to help configure clinical decision-support tools as well computerized physician order entry systems—all qualities that are necessary for a successful stewardship program. The SHM supports such efforts, in conjunction with the CDC and other organizations (G.M. Maynard, SHM, California 2014, personal communication). Antimicrobial stewardship programs, once revised or formally established, also need to undergo continuous evaluation, and to be able to evolve or change depending on the institutional and or national climate; once again, these are areas in which hospitalists can and should play a vital role.

References

Albanto, P., Haneley, J., and Costillo, J. (2009) Prevalence and use of antibiotics in community associated methicillin-resistant *Staphylococcus aureus* (CA-MRSA) infections in a community hospital. Presentation to: SHM (Society of Hospital Medicine) Annual Meeting, May 14–17, 2009 in Chicago, Illinois, Abstract 1. Available at: http://www.shmabstracts.com/abstract/prevalence-and-use-of-antibiotics-in-community-associated-methicillinresistant-staphylococcus-aureus-infections-camrsa-in-a-community-hospital/ (accessed 10 June 2016).

AHA (2013) AHA Annual Survey Database™ Fiscal Year 2014. American Hospital Association, Chicago, Illinois and Washington, DC. Available at: http://www.ahadataviewer.com/book-cd-products/aha-survey/ (accessed 19 January 2014).

Dellit, T.H., Owens, R.C., McGowan, J.E., Jr., Gerding, D.N., Weinstein, R.A., Burke, J.P., Huskins, W.C., Paterson, D.L., Fishman, N.O., Carpenter, C.F. *et al.* (2007) Infectious Diseases Society of America and the Society for Healthcare Epidemiology of America guidelines for developing an institutional program to enhance antimicrobial stewardship. *Clinical Infectious Diseases* 44, 159–177.

Gordon, S.M., Nabin, S.K., and Rehm, S.J. (2011) Transitioning antimicrobial stewardship beyond the hospital: the Cleveland Clinic's community based parenteral anti-infective therapy (CoPAT) program. *Journal of Hospital Medicine* 6(Suppl 1) S24–S30.

Hartley, S., Valley, S., Kuhn, L., Washer, L., Gandhi, T., Malani, A., Chenoweth, C., Saint, S., and Flanders, S. (2011) Inappropriate testing for urinary tract infection in hospitalized patients: an opportunity for improvement. Presented at: SHM (Society of Hospital Medicine) Annual Meeting, May 10–13, 2011 in Dallas, Texas, Abstract 55. Available at: http://www.shmabstracts.com/abstract/inappropriate-testing-for-urinary-tract-infection-in-hospitalized-patients-an-opportunity-for-improvement/ (accessed 10 June 2016).

Herzke, C., Chen, L., Brotman, D., Anderson, D., Choi, Y., Sexton, D., and Kaye, K. (2011) Factors associated with inappropriate definitive antimicrobial therapy for bloodstream infection from methicillin resistant *Staphylococcus aureus*. Presented at: SHM (Society of Hospital Medicine) Annual Meeting, May 10–13, 2011 in Dallas, Texas, Abstract 56. Available at: http://www.shmabstracts.com/abstract/factors-associated-with-inappropriate-definitive-antimicrobial-therapy-for-bloodstream-infection-from-methicillinresistant-staphylococcus-aureus/ (accessed 10 June 2016).

Kisuule, F., Wright, S., Barreto, J., and Zenilman, J. (2008) Improving antibiotic utilization among hospitalists: a pilot academic detailing project with a public health approach. *Journal of Hospital Medicine* 3, 64–70.

Ohi, C.A. and Luther, V.P. (2011) Antimicrobial stewardship for inpatient facilities. *Journal of Hospital Medicine* 6(Suppl 1), S4–S15.

Pile, J.C. (2011) Antimicrobial stewardship: optimizing antibiotic use in an era of increasing resistance and rising costs. *Journal of Hospital Medicine* 6(Suppl 1), S1–S3.

Rhode, J., Jacobsen, D. and Rosenberg, D.J. (2013) Role of the hospitalist in antimicrobial stewardship: a review of work completed and description of a multisite collaborative. *Clinical Therapeutics* 35, 751–757.

SHM (2009) Definition of a Hospitalist and Hospital Medicine. Society of Hospital Medicine, Philadelphia, Pennsylvania. Available at: http://www.hospitalmedicine.org/AM/Template.cfm?Section=Hospitalist_Definition&Template=/CM/HTMLDisplay.cfm&ContentID=24835 (accessed 17 January 2014).

Srinivasan, A. (2011) Engaging hospitalists in antimicrobial stewardship: the CDC perspective. *Journal of Hospital Medicine* 6(Suppl 1), S31–S33.

Wachter, R.M. and Goldman, L. (1996) The emerging role of "hospitalists" in the American health care system. *The New England Journal of Medicine* 335, 514–516.

Index

Page numbers in **bold** type refer to figures and tables.